D1348462

30116 003379359

Neural and Endocrine Peptides and Receptors

GWUMC Department of Biochemistry Annual Spring Symposia

Series Editors:
Allan L. Goldstein, Ajit Kumar, George V. Vahouny, and J. Martyn Bailey
The George Washington University Medical Center

DIETARY FIBER IN HEALTH AND DISEASE
Edited by George V. Vahouny and David Kritchevsky

EUKARYOTIC GENE EXPRESSION
Edited by Ajit Kumar

NEURAL AND ENDOCRINE PEPTIDES AND RECEPTORS
Edited by Terry W. Moody

PROSTAGLANDINS, LEUKOTRIENES, AND LIPOXINS
Biochemistry, Mechanism of Action, and Clinical Applications
Edited by J. Martyn Bailey

THYMIC HORMONES AND LYMPHOKINES
Basic Chemistry and Clinical Applications
Edited by Allan L. Goldstein

Neural and Endocrine Peptides and Receptors

Edited by
Terry W. Moody
The George Washington University School of Medicine and Health Sciences
Washington, D.C.

Plenum Press • New York and London

*

Library of Congress Cataloging in Publication Data

Neural and endocrine peptides and receptors.
(GWUMC Department of Biochemistry annual spring symposia)
Includes bibliographies and index.
1. Peptide hormones—Congresses. 2. Peptide hormones—Receptors—Congresses. 3. Neuropeptides—Congresses. 4. Neuropeptides—Receptors—Congresses. I. Moody, Terry W. II. Series.
QP572.P4N48 1986 612′.814 86-20462
ISBN 0-306-42300-6

A Division of Plenum Publishing Corporation
233 Spring Street, New York, N.Y. 10013

Printed in the United States of America

Preface

The Fifth Annual Washington Spring Symposium on Health Sciences attracted over 400 scientists from 20 countries. It was held at the Lisner Auditorium of the George Washington University in Washington, D.C. The theme of the meeting was neural and endocrine peptides and receptors.

The meeting emphasized basic and clinical research on neural and endocrine peptides and receptors. The six plenary sessions emphasized pituitary peptides, releasing factors, brain peptides, growth factors, peripheral peptides, and clinical applications. The chapters in this volume are derived from each of these six scientific sessions plus the poster and special sessions.

The Abraham White Distinguished Scientist Award was presented to Dr. Julius Axelrod for his numerous contributions to the field of neurochemistry. He presented the keynote address, which was entitled "The Regulation of the Release of ACTH." Dr. Axelrod discussed numerous factors, such as the peptides CRF, VIP, and somatostatin, that regulate hormone secretion from pituitary cells.

The Distinguished Public Service Award was presented to Senator Lowell Weicker, Jr., in recognition of his leadership and outstanding achievements in the United States Senate and for his legislative support for biomedical research and education. In the symposium banquet address, Senator Weicker stressed the need for continued federal support of biomedical science research.

The field of peptide research has been active for over three decades. It has been richly endowed with Nobel Laureates, including Drs. Vincent DuVigneaud (the chemical synthesis of peptides), Bruce Merrifield (the chemical sequencing of peptides), Rosalyn Yalow (the development of the radioimmunoassay), Roger Guillemin (the isolation of TRH), and Andrew Schally (the isolation of somatostatin). The symposium emphasized that important contributions have been and continue to be made in this area of biological research.

In its infancy, this field was concerned with the sequencing of neural and endocrine peptides isolated and purified from various organs. Subsequently, radioimmunoassays were developed to determine the peptide content of various biological specimens. Now, with recent advances in biochemistry, numerous peptide analogues can rapidly be synthesized and tested for biological activity in receptor binding and biological assays. In addition, progress in molecular biology has made possible the cloning of genes that contain biologically important peptides.

It has long been accepted that peptides are biologically important both in the nervous system, where peptides such as neurotensin alter behavior, and in the gastrointestinal tract, where peptides such as cholecystokinin regulate food intake and digestion. Considerable progress has been made, however, in establishing the facts that peptides such as β-endorphin modify immune responses and that peptides such as bombesin alter the growth of normal and malignant cells. It also has been determined that releasing factors such as CRF play an important role in hormone secretion from the pituitary.

Rapid advances are being made in the clinical application of neural and endocrine peptides. New substance P receptor antagonists continue to be synthesized and their biological potency evaluated. Also, somatostatin and LHRH agonists are being used as therapeutic agents in the treatment of certain types of cancer. Monoclonal antibodies may serve as valuable treatment agents in disease associated with excessive peptide production and secretion.

In summary, although much progress has been made much more remains to be learned concerning the chemistry and biology of neural and endocrine peptides and receptors. This symposium provided an international forum for the exchange of new ideas, and this volume will be of interest to neuroscientists, endocrinologists, gastroenterologists, psychiatrists, oncologists, and others interested in acquiring state-of-the-art information about neural and endocrine peptides and receptors.

Terry W. Moody

Contents

PART I—PITUITARY PEPTIDES

PART II—PEPTIDE-RELEASING FACTORS

PART VI—CLINICAL APPLICATIONS OF NEURAL AND ENDOCRINE PEPTIDES

I

Pituitary Peptides

1

An Endogenous Peptide Ligand for the PCP/σ-Opiod Receptor

PATRICIA C. CONTRERAS, DEBORA A. DiMAGGIO, RÉMI QUIRION, and THOMAS L. O'DONOHUE

1. INTRODUCTION

1.1. Classification of Opioid Receptors

Opioid receptors comprise a heterogeneous group that can be divided into at least four biochemically and topographically distinct subtypes, designated μ, κ, σ, and δ. Martin *et al.* (1976) proposed the existence of μ-, κ-, and σ-opioid receptors based on observed differences in pharmacological profiles of drugs seen with variably selective opioid agonists and antagonists. Other groups have provided evidence for the δ subtype (Hughes *et al.*, 1975).

The μ-opioid receptor has been extensively characterized (Lord *et al.*, 1977; Chang *et al.*, 1979; Chang and Cuatrecasas, 1981). Classical opioid effects, such as analgesia induced by morphine and its congeners, are thought to be mediated by the μ-opioid receptor, which preferentially binds the levorotatory isomer. δ-Opioid receptors, which have been well characterized (Simantov *et al.*, 1978), are believed to be involved in reward processes and seizure. From the evidence available on κ-opioid receptors (Kosterlitz and Paterson, 1980; Chang and Cuatrecasas, 1979), it appears that these sites are involved in mediating analgesia and sedation. Finally, it has been postulated that σ-opioid receptors are involved in mediating the psy-

PATRICIA C. CONTRERAS • Experimental Therapeutics Branch, National Institute of Neurological and Communicative Disorders and Stroke, National Institutes of Health, Bethesda, Maryland 20205, and Department of Medicine, Howard University, Washington, D.C. 20059. DEBORA A. DiMAGGIO and THOMAS L. O'DONOHUE • Experimental Therapeutics Branch, National Institute of Neurological and Communicative Disorders and Stroke, National Institutes of Health, Bethesda, Maryland 20205. RÉMI QUIRION • Douglas Hospital Research Centre and Department of Psychiatry, Faculty of Medicine, McGill University, Verdun, Quebec H4H 1R3, Canada.

chotomimetic actions seen with some of the benzomorphans such as cyclazocine, N-allylnormetazocine (SKF 10,047), phencyclidine and phencyclidine analogues (Quirion *et al.*, 1981b).

1.2. Endogenous Opioid Ligands

The μ-, κ-, and δ-opioid receptors have been found to interact with endogenous peptide ligands that share certain pharmacological properties with opioid drugs. These peptide ligands are derived from at least three different prohormones located in both the central and peripheral nervous sytems and also in the endocrine system.

δ-Opioid receptors are relatively selective for two related pentapeptides, methionine enkephalin and leucine enkephalin (Met- and Leu-enkephalin), which were originally isolated from porcine brain (Hughes, 1975). Both Met- and Leu-enkephalin inhibit electrically induced contractions of guinea pig ileum, an effect that mimics that of opioid drugs and is naloxone reversible. The enkephalins are processed posttranslationally from proenkephalin and secreted from central and peripheral neurons and endocrine cells in the adrenal medulla.

Neurons in the brain and spinal cord secrete peptides derived from prodynorphin. Some of these peptides appear to be endogenous ligands for κ-opioid receptors. Dynorphin$_{1-13}$, isolated from porcine pituitary by Goldstein *et al.* (1979), contains within its sequence Leu-enkephalin, which appears to be one of the products of posttranslational processing (Zamir *et al.*, 1984; Palkovits *et al.*, 1983). Dynorphin$_{1-13}$ is 700 times more potent than the enkephalins in inhibiting electrical contractions of the guinea pig ileum longitudinal muscle (Goldstein *et al.*, 1979).

The third prohormone from which opioid peptides are derived is proopiomelanocortin, which yields a number of nonopioid and opioid peptide products (O'Donohue and Dorsa, 1982). Of these products, β-endorphin, an untriakontapeptide isolated from camel pituitary gland by Li and Chung (1976), is thought to interact primarily with μ- and δ-opioid receptors.

1.3. Evidence for Specific PCP/σ-Opioid Receptors

In recent years, several investigators have reported the presence of a class of high-affinity binding sites for phencyclidine [1-(1-cyclohexylphenyl)piperidine, PCP], a dissociative anesthetic with psychotomimetic properties. This binding was shown to be saturable, reversible, and selective (Vincent *et al.*, 1979; Zukin and Zukin, 1979; Quirion *et al.*, 1981a; Vignon *et al.*, 1982) as well as stereoselective (Quirion *et al.*, 1981a). Also, the binding of [^{3}H]PCP was rapidly inactivated by heat and destroyed by proteases, indicating that the PCP receptor is a protein (Vignon *et al.*, 1982). Receptor densities using [^{3}H]PCP (Quirion *et al.*, 1981b) and [^{3}H]TCP {N-[1-(2-thienyl)cyclohexyl[3,4[3,4[^{3}H]piperidine} (Contreras *et al.*, 1985a) to label the binding sites were reported to be highest in cortical regions and hippocampus, indicating that the distribution of PCP binding sites correlates well with the psychotomimetic properties of the drug. [^{3}H]Phencyclidine appeared to label PCP binding sites specifically, as serotonin, LSD, benzodiazepines, and dopaminergic and adrenergic agonists and antagonists did not inhibit the binding of [^{3}H]PCP. Although PCP appears to bind to muscarinic and μ-opioid receptors with very low affinity (Vincent *et al.*, 1979), its action on these receptors is not compatible with its pharmacological profile (Domino, 1981).

The relevance of PCP receptors is supported by the finding that there is a good correlation between the ability of PCP analogues to bind to PCP receptors and to produce ataxia (Vincent *et al.*, 1979; Vignon *et al.*, 1982; Vaupel *et al.*, 1984; Contreras *et al.*, 1962), stereotyped behavior (Contreras *et al.*, 1985a, 1986a), and a PCP-like stimulus in drug discrimination paradigms (Shannon, 1981; Holtzman, 1980; Brady and Balster, 1981).

Only drugs that produce PCP-like psychotomimetic effects, PCP analogues, dexoxadrol, and σ opioids, inhibited the binding of [^{3}H]PCP. Conversely, PCP inhibits the binding of [^{3}H]cyclazocine (Zukin and Zukin, 1981). Also, in drug discrimination paradigms, σ opioids generalize to PCP stimulus (Shannon, 1981; Brady *et al.*, 1981), and PCP generalizes to σ opioids (Teal and Holtzman, 1980; Shannon, 1983). Because σ opioids and PCP produce many of the same effects *in vitro* and *in vivo,* it has been suggested that PCP and σ opioids act through the same binding site. However, recent work by a number of investigators (Su, 1982; Tam, 1983; Martin *et al.*, 1984; Contreras *et al.*, 1985a) suggests that PCP binding sites and σ-opioid sites may be distinct sites because of differences in drug selectivity and regional distribution.

2. AN ENDOGENOUS LIGAND FOR THE PCP/σ-OPIOID RECEPTOR

The presence of highly specific and selective binding sites for PCP/σ opioids in brain strongly suggested the presence of an endogenous ligand for these receptors, such has been found to be case for μ-, κ-, and σ-opioid receptors. Quirion *et al.* (1984) and O'Donohue *et al.* (1983) have reported the isolation of an endogenous factor from preparative scale porcine brain acid extracts that inhibits the binding of [^{3}H]PCP. This endogenous factor has been purified to homogeneity and has been given the name α-endopsychosin.

2.1. Extraction and Purification of α-Endopsychosin

Porcine brains (200) were homogenized at 4°C in 5 volumes of acid solution (Bennett *et al.*, 1979) consisting of trifluoroacetic acid, formic acid, hydrochloric acid, and sodium chloride. Following centrifugation, the homogenate was extracted with petroleum ether. The resulting supernatant was fractionated using ultrafiltration and chromatographed in a series of runs using preparative and semipreparative liquid chromatography followed by several analytical reverse-phase high-pressure liquid chromatography steps. Throughout the purification, the binding of [^{3}H]PCP to rat brain membranes was used to assess the progress of this separation procedure.

2.2. Chemical Nature of α-Endopsychosin

Supernatants were chromatographed over a column of Sephadex G-10, G-25, and G-50, and aliquots of collected fractions were assessed for their ability to inhibit the binding of [^{3}H]PCP. The PCP-like activity eluted near the included volume of a Sephadex G-50 column, in the void volume of a Sephadex G-10 column, and in the fractionation range of a Sephadex G-25 column. These results indicated that the endogenous material has a molecule weight of about 3000.

The effects of various enzymes on the activity of HPLC fractions that inhibited the binding of [^{3}H]PCP was investigated to determine whether the PCP-like activity resided in a peptide or protein. Pronase, carboxypeptidase A, and trypsin markedly decreased the potency of 10 nanounits of PCP-like activity. No significant change in activity was seen when fractions were incubated with α-chymotrypsin. Boiled enzymes did not alter the ability of active fractions to inhibit the binding of [^{3}H]PCP. These data suggest that the endogenous material is a protein or peptide.

2.3. Structural Analysis of α-Endopsychosin

An aliquot of the most active PCP-like material was hydrolyzed in acid, and the amino acid composition was determined using OPA detection. It was determined that the peptide contained approximately 26 amino acids, in close agreement with the molecular weight predicted by Sephadex gel filtration studies. N-terminal analysis revealed that the peptide was blocked at this site. The nature of this blockade is yet to be determined.

2.4. Selectivity and Specificity of Active Fractions

A comparative dose of 10 nanounits of PCP-like activity inhibited the binding of [^{3}H]PCP in rat brain membranes but did not inhibit binding of [^{3}H]dihydromorphine, [^{3}H]D-Ala^{2}-D-Leu^{5}-enkephaline, [^{3}H]ethylketocyclazocine, [^{3}H]diazepam, or [^{3}H]neurotensin. These results indicate that the PCP-like active material selectively binds to PCP receptors, as it did not appear to bind to μ-, δ-, or κ-opioid receptors or to benzodiazepine or neurotensin receptors.

Aliquots of fractions from an intermediate chromatography step were assayed for their ability to inhibit the binding of [^{3}H]PCP. The endogenous material inhibited binding of [^{3}H]PCP to rat brain membranes and did so in a dose-related fashion but did not inhibit the binding of [^{3}H]haloperidol. Thus, α-endopsychosin appears to bind preferentially to binding sites labeled by [^{3}H]PCP.

2.5. Regional Distribution of Active Material in Rat Brain

The distribution of the active material that inhibited the binding of [^{3}H]PCP correlated well with the distribution of PCP receptors. The highest concentrations of endogenous material were found in hippocampus and cortex, whereas relatively little material was detectable in striatum and brainstem.

2.6. *In Vivo* Effects of the PCP-like Fractions

Active fractions tested for electrophysiological actions on hippocampal and cortical cells mimicked the actions of PCP (Quirion *et al.*, 1984). Iontophoresis of PCP inhibited spontaneous cortical and hippocampal cell firing, as did micropressure ejection of the PCP-like material. Fractions that did not possess PCP-like actions in the binding assays had no effect on spontaneous neuronal activity.

Similarly, in a behavioral paradigm in which unilateral injection of PCP into substantia nigra induces contralateral turning behavior (Quirion *et al.*, 1984), PCP-like fractions also mimicked the actions of PCP, producing significant rotational behavior contralateral to the injection site. Once again, inactive fractions elicited no response in this test. These two experiments *in vivo* demonstrate that the endogenous material is a PCP-like agonist.

3. Antagonism of PCP in Behavioral and Binding Assays by Metaphit

In order to demonstrate that α-endopsychosin and PCP exert similar effects by interacting with the same receptor, a specific PCP antagonist is required. Furthermore, the physiological role of the endogenous PCP-like system could be predicted by antagonism of α-endopsychosin much as naloxone has been used to predict the role of endogenous opioids. Along this line of investigation, we began to study and characterize the *in vivo* effects of metaphit (Contreras *et al.*, 1985a,b). Metaphit {1-[1-(3-isothiocyanatophenyl)cyclohexyl]piperidine} is a PCP analogue that specifically acylates PCP receptors *in vitro* (Rafferty *et al.*, 1985). Additional experiments were performed to determine whether metaphit could acylate PCP receptors *in vivo* and antagonize PCP induction of stereotyped behavior and ataxia. If metaphit is found to antagonize the behavioral effects of PCP by acylating PCP receptors, then metaphit will be useful in determining whether α-endopsychosin produces PCP-like effects by binding to PCP receptors.

3.1. Effect of Metaphit on the Binding of [^{3}H]PCP

Pretreatment of rats with 1 or 2 μmol/rat of metaphit i.c.v. 24 hr prior to sacrifice resulted in a 25% or 40% decrease in the B_{max} but no change in the K_d of binding of [^{3}H]PCP. Metaphit pretreatment did not alter the binding of [^{3}H]etorphine or [^{3}H]spiroperidol. Since the preparation of the brain homogenates for the binding assays consisted of two washes and was done 24 hr after metaphit pretreatment, these results indicate that metaphit specifically binds irreversibly to PCP receptors after *in vivo* administration.

3.2. Effect of Metaphit on PCP Induction of Stereotyped Behavior and Ataxia

Metaphit administered alone at doses up to 1 μmol/rat did not produce any significant behavioral effects. However, at doses of 2 μmol/rat and larger, metaphit produced PCP-like stereotyped behavior and ataxia. Thus, metaphit is a very weak PCP agonist. In addition to acute effects, metaphit produced convulsions, which were evident between 5 and 24 hr after intracerebroventricular (i.c.v.) adminstration of 2 μmol/rat.

Metaphit administered i.c.v. prior to PCP administered i.c.v. antagonized PCP induction of stereotyped behavior and ataxia up to 5 days after metaphit pretreatment. The antagonism of the behavioral effects of PCP by metaphit was dose dependent. Furthermore, this antagonism by metaphit is specific, as metaphit pretreatment i.c.v. did not antagonize amphetamine-induced stereotyped behavior and could be prevented by pretreating rats with PCP just prior to metaphit administration. These results indicate that acylation of PCP receptors results in decreased ability of PCP to induce stereotyped behavior.

Since metaphit appears specifically to acylate PCP receptors, metaphit is a useful tool to study the physiological role of PCP receptors. When metaphit was administered i.c.v. prior to i.p. administration of PCP, metaphit antagonized only the ability of PCP to induce sterotyped behavior but not ataxia. Thus, it appears that ataxia is mediated by both central and peripheral mechanisms. It is unlikely that the peripheral effect of PCP in induction of ataxia is mediated by PCP receptors, as 20 mg/kg of metaphit administered i.v. only antagonized PCP-induced stereotyped behavior when PCP was also administered peripherally.

3.3 Effect of Metaphit on the Behavioral Effects of (−)Cyclazocine

(−)Cyclazocine and PCP probably do not induce stereotyped behavior and ataxia through an interaction with the same receptor, as metaphit did not antagonize the behavioral effects of (−)cyclazocine. Yet, (−)cyclazocine was able to inhibit the binding of [^{3}H]PCP. These findings are consistent with the findings that metaphit can only bind irreversibly to about 50% of the binding sites labeled by [^{3}H]PCP (Rafferty *et al.*, 1985), which indicates that [^{3}H]PCP binds to more than one type of binding site. Thus, PCP and (−)cyclazocine probably exert their effects through different receptors.

4. DISCUSSION

The existence of high-affinity and stereoselective binding sites for PCP, a dissociative anesthetic with psychotomimetic properties, has been demonstrated in brain by a number of groups (Vincent *et al.*, 1979; Zukin and Zukin, 1979; Quirion *et al.*, 1981a; Vignon *et al.*, 1982). The identification of these PCP receptors in central nervous tissue suggested the existence of an endogenous ligand specific for these pharmacologically relevant binding sites. We have identified what appears to be a selective peptide ligand for the PCP receptor from porcine brain acid extracts and have named this compound α-endopsychosin. We have assigned the α designation to this peptide since activity profiles of fractionated porcine brain show that other fractons are also PCP-active, although these do not appear to be as potent as the fraction characterized. This putative endogenous ligand for the PCP receptor, which is regionally distributed throughout the CNS and inhibits binding of [^{3}H]PCP, appears to be an agonist relative to PCP, as indicated by its actions in electrophysiological and behavioral tests. The peptide appears to be selective for PCP receptors, since it does not interact with μ-, κ-, δ-opioid receptors; nor does it interact with a number of other peptide binding sites. Furthermore, the distribution of this peptide is compatible with PCP receptor densities, with highest concentrations of both ligand and receptor in areas that could be relevant to the psychotomimetic properties of PCP. Amino acid analysis of an aliquot of the purified fraction shows the peptide to contain at least 26 residues. Because the peptide is blocked at the N terminus, enzyme fragments have been generated and purified over HPLC for sequence determination.

Without a specific PCP antagonist, it could not be stated unequivocably that the endogenous ligand and PCP act at the same receptor to induce specific electrophysiological and behavioral actions. Clearly, the discovery of specific PCP antagonists is necessary for complete characterization of the PCP receptor/endogenous ligand system. Along this line of investigation, Rafferty *et al.* (1985) have reported that metaphit, a PCP analogue, specifically acylates PCP receptors *in vitro*. In our own studies, we have shown that metaphit specifically binds irreversibly to PCP receptors after *in vivo* administration. Furthermore, metaphit, when administered i.c.v. in rats prior to administration of PCP, blocks stereotyped behavior, an effect that we have shown to be the result of PCP receptor binding. Metaphit did not antagonize induction of ataxia following peripheral administration, so it is likely that ataxia is mediated by both central and peripheral mechanisms. Additionally, at high doses, metaphit appears to be a weak PCP agonist. It will be interesting to study the effects of this specific acylating agent on the interaction of the endogenous ligand with PCP receptors.

Although PCP affects several ion channels (Aguayo *et al.*, 1982; Alburquerque *et al.*, 1980) and numerous neurotransmitter systems, it is believed to exert its psychotomimetic effects via interaction with a specific receptor. It has been suggested that PCP receptor sites are actually σ-opioid sites, since SKF 10,047 and cyclazocine, σ-opioid psychotomimetics, inhibit the binding of [^{3}H]PCP to membranes *in vitro*. Phencyclidine analogues have been shown to inhibit binding of [^{3}H]cyclazocine (Zukin and Zukin, 1981). Furthermore, σ opioids generalize to PCP stimulus in drug discrimination paradigms (Brady *et al.*, 1982; Shannon, 1981), and PCP has been shown to generalize to σ-opioid stimulus (Shannon, 1983). However, metaphit inhibits PCP-induced stereotypy but does not inhibit sterotyped behavior typically seen with cyclazocine treatment. In addition, evidence exists that suggests that PCP,SKF 10,047, and cyclazocine do not bind to a homogenenous population of receptors (Contreras *et al.*, 1985a). Recent work presented by a number of investigators (Su, 1982; Tam, 1983; Martin *et al.*, 1984; Contreras *et al.*, 1985a, 1986b) also indicate that because of differences in regional distribution or drug selectivity, PCP receptors and σ-opioid sites represent different binding sites. Largent *et al.* (1984) have shown that sites labeled with [^{3}H]PPP [3-(3-hydroxyphenyl)-N-(1-propyl)piperidine] are specific for sites that are distinct from PCP sites. Finally Pilapil *et al.* (1985) have delineated a binding profile for dexoxadrol, a putative σ agonist, that is distinct from PCP binding. In this light, it is particularly interesting to note that the PCP-like peptide inhibits binding of [^{3}H]PCP but not [^{3}H](+)SKF 10,047, [^{3}H]dexoxadrol, or [^{3}H]haloperidol. The endogenous ligand for the PCP receptor may therefore be a far more selective ligand than any of the synthetic compounds. This also raises the possibility that there may be other endogenous ligands for the PCP/σ-opioid receptors.

A controversial issue concerns whether PCP/σ opioids are in fact opioids at all, as their "σ" psychotomimetic actions are not antagonized by naloxone. The three known opioid peptide precursors share marked structural homology in that they all contain multiple cysteine residues in the N-terminal region, they are all approximately the same size, and they all contain an enkephalin sequence. It will be interesting to see whether α-endopsychosin is the peptide product of a distinct yet similar prohormone. If so, then perhaps α-endopsychosin and PCP binding sites may belong to the superfamily of opioid peptides and receptors.

The functions of endogenous PCP-like peptides and receptors are, of course, unknown. However, the high density of α-endopsychosin and PCP receptors and the electrophysiological effects of PCP and α-endopsychosin in the hippocampus and cerebral cortex suggest a possible involvement of an endogenous PCP-like system in higher cognitive function. A role in memory processes may be indicated by reports of amnesia in humans after PCP intoxication. Roles for the endogenous PCP-like system in mediating higher cortical processes such as those controlling arousal, emotion, rationality, and aggression are also suggested by the clinical effects of PCP. It is of paramount importance to determine if pathologies of the endogenous PCP-like system are involved in cognitive dysfunctions such as psychoses.

ACKNOWLEDGMENT. This work was supported in part by the Chemical Research and Development Center, U.S. Army.

REFERENCES

Arguayo, L. G., Warnick, J. E., Maayani, S., Glick, S. D., Weinstein, H., and Alburquerque, E. X., 1982, Interaction of phencyclidine and its analogues on ionic channels of the electrically excitable membrane and nicotinic receptor: Implication for behavioral effects, *Mol. Pharmacol.* **21:**637–647.

Alburquerque, E. X., Tsai, M. C., Aronstam, R., Witkop, B., Eldefrawi, A. R., and Eldefrawi, M. E., 1980, Phencyclidine interactions with the ionic channel of the acetylcholine receptor and electrogenic membrane, *Proc. Natl. Acad. Sci. U.S.A.* **77:**1224–1228.

Bennett, H. P. J., Browne, C. A., Goltzman, D., and Solomon S., 1979, Isolation of peptide hormones by reverse-phase high pressure liquid chromatography, in: *Peptides: Structure and Biological Function* (E. Gross and J. Meienhofer, eds.), Pierce Chemical Co., Rockford, IL, pp. 121–125.

Brady, K. T., and Balster, R. L., 1981, Discriminative stimulus properties of phencyclidine and five analogues in the squirrel monkey, *Pharmacol. Biochem. Behav.* **14:**213–218.

Brady, K. T., Balster, R. L., and May, E. L., 1982, Stereoisomers of N-allylnormetazocine: Phencyclidine-like behavioral effects in squirrel monkeys and rats, *Science* **215:**178–179.

Chang, K. J., and Cuatrecasas, P., 1979, Novel opiate binding sites selective for benzomorphan drugs, *J. Biol. Chem.* **254:**2610–2618.

Chang, K. J., and Cuatrecases, P., 1981, Heterogeneity and properties of opiate receptors, *Fed. Proc.* **40:**2729–2734.

Chang, K. J., Cooper, B. R., Hazum, E., and Cuatrecasas, P., 1979, Multiple opiate receptors: Different regional distribution in the brain and differential binding of opiates and opioid peptides, *Mol Pharmacol.* **16:**91–104.

Contreras, P. C., Quirion, R., and O'Donohue, T. L., 1985a, Agonistic and antagonistic effects of PCP-derivatives and sigma opioids in PCP behavioral and receptor assays, *NIDA Monogr.* **64:**80–94.

Contreras, P. C., Rafferty, M. F., Lessor, R. A., Rice, K. C., Jacobson, A. E., and O'Donohue, T. L., 1985b, A specific acylating ligand for phencyclidine (PCP) receptors antagonizes PCP behavioral effects, *Eur. J. Pharmacol.,* **111:**405–406.

Contreras, P. C., Rice, K. C., Jacobson, A. E., and O'Donohue, T. L., 1986a, Stereotyped behavior correlates better than ataxia with phencyclidine–receptor interactions *Eur. J. Pharmacol.* **121:**9–18.

Contreras, P. C., Quirion, R., and O'Donohue, T. L., 1986b, Autoradiographic distribution of phencyclidine receptors in the rat brain using (3H)TCP *Neurosci. Lett.* (in press).

Domino, E. F., 1981, *PCP: Historical and Current Perspectives,* NPP Books, Ann Arbor.

Goldstein, A., Tachibana, S., Lowney, L. I., Hunkapiller, M., and Hood, L., 1979, Dynorphin-(1–13), an extraordinarily potent opioid peptide, *Proc. Natl. Acad. Sci. U.S.A.* **76:**6666–66770.

Holtzman, S. G., 1980, Phencyclidine-like discriminative effects of opioids in the rat, *J. Pharmacol. Exp. Ther.* **214:**614–620.

Hughes, J., 1975, Isolation of an endogenous compound from the brain with pharmacological properties similar to morphine, *Brain Res.* **88:**295–308.

Hughes, J., Smith, T. W., Kosterlitz, H. W., Fotherfill, L. A., Morgen, B. A., and Morris, H. R., 1975, Identification of two related pentapeptides from the brain with potent opiate agonist activity, *Nature* **258:**577–579.

Kosterlitz, H. W., and Paterson, S. J., 1980, Characterization of opioid receptors in nervous tissue, *Proc. R. Soc. Lond. [Biol.]* **210:**113–126.

Largent, B. L., Gundlach, A. L., and Synder, S. H., 1984, Psychotomimetic opiate receptors labeled and visualized with (±)-(^{3}H)3-(3-hydroxyphenyl)-N-(1-propyl)piperidine, *Proc. Natl. Acad. Sci. U.S.A.* **81:**4983–4987.

Li, C. H., and Chung, D., 1976, Isolation and structure of an untriakontapeptide with opiate activity from camel pituitary glands, *Proc. Natl. Acad. Sci. U.S.A.* **73:**1145–1148.

Lord, J., Waterfield, A., Hughes, J., and Kosterlitz, H., 1977, Endogenous opioid peptides: Multiple agonists and receptors, *Nature* **267:**495–500.

Martin, B. R., Katzen, J. S., Woods, J. A., Tripathi, H. L., Harris, L. S., and May, E. L., 1984, Stereoisomers of (^{3}H)-N-allylnormetazocine bind to different sites in mouse brain, *J. Pharmacol. Exp. Ther.* **231:**539–544.

Martin, W. R., Eades, C. G., Thompson, J. A. Huppler, R. E., and Gilbert, P. E., 1976, The effects of morphine- and nalorphine-like drugs in the non-dependent and morphine-dependent chronic spinal dog, *J. Pharmacol. Exp. Ther.* **197:**517–532.

O'Donohue, T. L., and Dorsa, D. M., 1982, The opiomelanotropinergic neuronal and endocrine sytems, *Peptides* **3:**353–395.

O'Donohue, T. L., Pert, C. B., French, E., Pert, A., DiMaggio, D. A., Everist, H., and Quirion, R., 1983, Evidence for an endogenous CNS ligand for the PCP receptor, in: *Peptides, Structure and Function* (V. J. Hruby and D. H. Rich, eds.), Pierce Chemical Co., Rockford, IL, pp. 433–437.

Palkovits, M., Brownstein, M. J., and Zamir, N., 1983, Immunoreactive dynorphin and alpha-neo-endorphin in rat hypothalamo–neurohypophyseal system, *Brain Res.* **278:**258–278.

Pilapil, C., Contreras, P. C., O'Donohue, T. L., and Quirion, R., 1985, Autoradiographic distribution of (^{3}H)dexoxadrol, a phencyclidine-related ligand, binding sites in rat and human brain, *Neurosci. Lett.* **56:**1–6.

Quirion, R., Rice, K. C., Skolnick, P., Paul, S., and Pert, C. B., 1981a, Stereospecific displacement of ^{3}H phencyclidine (PCP) receptor binding by an enantiomeric pair of analogs, *Eur. J. Pharmacol.* **74:**107–108.

Quirion, R., Hammer, R. P., Jr, Herkenham, M., and Pert, C. B., 1981b, Phencyclidine (angel dust)/sigma "opiate" receptor: Visualization by tritium-sensitive film, *Proc. Natl. Acad. Sci. U.S.A* **78:**5881–5885.

Quirion, R., DiMaggio, D. A., French, E. D., Contreras, P. C., Shiloach, J., Pert, C. B., Everist, H., Pert, A., and O'Donohue, T. L., 1984, Evidence for an endogeous peptide ligand for the phencyclidine receptor, *Peptides* **15:**967–973.

Rafferty, M. F., Mattson, M. Jacobson, A. E., and Rice, K. K, 1985, A specific acylating agent for the [^{3}H]phencyclidine receptors in rat brain, *FEBS Lett.* **181:**318–322.

Shannon, H. E., 1981, Evaluation of phencyclidine analogs on the basis of their discriminative stimulus properties in the rat, *J. Pharmacol. Exp. Ther.* **216:**543–552.

Shannon, H. E., 1982, Pharmacological analysis of the phencyclidine-like discriminative stimulus properties of narcotic derivatives in rats, *J. Pharmacol. Exp. Ther.* **222:**146–151.

Shannon, H. E., 1983, Pharmacological evaluation of N-allylnormetazocine (SKF-10,047) on the basis of its discriminative stimulus properties in rat, *J. Pharmacol. Exp. Ther.* **225:**144–150.

Simantov, R., Childers, D., and Snyder, S., 1978, The opiate receptor binding interactions of ^{3}H-methionine enkephalin, an opioid peptide, *Eur. J. Pharmacol.* **47:**319–331.

Su, T. P., 1982, Evidence for sigma-opioid receptor: Binding of (^{3}H)-SKF-10,047 to etorphine-inaccessible sites in guinea pig brain, *J. Pharmacol. Exp. Ther.* **223:**284–290.

Tam, S. W., 1983, Naloxone-inaccessible sigma receptor in rat central nervous system, *Proc. Natl. Acad. Sci. U.S.A.* **80:**6703–6707.

Teal, J. J., and Holtzman, S. G., 1980, Discriminative stimulus effects of cyclazocine in the rat, *J. Pharmacol. Exp. Ther.* **212:**368–376.

Vaupel, D. B., McCoun, D., and Cone, E. J., 1984, Phencyclidine analogs and precursors: Rotarod and lethal dose studies in the mouse, *J. Pharmacol. Exp. Ther.* **230:**20–27.

Vignon, J., Vincent, J. P., Bidard, J. N., Kamenka, J. M., Geneste, P., Monier, S., and Lazdunski, M., 1982, Biochemical properties of the brain phencyclidine receptor, *Eur. J. Pharmacol.* **81:**531–542.

Vincent, J. P., Cavey, D., Kamenka, J. M., Geneste, P., and Lazdunski, M., 1978, Interaction of phencyclidine with the muscarinic and opiate receptors in the central nervous system, *Brain Res.* **152:**176–182.

Vincent, J. P., Kartalovski, P., Geneste, J. M., and Lazdunski, M., 1979, Interaction of phencyclidine ("angel dust") with a specific receptor in rat brain membranes, *Proc. Natl. Acad. Sci. U.S.A.* **76:**4678–4682.

Zamir, N., Zamir, D., Eiden, L. E., Palkovits, M., Brownstein, M. J., Eskay, R. L., Weber, E., Faden, A. I., and Feuerstein, G., 1985, Methionine and leucine enkephalin in rat neurohypophysis: Different responses to osmotic stimuli and T2 toxin, *Science* **228:**605–608.

Zukin, S. R., and Zukin, R. S., 1979, Specific ^{3}H phencyclidine binding in rat central nervous system, *Proc. Natl. Acad. Sci. U.S.A.* **76:**5372–5376.

Zukin, R. S., and Zukin, S. R., 1981, Demonstration of (^{3}H)cyclazocine binding to multiple opiate receptor sites, *Mol. Pharmacol.* **20:**246–254.

2

β-Lipotropin

CHOH HAO LI

1. INTRODUCTION

Pituitary extracts have been known for nearly 50 years to contain fat-mobilizing or lipolytic activity. Now that pituitary hormones have been isolated and their primary structures are known, it is possible to conclude that somatotropin, thyrotropin, corticotropin (ACTH), and melanotropins (α- and β-MSH) are active as lipolytic agents. A new pituitary lipolytic factor, β-lipotropin (β-LPH), whose chemical characteristics are completely different from known pituitary hormones, was discovered in 1964 during the course of improving the procedure for the isolation of ACTH from sheep pituitary glands (Li, 1964; Birk and Li, 1964). β-Lipotropin has subsequently been isolated from pituitary glands of cow, pig, rat, whale, turkey, and ostrich (Li, 1982; Naudé and Oelofsewn, 1981). The complete amino acid sequences of human, sheep, pig, cow and ostrich (see Fig. 1) have been proposed (Li, 1982; Naudé *et al.*, 1981). β-Lipotropin is the prohormone for β-MSH and β-endorphin (β-EP). It is initially biosynthesized in a large precursor, proopiomelanocortin (POMC), which is also the precursor for ACTH (Mains *et al.*, 1977; Nankanishi *et al.*, 1979).

2. SHEEP β-LPH

β_s-Lipotropin consists of 91 amino acids with Glu as the NH_2-terminal amino acid. The primary structure of β_s-LPH was initially proposed in 1965 (Li *et al.*, 1965) with minor revision in 1973 (Gráf and Li, 1973). The single Trp residue is in position 52, and four other amino acids occupy the following positions: His 49, 87; Val 31, 75; Met 47, 65; Ile 82, 83. It should be noted that the peptide segment

CHOH HAO LI • Laboratory of Molecular Endocrinology, University of California, San Francisco, California 94143.

```
                         5                   10                  15                  20
  Human:  H-Glu-Leu-Thr-Gly-Gln-Arg-Leu-Arg-Glu-Gly-Asp-Gly-Pro-Asp-Gly-Pro-Ala-Asp-Asp-Gly-
  Ovine:  H-Glu-Leu-Thr-Gly-Glu-Arg-Leu-Glu-Gln-Ala-Arg-Gly-Pro-Glu-Ala-Gln-Ala-Glu-Ser-Ala-
 Bovine:  H-Glu-Leu-Thr-Gly-Glu-Arg-Leu-Glu-Gln-Ala-Arg-Gly-Pro-Glu-Ala-Gln-Ala-Glu-Ser-Ala-
Porcine:  H-Glu-Leu-Ala-Gly-Ala-Pro-Pro-Glu-Pro-Ala-Arg-Asp-Pro-Glu-Ala-Pro-Ala-Glu-Gly-Ala-
Ostrich:  H-Ala-Leu-Pro-Pro-Ala-Ala-Met-Leu-Pro-Ala-Ala-Ala-Glu-Glu-Glu-Glu-Gly-Glu-Glu-Glu-

                      25                  30                                      35
            Ala-Gly-Ala-Gln-Ala-Asp-Leu-Glu-His-Ser-Leu-Leu-Val(                 )Ala-Ala-
            Ala-Ala-Arg-Ala(   )Glu-Leu-Glu-Tyr-Gly-Leu-Val-Ala-Glu(      )Ala-Glu-Ala-Ala-
            Ala-Ala-Arg-Ala(   )Glu-Leu-Glu-Tyr-Gly-Leu-Val-Ala-Glu-Ala-Glu-Ala-Glu-Ala-Ala-
            Ala-Ala-Arg-Ala(   )Glu-Leu-Glu-His-Gly-Leu-Val-Ala-Glu(      )Ala-Gln-Ala-Ala-
            Glu-Glu-Gly-Glu(                                                         )Ala-

                      40                  45                  50                  55
            Glu-Lys-Lys-Asp-Glu-Gly-Pro-Tyr-Arg-Met-Glu-His-Phe-Arg-Trp-Gly-Ser-Pro-Pro-Lys-
            Glu-Lys-Lys-Asp-Ser-Gly-Pro-Tyr-Lys-Met-Glu-His-Phe-Arg-Trp-Gly-Ser-Pro-Pro-Lys-
            Glu-Lys-Lys-Asp-Ser-Gly-Pro-Tyr-Lys-Met-Glu-His-Phe-Arg-Trp-Gly-Ser-Pro-Pro-Lys-
            Glu-Lys-Lys-Asp-Glu-Gly-Pro-Tyr-Lys-Met-Glu-His-Phe-Arg-Trp-Gly-Ser-Pro-Pro-Lys-
            Glu-Lys-Glu-Asp-Gly-Gly-Ser-Tyr-Arg-Met-Arg-His-Phe-Arg-Trp-Gln-Ala-Pro-Leu-Lys-

                      60                  65                  70                  75
            Asp-Lys-Arg-Tyr-Gly-Gly-Phe-Met-Thr-Ser-Glu-Lys-Ser-Gln-Thr-Pro-Leu-Val-Thr-Leu-
            Asp-Lys-Arg-Tyr-Gly-Gly-Phe-Met-Thr-Ser-Glu-Lys-Ser-Gln-Thr-Pro-Leu-Val-Thr-Leu-
            Asp-Lys-Arg-Tyr-Gly-Gly-Phe-Met-Thr-Ser-Glu-Lys-Ser-Gln-Thr-Pro-Leu-Val-Thr-Leu-
            Asp-Lys-Arg-Tyr-Gly-Gly-Phe-Met-Thr-Ser-Glu-Lys-Ser-Gln-Thr-Pro-Leu-Val-Thr-Leu-
            Asp-Lys-Arg-Tyr-Gly-Gly-Phe-Met-Ser-Ser-Glu-Arg-Gly-Arg-Ala-Pro-Leu-Val-Thr-Leu-

                      80                  85              89           Total Residues
            Phe-Lys-Asn-Ala-Ile-Ile-Lys-Asn-Ala-Tyr-Lys-Lys-Gly-Glu-OH           89
            Phe-Lys-Asn-Ala-Ile-Ile-Lys-Asn-Ala-His-Lys-Lys-Gly-Gln-OH           91
            Phe-Lys-Asn-Ala-Ile-Ile-Lys-Asn-Ala-His-Lys-Lys-Gly-Gln-OH           93
            Phe-Lys-Asn-Ala-Ile-Val-Lys-Asn-Ala-His-Lys-Lys-Gly-Gln-OH           91
            Phe-Lys-Asn-Ala-Ile-Val-Lys-Ser-Ala-Tyr-Lys-Lys-Gly-Gln-OH           79
```

FIGURE 1. Amino acid sequences of β-LPH from various species.

corresponding to residues 41–58 is identical to that of bovine β-MSH (Geschwind *et al.*, 1957).

β_s-Lipotropin exists in two polymorphic forms by partition chromatography on agarose gel (Yamashiro and Li, 1976). The difference between the primary structures of these two forms resides in the NH_2-terminal residue: the major component has glutamic acid and the other pyroglutamic acid. It is unlikely that either <Glu or Gln could give rise to the NH_2-terminal Glu of β_s-LPH. Thus, two forms of β_s-LPH are probably synthesized by two different genes in the pituitary gland.

Chemical synthesis of β_s-LPH has been accomplished (Yamashiro and Li, 1978) using the solid-phase method. The synthetic product was shown to be identical to natural hormone by paper electrophoresis at two pH values, thin-layer chromatography, amino acid analyses of acid or enzymic hydrolysis, NH_2-terminal residue analysis, peptide map of tryptic digests, isoelectric focusing, circular dichroism spectra, lipolytic activity in isolated rabbit fat cells, radioimmunoassay, and complement fixation.

In high doses, synthetic β_s-LPH stimulates both corticosterone production in rat adrenal decapsular cells and aldosterone production in capsular cells (Li *et al.*, 1982). This confirms the earlier finding that β-LPH may be an aldosterone-stimulating factor (Matsuoka *et al.*, 1980, 1981).

3. HUMAN β-LPH

Human β-LPH was first isolated by Cseh *et al.* (1968) in 1968 and later by others (Scott and Lowry, 1974; Chrétien *et al.*, 1976; Li and Chung, 1981). The amino acid sequence of the hormone was proposed in 1981 (Li and Chung, 1981; Takahashi *et al.*, 1981). It consists of 89 amino acids with the single Trp residue in position 50 (see Fig. 1).

Chemical synthesis of β_h-LPH was achieved by the segment-coupling method (Blake, 1981) in aqueous solution (Blake and Li, 1983). This involves the use of thiol acids for coupling segments in aqueous media as shown in the following reaction:

$$R_1CO\text{-}S^- + 2\ Ag^+ + H_2N\text{-}R_2 \rightarrow R_1CONHR_2 + Ag_2S + H^+$$

where R_1 and R_2 are peptide segments that may be synthesized by the solution or solid-phase method.

The strategy for the total synthesis of β_h-LPH (Blake and Li, 1983) was to synthesize peptide corresponding to β_h-LPH_{1-60} and β_h-LPH_{61-89}, which could be coupled together to give the Gly^{60}-Gly^{61} peptide bond. In order to obtain the desired peptide $[Lys(Cit)^{67,77,82,86,87}]$-$\beta_h$-$LPH_{61-89}$ (Ia) by conventional procedures of solid-phase method, it was necessary to block the terminal α-amino function with a group that would be stable to HF and thus allow selective protection of the lysine amino groups. Previously, the 9-(fluorenyl)methyloxycarbonyl group was useful for this purpose (Blake and Li, 1981), but in this case the decrease in aqueous solubility conferred by this group was unacceptable. It was decided to take advantage of the absence of arginine from β_h-LPH_{61-89}. Accordingly, Ac-Arg-β_h-LPH_{61-89} (I) was synthesized and readily purified by chromatography on CM-cellulose and partition chromatography on Sephadex G-50. Reaction of peptide I with citraconic anhydride followed by digestion with trypsin to remove the acetylarginyl residue gave Ia.

BOC-Gly-S resin (Blake and Li, 1981) was used for synthesis of [$GlyS^{60}$]-β_h-LPH_{1-60} (II) by the solid-phase method. The crude product obtained from HF cleavage was chromatographed on CM-cellulose, deformylated by brief treatment with aqueous alkali, and rechromatographed on CM-cellulose. Partition chromatography gave the highly purified product, which was treated with citraconic anhydride to give Cit-[$Lys(Cit)^{37,38,55,57}$, $GlyS^{60}$]-β_h-LPH_{1-60} (IIa).

Peptides Ia and IIa were coupled together by reaction with silver nitrate/N-hydroxysuccinimide in 50% dimethylformamide. Overnight reaction with 25% acetic acid gave the crude product that was purified by chromatography on CM-cellulose. The material in the peak tubes that had the same elution position as native β_h-LPH was shown to be the desired peptide by amino acid analysis. Further purification was necessary by gel chromatography on Sephadex G-50 and rechromatography on CM-cellulose. The final yield of β_h-LPH was 10%. The synthetic product was identical to the natural hormone in its electrophoretic mobility on paper, behavior on high-performance liquid chromatography, and position in isoelectric focusing gels. Assay for lipolytic activity in isolated rabbit fat cells and radioimmunoreactivity showed the synthetic product to be equipotent with the native hormone. Thus, the synthetic and natural β_h-LPH are identical, and the proposed amino acid sequence is correct.

Synthetic β_h-LPH was found to stimulate aldosterone but not cortisol secretion in the autoperfused dog adrenal gland (H.-G. Güllner, unpublished results). The synthetic peptide was administered in a dose of 100 ng/min into the *in situ* perfused adrenal gland of hypophysectomized and nephrectomized dogs. This experimental model has been described previously (Güllner and Gill, 1983). Aldosterone and cortisol were measured by radioimmunoassay. Figure 2 summarizes the results. Synthetic human β-LPH stimulated aldosterone secretion but had virtually no effect on cortisol secretion. The time course of the effect on aldosterone was similar to that of ACTH. Compared to synthetic human β-EP (Güllner and Gill, 1983), at least three times higher doses of β-LPH were needed to cause a comparable increase in aldosterone secretion. Thus, these data indicate that synthetic β_h-LPH selectively simulates *in vivo* aldosterone production.

4. β_s-$LIPOTROPIN_{61-91}$

β_s-$Lipotropin_{61-91}$ (see Fig. 1) is known as β-EP (Li and Chung, 1976). An intracellular protease in the pituitary causes the cleavage of the Arg-Tyr bond in positions 60 and 61 in the β-LPH structure to generate the opioid peptide (Gráf and Kenessey, 1981). β-Endorphin has been isolated and sequenced from pituitary

Birk, Y., and Li, C. H., 1964, Isolation and properties of a new, biologically active peptide from sheep pituitary glands, *J. Biol. Chem.* **239:**1048.

Blake, J., 1981, Peptide segment coupling in aqueous medium: Silver ion activation of the thiolcarboxyl group, *Int. J. Peptide Protein Res.* **27:**273–274.

Blake, J., and Li, C. H., 1981, New segment-coupling method for peptide synthesis in aqueous solution: Application to synthesis of human [Gly^{17}]-β-endorphin, *Proc. Natl. Acad. Sci. U.S.A.* **78:**4055–4058.

Blake, J., and Li, C. H., 1983, Total synthesis of human β-lipotropin, *Proc. Natl. Acad. Sci. U.S.A.* **80:**1556–1559.

Chrétien, M., Gillardeau, C., Seidah, N., and Lis, M., 1976, Purification and partial chemical characterization of human pituitary lipolytic hormone, *Can. J. Biochem.* **54:**778–782.

Cseh, G., Gráf, L., and Goth, E., 1968, Lipotropic hormone obtained from human pituitary gland, *FEBS Lett.* **2:**42–44.

Geschwind, I. I., Li, C. H., and Barnafi, L., 1957, The isolation and structure of a melanocyte-stimulating hormone from bovine pituitary gland, *J. Am. Chem. Soc.* **79:**1003.

Gráf, L., and Kenessey, A., 1981, Characterization of proteinases involved in the generation of opioid peptides from β-lipotropin, in: *Hormonal Proteins and Peptides,* Vol. X (C. H. Li, ed.), Academic Press, New York, pp. 35–63.

Gráf, L., and Li, C. H., 1973, Action of plasmin on ovine β-lipotropin: Revision of the carboxyl terminal sequence, *Biochem. Biophys. Res. Commun.* **53:**1304–1309.

Güllner, H.-G., and Gill, J. R., 1983, Beta endorphin selectively stimulates aldosterone secretion in hypophysectomized, nephrectomized dogs, *J. Clin. Invest.* **71:**124–128.

Hammonds, R. G., Jr., Nicolas, P., and Li, C. H., 1984, β-Endorphin-(1–27) is an antagonist of β-endorphin analgesia, *Proc. Natl. Acad. Sci. U.S.A.* **81:**1389–1390.

Lee, C. Y., McPherson, M., Licko, V., and Ramachandran, J., 1980, Pituitary corticotropin-inhibiting peptide: Properties and use in study of corticotropin action, *Arch. Biochem. Biophys.* **201:**411–419.

Li, C. H., 1964, Lipotropin, a new active peptide from pituitary glands, *Nature* **201:**924.

Li, C. H., 1982, The lipotropins, in: *Biochemical Actions of Hormones,* Vol. IX (G. Litwack, ed.), Academic Press, New York, pp. 1–41.

Li, C. H., and Chung, D., 1976, Isolation and structure of an untriakontapeptide with opiate activity from camel pituitary glands, *Proc. Natl. Acad. Sci. U.S.A.* **73:**1145–1148.

Li, C. H., and Chung, D., 1981, Isolation, characterization and amino sequence of β-lipotropin from human pituitary glands, *Int. J. Peptide Protein Res.* **17:**131–142.

Li, C. H., Barnafi, L., Chrétien, M., and Chung, D., 1965, Isolation and amino-acid sequence of β-LPH from sheep pituitary glands, *Nature* **208:**1093–1094.

Li, C. H., Chung, D., Yamashiro, D., and Lee, C. Y., 1978, Isolation, characterization, and synthesis of a corticotropin-inhibiting peptide from human pituitary glands, *Proc. Natl. Acad. Sci. U.S.A.* **75:**4306–4309.

Li, C. H., Ng, T. B., and Cheng, C. H. K., 1982, Melanotropins: Aldosterone- and corticosterone-stimulating activity in isolated rat adrenal cells, *Int. J. Peptide Protein Res.* **19:**361–365.

Mains, R. E., Eipper, R. A., and Ling, N., 1977, Common precursor to corticotropins and endorphins, *Proc. Natl. Acad. Sci. U.S.A.* **74:**3014–3018.

Matsuoka, H., Mulrow, P. J., and Li, C. H., 1980, β-Lipotropin: A new aldosterone-stimulating factor, *Science* **209:**307–308.

Matsuoka, H., Mulrow, P. J., Franco-Saenz, R., and Li, C. H., 1981, Effects of β-lipotropin and β-lipotropin-derived peptides on aldosterone production in the rat adrenal gland, *J. Clin. Invest.* **68:**752–759.

Nakanishi, S., Inoue, A., Kita, T., Nakamura, M., Chang, A. C. Y., Cohen, S. N., and Numa, S., 1979, Nucleotide sequence of cloned cDNA for bovine corticotropin–β-lipotropin precursor, *Nature* **278:**423–427.

Naudé, R. J., and Oelofsen, W., 1981, Isolation and characterization of β-lipotropin from the pituitary gland of the ostrich, *Struthio camelus, Int. J. Peptide Protein Res.* **18:**135–137.

Naudé, R. J., Chung, D., Li, C. H., and Oelofsen, W., 1981, β-Lipotropin: Primary structure of the hormone from the ostrich pituitary gland, *Int. J. Peptide Protein Res.* **18:**138–147.

Ng, T. B., Chung, D., and Li, C. H., 1981, Isolation and properties of β-endorphin-(1–27), N^{α}-acetyl-β-endorphin, corticotropin, gamma-lipotropin and neurophysin from equine pituitary glands, *Int. J. Peptide Protein Res.* 18:443–450.

Ng, T. B., Chung, D., and Li, C. H., 1982, Isolation and properties of β-endorphin-(1–27)-like peptide from bovine brains, *Int. J. Peptide Protein Res.* **19:**343–347.

Nicolas, P., and Li, C. H., 1985, β-Endorphin-(1–27) is a naturally occurring antagonist to etorphine-induced analgesia, *Proc. Natl. Acad. Sci. U.S.A.* **82:**3178–3181.

Scott, A. P., and Lowry, P. J., 1974, Adrenocorticotrophic and melanocyte-stimulating peptides in the human pituitary, *Biochem. J.* **139:**593–602.

Smyth, D. G., 1983, β-Endorphin and related peptides in pituitary, brain, pancreas and antrum, *Br. Med. Bull.* **39:**25–30.

Takahashi, H., Teranishi, Y., Nakanishi, S., and Numa, S., 1981, Isolation and structural organization of the human corticotropin–β-lipotropin precursor gene, *FEBS Lett.* **135:**97–102.

Yamashiro, D., and Li, C. H., 1976, Isolation and properties of ovine [1-pyroglutamic acid]-β-lipotropin, *Biochim. Biophys. Acta* **451:**124–132.

Yamashiro, D., and Li, C. H., 1978, Total synthesis of ovine β-lipotropin by the solid-phase method, *J. Am. Chem. Soc.* **100:**5174–5179.

Yamashiro, D., and Li, C. H., 1984, β-Endorphin: Structure and activity, in: *The Peptides, Analysis, Synthesis, Biology,* Vol. 6 (S. Udenfriend and J. Meienhofer, eds.), Academic Press, New York, pp. 191–217.

3

Proopiomelanocortin-Derived Peptides and Their Receptors in the Immune System

KENNETH L. BOST, ERIC M. SMITH, and J. EDWIN BLALOCK

1. INTRODUCTION

It has been known for many years that stressful situations can be a contributing factor in the development of a variety of bacterial, viral, and neoplastic diseases (Riley, 1981; Stein, 1985). Specifically, decreased immunocompetence seems to account for the increased susceptibility of stressed hosts for these disease states. An increasing interest in the biochemical basis for this stress-induced immune suppression has resulted in the emergence of a new discipline (Holden, 1980). This field has been referred to as psychoimmunology, neuroimmunoendocrinology, neuroimmunomodulation, and behavioral medicine. Regardless of the title given to this new field, the question being addressed is the same: How are the nervous, endocrine, and immune systems interrelated?

Much evidence has accumulated to suggest that the neuroendocrine system can influence immune function. For example, immune responses can be regulated using classical Pavlovian conditioning (Ader, 1981). Furthermore, destructive lesioning of brain hypothalamic regions has been shown to have a variety of immunologic effects, including (1) decreasing the number of splenocytes and thymocytes (Brooks *et al.*, 1982), (2) lowering antibody production (Tyrey and Nalbandov, 1972; Goldstein, 1978), (3) suppressing cellular immunity (Macris *et al.*, 1970; Jankovic and Isakovic, 1973), and (4) decreasing proliferative responses of lym-

KENNETH L. BOST, ERIC M. SMITH, and J. EDWIN BLALOCK • Department of Microbiology, University of Texas Medical Branch, Galveston, Texas 77550. *Present address of K.L.B. and J.E.B.:* Department of Physiology and Biophysics, University of Alabama, Birmingham, Alabama 35294.

phocytes to concanavalin A (Cross *et al.*, 1980). Conversely, lesioning in the hippocampus, mammillary bodies, or the amygdaloid complex resulted in increased proliferation of lymphoid cells to concanavalin A (Brooks *et al.*, 1982). Taken together with the well-known effects of stress on immune function, these studies have provided a basis for postulating that the nervous system can influence immune responsiveness. However, the molecular mechanisms for this nervous-system-induced immune regulation have only recently been investigated. Studies conducted in our laboratory and by other investigators have shown that many neuroendocrine hormones can be made by cells of the immune system. Furthermore, receptors for these neuroendocrine hormones can also be found on immunocytes. Therefore, it seems logical to suggest that the ability of the nervous system to influence immune responsiveness may be the result of ligands and receptors that are common to both systems. Here we summarize some of the work that supports the ability of the nervous system to influence immune responsiveness and also some recent data to suggest that the immune system can also influence neuroendocrine function.

2. PRODUCTION OF NEUROENDOCRINE PEPTIDES BY CELLS OF THE IMMUNE SYSTEM

One mechanism by which the nervous and immune systems may influence one another's function is the ability of both systems to produce similar, if not identical, ligands. It is becoming apparent that cells of the immune system are capable of producing a variety of neuroendocrine peptide hormones. Our laboratory first demonstrated that lymphocytes have the ability to produce immunoreactive (ir) ACTH and endorphins and that these lymphocyte-derived hormones were identical to their pituitary-derived counterparts by several criteria, including antigenicity, molecular weight, and retention time on reversed-phase high-pressure liquid chromatography (Smith and Blalock, 1981). The lymphocyte-derived ACTH and endorphins also had the appropriate biological activities of inducing steroidogenesis in adrenal cells and producing analgesia and catatonia, respectively. Since the ACTH and endorphins were intrinsically radiolabeled with amino acids during their *in vitro* induction, there is no doubt that these peptides were synthesized by the lymphocytes *de novo* and not just passively acquired. Studies in other laboratories have extended our initial observations. Lolait *et al.* (1984) demonstrated that a subset of splenic macrophages had the ability to produce ir β-endorphin, ACTH, and a putative precursor. Thus, it seems that lymphocytes are capable of *de novo* synthesis of peptides derived from proopiomelanocortin (i.e., ACTH and endorphins) and that these lymphocyte-derived peptides appear to be identical to their neuroendocrine counterparts.

Hormone production by cells of the immune system is not limited to ACTH or endorphins, though these have been the best characterized. Vasoactive intestinal peptide was demonstrated in unstimulated polymorphonuclear leukocytes (O'Dorisio *et al.*, 1980), in mononuclear leukocytes and polymorphonuclear leukocytes (Lygren *et al.*, 1984), and in mast cells (Cutz *et al.*, 1978; Giachetti *et al.*, 1978).

Other neuroendocrine hormones, including somatostatin (Lygren *et al.*, 1984) and thyrotropin (Smith *et al.*, 1983), have been detected in mononuclear leukocytes. The *de novo* synthesis of thyrotropin by lymphocytes was demonstrated by intrinsically radiolabeling the hormone during its production *in vitro*. Preliminary immunologic and biochemical data from our laboratory also suggest that human and mouse mononuclear leukocytes can produce immunoreactive chorionic gonadotropin, growth hormone, follicle-stimulating hormone, and luteinizing hormone given the appropriate stimulus (E. M. Smith, D. V. Harbour-McMenamin, K. L. Bost, and J. E. Blalock, unpublished observations). Thus, it appears that cells of the immune system have the potential to produce a wide variety of neuroendocrine peptide hormones.

3. FACTORS STIMULATING THE PRODUCTION OF MONONUCLEAR-LEUKOCYTE-DERIVED HORMONES

Different stimuli have been shown to induce mononuclear leukocytes to produce peptide hormones. Interestingly, the type of stimulus seems to determine the particular hormone that is produced. For example, the B-cell mitogen lipopolysaccharide (LPS) stimulates mononuclear leukocytes to produce ACTH and endorphinlike peptides that are antigenically, structurally, and functionally similar to their proopiomelanocortin-derived neuroendocrine peptides (Harbour-McMenamin *et al.*, 1985). These data are interesting in light of the observation that endorphins have been implicated as possible mediators of some pathophysiological changes induced during endotoxic shock (Holaday and Faden, 1978). In contrast, a T lymphocyte mitogen (staphylococcal enterotoxin A) stimulated human lymphocytes to produce thyrotropin but not ACTH (Smith *et al.*, 1983).

In addition to mitogens, viral, bacterial, or tumor cell interactions with mononuclear leukocytes stimulate the production of some neuroendocrine peptides. Specifically, Newcastle disease virus-infected lymphocytes synthesized ACTH and endorphins in response to the virus (Smith and Blalock, 1981). Similar results were obtained in the same study when lymphocytes were exposed to tumor cells. Recently, leukocytes from chickens injected with *Salmonella pullorum* antigen were shown to synthesize an ACTH-like substance (Siegel *et al.*, 1985). This study is interesting since a bacterial antigen stimulated the production of ACTH-like substances by leukocytes from the nonmammalian class *Aves*. From the studies just described, it appears that different stimuli induce the production of different peptide hormones. There are probably two major variables that determine which peptide hormones are produced: (1) the nature of the stimulus and (2) the specific cell type that is stimulated.

Although it is apparent from the studies cited above that classical lymphocyte stimuli (i.e., mitogens and antigens) can result in peptide hormone production, our laboratory has recently reported that hypothalamic releasing factors can also stimulate release of hormones from mononuclear leukocytes. Specifically, corticotropin-releasing factor was observed to cause the *de novo* synthesis and release of mononuclear-leukocyte-derived ACTH and β-endorphin (Smith *et al.*, 1986). At approximately a tenfold higher concentration, arginine vasopressin was also observed to have intrinsic corticotropin-releasing activity. At concentrations frequently used to induce cultured pituitary cells, corticotropin-releasing factor and arginine vasopressin acted together in an additive fashion to induce ACTH and β-endorphin. This induction was blocked by dexamethasone. Thus, mononuclear leukocytes seem quite similar to anterior pituitary cells with respect to the ability to respond to a positive hypothalamic signal and to have that signal inhibited by a synthetic glucocorticoid hormone. In summary, immunostimulants as well as hypothalamic releasing factors induce peptide hormone production in mononuclear leukocytes. It will be interesting to determine which of these stimuli are most important with respect to immunologic responses *in vivo*.

4. PROCESSING OF PROOPIOMELANOCORTIN BY MONONUCLEAR LEUKOCYTES

ACTH, the endorphins, the melanocyte-stimulating hormones (MSH), and the lipotropins (LPH) are all processed from a common polyprotein precursor called proopiomelanocortin (POMC) (Nakanishi *et al.*, 1979). Thus, one might think that expression of the POMC mRNA would always result in the expression of all four peptides. However, this does not seem to be the case, since differential processing of the POMC polyprotein results in preferential expression of varying peptides. For example, in the anterior pituitary, the predominant products from the POMC polyprotein are ACTH and β-LPH, whereas in the intermediate lobe, α-MSH and β-endorphin predominate (Krieger, 1983). Thus, differential processing of the POMC protein allows for the expression of different activities derived from the same precursor molecule.

Interestingly, studies in our laboratory have shown that mononuclear leukocytes may also have differential processing pathways for POMC. For example, Newcastle disease virus-infected lymphocytes or lymphocytes exposed to corticotropin-releasing factor produced peptides corresponding to $ACTH_{1-39}$ and β-endorphin (Smith and Blalock, 1981; Smith *et al.*, 1985). In contrast, although lipopolysaccharide induces ACTH and endorphin production, these peptides have molecular weights that correspond to $ACTH_{1-24}$ to $ACTH_{1-26}$ and α- or γ-endorphin (Harbour-McMenamin *et al.*, 1985). The induction of an ACTH molecule with a molecular weight of approximately 2900 suggests a novel processing of the POMC precursor that has not previously been described. There are two likely possibilities for the differential processing observed in mononuclear leukocytes. One possibility is that different stimuli (i.e., LPS versus virus or CRF) may affect the way in which POMC is processed. A second possibility is that different cell subpopulations (i.e., B versus T versus macrophages) responding to the varying stimuli process POMC in different ways. Another point that should be considered is that these mononuclear leukocytes differ from virtually all other extrapituitary tissues that have been looked at in terms of processing POMC (Kreiger, 1983). For example, though ACTH and β-endorphin have been detected, we have yet to observe the production of α-MSH, as is seen in most other POMC-producing tissues. Such novel processing of POMC by cells of the immune system suggests that the expression of different POMC peptides with different stimuli may have important implications for regulation of immune responses.

5. RECEPTORS FOR NEUROENDOCRINE HORMONES ON MONONUCLEAR LEUKOCYTES

In addition to producing peptide hormones, cells of the immune system have also been shown to possess receptors for many neuroendocrine peptides. Johnson *et al.* (1982) demonstrated that mouse spleen mononuclear cells have both high- and low-affinity receptors for ACTH (K_d = 0.1 and 4.8 nm, respectively) with approximately 3000 high- and 50,000 low-affinity sites per cell. Interestingly, ACTH receptors on rat adrenal cells possessed similar high- and low-affinity binding sites (K_d = 0.25 and 10.0 nm, respectively) with 3000 high- and 30,000 low-affinity sites per cell (McIlhinny and Schulster, 1975). These initial studies showed that ACTH receptors on leukocytes and adrenal cells had similar binding characteristics, suggesting that ACTH receptors on these two cell popoulations may have structural similarities.

To address this possibility directly, the molecular characteristics of ACTH receptors have recently been explored in our laboratory using a new technology (Bost *et al.*, 1985). ACTH receptors were immunoaffinity purified from mouse Y-1 adrenal cells, and the molecular weight and subunit structure were determined using polyacrylamide gel electrophoresis. To summarize our results, the total molecular weight of the adrenal ACTH receptor was 225,000, and it was composed of four polypeptide chains of 83,000, 64,000, 52,000, and 22,000 daltons. The 83,000- and 52,000-dalton chains were shown to be disulfide linked, with the 83,000-dalton chain containing the ACTH binding site (K. L. Bost and J. E. Blalock, unpublished data). When ACTH receptors were isolated from mouse splenic or human peripheral blood mononuclear cells, similar if not identical molecular weights were found for the whole receptor and the four subunits. Thus, at a molecular level, ACTH receptors from adrenal and immune cells seem to be identical. Therefore, the neuroendocrine and immune systems are able to communicate bidirectionally by sharing common signal molecules (e.g., ACTH) and common receptors (e.g., ACTH receptors).

In addition to ACTH receptors, opiate binding sites on leukocytes have been demonstrated by a number of investigators. Wybran *et al.* (1979) published indirect evidence that human peripheral blood T cells possessed opiate receptors by demonstrating that morphine and methionine enkephalin inhibited and enhanced, respectively, T-cell rosetting of sheep red blood cells. This inhibition and enhancement could be reversed by the opiate antagonist naloxone but not by the inactive enantimere levomoramide. Johnson *et al.* (1982) demonstrated methionine enkephalin receptors on mouse mononuclear leukocytes more directly using competitive binding experiments with ^{3}H-labeled methionine enkephalin. These studies demonstrated a single high-affinity binding site of approximately 0.6 nm. Lopker *et al.* (1980) demonstrated specific binding of ^{3}H-dihydromorphine to granulocytes and monocytes with apparent K_ds of 10 nm and 8 nm, respectively, and 3000 and 4000 binding sites per cell, respectively. Hazum *et al.* (1979) demonstrated high- and low-affinity β-endorphin binding sites on cultured human lymphocytes. Interestingly, these receptors did not appear to be classical opiate receptors, since the binding was not affected by naloxone, morphine, enkephalins, or α-endorphin. Since α-endorphin did not inhibit the β-endorphin binding, it was suggested that the carboxyl end of β-endorphin (which is absent in α-endorphin) is essential for binding to this receptor. More recently, Mehrishi and Mills (1983) have demonstrated classical opioid receptors on human lymphocytes using radiolabeled morphine and naloxone. Thus, leukocytes have been shown to possess opiate receptors, and it appears that lymphocytes may have two different types of opiate receptors (classical and nonclassical endorphin receptors).

Leukocytes have also been shown to possess receptors for other neuroendocrine peptides. Human peripheral blood lymphocytes were shown to possess vasoactive intestinal peptide (VIP) receptors with a K_d of 0.47 nm and approximately 1700 binding sites per cell (Danek *et al.*, 1983). Ottaway *et al.* (1983) demonstrated similar VIP binding to human circulating mononuclear cells. Mouse T lymphocytes were also recently shown to have specific receptors for VIP (Ottaway and Greenberg, 1984). Beed *et al.* (1983) have shown that the VIP receptor on Molt 4 cells (a T-lymphoblastoid cell line) is coupled to the adenylate cyclase system. Substance P receptors have been identified on human T lymphocytes and on IM-9 lymphoblasts (Payan *et al.*, 1984a,b). Neurotensin receptors have been demonstrated on macrophages (Bar-Shavit *et al.*, 1982). Also, the presence of corticotropin-releasing factor and vasopressin receptors on mononuclear leukocytes has been suggested (Smith *et al.*, 1986). Thus, it seems that cells of the immune and nervous systems have many types of receptors in common.

6. EFFECTS OF NEUROENDOCRINE HORMONES ON IMMUNE RESPONSES

Since it is clear that cells functioning in immune responses produce peptide hormones and have their specific receptors, these receptors must have a function. Therefore, a major question to be addressed is: What role do the peptide hormones play in regulating immune responses? Although recent experiments have only begun to shed some light on this question, these studies have produced some exciting and intriguing results. For example, although the classical role of ACTH is to induce adrenal steroidogenesis, Johnson *et al.* (1982) found that ACTH is a potent inhibitor of antibody production. In an *in vitro* plaque-forming cell assay, $ACTH_{1-39}$ was shown to inhibit antibody responses to both T-lymphocyte-dependent (sheep red blood cells) and T-independent (dinitrophenol–Ficoll) antigens. Inhibition of the T-dependent antibody response required only one-fourth the amount of ACTH necessary to achieve the same level of inhibition of the T-independent response. In addition, although $ACTH_{1-39}$ was suppressive, $ACTH_{1-24}$ was not (Johnson *et al.*, 1984). This is in contrast to the steroidogenic activity, in which $ACTH_{1-39}$ and $ACTH_{1-24}$ are equally active (McIlhinney and Schulster, 1975). $ACTH_{1-39}$ was also found to suppress *in vitro* production of interferon-γ by lymphocytes, whereas $ACTH_{1-24}$ and $ACTH_{1-13}$ did not (Johnson *et al.*, 1984). Again, there seems to be little correlation between ACTH's steroidogenic and immunomodulatory activities, since $ACTH_{1-24}$ is not immunosuppressive *in vitro* but does induce steroidogenesis in adrenal cells.

Whereas ACTH was shown to decrease antibody responses, thyrotropin enhanced the *in vitro* antibody response to sheep red blood cells. This effect required the addition of the hormone during the early phase of the response (1–2 days) (Blalock *et al.*, 1985). Interestingly, two different hormones (TSH and ACTH) that are produced by mononuclear leukocytes in response to two different stimuli (enterotoxin A and virus infection, respectively) have opposite effects on antibody production.

Endogenous opiates have also been evaluated for immunoregulatory activity and have been shown to affect antibody production, lymphocyte proliferation, and natural killer cell activity (Wybran, 1985). Gilman *et al.* (1982) showed that in response to the T-cell mitogens concanavalin A and phytohemagglutinin, β-endorphin enhanced rat splenic lymphocyte proliferation in a dose-dependent manner. However, this inhibition was not blocked by the opiate antagonist naloxone, suggesting that β-endorphin does not act through classical opiate receptors in this system (Simon and Miller, 1981). In contrast to the effects of β-endorphin, α-endorphin and [D-Ala[2]-Met[5]]enkephalin did not alter mitogen responsiveness (Gilman *et al.*, 1982). McCain *et al.* (1982) found that human T-cell mitogenesis was suppressed by β-endorphin. Although these results seem to conflict with those of Gilman *et al.* (1982), the differences may be explained by the different source of lymphocytes or by the higher concentrations of β-endorphin used by McCain *et al.* (1982). Opiates frequently display opposite activities between high and low concentrations (Faith *et al.*, 1986). Finally, methionine and leucine enkephalins were shown to enhance phytohemagglutinin response of human lymphocytes (Plotnikoff and Miller, 1983). Thus, opiates have the ability to modulate lymphocyte proliferation in response to nonspecific mitogen stimulation.

Although mitogen responsiveness may suggest an immunomodulatory role for the opiates, more direct methods of assessing the effect of opiates on immune responses have been used. Johnson *et al.* (1982) demonstrated that the *in vitro* antibody response of mouse splenic leukocytes to sheep red blood cells was highly suppressed (80% inhibition) by α-endorphin. Methionine and leucine enkephalin also inhibited this antibody response (approximately 60% suppression), whereas γ- and β-endorphins were not suppressive. The extra amino acids on the carboxyl end of γ- and β-endorphin that are not present on α-endorphin may be the reason for the differing effects of these endorphins on *in vitro* antibody production. It should also be noted that naloxone and β-endorphin could block the inhibitory effects of α-endorphin. This result suggests that opiatelike receptors were involved and that merely binding to opiate receptors (which β-endorphin does) was not sufficient for inhibition of antibody production.

Opiates have also been shown to regulate natural killer (NK) cell activity. Mathews *et al.* (1983) and Kay *et al.* (1984) found that β-endorphin enhanced NK activity 20 to 55% in a dose-dependent fashion. The augmented NK activity was blocked by naloxone. Interestingly, α-endorphin, leucine enkephalin, or morphine had very little effect on NK activity, whereas γ-endorphin demonstrated some NK activity (Mathews *et al.*, 1983; Kay *et al.*, 1984). More recent studies (Wybran, 1985) have shown that morphine and leucine and methionine enkephalins increased NK activity (approximately 30%) and that this enhancement could be blocked by naloxone. Thus, these results demonstrate that NK activity can be enhanced by a variety of endogenous opiates.

One final effect of opiates on immune cell function has been described by Van Epps and Saland (1984). They have observed that mononuclear leukocyte chemotaxis is stimulated by both β-endorphin and methionine enkephalin. Both opiates were active at very low concentrations (10^{-14} M), and the enhanced cellular migration was blocked by naloxone.

Recent studies using a variety of peptides derived from the nervous system have shown that these peptides can also have immunomodulatory effects. Payan *et al.* (1983) found that human peripheral blood T lymphocytes were stimulated to proliferate by substance P. Substance P also enhanced the mitogenic response of these cells to phytohemaglutinin. Ottaway and Greenberg (1984) found that vasoactive intestinal peptide could inhibit lymphocyte mitogenesis to concanavalin A and phytohemagglutinin, but the response to lipopolysaccharide was not affected. Somatostatin was shown to inhibit the proliferation of Molt-4 lymphoblasts and phytohemagglutinin-stimulated T lymphocytes (Payan *et al.*, 1984b). It is evident from these studies that many peptides functioning in the nervous system also seem to have effects on immune cells. However, a major question that needs to be addressed is the importance of these peptides in regulating immune responses *in vivo*.

7. EFFECTS OF MONONUCLEAR-LEUKOCYTE-DERIVED ACTH AND ENDORPHINS ON NEUROENDOCRINE FUNCTION

Whereas evidence has accumulated to demonstrate that the nervous system can modulate immune responses, the ability of the immune system to modulate nervous system responses has not been as well studied. Our laboratory has suggested the existence of a lymphoid–adrenal axis (Blalock and Smith, 1985), which is based on the observation that lymphocytes can produce ACTH that has steroidogenic activity. Specifically, when hypophysectomized mice were infected with Newcastle disease virus (which induces lymphocyte ACTH *in vitro*), a time-dependent increase in corticosterone production was observed. The corticosterone production was inhibited by dexamethasone (Smith *et al.*, 1982). Thus, it is likely in these experiments that the increased corticosteroid levels were evoked by lymphocyte-derived ACTH. Studies by other investigators also support the notion that lymphocyte-derived ACTH can result in a steroidogenic response. Leukocytes from chickens injected with *Salmonella pullorum* antigen were shown to synthesize an ACTH-like substance (Siegel *et al.*, 1985). The authors went on to show that this leukocyte-derived substance could cause a corticosteroid response when placed on isolated adrenal cells.

Support for a lymphoid adrenal axis has also come from human studies. For example, bacterial lipopolysaccharide (Piromen®) induced cortisol responses in seven out of eight patients who underwent pituitary stalk sectioning (Van Wyk *et al.*, 1960). More recently, a patient was observed to have clinical and laboratory features found in ectopic ACTH syndrome. However, no ACTH-producing tumor could be found, and the absence of a gradient in ACTH concentration between the inferior petrosal sinuses and the periphery argued against pituitary overproduction. Thus, as a last resort, a bilateral adrenalectomy was performed to correct the hypercortisolism. Six months later, a pseudotumor containing only fat cells and inflammatory tissue was observed in the patient. On removal of the inflammatory tissue, plasma ACTH levels immediately returned to normal. It was found that leukocytes within the inflammatory tissue stained positively for ACTH (Dupont *et al.*, 1984). Taken together, these studies leave little doubt that cells of the immune system have the ability to produce ACTH and to affect adrenal function directly.

Another example in which the immune system can influence nervous system responses may be that of gram-negative sepsis and endotoxic shock. The production of endorphins has been linked to some of the pathophysiological effects associated with these trauma, since naloxone could alleviate endotoxin-induced hypotension and hyperthermia (Reynolds *et al.*, 1980). Carr *et al.* (1982) suggested that following endotoxin administration, endorphins may have been produced from the immune system. Furthermore, our laboratory has found that lipopolysaccharide-treated mononuclear leukocytes make molecules that are antigenically, structurally, and functionally related to ACTH and γ-endorphin (Harbour-McMenamin *et al.*, 1985). Since lipopolysaccharide can induce leukocyte-derived endorphin, we suggest that the most likely source of opiates during gram-negative sepsis and endotoxin shock may be cells of the immune system. Support for this idea comes from the observation that lymphocyte depletion can abrogate a number of endotoxin-induced cardiopulmonary changes (Bohs *et al.*, 1979).

8. CONCLUSIONS

From the studies described above, there can be little doubt that cells of the immune system can synthesize biologically active neuroendocrine peptide hormones. Furthermore, immunocytes also possess receptors for many of these peptides, suggesting that these hormones can influence immune responsiveness. A major question to be addressed is why do cells of the immune system synthesize and respond to neuroendocrine hormones? We have postulated that the immune system can serve as a sensory organ for external stimuli that cannot be detected by the nervous system (Blalock, 1984a,b). For example, whereas the nervous system detects classical sensory stimuli (physical, visual, etc.), the immune system recognizes stimuli such as viruses, bacteria, and tumors. Thus, production of neuroendocrine hormones by immunocytes in response to these noncognitive stimuli could induce physiological changes (e.g., lymphocyte-derived ACTH inducing steroidogenesis). To complete the immune–adrenal axis, these neuroendocrine hormones could also influence immune function (e.g., ACTH decreasing *in vitro* antibody responses). Since mononuclear leukocytes can produce hormones and also have receptors for the same hormones (e.g., ACTH and ACTH receptors), it is possible that these immunocytes may also influence their own function in an autocrinelike fashion (Heldin and Westermark, 1984).

ACKNOWLEDGMENTS. This work was supported in part by NIH Grant R01 AM 33839-O1A1, the Office of Naval Research (N0014-K-0486), and Triton Biosciences, Inc.

REFERENCES

Ader, R., ed., 1981, *Psychoneuroimmunology*, Academic Press, New York.

Bar-Shavit, Z., Terry, S., Blumberg, S., and Goldman, R., 1982, Neurotensin–macrophage interaction: Specific binding and augmentation of phagocytosis, *Neuropeptides* **2:**325–335.

Beed, E. A., O'Dorisio, S., O'Dorisio, T. M., and Gaginella, T. S., 1983, Demonstration of a functional receptor for vasoactive intestinal polypeptide on Molt 4b T lymphoblasts, *Regul. Peptides* **6:**1–12.

Blalock, J. E., 1984a, Relationships between neuroendocrine hormones and lymphokines, *Lymphokines* **9:**1–13.

Blalock, J. E., 1984b, The immune system as a sensory organ, *J. Immunol.* **132:**1067–1070.

Blalock, J. E., and Smith, E. M., 1985, A complete regulatory loop between the immune and neuroendocrine systems, *Fed. Proc.* **44:**108–111.

Blalock, J. E., Johnson, H. M., Smith, E. M., and Torres, B. A., 1985, Enhancement of the *in vitro* antibody response by thyrotropin, *Biochem. Biophys. Res. Commun.* **125:**30–34.

Bohs, C. T., Fish, J. C., Miller, T. H., and Traber, D. L., 1979, Pulmonary vascular response to endotoxin in normal and lymphocyte depleted sheep, *Circ. Shock* **6:**13–21.

Bost, K. L., Smith, E. M., and Blalock, J. E., 1985, Similarity between the corticotropin (ACTH) receptor and a peptide encoded by an RNA that is complementary to ACTH mRNA, *Proc. Natl. Acad. Sci. U.S.A.* **82:**1372–1375.

Brooks, W. H., Cross, R. J., Roszman, T. L., and Markesbery, W. R., 1982, Neuroimmunomodulation: Neural anatomical basis for impairment and facilitation, *Ann. Neurol.* **12:**56–61.

Carr, D. B., Bergland, R., Hamilton, A., Blume, H., Kasting, N., Arnold, M., Martin, M. B., and Rosenblatt, M., 1982, Endotoxin stimulated opioid peptide secretion: Two secretory pools and feedback control *in vivo*, *Science* **217:**845–848.

Cross, R. J., Markesbery, W. R., Brooks, W. H., and Roszman, T. L., 1980, Hypothalamic–immune interactions. I. The acute effect of anterior hypothalamic lesions on the immune response, *Brain Res.* **196:**79–87.

Cutz, E., Chan, W., Track, N., Goth, A., and Said, S., 1978, Release of vasoactive intestinal peptide in mast cells by histamine liberators, *Nature* **275:**661–662.

Danek, A., O'Dorisio, M. S., O'Dorisio, T. M., and George, J. M., 1983, Specific binding sites for vasoactive intestinal polypeptide on nonadherent peripheral blood lymphocytes, *J. Immunol.* **131:**1173–1177.

Dupont, A. G., Somers, G., Van Steirteghen, A. C., Warson, F., and Vanhaelst, L., 1984, Ectopic adrenocorticotropin production: Disappearance after removal of inflammatory tissue, *J. Clin. Endocrinol. Metab.* **58:**654–658.

Faith, R. E., Plotnikoff, N. P., and Murgo, A. J., 1986, Effects of opiates and enkephaline-endorphins on immune functions, in: *Proceedings, National Institute of Drug Abuse Technical Meeting on Mechanisms of Tolerance and Dependence* (in press).

Giachetti, A., Goth, A., and Said, S. I., 1978, Vasoactive intestinal polypeptide (VIP) in rabbit platelets and rat mast cells, *Fed. Proc.* **37:**657.

Gilman, S. C., Schwartz, J. M., Milner, R. J., Bloom, F. E., and Feldman, J. D., 1982, β-Endorphin enhances lymphocyte proliferative responses, *Proc. Natl. Acad. Sci. U.S.A.* **79:**4226–4230.

Goldstein, M. M., 1978, Antibody-forming cells of the rat spleen after injury to the midbrain, *Bull. Exp. Biol. Med.* **82:**183–187.

Harbour-McMenamin, D. V., Smith, E. M., and Blalock, J. E., 1985, Bacterial lipopolysaccharide induction of leukocyte derived ACTH and endorphins, *Infect. Immun.* **48:**813–817.

Hazum, E., Chang, K., and Cuatrecasas, P., 1979, Specific nonopiate receptors for β-endorphin, *Science* **205:**1033–1035.

Heldin, C. H., and Westermark, B., 1984, Growth factors: Mechanism of action and relation to oncogenes, *Cell* **37:**9–20.

Holaday, J. W., and Faden, A. I., 1978, Naloxone reversal of endotoxin hypotension suggests role of endorphin in shock, *Nature* **275:**450–451.

Holden, C., 1980, Behavioral medicine: An emergent field, *Science* **209:**479–481.

Jankovic, B. D., and Isakovic, K., 1973, Neuroendocrine correlates of immune response, *Int. Arch. Allergy Appl. Immunol.* **45:**360–372.

Johnson, H. M., Smith, E. M., Torres, B. A., and Blalock, J. E., 1982, Neuroendocrine hormone regulation of *in vitro* antibody production, *Proc. Natl. Acad. Sci. U.S.A.* **79:**4171–4174.

Johnson, H. M., Torres, B. A., Smith, E. M., Dion, L. D., and Blalock, J. E., 1984, Regulation of lymphokine (γ-interferon) production by corticotropin, *J. Immunol.* **132:**246–250.

Kay, N., Allen, J., and Morley, J. E., 1984, Endorphins stimulate normal human peripheral blood lymphocyte natural killer activity, *Life Sci.* **35:**53–59.

Krieger, D. T., 1983, Brain peptides: What, where and why? *Science* **222:**975–985.

Lolait, S. J., Lim, A. T. W., Toh, B. H., and Funder, J. W., 1984, Immunoreactive β-endorphin in a subpopulation of mouse spleen macrophages, *J. Clin. Invest.* **73:**277–280.

Lopker, A., Abood, L. G., Hoss, W., and Lionetti, F. J., 1980, Stereoselective muscarinic acetylcholine and opiate receptors in human phagocytic leukocytes, *Biochem. Pharmacol.* **29:**1361–1365.

Lygren, I., Revhaug, A., Burhol, P. G., Giercksky, K. E., and Jenssen, T. G., 1984, Vasoactive intestinal peptide and somatostatin in leukocytes, *Scand. J. Clin. Lab. Invest.* **44:**347–351.

Macris, N. T., Schiavi, R. C., Camerino, M. S., and Stein, M., 1970, Effects of hypothalamic lesions on immune processes in the guinea pig. *Am. J. Physiol.* **219:**1205–1209.

Mathews, P. M., Droelich, C. J., Sibbitt, W. L., Jr., and Bankhurst, A. D., 1983, Enhancement of natural cytotoxicity by β-endorphin, *J. Immunol.* **130:**1658–1662.

McCain, H. W., Lamster, I. B., Bozzone, J. M., and Grbic, J. T., 1982, β-endorphin modulates human immune activity via non-opiate receptor mechanisms, *Life Sci.* **31:**1619–1624.

McIlhinney, R. A. J., and Schulster, D., 1975, Studies on the binding of ^{125}I-labelled corticotropin to isolated rat adrenocortical cells, *J. Endocrinol.* **64:**175–184.

Mehrishi, J. N., and Mills, I. H., 1983, Opiate receptors on lymphocytes and platelets in man, *Clin. Immunol. Immunopathol.* **27**:240–249.

Nakanishi, S., Inoue, A., Kita, T., Nakamura, M., Chang, A. C. Y., Cohen, S. N., and Numa, S., 1979, Nucleotide sequence of cloned cDNA for bovine corticotropin–β-lipotropin precursor, *Nature* **278**:423–427.

O'Dorisio, M. S., O'Dorisio, T. M., Cataland, S., and Balcerzak, S. P., 1980, Vasoactive intestinal peptide as a biochemical marker for polymorphonuclear leukocytes, *J. Lab. Clin. Med.* **96**:666–670.

Ottaway, C. A., and Greenburg, G. R., 1984, Interaction of vasoactive intestinal peptide with mouse lymphocytes, *J. Immunol.* **132**:417–423.

Ottaway, C. A., Bernaerts, C., Chan, B., and Greenberg, G. R., 1983, Specific binding of vasoactive intestinal peptide to human circulating mononuclear cells, *Can. J. Physiol. Pharmacol.* **61**:664–671.

Payan, D. G., Brewster, D. R., and Goetzl, E. J., 1983, Specific stimulation of human T lymphocytes by substance P, *J. Immunol.* **131**:1613–1615.

Payan, D. G., Brewster, D. R., and Goetzl, E. J., 1984a, Stereospecific receptors for substance P on cultured human IM-9 lymphoblasts, *J. Immunol.* **133**:3260–3265.

Payan, D. G., Hess, C. A., and Goetzl, E. J., 1984b, Inhibition of somatostatin of the proliferation of T-lymphocytes and Molt-4 lymphoblasts, *Cell. Immunol.* **84**:433–438.

Plotnikoff, N., and Miller, G. C., 1983, Enkephalins as immunomodulators, *Int. J. Immunopharmacol.* **5**:437–441.

Reynolds, D. G., Guill, N. V., Vargish, T., Hechner, R. B., Fader, A. I., and Holaday, J. W., 1980, Blockade of opiate receptors with naloxone improves survival and cardiac performance in canine endotoxic shock, *Circ. Shock* **7**:39–48.

Riley, V., 1981, Psychoneuroendocrine influences on immunocompetence and neoplasia, *Science* **212**:1100–1109.

Siegel, H. S., Gould, N. R., and Lattimer, J. W., 1985, Splenic leukocytes from chickens injected with *Salmonella pullorum* antigen stimulate production of corticosteroids by isolated adrenal cells, *Soc. Exp. Biol. Med.* **178**:523–530.

Simon, E. J., and Miller, J. M., 1981, Opioid peptides and opiate receptors, in: *Basic Neurochemistry* (G. J. Seigel, R. W. Albers, B. W. Agaranoff, and R. Katzman, eds.), Little, Brown, Boston, p. 255.

Smith, E. M., and Blalock, J. E., 1981, Human lymphocyte production of ACTH and endorphin-like substances: Association with leukocyte interferon, *Proc. Natl. Acad. Sci. U.S.A.* **78**:7530–7534.

Smith, E. M., Meyer, W. J., and Blalock, J. E., 1982, Virus-induced increases in corticosterone in hypophysectomized mice: A possible lymphoid–adrenal axis, *Science* **218**:1311–1312.

Smith, E. M., Phan, M., Coppenhaver, D., Kruger, T. E., and Blalock, J. E., 1983, Human lymphocyte production of immunoreactive thyrotropin, *Proc. Natl. Acad. Sci. U.S.A.* **80**:6010–6013.

Smith, E. M., Morrill, A. C., Meyer, W. J., and Blalock, J. E., 1986, Corticotropin releasing factor induction of leukocyte derived immunoreactive ACTH and endorphins, *Nature* (in press).

Stein, M., 1985, Bereavement, depression, stress and immunity, in: *Neural Modulation of Immunity* (R. Guillemin, M. Cohn, and T. Melnechuk, eds.), Raven Press, New York, pp. 29–41.

Tyrey, L., and Nalbandov, A. V., 1972, Influence of anterior hypothalamic lesions on circulating antibody titers in the rat, *Am. J. Physiol.* **222**:179–185.

Van Epps, D., and Saland, L., 1984, β-Endorphin and met-enkephalin stimulate human peripheral blood mononuclear cell chemotaxis, *J. Immunol.* **132**:3046–3053.

Van Wyk, J. J., Dugger, G. S., Newsome, J. F., and Thomas, P. Z., 1960, The effect of pituitary stalk section in the adrenal function of women with cancer of the breast, *J. Clin. Endocrinol. Metab.* **20**:157–172.

Wybran, J., 1985, Enkephalins and endorphins as modifiers of the immune system, *Fed. Proc.* **44**:92–94.

Wybran, J., Appelbroom, T., Famaey, J.-P., and Govaerts, A., 1979, Suggestive evidence for receptors for morphine and methionine-enkephalin on normal blood T lymphocytes, *J. Immunol.* **123**:1068–1070.

4

Prohormone-Converting Enzymes

D. C. PARISH

It is well established that peptide hormones are synthesized as larger peptide precursors. Initially this was demonstrated by pulse-chase studies, *in vitro* synthesis, and immunoprecipitation (Docherty and Steiner, 1982; Loh and Gainer, 1983, for reviews), but molecular biology has now made the sequences of a large number of hormone precursors available, for example, proopiomelanocortin (Nakanishi *et al.*, 1979), proenkephalin (Noda *et al.*, 1982), provasopressin (Land *et al.*, 1982), and prooxytocin (Land *et al.*, 1983). These sequences show certain common features (Fig. 1); in particular, the hormone sequences are bracketed by pairs of basic amino acid residues such as Lys-Arg, Lys-Lys, or Arg-Arg. The precursor forms of these hormones generally lack biological activity, so that limited proteolysis at paired basic residues is necessary to generate the hormones, followed by exopeptidase activity to remove the Lys and Arg terminal residues, which are not found in the mature peptide hormones. Carboxypeptidases (Fricker and Snyder, 1983; Hook and Loh, 1984) and aminopeptidases (Gainer *et al.*, 1984) have been described that are capable of such modifications. However, this discussion is confined to the initial proteolysis by endopeptidases. Cleavage may be followed by other modifications such as acetylation and amidation in order to produce biological potency (Loh and Gainer, 1983, for review).

Identification and purification of an enzyme or family of enzymes responsible for this limited proteolysis has been the target of a number of laboratories in this field, but this has been no easy task, in part because of the large number of proteases in the cell. Several criteria have been suggested (Docherty and Steiner, 1982) for the identification of physiologically relevant proteases. These criteria include (1) that the enzyme should cleave the prohormone *in vitro* to generate the products seen *in vivo*, (2) that the enzyme should be located in the subcellular compartment where cleavage takes place, and (3) that the enzyme should function under the conditions found in that subcellular compartment.

D. C. PARISH • Laboratory of Neurochemistry and Neuroimmunology, National Institute of Child Health and Human Development, National Institutes of Health, Bethesda, Maryland 20205.

The subcellular organization of posttranslational precursor processing is probably best understood in the neural lobe of the pituitary. The magnocellular cells of the neural lobe have their cell bodies in the synaptic and paraventricular nuclei of the hypothalamus, and their axons course down through the median eminence to the neural lobe. These cells synthesize the hormones oxytocin and vasopressin, together with binding proteins called neurophysins, as larger-molecular-weight precursors (Fig. 1). Pulse-labeling studies showed that proteolytic cleavage of those precursors occurred in the median eminence during axonal transport (Gainer *et al.*, 1977), when the precursor is contained in secretory vesicles. Therefore, the paired basic residue-cleaving enzyme must also be localized in the neurosecretory vesicles. There is also evidence that precursor processing occurs within secretory vesicles for proinsulin (Fletcher *et al.*, 1980) and proopiomelanocortin (Loh and Gritch, 1981).

Therefore, isolation of secretory vesicles by density gradient centrification offers a powerful and rational purification step for isolating the processing enzymes (Loh *et al.*, 1984). This paradigm has been used in examining putative prohormone-converting enzymes from a number of tissues (Table I), including adrenal medulla

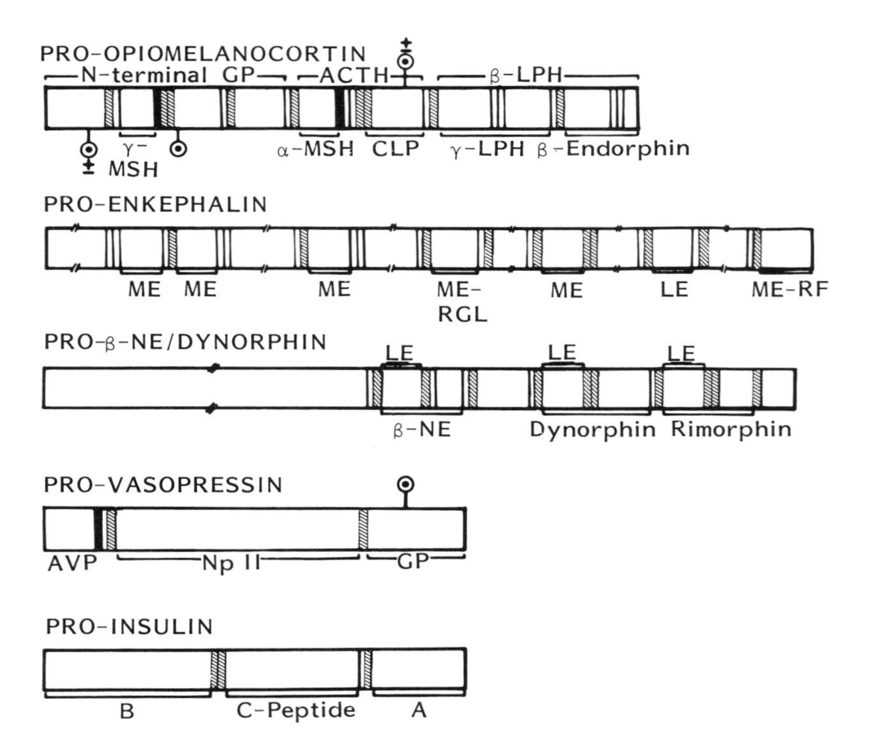

FIGURE 1. Prohormone sequences determined from the cDNA sequence by molecular biology. □ = Lys, ▧ = Arg, ■ = Gly.

TABLE I. Putative Proprotein-Converting Enzymes in Different Tissues

Source	Substrate	Specificity	Subcellular localization	Reference
Bovine adrenal medulla	Peptide E	Lys-Arg	Chromaffin granules	Lindberg *et al*. (1984)
	Peptide F	Arg-Arg		
		Lys-Lys		
Bovine adrenal medulla	Tosyl arginine methyl ester		Chromaffin granules	Evangelista *et al*. (1982)
Bovine adrenal medulla	Peptide E	Lys-Arg	Chromaffin granules	Troy and Musacchio (1982)
		Arg-Arg		
Bovine adrenal medulla	Leu-Enkephalin-LysArgPheAlaNH_2	Lys-Arg	Chromaffin granules	Mizuno and Matsuo (1985); Mizuno *et al*. (1985)
Yeast	Leu-Enkephalin-related synthetic peptides	Lys-Arg		Mizuno and Matsuo (1984)
		Arg-Arg	Unknown	
		Arg-Lys		
Yeast	BOCGlnArgArgMCA	Arg-Arg	Unknown	Julius *et al*. (1984)
	BOCGluLysLysMCA	Lys-Lys		
Anglerfish pancreatic islets	Proinsulin	Lys-Arg	Secretory vesicles and microsomes	Fletcher *et al*. (1980, 1981); Noe *et al*. (1984)
	Proglucagon	Arg-Arg		
	Prosomatostatin			
Rat pancreas	Proinsulin	Arg-Arg	Secretory vesicles	Kemmler *et al*. (1973)
Rat pituitary intermediate lobe	Proopiomelanocortin	Lys-Arg	Secretory vesicles	Chang and Loh (1984)
Bovine pituitary intermediate lobe	Proopiomelanocortin	Lys-Arg	Secretory vesicles	Loh *et al*. (1985)
Rat neurointermediate lobe	N-AcetylLysArgTyrAsnLeuNH_2	Lys-Arg	Unknown	Pelaprat *et al*. (1984)
Porcine neurointermediate lobe	$ACTH_{11-24}$	Lys-Lys	Unknown	Cromlish *et al*. (1985)
		Lys-Arg		
Porcine pituitary	LysAspLysArgTyrGly	Lys-Arg	Secretory granules	Smyth *et al*. (1977)
Bovine pituitary neural lobe	Proopiomelanocortin	Lys-Arg	Secretory vesicles	Chang *et al*. (1982)
Rat anterior pituitary	Proopiomelanocortin	Lys-Arg	Secretory vesicles	Chang and Loh (1983)
Bovine parathyroid	Proparathyroid hormone	Lys-Arg	Microsomes	MacGregor *et al*. (1976, 1978)
Rat liver	Proalbumin	Arg-Arg	Large granules	Quinn and Judah (1978)

(Lindberg *et al.*, 1984; Mizuno and Matsuo, 1985), pituitary intermediate (Loh *et al.*, 1985) and neural lobe (Chang *et al.*, 1982), and pancreas (Fletcher *et al.*, 1980, 1981; Noe *et al.*, 1984). Since cleavage occurs within the secretory vesicle, candidates for the prohormone-converting enzyme should be active *in vitro* under conditions that mimic those in the vesicle. Some of these conditions are still unclear; however, it is known that these vesicles have an acidic internal pH (Loh *et al.*, 1984; Russell, 1984; Gainer *et al.*, 1985).

A large number of paired basic residue-specific putative processing enzymes have been described that meet some or all of these criteria (Table I; Loh and Gainer, 1983). In the adrenal chromaffin granules, one such enzyme has been partially purified by affinity chromatography on a soybean trypsin inhibitor column (Mizuno and Matsuo, 1985; Mizuno *et al.*, 1985). This enzyme is a serine protease with a basic pH optimum and generates leucine enkephalin from leucine enkephalin-LysArgPheAla-amide, a peptide sequence corresponding to a sequence within proenkaphalin. This cleavage occurs exclusively between the Lys and Arg residues; however, other workers (Lindberg *et al.*, 1984) have described a serine protease prepared from adrenal chromaffin granules by the same affinity chromatography method that can cleave peptide E and peptide F (intermediates from proenkephalin) between or to the carboxy side of the paired basic residues. An acid thiol protease has also been described in chromaffin granules, measured by its ability to cleave tosyl arginine methyl ester (Evangelista *et al.*, 1982) or peptide E (Troy and Musacchio, 1982).

Yeast synthesize a precursor to α mating factor that is cleaved at paired basic residues (Kurjan and Hershowitz, 1982). An enzyme activity has been prepared from yeast (Mizuno and Matsuo, 1984) that is similar to the basic serine protease isolated from chromaffin granules. This enzyme also binds to soybean trypsin inhibitor affinity columns and generates leucine enkephalin from synthetic precursor sequences. However, another enzyme has also been suggested as pro-α factor-converting enzyme based on a different paradigm (Julius *et al.*, 1984). Yeast mutants were selected that were deficient in the processing of pro-α factor, and from mapping studies with the wild type, a putative structural gene for the pro-α mating factor-converting enzyme was isolated. Introduction of this gene into the mutants restored the biological activity associated with processing of pro-α factor, suggesting that this gene coded for the appropriate processing enzyme. The gene appears to code for an endopeptidase that is inhibited by Zn^{2+} and sulfhydryl group reagents but not by serine protease inhibitors. This enzyme activity was measured by the cleavage of paired-basic-residue-containing synthetic peptides. The use of a mutant incapable of processing prohormones was proposed as one of the criteria necessary to identify a prohormone-processing enzyme (Docherty and Steiner, 1982), and this is the first such demonstration.

An enzyme activity has been described in anglerfish pancreatic islet cells (Fletcher *et al.*, 1980; Noe *et al.*, 1984) that cleaves proinsulin, proglucagon, and prosomatostatin to form insulin, glucagon, and somatostatin. This enzyme is an acid thiol protease that is found in both vesicles and microsomes, although the converting activity for prosomatostatin I (but not prosomatostatin II) was confined to the secretory granules, suggesting that this activity is the product of more than one enzyme. A serine protease has been reported in rat pancreatic islet granules that cleaves proinsulin to insulin (Kemmler *et al.*, 1973).

Proopiomelanocortin (POMC) is the precursor in the intermediate lobe of the pituitary and contains the sequences of ACTH, β-endorphin, α-MSH,β-LPH, and a glycopeptide. An enzyme activity has not been described in rat (Chang and Loh, 1984) and bovine (Loh *et al.*, 1985) secretory vesicles that cleaves labeled POMC at the paired basic residues to yield 21K and 23K ACTH, 13K ACTH, 4.5K ACTH, β-LPH, β-endorphin, and a 16K glycopeptide, products that are seen by pulse labeling *in situ*. The enzyme activity in bovine intermediate lobe vesicles was purified to apparent homogeneity (as judged by SDS gel electrophoresis) by gel filtration on Sephadex G-75 followed by affinity chromatography on concanavalin A-Sepharose and preparative polyacrylamide gel electrophoresis. The enzyme was characterized as a glycoprotein with an apparent molecular weight of 10,000 and an apparent isoelectric point between 3.5 and 4.0. It had a pH optimum between 4 and 5 and was inhibited by pepstatin A. Cleavages made by this enzyme occurred both on the carboxy side of the Arg and in between the Lys-Arg pairs of POMC. The enzyme was also capable of cleaving [^{125}I]proinsulin to products that comigrated with the insulin A and B chains on SDS polyacrylamide gels. In the intermediate lobe *in vivo,* ACTH is further processed to produce α-MSH; however, this enzyme did not convert ACTH *in vitro,* suggesting that there is a second enzyme in the intermediate lobe responsible for cleaving ACTH.

This is the first putative prohormone-converting enzyme that has been purified to apparent homogeneity. Other proteases have been described in the pituitary that are potential candidates for an intermediate lobe prohormone-converting enzyme. Four thiol proteases have been described in rat neurointermediate lobe that could cleave model peptides such as N-acetyl-Lys-Arg-Tyr-Asn-Leu-amide (Pelaprat *et al.*, 1984). However, this activity did not seem to be localized in the secretory granule fraction. A serine protease activity was demonstrated in porcine neurointermediate lobe that cleaved $ACTH_{11-24}$ at paired basic residues (Cromlish *et al.*, 1985), although a subcellular localization was not attempted. Another activity has been described in whole pituitary capable of cleaving synthetic hexapeptides containing Lys-Arg, with a basic pH optimum (Smyth *et al.*, 1977).

In the neural lobe of the pituitary, the hormone precursors are prooxytocin and provasopressin. Each of these precursors contains a specific neurophysin that is associated with only one of the peptide hormones, and monoclonal antibodies are now available for these different neurophysins (Ben-Barak *et al.*, 1984). Pulse labeling and immunoprecipitation of hypothalamic extracts demonstrate the existence of a number of intermediates between the precursor and neurophysins that can be separated by isoelectric focusing (Fig. 2). A putative neural lobe provasopressin-processing enzyme should convert the prohormone into these intermediates seen *in vivo*. An enzyme activity was described in this laboratory from bovine neural lobe secretory granules (Chang *et al.*, 1982) that converted POMC to the

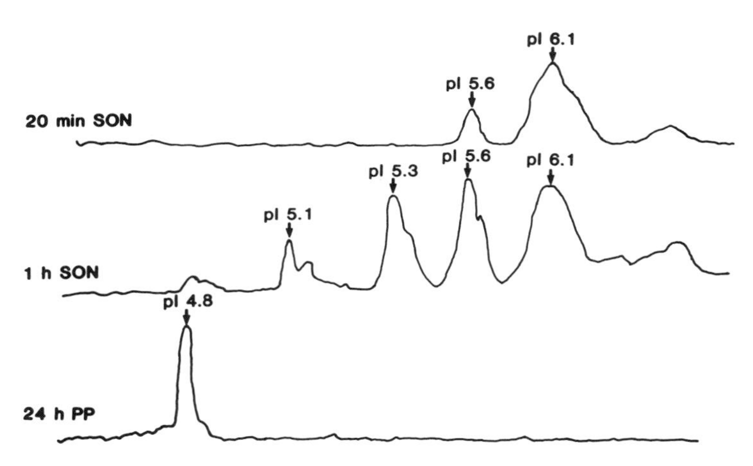

FIGURE 2. Rat provasopressin and intermediates found in the supraoptic nucleus (SON) and pituitary after injection of [^{35}S]cysteine into the nucleus. The SON was taken at 20 min or 1 hr after injection, and the posterior pituitary after 24 hr. The proteins were immunoprecipitated with specific monoclonal antibodies and separated on slab isoelectric focusing. Autoradiographs of the gels were densitometrically scanned.

products seen in the pituitary *in vivo*. Although heterologous, this assay demonstrated that the enzyme was acting at paired basic residues. It was inhibited by pepstatin A and had an acid pH optimum. This prohormone-converting enzyme has now been further purified, to apparent homogeneity as judged by SDS gel electrophoresis, by concanavalin A and preparative polyacrylamide gel electrophoresis. The enzyme is a glycoprotein with an apparent molecular weight of 70,000. These characteristics are very similar to those of the intermediate lobe prohormone-converting enzyme already described, and this may be a closely related or possibly identical enzyme.

In order to test this enzyme activity against the precursor present in the neural lobe, [^{35}S]provasopressin was prepared by pulse labeling rat supraoptic nuclei and immunoprecipitation with vasopressin-neurophysin-specific monoclonal antibodies. This procedure produces a mixture of provasopressin and intermediates that contain vasopressin-associated neurophysin. These can be separated by isoelectric focusing electrophoresis, in which provasopressin has a pI of 6.1 (Fig. 3). This substrate was then incubated with concanavalin A affinity-purified neural lobe prohormone-converting enzyme. Incubation completely cleaved the prohormone and increased the proportion of 5.1 and 5.3 pI peak seen on the IEF profile (Fig. 3). The intermediates from the substrate preparation act as internal markers, showing that the products of incubation comigrated with the naturally occurring intermediates. This incubation also generated a TCA-soluble vasopressinlike labeled peptide. Studies are in progress with more highly purified forms of provasopressin and prooxytocin that should yield unambiguous cleavage patterns, but these data strongly suggest that the putative neural lobe converting enzyme is capable of cleaving the appropriate prohormone to produce the peptides seen *in vivo*.

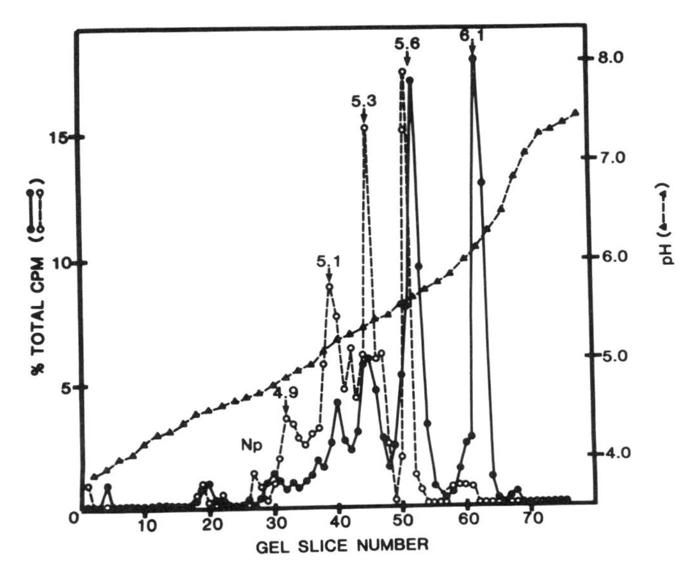

FIGURE 3. [^{35}S]Provasopressin and its intermediates were prepared as in Fig. 2 and separated on the tube IEF gels before (○—○) and after (●—●) incubation with concanavalin A-purified neural lobe prohormone-converting enzyme. The gels were sliced, and radioactivity in the slices was measured by liquid scintillation. The pH gradient was determined by eluting ampholytes from another gel run in parallel with these gels.

The criteria, discussed earlier in this chapter, for identifying a possible prohormone-converting enzyme have now become widely accepted. For example, many of the enzymes mentioned here were prepared from secretory vesicles rather than from whole tissue. Assays using complete prohormones as substrates have proved successful, though cumbersome, in guiding converting enzyme purification. It should also be noted that using a whole prohormone presents a wide variety of peptide sequences to an enzyme, so that if it has a specificity outside of the paired basic residues, then this will rapidly become apparent.

Once prohormone-converting enzymes from various tissues have been purified in sufficient quantity to sequence them and raise antibodies, then a whole new set of questions can be asked. For example, are these enzymes separately regulated from the precursors they cleave? Several precursors contain more than one active hormone sequence (e.g., POMC and the other opiate peptides), and the pattern of products could presumably be drastically altered by a variation in the type or amount of prohormone-converting enzyme packaged into secretory granules. There is also the basic question of how enzyme and substrate are colocalized in the same granules. Future work in this field should provide answers to these questions.

ACKNOWLEDGMENTS. I would like to thank Dr. Y. P. Loh and Dr. H. Gainer for their generous assistance, guidance, and support and Dr. M. J. Brownstein for performing the stereotactic injections to label all the prooxytocin and provasopressin used in these studies.

REFERENCES

Ben-Barak, Y., Russell, J. T., Whitnall, M. H., Ozato, K., and Gainer, H., 1984, Neurophysin in the hypothalamo–neurohypophysial system. I. Production and characterization of monoclonal antibodies, *J. Neurosci.* **5:**81–97.

Chang, T.-L., and Loh, Y. P., 1983, Characterization of proopiocortin converting activity in rat anterior pituitary secretory granules, *Endocrinology* **112:**1832–1839.

Chang, T.-L., and Loh, Y. P., 1984, *In vitro* processing of proopiocortin by membrane associated and soluble converting enzyme activities from rat intermediate lobe secretory granules, *Endocrinology* **114:**2092–2099.

Chang, T.-L., Gainer, H., Russell, J. T., and Loh, Y. P., 1982, Proopiocortin-converting enzyme activity in bovine neurosecretory granules, *Endocrinology* **111:**1607–1614.

Cromlish, J. A., Seidah, N. G., and Chretien, M. N., 1985, Isolation and characterization of four proteases from porcine pituitary neurointermediate lobes: Relationship to the maturation enzyme of prohormones, *Neuropeptides* **5:**493–496.

Docherty, K., and Steiner, D., 1982, Post-translational proteolysis in polypeptide hormone biosynthesis, *Annu. Rev. Physiol.* **44:**625–638.

Evangelista, R., Ray, P., and Lewis, R. V., 1982, A "trypsin-like" enzyme in adrenal chromaffin granules: A proenkephalin processing enzyme, *Biochem. Biophys. Res. Commun.* **106:**895–902.

Fletcher, D. J., Noe, B. D., Bauer, G. E., and Quigley, J. P., 1980, Characterization of the conversion of a somatostatin precursor to somatostatin by islet secretory granules, *Diabetes* **29:**593–599.

Fletcher, D. J., Noe, B. D., Bauer, G. E., and Quigley, J. P., 1981, Characterization of proinsulin and proglucagon converting activities in isolated islet secretory granules, *J. Cell Biol.* **90:**312–322.

Fricker, L. D., and Snyder, S. H., 1983, Purification and characterization of enkephalin convertase, an enkephalin-synthesizing carboxypeptidase, *J. Biol. Chem.* **258:**1050–1055.

Gainer, H., Same, Y. L., and Brownstein, M. J., 1977, Biosynthesis and axonal transport of rat neurohypophysial proteins and peptides, *J. Cell Biol.* **73:**366–381.

Gainer, H., Russell, J. T., and Loh, Y. P., 1984, An aminopeptidase activity in bovine pituitary secretory vesicles that cleaves the N-terminal arginine from β-lipotropin$_{60-65}$, *FEBS Lett.* **175:**135–139.

Gainer, H., Russell, J. T., and Loh, Y. P., 1985, The enzymology and intracellular organization of peptide precursor processing: The secretory vesicle hypothesis, *Neuroendocrinology* **40:**171–184.

Hook, V. Y. H., and Loh, Y. P., 1984, Carbopeptidase B-like converting enzyme activity in secretory granules of rat pituitary, *Proc. Natl. Acad. Sci. U.S.A.* **81:**2776–2780.

Julius, D., Brake, A., Blair, L., Kunisawa, R., and Thomer, J., 1984, Isolation of the putative structural gene for the lysine arginine cleaving endopeptidase required for processing of prepro α factor, *Cell* **36:**1075–1089.

Kemmler, W., Steiner, D. F., and Borg, J., 1973, Studies on the conversion of proinsulin to insulin, *J. Biol. Chem.* **248:**4544–4551.

Kurjan, J., and Herskowitz, I., 1982, Structure of a yeast pheromone gene (MF α) a putative α factor precursor contains four tandem repeats of mature α factor, *Cell* **30:**933–942.

Land, H., Schutz, G., Schmale, H., and Richter, D., 1982, Nucleotide sequence of cloned cDNA encoding bovine arginine vasopressin-neurophysin II precursor, *Nature* **295:**299–303.

Land, H., Grez, M., Ruppert, S., Schmale, H., Rahbein, M., Richter, D., and Schutz, G., 1983, Deduced amino acid sequence from the bovine oxytocin-neurophysin I precursor DNA, *Nature* **302:**342–344.

Lindberg, I., Yang, M.-Y. T., and Costa, E., 1984, Further characterization of an enkephalin-generating enzyme from adrenal-medullary chromaffin granules, *J. Neurochem.* **42:**1411–1419.

Loh, Y. P., and Gainer, H., 1983, Biosynthesis and processing of neuropeptides, in: *Brain Peptides* (D. Krieger, M. Brownstein, and J. Martin, eds.), John Wiley & Sons, New York, pp. 79–116.

Loh, Y. P., and Gritsch, H. A., 1981, Evidence for intragranular processing of pro-opiocortin in the mouse pituitary intermediate lobe, *Eur. J. Cell Biol.* **26:**177–183.

Loh, Y. P., Tam, W. H. H., and Russell, J. T., 1984, Measurement of ΔpH and membrane potential in secretory vesicles isolated from bovine pituitary intermediate lobe, *J. Biol. Chem.* **259:**8238–8245.

Loh, Y. P., Parish, D. C., and Tuteja, R., 1985, Purification and characterization of a paired basic residue-specific proopiomelanocortin converting enzyme from bovine pituitary intermediate lobe vesicles, *J. Biol. Chem.* **260:**7194–7205.

MacGregor, R. R., Chu, L. L. M., and Cohn, D. V., 1976, Conversion of proparathyroid hormone to parathyroid hormone by a particulate enzyme of the parathyroid gland, *J. Biol. Chem.* **251:**6711–6716.

MacGregor, R. R., Hamilton, J. W., and Cohn, D. V., 1978, The mode of conversion of proparathormone to parathormone by a particulate converting enzymic activity of the parathyroid gland, *J. Biol. Chem.* **253:**2012–2017.

Mizuno, K., and Matsuo, H., 1984, A novel protease from yeast with specificity towards paired basic residues, *Nature* **309:**558–560.

Mizuno, K., and Matsuo, H., 1985, Proenkephalin processing enzyme with specificity towards paired basic residues purified from bovine adrenal chromaffin granules, *Neuropeptides* **5:**489–492.

Mizuno, K., Kojima, M., and Matsuo, H., 1985, A putative prohormone processing protease in bovine adrenal medulla specifically cleaving in between Lys-Arg sequences, *Biochem. Biophys. Res. Comm.* **128:**884–891.

Nakanishi, S., Ionue, A., Kita, T., Nakamura, M., Chang, A. C. Y., Cohen, S. N., and Numa, S., 1979, Nucleotide sequence of cloned cDNA for bovine corticotropin–β-lipotropin precursor, *Nature* **278:**423–427.

Noda, M., Taranishi, Y., Takahashi, H., Toyosato, M., Natake, M., Nakanishi, S., and Numa, S., 1982, Isolation and structural organization of the human preproenkephalin gene, *Nature* **297:**432–434.

Noe, B. D., Debo, G., and Speiss, J., 1984, Comparison of prohormone-processing activities in islet microsomes and secretory granules: Evidence for distinct converting enzymes for separate islet prosomatostatins. *J. Cell Biol.* **99:**578–587.

Pelaprat, D., Sidah, N. G., Sikstrom, R. A., Lambelin, P., Hamelin, J., Lazure, C., Cromlish, J. A., and Chretien, M., 1984, Subcellular fractionation of pituitary neurointermediate lobes: Revelation of various basic proteases, *Endocrinology* **115:**581–590.

Quinn, P. S., and Judah, J. D., 1978, Calcium dependent Golgi-vesicle fusion and cathepsin B in the conversion of proalbumin into albumin in rat liver, *Biochem. J.* **172:**301–309.

Russell, J. T., 1984, ΔpH, H^+ diffusion potentials, and Mg^{2+} ATPase in neurosecretory vesicles isolated from bovine neurohypophyses, *J. Biol. Chem.* **259:**9496–9507.

Smyth, D. G., Austen, B. M., Bradbury, A. F., Geison, M. J., and Small, C. R., 1977, Biogenesis and metabolism of opiate active peptides, in: *Centrally Acting Peptides* (J. Hughes, ed.), Macmillan, London, pp. 231–239.

Troy, C. M., and Musacchio, J. M., 1982, Processing of enkephalin precursors by chromaffin granule enzymes, *Life Sci.* **31:**1717–1720.

5

α-Melanotropin Analogues for Biomedical Applications

MAC E. HADLEY, VICTOR J. HRUBY,
DHIRENDRA N. CHATURVEDI, AND TOMI K. SAWYER

1. INTRODUCTION

Until recently, the melanotropins were of interest mainly to comparative physiologists, since their role as hormones was believed to be restricted mainly to the control of integumental color changes in lower vertebrates. It is now realized that the melanotropic peptides are also found localized to the nervous system (O'Donohue and Dorsa, 1982), where as neurohormones they may serve a number of central nervous system functions (Beckwith and Sandman, 1982). Furthermore, localization of melanotropins to peripheral organs suggests that these peptides, possibly functioning as local hormones, subserve other as yet undefined physiological roles (Chen *et al.*, 1984).

2. SOURCE, SYNTHESIS, AND MECHANISMS OF ACTION

The melanotropins are a family of structurally related peptides, derived from a precursor protein, proopiomelanocortin. This precursor is acted on by cell-specific enzymes to release the hormonal peptide characteristic of the particular cell. In the pars distalis, for example, corticotropin (ACTH) is the hormonal peptide secreted by the corticotroph, whereas in the melanotrophs of the pars intermedia, α-melanotropin (α-melanocyte-stimulating hormone, α-MSH) is the hormonal secretory

MAC E. HADLEY • Department of Anatomy, University of Arizona, Tucson, Arizona 85724.
VICTOR J. HRUBY • Department of Chemistry, University of Arizona, Tucson, Arizona 85721.
DHIRENDRA N. CHATURVEDI • Vega Biotechnologies, Inc., Tucson, Arizona 85706.
TOMI K. SAWYER • Biopolymer Chemistry/Biotechnology, The Upjohn Company, Kalamazoo, Michigan 49001.

product. Although other melanotropic peptides are components of proopiomelanocortin, none of these molecules (e.g., β-lipotropin, γ-MSH, β-MSH) are presently candidates as either neurohormones or as systemic hormones.

The mechanisms of action of melanotropic peptides have been documented in several bioassays. The *in vitro* frog (e.g., *Rana pipiens*) and lizard (*Anolis carolinensis*) skin bioassays (Castrucci *et al.*, 1984a) have proved particularly useful in characterizing the nature of melanotropin receptors through classical structure–activity studies (Hruby *et al.*, 1984). In the frog and lizard skin bioassays, it has been determined that there is a receptor-specific calcium requirement for signal transduction and adenylate cyclase activation. Cultured mouse melanoma cells have provided the relevant information on the intracellular events associated with melanotropin receptor activation. Melanotropins activate plasmalemmal adenylate cyclase and the production of cAMP, a postulated intracellular second messenger of hormone action (Pawelek and Körner, 1982). The immediate actions of melanotropins on the production of cAMP can be documented by the melanoma adenylate cyclase assay, which utilizes melanoma cell membranes (Bregman *et al.*, 1980). In the melanoma tyrosinase bioassay, the long-term actions of melanotropins on melanoma cells grown in tissue culture reveal both transcriptional and translational requirements for the synthesis and/or activation of tyrosinase, the rate-limiting enzyme in melanin formation (Fuller and Hadley, 1978; Pawelek and Körner, 1982). Important reviews on the subject provide further information on the distribution, cellular sources, chemistry, biological roles, and mechanisms of action of the melanotropins (Hadley, 1984; Hruby *et al.*, 1984).

3. MELANOTROPIN STRUCTURE–FUNCTION STUDIES

We have determined that the structural requirements of melanotropins for receptor activation in the frog and lizard skin bioassays differ considerably (Hruby *et al.*, 1984). In general, the structural requirements for receptor activation by linear fragments of α-MSH are similar in both bioassays. In the lizard only the 4–11 melanotropin sequence is needed for near equipotency to α-MSH, whereas in the frog the 4–12 sequence is required. When substitutions of individual amino acids within the 4–9 sequence are made, differences in receptor recognition by the two species become clearly evident. Certain cyclic melanotropins, for example, are superagonists in the frog skin bioassay, whereas in the lizard skin bioassay, the cyclic fragment analogues are sometimes less active than their linear counterparts (Cody *et al.*, 1984a).

In the design of melanotropins that might prove useful for biomedical applications, it will be important to understand the comparative endocrinology of melanotropin receptors. Our present structure–activity studies indicate that S91 mouse melanoma melanotropin receptors are similar, if not identical, to lizard (*Anolis*) melanocyte receptors. It is, however, unknown whether S91 mouse melanoma receptors are similar to those of humans. Almost nothing is known about human cutaneous melanocyte or melanoma receptors. Normal human melanocytes, and possibly melanoma cells, are responsive to melanotropins. In the design of melanotropins that might be most useful for human biomedical applications, it will be important to characterize human melanotropin receptors through the use of classical structure–activity studies.

3.1. The Message Sequence

α-Melanotropin is a tridecapeptide, Ac-Ser-Tyr-Ser-Met-Glu-His-Phe-Arg-Trp-Gly-Lys-Pro-Val-NH_2, in which the central 4–10 sequence, which is also present in other members of the melanotropin family of peptides, is considered the active site or so-called message sequence. We have determined in several melanocyte bioassays that the N-terminal tripeptide sequence, -Ser-Tyr-Ser-, is rather unessential for full biological activity. The C-terminal valine is also only minimally important for melanotropic activity in the frog and lizard skin bioassays.

3.2. Superpotent Melanotropin Analogues Exhibiting Prolonged Activity

Heat–alkali treatment (racemization) of α-MSH leads to a mixture of peptides that exhibit prolonged melanotropic activity (Bool *et al.*, 1981). That is, after removal of the melanotropins from the bathing or culture media, there still remains residual activity in the "absence" of the peptides. We determined the D–L ratios of the individual amino acids present within the racemized peptides (Sawyer *et al.*, 1983). Based on this information, we synthesized an α-MSH analogue in which D–Phe was substituted for the L-enantiomer at position 7. In addition, we also substituted norleucine for methionine at position 4 of α-MSH. This α-MSH analogue, [Nle^4, D-Phe^7]-α-MSH, exhibited superpotent and prolonged (residual) activity in each of our bioassays (Sawyer *et al.*, 1980). Other fragment analogues of α-MSH, Ac-[Nle^4, D-Phe^7]-α-MSH_{4-11}-NH_2 and Ac-[Nle^4, D-Phe^7]-α-MSH_{4-10}-NH_2, also possessed the unique biological characteristics of the parent analogue (Sawyer *et al.*, 1982a). Many other α-MSH analogues have subsequently been synthesized that also exhibit superpotency and residual activity (Hruby *et al.*, 1984).

3.3. Superagonist Cyclic Melanotropin Analogues

Perhaps because of the postulated inherent conformational flexibility of α-MSH, the three-dimensional topochemical properties that may be important for its biologically active conformation at the melanotropin receptor had not been experimentally examined until we began our studies. The various conformational requirements in a peptide hormone that are important for biological activity can be defined if semirigid synthetic analogues can be prepared that exhibit differential biological activities (superagonism, antagonism, etc.). Based on our demonstration of the exceptional *in vitro* (Sawyer *et al.*, 1980) and *in vivo* (Hadley, 1981) biological properties of [Nle^4, D-Phe^7]-α-MSH on both normal and transformed melanocytes, a knowledge of the known conformational effects of introducing D-amino acids into peptides, and examination of three-dimensional molecular models of this stereoisomeric analogue, we concluded that a β-turn or other peptide chain-reversal region within the central active site (His-Phe-Arg-Trp) of α-MSH might be functionally related to its biologically active conformation.

To evaluate the effect of covalently locking α-MSH into a reverse-turn conformation, we substituted cysteine residues for the methionine-4 and glycine-10 residues (essentially an isosteric replacement). We have reported the synthesis of [Cys^4, Cys^{10}]-α-MSH which possesses superagonist potency in the frog skin bioassay (Sawyer *et al.*, 1982b). The increased biological potency of [Cys^4, Cys^{10}]-α-MSH in certain melanocyte bioassays provides evidence for the possible existence of a pseudocyclic biologically active conformation of the linear native hormone. The structurally related 4–13, 4–12, 4–11, and 4–10 fragment analogues were also considerably more active than their linear counterparts in the frog skin bioassay (Knittel *et al.*, 1982; Cody *et al.*, 1984a). We also have synthesized [Cys^4, D-Phe^7, Cys^{10}]-α-MSH_{4-13}-NH_2 as well as the shorter 4–12, 4–11, and 4–10 analogues. All of these analogues are superagonists in some bioassays, and the 4–13 and 4–12 analogues exhibit ultraprolonged biological activity (Cody *et al.*, 1984b).

3.4. Enzyme-Resistant Melanotropin Analogues

Normal α-MSH has a very short half-life *in vivo,* about 2 min or less in man and rodents. An important attribute of the melanotropins we have developed is their stability *in vivo* or under *in vitro* culture conditions. Analogues of α-MSH have been synthesized and studied relative to their resistance to inactivation by sera and purified proteolytic enzymes (α-chymotrypsin, trypsin). When incubated with serum, the melanotropic activity of α-MSH was lost after a 4-hr incubation, whereas β-MSH retained about one-half of its original biological activity during this same time period. Both α- and β-MSH were rapidly inactivated by α-chymotrypsin and trypsin. In contrast, [Nle^4, D-Phe^7]-α-MSH and the 4–11 and 4–10 fragment analogues were totally resistant to biological inactivation by rat serum, α-chymotrypsin, or trypsin. The conformationally restricted melanotropin, [Cys^4, Cys^{10}]-α-MSH as well as certain other cyclic fragment analogues were also totally resistant to inactivation by enzymes (Castrucci *et al.*, 1984b). These [Nle^4, D-Phe^7]-substituted or [Cys^4, Cys^{10}]-substituted melanotropins may provide useful biologically stable probes for *in vitro* and *in vivo* studies of melanotropin target tissues and also for biomedical applications as described below.

4. MELANOTROPIN ANALOGUES FOR DIAGNOSTIC APPLICATIONS

As noted above, [Nle^4, D-Phe^7]-substituted α-melanotropins exhibit prolonged biological activity. If this residual activity results from irreversible binding of the peptides with melanotropin receptors or to some component of the transduction process, then it should be possible to deliver drugs or other ligands to target tissues by a site-specific, receptor-directed mechanism.

4.1. Radiolabeled Melanotropins

Radiolabeled (tritium, ^{3}H; ^{125}I; ^{131}I) hormones have been utilized by many investigators to determine number and distribution of cellular hormone receptors. We previously reported that α-MSH and β-MSH are inactivated by the methods generally used to radioiodinate peptide hormones. We presumed from previous studies that inactivation of the melanotropins resulted from oxidation of methionine at position 4 or 7, respectively, of these melanotropins. We therefore substituted norleucine (Nle) at these positions and synthesized $[Nle^{4}]$-α-MSH and $[Nle^{7}]$-β-MSH, which were found to be resistant to the oxidative methods used to iodinate the peptides and were also more potent than the native hormones (Heward *et al.*, 1979a,b). We successfully iodinated $[Nle^{4}]$-α-MSH by the iodogen method (Heward *et al.*, 1981). Nevertheless, not all iodinations are successful, apparently because of interaction of iodine with amino acid residues at other sites within the melanotropin. It is hoped that further improvements in iodination techniques and suitable amino acid substitutions at nonessential sites within melanotropins (e.g., substitutions for His^{5}, Arg^{8}, or Trp^{9}) that do not lead to loss of potency may eliminate the problem of radioinactivation and provide melanotropins even more useful for biomedical studies.

We have reported the successful synthesis of a tritium-labeled melanotropin (Wilkes *et al.*, 1983): Ac-{$[^{3}H]Nle^{4}$, D-Phe^{7}}-α-MSH_{4-11}-NH_2, an octapeptide, was prepared by direct incorporation of ^{3}H-labeled norleucine into the peptide. The tritium-labeled peptide is a superpotent agonist of frog and lizard skin melanocytes and mouse S91 melanoma cells. This melanotropin possessed ultraprolonged activity on melanocytes both *in vitro* and *in vivo,* and the peptide was resistant to inactivation by serum and purified proteolytic enzymes. The melanotropic activity of the labeled peptide was identical to that of the unlabeled analogue. This analogue or other radiolabeled [Nle^{4}, D-Phe^{7}]-substituted melanotropins should provide useful probes for determining the presence of melanotropin receptors on normal and abnormal (melanoma) melanocytes and on other putative melanotropin target tissues. Since [Nle^{4}, D-Phe^{7}]-substituted melanotropins may irreversibly interact with melanoma cells (Abdel Malek *et al.*, 1985), then attached to a radioisotope (^{125}I, ^{131}I), these peptides might provide a method for melanoma tumor localization by external scintography.

4.2. Biotin–Melanotropin Conjugates

We have prepared biocytin (Bct) derivatives of [Nle^{4}, D-Phe^{7}]-α-MSH (Chaturvedi *et al.*, 1984): [N^{α}-Bct-Ser^{1}, Nle^{4}, D-Phe^{7}]-α-MSH and [N^{12}-Bct-N^{α}-dodecanoyl-Ser^{1}, Nle^{4}, D-Phe^{7}]-α-MSH were synthesized by solid-phase techniques, and the coupling of biotin and 12-aminododecanoic acid was achieved through their succinimido esters. These melanotropins possessed almost identical actions to their parent analogue, [Nle^4, D-Phe^7]-α-MSH. Both biocytin derivatives were superagonists, exhibited ultraprolonged biological activity, and were equipotent to or more potent than α-MSH as determined by several melanotropin melanocyte and melanoma bioassays. The melanotropins were also resistant to inactivation by proteolytic enzymes.

4.3. Fluorescein–Melanotropin Conjugate

We have synthesized a fluorescein-labeled [Nle^4, D-Phe^7]-α-MSH analogue using dichlorotriazinylamino fluorescein. Like the parent peptide, [Nle^4, D-Phe^7]-α-MSH, the fluorescein analogue was a superagonist, was nonbiodegradeable, and possessed ultraprolonged biological activity (Chaturvedi *et al.*, 1985). This fluorescein conjugate together with the biotin-labeled melanotropins should provide useful probes for answering basic biological questions relating to the nature and distribution of melanotropin receptors.

5. MELANOTROPIN ANALOGUES FOR MELANOMA CHEMOTHERAPY

Melanoma is notorious for its difficulty of early detection and refractoriness to chemotherapy. Worldwide statistics indicate that the incidence of melanoma is on the increase. The cutaneous melanomas are a group of malignancies in which there is a premium on diagnosis at the *in situ* and early invasive phases of growth. Since mouse melanoma cells as well as human melanocytes possess melanotropin receptors, it may be possible to utilize melanotropins to understand the basic molecular mechanism involved in melanoma growth and proliferation and to deliver cytotoxic (anticancer) drugs to these cancer cells by a receptor-directed mechanism.

5.1. Melanotropins and Melanoma Melanotropin Receptors

A number of mouse melanoma lines have been utilized to unravel basic questions related to melanogenesis and cellular proliferation. We have determined that [Nle^4, D-Phe^7]-α-MSH and structurally related 4–11 and 4–10 fragment analogues are 100–1000 times more active than α-MSH in stimulating tyrosinase activity in Cloudman S91 melanoma cells (Marwan *et al.*, 1985).

We previously demonstrated that [Nle^4, D-Phe^7]-α-MSH causes an ultraprolonged *in vitro* (hours) and *in vivo* (days–weeks) dispersion of melanin granules (melanosomes) within reptilian and amphibian integumental melanophores (Hadley *et al.*, 1981, 1985). These melanophores are highly differentiated cells in which melanosome movements are rapidly activated events. Melanoma cells, on the other hand, may be viewed as relatively undifferentiated in that phenotypic expression (tyrosinase activation and melanin production) involves transcriptional and translational events in which new protein synthesis is initiated. We have demonstrated that certain melanotropin analogues can turn on these genomic events almost indefinitely (Hadley *et al.*, 1985; Abdel Malek *et al.*, 1985).

After an exposure of S91 melanoma cells to [Nle^4, D-Phe^7]-α-MSH for only 4 hr or less, genomic mechanisms involving both messenger RNA synthesis and protein synthesis that lead to increased tyrosinase activity are turned on for at least 6 days subsequent to "removal" of the peptide and in spite of continuing cell division. This observation raises a number of important questions. Either the melanotropin analogue irreversibly activates plasma membrane processes related to adenylate cyclase activity, which are manifested even in the absence of the analogue from the culture medium, or, alternatively, the analogue may be irreversibly bound to melanoma cell melanotropin receptors or to some component of the transduction process between receptors and adenylate cyclase (and retained there through subsequent cell divisions).

The first suggestion is without experimental precedent, whereas the second suggestion is supported by the observation that cholera toxin is able to irreversibly bind and activate adenylate cyclase in many cell types. The data suggest that S91 melanoma cells possess spare receptors and that more receptors become occupied initially by [Nle^4, D-Phe^7]-α-MSH than are needed to activate fully adenylate cyclase and thereafter tyrosinase. The tyrosinase response to a 4-hr treatment with the melanotropin analogue supports this contention, since it was three cell divisions later that a maximal response was attained (Abdel Malek *et al.*, 1985). The initial activation of tyrosinase in response to α-MSH, although equal in magnitude to that by [Nle^4, D-Phe^7]-α-MSH, was rapidly diminished. This observation argues against the residual response being caused by the production of a long-lived intracellular messenger, since both α-MSH and the analogue should stimulate the production of the same second messenger. During subsequent cell divisions, these receptors would be randomly distributed to the daughter cells. During continuous cell division, it appears that some receptors remain occupied, and this results in enhanced tyrosinase activity for almost 6 consecutive days (about six cell divisions). If the melanotropin analogue is irreversibly sequestered within the melanoma cell membrane, it would also be expected that the analogue would eventually be equally distributed to daughter cells. In time, one would anticipate that the number of receptors occupied would become diminished through cell division and that tyrosinase activity would decline, as shown in our recent experiments. These interpretations of our results have important implications for cancer research and for cell biology in general.

5.2. Melanotropin–Antitumor Drug Conjugates

Clinical cancer chemotherapy is hindered by the lack of selective toxicity of most antitumor drugs for tumor cells relative to normal cells. There is an urgent need for the development of effective delivery systems for altering and controlling the absorption, blood levels, metabolism, organ distribution, and cellular uptake of antitumor agents. The ultimate form of controlled drug delivery would be the realization of Paul Ehrlich's "magic bullet" concept, the development of drug–carrier conjugates that deliver drugs exclusively to a particular target cell type. Many potent cytotoxic agents have been used with a variety of carriers with the hope of improving their specificity for a particular tumor.

Melanotropins and related analogues offer a unique approach as potential drug carriers since they possess a very high degree of specificity and affinity for melanoma cell melanotropin receptors. Similar to the concept of hormone action, melanotropin–drug conjugate targeting would also involve a recognition event between the drug conjugate and specific receptors at the cell surface. The drug complex could then be internalized, as reported by some investigators, and the antitumor drug released for action inside the cell. The antitumor drug might also exert its cytotoxicity action by interaction at the cell surface, as has been suggested for adriamycin, retinoic acid, and some other drugs. Analogues of α-MSH that are superpotent, nonbiodegradeable, and exhibit ultraprolonged biological activity should provide the ideal carriers for targeting cytotoxic agents to melanoma cells.

6. MELANIN PIGMENTATION OF THE SKIN

Pelage pigmentation in some mammals may be regulated by melanotropins (Thody, 1980). In a mouse model, follicular melanogenesis can be induced by melanotropins. We have determined that [Nle^4, D-Phe^7]-α-MSH is a superpotent agonist of follicular melanogenesis in this strain (C57BL/6JA^y) of mouse (Upton *et al.*, 1985). This analogue is about 10,000 times more active than α-MSH in stimulating follicular melanogenesis. The increased potency difference between the analogue and α-MSH may result from the fact that the analogue, unlike the native hormone, is resistant to inactivation by serum proteolytic enzymes. These results demonstrate the potential utility of certain melanotropins for *in vivo* studies of melanotropin target tissues.

Melanin synthesis within hair follicles only occurs during the growth phase (anagen) of the hair cycle. By the use of melanotropin analogues it is possible to demonstrate patterns of follicular melanogenesis that indicate those areas of the skin wherein hairs are actively growing. Melanin deposition within hair is an indelible marker of specific temporal biosynthetic activities and cellular interactions within the follicle. For the developmental biologist, melanotropin analogues may provide useful tools to answer questions related to cutaneous cellular proliferation, particularly as it relates to hair growth.

7. MELANOTROPINS AND NERVOUS SYSTEM FUNCTION

Melanotropins or related peptides have been localized to specific neurons within the brain. As putative neurohormones, these melanotropins might function as neurotransmitters and/or neuromodulators. Systemic or intracerebroventricular injections of MSH/ACTH peptides into rodents stimulate certain behavioral activities (O'Donohue and Dorsa, 1982). [Cys[4], Cys[10]]-α-Melanotropin and related cyclic fragment analogues are potent inducers of excessive grooming behavior in the rat (Kobobun *et al.*, 1983; Hirsch *et al.*, 1984). The Ac-[Cys[10], Cys[10]]-α-MSH_{4-10}-NH_2 analogue was the most active of the analogues studied, whereas the linear $ACTH_{4-10}$ and Ac-[Nle[4], D-Phe[7]]-α-MSH_{4-10}-NH_2 peptides were inactive in the bioassay at the concentrations employed.

We have determined that certain [D-Phe[7]]-substituted linear and cyclic analogues of α-MSH are resistant to inactivation by rat brain homogenates, whereas the L-Phe[7] enantiomeric peptides are rapidly inactivated (Akiyama *et al.*, 1980). If melanotropins function as neurohormones, then their actions must be mediated through melanotropin receptors. Certain melanotropin analogues resistant to inactivation by serum and brain enzymes may provide useful ligands for the localization and characterization of brain melanotropin receptors.

8. ANTIMELANOTROPINS AND MELANIN-CONCENTRATING HORMONE

Any hypothesis regarding the relationship(s) of the structural, conformational, and dynamic properties of a peptide hormone to biological activity should embody some hypothesis for design of inhibitors, that is, the features that distinguish the biological message (signal transduction) from the binding message. The antagonist can provide direct evidence of those chemicophysical properties important to binding activity and indirectly (by comparison with agonists) provide insight into those elements of structure and conformation important to the transduction process (biological activity message). This information can be very useful in the design of more potent, longer-lasting hormone analogues. In addition, if the antagonist is a true competitive inhibitor, it can be used in a dose–response assay to determine the dissociation constant for the hormone–receptor interaction. Obviously, it also can be used both *in vitro* and *in vivo* to block specifically the biological activity caused by a particular hormone agonist so that other hormones or other agonists that utilize separate receptors can be studied independently. These and other studies can provide important insights into mechanisms of hormone action. In *in vivo* studies, antagonists also provide the ability to interact with receptors without producing a biological response. In the case of melanotropins, this could be very valuable in tumor localization or related studies, especially if it turns out that melanotropins might also affect (adversely) behavior or other nervous system or physiological functions.

A putative melanin-concentrating hormone (MCH) was isolated from salmon pituitaries, and its primary structure was determined (Kawauchi *et al.*, 1983). Our initial interest in the reported primary structure of MCH related to the possibility that the peptide might antagonize the actions of α-MSH. The existence of such an antagonist would be important in studies on mechanisms of α-MSH action. We synthesized the heptadecapeptide, Asp-Thr-Met-Arg-Cys-Met-Val-Gly-Arg-Val-Tyr-

Arg-Pro-Cys-Trp-Glu-Val (Wilkes *et al.*, 1984a,b), and it may be possible to synthesize structural analogues of MCH that possess more or less MCH- or MSH-like activity. It will be important to determine the essential structure and conformational features of the peptide that are required for receptor binding and signal transduction that lead to melanosome translocation within melanophores. This will require classical structure–activity studies, which may provide important insights into the evolution of melanotropins, their receptors, and the target tissues that they regulate.

9. CONCLUSION

We have developed α-MSH analogues that are superpotent, enzyme-resistant, and that exhibit ultraprolonged biological activity. These analogues should prove useful for many studies in physiology and medicine. It is likely that these melanotropins could be utilized as radiolabeled or fluorescent-labeled ligands for the detection of melanoma in humans. These molecules may also serve as proverbial "magic bullets" for the site-specific delivery of antitumor drugs to melanoma. A developing area of potential promise for these analogues is for the treatment of problems in learning and behavior. Further research in these areas should have a significant impact on drug development studies for a variety of biomedical applications.

ACKNOWLEDGMENTS. The research reported here was supported, in part, by U.S. Public Health Service grant AM17420 and by National Science Foundation grants PCM-8112200 and PCM 8110708. We would like to acknowledge the research contributions of Wayne L. Cody and Brian C. Wilkes to these studies.

REFERENCES

Abdel Malek, Z. A., Kreutzfeld, K. L., Marwan, M. H., Hadley, M. E., Hruby, V.J., and Wilkes, B.C., 1985, Prolonged stimulation of S91 melanoma tyrosinase by [Nle^4 D-Phe^7]-substituted α-melanotropins, *Cancer Res.* **45:**4735–4740.

Akiyama, K., Yamamura, H. I., Wilkes, B. C., Cody, W. L., Hruby, V. J., Castrucci, A. M. de L., and Hadley, M. E., 1984, Relative stability of α-melanotropin and related analogues to rat brain homogenates, *Peptides* **5:**1191–1195.

Beckwith, B. E., and Sandman, C. A., 1982, Central nervous system and peripheral effects of ACTH, MSH and related neuropeptides, *Peptides* **3:**411–420.

Bool, A. M., Gray, G. H., II, Hadley, M. E., Heward, C. B., Hruby, V. J., Sawyer, T. K., and Yang, Y. C. S., 1981, Racemization effects on melanocyte stimulating hormones and related peptides, *J. Endocrinol.* **88:**57–65.

Bregman, M. D., Sawyer, T. K., Hadley, M. E., and Hruby, V. J., 1980, Adenosine and divalent cation effects on S-91 melanoma adenylate cyclase, *Arch. Biochem. Biophys.* **200:**1–7.

Castrucci, A. M. de L., Hadley, M. E., and Hruby, V. J., 1984a, Melanotropin bioassays: *In vitro* and *in vivo* comparisons, *Gen. Comp. Endocrinol.* **55:**104–111.

Castrucci, A. M. de L., Hadley, M. E., Sawyer, T. K., and Hruby, V. J., 1984b, Enzymological studies of melanotropins, *Comp. Biochem. Physiol.* **78B:**519–524.

Chaturvedi, D. N., Knittel, J. J., Hruby, V. J., Castrucci, A. M. de L., and Hadley, M. E., 1984, Highly potent peptide hormone analogues: Synthesis and biological actions of biotin-labeled melanotropins, *J. Med. Chem.* **27:**1406–1410.

Chaturvedi, D. N., Hruby, V. J., Castrucci, A. M. de L., Kreutzfeld, K. L., and Hadley, M. E., 1985, Synthesis and biological evaluation of the superagonist [N^{α}-chlorotriazinylamino-fluorescein-Ser^{1}, Nle^{4}, D-Phe^{7}]-α-MSH, *J. Pharm. Sci.* **74:**237–240.

Chen, C.-L. C., Mather, J. P., Morris, P. L., and Bardin, C. W., 1984, Expression of pro-opiomelanocortin-like gene in the testis and epididymis, *Proc. Natl. Acad. Sci. U.S.A.* **81:**5672–5675.

Cody, W. L., Wilkes, B. C., Muška, B. J., Hruby, V. J., Castrucci, A. M. de L., and Hadley, M. E., 1984a, Cyclic melanotropins, Part VI: Importance of the C-terminal tripeptide (Lys-Pro-Val), *J. Med. Chem.* **27:**1186–1190.

Cody, W. L., Mahoney, M., Knittel, J. J., Hruby, V. J., Castrucci, A. M. de L., and Hadley, M. E., 1984b, Cyclic melanotropins, Part IX: 7-D-Phenylalanine analogues of the active site sequence, *J. Med. Chem.* **28:**583–588.

Fuller, B. B., and Hadley, M. E., 1978, Transcriptional and translational requirements for MSH action on melanoma cells, *Pigment Cell* **4:**97–104.

Hadley, M. E., 1984, *Endocrinology,* Prentice-Hall, Englewood Cliffs, New Jersey.

Hadley, M. E., Anderson, B., Heward, C. B., Sawyer, T. K., and Hruby, V. J., 1981, Calcium-dependent prolonged effects on melanophores of [4-norleucine, 7-D-phenylalanine]-melanotropin, *Science* **213:**1025–1027.

Hadley, M. E., Abdel Malek, Z. A., Marwan, M. M., Kreutzfeld, K. L., and Hruby, V. J., 1984, [Nle^{4}, D-Phe^{7}]-α-MSH: A superpotent melanotropin that "irreversibly" activates melanoma tyrosinase, *Endocrinol. Res. Commun.* **11:**157–170.

Hadley, M. E., Mieyr, J. H., Martin, B. E., Castrucci, A. M. de L., Hruby, V. J., Sawyer, T. K., Powers, E. A., and Ranga Rao, K., 1985, [Nle^{4}, D-Phe^{7}]-α-MSH: A superpotent melanotropin with prolonged action on vertebrate chromatophores, *Comp. Biochem. Physiol.* **81A:**1–6.

Heward, C. B., Yang, Y. C. S., Ormberg, J. F., Hadley, M. E., and Hruby, V. J., 1979a, Effects of chloramine T and iodination on the biological activity of α-melanotropin, *Hoppe Seylers Z. Physiol. Chem.* **360:**1851–1859.

Heward, C. B., Yang, Y. C. S., Sawyer, T. K., Bregman, M. D., Fuller, B. B., Hruby, V. J., and Hadley, M. E., 1979b, Iodination associated inactivation of α-melanocyte stimulating hormone, *Biochem. Biophys. Res. Commun.* **88:**266–273.

Heward, C. B., Kreutzfeld, K. L., Hadley, M. E., Larsen, B., Sawyer, T. K., and Hruby, V. J., 1981, Preparation of radiolabeled melanotropin suitable for use as a tracer in a radioreceptor assay: [^{125}I-Tyr^{2}, Nle^{4}]-α-MSH, in: *Pigment Cell 1981* (M. Seiji, ed.), University of Tokyo Press, Tokyo, pp. 339–346.

Hirsch, M. D., O'Donohue, T. L., Wilson, R., Sawyer, T. K., Hruby, V. J., Cody, W. L., Knittel, J. J., and Crawley, J. N., 1984, Structural–conformational modification of α-MSH/$ACTH_{4-10}$ analogues with highly potent behavioral activities, *Peptides* **5:**1197–1201.

Hruby, V. J., Knittel, J. J., Mosberg, H. I., Rockway, T. W., Wilkes, B. C., and Hadley, M. E., 1983, Conformational considerations in the design of highly potent and long-acting peptide hormone agonists and antagonists, in: *Peptides 1982* (K. Blaha and P. Melon, eds.), Walter de Gruyter, Berlin, pp. 19–30.

Hruby, V. J., Wilkes, B. C., Cody, W. L., Sawyer, T. K., and Hadley, M. E., 1984, Melanotropins: Structural, conformational and biological considerations in the development of superpotent and superprolonged analogs, in: *Peptide Protein Review,* Vol. 3 (M. T. W. Hearn, ed.), Marcel Dekker, New York, pp. 1–64.

Kawauchi, H., Kawazoe, M., Tsubokawa, M., Kishida, M., and Baker, B. I., 1983, Characterization of melanin-concentrating hormone in chum salmon pituitaries, *Nature* **305:**321–323.

Knittel, J. J., Sawyer, T. K., Hruby, V. J., and Hadley, M. E., 1982, Structure–activity studies of highly potent cyclic [Cys^4, Cys^{10}]-α-melanotropin analogues, *J. Med. Chem.* **28:**125–129.

Kobobun, K., O'Donohue, T. H., Adelmann, G. E., Sawyer, T. K., Hruby, V. J., and Hadley, M. E., 1983, Behavioral effects of 4-norleucine, 7-D-phenylalanine-α-melanocyte stimulating hormone, *Peptides* **4:**721–724.

Marwan, M. M., Abdel Malek, Z. A., Kreutzfeld, K. L., Hadley, M. E., Wilkes, B. C., Hruby, V. J., and Castrucci, A. M. de L., 1985, Stimulation of S91 melanoma tyrosinase activity by superpotent α-melanotropins, *Mol. Cell. Endocrinol.* **41:**171–177.

O'Donohue, T. L., and Dorsa, M. D., 1982, The opiomelanotropinergic neuronal and endocrine systems, *Peptides* **3:**353–395.

Pawelek, J. M., and Körner, A. M., 1982, The biosynthesis of mammalian melanin, *Am. Scientist* **70:**136–145.

Sawyer, T. K., Sanfilippo, P. J., Hruby, V. J., Engel, M. H., Heward, C. B., Burnett, J. B., and Hadley, M. E., 1980, [Nle^4, D-Phe^7]-α-Melanocyte stimulating hormone: A highly potent α-melanotropin with ultralong biological activity, *Proc. Natl. Acad. Sci. U.S.A.* **77:**5754–5758.

Sawyer, T. K., Hruby, V. J., Darman, P. S., and Hadley, M. E., 1982a, [4-Half-cystine, 10-half-cystine]-α-melanocyte stimulating hormone: A cyclic α-melanotropin exhibiting superagonist biological activity, *Proc. Natl. Acad. Sci. U.S.A.* **79:**1751–1755.

Sawyer, T. K., Hruby, V. J., Wilkes, B. C., Draelos, M. T., Hadley, M. E., and Bergsneider, M., 1982b, Comparative biological activities on highly potent analogues of α-melanotropin$_{4-10}$, *J. Med. Chem.* **25:**1022–1027.

Sawyer, T. K., Hruby, V. J., Hadley, M. E., and Engel, M. H., 1983, α-Melanocyte stimulating hormone: Chemical nature and mechanism of action, *Am. Zool.* **23:**529–540.

Thody, A. J., 1980, *The MSH Peptides,* Academic Press, New York.

Upton, J. L., Woods, R., Jessen, G. L., Castrucci, A. M. de L., Hadley, M. E., Wilkes, B. C., and Hruby, V. J., 1985, Superpotent melanotropins and mouse pelage melanogenesis, *Pigment Cell* **1985:**139–143.

Wilkes, B. C., Hruby, V. J., Yamamura, H. I., Akiyama, K., Castrucci, A. M. de L., Hadley, M. E., Andrews, J. R., and Wan, Y.-P., 1983, Synthesis of tritium labeled Ac-[Nle^4, D-Phe^7]-α-MSH$_{4-11}$-NH$_2$: A superpotent melanotropin with prolonged biological activity, *Life Sci.* **34:**977–984.

Wilkes, B. C., Hruby, V. J., Castrucci, A. M. de L., Sherbrooke, W. C., and Hadley, M. E., 1984a, Synthesis of a cyclic melanotropic peptide exhibiting both melanin concentrating and dispersing activities, *Science* **224:**111–113.

Wilkes, B. D., Hruby, V. J., Sherbrooke, W. C., Castrucci, A. M. de L., and Hadley, M. E., 1984b, Synthesis and biological actions of melanin concentrating hormone, *Biochem. Biophys. Res. Commun.* **122:**613–619.

6

β-Endorphin

Processing and Regulation

J. HAM and D. G. SMYTH

1. INTRODUCTION

β-Endorphin, a 31-residue peptide with potent analgesic activity, was first isolated from pituitary glands when it was found to coexist with a related peptide, β-endorphin$_{1-27}$, which possessed only slight analgesic activity (Bradbury *et al.*, 1975; Smyth *et al.*, 1978). This finding, that a hormone can exist in inactive as well as active forms, did not appear to have a precedent, and it was compounded when the two peptides were seen to occur also in α,N-acetylated forms that were devoid of opiate activity (Smyth *et al.*, 1979). Subsequently, it was found that β-endorphin is derived from a complex prohormone that is the precursor of several biologically active peptides (Roberts and Herbert, 1977), and this suggested that the specific processing events that give rise to the multiple forms of β-endorphin might be related to the polyfunctional nature of the prohormone.

Our continuing studies have revealed that β-endorphin exhibits different patterns of processing in different tissues (Smyth and Zakarian, 1980; Zakarian and Smyth, 1982a;b), and it is of particular interest that the reactions have been found to be influenced by pharmacological and physiological signals (Ham and Smyth, 1984b, 1985, 1986). In addition, recent *in vitro* studies of the binding of β-endorphin and its related peptides to brain membranes have indicated that preferential receptors may exist for these peptides that mediate different neuroregulatory actions (Toogood *et al.*, 1984). We present here the results of experiments that suggest that the control of processing is a new and important level for the regulation of bioactivity.

J. HAM and D. G. SMYTH • National Institute for Medical Research, The Ridgeway, Mill Hill, London NW7 1AA, England.

2. β-ENDORPHIN-RELATED PEPTIDES: IDENTITY AND BIOLOGICAL ACTIVITY

β-Endorphin occurs in high concentration in the pituitary gland, from which it is released into the circulation as a hormone; it also occurs in brain, where it is distributed in axons and terminals in the manner of a classical neurotransmitter. In addition, this potent analgesic peptide occurs at low concentration in a variety of peripheral tissues such as pancreas, antrum, testis, and placenta. In most cases the 31-residue form of β-endorphin is accompanied by a series of related peptides that appear to be formed by specific proteolysis in the C-terminal region of its peptide chain. Principal among the shortened forms are β-endorphin$_{1-27}$ (Smyth *et al.*, 1978) and β-endorphin$_{1-26}$ (Smyth *et al.*, 1981a); small amounts of γ-endorphin (β-endorphin$_{1-17}$) and α-endorphin (β-endorphin$_{1-16}$) (Ling *et al.*, 1976) have also been identified. The existence of this range of peptides raises the possibility that each one may possess a distinctive biological activity, since they all contain the sequence of methionine enkephalin at their NH_2 terminus but have C-terminal extensions of different length.

In certain cells these peptides exist principally in an α,N-acetylated form, a modification that is of particular significance since the acetylated peptides are devoid of opiate activity (Deakin *et al.*, 1980). Thus, the principal naturally occurring forms of β-endorphin are the 31-, 27-, and 26-residue peptides and their corresponding N-acetylated derivatives (Fig. 1).

Although the formation of β-endorphin$_{1-27}$ would seem to require the action of an enzyme with specificity of the paired lysine residues present at positions 28 and 29 of the β-endorphin sequence, and the formation of β-endorphin itself involves the action of an enzyme with specificity for lysylarginine, there is evidence that

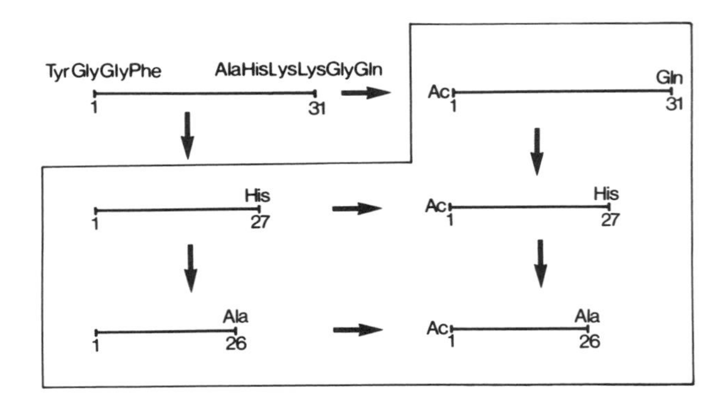

FIGURE 1. β-Endorphin-related peptides identified in pituitary and brain. The peptides enclosed by the box are formed by specific processing of β-endorphin$_{1-31}$ and lack signficant analgesic properties.

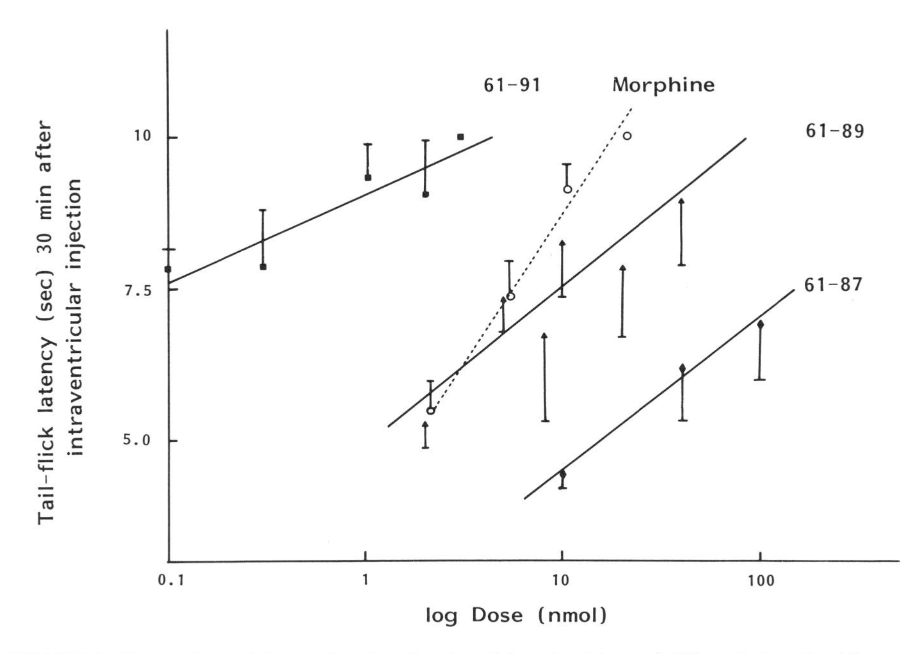

FIGURE 2. Comparison of the analgesic potencies of β-endorphin$_{1-31}$ (LPH$_{61-91}$), β-endorphin$_{1-29}$ (LPH$_{61-89}$), β-endorphin$_{1-27}$ (LPH$_{61-87}$), and morphine. Analgesic activity was determined by the rat tail-flick assay.

the endopeptidases involved in the generation of these peptides may differ from each other. Similarly, it appears that the formation of β-endorphin$_{1-26}$ may require a different enzyme from that involved in the biogenesis of β-endorphin$_{1-31}$, and the 26-residue peptide may not be formed via β-endorphin$_{1-27}$ as an intermediate. Thus, certain cells are known to contain β-endorphin$_{1-31}$ and little or no β-endorphin$_{1-27}$ or β-endorphin$_{1-26}$ (Smyth *et al.*, 1981b). That the shorter forms of β-endorphin appear to be generated independently and have characteristic biological activities emphasizes the importance of studying the mechanisms involved in their formation.

With respect to analgesic activity, β-endorphin$_{1-31}$ is unique in that it possesses a very high potency. Administered directly into the ventricles, it produces profound long-lasting analgesia. On a molar basis, the activity of β-endorphin$_{1-31}$ in the cat is about 100 times greater than that of morphine (Feldberg and Smyth, 1977). In the rat hot-plate assay, its potency is approximately 50 times greater than that of morphine (Bradbury *et al.*, 1977), and in the tail-flick assay values of 19 (Loh *et al.*, 1976) to 100 (Geisow *et al.*, 1977) have been reported. The analgesia produced by β-endorphin, and most of its other actions, is reversible by naloxone, which indicates that the effects are mediated at opiate receptors.

β-Endorphin$_{1-27}$, in contrast, possesses approximately 500 times less analgesic activity than the parent peptide (Fig. 2), but its action has a similar duration. The four C-terminal residues of β-endorphin$_{1-31}$, Lys-Lys-Gly-Gln, are thus important for potency but not for the persistence of analgesia. The tetrapeptide, of course, could have a biological activity of its own (Carter *et al.*, 1979), or it could serve as the precursor of glycylglutamine, which has been shown to have an inhibitory action on the firing of neurons in the reticular formation of the brainstem (Parish *et al.*, 1983). In this way, the loss of the analgesic activity of β-endorphin associated with its C-terminal processing may be accompanied by the appearance of a new activity, and since β-endorphin and glycylglutamine are formed in the same cell, the possibility should be considered that the two peptides may function synergistically.

The α,N-acetylated derivatives of β-endorphin possess no detectable analgesic activity; indeed, they are essentially devoid of all opiate properties. However, there have been reports that acetyl-γ-endorphin is active in certain behavioral tests (Van Ree *et al.*, 1984; Wiegant *et al.*, 1985), and there is good evidence that the acetylated forms of β-endorphin influence the proliferation of lymphocytes (Hazum *et al.*, 1979). Thus, although the acetylated and shorter derivatives of β-endorphin lack analgesic activity, they appear to possess neuronal and peripheral activities of their own.

3. TISSUE-SPECIFIC PROCESSING OF β-ENDORPHIN

Immunohistochemical studies in a range of species have shown that β-endorphin related peptides occur in specific cells in the anterior pituitary and in all the cells of the pars intermedia (Bloom *et al.*, 1978; Zakarian and Smyth, 1979). Furthermore, because of their origin from a common prohormone, β-endorphin and ACTH are elaborated in the same cells (Watson *et al.*, 1977). The nature of the peptides revealed by immunocytochemistry, however, is not precisely defined, since the antibodies employed generally react with the peptide sequences when they form part of the structure of their various biosynthetic precursors. In studies of β-endorphin, the antibodies that have been used also reacted with the acetylated and shortened forms of the peptide. To differentiate among the various derivatives, the endogenous peptides have been extracted from tissues, resolved chromatographically, and their concentrations determined by radioimunoassay (RIA) (Zakarian and Smyth, 1982b).

In many species the anterior region of the pituitary has been found to contain approximately equal concentrations of β-endorphin and lipotropin, its immediate precursor. In contrast, the pars intermedia contains almost exclusively peptides that have the approximate molecular size of the 31-residue peptide, and it contains only small amounts of lipotropin (Fig. 3). Thus, the degree of proteolysis undergone by the β-endorphin prohormone differs markedly in these two tissues. Further examination of the peptides (Fig. 4) has revealed that all six of the β-endorphin-

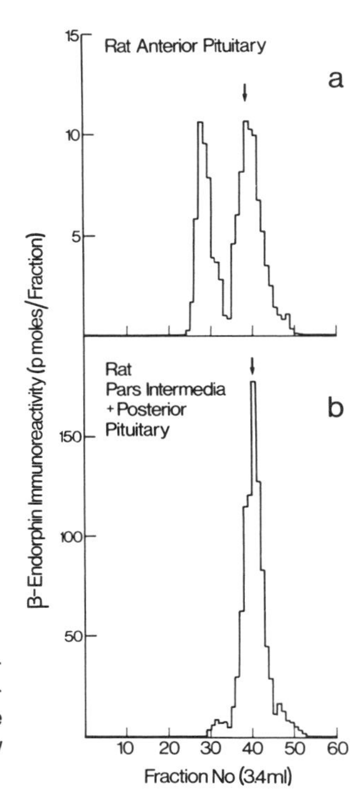

FIGURE 3. Gel-exclusion chromatography of β-endorphin-related peptides from rat anterior pituitary (a) and pars intermedia (b). Note that lipotropin is a major component in the anterior pituitary and not in the pars intermedia. The arrow indicates the elution position of [^{125}I]β-endorphin$_{1-31}$.

related peptides shown in Fig. 1 are present in the pars intermedia, whereas the predominant peptide in the anterior pituitary is the NH_2 form of β-endorphin$_{1-31}$. In the brain, immunocytochemical studies have shown that β-endorphin-related peptides occur principally in the hypothalamus and midbrain, but significant amounts of the peptide can occur in the amygdala, hippocampus, colliculae, and brainstem (Zakarian and Smyth, 1982b). On extraction and analysis of the peptides, a variety of different processing patterns are seen. The main peptide in the hypothalamus, and also in the midbrain and amygdala, is β-endorphin$_{1-31}$, but it is accompanied by the 26- and 27-residue peptides. Notably, these regions of brain contain very little of the acetylated forms. In regions distal from the hypothalamus, the degree of processing of the β-endorphin prohormone is more extensive, and the acetylated, shortened forms of β-endorphin can be the major peptides (Fig. 5).

It should be noted that the patterns can exhibit a degree of variation, and in tissues where the concentration of stored β-endorphin is low, it may not be possible to assign a definitive processing pattern, particularly if the release of peptides from the cells is under active regulation. In general, however, the observed processing patterns are broadly characteristic of each tissue.

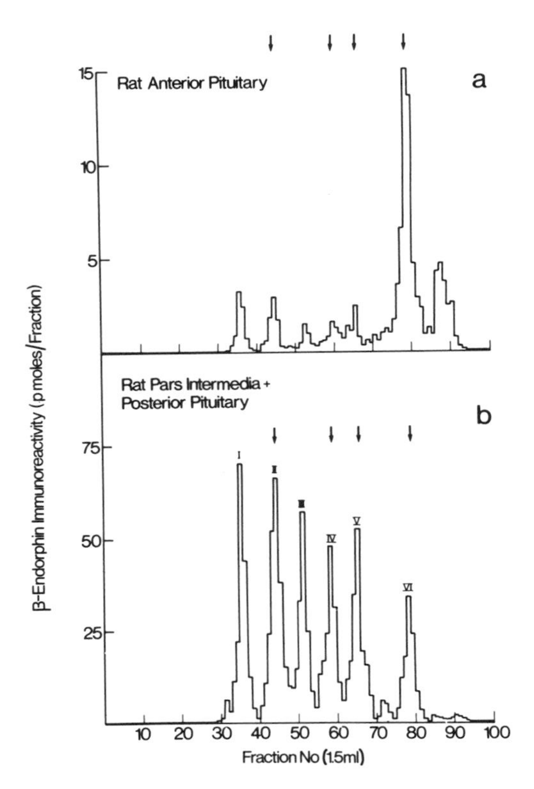

FIGURE 4. Resolution of β-endorphin-related peptides in rat anterior (a) and pars intermedia (b) by ion-exchange chromatography. Peptides with the approximate molecular size of β-endorphin were first obtained by gel exclusion as in Fig. 3. The arrows from left to right indicate the elution positions of α,N-acetyl-β-endorphin$_{1-27}$ β-endorphin$_{1-27}$, α,N-acetyl-β-endorphin $_{1-31}$ and β-endorphin $_{1-31}$, and β-endorphin$_{1-31}$.

4. INFLUENCE OF PHARMACOLOGICAL AND PHYSIOLOGICAL AGENTS ON β-ENDORPHIN PROCESSING

It is generally accepted that prohormone processing is "tissue specific," but it has not been established whether the patterns of the peptides produced are completely fixed and immutable. If a cell were undergoing dynamic secretion, for example, it might be expected that the intracellular peptides would be less exposed to the processing enzymes and the products might contain a higher proportion of "immature" precursors. A cell undergoing basal secretion, on the other hand, might be expected to accumulate "end products" of processing in which the extensively processed fragments would predominate.

Experiments have been carried out to examine whether factors that affect secretion might also influence the mode and degree of processing. Haloperidol (1 mg/kg) administered daily to rats over a period of 12 days has been shown to induce a significant change in the nature of the β-endorphin-related peptides in the pars intermedia (Ham and Smyth, 1985). Under the influence of this dopamine antagonist, the content of the peptides increased substantially, although haloperidol is known to stimulate secretion from the pars intermedia (Höllt and Bergmann, 1982). Of particular interest was the finding that in comparison with control animals the degree of proteolysis of β-endorphin, and of its acetylation, was enhanced: The concentrations of the acetylated forms of β-endorphin$_{1-27}$ and β-endorphin$_{1-26}$ were elevated relative to the NH_2 forms of these peptides, and the degree of acetylation of β-endorphin$_{1-31}$ was also greater than in the control animals (Fig. 6). The experiments have shown that haloperidol not only stimulates secretion but has a profound influence on the form of the peptide produced. It is implicit that the processing of β-endorphin is dynamic, and a relationship appears to exist between processing and secretion. It is somewhat surprising that the observed effects were in the opposite direction from those anticipated; the increased secretion rates were in fact accompanied by increased and not decreased degrees of processing. The processing reactions therefore do not appear to be determined by the length of time a peptide is present in a cell.

More direct information on a possible relationship between processing and

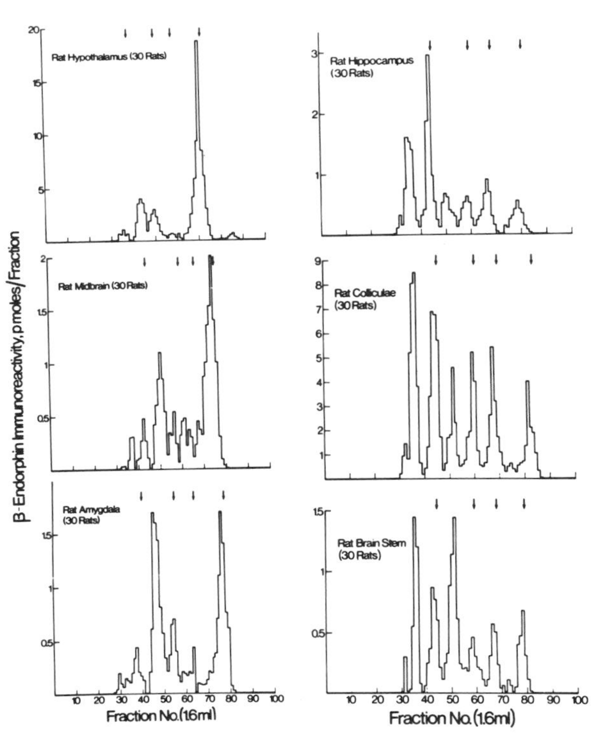

FIGURE 5. Resolution of β-endorphin-related peptides present in regions of rat brain. The peptides were resolved as in Fig. 4.

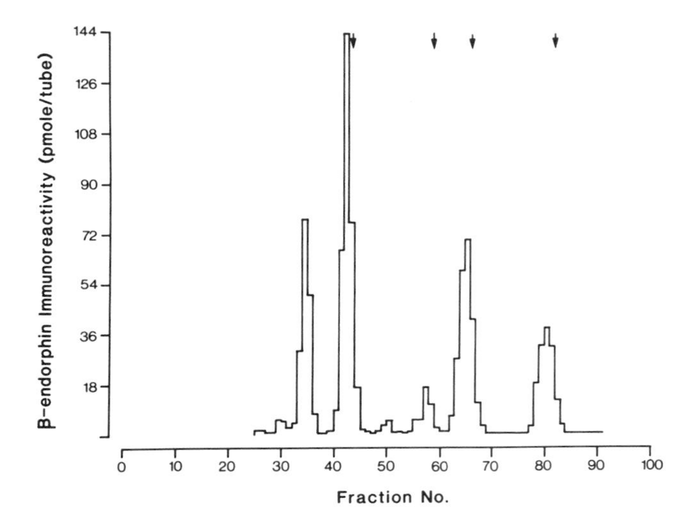

FIGURE 6. Ion-exchange chromatography of β-endorphin-related peptides present in the pars intermedia of rats treated with haloperidol. The marker peptides are as indicated in Fig. 4.

secretion has been obtained in studies of rat pars intermedia cells grown in monolayer culture. Under control conditions, in which dopaminergic innervation was absent, the pattern of β-endorphin-related peptides in the cells was clearly different from the corresponding pattern *in vivo* (Figs. 4 and 7). The major peptide in the cultured cells was the opiate active peptide β-endorphin$_{1-31}$, whereas *in situ* the major

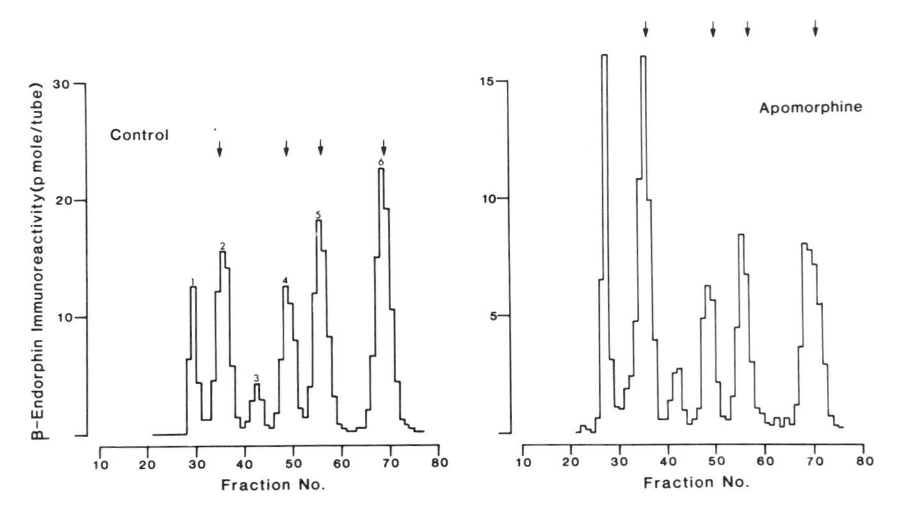

FIGURE 7. Intracellular content of β-endorphin-related peptides in monolayer cultures of untreated and apomorphine-treated anterior pituitary cells. The peptides were resolved as in Fig. 4.

peptides are the acetylated forms of β-endorphin$_{1-27}$ and β-endorphin$_{1-26}$. In the monolayer cultures, the cells appear to undergo uninhibited secretion, and the processing reactions generate a high proportion of the bioactive peptide. However, when the cultures are incubated for 3 hr in the presence of dopamine agonists such as apomorphine (10^{-5} M) or ergocryptine (10^{-6} M), the intracellular processing patterns revert toward the typical pattern seen *in situ,* with the acetylated and shorter forms of β-endorphin predominating (Fig. 7). These findings seem likely to have physiological relevance since with anterior pituitary cells no dopaminergic effect on β-endorphin processing would be anticipated, and none was observed.

A series of experiments using primary cultures of rat anterior pituitary cells has provided strong evidence for the existence of different "pools" of peptides, which can be secreted under different conditions (Ham and Smyth, 1986). Under basal conditions, in the absence of added secretagogue, the principal peptides released were lipotropin and β-endorphin$_{1-31}$, and this pattern remained unchanged over a period of 48 hr. The corresponding intracellular patterns, however, showed that the predominant peptide was β-endorphin$_{1-31}$, and lipotropin was a minor component. When the cells were stimulated acutely with 10^{-7} M corticotropin-releasing factor (CRF) for up to 1 hr, the released peptides reflected the average intracellular content, with bioactive β-endorphin as the main peptide; during pro-

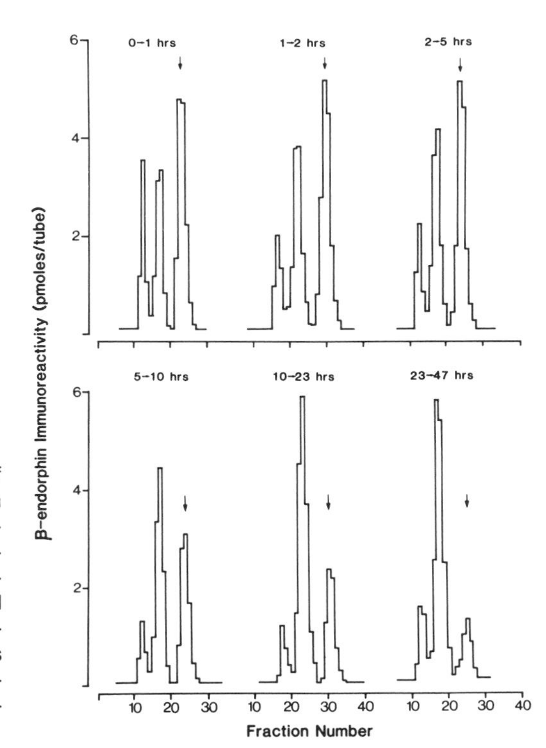

FIGURE 8. Time-dependent release of β-endorphin-related peptides from monolayer cultures of rat anterior pituitary cells treated with corticotropin-releasing factor (CRF). The culture media were removed at each time interval and replaced by fresh medium containing 10^{-7} M CRF. The secreted peptides were analyzed by gel filtration on Sephadex G75. The arrow indicates the elution position of [^{125}I]β-endorphin.

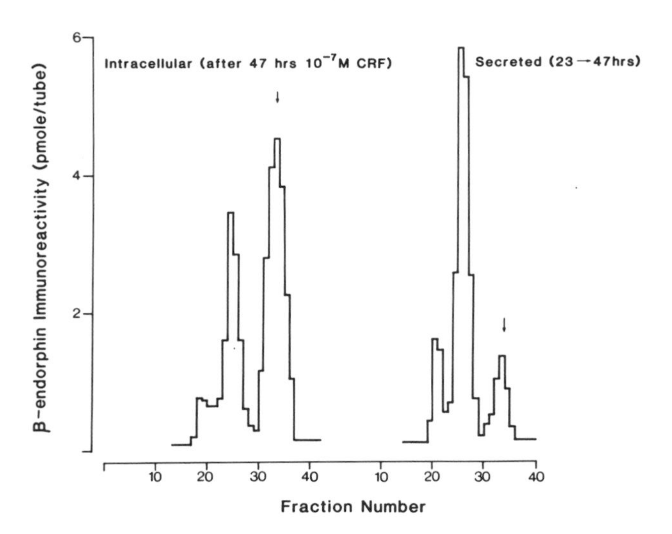

FIGURE 9. Comparison of intracellular with secreted β-endorphin-related peptides after chronic stimulation with 10^7 M CRF. The peptides were resolved as in Fig. 8.

longed stimulation, however, the proportion of secreted lipotropin to β-endorphin increased progressively, to the point that by 24 hr lipotropin was essentially the sole component (Fig. 8).

It is important to note that the composition of the peptides released under chronic stimulation was different from the average intracellular composition of the cells, which contained β-endorphin$_{1-31}$ as the principal peptide (Fig. 9). Thus, even after chronic stimulation of secretion, the pituitary cells still contained maturely processed products, and the less-processed peptides that were secreted under these conditions could not be attributed to an inability of the cell machinery to maintain the normal rate of processing.

The results point to the existence of multiple populations of granules within a single cell or to the existence of more than one cell type from which different segments of the prohormone sequence can be secreted. In either case, it is clear that secretion can occur from different pools, which may respond to specific releasing factors. The presence of approximately equal amounts of lipotropin and β-endorphin in rat anterior pituitary *in situ* (Fig. 3) could be attributed to the presence of certain granules (or cells) that process the prohormone only as far as lipotropin, together with other granules that contain an additional processing enzyme and produce β-endorphin as the end product. The experiments with the cultured pituitary cells indicate that specific signals may be capable of recruiting different peptides to enter the secretory pathway, thus providing a degree of flexibility in the control mechanisms. It should perhaps be restated that the secretion of distinctive packages of peptides could be interpreted in terms of cellular heterogeneity, or it could take place from multiple secretory granules.

5. BIOLOGICAL SIGNIFICANCE OF FLEXIBILITY IN PROHORMONE-PROCESSING REACTIONS

Evidence is now accumulating that prohormone-processing reactions respond to physiological and pharmacological signals. Thus, the peptides released from a cell may differ markedly from the average intracellular composition. On this basis, when a series of biologically active peptides is generated from a common prohormone, mechanisms appear to exist that permit the secretion of the various activities in different combinations. Specific triggering of a cell may thus lead to the release of intricate patterns of neuroactive peptides, which can be varied to meet the environmental requirements.

Recent studies with [^{125}I] β-endorphin have shown that the radiolabeled peptide administered intravenously in the rat appears to "home" on to the gastric antrum and small intestine (Ham *et al.*, 1984). If this relatively specific binding is indicative of biological function, a possible role for β-endorphin released from the pituitary might be to inhibit contraction of the gastrointestinal tract during stress, whereas the ACTH that is released concomitantly acts on the adrenal gland. Our finding that processing reactions are dynamic suggests the possibility that ACTH can be accompanied by differing proportions of β-endorphin according to the need for each peptide. By analogy, the formation of bioactive peptides from the adrenal prohormone of methionine enkephalin may also be influenced by physiological signals. This prohormone contains at least seven different forms of enkephalin, each one comprising an identical pentapeptide sequence but followed by a different "address" sequence (Comb *et al.*, 1982), and it appears that these peptides have different opiate specificities (Lord *et al.* 1977). Variations in the processing of the enkephain prohormone would lead to changes in the proportions of the different opiate peptides produced, and it seems likely that the resulting patterns will be found to be associated with the requirement for different neuronal activities.

A further and potentially exciting aspect in the production of a number of bioactivities from a single prohormone is that certain peptides may act at different receptors in a synergistic manner, functioning to produce cooperative effects. The flexibility that can take place in prohormone processing, shown in our experiments, suggests that a new level may exist at which the regulation of these multiple activities is attained.

REFERENCES

Bloom, F., Battenberg, E., Rossier, J., Ling, N., and Guillemin, R., 1978, Neurons containing β-endorphin in rat brain exist separately from those containing enkephalin: Immunocytochemical studies, *Proc. Natl. Acad. Sci. U.S.A.* **75:**1591–1595.

Bradbury, A. F., Smyth, D. G., and Snell, C. R., 1975, Biosynthesis of α-MSH and ACTH, in: *Peptides, Chemistry, Structure and Biology* (R. Walter and J. Meienhofer, eds.), Ann Arbor Press, Ann Arbor, pp. 609–615.

Bradbury, A. F., Smyth, D. G., Snell, C. R., Deakin, J. F. W., and Wendlandt, S., 1977, Comparison of the analgesic properties of lipotropin C-fragment and stabilized enkephalins in the rat, *Biochem. Biophys. Res. Commun.* **64:**748–753.

Carter, R. U., Shuster, S., and Morley, J. S., 1979, Melanotropin potentiating factor in the C-terminal tetrapeptide of human β-lipotropin, *Nature* **279:**74–75.

Comb, M., Seeberg, P. H., Adelman, J., Eiden, L., and Herbert, E., 1982, Primary structure of the human Met- and Leu-enkephalin precursor and its mRNA, *Nature* **295:**663–666.

Deakin, J. F. W., Doströvsky, J. O., and Smyth, D. G., 1980, Influence of NH_2-terminal acetylatin and COOH-terminal proteolysis on the analgesic activity of β-endorphin, *Biochem. J.* **189:**501–506.

Feldberg, W. S., and Smyth, D. G., 1977, C-Fragment of lipotropin—an endogenous potent analgesic peptide, *Br. J. Pharmacol.* **60:**445–454.

Geisow, M. J., Doströvsky, J. O., and Smyth, D. G., 1977, Analgesic activity of lipotropin C-fragment depends on carboxyl terminal tetrapeptide, *Nature* **269:**167–168.

Ham, J., and Smyth, D. G., 1984, Regulation of bioactive β-endorphin in rat pars intermedia, *FEBS Lett.* **175:**407–411.

Ham, J., and Smyth, D. G., 1985, β-Endorphin processing in pituitary and brain is sensitive to haloperidol stimulation, *Neuorpeptides* **5:**497–500.

Ham, J., and Smyth, D. G., 1986, β-Endorphin and ACTH related peptides in primary cultures of rat anterior pituitary cells: Evidence for different intracellular pools, *FEBS Lett.* **190:**253–258.

Ham, J., McFarthing, K. G., Toogood, C. I. A., and Smyth, D. G., 1984, Influence of dopaminergic agents on β-endorphin processing in rat pars intermedia, *Biochem. Soc. Trans.* **12:**927–929.

Hazum, E., Chang, K. J., and Cuatrecasas, P., 1979, Specific nonopiate receptors for β-endorphin, *Science* **205:**1033–1035.

Höllt, V., and Bergmann, M., 1982, Effects of acute and chronic haloperidol treatment on the concentrations of immunoreactive β-endorphin in plasma, pituitary and brain of rats, *Neuorpharmacology* **21:**147–154.

Ling, N., Burgus, R., and Guillemin, R., 1976, Isolation, primary structure and synthesis of α-endorphin and γ-endorphin, two peptides of hypothalamus–hypophysial origin with morphinomimetric activity, *Proc. Natl. Acad. Sci. U.S.A.* **73:**3942–3946.

Loh, H. H., Tseng, L. F., Wei, E., and Li, C. H., 1976, β-Endorphin is a potent analgesic agent, *Proc. Natl. Acad. U.S.A.* **73:**2895–2898.

Lord, J. A. H., Waterfield, A. A, Hughes, J., and Kosterlitz, H. W., 1977, Endogenous opioid peptides: Multiple agonists and receptors, *Nature* **267:**495–499.

Parish, D. C., Smyth, D. G., Normanton, J. R., and Wolstencroft, J. H., 1983, Glycylglutamine, an inhibitory neuropeptide derived from β-endorphin, *Nature* **306:**267–270.

Roberts, J. L., and Herbert, E., 1977, Characterization of a common precursor to corticotropin and β-lipotropin: Cell free synthesis of the precursor and identification of corticotropin peptides in the molecule, *Proc. Natl. Acad. Sci. U.S.A.* **74:**4826–4830.

Smyth, D. G., and Zakarian, S., 1980, Selective processing of β-endorphin in regions of porcine pituitary, *Nature* **288:**613–615.

Smyth, D. G., Massey, D. E., Zakarian, S., and Finnie, M. D. A., 1979, Endorphins are stored in biologically active and inactive forms; isolation of α,N-acetyl peptides, *Nature* **279:**252–254.

Smyth, D. G., Snell, C. R., and Massey, D. E., 1978, Isolation of lipotropin C-fragment from porcine pituitary and C-fragment from brian, *Biochem. J.* **175:**261–270.

Smyth, D. G., Smith, C. C. F., and Zakarian, S., 1981a, Isolation and identification of two new peptides related to β-endorphin, in: *Advances in Endogenous and Exogenous Opioids* (H. Takagi, ed.), Kodansha–Elsevier, Tokyo–Amsterdam, pp. 145–148.

Smyth, D. G., Zakarian, S., Deakin, J. F. W., and Massey, D. E., 1981b, β-Endorphin related peptides in the pituitary gland: Isolation, identification and distribution, in: *Peptides of the Pars Intermedia* (G. Lawrenson and D. C. Evered, eds.), Pitman Medical, London, pp. 79–100.

Toogood, C. I. A., McFarthing, K. G., Hulme, E. C., and Smyth, D. G., 1984, [^{125}I-Tyr^{27}]β-Endorphin, a radiolabelled derivative of β-endorphin for study of opiate receptors, *Neuropeptides* **5:**121–124.

Van Ree, J. M., Gaffori, O., and Kiraly, I., 1984, β-Endorphin and N-acetyl-β-endorphin interfere with distinct dopaminergic systems in the nucleus accumbens via opioid and non-opioid mechanisms, *Life Sci.* **34:**1317–1324.

Watson, S. J., Verhoef, J., Burbach, J. P. H., van Amerogen, A., Gaffori, O., Sitsen, J. M. A., and de Wied, D., 1985, N^{α}-acetyl-γ-endorphin is an endogenous non-opioid neuropeptide with biological activity, *Life Sci.* **36:**2277–2285.

Zakarian, S., and Smyth, D. G., 1979, Distribution of active and inactive forms of endorphins in the pituitary and brain of the rat, *Proc. Natl. Acad. Sci. U.S.A.* **76:**5972–5976.

Zakarian, S., and Smyth, D. G., 1982a, β-Endorphin is processed differently in specific regions of rat pituitary and brain, *Nature* **296:**250–253.

Zakarian, S., and Smyth, D. G., 1982b, Distribution of β-endorphin related peptides in rat pituitary and brain, *Biochem. J.* **202:**561–571.

II

Peptide-Releasing Factors

7

Endocrine, Gastrointestinal, and Antitumor Activity of Somatostatin Analogues

ANDREW V. SCHALLY, REN-ZHI CAI, I. TORRES-ALEMAN, TOMMIE W. REDDING, BALAZS SZOKE, DADIN FU, MARION T. HIEROWSKI, JOHN COLALUCA, and STANISLAW KONTUREK

1. INTRODUCTION

Since recent studies with somatostatin were carefully reviewed by Reichlin (1983) and by a variety of authors in the *2nd International Symposium on Somatostatin,* which was held in Athens in 1981 but published with updates in 1984 (Raptis *et al.*, 1984), no attempts are made to cover new developments with somatostatin-14 and somatostatin-28. Instead, this chapter focuses on the recent work in the field of somatostatin analogues and the investigations of their endocrine, gastrointestinal, and antitumor activities.

2. BACKGROUND

Somatostatin is of little therapeutic value since it has a wide spectrum of biological actions and a short half-life (Pfeiffer and Schusdziarra, 1984; Schally *et al.*, 1978). Most somatostatin analogues have therefore been designed for selective, enhanced, and prolonged activities. The analogues also provide insight into struc-

ANDREW V. SCHALLY, REN-ZHI CAI, I. TORRES-ALEMAN, TOMMIE W. REDDING, BALAZS SZOKE, DADIN FU, MARION T. HIEROWSKI, and JOHN COLALUCA • Veterans Administration Medical Center and Tulane University School of Medicine, New Orleans, Louisiana 70146. STANISLAW KONTUREK • Medical Academy, Krakow, Poland.

ture–function relationships of somatostatin. Many analogues have been synthesized and screened for multiple activities (Schally *et al.*, 1978, 1984b; Gottesman *et al.*, 1982). Several investigators have examined the systematic incorporation of D-amino acids into the somatostatin backbone and obtained some analogues with greatly increased activities (Rivier *et al.*, 1975, 1977; Brown *et al.*, 1977; Meyers *et al.*, 1977). D-Amino acid substitutions offer resistance to degradative enzymes and thereby prolong activity. Other successful strategies in designing analogues of peptide hormones have included derivatization of various functional groups and deletion, addition, or exchange of amino acids (Grant *et al.*, 1976; Meyers *et al.*, 1978). These investigations have been reviewed previously (Schally *et al.*, 1978, 1984b; Gottesman *et al.*, 1982), and only recent studies on antitumor activities of older analogues are mentioned.

3. SYNTHESIS AND ENDOCRINE EVALUATION OF MODERN ANALOGUES OF SOMATOSTATIN

3.1. Work by Other Groups

Veber *et al.* (1979, 1981) carried out conformational analysis and designed several analogues by replacing nine of the 14 amino acids of somatostatin with a single proline residue. Some of the resulting hexapeptide analogues, particularly cyclo(-Pro-Phe-D-Trp-Lys-Thr-Phe), were reported to be highly active in the inhibition of growth hormone, insulin, and glucagon after parenteral or oral administration (Veber *et al.*, 1981). More recently, the same group reported the synthesis of another series of analogues. Among a large number of cyclic hexapeptides synthesized, the analogue cyclo(N-Me-Ala-Tyr-D-Trp-Lys-Val-Phe) was stated to be 50–100 times more potent than somatostatin in inhibiting growth hormone, insulin, and glucagon (Veber *et al.*, 1984). Studies in diabetic animals showed improved control of postprandial hyperglycemia when this analogue was given in combination with insulin.

Bauer *et al.* (1982) synthesized another series of highly potent octapeptide analogues of somatostatin. They retained the sequence 7–10 of somatostatin, Phe-Trp-Lys-Thr, originally proposed as essential by Veber *et al.* (1981), and incorporated this sequence with the Trp residue in the D configuration as in the original Veber *et al.* (1981) analogue into a series of cystine-bridged analogues of which D-Phe-Cys-Phe-D-Trp-Lys-Thr-Cys-Thr-OL containing a C-terminal amino alcohol was the most active (Bauer *et al.*, 1982). This analogue was reported to be long acting and 45–70 times more potent than somatostatin in tests on inhibition of growth hormone secretion in rats and monkeys. Moreover, this analogue suppressed GH secretion more selectively than that of the islet hormones, its potency for the inhibition of insulin or glucagon being three and 23 times greater, respectively, than that for somatostatin. This analogue was active parenterally and orally and was subjected to careful clinical evaluation (Marbach *et al.*, 1984).

3.2. Work in Our Laboratory

In our early studies in various animal tumor models, we used analogues of somatostatin-14 and detected a significant antitumor activity for some analogues that contained disulfide linkages (Redding and Schally, 1983, 1984; Schally *et al.*, 1984a,c; De Quijada *et al.*, 1983). In some of these tests the Veber *et al.* (1981) analogue was devoid of antitumor activity. Since the Bauer (1982) analogue was about ten times more potent for inhibition of GH release than the Veber analogue and had disulfide linkages, which we speculatively considered necessary for antitumor activity, we decided to concentrate on the syntheses of a series of analogues related to the Bauer *et al.* (1982) octapeptide but all containing a C-terminal amide. Having had much experience with the successful use of microcapsules of D-Trp-6-LH-RH in poly(DL-lactide-coglycolide) for once-a-month intramuscular administration (Redding *et al.*, 1984), we planned to encapsulate our somatostatin analogues and were thus not too interested in oral activity.

Thus, nearly 200 analogues of somatostatin were synthesized by solid-phase methods (Merrifield, 1967), employing Beckman model 990 automatic peptide synthesizers or the usual glass reaction vessels (Cai *et al.*, 1985). Synthetic peptides were purified by gel filtration on Sephadex, MPLC, and HPLC. After purification and before antitumor tests, the analogues were tested biologically for their inhibitory effect *in vivo* on GH, glucagon, and insulin release and also for inhibition of gastric acid, secretin, and CCK. Inhibition of release of GH was measured in adult male rats anesthetized with sodium pentobarbital. Blood samples were taken from the jugular vein 15 min after peptide injection, and serum GH was assayed by RIA. For some analogues, doses as low as 0.005–0.02 μg/100 g were active. Inhibition of release of insulin and glucagon was determined in rats fasted for 30 hr and anesthetized with pentobarbital. Somatostatin analogues were injected into the jugular vein, and 5 min later blood was collected from the hepatic portal vein for insulin and glucagon determinations. In all assays the responses were compared to those elicited by standard doses of somatostatin-14.

The analogues Ac-Phe-Cys-Phe-D-Trp-Lys-Thr-Cys-Thr-NH_2 (RC-101-I) and D-Phe-Cys-Phe-D-Trp-Lys-Thr-Cys-Trp-NH_2 (RC-95-I) were 71 times and 53 times more potent, respectively, than somatostatin (Cai *et al.*, 1985). In many cases, the activity of somatostatin analogues could be enhanced by incorporation of Tyr and Val in positions corresponding to residues 7 and 10, respectively, of somatostatin-14 (SS-14). Thus, the analogues D-Phe-Cys-Tyr-D-Trp-Lys-Val-Cys-Thr-NH_2 (RC-121) and D-Phe-Cys-Tyr-D-Trp-Lys-Val-Cys-Trp-NH_2 (RC-160) were 177 times (17700%) and 113 times (11300%) more potent, respectively, than SS-14 in tests for inhibition of growth hormone release, based on four-point assays (Cai *et al.*, 1985). This activity is two to four times greater than that reported for the analogue of Bauer *et al.* (1982). A prolonged duration of action of at least 3 hr was also observed. Both series of these analogues (Cai *et al.*, 1985) are selective for suppressing GH release, the inhibitory potencies for insulin and glucagon in rats being much smaller. Thus, the insulin release inhibitory activity of analogues Ac-Phe-

Cys-Phe-D-Trp-Lys-ThrCys-Thr-NH_2 (RC-101-I), D-Phe-Cys-Phe-D-Trp-Lys-Thr-Cys-Trp-NH_2 (RC-95-I), Ac-Phe-Cys-Tyr-D-Trp-Lys-Val-Cys-Thr-NH_2 (RC-121), and D-Phe-Cys-Tyr-D-Trp-Lys-Val-Cys-Trp-NH_2 (RC-160) was only 23.4, 34.5, 8.7, and 6.2 times greater, respectively, than that of somatostatin-14. Inhibition of glucagon release for analogues RC-121 and RC-160 was several times greater than that for insulin. Since these analogues block GH response and suppress glucagon more than insulin release, this selectivity of action could make them useful for the treatment of diabetic retinopathy and diabetes mellitus type II (not dependent on insulin). Such analogues could also possibly serve as adjuncts to insulin in type I (insulin-dependent) diabetes.

4. EFFECT OF SOMATOSTATIN ANALOGUES ON GASTRIC AND PANCREATIC SECRETION

In previous studies it was demonstrated that analogues obtained by substitutions of certain residues in somatostatin-14 possess a spectrum of biological actions on gastrointestinal secretions similar to somatostatin (Schally *et al.*, 1984b; Konturek *et al.*, 1977; Brown *et al.*, 1978). Thus, D-Trp^8-somatostatin and D-Cys^{14}-somatostatin and other analogues were tested for inhibition of pentagastrin-induced gastric secretion *in vivo* in cats (Brown *et al.*, 1978; Hirst *et al.*, 1980) and dogs (Konturek *et al.*, 1977). For some analogues, the gastric acid and pepsin inhibitory activity was smaller than their potency for suppressing GH release *in vitro*. D-Trp^8-somatostatin and D-Cys^{14}-SS also inhibited pancreatic bicarbonate secretion induced by secretin but were essentially equipotent to somatostatin in gastric and pancreatic secretion (Konturek *et al.*, 1977).

However, there is much less information on the effect of modern analogues of somatostatin on gastric and pancreatic secretion. Recently, Hirst *et al.* (1984) reported that the gastric acid and pepsin inhibitory activity of the analogue of Veber *et al.* (1981), cyclo(Pro-Phe-D-Trp-Lys-Thr-Phe), in cats with gastric fistulae was reduced compared with somatostatin. Moreover, the inhibition was not dose related in the range 10–50 μg/kg per hr.

Two different methods for measuring inhibition of gastrin effects have been used in our laboratory. The first is inhibition of pentagastrin- or desglugastrin-induced gastric acid secretion in fasted dogs with gastric fistulae (Konturek *et al.*, 1977). Inhibitory responses obtained with the analogue are compared to those induced by SS-14. Similar potencies were obtained if SS-14 or analogues are infused in saline or 0.5% dog serum albumin. The second method involves inhibition of gastric acid in adult male rats equipped with chronic gastric fistulae (Gibbs *et al.*, 1973). Des-Glu-gastrin, 5 μ/kg (s.c.), was used as a stimulant of gastric acid secretion. Thirty minutes after the des-Glu-gastrin is injected, SS-14 or the analogues are given s.c., and gastric juice samples are collected.

Using these methods we demonstrated that the gastric acid inhibition activity of cyclo(Pro-Phe-D-Trp-Lys-Thr-Phe) in dogs was 45% that of somatostatin-14 in dogs on a weight basis. Good dose responses were obtained in the range of 2.5–5 μg/kg per hr. In rats the gastric acid inhibitory activity of the cyclic hexapeptide was 132% that of somatostatin-14 on a weight basis.

We have also investigated the effects of cyclo(Pro-Phe-D-Trp-Lys-Thr-Phe) (cyclohexapeptide) of Veber *et al.* (1981) on the pancreatic response to endogenous stimulants (meal and duodenal acidification) and exogenous secretin and CCK in dogs with esophageal, gastric, and pancreatic fistulae (Konturek *et al.*, 1985). Cyclohexapeptide given i.v. in graded doses against a constant background stimulation with secretin inhibited pancreatic HCO_3 and protein secretion, being about twice as potent as somatostatin-14 (SS-14). Cyclohexapeptide also caused an inhibition of pancreatic secretion induced by feeding a meat meal, sham feeding, and duodenal acidification, or infusion of CCK. The inhibition of pancreatic secretion by cyclohexapeptide resulted in part from direct inhibitory action on the exocrine pancreas as well as from the suppression of the release of secretin, insulin, and pancreatic polypeptide. It was concluded that in dogs cyclohexapeptide is a more potent inhibitor of pancreatic secretion than SS-14 (Konturek *et al.*, 1985).

The activity of the original Bauer *et al.* (1982) analogue SMS 201-995 and several analogues related to it on des-Glu-gastrin-stimulated gastric secretion was also investigated in dogs. The potency of the Bauer *et al.* (1982) analogue on the inhibition of gastric secretion was nearly four times greater (367%) on a weight basis than that of somatostatin-14. On a molar basis (since the mol. wt. of the analogue is 1019 versus 1637 for SS-14), the gastric acid inhibitory activity would be only 228% higher. This activity is much smaller than that for inhibition of growth hormone *in vivo* and corresponds roughly to the values of the insulin inhibitory potency. This finding indicates that residues 1–6 and/or 12–14 in the amino acid chain of somatostatin are important for gastric acid inhibitory activity. The same range of gastric acid inhibitory potencies was found for analogues *p*-Cl-D-Phe-Cys-Tyr-DTrp-Lys-Val-Cys-Thr-NH_2 (RC-88-II), D-Phe-Cys-Tyr-D-Trp-Lys-Val-Cys-Thr-NH_2 (RC-121), and D-Phe-Cys-Tyr-D-Trp-Lys-Val-Cys-Trp-NH_2 (RC-160), which were 216%, 475%, and 433% more potent, respectively, than SS-14. Thus, all these analogues had a much smaller potency for inhibition of gastric acid than of GH *in vivo*, showing a marked selectivity in their activity. Nevertheless, the ability to inhibit plasma gastrin, gastric acid, and pepsin release and the secretion of pancreatic enzymes could make them potentially useful therapeutic adjuncts for the treatment of ulcers and acute pancreatitis (Konturek, 1977; Dollinger *et al.*, 1976; Usadel *et al.*, 1980; 1984).

5. ANTITUMOR ACTIVITY OF SOMATOSTATIN ANALOGUES: STUDIES IN ANIMAL MODELS

A number of analogues of somatostatin-14 as well as the cyclic hexapeptide of Veber *et al.* (1981) and several analogues related to the octapeptide of Bauer *et al.* (1982) were tested in various animal tumor models.

5.1. Prostate Cancer

It is possible that some somatostatin analogues could be used as adjuncts together with LH-RH agonists in the treatment of androgen-dependent prostatic carcinoma. This hypothesis is based on the evidence that prolactin may be a promoter of prostate growth and could be involved in prostate cancer as a cofactor (Schally *et al.*, 1984a,c). The reduction in prolactin levels produced by the administration of a suitable somatostatin analogue, combined with the decrease in testosterone values that results from chronic treatment with LH-RH agonists, may lead to a greater inhibition of prostate tumors than that which can be obtained with LH-RH agonists alone. The administration of somatostatin analogues also decreases GH levels, and this might also contribute to an additional inhibition of tumor growth.

Preliminary evidence obtained in animal models of prostate tumors is in agreement with this view. In the Dunning R3327H model of prostate adenocarcinoma in Copenhagen-Fisher rats, twice daily s.c. administration of 25μg D-Trp^6-LH-RH or L-5F-Trp^8-D-Cys^{14}-somatostatin significantly reduced the percentage change in tumor volume. The combination of somatostatin analogue with the LH-RH agonist resulted in an even greater decrease in tumor volume than that obtained with either peptide given alone. The investigation of the effectiveness of various modern analogues of somatostatin given alone or combined with injectable microcapsules of D-Trp^6-LH-RH, designed for a constant, controlled release of this analogue over a 30-day period, is in progress in the Dunning model and in the human lines of prostate carcinoma transplanted into nude mice. The aim of these studies is to determine whether the combination of LH-RH agonists with somatostatin analogue could result in an increase in the therapeutic response in prostate cancer.

5.2. Breast Cancer

It has been established that approximately 30–40% of all breast cancers are estrogen dependent and can show a regression following endocrine manipulations such as the antiestrogen tamoxifen or LH-RH agonist therapy. Various studies support the hypothesis that prolactin may play a role in the growth of breast neoplasms, not only in rodents but also in humans (Schally *et al.*, 1984a). The presence of prolactin receptors in mammary tumors; including human breast cancer cells, also supports the concept that a certain proportion of mammary tumors can be prolactin dependent. In addition, it has been suggested that growth hormone and somatomedins may be involved in the growth of human breast cancer. Thus, reduction in prolactin and GH levels induced by administration of a suitable somatostatin analogue, combined with the decrease in estrogen values that results from chronic treatment with LH-RH agonist or antagonist, could lead to a greater inhibition of mammary tumors than can be obtained with LH-RH analogue alone. If this view is correct, somatostatin analogues could be tried as adjuncts together with agonistic or antagonistic analogues of LH-RH in the treatment of breast cancer. Rose *et al.* (1983) showed that in female rats bearing N-nitrosomethylurea (NMU)-

induced mammary carcinomas, treatment with pergolide mesylate or with somatostatin analogue cyclo(ω-aminoheptanoic acid-Cys-Phe-D-Trp-Lys-Thr-Cys) did not induce tumor regression when these drugs were given alone. However, when the analogue of somatostatin was combined with pergolide, it produced tumor regression. The results were interpreted that simultaneous inhibition of growth hormone and prolactin levels can induce the regression of mammary carcinoma in the NMU-model (Rose *et al.*, 1983).

Our own studies in Wistar/Furth rats bearing estrogen- and prolactin-dependent MT/W9A mammary adenocarcinoma showed that once-a-month administration of microcapsules with D-Trp^6-LH-RH or daily injections of somatostatin analogue Ac-*p*-Cl-D-Phe-Cys-Phe-D-Trp-Lys-Thr-Cys-Thr-NH_2 (RC-15) can inhibit the growth of this tumor. However, the first study conducted over a period of only 27 days did not demonstrate any significant synergism between the LH-RH agonists and the somatostatin analogue. Other studies with somatostatin analogues and D-Trp^6-LH-RH agonist are in progress in mice with MXT mammary tumors and in rats with MT/W9A breast carcinoma. If these experimental studies are successful, in future clinical trials, breast tissue specimens from women with breast cancer could be screened not only for levels of receptors for estrogen and progesterone, which predict the response to estrogen deprivation, but also for the presence of membrane binding sites for prolactin. Consequently, a more rational therapy could be initiated based on the status of the receptors.

5.3. Pituitary Tumors

Pituitary tumors are presently treated by neurosurgery or dopamine agonists such as 2-bromo-α-ergocriptine (bromocriptine). Bromocriptine is more effective for prolactinomas than for tumors associated with acromegaly. Various studies in animals and in man have demonstrated that somatostatin-14 and its analogues can decrease the release of GH, TSH, prolactin, and ACTH from the pituitary (Hall and Gomez-Pan, 1976; Schally *et al.*, 1984b), but except for isolated trials, these findings have not been applied therapeutically (see also Section 6). More work may be needed in animal models of pituitary tumors.

We have previously demonstrated that chronic administration of D-5-methoxy-Trp^8-somatostatin inhibited the growth of the prolactin- and ACTH-secreting pituitary tumor 7315a in female Buffalo rats (De Quijada *et al.*, 1983). Lamberts *et al.* (1984) also showed that the Bauer *et al.* (1982) analogue inhibits the growth of rat pituitary tumor 7315a by 40%. However, the treatment with the analogues cyclo(Pro-Phe-D-Trp-Lys-Thr-Phe) and D-Phe-Cys-Phe-D-Trp-Lys-Thr-Cys-Thr-NH_2 appeared to enhance the weight and volume of the GH- and prolactin (PRL)-secreting pituitary GH_3 tumor in rats. Prolactin levels were increased during the treatment (Torres-Aleman *et al.*, 1985). The growth of both 7315a and GH_3 pituitary tumor was markedly inhibited by treatment with D-Trp^6-LH-RH. Serum GH and PRL levels were also reduced. This unexplained stimulation of the growth of the GH_3 tumor by the two somatostatin analogues may be related to the dosage administered

(5 μg b.i.d.) or to the experimental conditions. Additional investigations, experimental and clinical, are needed to determine whether certain somatostatin analogues, LH-RH agonists, or their combination would be useful under clinical conditions for the management of patients with pituitary tumors.

5.4. Tumors of Bone and Cartilage

The incidence of osteosarcomas is a function of the amount of cellular activity in the bones and may be influenced by growth hormone. Hormonal factors also play an important role in the growth and differentiation of normal and malignant cartilage tissue. Swarm rat chondrosarcoma is a malignant, transplantable, hormone-responsive tumor, similar to human tumors, whose growth is dependent on GH, somatomedins, glucocorticoids, and insulin (McCumbee *et al.*, 1980; Salomon *et al.*, 1979). This growth hormone dependence of rat chondrosarcomas prompted us to test experimentally the hypothesis that somatostatin analogues, by inhibiting the GH secretion, could have potential therapeutic implications in treatment of this neoplasm (Redding and Schally, 1983). D-Trp^6-LH-RH was also tried because it causes inhibition of pituitary gonadotropin release and sex-hormone deprivation. In male Sprague–Dawley rats bearing transplanted swarm chondrosarcoma, chronic administration of *p*NH$_2$-Phe^4-SS-14, D-5F-Trp^8-SS-14, or D-5-methoxy-Trp^8-SS-14 significantly reduced tumor weight and/or volume.

Administration of 30 μg/day for 30 days of *p*NH$_2$-Phe^4-SS-14 decreased tumor weights by 38% compared to controls. Treatment of tumor rats with 30 μg b.i.d. of D-5F-Trp^8-SS-14 for 14 to 21 days decreased tumor weights by 32–46% ($P < 0.01$) and tumor volume by 50–60% versus controls. D-Methoxy-Trp^8-SS-14 (30 μg b.i.d.) decreased tumor volume by 27% ($P < 0.05$) as compared to controls. In three different experiments, D-Trp^6-LH-RH in doses of 30 μg/day administered alone or with somatostatin analogues also reduced the weight and/or volume of chondrosarcomas. Growth hormone and prolactin levels were significantly decreased in rats treated with D-5F-Trp^8-SS-14 or with D-Trp^6-LH-RH. There appeared to be a greater suppression of serum GH and prolactin levels when D-5F-Trp^8-SS was administered together with D-Trp^6-LH-RH.

Several experiments were also carried out in mice with Dunn osteosarcomas. In the first experiment, L-5F-Trp^8-D-Cys^{14}-SS-14 prolonged the survival rate by 86%, and D-Trp^6-LH-RH by 29% (Schally *et al.*, 1984a,c). In the second experiment, the prolongation of survival of mice with Dunn osteosarcomas was 37% in the case of D-Trp^6-LH-RH and 73% for L-5Br-Trp^8-D-Cys^{14}-SS-14 (Schally *et al.*, 1984a,c).

The analogues Ac-*p*-Cl-Phe-Cys-Phe-D-Trp-Lys-Thr-Cys-Thr-NH$_2$ (RC-15) and D-Phe-Cys-Tyr-D-Trp-Lys-Val-Cys-Trp-NH$_2$ (RC-160-2H) in doses of 2–5 μg b.i.d. appeared to have antitumor activities as shown by an increased survival rate in mice bearing the Dunn osteosarcoma.

The inhibitory effect of analogues of somatostatin on the growth of chondrosarcomas can most likely be explained by suppression of GH levels. However, for some analogues, there is no correlation between GH release-inhibiting activity and antitumor activity. Cyclic hexapeptide (Veber *et al.*, 1981), which is five times more active than somatostatin-14 in inhibiting GH release, did not prolong the survival of mice with osteosarcomas. The analogues L-5-Br-Trp^{8}-D-Cys^{14}-SS or L-5-F-Trp^{8}-D-Cys^{14}-SS, which are only about twice as potent as somatostatin-14 in inhibiting GH release *in vivo,* were most effective in suppressing Dunn osteosarcomas. It is thus possible that some somatostatin analogues have a direct antiproliferative effect on tumor tissue by preventing metastases to the lungs. Several classes of somatostatin analogues must be investigated in order to find a correlation between structure and antitumor activity. The mechanism of suppression of these tumors by D-Trp^{6}-LH-RH is not clear, but the state of deprivation of sex hormones induced by therapy with D-Trp^{6}-LH-RH may affect estrogen- or testosterone-dependent proteins in bone and cartilage.

Much additional experimental work is needed in animal models of these tumors to extend our observation. Inhibition of the swarm chondrosarcomas and Dunn osteosarcomas by analogues of somatostatin and LH-RH agonists raises some hope that they might lead to a new endocrine therapy for chondrosarcomas, osteosarcomas, and related hormone-dependent neoplasms. Treatment with derivatives of hypothalamic hormones could be of value in patients with osteogenic malignancies in whom conventional therapy has failed.

5.5. Pancreatic Carcinoma

Malignant exocrine tumors arise most frequently from the ducts, and thus most carcinomas are in the head of the pancreas (Snodgrass, 1977). Ductal types constitute about 80–90% of cases, and acinar 10–15%. Carcinoma of the head of the pancreas has a very poor prognosis.

Tumors of the endocrine pancreas (islet cells tumors) may be benign or malignant, and they may be functioning or nonfunctioning. Various experimental and clinical findings suggest that it might be possible to develop a new hormonal therapy for some malignant tumors of the pancreas based on somatostatin analogues. Somatostatin and some of its analogues inhibit secretion and/or action of gastrin, secretin, cholecystokinin (CCK), and vasoactive intestinal polypeptide (VIP) (Raptis *et al.*, 1984; Schally *et al.*, 1978). Gastrin, cholecystokinin, and secretin promote the growth of the exocrine pancreas and increase DNA, RNA, and protein content. This ability to produce hyperplasia and hypertrophy of the exocrine pancreas is now a well-established action of gastrin, CCK, and secretin (Dembinski and Johnson, 1980). It is likely that gastrointestinal hormones influence the growth of the malignant cells of the pancreas as well. Cholecystokinin, secretin, and gastrin can stimulate the growth of pancreatic adenocarcinoma cells, duodenal cells, gastric mucosa cells, and rat stomach cancer cells in tissue cultures. Townsend *et al.* (1981) have shown that cerulein, which is structurally closely related to CCK, given together with secretin, stimulated the *in vivo* growth of pancreatic ductal adenocarcinoma H-2-T in golden hamsters.

Clinical studies have demonstrated that somatostatin or its analogues can inhibit the secretion of insulin in patients with insulinomas, glucagon in cases of glucagonoma, as well as the secretion of ectopic endocrine tumors of the pancreas such as gastrin-producing tumors in Zollinger–Ellison syndrome and VIP in Verner–Morrison syndrome (Bloom *et al.*, 1974; Mortimer *et al.*, 1974; Adrian *et al.*, 1981; Long *et al.*, 1979). However, neither somatostatin nor its analogues have been tried chronically in patients with carcinoma of the pancreas.

Using animal models of pancreatic cancer acinar and ductal type, we investigated the effect of analogues of hypothalamic hormones on the growth of pancreatic tumors (Redding and Schally, 1984). In Wistar/Lewis rats bearing the acinar pancreatic tumors DNCP-322, chronic administration of L-5Br-Trp^8-somatostatin significantly decreased tumor weight and volume. The cyclic hexapeptide of somatostatin, cyclo(Pro-Phe-D-Trp-Lys-Thr-Phe), failed to influence the growth of this tumor (Redding and Schally, 1984); D-Trp^6-LH-RH also significantly decreased tumor weight and volume. In Syrian hamsters with a ductal form of pancreatic cancer, chronic administration of L-5-Br-Trp^8-SS at a dose of 20 μg b.i.d. for 21–30 days diminished tumor weight and volume. Percentage change in tumor volume was significantly decreased compared to control animals.

In one experiment cyclic hexapeptide of somatostatin also inhibited the growth of this tumor; D-Trp^6-LH-RH, given twice daily or injected in the form of constant-release microcapsules, significantly decreased tumor weight and volume. In these animals the testes were significantly reduced, and serum levels of testosterone were suppressed. Hamsters castrated 4 days after transplantation of the pancreatic tumor also showed a significant decrease in tumor weight and volume. This indicates that pancreatic cancers may be at least in part androgen sensitive. D-Trp^6-LH-RH might decrease the growth of pancreatic carcinomas by creating a state of sex hormone deprivation. The presence of specific hormone receptors for estrogen and androgen in the pancreas indicated that sex hormones could influence neoplastic cell processes in this tissue. These findings suggest that pancreatic adenocarcinoma may be sensitive to both gastrointestinal and sex hormones. Somatostatin analogues reduce the growth of pancreatic ductal and acinar cancers, probably by inhibiting the release and/or stimulatory action of gastrointestinal hormones on tumor cells. However, only scanty information exists on inhibition of secretin and CCK by SS analogues.

It is apparent that for some somatostatin analogues there is no correlation between GH and gastrin release-inhibiting activity and antitumor activity. Some analogues such as the cyclic hexapeptide of Veber *et al.* (1981), which is five times more active than somatostatin-14 in inhibiting GH and equipotent on gastrin release, did not inhibit the growth of rat acinar pancreatic tumors. Somatostatin-28 also had no effect on this tumor, although it is more potent and longer acting than SS-14 on insulin and GH and essentially equipotent on gastrin. On the other hand, L-5-Br-Trp^8-SS, which is only slightly more potent as somatostatin-14 in inhibiting GH and gastrin release *in vivo,* was very effective in suppressing rat pancreatic tumor. It is thus possible that some somatostatin analogues act directly on tumor tissue.

In any event, several classes of somatostatin analogues must be carefully investigated in order to find a correlation between structure and antitumor activity in pancreatic carcinoma models. Some of these analogues of somatostatin would also be useful in treatment of gastric and colon cancer. A combined administration of a suitable somatostatin analogue with an LH-RH agonist could inhibit cancers of the pancreas more effectively than somatostatin analogues alone. The inhibition of the growth of ductal and acinar pancreatic tumors by D-Trp^6-LH-RH or analogues of somatostatin obtained in our laboratory appears to be the first attempt at endocrine management of these tumors. These observations could be of clinical significance. Inhibition of animal models of pancreatic tumors in rats and hamsters in response to chronic administration of somatostatin analogues and D-Trp^6-LH-RH suggests that these compounds should be considered for the development of a new hormonal therapy for cancer of the pancreas.

5.6. New Hypothesis on the Possible Mechanism of Action of Somatostatin on Tumors

Somatostatin was shown to be a powerful growth inhibitor in a variety of systems, e.g., HeLa cells (Mascardo and Sherline, 1982). Possible biochemical mechanisms by which somatostatin may inhibit the growth of human pancreatic cancers were investigated in MIA PaCa-2 human pancreatic carcinoma line (Hierowski *et al.*, 1985). Binding of somatostatin to the membrane receptor in MIA PaCa-2 cell line activated dephosphorylation of a phosphotyrosyl membrane protein whose phosphorylation was promoted by epidermal growth factor (EGF). The EGF receptor consists of three regions, an EGF binding site located on the external cell surface, the transmembrane portion, and cytoplasmic tyrosine kinase domain. Phosphorylation of EGF receptor at tyrosine residues plays an important role in regulation of cellular proliferation and growth of normal tissue. Somatostatin nullifies the growth stimulation produced by EGF (Mascardo and Sherline, 1982). Our work shows that somatostatin reversed the stimulatory effect of EGF on phosphorylation of the tyrosine kinase portion of the EGF receptor in the MIA PaCa-2 pancreatic cancer line (Hierowski *et al.*, 1985).

It has been reported by Waterfield and his collaborators that the product of viral oncogene *erbB* is a truncated receptor for EGF (Downward *et al.*, 1984). The sequence encoded in *erbB* corresponds to the transmembrane and cytoplasmic portions of the EGF receptor (Downward *et al.*, 1984; Yamamoto *et al.*, 1983). Although the *erbB* oncogene may have been derived from the cellular gene for the epidermal growth factor, the gene segment coding for the EGF binding region was apparently not acquired by the virus (Downward *et al.*, 1984). No tyrosine kinase activity has been demonstrated so far for the product of the *erbB* oncogene, but other viral oncogenes are known to possess it. It has been hypothesized that certain oncogenes or some of their products such as growth factors or aberrant receptors may stimulate cell division. In the case of the *erbB* oncogene, since its products lack the binding region for growth factors, they could create a condition whereby the cells are maintained in a continuous state of division (Downward *et al.*, 1984).

At present, it is still not known whether the uncontrollable growth of cancer cells could be the result of the interaction among growth factors, oncogenes or their products, and the receptors (Heldin and Westermark, 1984). However, it is interesting to speculate whether somatostatin or its analogues, by virtue of their antiproliferative properties, possibly through activation of dephosphorylation (Hierowski *et al.*, 1985), could be used to inhibit the growth of pancreatic and other cancers. Manipulations based on growth inhibitors such as powerful analogues of somatostatin could be one of the ways to suppress multifactorial process of malignant growth.

6. CLINICAL TRIALS WITH SOMATOSTATIN ANALOGUE

No clinical studies have been published on the original Veber *et al.* (1981) analogue, cyclo(Pro-Phe-D-Trp-Lys-Thr-Phe). The Veber *et al.* (1984) analogue was chosen for clinical evaluation of the inhibition of glucagon release as an adjunct to insulin in the treatment of patients with diabetes mellitus.

Various interesting clinical trials were carried out with the Bauer *et al.* (1982) analogue D-Phe-Cys-Phe-D-Trp-Lys-Thr-Cys-Thr-OL (SMS 201-995). The original assumptions were that this analogue might be useful as an adjunct in therapy in diabetes and for the treatment of acromegaly, diabetic retinopathy, and various gastrointestinal disorders (Bauer *et al.*, 1982).

Plewe *et al.* (1984) showed that a simple subcutaneous injection of 50 μg of SMS-201-995 markedly reduced serum GH levels in six out of seven acromegalics, the effect lasting for at least 9 hr. Some reduction in serum glucagon but no side effects were observed. Lamberts *et al.* (1984) reported similar results in five acromegalic patients. Previously, various analogues of somatostatin such as Phe^4-somatostatin, p-NH_2-Phe^4-D-Trp^8-somatostatin and Ala^2-D-Trp^8-Cys^{14}-somatostatin, and L-5F-Trp^8-somatostatin were administered to normal and acromegalic patients and shown to inhibit the basal or arginine-stimulated GH levels, but the effects were short-lived and persisted only during the infusion of these analogues (Wajchenberg *et al.*, 1983; Gonzalez-Barcena *et al.*, 1984; Szilagyi *et al.*, 1984; Charro *et al.*, 1984). These analogues appeared to suppress insulin and glucagon less than GH (Wajchenberg *et al.*, 1983; Gonzalez-Barcena *et al.*, 1984), and generally all of these hormones showed a rebound after the infusion. The Bauer octapeptide appears to be the first analogue suitable for the clinical management of acromegaly. This analogue was also used for the treatment of a 15-year-old female patient with a GH-releasing factor producing intestinal tumor and evidence of GH hypersecretion (Von Verder *et al.*, 1984). Twice-daily subcutaneous administration of the analogue reduced levels of GH and also caused some decrease in insulin but no disturbances in the carbohydrate metabolism (Von Verder *et al.*, 1984). A possible application of this analogue in diabetes may be indicated by the finding that it improved glycemic stability and decreased insulin requirement in diabetic patients (Marbach *et al.*, 1984).

Previously, somatostatin and some of its analogues were tested in patients with endocrine pancreatic tumors and tumors of the gastrointestinal tract (Mortimer *et al.*, 1974; Bloom *et al.*, 1984; Schally *et al.*, 1984b; Adrian *et al.*, 1981; Long *et al.*, 1979). The octapeptide analogue des-$AA^{1,2,4,5,12,13}$-D-Trp^{8}-somatostatin was used in patients with pancreatic endocrine tumors. The analogue given subcutaneously suppressed the tumor-derived hormones in patients with insulinomas, glucagonomas, and gastrinomas. A prolonged action was obtained only after s.c. injection in 10% human serum albumin but not after i.v. administration (Adrian *et al.*, 1981; Long *et al.*, 1979). Recently, it was reported that SMS 201-995 octapeptide used for a period of 8 months in a patient with VIPoma not only reduced VIP secretion and prevented torrential diarrhea but also led to a reduction in the size of hepatic metastases (Kraenzlin *et al.*, 1983, 1984). Maton *et al.* (1985) have also used the Bauer *et al.* (1982) octapeptide to treat a patient with pancreatic cholera and a VIP-producing metastatic non-β islet cell tumor. They were able to demonstrate a decrease in diarrhea and a reduction in the plasma levels of VIP, pancreatic polypeptide, and other gastrointestinal hormones. However, prolactin levels rose significantly, and there was no change in tumor size after 3 months of treatment.

In conclusion, these clinical results and our experimental work in animal tumor models (Redding and Schally, 1983, 1984; Schally *et al.*, 1984a) suggest that somatostatin analogues may be useful not only for the treatment of endocrine-related diseases and disorders of the gastrointestinal tract but also in the field of hormone-sensitive tumors.

REFERENCES

Adrian, T. E., Barnes, A. J., Long, R. G., O'Shaughnessy, D. J., Brown, M. R., Rivier, J., Vale, W., Blackburn, A. M., and Bloom, S. R., 1981, The effect of somatostatin analogs on secretion of growth, pancreatic, and gastrointestinal hormones in man, *J. Clin. Endocrol. Metab.* **53:**675–681.

Bauer, W., Briner, U., Doepfner, W., Haller, R., Huguenin, R., Marbach, P., Petcher, T. J., and Pless, J., 1982, SMS-201-995: A very potent and selective octapeptide analogue of somatostatin with prolonged action, *Life Sci.* **31:**1133–1140.

Bloom, S. R., Thorner, M. O., Besser, G. M., Hall, R., Gomez-Pan, A., Roy, V. M., Russell, R. C. G., Coy, D. H., Kastin, A. J., and Schally, A. V., 1974, Inhibition of gastrin and gastric-acid secretion by growth-hormone release-inhibiting hormone, *Lancet* **2:**1106.

Brown, M. E., Rivier, J. E., and Vale, W. W., 1977, Somatostatin analogs with selected biological activities, *Science* **196:**1467–1468.

Brown, M. P., Coy, D. H., Gomez-Pan, A., Hirst, B. H., Hunter, M., Meyers, C., Reed, J. D., Schally, A. V., and Shaw, B., 1978, Structure–activity relationships of eighteen somatostatin analogues on gastric secretion, *J. Physiol. (Lond.)* **277:**1–14.

Cai, R.-Z., Szoke, B., Fu, D., Redding, T. W., Colaluca, J., Torres-Aleman, I., and Schally, A. V., 1986, in: *Proceedings of the Ninth American Peptide Symposium June 1985* (V. J. Hruby, K. D. Kopple, and C. M. Deber, eds.), Pierce Chemical Co., Rockford, Il, pp. 627–630.

Charro, A. L., Cabranes, J. A., Lafuente, A., Diaz-Salgado, C., Cordero, M., Almoguera, E., Comaru-Schally, A. M., Meyers, C. A., Coy, D. H., and Schally, A. V., 1984, Inhibition of growth hormone and glucagon release by Phe-4-somatostatin in patients with acromegaly, *Res. Commun. Chem. Pathol. Pharmacol.* **45:**432–444.

Dembinski, A. B., and Johnson, L. R., 1980, Stimulation of pancreatic growth by secretin, caerulein, and pentagastrin, *Endocrinology* **106:**323–328.

De Quijada, M. G., Redding, T. W., Coy, D. H., Torres-Aleman, I., and Schally, A. V., 1983, Inhibition of growth of a prolactin-secreting pituitary tumor in rats by analogs of luteinizing hormone-releasing hormone and somatostatin, *Proc. Natl. Acad. Sci. U.S.A.* **80:**3485–3488.

Dollinger, H. C., Raptis, S., and Pfeiffer, E. F., 1976, Effects of somatostatin on exocrine and endocrine pancreatic function stimulated by intestinal hormones in man, *Horm. Metab. Res.* **8:**74–78.

Downward, J., Yarden, Y., Mayes, E., Scrace, G., Totty, N., Stockwell, P., Ullrich, A., Schlessinger, J., and Waterfield, M. D., 1984, Close similarity of epidermal growth factor receptor and v-*erb*-B oncogene protein sequences, *Nature* **307:**521–527.

Gibbs, J., Young, R. C., and Smith, G. P., 1973, Cholecystokinin elicits satiety in rats with open gastric fistulas, *Nature* **245:**323–325.

Gonzalez-Barcena, D., Mendoza, F., Medina, R., Coy, D. H., Murphy, W. A., and Schally, A. V., 1984, Effect of Ala-2-D-Trp-8-D-Cys-14-somatostatin on the arginine induced release of insulin, GH and glucagon in normal men, *Endocrine Res.* **10:**39–48.

Gottesman, E. S., Mandarino, L. J., and Gerich, J. E., 1982, Somatostatin, in: *Special Topics in Endocrinology and Metabolism,* Vol. 4 (M. Cohen, ed.), Alan R. Liss, New York, pp. 177–243.

Grant, N., Clark, D., Garsky, V., Juanakais, E., McGregor, W., and Sarantakis, D., 1976, Dissociation of somatostatin effects. Peptides inhibiting the release of growth hormone but not glucagon or insulin in rats, *Life Sci.* **19:**629–632.

Hall, R., and Gomez-Pan, A., 1976, The hypothalamic regulatory hormones and their clinical applications, *Adv. Clin. Chem.* **18:**173–212.

Heldin, C.-H., and Westermark, B., 1984, Growth factors: Mechanism of action and relation to oncogenes, *Cell* **37:**9–20.

Hierowski, M. T., Liebow, C., du Sapin, K., and Schally, A. V., 1985, Stimulation by somatostatin of dephosphorylation of membrane proteins in pancreatic cancer MIA PaCa-2 cell line, *FEBS Lett.* **179:**252–256.

Hirst, B. H., Shaw, B., Meyers, C. A., and Coy, D. H., 1980, Structure–activity studies with somatostatin: The role of tryptophan in position 8, *Regul. Peptides* **1:**97–113.

Hirst, B. H., Arilla, E., Coy, D. H., and Shaw, B., 1984, Cyclic hexa- and pentapeptide somatostatin analogues with reduced gastric inhibitory activity, *Peptides* **5:**857–860.

Konturek, S. J., Tasler, J., Krol, R., Dembinski, A., Coy, D. H., and Schally, A. V., 1977, Effect of somatostatin analogs on gastric and pancreatic secretion, *Proc. Soc. Exp. Biol. Med.* **155:**519–522.

Konturek, S. J., Cieszkowski, M., Bilski, J., Konturek, J., Bielanski, W., and Schally, A. V., 1985, Effects of cyclic hexapeptide analog of somatostatin on pancreatic secretion in dogs, *Proc. Soc. Exp. Biol. Med.* **178:**68–72.

Kraenzlin, M. E., Ch'ng, J. C., Wood, S. M., and Bloom, S. R., 1983, Can inhibition of hormone secretion be associated with endocrine tumour shrinkage? *Lancet* **2:**1501.

Kraenzlin, M. E., Ch'ng, J. L. C., Wood, S. M., and Bloom, S. R., 1984, Remission of symptoms and shrinkage of metastasis with long term treatment with somatostatin analogue, *Gut* **25:**A576.

Lamberts, S. W. J., Uitterlinden, P., Zuiderwijk, J., Verleun, T., Neufeld, M., and Del Pozo, E., 1984, The effects of a long-acting somatostatin analog on pituitary tumor growth and hormone secretion in rat and man, *Excerpta Med.* **652:**847.

Long, R. G., Barnes, A. J., Adrian, T. E., Mallinson, C. N., Brown, M. R., Vale, W., Rivier, J. E., Christopides, N. D., and Bloom, S. R., 1979, Suppression of pancreatic endocrine tumour secretion by long-acting somatostatin analogue, *Lancet* **2:**764–767.

Marbach, M., Neufeld, M., and Pless, J., 1984, Clinical applications of somatostatin analogs, in: *Proceedings of the Symposium on Somatostatin, Montreal, July 1984* (Y. C. Patel and G. S. Tannenbaum, eds.), Elsevier, Amsterdam, pp. 339–345.

Mascardo, R. N., and Sherline, P., 1982, Somatostatin inhibits rapid centrosomal separation and cell proliferation induced by epidermal growth factor, *Endocrinology* **111:**1394–1396.

Maton, P. N., O'Dorisio, T. M., Howe, B. A., McArthur, K. E., Howard, J. M., Cherner, J. A., Malarkey, T. B., Collen, M. J., Gardner, J. D., and Jensen, R. T., 1985, Effect of a long-acting somatostatin analogue (SMS 201-995) in a patient with pancreatic cholera, *N. Engl. J. Med.* **312:**17–21.

McCumbee, W. D., McCarty, D. S., Jr., and Lebovitz, H. E., 1980, Hormone responsiveness of a transplantable rat chondrosarcoma. II. Evidence for *in vivo* hormone dependence, *Endocrinology* **106:**1930–1940.

Merrifield, R. B., 1967, Solid phase peptide synthesis. III. An improved synthesis of bradykinin, *Recent Prog. Horm. Res.* **23:**451–482.

Meyers, C., Arimura, A., Gordin, A., Fernandez-Durango, R., Coy, D. H., Schally, A. V., Drouin, J., Ferland, L., Beaulieu, M., and Labrie, F., 1977, Somatostatin analogs which inhibit glucagon and growth hormone more than insulin release, *Biochem. Biophys. Res. Commun.* **74:**630–636.

Meyers, C. A., Coy, D. H., Huang, W. Y., Schally, A. V., and Redding, T. W., 1978, Highly active position eight analogues of somatostatin and separation of peptide diastereomers by partition chromatography, *Biochemistry* **17:**2326–2331.

Mortimer, C. H., Tunbridge, W. M. G., Carr, D., Yeomans, L., Lind, T., Coy, D. H., Bloom, S. R., Kastin, A., Mallinson, C. N., Besser, G. M., Schally, A. V., and Hall, R., 1974, Effects of growth-hormone release-inhibiting hormone on circulating glucagon, insulin, and growth hormone in normal, diabetic, acromegalic, and hypo-pituitary patients, *Lancet* **1:**697–701.

Pfeiffer, E. F., and Schusdziarra, V., 1984, What can a clinician extract from the development of somatostatin for therapy? in: *2nd International Symposium on Somatostatin, Athens (Greece)* (S. Raptis, J. Rosenthal, and J. E. Gerich, eds.), Attempto Verlag, Tübingen, pp. 222–231.

Plewe, G., Beyer, J., Krause, U., Neufeld, M., and Del Pozo, E., 1984, Long-acting and selective suppression of growth hormone secretion by somatostatin analogue SMS 201-995 in acromegaly, *Lancet* **2:**782–784.

Raptis, S., Rosenthal, J., and Gerich, J. E., eds., 1984, *2nd International Symposium on Somatostatin, Athens (Greece),* Attempto Verlag, Tübingen.

Redding, T. W., and Schally, A. V., 1983, Inhibition of growth of the transplantable rat chondrosarcoma by analogs of hypothalamic hormones, *Proc. Natl. Acad. Sci. U.S.A.* **80:**1078–1082.

Redding, T. W., and Schally, A. V., 1984, Inhibition of growth of pancreatic carcinomas in animal models by analogs of hypothalamic hormones, *Proc. Natl. Acad. Sci. U.S.A.* **81:**248–252.

Redding, T. W., Schally, A. V., Tice, T. R., and Meyers, W. E., 1984, Long-acting delivery systems for peptides: Inhibition of rat prostate tumors by controlled release of [D-Trp-6]luteinizing hormone-releasing hormone from injectable microcapsules, *Proc. Natl. Acad. Sci. U.S.A.* **81:**5845–5848.

Reichlin, S., 1983, Somatostatin, *N. Engl. J. Med.* **309:**1495–1501, 1556–1563.

Rivier, J., Brown, M., and Vale, W., 1975, D-Trp-8-Somatostatin: An analog of somatostatin more potent than the native molecule, *Biochem. Biophys. Res. Commun.* **65:**746–751.

Rivier, J., Brown, M., Rivier, C., Ling, N., and Vale, W., 1977, Hypothalamic hypophysiotropic hormones. Review on the design of synthetic analogs, in: *Peptides 1976* (A. Loffet, ed.), l'Université de Bruxelles, Brussels, pp. 427–449.

Rose, D. P., Gottardis, M., and Noonan, J. J., 1983, Rat mammary carcinoma regressions during suppression of serum growth hormone and prolactin, *Anticancer Res.* **3:**323–326.

Salomon, D. S., Paglia, L. M., and Verbruggen, L., 1979, Hormone-dependent growth of a rat chondrosarcoma *in vivo, Cancer Res.* **39:**4387–4395.

Schally, A. V., Coy, D. H., and Meyers, C. A., 1978, Hypothalamic regulatory hormones, *Annu. Rev. Biochem.* **47:**89–128.

Schally, A. V., Comaru-Schally, A. M., and Redding, T. W., 1984a, Antitumor effects of analogs of hypothalamic hormones in endocrine-dependent cancers, *Proc. Soc. Exp. Biol. Med.* **175:**259–281.

Schally, A. V., Coy, D. H., Murphy, W. A., Redding, T. W., Comaru-Schally, A. M., Hall, R., Rodriguez-Arnao, M. D., Gomez-Pan, A., Gonzalez-Barcena, D., Millar, R. P., and Konturek, S., 1984b, Physiological and clinical studies with somatostatin analogs and pro-somatostatin, in:

2nd International Symposium on Somatostatin, Athens (Greece) (S. Raptis, J. Rosenthal, and J. E. Gerich, eds.), Attempto Verlag, Tübingen, pp. 188–196.

Schally, A. V., Redding, T. W., and Comaru-Schally, A. M., 1984c, Potential use of analogs of luteinizing hormone-releasing hormones in the treatment of hormone-sensitive neoplasms, *Cancer Treat. Rep.* **68:**281–289.

Snodgrass, P. J., 1977, Section 7, Disorders of the pancreas, in: *Harrison's Principles of Internal Medicine,* 8th ed. (G. W. Thorn, R. D. Adams, E. Braunwald, K. J. Isselbacher, and R. G. Petersdorf, eds.), McGraw-Hill, New York, pp. 1643–1645.

Szilagyi, G., Irsy, G., Goth, M., Szabolcs, I., Coy, D. H., Comaru-Schally, A. M., and Schally, A. V., 1984, Effect of a somatostatin analogue on trophic hormone levels in acromegalic patients with elevated hGH after adrenomectomy and treatment with bromocriptine, *Exp. Clin. Endocrinol.* **84:**190–196.

Torres-Aleman, I., Redding, T. W., and Schally, A. V., 1985, Inhibition of growth of a prolactin and growth hormone-secreting pituitary tumor in rats by D-tryptophan-6 analog of luteinizing hormone-releasing hormone, *Proc. Natl. Acad. Sci. U.S.A.* **82:**1252–1256.

Townsend, C. M., Jr., Franklin, R. B., Watson, L. C., Glass, E. J., and Thompson, J. C., 1981, Stimulation of pancreatic cancer growth by caerulein and secretin, *Surg. Forum* **34:**228–229.

Usadel, K. H., Leuschner, U., and Uberla, K. K., 1980, Treatment of acute pancreatitis with somatostatin: A multicenter double-blind trial, *N. Engl. J. Med.* **303:**999–1000.

Usadel, K. H., Leuschner, U., and Uberla, K. K., 1984, Report on a randomized double-blind European multicenter study on the use of somatostatin in acute pancreatitis (APTS) in: *2nd International Symposium on Somatostatin, Athens (Greece)* (S. Raptis, J. Rosenthal, and J. E. Gerich, eds.), Attempto Verlag, Tübingen, pp. 236–243.

Veber, D. F., Holly, F. W., Nutt, R. F., Bergstrand, S. J., Brady, St. F., Hirschmann, R., Glitzer, M. S., and Saperstein, R., 1979, Highly active cyclic and bicyclic somatostatin analogs of reduced ring size, *Nature* **280:**512–514.

Veber, D. F., Freidinger, R. M., Schwenk-Perlow, D., Paleveda, W. J., Jr., Holly, R. W., Strachan, R. G., Nutt, R. F., Arison, B. H., Homnick, C., Randall, W. C., Glitzer, M. S., Saperstein, R., and Hirschmann, R., 1981, A potent cyclic hexapeptide analogue of somatostatin, *Nature* **292:**55–58.

Veber, D. F., Saperstein, R., Nutt, R. F., Freidinger, R. M., Brady, S. F., Curley, P., Perlow, D. S., Paleveda, W. J., Colton, C. D., Zacchei, A. G., Tocco, D. J., Hoff, D. R., Vandlen, R. L., Gerich, J. E., Hall, L., Mandarino, L., Cordes, E. H., Anderson, P. S., and Hirschmann, R., 1984, A super active cyclic hexapeptide analog of somatostatin, *Life Sci.* **34:**1371–1378.

Von Werder, K., Losa, M., Muller, O. A., Schweiberer, L., Falhbusch, R., and Del Pozo, E., 1984, Treatment of metastasising GRF-producing tumour with a long-acting somatostatin analogue, *Lancet* **2:**282–283.

Wajchenberg, B. L., Cesar, F. P., Leme, C. E., Borghi, V. C., Souza, I. T. T., Neto, D. G., Germek, O. A., Coy, D. H., Comaru-Schally, A. M., and Schally, A. V., 1983, Dissociated effects of somatostatin analogs on arginine-induced insulin, glucagon and growth hormone release in acromegalic patients, *Horm. Metab. Res.* **15:**471–474.

Yamamoto, T., Nishida, T., Miyajima, N., Kawai, S., Ooi, T., and Toyoshima, K., 1983, The *erbB* gene of avian erythroblastosis virus is a member of the *src* gene family, *Cell* **35:**71–78.

8

The Role of Brain Peptides in the Control of Anterior Pituitary Hormone Secretion

S. M. McCANN, W. K. SAMSON, M. C. AQUILA, J. BEDRAN de CASTRO, N. ONO, M. D. LUMPKIN, and O. KHORRAM

1. INTRODUCTION

A host of peptides act within the hypothalamus and also directly on the pituitary to control the secretion of anterior pituitary hormones. The classical releasing and inhibiting hormones act directly on the pituitary to inhibit or stimulate the release of anterior pituitary hormones. They appear also to act within the brain to modulate their own release and that of other releasing factors. With the discovery of many additional brain peptides, most of which were found initially in the gastrointestinal tract, many of these have also been examined, and it is apparent that there are important hypothalamic actions of a number of these peptides to alter pituitary hormone secretion (McCann *et al.*, 1984b). In this chapter, we review some of our recent work in this area beginning with an examination of possible additional peptidic releasing or inhibiting factors that may directly alter pituitary hormone secretion and continuing with an examination of the intrahypothalamic action of the various peptides to alter their own release and that of pituitary hormones.

2. EVIDENCE FOR AN FSH-RELEASING FACTOR

Since LH-releasing hormone (LHRH) will also release FSH, albeit to a lesser extent, the concept has developed that there is only one gonadotropin-releasing

---

S. M. McCANN, W. K. SAMSON, M. C. AQUILA, J. BEDRAN de CASTRO, N. ONO, M. D. LUMPKIN, and O. KHORRAM • Department of Physiology, The University of Texas Health Science Center at Dallas, Southwestern Medical School, Dallas, Texas 75235.

hormone (GnRH), which controls the release of FSH and LH and is the decapeptide LHRH (Schally *et al.*, 1971). Overwhelming evidence now supports the concept that there must be a separate FSH-releasing factor, which is yet to be isolated. The evidence for this concept is the following. Electrochemical (Kalra *et al.*, 1971; Chappel and Barraclough, 1976) or prostaglandin E_2 (Ojeda *et al.*, 1977a,b) stimulation of the dorsal anterior hypothalamic area and regions extending ventrally and caudally from this area, particularly the caudal median eminence, evoked selective FSH secretion, whereas stimulations in the preoptic area evoked selective LH release (Kalra *et al.*, 1971). Lesions in the dorsal anterior hypothalamic area produced a partial blockade of FSH secretion in the castrate and in the steroid-primed animal (Lumpkin and McCann, 1984) and eliminated pulsatile FSH secretion in the castrate, whereas LH pulsations proceeded apace (Lumpkin *et al.*, 1983). Conversely, lesions in the suprachiasmatic–preoptic area blocked the steroid-induced LH but not FSH release (Bishop *et al.*, 1972). These observations are consistent with the concept that the LH-controlling area is in the preoptic area, whereas the FSH-controlling area is, at least in part, localized to the dorsal anterior hypothalamic area.

Antisera against LHRH have recently been reported to inhibit selectively LH but not FSH release acutely (Culler and Negro-Vilar, 1985). Intragastric administration of alcohol in the castrate rat produced a complete blockade of LH pulsations, and the basal level of plasma LH fell, whereas pulsations of FSH and the basal level of that hormone were maintained (Dees *et al.*, 1985).

Extracts of the posterior median eminence contained more FSH-releasing activity than could be accounted for by the content of LHRH (Mizunuma *et al.*, 1983), and an even more pronounced exaggeration of FSH-releasing activity was found in extracts of the organum vasculosum lamina terminalis (Samson *et al.*, 1980). Recently, we have completed experiments employing gel filtration on Sephadex G25 of sheep hypothalamic extracts and observed elution of FSH-releasing activity measured by an *in vivo* assay in ovariectomized, steroid-blocked rats in the same tubes in which we observed it 20 years ago by bioassay of FSH (Lumpkin *et al.*, 1986). These tubes contained no detectable radioimmunoassayable LHRH and had no bioassayable LHRH activity. Following these tubes, LHRH was eluted in the region previously found as measured by radioimmunoassay of the peptide and also by bioassay of LH-releasing activity. This LH-releasing zone contains some FSH-releasing activity, probably attributable to the LHRH present. With the thought that the LHRH recently isolated from chick and fish hypothalami might possibly represent the elusive FSH-releasing factor, we have assayed these peptides but have not found any selective FSH-releasing activity (Lumpkin *et al.*, 1986). The FSH-releasing factor remains to be isolated and synthesized; however, the weight of evidence is almost overwhelmingly in favor of its existence.

In addition, several peptides other than LHRH have been found capable of stimulating gonadotropin release from dispersed pituitary cells perifused *in vitro*. Examples of this are provided by gastric inhibitory polypeptide (Ottlecz *et al.*, 1985) and neuropeptide Y (McDonald *et al.*, 1985a). Whether these effects are of physiological significance remains to be determined.

3. EVIDENCE FOR A PEPTIDIC PROLACTIN-INHIBITING FACTOR

It has now become apparent that many factors can act directly on the lactotroph to alter prolactin secretion (McCann *et al.*, 1984a). A number of inhibitors have been discovered, including dopamine, γ-aminobutyric acid (GABA), and acetylcholine. The evidence is conclusive that dopamine exercises a physicological role in this regard. Early evidence suggested the presence of a peptidic prolactin-inhibiting factor in hypothalamic extracts (Dhariwal *et al.*, 1968). We have recently examined the fractions from the Sephadex column for prolactin release-inhibiting activity as measured by the inhibition of prolactin release from dispersed anterior pituitary cells of adult male or estrogen-primed, ovariectomized rats. With this system, prolactin-inhibiting activity was detected in the same tubes in which we had found it many years ago assaying for the activity by *in vivo* bioassay. These fractions also contained LHRH and somatostatin; however, these peptides had no prolactin-inhibiting activity in the quantities present. Neither dopamine nor GABA was detected in the active fractions by radioenzymatic and fluorophotoenzymatic assay, respectively. In addition, receptor blockers for dopamine or GABA did not interfere with the PIF activity (Mizunuma *et al.*, 1985). In other experiments we have found a pronase-labile PIF activity in crude rat hypothalamic extracts (Khorram *et al.*, 1984b). Taken together, these findings indicate that this PIF cannot be either dopamine or GABA and is presumably a peptide. Recently, Adelman *et al.* (1986) have detected PIF activity of a 58-amino-acid fragment of the prepro-LHRH molecule. This fragment is presumably larger than the molecule that has been shown to have PIF activity in the present experiments. It is possible that PIF may represent a fragment of this 58-amino-acid peptide. The evidence is clearly mounting that a peptidic PIF in fact exists and probably plays a physiologically significant role in the control of prolactin secretion.

In addition to multiple PIFs, there are multiple peptides with prolactin-releasing activity (McCann *et al.*, 1984a). These include thyrotropin-releasing factor, oxytocin, vasoactive intestinal polypeptide, peptide histidine isoleucine, neurotensin, substance P, and angiotensin II. In the case of oxytocin, the PRF activity is most readily demonstrated in perifused columns of pituitary cells (M.D. Lumpkin, W.K. Samson, and S.M. McCann, unpublished data), and it appears that oxytocin has a physiologically significant role in the suckling and proestrus-induced prolactin release on the basis of passive neutralization studies with antibodies directed against oxytocin (W.K. Samson, unpublished data).

4. INTRAHYPOTHALAMIC ACTIONS OF PEPTIDES TO ALTER GONADOTROPIN SECRETION

We have now evaluated the effects of a number of peptides on the secretion of gonadotropins (McCann, 1983). Space does not permit a complete elaboration of all of these effects. It is apparent that very minute doses of a number of these peptides can inhibit gonadotropin secretion in the ovariectomized rat. Larger doses are required in some cases. The important point is the physiological significance of these effects, which must be examined by the use of suitable antagonists to the peptides and/or antisera directed against the peptide.

Cholecystokinin (CCK) is an extremely potent inhibitor of LH release in the castrate female rat (Vijayan *et al.*, 1979a). The introduction of only 4 ng into the third ventricle of conscious ovariectomized rats was sufficient to inhibit LH but not FSH release in these animals. We recently tested the CCK antagonist proglumide and found that it has the opposite effect following its intraventricular injection; i.e., LH levels are elevated (E. Vijayan and S.M. McCann, unpublished data). Therefore, it appears that CCK may be a physiologically significant inhibitor of LH release in the castrate. By contrast, vasoactive intestinal polypeptide (VIP) was a powerful stimulant of LH release in the castrate (Vijayan *et al.*, 1979b) and released LHRH from hypothalamic synaptasomes incubated *in vitro*, suggesting that it may play a role in the stimulation of LH release in the castrate (Samson *et al.*, 1981); however, studies with antisera directed against the peptide have yet to be carried out. Third ventricular injection of both pancreatic polypeptide (McDonald *et al.*, 1985b) and neuropeptide Y (McDonald *et al.*, 1985a) also inhibited LH release in the castrate, and this was the opposite of the action of the latter peptide at the pituitary level to stimulate LH and FSH release by direct action on the gland in the perifusion system, as mentioned above (McDonald *et al.*, 1985a). These inhibitory actions of the peptides have been confirmed by others (Kalra and Crowley, 1984). Neuropeptide Y is a good example of a peptide that has different actions depending on the hormonal state of the animal. In the estrogen-primed animal, instead of inhibiting LH release, neuropeptide Y stimulates it (Kalra and Crowley, 1984).

It is now generally accepted that opiates inhibit LH release under most circumstances (McCann, 1983). β-Endorphin appears to play a physiological role in inhibition of LH release in the castrate female rat as determined by studies with intraventricular injection of β-endorphin antibodies (Bedran de Castro and McCann, 1985). The physiological significance of the endorphins is borne out by the ability of naloxone to elevate LH release under most conditions in the rat.

α-Melanocyte-stimulating hormone (α-MSH) appears to have a physiologically significant inhibitory action to decrease LH release in the castrate female rat, since intraventricular injection of minute doses of α-MSH lowers LH in the castrate (Khorram *et al.*, 1984a) and a significant elevation in LH levels in plasma follows the intraventricular injection of the globulin fraction of a highly specific α-MSH antiserium (Khorram and McCann, 1984). Similarly, α-MSH inhibits prolactin release, and antisera directed against the peptide elevate prolactin. These inhibitory actions of α-MSH on LH and prolactin are mediated by dopamine. Gastric inhibitory polypeptide has yet to be demonstrated in the brain; however, it has a unique ability to suppress FSH release following its intraventricular injection without altering LH release and, as already mentioned, has a stimulatory effect on both FSH and LH release from perifused pituitary cells *in vitro* (Ottlecz *et al.*, 1985). Its selective inhibitory effect on FSH release is shared by the gonadal peptide inhibin, which also inhibits only FSH and not LH release following its intraventricular injection in the face of normal responsiveness of the pituitary to LHRH (Lumpkin *et al.*, 1981a).

It has long been postulated that pituitary hormones have a short-loop feedback effect to suppress their own release by intrahypothalamic action (Piva *et al.*, 1979). We have found suggestive evidence for this in the case of LH, and prolactin clearly inhibits its own release rapidly by an intrahypothalamic action (Mangat and McCann, 1983). So far, we have not been able to determine any reproducible effect of FSH to modify its own release or that of other hormones in the ovariectomized rat.

5. ULTRASHORT-LOOP FEEDBACK OF PEPTIDES TO ALTER THEIR OWN RELEASE

Many years ago, Martini postulated ultrashort-loop feedback of releasing factors to suppress their own release (Piva *et al.*, 1979). Several years ago, we discovered what appears to be an ultrashort-loop feedback of somatostatin to suppress its own release. Following the intraventricular injection of small doses of the peptide in conscious ovariectomized rats, a paradoxical suppression of plasma growth hormone resulted, which was accompanied by a lowering of FSH, LH, and TSH levels. We postulated that the peptide was taken up from the third ventricle and then either inhibited its own release or stimulated the release of GRF or both and that this resulted in the suppression of GH. We postulated that the peptide also inhibited the release of LHRH and TRH to account for the lowering in the plasma concentrations of those two peptides (Lumpkin *et al.*, 1981b).

Subsequently, we searched for similar ultrashort-loop feedback actions of other releasing factors to alter their own release. We found that injection of GRF into the third ventricle elevated plasma GH until the dose was drastically reduced, at which point a lowering ensued. Presumably, following the higher doses of the peptide, the uptake was sufficient in portal vessels to deliver a quantity to the somatotrophs, which evoked release of GH. When the dose was cut, the intrahypothalamic concentration was sufficient to exert its inhibitory action, but insufficient quantities reached the pituitary to affect the somatotrophs directly (Lumpkin *et al.*, 1985a).

These experiments do not differentiate between an effect of the peptide on the secretion of GH-releasing factor and/or an effect on somatostatin release. Consequently, we have evaluated the effect of GRF on the release of somatostatin from small median eminence–stalk fragments of male rat hypothalami incubated *in vitro* (Aquila and McCann, 1985). We have observed a highly potent effect of GRF to enhance somatostatin release with a minimal effective concentration of 10^{-11} M. We have further determined the mechanism of this effect, since it is not blocked by a variety of synaptic receptor blockers but is blocked by naloxone, which should block certain opioid receptors. Consequently, it appears that the somatostatin-releasing action of GRF may be mediated via opioid peptides (Aquila and McCann, 1986). There are many β-endorphin terminals within this small fragment that may mediate this effect.

We have evaluated a possible ultrashort-loop feedback of LHRH to inhibit its own release in the castrate rat and have found that, as in the case of GRF, administration of relatively high doses of LHRH intraventricularly resulted in an elevation of LH, but if the dose was cut to 1 ng, a rapid and short-lived inhibition of LH release ensued. Presumably then, LHRH may have a physiologically significant ultrashort-loop negative feedback effect that may play a role in terminating the pulses of LH release that characterize the castrate condition (Bedran de Castro *et al.,* 1985).

Similarly, there appears to be an ultrashort-loop negative feedback of oxytocin to suppress its own release, since intraventricular injection of the peptide resulted in inhibition of prolactin release, whereas its direct action on the pituitary is to stimulate prolactin release (McCann *et al.,* 1984b).

With the determination of structure and synthesis of CRF, we also evaluated this peptide for a possible ultrashort-loop feedback action. Relatively high doses of CRF injected into the third ventricle (greater than 1 nmol) elevated plasma ACTH in a dose-related manner, which was attributed to uptake of the peptide from the ventricle, diffusion to the median eminence, and transport via the portal vessels to the pituitary with stimulation of the corticotrophs. Consequently, we diminished the dose of CRF, looking for a lowering of the already quite low plasma ACTH levels in these resting rats; however, no lowering was detected regardless of the dose injected. Therefore, we evaluated the effect of intraventricular injection of CRF in small doses on the release of ACTH during ether stress. To our surprise, employing doses 1000 times less than those required to elevate ACTH by an action on the pituitary, we found an augmentation of the stress-induced ACTH-release (Ono *et al.,* 1985a). We therefore postulate that in stress, CRF exerts an ultrashort-loop positive feedback to agument its own release. This may be of physiological significance, since one would wish to amplify the release of CRF in stress and not to suppress it, as might be the case under resting conditions. The situation is quite analogous to the stimulatory action of gonadal steroids around the time of the preovulatory release of LH, which is the classical example of positive feedback on the hypothalamus (McCann *et al.,* 1983).

Because the release of hormones other than ACTH is also altered by stress, we monitored the effects of the microinjection of ovine CRF into the third ventricle on the plasma levels of LH and GH. Lower doses of CRF had no effect on plasma LH; however, plasma LH levels were lowered by the highest dose of CRF injected (1 nmol), with the first significant lowering detectable at 30 min. Interestingly, the pattern of LH release following stress in the ovariectomized rat is an initial elevation followed by a lowering. This raises the possibility that release of CRF during stress could be responsible for the inhibition of LH release that supervenes. Microinjections of an even lower dose of 0.1 nmol of CRF lowered plasma GH levels within 15 min, and the suppression persisted for the duration of the experiment. There was no additional effect of the higher 1 nmol dose. Stress is followed by a lowering of GH in the rat, and it therefore seems reasonable to speculate that intrahypothalamic release of CRF following stress may mediate this lowering of GH (Ono *et al.*, 1985a). Similar results on both LH and GH have been obtained following the lateral ventricular injection of the peptide (Rivier and Vale, 1984).

It is already known that vasopressin can potentiate the release of ACTH induced by CRF by actions at both pituitary and hypothalamic levels (Yates *et al.*, 1971; Hedge *et al.*, 1966). To determine the physiological significance of arginine vasopressin in stress-induced release of anterior pituitary hormones, vasopressin antisera or normal rabbit serum was microinjected into the third ventricle of freely moving ovariectomized female rats. Because we had found this paradigm useful in prior experiments (Khorram and McCann, 1984), a single 3-μl injection was given, and 24 hr later the injection was repeated 30 min prior to application of ether stress for 1 min. Although arginine vasopressin antisera had no effect on basal plasma ACTH concentrations, the elevation of plasma ACTH induced by ether stress was lowered significantly (Ono *et al.*, 1985b). Plasma LH tended to increase following ether stress as expected; however, the level of LH following stress was significantly lower in the vasopressin antisera-treated group than in the group pretreated with normal rabbit serum. Ether stress lowered plasma GH levels as expected in the rat, and this lowering was slightly but significantly antagonized by vasopressin antisera. Ether stress also elevated plasma prolactin as expected, but these changes were not quite significantly modified by the antisera. As expected, vasopressin induced a dose-related release of ACTH in doses ranging from 10 ng (10 pM) to 10 μg per tube when incubated with dispersed anterior pituitary cells for 2 hr. There was no effect of any of these doses of arginine vasopressin on the release of LH or other anterior pituitary hormones *in vitro* except for a significant stimulation of TSH release at a high dose, which we have also found in other experiments.

These results indicate that vasopressin is involved in induction of ACTH and LH release during stress. The inhibitory action of the vasopressin antisera on ACTH release may occur intrahypothalamically by blocking the stimulatory action of vasopressin on CRF release. We have previously alluded to experiments indicating that microinjection of vasopressin into the hypothalamus could augment the action of CRF (Hedge *et al.*, 1966). Part of the effect of the vasopressin antisera could be related to direct blockade of the stimulatory action of vasopressin on the pituitary. Because of the minute amounts of antisera given, we are inclined to discount this possibility. The effects of vasopressin on LH release in stress are presumably brought about by blockade of a stimulatory action of vasopressin on the LHRH neural terminals, since there is no action of the peptide directly on the pituitary to alter LH release (Ono *et al.*, 1985b).

We have evaluated the effects of intravenous injection of antibodies directed against CRF. These antibodies nearly completely blocked the ACTH-releasing action of ether stress in agreement with previous findings of Rivier and Vale (1985). This is consistent with the statements made above, which indicate that CRF is the most important ACTH-releasing agent. Intraventricular injection of antibodies directed against CRF, following the same protocol just mentioned for vasopressin antibodies, resulted in no alteration in the resting levels of any of the pituitary hormones measured but largely blocked the ACTH release from ether anesthesia. In view of the minute amount of antibody given in these experiments, we speculate that the antibodies are taken up from the ventricle and block the ultrashort-loop positive feedback of CRF to augment its own release during stress (Ono *et al.*, 1985c). We cannot, however, completely exclude the possibility that some of the antibodies reached the pituitary and had a direct blocking action on the ACTH-releasing effects of CRF at the gland. The intraventricular injection of the antisera directed against CRF completely blocked the stress-induced lowering of GH, which suggests that CRF induces this lowering following its release during stress by an intrahypothalamic action. This, of course, is consistent with the previously described ability of intraventricular CRF to lower GH (Ono *et al.*, 1985a). Surprisingly, the release of LH was unaffected by the CRF antibodies, which could mean that the inhibition of LH release during stress does not involve CRF and/or that the passive neutralization was not complete, permitting the unneutralized CRF to exert its effect. The fact that it takes a higher dose of CRF given intraventricularly to lower LH than GH mitigates against this possibility.

In conclusion, these studies reveal a complex interplay of peptides within the hypothalamus and directly on the pituitary to modulate the secretion of anterior pituitary hormones. Additional studies with antibodies or antagonists against other peptides will be necessary to unravel the situation completely. Furthermore, the interactions of the peptides with the various monoaminergic systems already shown to play an important role in control of pituitary hormone secretion must be further studied.

6. SUMMARY

A host of peptides act within the hypothalamus and also directly on the pituitary to control the secretion of anterior pituitary hormones. The classical releasing and inhibiting hormones act directly on the gland to inhibit or stimulate the release of anterior pituitary hormones. In addition to those that have been structurally identified, strong evidence exists for the presence of an FSHRF and a peptidic prolactin-inhibiting factor within the hypothalamus. The releasing and inhibiting hormones appear also to act within the brain to modulate their own release and that of other releasing factors. For example, somatostatin acts centrally to stimulate GH secretion by altering the release of somatostatin and/or GH-releasing factor. The GH-releasing factor acts back to inhibit GH, probably by inhibiting its own release and by activating the release of somatostatin. This has been demonstrated by incubating the peptide with median eminence fragments *in vitro*. Luteinizing hormone-releasing hormone appears to inhibit its own release following its intraventricular injection into ovariectomized rats. Oxytocin, which has the capacity to stimulate prolactin release by direct action on the pituitary, appears to inhibit its own release following its injection into the third ventricle with a resultant decline in prolactin release. Corticotropin-releasing factor had no ability to suppress its own release in resting animals but appeared to exert an ultrashort-loop positive feedback to stimulate ACTH release to even higher levels following imposition of ether stress. At the same time, CRF is capable of inhibiting LH and GH secretion following its intraventricular injection, which suggests that it induces the stress pattern of release of hypothalamic peptides, which then brings about the stress pattern of pituitary hormone secretion. Similarly, somatostatin not only affects the release of GH but also suppresses the release of TSH and gonadotropins, indicative of its inhibitory effects on the release of other hypothalamic peptides.

Many other peptides affect the release of pituitary hormones via hypothalamic action. The most clearly identified example is the opioid peptides. β-Endorphin appears to have a tonic inhibitory effect on the release of LH and to be involved in the stress-induced release of prolactin as evidenced by studies with opioid receptor blockers and with antisera directed against β-endorphin. Cholecystokinin is very potent to alter the secretion of pituitary hormones. This action may have physiological significance in view of the opposite effects produced by the administration of the CCK antagonist proglumide. Melantropin has powerful effects to suppress prolactin and LH secretion, which may have physiological significance since they can be reversed by intraventricular injection of antisera directed against the peptide. These are but a few examples of the myriad interactions of peptides within the brain to alter pituitary hormone secretion.

REFERENCES

Adelman, J. P., Mason, A. J., Hayflick, J. S., and Seeburg, P. H., 1986, Isolation of gene and hypothalamic cDNA for common precursor of gonadotropin releasing hormone (GnRH) and prolactin release inhibiting factor (PIF) in human and rat, *Proc. Natl. Acad. Sci. U.S.A.* **83:**179–183.

Aquila, M. C., and McCann, S. M., 1985, Stimulation of somatostatin release *in vitro* by synthetic human growth hormone-releasing factor by a nondopaminergic mechanism, *Endocrinology* **117**(2):762–765.

Aquila, M. C., and McCann, S. M., 1986, Evidence that GRF stimulates SRIF release via *in vitro* β endorphin, *Endocrinology* (in press).

Bedran de Castro, J. C., and McCann, S. M., 1986, Role of opioid peptides in control of gonadotropin secretion, *Brain Res. Bull.* (submitted).

Bedran de Castro, J. C., Khorram, O., and McCann, S. M., 1985, Possible negative ultrashort-loop feedback of luteinizing hormone releasing hormone (LHRH) in the ovarectomized rat, *Proc. Soc. Exp. Biol. Med.* **179:**132–135.

Bishop, W., Kalra, P. S., Fawcett, C. P., Krulich, L., and McCann, S. M., 1972, The effects of hypothalamic lesions on the release of gonadotropins and prolactin in response to estrogen and progesterone treatment in female rats, *Endocrinology* **91:**1404–1410.

Chappel, S. C., and Barraclough, C. A., 1976, Hypothalamic regulation of pituitary FSH secretion, *Endocrinology* **98:**927–935.

Culler, M. D., and Negro-Vilar, A., 1985, Passive LHRH immunization suppresses pulsatile LH but not FSH secretion in the castrate male rat, in: *Program of the 67th Annual Meeting of the Endocrine Society,* Endocrine Society, Baltimore, p. 197.

Dees, L., Rettori, V., Kozlowski, J., and McCann, S. M., 1985, Ethanol and the pulsatile release of luteinizing hormone, follicle stimulating hormone and prolactin in ovariectomized rats, *Alcohol* **2:**641–646.

Dhariwal, A. P. S., Antunes-Rodrigues, J., Grosvenor, C., and McCann, S. M., 1968, Purification of ovine prolactin-inhibiting factor (PIF), *Endocrinology* **82:**1236–1241.

Hedge, G. A., Yates, M. B., Marcus, R., and Yates, F. E., 1966, Site of action of vasopressin in causing corticotropin release, *Endocrinology* **79:**328–340.

Kalra, S. P., and Crowley, W. R., 1984, Differential effects of pancreatic polypeptide on luteinizing hormone release in female rats, *Neuroendocrinology* **38:**511–513.

Kalra, S. P., Ajika, K., Krulich, L., Fawcett, C. P., Quijada, M., and McCann, S. M., 1971, Effects of hypothalamic and preoptic electrochemical stimulation on gonadotropin and prolactin release in proestrous rats, *Endocrinology* **98:**1150–1158.

Khorram, O., and McCann, S. M., 1984, Physiological role of alpha melanocyte-stimulating hormone in modulating the secretion of prolactin and luteinizing hormone in the female rat, *Proc. Natl. Acad. Sci. U.S.A.* **81:**8004–8008.

Khorram, O., DePalatis, L. D., and McCann, S. M., 1984a, The effect and possible mode of action of αMSH on gonadotropin release in the ovariectomized rat, *Endocrinology* **114:**227–233.

Khorram, O., DePalatis, L. D., and McCann, S. M., 1984b, Hypothalamic control of prolactin secretion during the perinatal period in the rat, *Endocrinology* **115:**1698–1704.

Lumpkin, M. D., and McCann, S. M., 1984, Effect of destruction of the dorsal anterior hypothalamus on follicle-stimulating hormone secretion in the rat, *Endocrinology* **115:**2473–2480.

Lumpkin, M. D., Negro-Villar, A., Franchimont, B., and McCann, S. M., 1981a, Evidence for a hypothalamic site of action of inhibin to suppress FSH release, *Endocrinology* **108:**1101–1104.

Lumpkin, M. D., Samson, W. K., and McCann, S. M., 1981b, Paradoxical elevation of growth hormone by intraventricular somatostatin: Possible ultrashort-loop feedback, *Science* **211:**1072–1074.

Lumpkin, M. D., Samson, W. K., McDonald, J. E., and McCann, S. M., 1983, Lesions of the paraventricular region suppress pulsatile FSHJ but not LH release, *Soc. Neurosci. Abstr.* **13:**708.

Lumpkin, M. D., Samson, W. K., and McCann, S. M., 1985, Effects of intraventricular growth hormone-releasing factor on growth hormone release: Further evidence for ultrashort-loop feedback, *Endocrinology* **116:**2070–2074.

Lumpkin, M. D., Moltz, J. H., Yu, W., Fawcett, C. P., and McCann, S. M., 1986, Purification of sheep FSH-releasing factor and detection of its activity by *in vivo* bioassay (submitted).

Mangat, H. K., and McCann, S. M., 1983, Acute inhibition of prolactin and TSH secretion after intraventricular injection of prolactin in ovariectomized rats, *Am. J. Physiol.* **224:**E31–E36.

McCann, S. M., 1983, Present status of LHRH: Its physiology and pharmacology, in: *Role of Peptides and Proteins in Control of Reproduction* (S. M. McCann and D. S. Dhindsa, eds.), Elsevier, New York, pp. 3–26.

McCann, S. M., Lumpkin, M. D., Mizunuma, H., Khorram, O., Ottlecz, A., and Samson, W. K., 1984a, Peptidergic and dopaminergic control of prolactin release, *Trends Neurosci.* **7:**127–131.

McCann, S. M., Lumpkin, M. D., Ono, N., Khorram, O., Ottlecz, A., and Samson, W. K., 1984b, Interactions of brain peptides within the hypothalamus to alter anterior pituitary hormone secretion, in: *Endocrinology* (F. Labrie and L. Proulx, eds.), Excerpta Medica, Amsterdam, pp. 185–190.

McDonald, J. K., Lumpkin, M. D., Samson, W. K., and McCann, S. M., 1985a, Neuropeptide Y affects secretion of luteinizing hormone and growth hormone in ovariectomized rats, *Proc. Natl. Acad. Sci. U.S.A.* **82:**561–564.

McDonald, J. K., Lumpkin, M. D., Samson, W. K., and McCann, S. M., 1985b, Pancreatic polypeptides affect LH and growth hormone secretion in rats, *Peptides* **6:**79–84.

Mizunuma, H., Samson, W. K., Lumpkin, M. D., and McCann, S. M., 1983, Evidence for an FSH-releasing factor in the posterior portion of the rat median eminence, *Life Sci.* **33:**2003–2009.

Mizunuma, H., Khorram, O., and McCann, S. M., 1985, Purification of a nondopaminergic and non-GABAergic prolactin release-inhibiting factor (PIF) in sheep stalk–median eminence, *Proc. Soc. Exp. Biol. Med.* **178:**114–120.

Ojeda, S. R., Jameson, H. E., and McCann, S. M., 1977a, Hypothalamic areas involved in prostaglandin (PG)-induced gonadotropin release. I. Effects of PGE_2 and PGF implants on LH release, *Endocrinology* **100:**1585–1594.

Ojeda, S. R., Jameson, H. E., and McCann, S. M., 1977b, Hypothalamic areas involved in prostaglandin (PG)-induced gonadotropin release. II. Effects of PGE_2 and PGF implants on follicle stimulating hormone release, *Endocrinology* **100:**1595–1603.

Ono, N., Bedran de Castro, J. C., and McCann, S. M., 1985a, Possible negative ultrashort-loop feedback of luteinizing hormone-releasing hormone (LHRH) in the ovariectomized rat, *Proc. Soc. Exp. Biol. Med.* **82:**3528–3531.

Ono, N., Bedran de Castro, J., Khorram, O., and McCann, S. M., 1985b, Role of arginine vasopressin in control of ACTH and LH release during stress, *Life Sci.* **36:**1779–1786.

Ono, N., Samson, W. K., McDonald, J. K., Lumpkin, M. D., Bedran de Castro, J. C., and McCann, S. M., 1985c, The effects of intravenous and intraventricular injection of antisera directed against corticotropin releasing factor (CRF) on the secretion of anterior pituitary hormones, *Proc. Natl. Acad. Sci. U.S.A.* **82:**7787–7790.

Ottlecz, A., Samson, W. K., and McCann, S. M., 1985, The effects of gastric inhibitory polypeptide (GIP) on the release of anterior pituitary hormones, *Peptides* **6:**115–119.

Piva, F., Motta, M., and Martini, L., 1979, Regulation of hypothalamic and pituitary function: Long, short and ultrashort-loops feedback, in: *Endocrinology* (L. J. DeGroot, G. F. Cahill, Jr., and L. Martini, eds.), Grune & Stratton, New York, pp. 21–33.

Rivier, C., and Vale, W., 1984, Corticotropin-releasing factor (CRF) acts centrally to inhibit growth hormone secretion in the rat, *Endocrinology* **114:**2409–2411.

Rivier, C., and Vale, W. W., 1985, Effects of corticotropin releasing factor, neurohypophyseal peptides and catecholamines on pituitary function, *Fed. Proc.* **44:**189–196.

Samson, W. K., Snyder, G. D., Fawcett, C. P., and McCann, S. M., 1980, Chromatographic and biological analysis of ME and OVLT LHRH, *Peptides* **1:**97–102.

Samson, W. K., Burton, K. P., Reeres, J. P., and McCann, S. M., 1981, Vasoactive intestinal peptides stimulates LHRH release from median eminence synaptosomes, *Regul. Peptides* **2:**253–264.

Schally, A. V., Arimura, A., Kastin, A. J., Matsuo, H., Baba, R., Redding, T. W., Nair, R. W. G., Debeljuk, L., and White, W. F., 1971, The gonadotropin-releasing hormone: A single hypothalamic peptide regulates the secretion of both LH and FSH, *Science* **173:**1036–1038.

Vijayan, E., Samson, W. K., and McCann, S. M., 1979a, *In vivo* and *in vitro* effects of cholecystokinin on gonadotropin, prolactin, and growth hormone and thyrotropin release in the rat, *Brain Res.* **172:**295–302.

Vijayan, E., Samson, W. K., Said, S. I., and McCann, S. M., 1979b, Vasoactive intestinal peptide (VIP): Evidence for a hypothalamic site of action to release growth hormone, luteinizing hormone and prolactin in conscious ovariectomized rats, *Endocrinology* **104:**53–57.

Yates, F. E., Russel, S. M., Dallman, M. F., Hedge, G. A., McCann, S. M., and Dhariwal, A. P. S., 1971, Potentiation by vasopressin of corticotropin-release induced by corticotropin-releasing factor, *Endocrinology* **88:**3–15.

9

Brain Noradrenergic Neurons, Corticotropin-Releasing Factor, and Stress

RITA J. VALENTINO and STEPHEN L. FOOTE

1. THE ROLE OF CORTICOTROPIN-RELEASING FACTOR IN STRESS

1.1. Introduction

The hallmark of the organismic state of stress, and the endpoint that has been utilized to determine whether a stimulus is stressful, is the release of adrenocorticotropin (ACTH) from the anterior pituitary. It has long been assumed that ACTH release is controlled by a hypothalamic releasing factor. Such a substance, corticotropin-releasing factor (CRF, a 41-amino-acid peptide), has recently been isolated, characterized, and synthesized by Vale and colleagues (1981). Substantial evidence, reviewed in this chapter, now suggests that in addition to its function as a releasing factor, CRF may act as a neurotransmitter. CRF-containing cell bodies and fibers have been localized in extrahypothalamic regions, and administration of CRF elicits physiological and behavioral effects that do not depend on the integrity the hypothalamic-pituitary axis. In these dual roles, CRF may act synergistically as a hypothalamo-pituitary releasing factor and as a neurotransmitter in extrahypothalamic circuits to mobilize the organism to respond to "stressful" situations.

One extrahypothalamic site where CRF has been visualized is the noradrenergic nucleus locus coeruleus (LC). The physiological characteristics of LC cells during different behavioral states, and the changes observed in norepinephrine (NE) turnover in LC target areas during stress, suggest that these cells are activated by stressful stimuli. Moreover, the massive divergent LC efferent projections and NE

RITA J. VALENTINO • Department of Pharmacology, George Washington University Medical Center, Washington, D.C. 20037. STEPHEN L. FOOTE • Department of Psychiatry, University of California, San Diego, La Jolla, California 92093.

effects produced on target neurons as a consequence of LC activation provide a potential for altering activity in areas of the CNS involved in behavioral and autonomic regulation.

In this chapter, evidence is reviewed that is relevant to the hypothesis that CRF is released within the LC in response to stressful stimuli, activating LC neurons, and that the LC then serves to initiate and coordinate certain behavioral and physiological aspects of the organism's stress response.

1.2. Endocrine Effects

In order to demonstrate that a factor is directly involved in endocrine responses to stress, it is not sufficient to demonstrate that it elicits ACTH release *in vivo* and *in vitro*. In addition, it must be demonstrated that ACTH release by different stressors is blocked or attenuated by antagonists of that factor or by modifications that alter the action of that factor. These two criteria have been satisfied by CRF. The list of endogenous substances that release ACTH includes the peptides arginine vasopression (AVP) (Vale *et al.*, 1978; Spinedi and Negro-Villar, 1983), oxytocin (Vale *et al.*, 1983), and angiotensin (Spinedi and Negro-Villar, 1983, 1984; Vale *et al.*, 1983) as well as the biogenic amines norepinephrine and epinephrine (Raymond and Negro-Villar, 1984). However, none of these compounds is as potent or as efficacious as CRF. In addition, many of these substances act by potentiating CRF. For example, the release of ACTH by vasopressin, oxytocin, and the catecholamines can be reduced markedly by administration of either anti-CRF antiserum or a CRF antagonist or by blocking CRF release with chlorpromazine–pentobarbital–morphine pretreatment (Rivier and Vale, 1985).

Animals subjected to various stressors demonstrate plasma ACTH increases that are blocked by previous administration of CRF antiserum. For example, ACTH release elicited by tail restraint, hemorrhage, hypoglycemia, ether stress, or adrenalectomy is markedly decreased by CRF antiserum (Tilders *et al.*, 1985; Plotsky, 1985; Rivier *et al.*, 1982). Although these data demonstrate an integral role of CRF in stress-induced ACTH release, they do no rule out the importance of the potentiating effects of other endogenous substances. During ether-induced stress, for example, chlorisondamine, a ganglionic blocker, decreases the rise in plasma ACTH by 40–60%. The combination of both chlorisondamine and CRF antiserum can completely abolish ACTH release in these animals (Rivier and Vale, 1983). Thus, although CRF plays a primary role in ACTH release during stress, other neuropeptides or transmitters modulate this response, and the degree of modulation may depend on the specific stressor involved.

1.3. Anatomic Distribution

As has been demonstrated for other releasing factors, CRF has a wide distribution in the brain, suggesting that it serves as a neurotransmitter as well as a releasing factor. This raises the possibility that CRF neurons projecting to several brain regions are coordinately activated in response to stressful stimuli, simultaneously initiating ACTH release, autonomic aspects of the stress response, and appropriate behavioral changes. The distribution of CRF has been determined by a number of investigators using both radioimmunoassay (Fischman and Moldow, 1982; Palkovits *et al.*, 1983, 1985) and immunohistochemistry (Bloom *et al.*, 1982; Bugnon *et al.*, 1982; Cummings *et al.*, 1983; Kawata *et al.*, 1982; Merchenthaler *et al.*, 1982; Olschowka *et al.*, 1982; Swanson *et al.*, 1983), with the two approaches yielding compatible results. The hypothalamic distribution of CRF reflects its endocrine role in releasing ACTH from the anterior pituitary. A cluster of CRF-immunoreactive neuronal perikarya is localized to the parvocellular region of the paraventricular nucleus (PVN). The highest level of CRF by RIA is found in the median eminence, where CRF-containing fibers and terminals are found mainly in the external layer. This is presumed to be the source of the CRF that mediates ACTH release. In the anterior magnocellular division of the PVN, CRF may be colocalized with oxytocin (Sawchenko *et al.*, 1984). Interestingly, the PVN is an area of the hypothalamus that projects to brainstem and spinal cord nuclei involved in autonomic function. Thus, CRF release from axons of this cell group could mediate autonomic responses to stressful stimuli.

The extrahypothalamic distribution of CRF includes nuclei in the basal forebrain and brainstem, many of which have been implicated in autonomic functions. CRF-immunoreactive neurons have been visualized in the central nucleus of the amygdala, the bed nucleus of the stria terminalis, lateral septal nucleus, nucleus of the diagonal band, the dorsal raphe, parabrachial nucleus, and the LC; CRF has also been identified in the spinal cord in laminae I, II, and X, where it may play a role in nociception, and in laminae V–VII, where is extends over the intermediolateral column (Merchenthaler *et al.*, 1983; Schipper *et al.*, 1983). This extrahypothalamic distribution of CRF provides an anatomic substrate for autonomic aspects of stress responses. Finally, CRF-containing cells have been visualized in the cerebral cortex. These cells may play a role in behavioral responses associated with stress.

The case of CRF as a neurotransmitter would be considerably strengthened by the demonstration of localized specific binding sites for this peptide. Unfortunately, there are few investigations reporting such localization. However, homogenates of anterior pituitary cells exhibit high-affinity binding of CRF (Wynn *et al.*, 1983), and autoradiographic studies have demonstrated binding in the neocortex, cerebellar cortex, olfactory tract, bed nucleus of the stria terminalis, interpeduncular nucleus, amygdala, and external layer of the median eminence (DeSouza *et al.*, 1984; Wynn *et al.*, 1984). All of these areas also contain fibers that are immunoreactive for CRF.

Information about afferent inputs to major CRF-containing areas would help elucidate how information about stressful stimuli might be relayed to these areas. For example, autonomic information may reach the PVN by way of the nucleus of the solitary tract (NTS), which projects to this area. The A_2/C_2 cell group within the NTS, and A_1/C_1 cell group in the ventral medulla, and the LC also project to the area of the PVN where CRF has been localized (Sawchenko and Swanson, 1982, 1983; Swanson and Sawchenko, 1983). The parabrachial nucleus, which is thought to be involved in respiration and in which dense CRF fibers have been visualized, also projects to the parvocellular region of the PVN (Saper and Lowey, 1980). Additionally, other hypothalamic nuclei project to CRF-containing cells of the PVN. These afferent projections may provide a means by which autonomic information can control CRF release. Finally, since many nuclei that receive CRF innervation also project to CRF-containing cells in the PVN, there may be neural feedback mechanisms acting on CRF release.

1.4. Autonomic Effects

A neurotransmitter function for CRF is supported by studies demonstrating CRF effects that are distinct from, and do not require, ACTH release. Among these are autonomic, metabolic, and gastric effects that are generally associated with stress. Following intracerebroventricular (i.c.v.) injection of CRF, plasma levels of glucose, norepinephrine, and epinephrine are increased, oxygen consumption rises, blood pressure and heart rate become elevated (Brown *et al.*, 1982a,b; Fisher *et al.*, 1982), and gastric acid secretion is inhibited (Tache *et al.*, 1983, 1984). These actions can be produced in hypophysectomized or adrenalectomized animals and in animals that have been administered anti-CRF antiserum peripherally. However, administration of a ganglionic blocker, chlorisondamine, attenuates these effects. Thus, these effects appear to be central actions of CRF to increase sympathetic tone and decrease parasympathetic tone.

The cardiovascular effects produced by CRF are of interest because total peripheral resistance is decreased while blood pressure is increased. Blood pressure elevation is thought to result from increased cardiac output, which results in part from catecholamine-induced effects on contractility as well as decreased vagal tone. The latter has been suggested because CRF has also been demonstrated to block reflex bradycardia produced by phenylephrine (Brown and Fisher, 1985). The net effect of increased cardiac output, increased blood pressure, and decreased total peripheral resistance would be increased tissue perfusion, an effect that may be adaptive during certain conditions. Thus, administration of exogenous CRF produces autonomic effects that could be adaptive in certain demanding situations.

In order to evaluate the possible physiological role of CRF in autonomic responses to stress, manipulations of endogenous CRF levels or availability have been used prior to stressing animals. Dexamethasone, which inhibits CRF release, attenuates increases in plasma epinephrine and norepinephrine in ether-stressed rats (Brown *et al.*, 1983). Similarly, adminstration of the CRF antagonist α-helical CRF_{9-41} prevents the rise in epinephrine associated with ether stress (Brown *et al.*, 1984).

The finding that exogenously administered CRF can produce autonomic effects similar to those observed in stress and that alterations in endogenous CRF function can alter some autonomic responses during stress is compatible with the hypothesis that, apart from its endocrine role, CRF has a wide range of CNS actions including an autonomic component that is elicited by stressful stimuli. The anatomic distribution of CRF fibers to CNS centers involved in control of these functions is also consistent with this hypothesis.

1.5. Behavioral Effects

In addition to endocrine and autonomic aspects of stress, there are also behavioral changes exhibited by animals in this state. The adaptive response is one of enhanced alertness and arousal. Studies of administration of exogenous CRF in rats and primates suggest that brain CRF may mediate these responses apart from its endocrine effects.

Low doses of CRF (15–150 pmol) produce long-lasting EEG activation and interictal spikes in hippocampus and amygdala (Ehlers *et al.*, 1983). Higher doses result in afterdischarges in the amygdala and, in some animals, produce motor seizures. Behaviors associated with EEG activation at the lower doses (15–150 pmol) include locomoter activation, increased grooming, rearing, and sniffing, whereas higher doses are associated with more bizarre behaviors. These behavioral effects have been demonstrated in hypophysectomized rats and so are distinct from ACTH release produced by CRF.

CRF-induced behaviors are dependent on situation as well as dose. Whereas rats tested in photocell cages exhibit increased locomotor activity (Sutton *et al.*, 1982), rats tested in a novel open-field environment show decreases in locomotion, which is consistent with increased aversiveness or fear of the environment (Sutton *et al.*, 1982). In open-field conflict paradigms in which food-deprived rats are placed in an open-field with a secured food pellet in the center, rats injected with CRF do not approach the food as often as and eat less than untreated rats (Britton *et al.*, 1982). In addtion, CRF-treated rats exhibit increased grooming. These behaviors have been interpreted as neophobic responses and are opposite to those observed in rats administered anxiolytic agents (Britton and Britton, 1981).

Investigations of the behavioral effects of CRF in primates also show variability depending on the environment. Chair-restrained monkeys administered CRF (i.c.v.) show behaviors indicative of arousal such as struggling, whereas behavior appears to be inhibited when the animals are unrestrained in their home cages (Kalin *et al.*, 1983). Huddling and lying down in the home cage were also prominant after CRF administration. Other effects of CRF have been reported on social behavior: CRF increases mounting and aggression in male rats (Sutton *et al.*, 1982) and decreases sexual receptivity in female rats (Sirinathsingji *et al.*, 1983).

As is the case for the possible mediation of autonomic effects by CRF, putative anatomic substrates exist for CRF mediation of the behavioral concomitants of stress. However, the anatomic substrates for specific behaviors are not well characterized, nor are the behaviors associated with specific stressors or with various doses of CRF. These must be elucidated in order to determine the way in which extrahypothalamic CRF may mediate some or all of these behavioral responses.

1.6. Electrophysiological Effects

As a putative neurotransmitter, CRF should produce direct electrophysiological effects on cells that are involved in the proposed function of the peptide; i.e., CRF should alter the activity of cells that can potentially mediate behavioral or autonomic effects associated with stress. A preliminary study of iontophoretic effects of CRF demonstrated that direct application of CRF to cells of the cerebral cortex and hypothalamus activated these neurons, whereas application to cells in the thalamus and lateral septal area resulted in inhibition (Eberly *et al.*, 1983). As mentioned in Section 1.3, CRF has been visualized in these areas, and binding sites for CRF have been demonstrated in some of these sites.

The electrophysiological effects of CRF have also been investigated *in vitro* in slice preparations (Aldenhoff *et al.*, 1983; Siggins *et al.*, 1985), which offer the advantage of greater ease of intracellular recording and the potential for elucidating ionic mechanisms by which drugs and hormones act. In the hippocampal slice preparation, superfusion of CRF (10–200 nM) decreased afterhyperpolarizations in CA1 and CA3 pyramidal cells. These effects occurred without significant changes in membrane potential or input resistance. Higher concentrations (>250 nM) depolarized cells and increased input resistance, resulting in enhanced spontaneous discharge rates (Aldenhoff *et al.*, 1983). These excitatory effects are consistent with the EEG activation observed in freely moving rats following CRF administration and may play a role in CRF-induced behaviors.

In contrast to hippocampal neurons, cells in hypothalamic slices are more consistently inhibited by CRF (Siggins *et al.*, 1985). These effects have been studied primarily in the magnocellular region of the PVN.

One nucleus that receives CRF innervation and that has been implicated in behavioral arousal and autonomic function, and in which the electrophysiological effects of CRF have been studied, is the LC. The following sections outline the role of the LC in stress and the physiological effects of CRF on LC neurons.

2. THE ROLE OF THE NUCLEUS LOCUS COERULEUS IN STRESS

2.1. Introduction

There has long been evidence that LC–NE projections play a role in producing brain-mediated aspects of stress responses. Stress enhances NE turnover in several brain regions including some that receive their entire NE input from the LC. Since the LC is the largest NE cell group in the brain, and it has massive divergent axonal projections, it has the potential to influence activity throughout the CNS. Locus coeruleus activation produces demonstrable electrophysiological changes in target neurons, and the activity of LC neurons is affected by certain stressful stimuli. Thus, this nucleus has been postulated to respond to stress and consequently alter activity in other brain regions. The following sections outline the anatomic, physiological, and neurochemical evidence supporting the role of the LC in stress.

2.2. Anatomy of Locus Coeruleus Efferent System

The anatomic distribution of LC efferents is important as an indication of the regions possibly influenced by stress-induced LC activity. The efferent projections of this noradrenergic nucleus LC make up a widely divergent system that innervates every major region of the neuraxis including the spinal cord, cerebellum, and paleo- and neocortex (Grzanna and Molliver, 1980; Swanson, 1976; Swanson and Hartman, 1976). Thus, nuclei involved in endocrine, autonomic, and behavioral functions receive an NE innervation from this nucleus. For example, the LC innervates areas of the CNS involved in cardiovascular function including the hypothalamus (Jones and Moore, 1977; Kobayashi *et al.*, 1974; Moore and Bloom, 1979; Swanson, 1976, 1977), nucleus tractus solitarius (Loizou, 1969; McBride and Suttin, 1976; Olson and Fuxe, 1972; Swanson, 1977), dorsal motor nucleus of the vagus (Jacobowitz and Kostrzewa, 1971; Loizou, 1969; McBride and Suttin, 1976; Olson and Fuxe, 1972), and nucleus raphe dorsalis (Conard *et al.*, 1974; Pickel *et al.*, 1977). In addition, essentially all of the NE found in the cortex and hippocampus is contained within LC projections (Jones *et al.*, 1978; Kobayashi *et al.*, 1974). This divergent efferent system provides a substrate whereby the LC could coordinate effects of drugs or hormones to alter endocrine, autonomic, and behavioral responses.

It is of interest that the LC sends projections to the subdivision of the PVN where CRF-containing neurons have been localized (Sawchenko and Swanson, 1982). Since CRF fibers have also been localized in the LC, reciprocal innervation is present, which may function as a feedback loop.

2.3. Physiology of Locus Coeruleus Neurons

The physiological characteristics of LC-NE neurons and the effects of stimulation of the LC on target neurons are consistent with a role of the LC in stress. Electrical stimulation of the LC produces effects on target cells that are mimicked by iontophoresis of its neurotransmitter, norepinephrine (reviewed by Foote *et al.*, 1983). In general, this involves enhancement or synaptically evoked activity, whereas spontaneous activity is either decreased of unaffected (reviewed by Foote *et al.*, 1983). Enhancement of the effects of other neurotransmitters (i.e., GABA and acetylcholine) has also been reported after stimulation of the LC or concurrent iontophoresis of norepinephrine (reviewed by Foote *et al.*, 1983). Most of these effects have been observed in the neocortex (Foote *et al.*, 1975; Waterhouse *et al.*, 1980, 1982), hippocampus (Segal and Bloom, 1974a,b, 1976), and cerebellum

(Freedman *et al.*, 1977; Moises *et al.*, 1979, 1981; Moises and Woodward, 1980), although similar effects have been described in the lateral geniculate nucleus (Kayama *et al.*, 1982; Rogawski and Aghajanian, 1980), spinal cord (White and Neuman, 1980), and olfactory bulb (Jahr and Nicoll, 1982). These physiological consequences of LC activation and NE release have led to the hypothesis that activation of the LC system functions to cause enhancement of synaptic messages to target neurons with respect to background activity (Foote *et al.*, 1975; Woodward *et al.*, 1979).

Hypotheses of LC function must also consider the physiological characteristics of these cells under different behavioral conditions. These characteristics differ between anesthetized and unanesthetized animals. For example, LC activity of unanesthetized animals is very labile compared to that in anesthetized animals, where cells discharge at a slow regular rate (1–5 Hz) (Foote *et al.*, 1980, 1983; Foote and Bloom, 1979; Aston-Jones and Bloom, 1981a,b). Locus coeruleus cells of anesthetized animals generally respond only to noxious stimuli such as foot shcok (Aston-Jones *et al.*, 1982; Cederbaum and Aghajanian, 1978; Takigawa and Mogenson, 1977), whereas cells of unanesthetized animals are sensitive to a variety of nonnoxious sensory stimuli (Foote *et al.*, 1980; Aston-Jones and Bloom, 1981b). Responses to sensory stimuli consist of a brief period of activation followed by a longer period of suppression prior to a return to spontaneous levels of activity. Habituation of LC responses to sensory stimuli has also been reported to occur in monkeys (Foote *et al.*, 1980), suggesting that stimulus novelty is an important determinant of the vigor of LC neuronal activity. Combined behavioral and physiological studies in unanesthetized animals have described fluctuations of LC activity that are dependent on level of arousal. For example, LC activity is greatest when the animal is attending to stimuli in the external environment, decreased during consumptive behaviors such as grooming, eating, or slow-wave sleep, and essentially absent during REM sleep (Aston-Jones and Bloom, 1981a,b; Foote *et al.*, 1980). Finally, increases in LC activity anticipate behaviors associated with arousal or orientation to the external environment.

These characteristics of LC neuronal activity have led to the hypothesis that LC activation maintains a state of vigilance and arousal such that the animal's behavior is oriented towards the external environment and analysis of information, enhancing the performance of appropriate behavioral responses. Thus, LC cells are activated by external stimuli through sensory afferent projections, which results in enhanced signal-to-noise characteristics in target areas, in turn producing behavioral or autonomic responses that mobilize the animal to cope more effectively with threatening or noxious external stimuli.

2.4. Norepinephrine and Stress

The behavioral state of heightened vigilance and arousal associated with increased LC activity may be important in coping with stressful stimuli. The effects of different stressors on NE metabolism and turnover have been described, and, in general, NE turnover in the CNS increases during stress (Thierry *et al.*, 1968; Korf *et al.*, 1973; Cassens *et al.*, 1980, 1981). Thierry *et al.* (1968) have demonstrated a selective increase in NE turnover in the brainstem, mesencephalon, and spinal cord in animals exposed to footshock. Increases in NE turnover after footshock in rates have also been observed in cerebral cortex and hippocampus, two areas whose sole NE input is from the LC. Unilateral LC lesions abolished the increased turnover on the ipsilateral side. Increases in NE turnover can also be conditioned or elicited by previously neutral stimuli that have become associated with shock (Cassens *et al.*, 1980, 1981). Finally, previous experience with shock does not result in tolerance to effects on NE turnover. Taken together, these biochemical and electrophysiological studies suggest that footshock activates the LC, resulting in increased NE turnover in major LC target areas.

The results outlined above provide indirect evidence that LC activation is an integral component of stress. More direct evidence comes from studies in which LC activity is recorded during the presentation of stressful stimuli. Although most sensory stimuli do not alter LC discharge rate in anesthetized animals, noxious stimuli such as footshock have been demonstrated to produce a 50- to 100-msec period of activation followed by a 300- to 400-msec period of inhibition. Certain cardiovascular stimuli that may be considered stressful also affect LC activity in anesthetized animals. For example, blood volume loading decreases discharge rates of LC cells, whereas decreases in blood volume, as would occur during hemorrhage, increase discharge rate (Elam *et al.*, 1984; Svensson and Thoren, 1979). Likewise, hypercapnea activates LC neurons and increases brain NE turnover (Elam *et al.*, 1981; Garcia de Yebenes Prous *et al.*, 1977). Thus, the LC is activated by certain cardiovascular and metabolic stressors.

It is of interest that some stimuli that activate LC neurons (i.e., hemorrhage and footshock) also increase ACTH release, whereas some that decrease neuronal activity (i.e., blood volume loading, Plotsky, 1985) decrease release of ACTH (Plotsky, 1985). The changes in ACTH release produced by these stressors have been demonstrated to depend on CRF (see Section 1.2). Thus, strong parallels exist among CRF release, LC activity, and NE turnover. Since CRF fibers have been visualized in the LC, we have begun to test the hypothesis that certain stressful stimuli elicit extrahypothalamic release of CRF, which, in turn, increases LC activity. The LC, through its vast efferent system, could then serve as an amplification mechanism for CRF, acting to mobilize many different brain systems in response to stressful stimuli. In order to test this hypothesis, CRF-induced activation of LC neurons must be demonstrated. In addition, these changes in neuronal activity must be correlated with behavioral and/or autonomic responses associated with CRF mobilization or administration. The following section deals with the effects of CRF on spontaneous and sensory evoked activity of LC neurons in anesthetized and unanesthetized rats.

3. CRF ACTIONS ON LOCUS COERULEUS NEURONS

3.1. Anesthetized Rats

3.1.1. Spontaneous Activity

To test the hypothesis that the LC may amplify and coordinate the effects of CRF released as a neurotransmitter in response to stress, it was first necessary to demonstrate that CRF is capable of altering the activity of LC neurons (Valentino *et al.*, 1983a). Intracerebroventricular (i.c.v.) injection of CRF (3.0 μg) in halothane-anesthetized rats increased the discharge rates in all 12 LC cells tested by 65 ± 17%, the effect becoming maximal approximately 7 min after injection (Fig. 1; Valentino *et al.*, 1983a). Like other physiologically active peptides, CRF is amidated at its carboxy terminus. The free acid analogue, CRF-OH, is less active than CRF in releasing ACTH (Vale *et al.*, 1981) and in producing behavioral effects (Sutton *et al.*, 1982). Likewise, this analogue did not cause a significant increase in discharge rates in any of five LC cells tested (Fig. 1; Valentino *et al.*, 1983a). Moreover, another less active analogue, Ala[14]-CRF, was also ineffective in increasing LC spontaneous discharge rates (Fig. 2; R. J. Valentino and S. L. Foote, unpublished data). The effective dose of CRF was within the range of doses that produce both autonomic and behavioral actions as described in Sections 1.4 and 1.5. Lower doses of CRF were not effective in anesthetized rats (Fig. 2; R. J. Valentino and S. L. Foote, unpublished data). These results demonstrate that central administration of CRF activates LC cells in a dose-dependent and structurally specific manner. Moreover, the structural specificity is similar to that shown for CRF effects on ACTH release, suggesting that the receptors responsible for LC activation are similar to those responsible for release of ACTH.

Intracerebroventricular administration of drugs or hormones could affect a

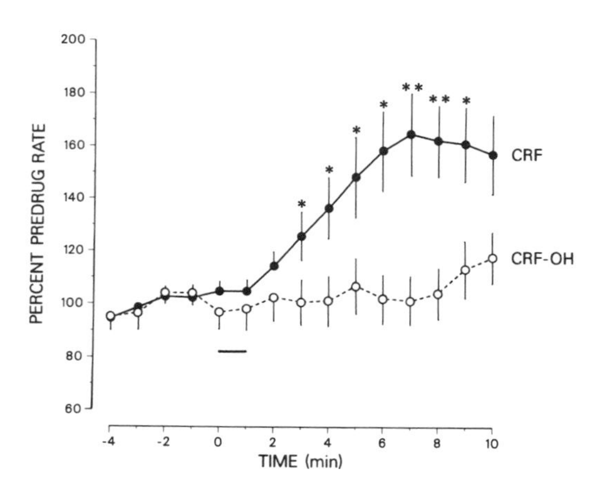

FIGURE 1. Average effect of CRF (solid circles) or CRF-OH (open circles) on spontaneous discharge rates of LC cells as a function of time before or after i.c.v. administration (at 0–1 min). Rates are expressed as a percentage of the mean predrug rate. Each point is the mean of 10 to 12 cells for CRF and three to five cells for CRF-OH; vertical lines indicate ± S.E.M. Statistical significance was determined by Student's *t*-test (two-tailed), $^{*}P < 0.05$, $^{**}P < 0.025$. (From Valentino *et al.*, 1983a, reproduced with permission.)

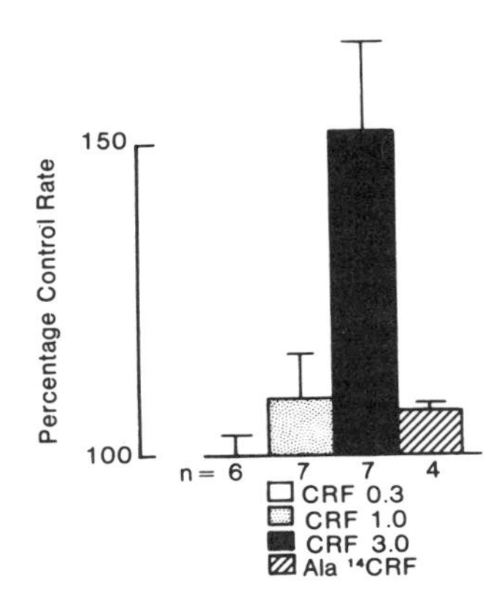

FIGURE 2. Dose dependence and structural specificity of CRF on LC activity. Bars indicate the LC spontaneous discharge rate after i.c.v. administration of different doses of CRF or Ala¹⁴-CRF (3.0 μg) expressed as a percentage of the mean rates prior to peptide administration. Vertical lines indicate ± S.E.M. Numbers below the bars indicate the number of cells tested. CRF caused a dose-dependent activation of LC spontaneous activity in halothane-anesthetized rats, and the effect is structurally specific since the less active analogue, Ala¹⁴-CRF, did not increase LC discharge rates.

number of afferents to the LC and thereby indirectly alter spontaneous activity. However, to demonstrate that CRF released from fibers innervating the LC can increase discharge rates, a direct effect of CRF on these cells must be demonstrated. The two approaches that were used for determining the direct actions of CRF on LC neurons were iontophoresis and pressure application of CRF onto the recorded neuron (Valentino *et al.*, 1983a). Iontophoresis of CRF to LC cells did not produce consistent alterations in activity. However, the application of small amounts of pressure to a micropipette containing CRF (0.1 mM), which released CRF directly onto recorded cells, resulted in enhanced discharge rates in nine of 14 cells (Valentino *et al.*, 1983a). The remaining five cells exhibited no change in discharge rates (see Fig. 3). In contrast to its effect on LC cells, CRF applied by this technique did not activate cerebellar Purkinje neurons and cells of the mesencephalic nucleus of the trigeminus. Interestingly, a dramatic activation was produced by CRF application to a neuron that was localized to the lateral parabrachial nucleus. This area is thought to be involved in respiration and is a pontine nucleus that projects

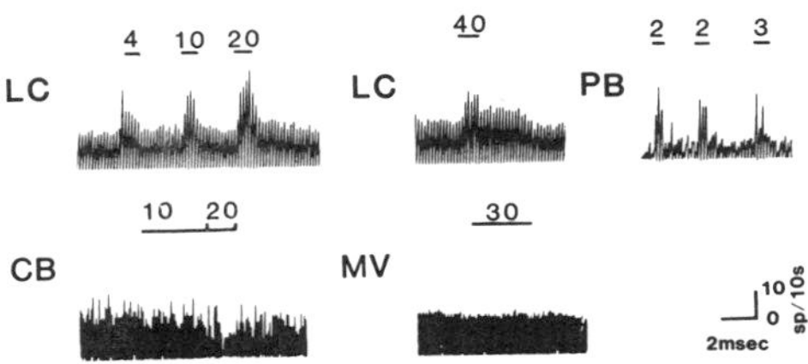

FIGURE 3. Effect of pressure application of CRF to different neurons. CRF was directly applied to neurons of the locus coeruleus (LC), cerebellar Purkinje cells (CB), neurons of the mesencephalic nucleus of the trigeminus (MV), or parabrachial cells (PB) for the time period indicated by bars above the records. Numbers above the bars indicate the amount of pressure in psi that was applied to the CRF-containing barrel of a three-barrel micropipette. The center barrel was used to record activity of the neurons. For CB and MV cells, the pipette used to administer CRF was the same as that used on LC cells in the corresponding records in the upper panel. Although CRF was excitatory on both LC cells and this PB neuron, it had no effect on CB or MV cells. (Reproduced with permission from Valentino *et al.*, 1983a.)

to CRF-containing cells of the PVN and receives CRF innervation. Thus, activation of LC neurons by CRF appears to be a direct effect and shows some regional specificity.

3.1.2. Sensory Responsiveness

Locus coeruleus cells of anesthetized animals are responsive to noxious stimuli. A brief shock to the paw excites contralateral LC neurons, and this is followed by a period of inhibition before base-line spontaneous activity returns (Fig. 4). In certain cases, this evoked activity is more sensitive to pharmacological agents than

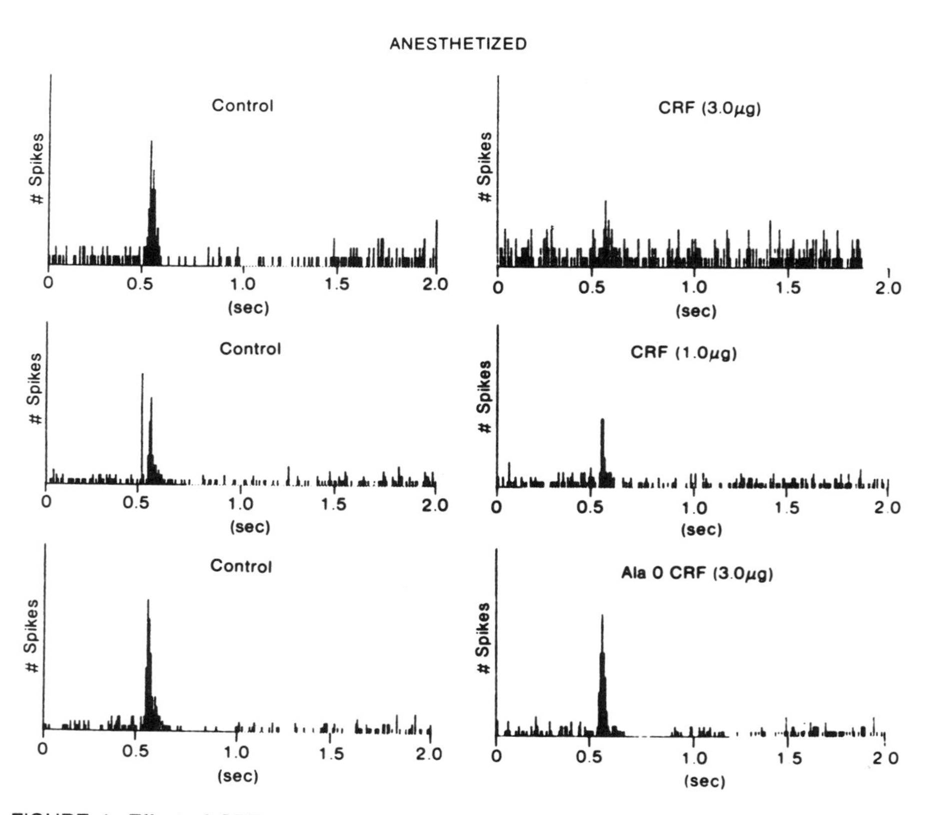

FIGURE 4. Effect of CRF on sensory evoked activity of LC neurons. The poststimulus time histograms shown indicate cumulative LC activity recorded during the presentation of 80–100 shocks to the contralateral paw. Shocks presented at the rate of 0.1 Hz occurred at 0.5 sec. The ordinate indicates the number of discharges of a single cell before and after stimulus presentation. The histograms shown illustrate the pattern of activity before and after i.c.v. administration of CRF (1.0 μg or 3.0 μg) or Ala^{14}-CRF (3.0 μg). CRF (3.0 μg) activated spontaneous or background activity and caused a relative decrease in evoked activity. Ala^{14}-CRF did not alter the pattern of LC activity in response to footshock.

is spontaneous activity (Aston-Jones *et al.*, 1982). When administered in doses that increase spontaneous discharge rates, CRF appeared to decrease the footshock response (Fig. 4). This effect was dose dependent; i.e., 0.3 μg of CRF had no effect, and 1.0 μg produced this effect in some cells (Fig. 4; R. J. Valentino and S. L. Foote, unpublished data). Although this action is difficult to interpret, it suggests that when the LC system is tonically activated, the magnitude of responses elicited by specific phasic sensory stimuli is decreased. The behavioral consequence of this electrophysiological phenomenon may be a tonically aroused animal whose level of arousal is not further phasically enhanced by discrete stimuli. However, until the behavioral manifestations of stress and of CRF administration are more fully characterized, the interpretation of CRF effects on sensory evoked LC activity must remain speculative.

3.2. Unanesthetized Rats

3.2.1. Spontaneous Activity

Compared to anesthetized animals, LC discharge activity in unanesthetized animals is more labile, changes with the level of arousal, and is sensitive to all modalities of nonnoxious sensory stimuli. The different characteristics of LC cells in the unanesthetized state suggest that sensitivity to CRF may also be altered. Additionally, the ultimate test of the hypothesis that the LC amplifies and coordinates behavioral and/or autonomic effects of CRF must be done in unanesthetized animals, where both behavior and LC activity may be monitored simultaneously. This led to an extension of our previous studies to the unanesthetized rat.

As in anesthetized rats, CRF increased the spontaneous discharge rates of LC cells in 12 of 15 determinations (Valentino *et al.*, 1983b). Interestingly, LC neurons in unanesthetized animals were sensitive to lower doses of CRF than those in anesthetized animals. For example, 1.0 μg, which does not cause significant activation of LC cells in anesthetized animals, produced a mean increase in discharge rate of 42 ± 9% in unanesthetized, sling-restrained rats. The latency of onset for the increase was variable. In addition, struggling of animals in the sling often followed increases in LC activity produced by CRF, interrupting the recording session. A higher dose of CRF (3.0 μg) produced this struggling more consistently. In four animals in which recording was not interrupted, the discharge rates of LC cells increased 50–200% during the 10- to 30-min period after injection. In animals that did not exhibit struggling, there was often vocalization, a behavior that was not evident prior to CRF administration. After the recording session, rats seemed sensitive to handling and vocalized when touched. This was not apparent in animals that were not administered CRF. Lower doses of CRF (0.3 μg) increased LC activity in two of four rats, and the less active analogue Ala^{14}-CRF had no effect in five rats (R. J. Valentino and S. L. Foote, unpublished data). Thus, these preliminary results suggest that central administration of CRF activates LC cells in unanesthetized rats in a dose-dependent, structurally specific manner and that these animals are more sensitive to CRF than anesthetized animals. Finally, the increase in LC discharge rates may precede struggling or signs of behavioral activation in rats.

3.2.2. Sensory Responsiveness

Locus coeruleus cells in unanesthetized animals are responsive to visual, auditory, and olfactory stimuli and respond in the same manner that LC cells in anesthetized animals respond to noxious stimuli (Aston-Jones and Bloom, 1981b; Foote *et al.*, 1980). The novelty of the stimulus may be an important characteristic in eliciting this response (Foote *et al.*, 1980). Since CRF has been thought to produce behaviors indicative of neophobia (Britton *et al.*, 1982), it might be expected to affect responsiveness of LC cells to sensory stimuli at doses that do not alter spontaneous activity. One might predict dose-dependent effects of CRF on sensory evoked activity such that low doses would enhance responsiveness whereas higher doses that increase arousal would produce relative decreases in responsiveness. This would be adaptive for an animal exposed to different intensities of stress.

Our initial studies of CRF in unanesthetized rats have utilized auditory stimuli to produce sensory responses in LC neurons (R. J. Valentino and S. L. Foote, unpublished data). As one might predict, the highest dose of CRF (3.0 μg) produced a substantial increase in spontaneous activity and a relative decrease in sensory evoked activity. Lower doses of CRF (0.3–1.0 μg) have produced inconsistent results, and the number of rats tested has been too small to characterize effects reliably. However, these results are consistent with those observed in anesthetized animals. In both preparations, CRF, at doses that substantially increase spontaneous discharge rates, decreases selective responsiveness to sensory stimuli. This may represent what occurs during high levels of arousal or in a very stressful situation. However, these interpretations must await a more complete characterization of the behavioral effects associated with stress responses and those associated with different doses of exogenously administered CRF.

3.3. Locus Coeruleus Activity and Blood Pressure

Our studies of LC activity in unanesthetized rats suggest that CRF activation of these cells precedes behavioral activation. Previous evidence indicates that CRF produces a number of autonomic as well as behavioral signs that are consistent with a role for mobilizing the organism to react to stressful stimuli. Since the LC receives input from nuclei involved in cardiovascular control and also sends efferent fibers to these centers, it is possible that the LC may coordinate the autonomic effects of CRF as well. A role for the LC in blood pressure regulation has been inferred from studies demonstrating that electrical stimulation of the LC produces a pressor effect (Chida *et al.*, 1983; Gurtu *et al.*, 1984; Kawamura *et al.*, 1978; Przuntek and Philippu, 1973; Ward and Gunn, 1976). Additionally, injections of aconitine (Perlman and Guideri, 1984) or morphine (Pant *et al.*, 1983) into the LC are associated with increases and decreases in blood pressure, respectively. Finally, changes in blood pressure and particularly blood volume have been shown to influence LC activity (Elam *et al.*, 1984). In order to determine whether the LC may mediate some of the autonomic effects of CRF, we have recorded both blood pressure and LC activity simultaneously in halothane-anesthetized rats during i.c.v. administration of CRF (3.0 μg). This dose produces LC activation in anesthetized rats and blood pressure elevation in unanesthetized rats. Interestingly, we found that although this dose increased LC discharge rates by 100% within 6 min of injection, no elevation of blood pressure occurred during this time (R. J. Valentino and S. L. Foote, unpublished data). These findings suggest that in anesthetized rats CRF is less effective in producing blood pressure changes and that LC activation by CRF may be dissociated from alterations in blood pressure. The two limitations of this interpretation are (1) that they are confounded by anesthesia—i.e., the LC may play a role in blood pressure effects induced by CRF in the absence of anesthesia (Brown and Fisher, 1985; Fisher *et al.*, 1982)—and (2) a greater activation of LC activity is necessary (>100%) to produce an effect on blood pressure. Extension of these studies to unanesthetized animals will be necessary to determine the role of the LC in autonomic effects of CRF.

4. SUMMARY

CRF may serve synergistic roles as a releasing factor and a neurotransmitter, acting simultaneously in both capacities to mobilize endocrine, physiological, and behavioral components of stress responses. Data compatible with this overall hypothesis have been generated by various approaches. The administration of exogenous CRF causes release of ACTH, behavioral activation, and a constellation of autonomic responses consistent with this hypothesis. The administration of CRF antiserum or antagonists to animals during stress can prevent some of these effects. Finally, the hypothalamic and extrahypothalamic distribution of CRF supports these roles. Likewise, noradrenergic neurons of the LC have been implicated in the generation of stress responses. The anatomic distribution of LC efferents, the physiological characteristics of these cells, and their effects on target neurons provide evidence for this role. This has been further supported by studies of NE turnover and activity of these cells during exposure to specific stressful stimuli.

Since CRF has been visualized in fibers innervating the LC, it may be postulated that stress elicits CRF release from these fibers, which then alters activity of LC cells. Through the divergent efferent system of the LC, these effects may be amplified to produce some of the physiological and behavioral aspects of stress responses. In testing this hypothesis, we have demonstrated that CRF directly alters LC spontaneous and sensory evoked activity in a dose-dependent manner that is both structurally and regionally specific. This alteration in LC activity appears to be associated with behavioral arousal more than with autonomic responses.

There are several limitations to the hypothesis that the LC plays an integral role in mediating CRF effects during stress. CRF-immunoreactive neurons have been visualized in other extrahypothalamic regions that are involved in behavior and autonomic function, and it is probable that CRF effects on these neurons contribute to the stress response. In addition, CRF innervation of the LC is relatively limited and has never been demonstrated ultrastructurally. Likewise, the density of CRF receptors in this nucleus is low. Although there is strong support for hypothalamic CRF release by stressful stimuli, it is not clear whether the same stimuli elicit release of CRF from extrahypothalamic neurons (however, see Britton *et al.*, 1984; Smith *et al.*, 1985). Moreover, the consequences of extrahypothalamic release have yet to be determined. We have cited investigations that imply a role for the LC in stress, but this role has not been thoroughly characterized. This requires determination of the effects of various stressors on LC activity and the characterization of the effect of LC activation on behavior and autonomic function. Thus, although existing evidence is consistent with a role for the LC in amplifying and coordinating the effects of CRF in stress, the contribution of the LC in the overall stress response has yet to be elucidated.

REFERENCES

Aldenhoff, J. B., Gruol, D. L., Rivier, J., Vale, W., and Siggins, G. R., 1983, Corticotropin releasing factor decreases postburst hyperpolarizations and excites hippocampal neurons, *Science* **221**:875–877.

Aston-Jones, G., and Bloom, F. E., 1981a, Activity of norepinephrine containing locus coeruleus neurons in behaving rats anticipates fluctuations in the sleep–waking cycle, *J. Neurosci.* **1**:876–886.

Aston-Jones, G., and Bloom, F. E., 1981b, Norepinephrine-containing neurons in behaving rats exhibit pronounced responses to non-noxious environmental stimuli, *J. Neurosci.* **1**:887–900.

Aston-Jones, G., Foote, S. L., and Bloom, F. E., 1982, Low doses of ethanol disrupt sensory responses of brain noradrenergic neurones, *Nature* **296**:857–860.

Bloom, F. E., Battenberg, E. L. F., Rivier, J., and Vale, W., 1982, Corticotropin releasing factor (CRF) immunoreactive neurons and fibers in rat hypothalamus, *Regul. Peptides* **4**:43–48.

Britton, D. R., and Britton, K. K., 1981, A sensitive open field measure of anxiolytic drug activity, *Pharmacol. Biochem. Behav.* **15**:577–582.

Britton, D. R., Koob, G. F., Rivier, J., and Vale, W., 1982, Intraventricular corticotropin releasing factor enhances behavioral effects of novelty, *Life Sci.* **31**:363–367.

Britton, K. T., Lyon, M., Vale, W., and Koob, G. F., 1984, Stress-induced secretion of corticotropin-releasing factor immunoreactivity in rat cerebrospinal fluid, *Soc. Neurosci. Abstr.* **10**(1):94.

Brown, M. R., and Fisher, L. A., 1985, Corticotropin-releasing factor: Effects on the autonomic nervous system and visceral systems, *Fed. Proc.* **44**:243–248.

Brown, M. R., Fisher, L. A., Rivier, J., Spiess, J., Rivier, C., and Vale, W., 1982a, Corticotropin-releasing factor: Effects on the sympathetic nervous system and oxygen consumption, *Life Sci.* **30**:207–210.

Brown, M. R., Fisher, L. A., Spiess, J., Rivier, C., Rivier, J., and Vale, W., 1982b, Corticotropin-releasing factor: Actions on the sympathetic nervous system and metabolism, *Endocrinology* **111**:928–931.

Brown, M. R., Fisher, L. A., and Vale, W. W., 1983, Sympathetic nervous system and adrenocortical interactions, *Soc. Neurosci. Abstr.* **9**:391.

Brown, M. R., Fisher, L. A., Vale, W. W., and Rivier, J. E., 1984, Corticotropin-releasing factor: Role in central nervous system regulation of the adrenal medulla, *Soc. Neurosci. Abstr.* **10**:1117.

Bugnon, C., Fellman, D., Gouge, A., and Cardot, J., 1982, Corticoliberin in rat brain: Immunocytochemical identification and localization of a novel neuroglandular system, *Neurosci. Lett.* **30**:25–30.

Cassens, G., Roffman, M., Kuruc, A., Orsulak, P. J., and Schildkraut, J. J., 1980, Alterations in brain norepinephrine metabolism induced by environmental stimuli previously paired with inescapable shock, *Science* **209:**1138–1139.

Cassens, G., Kuruc, A., Roffman, M., Orsulak, P. J., and Schildkraut, J. J., 1981, Alterations in brain norepinephrine metabolism and behavior induced by environmental stimuli previously paired with inescapable shock, *Behav. Brain Res.* **2:**387–407.

Cedarbaum, J. M., and Aghajanian, G. K., 1978, Activation of locus coeruleus meurons by peripheral stimuli: Modulation by a collateral inhibitory mechanism, *Life Sci.* **23:**1383–1392.

Chida, K., Kawamura, H., and Hatano, M., 1983, Participation of the nucleus locus coeruleus in DOCA-salt hypertensive rats, *Brain Res.* **273:**53–58.

Conard, L. C. A., Leonard, C. M., and Pfaff, D. W., 1974, Connections of the median and dorsal raphe nuclei in the rat: An autoradiographic and degeneration study, *J. Comp. Neurol.* **156:**179–206.

Cummings, S., Elde, R., Ells, J., and Lendall, A., 1983, Corticotropin-releasing factor immunoreactivity is widely distributed within the central nervous system of the rat: An immunohistochemical study, *J. Neurosci.* **3:**1355–1368.

DeSouza, E. B., Perrin, M. H., Insel, T. R., Rivier, J., Vale, W. W., and Kuhar, M. J., 1984, Corticotropin-releasing factor receptors in rat forebrain: Autoradiographic identification, *Science* **224:**1449–1451.

Eberly, L. B., Dudley, C. A., and Moss, R. L., 1983, Iontophoretic mapping of corticotropin-releasing factor (CRF) sensitive neurons in the rat forebrain, *Peptides* **4:**837–841.

Ehlers, C. L., Henriksen, S. J., Wang, M., Rivier, J., Vale, W. W., and Bloom, F. E., 1983, Corticotropin-releasing factor produces increases in brain excitability and convulsive seizures in the rat, *Brain Res.* **278:**332–336.

Elam, M., Yao, T., Thoren, P., and Svensson, T. H., 1981, Hypercapnia and hypoxia: Chemoreceptor-mediated control of locus coeruleus neurons and splanchnic, sympathetic nerves, *Brain Res.* **222:**373–381.

Elam, M., Yao, T., Svensson, T. H., and Thoren, P., 1984, Regulation of locus coeruleus neurons and splanchnic, sympathetic nerves by cardiovascular afferents, *Brain Res.* **290:**281–287.

Fischman, A. J., and Moldow, R. L., 1982, Extrahypothalamic distribution of CRF like immunoreactivity in the rat brain, *Peptides* **3:**149–153.

Fisher, L. A., Rivier, J., Rivier, C., Spiess, J., Vale, W. W., and Brown, M. R., 1982, Corticotropin-releasing factor (CRF): Central effects on mean arterial pressure and heart rate in rats, *Endocrinology* **110:**2222–2224.

Foote, S. L., and Bloom, F. E., 1979, Activity of norepinephrine-containing locus coeruleus neurons in the unanesthetized squirrel monkey, in: *Catecholamines: Basic and Clinical Frontiers* (E. Usdin, I. J. Kopin, and J. Barchas, eds.), Pergamon Press, New York, pp 625–627.

Foote, S. L., Freedman, R., and Oliver, A. P., 1975, Effects of putative neurotransmitters on neuronal activity in monkey auditory cortex, *Brain Res.* **86:**229–242.

Foote, S. L., Aston-Jones, G, and Bloom, F. E., 1980, Impulse activity of locus coeruleus neurons in awake rats and monkeys is a function of sensory stimulation and arousal, *Proc. Natl. Acad. Sci. U.S.A.* **77:**3033–3037.

Foote, S. L., Bloom, F. E., and Aston-Jones, G., 1983, Nucleus locus coeruleus: New evidence of anatomical and physiological specificity, *Physiol. Rev.* **63:**844–914.

Freedman, R., Hoffer, B. J., Woodward, D. J., and Riro, D., 1977, Interaction of norepinephrine with cerebellar activity evoked by mossy and climbing fibers, *Exp. Neurol.* **55:**269–288.

Garcia de Yebenes Prous, J., Carlson, A., and Mena Gomez, M. A., 1977, The effect of CO_2 on monamine metabolism in rat brain, *Naunyn Schmidebergs Arch. Pharmacol.* **301:**11–15.

Grzanna, R., and Molliver, M. E., 1980, The locus coeruleus in the rat: An immunohistochemical delineation, *Neuroscience* **5:**21–40.

Gurtu, C. G., Pant, K. K., Sinha, J. N., and Bhargava, K. P., 1984, An investigation into the mechanism of cardiovascular responses elicited by electrical stimulation of locus coeruleus and subcoeruleus in the cat, *Brain Res.* **39:**59–64.

Jacobowitz, D., and Kostrzewa, R., 1971, Selective action of 6-hydroxydopa on noradrenergic terminals: Mapping of preterminal axons of the brain, *Life Sci.* **10:**1329–1342.

Jahr, C. E., and Nicoll, R. A., 1982, Noradrenergic modulation of dendrodendritic inhibition in the olfactory bulb, *Nature* **297:**227–229.

Jones B. E., and Moore, R. Y., 1977, Ascending projections of the locus coeruleus in the rat. II. Autoradiographic study, *Brain Res.* **127:**23–53.

Jones B. E., Halaris, A. E., and Freeman, O. X., 1978, Innervation of forebrain regions by medullary noradrenaline neurons, a biochemical study in cats with central tegmental tract lesions, *Neurosci. Lett.* **10:**251–258.

Kalin, N. H. Shelton, S. E., Karemer, G. W., and McKenney, W. T., 1983, Corticotropin-releasing factor administered intraventriculary to rhesus monkeys, *Peptides* **4:**217–220.

Kawamura, H., Gunn, C. G., and Frohlich, E. D., 1978, Cardiovascular alteration by nucleus locus coeruleus in spontaneously hypertensive rat, *Brain Res.* **140:**137–147.

Kawata, M., Hashimoto, K., Takahara, J., and Sani, Y., 1982, Immunohistochemical demonstration of corticotropin-releasing factor containing neurons in the hypothalamus of mammals including primates, *Anat. Embryol.* **165:**303–313.

Kayama, Y., Negi, T., Sugitani, M., and Iwama, K., 1982, Effects of locus coeruleus stimulation on neuronal activities of dorsal lateral geniculate nucleus and perigeniculate reticular nucleus of rat, *Neuroscience* **7:**655–666.

Kobayashi, R. M., Palkovits, M., Kopin, I. J., and Jacobowitz, D. M., 1974, Biochemical mapping of noradrenergic nerves arising from the rat locus coeruleus, *Brain Res.* **77:**269–279.

Korf, J., Aghajanian, G. K., and Roth, R. H., 1973, Increased turnover of norepinephrine in the rat cerebral cortex during stress: Role of the locus coeruleus, *Neuropharmacology* **12:**933–938.

Loizou, L. A., 1969, Projections of the nucleus locus coeruleus in the albino rat, *Brain Res.* **15:**563–566.

McBride, R. L., and Suttin, J., 1976, Projections of the nucleus locus coeruleus and adjacent pontine tegmentum in the cat, *J. Comp. Neurol.* **165:**265–284.

Merchenthaler, I., Hynes, M. A., Vigh, S., Schally, A. V., and Petrusz, P., 1982, Immunocytochemical localizations of corticotropin-releasing factor (CRF) in rat brain, *Am. J. Anat.* **165:**383–396.

Merchenthaler, I., Hynes, M. A., Vigh, S., Schally, A. V., and Petrusz, P., 1983, Immunocytochemical localizations of corticotropin-releasing factor (CRF) in rat spinal cord, *Brain Res.* **275:**373–377.

Moises, H. C., and Woodward, D. J., 1980, Potentiation of GABA inhibitory action in cerebellum by locus coeruleus stimulation, *Brain Res.* **182:**327–344.

Moises, H. C., Woodward, D. J., Hoffer, B. J., and Freedman, R., 1979, Interactions of norepinephrine with Purkinje cell responses to putative amino acid neurotransmitters applied by microiontophoresis, *Exp. Neurol.* **64:**493–515.

Moises, H. C., Waterhouse, B. D., and Woodward, B. J., 1981, Locus coeruleus stimulation potentiates Purkinje cell responses to afferent input: The climbing fiber system, *Brain Res.* **222:**43–64.

Moore, R. Y., and Bloom, F. E., 1979, Central catecholamine neuron systems: Anatomy and physiology of the norepinephrine and epinephrine system, *Annu. Rev. Neurosci.* **2:**113–168.

Olschowka, J. A., O'Donahue, T. L., Mueller, G. P., and Jacobowitz, D. M., 1982, The distribution of corticotropin-releasing factor-like immunoreactive neurons in rat brain, *Peptides* **3:**995–1015.

Olson, L., and Fuxe, K., 1972, Further mapping out of central noradrenaline neuron systems: Projections of the "subcoruleus" area, *Brain Res.* **43:**289–295.

Palkovits, M., Brownstein, M. J., and Vale, W., 1983, Corticotropin releasing factor (CRF) immunoreactivity in hypothalamic and extrahypothalamic nuclei of sheep brain, *Neuroendocrinology* **37:**302–305.

Palkovits, M., Brownstein, M. J., and Vale, W., 1985, Distribution of corticotropin releasing factor in rat brain, *Fed. Proc.* **44:**215–219.

Pant, K. K., Gurtu, S., Sharma, D. K., Sinha, J. N., and Bhargara, K. P., 1983, Cardiovascular effects of microinjection of morphine into the nucleus locus coeruleus of the cat, *Jpn. J. Pharmacol.* **33:**253–256.

Perlman, R., and Guideri, G., 1984, Cardiovascular changes produced by the injection of aconitine at the area of the locus coeruleus in unanesthetized rats, *Arch. Int. Pharmacol.* **268:**202–215.

Pickel, V. M., Joh, T. H., and Reis, D. J., 1977, Serotonergic innervation of noradrenergic neurons in nucleus locus coeruleus: Demonstration by immunohistochemical localization of the transmitter specific enzyme tryosine hydroxylase and tryptophan hydroxylase, *Brain Res.*, **131:**197–214.

Plotsky, P. M., 1985, Hypophyseotropic regulation of adenohypophyseal adrenocorticotropin secretion, *Fed. Proc.* **344:**207–214.

Przuntek, H., and Philippu, A., 1973, Reduced pressor responses to stimulation of the locus coeruleus after lesion of the posterior hypothalamus, *Naunyn Schmiedebergs Arch. Pharmacol.* **276:**119–122.

Raymond, V., and Negro-Villar, A., 1984, Angiotensin II increases ACTH release in the absence of endogeneous arginine vasopression, *Life Sci.* **34:**721–729.

Rivier, C., and Vale, W., 1983, Modulation of stress-induced ACTH release by corticotropin releasing factor, catecholamines and vasopressin, *Nature* **305:**325–327.

Rivier, C., and Vale, W., 1985, Effects of corticotropin-releasing factor, neurohyphophyseal peptides, and catecholamines on pituitary function, *Fed. Proc.* **44:**189–195.

Rivier, C., Rivier, J., and Vale, W., 1982, Inhibition of adrenocorticotropic hormone secretion in the rat by immunoneutralization of corticotropin-releasing factor, *Science* **218:**377–378.

Rogawski, M. A., and Aghajanian, G. K., 1980, Activation of lateral geniculate neurons by norepinephrine: Mediation by an alpha-adrenergic receptor, *Brain Res.* **182:**345–359.

Saper, C. B., and Loewy, A. D., 1980, Efferent connections of the parabrachial nucleus in the rat, *Brain Res.* **194:**291–317.

Sawchenko, P. E., and Swanson, L. W., 1982, The organization of noradrenergic pathways from the brainstem to the paraventricular and supraoptic nuclei in the rat, *Brain Res. Rev.* **4:**285–325.

Sawchenko, P. E., and Swanson, L. W., 1983, The organization and biochemical specificity of afferent projections to the paraventricular and supraoptic nuclei, *Prog. Brain Res.* **60:**19–29.

Sawchenko, P. E., and Swanson, L. W., and Vale, W. W., 1984, Corticotropin-releasing factor: Coexpression within distance subsets of oxytocin-, vasopressin-, and neurotensin-immunoreactive neurons in the hypothalamus of the rat, *J. Neurosci.* **4:**1118–1129.

Schipper, J., Steinbusch, H. W. M., Vermes, I., and Tilders, F. J. H., 1983, Mapping of CRF-immunoreactive nerve fibers in the medulla oblongata and spinal cord of the rat, *Brain Res.* **267:**145–150.

Segal, M., and Bloom, F. E., 1974a, The action of norepinephrine in the rat hippocampus. I. Iontophoretic studies, *Brain Res.* **72:**79–97.

Segal, M., and Bloom, F. E., 1974b, The action of norepinephrine in the rat hippocampus. II. Activation of the input pathway, *Brain Res.* **72:**99–114.

Segal, M., and Bloom, F. E., 1976, The action of norepinephrine in the rat hippocampus. III. Hippocampal cellular responses to locus coeruleus stimulation in the awake rat, *Brain Res.* **107:**499–511.

Siggins, G. R., Gruol, D., Aldenhoff, J., and Pittman, Q., 1985, Electrophysiological actions of corticotropin-releasing factor in the central nervous system, *Fed. Proc.* **44:**237–242.

Sirinathsinghji, D. J. S., Rees, L. H., Rivier, J., and Vale, W., 1983, Corticotropin-releasing factor is a potent inhibitor of sexual receptivity in the female rat, *Nature* **305:**232–235.

Smith, M. A., Chappell, P. B., Bissette, G., Richie, J., Kilts, C. D., and Nemeroff, C. B., 1985, Stress induced alterations in corticotropin-releasing factor-like immunoreactivity (CRF-LI) in rat brain, *Soc. Neurosci. Abstr.* **11:**1269.

Spinedi, E., and Negro-Villar, A., 1983, Angiotensin II and ACTH release: Site of action and potency relative to corticotropin releasing factor and vasopressin, *Neuroendocrinology* **37:**446–453.

Spinedi, E., and Negro-Villar, A., 1984, Angiotensin II increases ACTH release in the absence of endogenous arginine vasopressin, *Life Sci.*, **34:**721–729.

Sutton, R. E., Koob, G. F., LeMoal, M., Rivier, J., and Vale, W., 1982, Corticotropin-releasing factor produces behavioral activation in rats, *Nature* **297:**331–333.

Svensson, T. H., and Thoren, P., 1979, Brain noradrenergic neurons in the locus coeruleus: Inhibition by blood volume load through vagal afferents, *Brain Res.* **172:**174–178.

Swanson, L. W., 1976, The locus coeruleus: A cytoarchitectonic, Golgi and immunohistochemical study in the albino rat, *Brain Res.* **110:**39–56.

Swanson, L. W., 1977, Immunohistochemical evidence for a neurophysin containing autonomic pathway arising in the paraventricular nucleus of the hypothalamus, *Brain Res.* **128:**346–353.

Swanson, L. W., and Hartman, B. K., 1976, The central adrenergic system. An immunoflourescence study of the location of cell bodies and their efferent connections in the rat utilizing dopamine-β-hydroxylase as a marker, *J. Comp. Neurol.* **163:**467–506.

Swanson, L. W., and Sawchenko, P. E., 1983, Hypothalamic integration: Organization of the paraventricular and supraoptic nuclei, *Annu. Rev. Neurosci.* **6:**275–325.

Swanson, L. W., Sawchenko, P. E., Rivier, J., and Vale, W., 1983, Organization of ovine corticotropin-releasing factor immunoreactive cells and fibers in the rat brain: An immunohistochemical study, *Neuroendocrinology* **36:**165–186.

Tache, Y., Goto, Y., Gunion, M., Vale, M., Rivier, J., and Brown, M., 1983, Inhibition of gastric acid secretion in rats by intracerebral injection of corticotropin-releasing factor (rCRF), *Science* **222:**935–937.

Tache, Y., Hamel, D., and Gunion, 1984, Inhibition of gastric acid secretion in rats by intracisternal or intrathecal injection of rat corticotropin-releasing factor (rCRF), *Dig, Dis. Sci.* **29:**86S.

Takigawa, M., and Mogenson, G. J., 1977, A study of inputs to antidromically indentified neurons of the locus coeruleus, *Brain Res.* **135:**217–230.

Thierry, A. M., Javoy, F., Glowinski, J., and Kety, S. S., 1968, Effects of stress on the metabolism of norepinephrine, dopamine and serotonin in the central nervous system of the rat. Modification of norepinephrine turnover, *J. Pharmacol. Exp. Ther.* **163:**163–171.

Tilders, F. J. H., Berkenbosch, E., Vernes, I., Linton, E. A., and Smith, P. G., 1985, Role of epinephrine and vasopressin in the control of the pituitary adrenal response to stress, *Fed. Proc.* **44:**155–160.

Vale, W., Rivier, C., Yang, L., Minick, S., and Guillemin, R., 1978, Effects of purified hypothalamic corticotropin-releasing factor and other substances on the secretion of adrenocorticotropin and β-endorphin like immunoreactivity *in vitro, Endocrinology* **103:**1910–1915.

Vale, W., Spiess, J., Rivier, C., and Rivier, J., 1981, Characterization of 41-residue ovin hypothalamic peptide that stimulates secretion of corticotropin and β-endorphin, *Science* **213:**1394–1397.

Vale, W., Vaughn, J., Smith, M., Yamenoto, G., Rivier, J., and Rivier, C., 1983, Effects of synthetic ovine corticotropin-releasing factor, glucocorticoids, catecholamines, neurohypophysical peptides and other substance on cultured corticotropin cells, *Endocrinology* **113:**1121–1131.

Valentino, R. J., Foote, S. L., and Aston-Jones, G., 1983a, Corticotropin-releasing factor activates noradrenergic neurons of the locus coeruleus, *Brain Res.* **270:**363–367.

Valentino, R. J., Foote, S. L., Aston-Jones, G., and Bloom, F. E., 1983b, Activation of locus coeruleus neurons by corticotropinreleasing factor (CRF), *Soc. Neurosci. Abstr.* **9:**140.

Ward, D. G., and Gunn, C. G., 1976, Locus coeruleus complex: Elicitation of a pressor response and brainstem region necessary for its occurence, *Brain Res.* **107:**401–406.

Waterhouse, B. D., Moises, H. C., and Woodward, D. J., 1980, Noradrenergic modulation of somatosensory cortical neuronal responses to iontophoretically applied putative neurotransmitters, *Exp. Neurol.* **69:**30–49.

Waterhouse, B. D., Moises, H. C., Yeh, H. H., and Woodward, D. J., 1982, Norepinephrine enhancement of inhibitory synaptic mechanisms in cerebellum and cerebral cortex: Mediation by β-adrenergic receptors, *J. Pharmacol. Exp. Ther.* **221:**495–506.

White, S. R., and Neuman, R. S., 1980, Facilitation of spinal motorneuron excitability by 5-hydroxytryptamine and noradrenaline, *Brain Res.* **188:**119–127.

Woodward, D. J., Moises, H. C., Waterhouse, B. D., Hoffer, B. J., and Freedman, R., 1979, Modulatory actions of norepinephrine in the central nervous system, *Fed. Proc.* **38:**2109–2116.

Wynn, P. C., Aguilera, G., Morell, J., and Catt, K. J., 1983, Properties and regulation of high-affinity pituitary receptors for corticotropin-releasing factor, *Biochem. Biophys. Res. Commun.* **110:**602–608.

Wynn, P. C., Hauger, R. L., Holmes, M. C., Millan, M. A., Catt, K. J., and Aguilera, G., 1984, Brain and pituitary receptors for corticotropin releasing factor: Localization and differential regulation after adrenalectomy, *Peptides* **5:**1077–1084.

10

Electrophysiological Studies into Circuitry Involved in LHRH-Mediated Sexual Behavior

CAROL A. DUDLEY and ROBERT L. MOSS

1. INTRODUCTION

Luteinizing hormone-releasing hormone (LHRH) has two well-documented neuroendocrinological functions; the decapeptide acts on the pituitary gland to stimulate the release of LH (Redding *et al.*, 1972; Ishikawa and Nagayama, 1973; Liu *et al.*, 1976; Blake, 1978), and it acts in the brain to enhance mating behavior in the female rat (Moss and McCann, 1973, 1975; Pfaff, 1973; Foreman and Moss, 1977, 1978a,b; Moss and Foreman, 1976; Sakuma and Pfaff, 1980; Riskind and Moss, 1979). Previous investigations indicated that the ability of LHRH to enhance mating behavior was not dependent on its effect on the pituitary–ovarian–adrenal axis. The decapeptide was shown to be capable of facilitating mating behavior in the ovariectomized, adrenalectomized (Moss and McCann, 1975) and ovariectomized, hypophysectomized female rat (Pfaff, 1973). Recent research efforts in our laboratory have been directed toward distinguishing the neuronal circuitry involved in the LHRH enhancement of mating behavior from the circuitry involved in LHRH-induced luteinizing hormone (LH) release. A combination of behavioral, endocrine, and electrophysiological techniques has been employed in an attempt to separate LHRH pathways participating in LH release from those contributing to the initiation of mating behavior.

CAROL A. DUDLEY and ROBERT L. MOSS • Department of Physiology, The University of Texas Health Science Center, Dallas, Texas 75235.

1.1. Localization of LHRH

1.1.1. Perikarya and Fibers

The major source of LHRH-containing cell bodies, as demonstrated by immunocytochemical studies, is the medial preoptic–septal–diagonal band of Broca area (Setalo *et al.,* 1976; Barry, 1979; Joseph *et al.,* 1981; Bennett-Clarke and Joseph, 1982). Perikarya containing LHRH are spread rather diffusely throughout this area. The LHRH fibers and terminal endings are concentrated most heavily in the organum vasculosum of the lamina terminalis and in the median eminence (ME) (Joseph *et al.,* 1981; Kozlowski and Hostetter, 1977), and less dense fiber concentrations are found in the arcuate ventromedial nucleus of the hypothalamus (ARC-VMH), the midbrain central gray (MCG), the subfornical organ, the corticomedial amygdala, the medial habenula, and the olfactory bulbs (Kozlowski and Hostetter, 1977; Shivers *et al.,* 1981, 1983; Witkin and Silverman, 1983; Bennett-Clarke and Joseph, 1982; Witkin *et al.,* 1982).

1.1.2. LHRH Pathways and the Control of LH Release

The role of various LHRH projection pathways in the control of LH release has been studied extensively. The bed of LHRH fibers in the ME is crucial for the stimulation of LH release. These LHRH fibers terminate on capillaries of the primary portal plexus of the pituitary (Hokfelt *et al.,* 1975), and the neurohormone is then transported to the pituitary gonadotrophs by the way of the portal vessels, where it acts to simulate the release of LH (Bergland and Page, 1978; Adams and Nett, 1979; Conn *et al.,* 1981). Complete deafferentation of the mediobasal hypothalamus, including the ARC-ME, has been shown to disrupt LH release (Blake and Sawyer, 1974; Arendash and Gallo, 1978; Kovacs *et al.,* 1983). Processes emanating from LHRH perikarya course along two or three well-defined pathways to reach the ME (Ibata *et al.,* 1979; Bennett-Clarke and Joseph, 1982; King *et al.,* 1982; Witkin *et al.,* 1982). Neurons situated in the medial aspect of the medial preoptic–septal–diagonal band continuum appear to adopt a medial course, following a ventricular route and arching through the ARC to enter the ME. The ventrolateral LHRH neurons appear to maintain a lateral course through the hypothalamus and to enter the lateral aspect of the ME. The relative contribution of these LHRH pathways to the release of LH is not yet clear. Some reports indicate that interruption of the lateral pathway markedly reduces LHRH staining in the ME (Jew *et al.,* 1984; Ibata *et al.,* 1979) and results in decreased LH concentrations (Phelps and Sawyer, 1977). Others, however, stress the importance of the medial LHRH pathway (for review see Kalra and Kalra, 1983).

1.1.3. LHRH Pathways and the Control of Mating Behavior

The LHRH pathways innervating brain sites that modulate mating behavior appear to be similar to those described above. Lesions of the VMH, MCG, the habenula (HAB), and the basolateral amygdaloid region have all been reported to reduce lordotic responding (Mathews and Edwards, 1977; Pfaff and Sakuma, 1979; Sakuma and Pfaff, 1979; Riskind and Moss, 1983; Modianos *et al.*, 1975; Rodgers and Schneider, 1979), whereas lesions of the MPOA have been reported to enhance, decrease, or have no effect on mating behavior (Powers and Valenstein, 1971; Popolow *et al.*, 1981; Singer, 1968; Gray *et al.*, 1978). All of these areas, as detailed above, contain immunocytochemically detectable LHRH fibers. Medial and/or lateral pathways originating from cell bodies in the septal–preoptic–retrochiasmatic area have been described (see Witkin *et al.*, 1982, for review). In addition, infusion of the decapeptide into the MPOA, ARC-VMH, and MCG has been demonstrated to increase lordotic behavior in ovariectomized estrogen-primed female rats (Foreman and Moss, 1977; Riskind and Moss, 1979; Sakuma and Pfaff, 1980).

1.2. Independence of Gonadotropic and Behavioral Regulation

1.2.1. Lesion Studies

The extent of functional overlap between the neuronal systems regulating mating behavior and those regulating the release of LH is not known. The proestrus surge of LH is temporally linked to the display of mating behavior; however, several investigators have demonstrated that the two systems can be differentiated. The first indication that the behavioral effect produced by LHRH was independent of its endocrine effects was the demonstration that the decapeptide induced lordotic behavior in ovariectomized, hypophysectomized rats (Pfaff, 1973). Lesions of the suprachiasmatic nuclei or the medial preoptic area have been reported to inhibit phasic LH release but to have no effect on mating behavior (Gray *et al.*, 1978). Dorsal preoptic area lesions produced an increase in lordotic behavior without significantly affecting ovarian function (Nance *et al.*, 1977). On the other hand, lesions of the HAB or basolateral amygdaloid region reduced mating behavior without affecting estrous cyclicity or spontaneous ovulation (Rodgers and Schneider, 1979). Lesions of the MCG also reduced lordotic responding without affecting basal levels of LH or estrogen and estrogen- plus progesterone-stimulated release of LH (Riskind and Moss, 1983). It has been proposed that the septal–medial preoptic–diagonal band pool of LHRH is common to both systems and that collateral fibers branching to the ME regulate the release of LH, whereas other fibers projecting to hypothalamic and extrahypothalamic sites are involved in the control of mating behavior (Moss and Dudley, 1984). Alternatively, the LHRH perikarya contributing to the release of LH and those modulating sexual behavior may reside at different sites within the medial preoptic–septal–diagonal band continuum.

1.2.2. Analogues of LHRH

The concept that LHRH control of gonadotropin release may be independent of its facilitatory effect on mating behavior has received support in recent studies in which administration of synthetic analogues of the decapeptide was demonstrated to have opposite effects on LH release and lordotic behavior. Some agonist analogues of LHRH that are more potent than the native hormone in inducing the release of LH (Rivier *et al.*, 1983a,b; Coy *et al.*, 1979; Vilchez-Martinez *et al.*, 1974) have been reported to inhibit mating behavior in the estrogen-primed female rat (Kastin *et al.*, 1980; Zadina *et al.*, 1981). In these same studies, some competitive antagonist analogues that inhibit the release of LH from the pituitary gland have been shown to enhance mating behavior in the same animal model. Electrophysiological experiments in which an LHRH antagonist analogue was iontophoretically applied to single cells in the VMH revealed that the neuronal membrane effects of the antagonist analogue were similar to those evoked by LHRH (Chan *et al.*, 1983). In addition, the membrane responsiveness to both LHRH and its antagonist analogue was modulated by systemically injected estrogen. These findings suggest that the structural requirements of the molecule necessary to produce maximal receptor binding, endocrine effects, and behavioral effects may be different.

1.2.3. Fragments of LHRH

Further evidence of a dissociation between the pituitary and behavioral effects of LHRH has been provided in experiments that examined the effect of fragments or portions of the LHRH molecule on both functions. Three fragments of LHRH, namely, des-Tyr^5-LHRH, des-Gly^{10}-LHRH, and Ac-$LHRH^{5-10}$, have been shown to facilitate lordotic responding when intraventricularly infused in ovariectomized, estrogen-primed rats (Dudley *et al.*, 1983). All three of these fragments are inactive in terms of *in vitro* and *in vivo* LH release (Arnold *et al.*, 1974; Rivier *et al.*, 1974; Vale *et al.*, 1979; Dudley *et al.*, 1983). Thus far, the smallest portion of the LHRH molecule that is behaviorally effective is Ac-$LHRH^{5-10}$; however, the finding that removal of the fifth- and tenth-position amino acids does not impair behavioral effectiveness suggests that the Ac-$LHRH^{5-10}$ fragment may be further reduced and still retain behavioral potentcy.

The physiological relevance of behavioral facilitation by fragments of the LHRH molecule is not yet clear. Endopeptidases with LHRH-degrading activity have been demonstrated in the brain (Koch *et al.*, 1974; Marks and Stern, 1974; Sundberg and Knigge, 1978). In hypothalamic–ME tissue, endopeptidase-catalyzed cleavage at the Tyr^5-Gly^6 bond has been reported to account for 80% of the LHRH degraded (Krause *et al.*, 1982). Total LHRH degradation in female rats has been shown to be physiologically relevant in that it decreases transiently prior to the first LH surge at puberty (Advis *et al.*, 1982). Reduction in peptidase activity has also been found to occur during the proestrus stage of the estrous cycle in intact female rats (O'Conner *et al.*, 1984). Changes in the rate of degradation of LHRH have been proposed to be a factor in the regulation of gonadotropin secretion (Advis *et al.*, 1982, 1983; Lapp and O'Conner, 1984; O'Conner *et al.*, 1984). The finding that fragments of the LHRH molecule can facilitate sexual behavior suggests that degradation of the decapeptide may also be important in this regard.

2. ELECTROPHYSIOLOGICAL EXPERIMENTS

Presumably, Ac-LHRH$^{5-10}$ modulates mating behavior by altering the neuronal activity of cells involved in the mating behavior circuit. To test this hypothesis, an electrophysiological experiment employing the techniques of single-unit extracellular recording, iontophoresis, and orthodromic identification was performed. The effect of iontophoretically applied Ac-LHRH$^{5-10}$ on the firing rate of neurons located in the VMH was assessed in ovariectomized animals that were primed or not primed with estradiol benzoate (EB). The effect of this LHRH fragment on neuronal firing rate was compared to the effect elicited by the whole LHRH molecule. An attempt to distinguish VMH neurons likely to participate in the control of mating behavior from those involved in other functions was made by using the orthodromic identification technique.

2.1. Method

2.1.1. Animals, Electrodes, and Testing Procedures

Thirty-seven female rats were ovariectomized and, 4 to 7 weeks later, 19 of the animals received an injection of EB (6 μg dissolved in corn oil) 24 hr prior to electrophysiological recording, whereas 18 animals remained untreated. On the day of recording, the experimental animal was anesthetized with urethane (1.2–1.4 g/kg) and placed in a stereotactic instrument. The electrophysiological recording setup is diagrammed in Fig. 1. After removal of the bone and dura dorsal to the mediobasal hypothalamus, a multibarreled glass microelectrode (tip diameter 2.5–3.0 μm) was manually lowered through the cortex to the dorsomedial hypothalamic area (DMH) where recording was begun. The electrode was further lowered through the DMH and the VMH with the aid of a hydraulic microdrive. Thus, the neurons recorded were located in the medial basal hypothalamus. The center barrel contained 4 M NaCl and was used to detect single-unit activity. Signals detected through the center barrel were amplified, displayed on an oscilloscope, processed through a window discriminator, and fed to a chart recorder for a permanent record. The outer barrels of the electrode were used for iontophoretic application of Ac-LHRH$^{5-10}$ (0.001 M; pH 6.0–6.5) or LHRH (0.001 M; pH 6.0–6.5). Another outer barrel contained 2 M NaCl, which was used to balance iontophoretic currents. Pontamine sky blue dye (2% in 0.5 M sodium acetate) was placed into another of the outer barrels and was ejected at the end of the experiment to mark the location of the electrode track. Ejection of peptides and balancing of iontophoretic currents was accomplished by a BH-2 iontophore unit (Medical Systems). Both peptides were ejected as cations. Between iontophoretic tests, a retaining current of 2.5 to 5.0 nA was applied to the outer barrels to prevent leakage. For a drug to be judged effective, a 30% change (increase or decrease) in mean firing frequency during drug application as compared to the mean frequency over the 30-sec period prior to drug application was required.

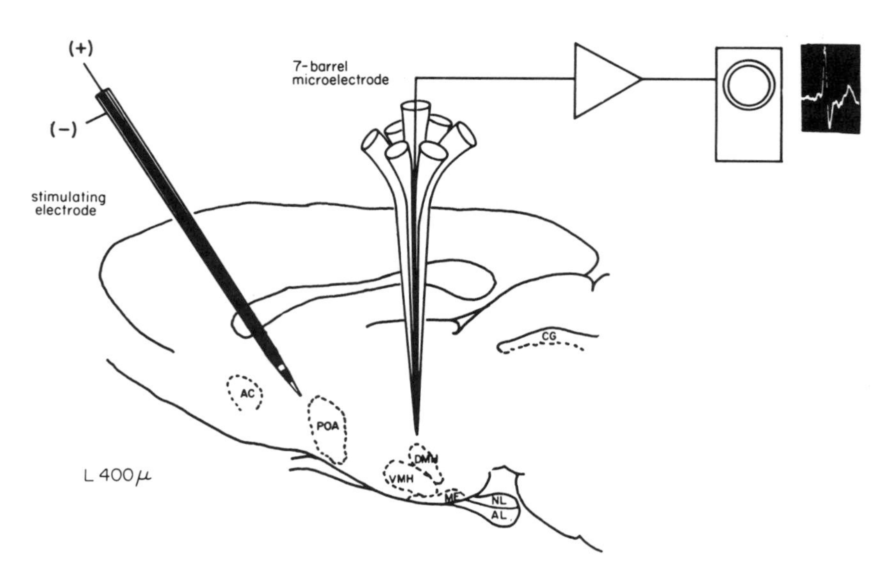

FIGURE 1. Diagram of the electrophysiological recording setup depicting the position of the recording and stimulating electrodes in the rat brain. Electrical activity of neurons in the dorsomedial hypothalamus (DMH) and ventromedial hypothalamus (VMH) was recorded through the center barrel of the microelectrode, amplified, and displayed on an oscilloscope. The stimulating electrode was placed in the medial septal area. POA, preoptic area; AC, anterior commisure; ME, median eminence; NL, neural lobe; AL, anterior lobe; CG, central gray.

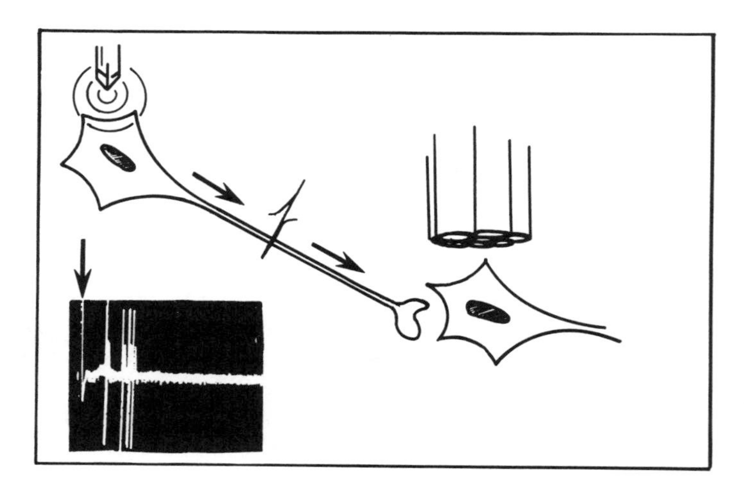

FIGURE 2. Schematic representation of the orthodromic identification technique. A stimulation pulse (upper left-hand corner) applied to a neuron in medial septal area causes generation of action potentials, which travel orthodromically down the axon and result in synaptic release of a transmitter substance. The released substance travels across the synapse and affects the firing rate of the basal hypothalamic neuron being recorded. The photograph in the lower left-hand corner was taken from a storage oscilloscope and shows the activity of a medial basal hypothalamic neuron generated in response to six stimulation pulses (superimposed at arrow). The neuron exhibited a variable short-latency (12–30 msec) increase in firing rate following stimulation. (Time from stimulation artifact to end of each sweep is 100 msec.)

2.1.2. Orthodromic Identification

Orthodromic identification was accomplished by placing a stimulating electrode in the medial septal area, which is a major source of LHRH-containing neurons. Biphasic stimulation pulses of 1–3 mA and 0.30-msec duration were delivered at a rate of 0.2 Hz to the septal area. As diagrammed in Fig. 2, action potentials generated as a result of this stimulation travel orthodromically down the axon to cause release of a substance (possibly LHRH) at the synapse. The effect of the released substance on the basal hypothalamic neuron under investigation was monitored through the recording electrode. If the substance released was excitatory, then the basal hypothalamic neuron displayed a variable, short-latency increase in activity in response to the stimulation pulse. On the other hand, if the substance released was inhibitory in nature, the basal hypothalamic neuron displayed a variable, short-latency decrease in firing rate. The photograph in Fig. 2 is an example of a VMH neuron displaying orthodromic excitation. If the basal hypothalamic neuron was not affected by the electrical stimulation, that neuron was considered nonidentified.

2.2. Results

2.2.1. Responsiveness to Ac-LHRH$^{5-10}$

Spontaneous activity was recorded from a total of 195 neurons. Their responsiveness to Ac-LHRH$^{5-10}$ is summarized in Table I. The majority of nonidentified neurons in both EB-primed and nonprimed animals were not responsive to Ac-LHRH$^{5-10}$. Of the neurons that were affected by Ac-LHRH$^{5-10}$, the predominant response was a decrease in firing rate. Estradiol benzoate priming tended to shift the response profile of nonidentified neurons from the "no effect" category to the inhibitory category. Typical examples of chart recordings demonstrating the inhib-

TABLE I. Responsiveness to Ac-LHRH$^{5-10}$

	Increase in firing rate		Decrease in firing rate		No change in firing rate	
	Percent	*n*	Percent	*n*	Percent	*n*
Nonprimed						
Nonidentified[a]	6	5	25	21	69	58
Orthodromically identified[a]	7	1	53	8	40	6
Estradiol-primed						
Nonidentified[a]	4	2	36	19	60	32
Orthodromically identified[a]	12	2	65	11	23	4

[a] Response profile of nonidentified and orthodromically identified neurons was significantly different in both nonprimed and EB-primed animals.

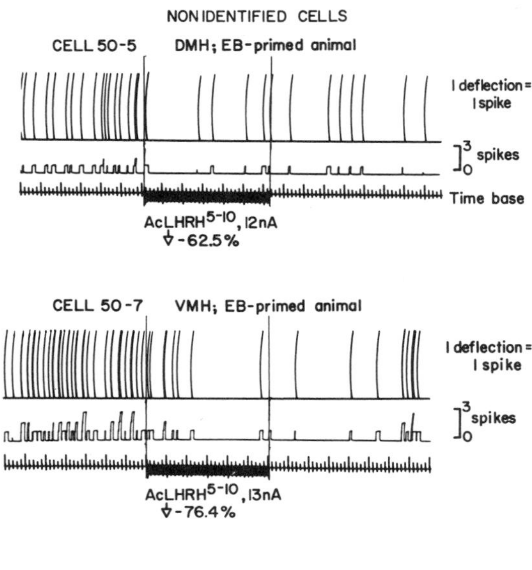

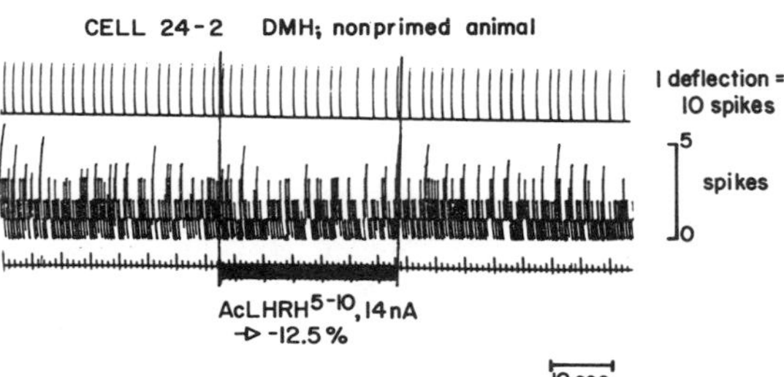

FIGURE 3. Three chart-recorder examples of neuronal response to iontophoresis of Ac-LHRH$^{5-10}$. The first line in each example represents the firing pattern of the cell; the middle line is a step-rate meter representation of the same data. The bottom line contains the time base, and the dark bars correspond to the length of time that the fragment was applied. The Ac-LHRH$^{5-10}$ inhibited the firing rate of the first two cells but was not effective in altering the firing rate of the third cell.

itory effect of Ac-LHRH$^{5-10}$ on nonidentified neurons in EB-primed animals are presented in Fig. 3. In contrast to nonidentified neurons, orthodromically identified units were predominantly responsive to the application of Ac-LHRH$^{5-10}$. The difference in responsiveness to Ac-LHRH$^{5-10}$ between orthodromically identified and nonidentified neurons was significant in both nonprimed ($\chi^2 = 5.99$; $df = 2$; $p = 0.05$) and EB-primed ($\chi^2 = 7.34$; $df = 2$; $P < 0.05$) animals. Thus, basal hypothalamic neurons receiving input from the septal area were more likely to be responsive to Ac-LHRH$^{5-10}$ than nonidentified neurons.

2.2.2. Responsiveness to LHRH

Neuronal responsiveness to LHRH is summarized in Table II. Again, as with the LHRH fragment, the majority of nonidentified neurons were not responsive to LHRH, and of the neurons that were affected by the decapeptide, the predominant response was a decrease in firing rate. Estradiol benzoate priming produced a

TABLE II. Responsiveness to LHRH

	Increase in firing rate		Decrease in firing rate		No change in firing rate	
	Percent	*n*	Percent	*n*	Percent	*n*
Nonprimed						
Nonidentified[a,b]	3	2	28	21	69	52
Orthodromically identified[a]	20	2	40	4	40	4
Estradiol-primed						
Nonidentified[b]	11	5	40	19	49	23
Orthodromically identified[a]	7	1	36	5	57	8

[a] Response profile of nonidentified and orthodromically identified neurons was significantly different in nonprimed animals.
[b] Response profile of nonidentified neurons was significantly different between EB-primed and nonprimed animals.

significant change in the response profile in these nonidentified neurons ($\chi^2 = 6.51$; $df = 2$; $P < 0.05$). More neurons were inhibited by LHRH and fewer were nonresponsive in EB-primed animals than in nonprimed animals. The response profile of orthodromically identified neurons was significantly different from that of nonidentified neurons in the nonprimed animal ($\chi^2 = 7.23$; $df = 2$; $P < 0.05$). This difference was caused by a shift from "no effect" to "inhibited" in the orthodromically identified neurons. This shift in responsiveness was not observed in orthodromically identified neurons recorded from EB-primed animals.

2.2.3. Responsiveness to Ac-LHRH$^{5-10}$ and LHRH

Nonidentified medial basal hypothalmic neurons that were tested with Ac-LHRH$^{5-10}$ and LHRH tended to exhibit the same response to both peptides (data not shown). In nonprimed animals, 86% of the neurons exhibited a similar response to the two peptides, although in EB-primed animals, the percentage of neurons responding similarly to Ac-LHRH$^{5-10}$ and LHRH was 58%. A smaller percentage of orthodromically identified neurons responded in the same manner to iontophoretic application of the two drugs (40% in the nonprimed animal and 50% in the EB-primed animal).

2.2.4. Orthodromic Identification

A total of 34 basal hypothalamic neurons were found to be orthodromically identified. Septal area stimulation resulted in an excitatory response in 30 of these neurons (latency range 5–30 msec; mean latency 12.8 msec). Of the four remaining neurons, three displayed a longer-latency (0.4–0.5 sec) decrease in excitability, and one neuron displayed short-latency excitation followed by a longer-latency inhibition. A correspondence between the orthodromic response produced by septal

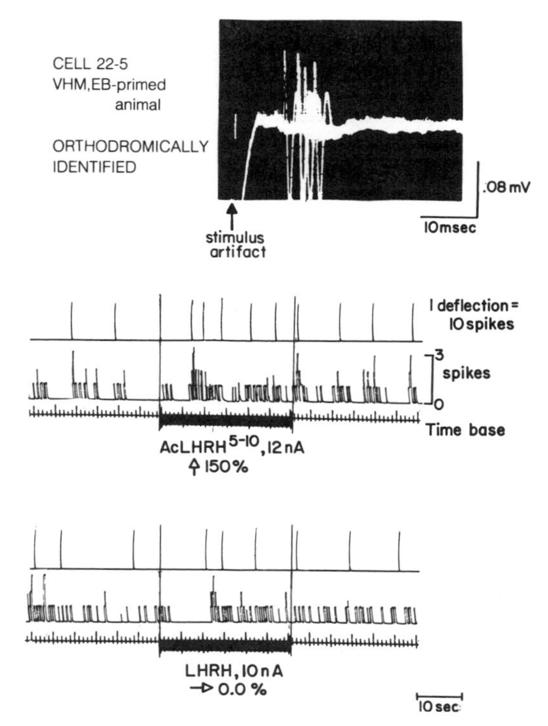

FIGURE 4. The effect of iontophoretically applied Ac-LHRH$^{5-10}$ and LHRH on an orthodromically identified neuron recorded in the ventromedial hypothalamus (VMH). The photograph at the top was taken from a storage oscilloscope following five stimulation pulses (superimposed at arrow). The neuron exhibited variable, short-latency (10–15 msec) excitation in response to the orthodromic stimulation and also demonstrated an increase in activity during iontophoresis of Ac-LHRH$^{5-10}$; LHRH was not effective in altering the firing rate of this VMH neuron.

area stimulation and the response elicited by iontophoretic application of LHRH or Ac-LHRH$^{5-10}$ was observed in eight neurons. In three of these neurons, the response to iontophoretically applied Ac-LHRH$^{5-10}$ was similar to the orthodromic response. An example of a VMH neuron that was orthodromically excited by septal area stimulation and iontophoretic application of Ac-LHRH$^{5-10}$ is shown in Fig. 4, and Fig. 5 presents a DMH neuron that was inhibited by both septal area stimulation and iontophoresis of Ac-LHRH$^{5-10}$. Application of LHRH did not affect the firing rate of either of these two neurons, however there were four cases in which the neuronal response to septal area stimulation was mimicked by the iontophoretic application of LHRH, and one case in which both Ac-LHRH$^{5-10}$ and LHRH increased the firing rate of an orthodromically excited neuron.

2.3. Summary

Acetyl-LHRH$^{5-10}$, a fragment of the LHRH molecule that is inactive in promoting LH release from the pituitary gland, was found to affect neuronal activity when iontophoretically applied to neurons in the medial basal hypothalamus. This finding indicates that Ac-LHRH$^{5-10}$ possesses CNS activity. The majority of responsive neurons exhibited a decrease in spontaneous activity during iontophoresis

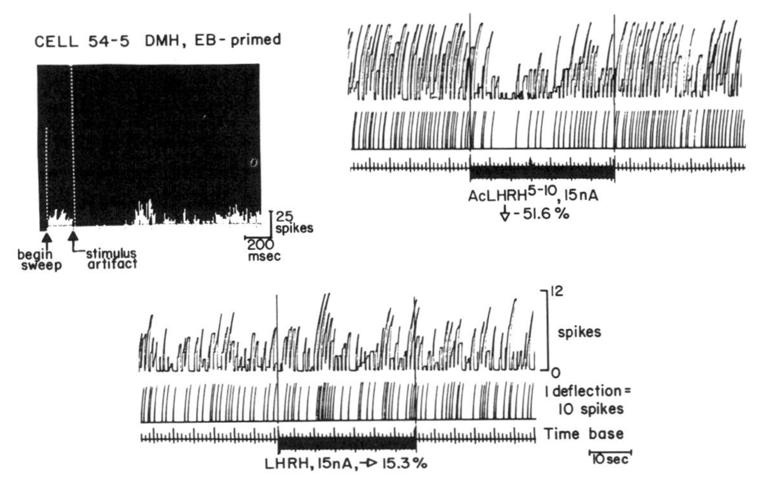

FIGURE 5. The effect of iontophoretically applied Ac-LHRH$^{5-10}$ and LHRH on an orthodromically identified neuron recorded from the dorsomedial hypothalamus (DMH). The photograph at the upper left is a peristimulus histogram showing neural activity before and after the simulation pulse. The histogram contains data from approximately 50 sweeps and demonstrates orthodromic inhibition, which begins 10 msec after the stimulation pulse and lasts for approximately 400 msec. Iontophoretically applied Ac-LHRH$^{5-10}$ also inhibited the firing of this neuron, whereas LHRH was without effect.

of Ac-LHRH$^{5-10}$. Estrogenic modulation of the neuronal responsiveness to Ac-LHRH$^{5-10}$ was observed in that the inhibitory effect of the fragment was more pronounced in EB-primed animals than in nonprimed rats. The effectiveness of Ac-LHRH$^{5-10}$ was particularly pronounced in a subpopulation of medial basal hypothalamic neurons. Basal hypothalamic neurons that were orthodromically responsive to stimulation of the septal area were more likely to be sensitive to Ac-LHRH$^{5-10}$ than nonidentified neurons. Iontophoretic application of LHRH produced essentially the same results as iontophoretic application of Ac-LHRH$^{5-10}$. It should be noted, however, that some neurons responded differentially to the two peptides. This finding provides evidence for a specific action of the LHRH fragment. In addition, a correspondence between the effect of septal area stimulation on neuronal firing rate and the effect of iontophoretically applied Ac-LHRH$^{5-10}$ or LHRH on neuronal firing rate was occasionally observed. Such a correspondence suggests that endogenous release of the two peptides may occur in the basal hypothalamic area.

3. DISCUSSION AND CONCLUSION

The finding that a fragment of LHRH is capable of changing the firing rate of medial basal hypothalamic neurons is consistent with the hypothesis that LHRH-degrading endopeptidases in hypothalamic–ME tissue produce biologically active fragments. Further support of this hypothesis was provided by the demonstration that Ac-LHRH$^{5-10}$ enhanced mating behavior in ovariectomized, estrogen-primed female rats (Dudley *et al.*, 1983). Stimulation applied to the medial septal area was found to elicit orthodromic responses in some medial basal hypothalamic neurons. This neuronal pathway from the septal area to the basal hypothalamus has previously been identified electrophysiologically (Blume *et al.*, 1982) and anatomically (Meiback and Siegel, 1977). Immunocytochemical studies indicate that projections from LHRH cell bodies located in the medial–preoptic–septal area course through the basal hypothalamus en route to the ME (Bennett-Clarke and Joseph, 1982; Witkin *et al.*, 1982). The correspondence between the neuronal response to medial preoptic–septal area stimulation and the response to iontophoretically applied LHRH obtained in the present study suggests that LHRH release was activated by the stimulation pulses. In addition, the similarity of effect of orthodromic stimulation and iontophoresis of Ac-LHRH$^{5-10}$ that was observed in the present study suggests that a fragment of LHRH structurally similar to Ac-LHRH$^{5-10}$ may also be released synaptically.

The concept that the endocrine and behavioral effects of LHRH are functionally independent is supported by the present investigation in that the Ac-LHRH$^{5-10}$ fragment does not promote LH release (Arnold *et al.*, 1974; Rivier *et al.*, 1974; Vale *et al.*, 1979; Dudley *et al.*, 1983) but is capable of modulating the firing rate of basal hypothalamic neurons. Fragments of the ACTH molecule and the vasopressin molecule have also been reported to exhibit a dissociation in behavioral and endocrine effects (De Wied, 1978; Walter *et al.*, 1980). This evidence suggests that further research at the electrophysiological level using fragments of various peptides may provide a way of distinguishing the neuronal systems involved in the control of behavior from those involved in the control of endocrine events.

ACKNOWLEDGMENTS. The authors express their appreciation to Deborah Aldridge for technical assistance and artwork and to June Roman for histological assistance. This work was supported by NIH grants NS 10434 and HD 11814.

REFERENCES

Adams, T. E., and Nett, T. M., 1979, Interaction of GnRH with anterior pituitary: Role of divalent cations, microtubules, and microfilaments in the GnRH activated gonadotroph, *Biol. Reprod.* **21:**1073–1086.

Advis, J. P., Krause, J. E., and McKelvy, J. F., 1982, Luteinizing hormone releasing hormone peptidase activities in discrete hypothalamic regions and anterior pituitary of the rat: Apparent regulation during the pre-pubertal period and first estrous cycle at puberty, *Endocrinology* **110:**1238–1245.

Advis, J. P., Krause, J. E., and McKelvy, J. F., 1983, Evidence that endopeptidase-catalyzed luteinizing hormone-releasing hormone cleavage contributes to the regulation of median eminence LHRH levels during positive steriod feedback, *Endocrinology* **112:**1147–1149.

Arendash, G., and Gallo, R. V., 1978, Apomorphine-induced inhibition of episodic LH release in ovariectomized rats with complete hypothalamic deafferentation, *Proc. Soc. Exp. Biol. Med.* **159:**121–125.

Arnold, W., Flouret, G., Morgan, R., Rippel, R., and White, W., 1974, Synthesis and biological activity of some analogs of gonadotropin releasing hormone, *J. Med. Chem.* **17**:314–319.

Barry, J., 1979, Immunohistochemistry of luteinizing hormone releasing hormone-producing neurons of the vertebrates, *Int. Rev. Cytol.* **60**:179–221.

Bennett-Clarke, C., and Joseph, S. A., 1982, Immunocytochemical distribution of LHRH neurons and processes in the rat: Hypothalamic and extrahypothalamic locations, *Cell Tissue Res.* **221**:493–504.

Bergland, R. M., and Page, R. B., 1978, Can the pituitary secrete directly to the brain? (Affirmative anatomical evidence), *Endocrinology* **102**:1325–1338.

Blake, C. A., 1978, Neurohumoral control of cyclic pituitary LH and FSH release, *Clin. Obstet. Gynecol.* **5**:305–327.

Blake, C. A., and Sawyer, C. H., 1974, Effects of hypothalamic deafferentation on the pulsatile rhythm in plasma concentrations of luteinizing hormone in ovariectomized rats, *Endocrinology* **94**:730–736.

Blume, H. W., Pittman, Q. J., Lafontaine, S., and Renaud, L. P., 1982, Lateral septum-medial hypothalamic connections: An electrophysiological study in the rat, *Neuroscience* **7**:2783–2792.

Chan, A., Dudley, C. A., and Moss, R. L., 1983, Action of prolactin, dopamine and LHRH on ventromedial hypothalamic neurons as a function of ovarian hormones, *Neuroendocrinology* **36**:397–403.

Conn, P. M., Marian, J., McMillan, M., Stern, J., Rogers, D., Hamby, M., Penna, A., and Grant, E., 1981, Gonadotropin-releasing hormone action in the pituitary: A three step mechanism, *Endocrinol. Rev.* **2**:174–185.

Coy, D. H., Seprodi, J., Vilchez-Martinez, J. A., Pedroza, E., Gardner, J., and Schally, A. V., 1979, Structure–function studies and prediction of conformational requirements for LH-RH, in: *Central Nervous System Effects of Hypothalamic Hormones and Other Peptides* (R. Collu, A. Barbeau, J. R. Ducharme, and J. G. Rochefort, eds.), Raven Press, New York, pp. 317–323.

De Wied, D., 1978, Effects of pituitary peptides on learning and memory processes, *Neurosci. Res. Prog. Bull.* **16**:321–328.

Dudley, C. A., Vale, W., Rivier, J., and Moss, R. L., 1983, Facilitation of sexual receptivity in the female rat by a fragment of the LHRH decapeptide, Ac-$^{5-10}$, *Neuroendocrinology* **36**:486–488.

Foreman, M. M., and Moss, R. L., 1977, Effects of subcutaneous injection and intrahypothalamic infusion of releasing hormone upon lordotic response to repetitive coital stimulation, *Horm. Behav.* **8**:219–234.

Foreman, M. M., and Moss, R. L., 1978a, Role of hypothalamic serotonergic receptors in the control of lordosis behavior in the female rat, *Horm. Behav.* **10**:97–106.

Foreman, M. M., and Moss, R. L., 1978b, Role of hypothalamic alpha and beta adrenergic receptors in the control of lordotic behavior in the ovariectomized, estrogen-primed rat, *Pharmacol. Biochem. Behav.* **9**:235–241.

Gray, G. D., Sodersten, P., Tallentire, D., and Davidson, J. M., 1978, Effects of lesions in various structures of the suprachiasmaticpreoptic region on LH regulation and sexual behavior in female rats, *Neuroendocrinology* **25**:174–191.

Hokfelt, T., Fuxe, K., Goldstein, M., Johansson, O., Fraser, H., and Jeffcoate, S., 1975, Immunofluorescence mapping of central monoamines and releasing hormone (LRH) systems, in: *Anatomical Neuroendocrinology* (W. E. Stumpf and L. D. Grant, eds.), S. Karger, Basel, p. 381.

Ibata, Y., Watanabe, K., Kinoshita, H., Kubo, S., and Sanyo, Y., 1979, The location of LH-RH neurons and their pathways to the median eminence, *Cell Tissue Res.* **198**:381–395.

Ishikawa, H., and Nagayama, T., 1973, The mechanism of stimulation of LH production by synthetic LH-releasing hormone (LH–RH) in tissue culture, *Biochem. Biophys. Res. Commun.* **55**:492–498.

Jew, J. Y., Leranth, C., Arimura, A., and Palkovits, M., 1984, Preoptic LH-RH and somatostatin in the rat median eminence, *Neuroendocrinology* **38**:169–175.

Joseph, S. A., Piekut, D. T., and Knigge, K. M., 1981, Immunocytochemical localization of luteinizing hormone-releasing hormone (LHRH) in vibratome-sectioned brain, *J. Histochem. Cytochem.* **29**:247–254.

Kalra, S. P., and Kalra, P. S., 1983, Neural regulation of luteinizing hormone secretion in the rat, *Endocrinol. Rev.* **4**:311–351.

Kastin, A. J., Coy, D. H., Schally, A. V., and Zadina, J. E., 1980, Dissociation of effects of LH-RH analogs on pituitary regulation and reproductive behavior, *Pharmacol. Biochem. Behav.* **13:**913–914.

King, J. C., Tobet, S. A., Snavely, F. L., and Arimura, A. A., 1982, LHRH immunopositive cells and their projections to the median eminence and organum vasculosum of the lamina terminalis, *J. Comp. Neurol.* **209:**287–300.

Koch, Y., Baram, T., Chobsieng, P., and Fridkin, M., 1974, Enzymatic degradation of luteinizing hormone-releasing hormone (LHRH) by hypothalamic tissue, *Biochem. Biophys. Res. Commun.* **61:**95–103.

Kovacs, M., Merchenthaler, I., Tima, L., Lengvari, I., and Setalo, G., 1983, Hypothalamic LHRH and plasma LH levels after electrical stimulation of the deafferented medial basal hypothalamus, *Cell Tissue Res.* **234:**209–217.

Kozlowski, G., and Hostetter, G., 1977, Cellular and subcellular localization and behavioral effects of gonadotropin-releasing hormone (Gn-RH) in the rat, in: *Brain Endocrine Interaction* III. *Neural Hormones and Reproduction* (D. E. Scott, G. P. Kozlowski, and A. Weindl, eds.), S. Karger, Basel, pp. 138–153.

Krause, J. E., Advis, J. P., and McKelvy, J. F., 1982, Characterization of the site of cleavage of luteinizing hormone-releasing hormone under conditions of measurement in which LHRH degradation undergoes physiologically related change, *Biochem. Biophys. Res. Commun.* **108:**1475–1481.

Lapp, C. A., and O'Conner, J. L., 1984, Peptidase activity in the hypothalamus of the rat: Utilization of leucine-*p*-nitroanilide to monitor the activity degrading luteinizing hormone releasing hormone, *Biol. Reprod.* **30:**848–854.

Liu, T.-C., Jackson, G. L., and Gorski, J., 1976, Effects of synthetic gonadotropin-releasing hormone on incorporation of radioactive glucosamine and amino acids into luteinizing hormone and total protein by rat pituitaries *in vitro, Endocrinology* **98:**151–163.

Marks, M., and Stern, F., 1974, Enzymatic mechanisms for the inactivation of luteinizing hormone releasing hormone, *Biochem. Biophys. Res. Commun.* **64:**1458–1463.

Mathews, D., and Edwards, D. A., 1977, The ventromedial nucleus of the hypothalamus and the hormonal arousal of sexual behaviors in the female rat, *Horm. Behav.* **8:**40–51.

Meiback, R. C., and Siegel, A., 1977, Efferent connections of the septal area in the rat: An analysis utilizing retrograde and anterograde transport methods, *Brain Res.* **119:**1–20.

Modianos, D. T., Hitt, J. C., and Popolow, H. B., 1975, Habenular lesions and feminine sexual behavior of ovariectomized rats: Diminished responsiveness to the synergistic effects of estrogen and progesterone, *J. Comp. Physiol. Psychol.* **89:**231–237.

Moss, R. L., and Dudley, C. A., 1984, The challenge of studying the behavioral effects of neuropeptides, in: *Handbook of Psychopharmacology,* Vol. 18 (L. L. Iversen, S. D. Iversen, and S. H. Snyder, eds.), Plenum Press, New York, pp. 397–454.

Moss, R. L., and Foreman, M. M., 1976, Potentiation of lordosis behavior by intrahypothalamic infusion of synthetic luteinizing hormone releasing hormone, *Neuroendocrinology* **20:**176–181.

Moss, R. L., and McCann, S. M., 1973, Induction of mating behavior in rats by luteinizing hormone-releasing factor, *Science* **181:**177–179.

Moss, R. L., and McCann, S. M., 1975, Action of luteinizing hormone-releasing factor (LRF) in the initiation of lordosis behavior in the estrone-primed, ovariectomized female rat, *Neuroendocrinology* **17:**309–318.

Nance, D. M., Christensen, L. W., Shryne, J. E., and Gorski, R. A., 1977, Modifications in gonadotropin control and reproductive behavior in the female rat by hypothalamic and preoptic lesions, *Brain Res. Bull.* **2:**307–312.

O'Conner, J. L., Lapp, C. A., and Mahesh, V. B. 1984, Peptidase activity in the hypothalamus and pituitary of the rat: Fluctuations and possible regulatory role of luteinizing hormone releasing hormone-degrading activity during the estrous cycle, *Biol. Reprod.* **30:**855–862.

Pfaff, D. W., 1973, Luteinizing hormone-releasing factor potentiates lordosis behavior in hypophysectomized ovariectomized female rats, *Science* **182:**1148–1149.

Pfaff, D. W., and Sakuma, Y., 1979, Deficit in the lordosis reflex of female rats caused by lesions in the ventromedial nucleus of the hypothalamus, *J. Physiol. (Lond.)* **288:**203–210.

Phelps, C. P., and Sawyer, C. H., 1977, Electrochemically stimulated release of luteinizing hormone and ovulation after surgical interruption of lateral hypothalamic connections in the rat, *Brain Res.* **131:**335–344.

Popolow, H. P., King, J. C., and Gerall, A. A., 1981, Rostral medial preoptic area lesions influence on female estrous processes and LHRH distribution, *Physiol. Behav.* **27:**855–861.

Powers, B., and Valenstein, E. S., 1971, Sexual receptivity: Facilitation by medial preoptic lesions in female rats, *Science* **175:**1003–1005.

Redding, T. W., Schally, A. V., Arimura, A., and Matsuo, H., 1972, Stimulation of release and synthesis of luteinizing hormone (LH) and follicle stimulating hormone (FSH) in tissue cultures of rat pituitaries in response to natural and synthetic LH and FSH releasing hormone, *Endocrinology* **90:**764–770.

Riskind, P., and Moss, R. L., 1979, Midbrain central gray: LHRH infusion enhances lordotic behavior in estrogen-primed ovariectomized rats, *Brain Res. Bull.* **4:**203–205.

Riskind, P., and Moss, R. L., 1983, Effects of lesions of putative LHRH-containing pathways and midbrain nuclei on lordotic behavior and luteinizing hormone release in ovariectomized rats, *Brain Res. Bull.* **11:**493–500.

Rivier, J., Amoss, M., Rivier, C., and Vale, W., 1974, Synthetic luteinizing homorne releasing factor. Short chain analogs, *J. Med. Chem.* **17:**230–233.

Rivier, C., Vale, W., and Rivier, J., 1983a, Effects of gonodatropin releasing hormone agonists and antagonists on reproductive functions, *J. Med. Chem.* **11:**1545–1550.

Rivier, C., Rivier, J., Perrin, M., and Vale, W., 1983b, Comparison of the effect of several gonadotropin releasing hormone antagonists on luteinizing hormone secretion, receptor binding, and ovulation, *Biol. Reprod.* **29:**374–378.

Rodgers, C. H., and Schneider, V. M., 1979, Facilitatory influences on mating behavior in the female rat affected by lesions of the habenula or the basolateral amygdaloid regions, *Psychoneuroendocrinology* **4:**237–244.

Sakuma, Y., and Pfaff, D. W., 1979, Mesencephalic mechanisms for integration of female reproductive behavior in ovariectomized rats, *Am. J. Physiol.* **5:**R285–R290.

Sakuma, Y., and Pfaff, D. W., 1980, LH–RH in the mesencephalic central gray can potentiate lordosis reflex of female rats, *Nature* **283:**566–567.

Setalo, G., Vigh, S., Schally, A., Arimura, A., and Flerko, B., 1976, Immunohistochemical study of the origin of LH-RH containing nerve fibers of the rat hypothalamus, *Brain Res.* **103:**597–602.

Shivers, B. D., Harlan, R. E., Morrell, J. I., and Pfaff, D. W., 1981, Immunocytochemical localization of LHRH in the male and the female rat brain: A quantitative comparison, *Soc. Neurosci. Abstr.* **7:**20.

Shivers, B. D., Harlan, R. E., Morrell, J. I., and Pfaff, D. W., 1983, Immunocytochemical localization of luteinizing hormone-releasing hormone in male and female rat brains, *Neuroendocrinology* **36:**1–12.

Singer, J. J., 1968, Hypothalamic control of male and female sexual behavior in female rats, *J. Comp. Physiol. Psychol.* **66:**738–742.

Sundberg, D. K., and Knigge, K. M., 1978, Luteinizing hormone releasing hormone (LH-RH) production and degradation by rat medial basal hypothalami *in vitro, Brain Res.* **139:**89–99.

Vale, W., Brown, M., Rivier, C., Perrin M., and Rivier, J., 1979, Development and application of analogs of LRF and somatostatin, in: *Brain Peptides: A New Endocrinology* (A. M. Gotto, Jr., E. J. Peck, Jr., and A. E. Boyd III, eds.), Elsevier/North Holland Biomedical Press, Amsterdam, pp. 71–88.

Vilchez-Martinez, J. A., Coy, D. H., Arimura, A., Coy, E. J., Hirotsu, Y., and Schally, A. V., 1974, Synthesis and biological properties of [D-Leu^6]-LH-RH and [D-Leu^6, Des Gly-$NH_2{}^{10}$]-LH-RH ethylamide, *Biochem. Biophys. Res. Commun.* **59:**1226–1232.

Walter, R., Rizmann, R. F., Tabakoff, B., Hoffman, P., and Flexner, L. B., 1980, Neurohypophyseal peptides and CNS adaptation, in: *The Role of Peptides in Neuronal Function* (J. L. Barker and T. G. Smith, Jr., eds.), Marcel Dekker, New York, pp. 654–666.

Witkin, J., and Silverman, A.-J., 1983, Luteinizing hormone-releasing hormone (LHRH) in rat olfactory systems, *J. Comp. Neurol.* **218:**426–432.

Witkin, J. W., Paden, C. M., and Silverman, A.-J., 1982, The luteinizing hormone-releasing hormone (LHRH) systems in the rat brain, *Neuroendocrinology* **35:**429–438.

Zadina, J. E., Kastin, A. J., Fabre, L. A., and Coy, D. H., 1981, Facilitation of sexual receptivity in the rat by an ovulation-inhibiting analog of LHRH, *Pharmacol. Biochem. Behav.* **15:**961–964.

11

Growth Hormone-Releasing Factor Analogues with Increased Receptor Affinity

JOHN J. NESTOR, Jr., TERESA L. HO, BARBARA M. DeLUSTRO, and ALAIN B. SCHREIBER

1. INTRODUCTION

After several erroneous earlier reports, the primary sequences of a family of related proteins exhibiting growth hormone-releasing factor (GRF) properties were reported (Guillemin *et al.*, 1982; Rivier *et al.*, 1982). The proteins were isolated from two different human pancreatic tumors that produced clinical signs of acromegaly, and these sequences differed only in chain length [37 (Esch *et al.*, 1983), 40 (River *et al.*, 1982; Esch *et al.*, 1983), or 44 (Esch *et al.*, 1983) residues]. Further studies with cloned human cDNA (Gubler *et al.*, 1983; Mayo *et al.*, 1983) or protein isolated from human hypothalami (Böhlen *et al.*, 1983; Ling *et al.*, 1984) suggest that the hypothalamic form of hGRF corresponds to the sequence (most likely the GRF_{1-44}-NH_2) isolated from the pancreatic tumors (Guillemin *et al.*, 1982). The shorter sequences may result from proteolysis. More recently, the sequences of GRFs (Fig. 1) were reported from the rat (Spiess *et al.*, 1983) and several domestic species (Brazeau *et al.*, 1984).

Synthetic replicates of the human pancreatic proteins, hpGRF_{1-44}-NH_2 (Guillemin *et al.*, 1982) and hpGRF_{1-40} (Rivier *et al.*, 1982), confirmed the biological activity of the proposed structures. Further syntheses demonstrated that the hp-

---

JOHN J. NESTOR, Jr., TERESA L. HO, BARBARA M. DeLUSTRO, and ALAIN B. SCHREIBER • Institutes of Bio-Organic Chemistry and Biological Sciences, Syntex Research, Palo Alto, California 94304. This is Contribution No. 220 from the Institute of Bio-Organic Chemistry, Syntex Research, Palo Alto, California.

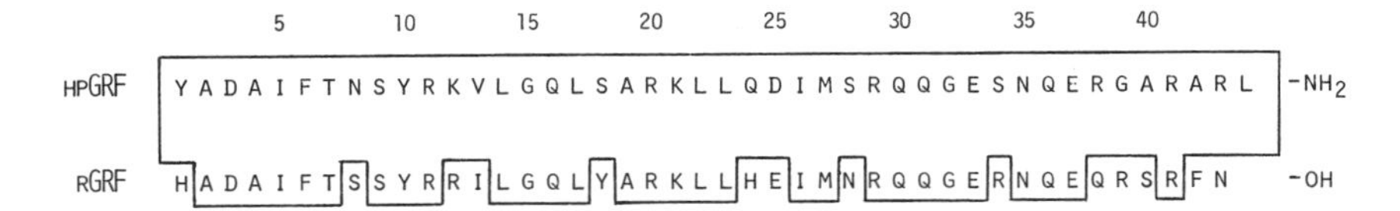

FIGURE 1. Amino acid sequences of human and rat GRFs. Homologies to hpGRF are indicated.

GRF_{1-27}-NH_2 sequence retained full intrinsic agonistic activity (20% potency), whereas hpGRF$_{1-29}$-NH_2 had full agonistic potency (Rivier *et al.,* 1982).

Sequence homologies (Guillemin *et al.,* 1982; Rivier *et al.,* 1982) were found between hpGRF and members of the secretin–glucagon family, especially PHI-27. The critical N-terminal residue, which is Tyr in all species examined except the rat, is His in the secretin family and in rGRF. Deletion of the N-terminal residue results in a loss of biological activity and pituitary receptor binding (Seifert *et al.,* 1985).

2. MEMBRANE INTERACTION

We have been examining substitutions in polypeptides that are designed to increase receptor affinity by a mechanism based on the hypothesis that ligand–membrane interaction precedes ligand–receptor interaction for membrane-bound receptors (Schwyzer *et al.,* 1983). Thus, the relatively unlikely interaction between a receptor binding site moving in two-dimensional space and a peptide with a random conformation tumbling in three-dimensional space (2D–3D) undergoes a "reduction in dimensionality" (to 2D–2D) on dissolution and concentration of the ligand in the membrane. It has also been shown that whereas calcitonin and adrenocorticotropin appear to have random conformations in aqueous solutions, they adopt a specific, perhaps the biologically active, conformation in detergents (Epand *et al.,* 1985) and hydrophobic solvents or on interaction with phospholipids (Gysin *et al.,* 1984). A sequence of events may therefore be envisioned wherein a peptide with a random conformation in the extracellular fluid penetrates the cell membrane, adopts its biologically active conformation, and is swept up by a receptor moving across the cell surface. If substitutions are made in a native ligand that facilitate this membrane interaction, an increased membrane-bound concentration of the ligand, and therefore an increased apparent receptor affinity, may result.

We have studied a series of unnatural amino acids that we designed (Nestor *et al.,* 1983, 1984) for phospholipid membrane interaction (Fig. 2). These new amino acids, the N^G,$N^{G'}$-dialkylhomoarginines [hArg(R_2)], should be able to interact with the phosphate head groups of phospholipids by a combination of electrostatic (guanidine–phosphate) and hydrophobic interactions. We have examined the effect of substitution by this class of amino acids in several polypeptide hormones

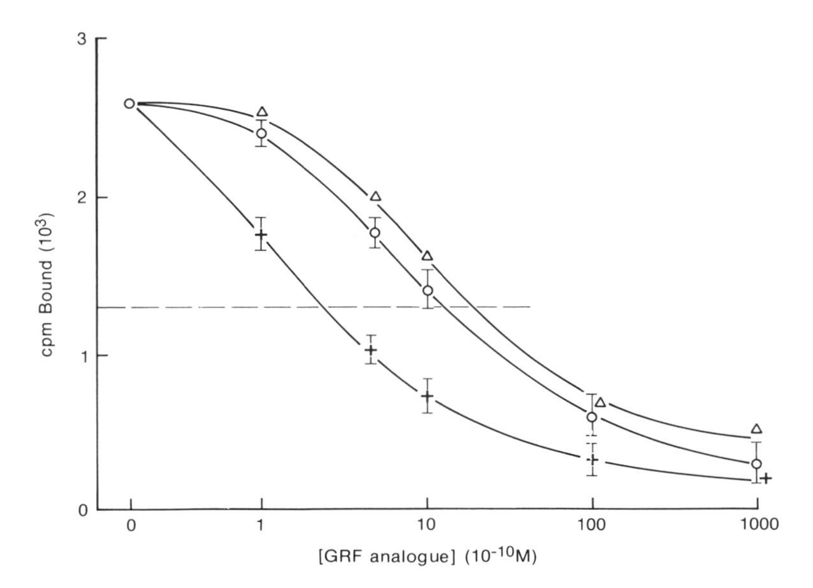

FIGURE 2. Displacement of binding of 2×10^{-9} M [^{125}I]hpGRF$_{1-44}$-NH$_2$ from GH3 cells by synthetic peptide analogues. The GH3 rat pituitary somatotrophic tumor cells (provided by Dr. D. Gospodarowicz, UCSF) were grown to confluence in 24-well trays (Costar) in Dulbecco's minimal essential medium (DMEM). Cells were washed with DMEM buffered at pH 7.4 (20 mM HEPES) containing 0.1% bovine serum albumin (DMEM-BSA). Cells were incubated at 4°C for 1 hr with 2×10^{-9} M [^{125}I]hpGRF$_{1-44}$-NH$_2$ in the presence of increasing concentrations of GRF analogues (250 μl final volume). Cells were washed three times with DMEM-BSA and lysed with 0.1 N NaOH, and the cell-associated radioactivity was determined in a counter. Nonspecific binding was assessed in the presence of 10^{-7} M synthetic hpGRF$_{1-44}$-NH$_2$ and did not exceed 15% of the specific binding. ○—○, hpGRF$_{1-44}$-NH$_2$; △—△, hpGRF$_{1-29}$-NH$_2$; +—+, [D-hArg(Et$_2$)^2]hpGRF$_{1-29}$NH$_2$.

(Nestor *et al.*, 1985a,b) and report here the increased receptor binding affinity that results when D-hArg(R$_2$) structures are incorporated into analogs of hpGRF.

3. BIOLOGICAL ASSAYS

The receptor binding affinities of our hpGRF analogues were assessed in GH3 cells and, for some analogues, in normal rat pituitary cells in monolayer culture (Vale *et al.*, 1972). We used [^{125}I]hpGRF$_{1-44}$ (20,000 cpm/ng protein) prepared with hpGRF$_{1-44}$-NH$_2$ (Peninsula Labs) by the lactoperoxidase method (Enzymobeads®, Bio-Rad) as the radioligand. GH3 cells are a subcloned rat pituitary cell line (provided by D. Gospodarowicz) that have GRF receptors on their surface but are abnormal in that they do not have GRF receptors on cytosolic secretory granules. Thus, although they bind hpGRF with an IC$_{50}$ virtually identical to that determined for normal rat pituitary monolayer cells (see Table II), they do not exhibit the prompt surge of GH release observed with normal cells. The comparison of receptor binding data between GH3 cells and normal rat pituitary cells presented here validates this assay and demonstrates that this GH3 cell line clone is a convenient system for measuring GRF receptor binding affinities (however, see Zeytin *et al.*, 1985).

The ability of the analogues to cause GH release was assessed on normal rat pituitary cells in monolayer culture (3 days) (Vale *et al.*, 1972). The culture supernatant was collected after a 3-hr incubation with various concentrations of test compound, and growth hormone (GH) present in the medium was assessed using a solid-phase radioimmunoassay kindly provided by Dr. S. Raiti of the NIH Pituitary Agency.

4. ANALOGUE STUDIES

Initial analogue studies in other laboratories suggested that the N-terminal 1–29 residue sequence was required for full agonistic potency (Rivier *et al.*, 1982). The N-terminal residue was required for receptor binding and GH release (Seifert *et al.*, 1985). Air oxidation of Met at position 27 was found to be facile and led to a loss of activity, so this residue was replaced by norleucine (Nle) with a resultant increase in potency ([Nle^{27}]hpGRF, twice hpGRF potency) (Rivier *et al.*, 1982). Substitution of His, the N-terminal residue of rGRF, into position 1 of hpGRF resulted in an increase in receptor binding affinity (fivefold) and GH release (3.2-fold) in assays on rat cells (Seifert *et al.*, 1985).

We began our analogue studies by combining the observations that potent GH-releasing structures were reportedly designed using Met-enkephalin (Tyr-Gly-Gly-Phe-Met-NH_2) as a starting point (Momany *et al.*, 1981) and that the N-terminal aromatic Tyr residue was important in both the enkephalin and GRF series. In the enkephalin studies, D-Ala substitution in position 2 gave substantial potency increases (Pert *et al.*, 1976) by protection from enzymatic degradation and possibly for conformational reasons. We therefore prepared [D-Ala^2]$hpGRF_{1-6}$-NH_2, [D-hArg$(Et_2)^2$]$hpGRF_{1-6}$-NH_2, and [N-Me-Tyr^1,D-hArg$(Et_2)^2$]$hpGRF_{1-6}$-NH_2. These analogues showed no receptor binding or GH release properties, however.

This substitution pattern was carried into a series of longer polypeptide analogues (Table I). We chose the 1–15 sequence next because it marks the end of a predicted secondary conformation domain (Chou–Fasman calculation, data not shown; however, see Coy *et al.*, 1985). Residues 3–14 are predicted to be in a β-pleated sheet conformation with a β-turn predicted for residues 7–10. Although $hpGRF_{1-15}$-NH_2 and [His^1]$hpGRF_{1-15}$-NH_2 exhibited no receptor binding and no GH release (Table I), [D-Ala^2]$hpGRF_{1-15}$-NH_2 did exhibit detectable binding (binding "potency" 0.5% relative to $hpGRF_{1-44}$). The analogue containing the amino acid designed for increased membrane affinity appeared to have significantly increased receptor affinity {[D-hArg$(Et_2)^2$]$hpGRF_{1-15}$-NH_2 binding potency ~3%}. Although this binding

TABLE I. Receptor Binding of $hpGRF_{1-15}-NH_2$ Analogues

No.	Compound	Receptor binding[a] to GH3 cells[b] Binding pot. (%)	IC$_{50}$ (10^{-9} M)
1	$hpGRF_{1-44}-NH_2$(standard)	100	2.4
2	$hpGRF_{1-5}-NH_2$	N.B.[c]	—
3	[His^1]$hpGRF_{1-15}-NH_2$	N.B.[c]	—
4	[D-Ala^2]$hpGRF_{1-15}-NH_2$	0.5	500
5	[D-hArg(Et_2)^2]$hpGRF_{1-15}-NH_2$	3	80

[a] IC_{50} values for competition with 2×10^{-9} M [^{125}I]$hpGRF_{1-44}-NH_2$. Values are ±0.1.
[b] GH3 rat pituitary cells are subcloned somatotrophic cell line provided by Dr. D. Gospodarowicz (U.C. San Francisco).
[c] No specific binding detected up to 5×10^{-5} M.

potency is low, it does suggest that analogues shorter than the commonly accepted 1–27 minimum size may exhibit significant GRF receptor binding.

Increased Receptor Affinity

Since $hpGRF_{1-29}-NH_2$ was reported to have full agonistic potency, we tested the D-hArg(R_2) substitution in this series. Although the $hpGRF_{1-29}-NH_2$ structure exhibited a slightly reduced receptor affinity (Table II), full receptor binding and GH release potency are exhibited by the [Nle^{27}]$hpGRF_{1-29}-NH_2$ analog, which cannot undergo air oxidation. The incorporation of D-hArg(Et_2), designed for increased membrane affinity, produced a remarkable 800% increase in receptor binding potency (**8,** Table II). Addition of the Nle substitution (**9**) led to a small decrease

TABLE II. Biological Activities of $hpGRF_{1-29}-NH_2$ Analogues

No.	Compound	Receptor binding[a]: GH3 cells Binding pot. (%)	Receptor binding[a]: GH3 cells IC$_{50}$ (10^{-9} M)	Receptor binding[a]: Rat pit. IC$_{50}$ (10^{-9} M)	GH release[b] from rat pit. cells Release (%)	GH release[b] from rat pit. cells EC$_{50}$ (10^{-10} M)
1	$hpGRF_{1-44}-NH_2$ (standard)	100	2.4	2.2	100	4
6	$hpGRF_{1-29}-NH_2$	60	4.1		100	4
7	[Nle^{27}]$hpGRF_{1-29}-NH_2$[c]	120	2.0	2.2	100	3
8	[D-hArg(Et_2)^2]$hpGRF_{1-29}-NH_2$	800	0.3	0.4	30	
9	[D-hArg(Et_2)^2,Nle^{27}]$hpGRF_{1-29}-NH_2$	600	0.4	0.3	35	

[a] IC_{50} values for competition with 2×10^{-9} M [^{125}I]$hpGRF_{1-44}-NH_2$. Values are ±0.1.
[b] Values are ±10%.
[c] Rivier *et al.* (1983).

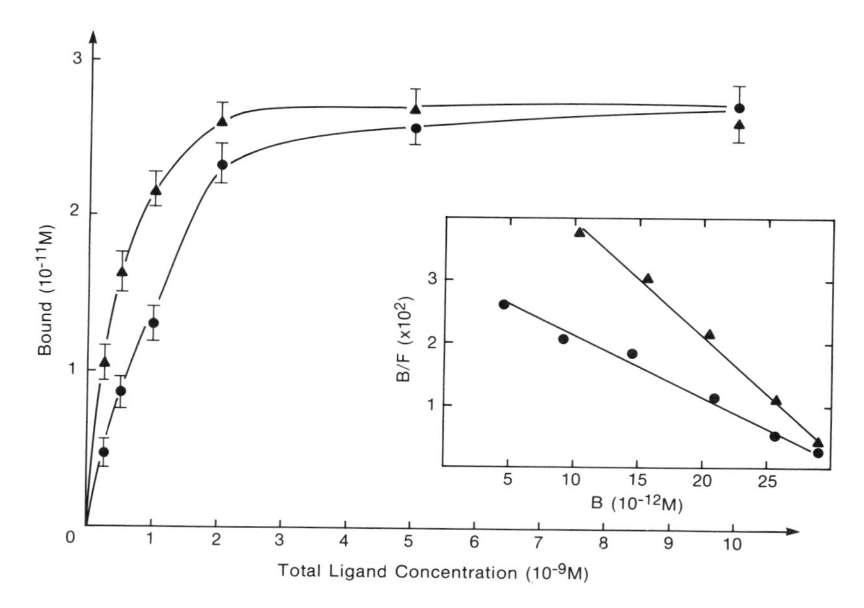

FIGURE 3. Saturation binding of radiolabeled GRF analogues to GH3 cells at 4°C. Binding assay conditions were the same as in Fig. 2. Insert: Scatchard plot derived from binding data. For both radiolabeled ligands, the number of binding sites is calculated to be close to 1.5×10^4 per cell. K_{app} for [^{125}I]hpGRF$_{1-44}$-NH$_2$ is 0.8×10^9 M^{-1} and for [^{125}I][D-hArg(Et$_2$)^2,Nle^{27}]hpGRF$_{1-29}$-NH$_2$, 0.6×10^{10} M^{-1}. ●—●, [^{125}I]hpGRF$_{1-44}$-NH$_2$; ▲—▲, [^{125}I]D-hArg(Et$_2$)^2,Nle^{27}hpGRF$_{1-29}$-NH$_2$.

in binding potency relative to **8** but still resulted in an analogue with a binding potency 600% greater than the standard.

The competition curve is parallel to that of the native ligand (Fig. 2) but is displaced to higher affinity. The K_D calculated for hpGRF$_{1-44}$-NH$_2$ to 0.9×10^{-9} M. This is comparable to the K_D ($\sim 0.4 \times 10^{-9}$ M) determined for hpGRF$_{1-40}$ on freshly dispersed rat pituitary cells (Seifert *et al.*, 1985). Analysis of the competition (Fig. 2) and Scatchard analysis (Fig. 3) suggest the presence of a single class of high-affinity receptors.

That the increased binding observed with [D-hArg(Et$_2$)^2,Nle^{27}]hpGRF$_{1-19}$-NH$_2$ is specific and competitive with native hpGRF is further confirmed by the finding that the IC$_{50}$ for competition of hpGRF$_{1-44}$-NH$_2$ with [^{125}I][D-hArg(Et$_2$)^2,Nle^{27}]hpGRF$_{1-29}$-NH$_2$ (2×10^{-9} M) is 11.6×10^{-9} M. The native ligand is therefore about sixfold less effective in competing with the hArg(Et$_2$)-containing radioligand than is cold [D-hArg(Et$_2$)^2,Nle^{27}]hpGRF$_{1-29}$-NH$_2$, which has an IC$_{50}$ of 2.1×10^{-9} M in this assay.

Altering the length of the side chain on the guanidine function significantly affected receptor affinity. Lengthening the chain from ethyl to propyl decreased binding potency to 340% (**10,** Table III). The decreased binding potency in this series may be caused by a steric hindrance to the electrostatic component of the

TABLE III. Biological Activities of $hpGRF_{1-29}$-NH_2 Analogues

No.	Compound	Receptor binding[a]: GH3 cells Binding pot. (%)	Receptor binding[a]: GH3 cells IC_{50} (10^{-9} M)	Receptor binding[a]: Rat pit. IC_{50} (10^{-9} M)	GH release[b] from rat pit. cells: Release (%)	GH release[b] from rat pit. cells: EC_{50} (10^{-10} M)
1	$hpGRF_{1-44}$-NH_2 (standard)	100	2.4	2.2	100	4
9	[D-hArg(Et_2)[2],Nle[27]]$hpGRF_{1-29}$-NH_2	600	0.4	0.3	35	
10	[D-hArg(Pr_2)[2], Nle[27]]$hPGRF_{1-29}$-NH_2	340	0.7		35	
11	[D-hArg(CH_2CH_2)[2], Nle[27]]$hpGRF_{1-29}$-NH_2	800	0.3			
12	[D-Arg[2], Nle[27]]$hpGRF_{1-29}$-NH_2	170	1.4		20	100
13	[hArg(Et_2)[2],Nle[27]]$hpGRF_{1-29}$-NH_2	480	0.5		20	
14	[D-hArg(Et_2)[1],Nle[27]]$hpGRF_{1-29}$-NH_2	270	0.9		15	
15	[D-Ala[2]]$hpGRF_{1-29}$-NH_2[d]	240	1.0	0.8	115	1

[a] IC_{50} values for competition with 2×10^{-9} M [^{125}I]$hpGRF_{1-44}$-NH_2. Values are ±0.1.
[b] Values are ±10%.
[c] Rivier *et al.* (1983).
[d] Lance *et al.* (1984).

proposed phospholipid–membrane interaction caused by the longer alkyl chains. Shortening the alkyl chain to an ethylene bridge restored the high binding potency {[D-Arg(CH_2CH_2)[2],Nle[27]]$hpGRF_{1-29}$-NH_2, 800% binding potency} observed in the hArg(Et_2) series. The corresponding analogue with the native amino acid with no alkyl groups on the side chain, [D-Arg[2],Nle[27]]$hpGRF_{1-29}$-NH_2, exhibits a further reduced potency compared to the hArg(R_2) class. This analogue does exhibit an increased binding potency relative to the standard, and this may reflect the ability of Arg to interact electrostatically with the cell membranes.

It is interesting to note that incorporation of the L form of hArg(Et_2) in position 2 (**13**) yields an analogue that binds almost as avidly as the corresponding D-containing analogue {[hArg(Et_2)[2],Nle[27]]$hpGRF_{1-29}$-NH_2, binding potency 480%}. A similar lack of rigid stereochemical requirements in this region of the molecule is illustrated by [D-hArg(Et_2)[1],Nle[27]]$hpGRF_{1-29}$-NH_2, which retains a relatively high receptor affinity although the D-amino acid is shifted to position 1.

Another analogue with increased receptor affinity is [D-Ala[2]]$hpGRF_{1-29}$-NH_2 (Lance *et al.*, 1984). Although [D–Ala[2]]$hpGRF_{1-29}NH_2$ was reported to have 50 times the $hpGRF_{1-29}NH_2$ potency *in vivo* in the pentobarbital rat model (Lance *et al.*, 1984) and *in vitro* (Heiman *et al.*, 1985), we found it to have only three- to fivefold increased potency in the rat model (C. Nerenberg, B. Rice, T. L. Ho, and J. J. Nestor, Jr., unpublished results). We also find that this analogue exhibits an increased receptor binding potency (2.4-fold) and GH release potency (four- to fivefold) *in vitro* (**15,** Table II).

The effect of the D-hArg(Et_2) substitution in full-length analogues was also

TABLE IV. Biological Activities of $hpGRF_{1-44}$-NH_2 Analogues

		Receptor binding[a]			GH release[b] from rat pit. cells	
		GH3 cells		Rat pit.		
No.	Compound	Binding pot. (%)	IC_{50} (10^{-9} M)	IC_{50} (10^{-9} M)	Release (%)	EC_{50} (10^{10} M)
1	$hpGRF_{1-44}$-NH_2 (standard)	100	2.4	2.2	100	4
16	[Leu^{27}]$hpGRF_{1-44}$-NH_2	77	3.1		85	
17	[D-hArg(Et_2)^2,Leu^{27}]$hpGRF_{1-44}$-NH_2	600	0.4	0.3	35	
18	[D-hArg(Et_2)^4,Leu^{27}]$hpGRF_{1-44}$-NH_2	90	2.6			
19	[D-Ala^2,D-hArg(Et_2)^4,Leu^{27}]$hpGRF_{1-44}$-NH_2	75	3.2		10	

[a] IC_{50} values for competition with 2×10^{-9} M [^{125}I]$hpGRF_{1-44}$-NH_2. Values are ±0.1.
[b] Values are ±10%.

studied (Table IV). In the 44-residue series, we employed a Leu substitution for the Met residue in position 27 and found it to be not as effective as Nle had been in the 29-residue series (compare **16** with **17**). Incorporation of D-hArg(Et_2) again produced an analogue with substantially increased receptor binding potency relative to the standard (**17,** 600% binding potency). This binding potency would be even higher (775%) if compared to the corresponding parent structure **16.** When the position of substitution is shifted to residue 4, the increased receptor binding effect is lost (**18, 19**).

5. APPROACH TO AN ANTAGONIST

Although all of the analogues containing D-hArg(Et_2) residues in position 2 exhibit substantially increased receptor binding, they are only partial agonists when GH release is studied in the dose range near the K_D (Fig. 4). Maximal release was in the range of 30% of that of the standard (compare **8** and **1,** Table II). The dose–response curve for GH release reflects both the increased binding (lower EC_{50} in the region below 1×10^{-10} M) and decreased maximal response (Fig. 4). We have recently extended the dose–response curve beyond the range originally investigated (10×10^{-10} M) and find an unusual response in that full agonism (at 10^{-7} M) is eventually achieved (Fig. 4). Whether this is because of GRF agonism or activation of an alternative receptor pathway (e.g., VIP receptors, see Laburthe *et al.*, 1983; Zeytin *et al.*, 1985) is not clear at this time.

Since [D-hArg(Et_2)^2,Nle^{27}]$hpGRF_{1-29}$-NH_2 exhibited very high receptor binding affinity but reduced intrinsic agonistic activity, it appeared to be an important lead toward the development of a GRF antagonist (Table V). The report that deletion of the N-terminal Tyr residue destroyed agonistic potency (Rivier *et al.*, 1983) led us to prepare the corresponding D-hArg(Et_2) analogue in the hope of decreasing the

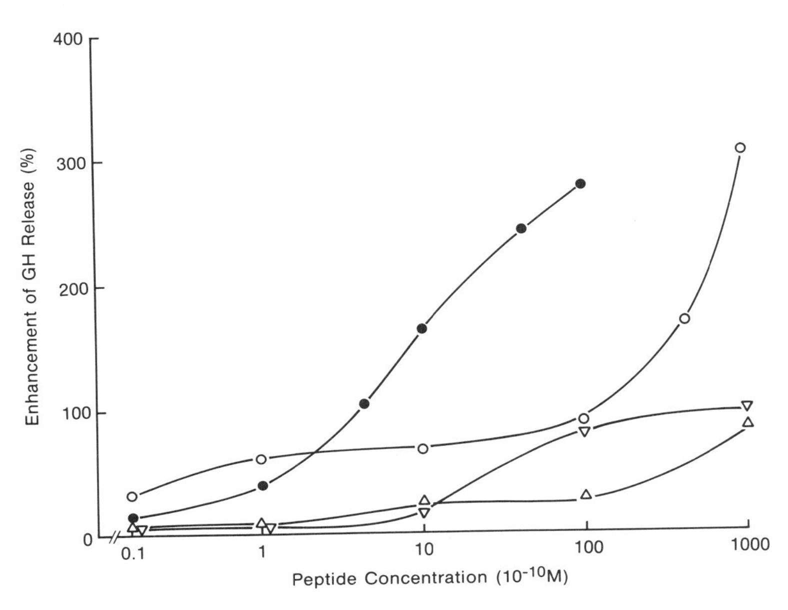

FIGURE 4. Induction of growth hormone release from normal rat pituitary cell cultures by GRF analogues. Normal rat anterior pituitary cells were plated in 24-well dishes (about 2×10^5 cells/well); GRF and GRF analogues were added to the cells in serum-free DMEM for various times at 37°C. The concentration of GH in the medium was determined by a solid-phase RIA using reagents from the NIH Pituitary Agency. Ninety-six-well polyvinylchloride plates were coated overnight at 4°C with rabbit anti-monkey Ig antibodies (Pel-Freez) at 0.1 mg/ml in phosphate-buffered saline (PBS). Plates were countercoated with a 3% bovine serum albumin solution in PBS for at least 1 hr at 37°C. Plates were washed and incubated with monkey anti-rat GH antiserum (1/10,000 final dilution) for 2 hr or more at 37°C. After several washes with PBS–0.1% BSA, plates were incubated at 37°C for 2 hr with 20 ng/ml [^{125}I]GH in the presence or absence of unlabeled GH or dilutions of cellular conditioned medium. After several washes, plates were dried and cut, and individual wells were counted in a γ counter. ●—●, hpGRF$_{1-44}$-NH$_2$; ○—○, [D-hArg(Et$_2$)^2,Nle^{27} hpGRF$_{1-29}$-NH$_2$; ▽—▽, [D-Arg^2, Nle^{27}],hpGRF$_{1-29}$-NH$_2$; △—△, [D-hArg(Et$_2$)^2, Nle^{27}] hpGRF$_{2-29}$-NH$_2$.

residual intrinsic agonistic activity in **9.** The resulting analogue, D-hArg(Et$_2$)^2,Nle^{27}]hpGRF$_{2-29}$-NH$_2$ (**20**), retains a receptor binding potency significantly higher than that of the native hormone (**1**), although the receptor affinity of the corresponding des-Tyr native sequence (**21**) is destroyed (Seifert *et al.*, 1985). This is a most dramatic demonstration of the effect of the D-hArg(Et$_2$) structure on receptor binding affinity.

Although compound **20,** when tested at doses near its K_D, displayed minimal agonistic activity, at very high concentrations *in vitro* it was found to be a very weak partial agonist (Fig. 4). This result may again be caused by activation of alternate receptors (see above).

TABLE V. Approach to an Antagonist of GRF

No.	Compound	Receptor binding[a] GH3 cells Binding pot. (%)	Receptor binding[a] GH3 cells IC_{50} (10^{-9} M)	Receptor binding[a] Rat pit. IC_{50} (10^{-9} M)	GH release[b] from rat pit. cells Release (%)	GH release[b] from rat pit. cells EC_{50} (10^{-10} M)
1	$hpGRF_{1-44}$-NH_2 (standard)	100	2.4	2.2	100	4
9	[D-hArg$(Et_2)^2$,Nle^{27}]$hpGRF_{1-29}$-NH_2	600	0.4	0.3	35	
12	[D-Arg^2,Nle^{27}]$hpGRF_{1-29}$-NH_2	170	1.4		20	100
20	[D-hArg$(Et_2)^2$,Nle^{27}]$hpGRF_{2-29}$-NH_2	240	1.0	1.2	5	100
21	des-Tyr^1-$hpGRF_{2-29}$-NH_2[c]	<1[c]				

[a] IC_{50} values for competition with 2×10^{-9} M [^{125}I]$hpGRF_{1-44}$-NH_2. Values are ±0.1.
[b] Values are ±10%.
[c] Seifert *et al.* (1985).

6. CONCLUSIONS

The incorporation of the hArg(Et_2) residue, an amino acid designed for phospholid–membrane interaction, has produced substantial increases (up to 800%) in receptor binding potency in the hpGRF series. The process by which the increased binding occurs does not appear to be sensitive to stereochemical factors. This result is consistent with the hypothesis that ligand–membrane interaction may play an important role in the receptor binding process and is in agreement with our results with analogues of transforming growth factor (Nestor *et al.*, 1985a) and luteinizing hormone-releasing hormone (Nestor *et al.*, 1985b).

The use of this GH3 tumor cell line clone for binding studies was validated by comparison with normal rat pituitary cells, and the resulting K_D values are consistent with a recent literature report (Seifert *et al.*, 1985). The use of this cell line offers several advantages, since large amounts of this homogeneous population of cells can be readily obtained. These cells do not, however, serve as the basis for a GH-release assay.

We have demonstrated that N-terminally shortened analogues of GRF that incorporate the hArg(Et_2) residue can retain binding affinity even greater than the native ligand, in contrast to the result when this residue is not present. The very reduced intrinsic agonistic activity that results offers hope that [D-hArg$(Et_2)^2$,Nle^{27}]$hpGRF_{1-29}$-NH_2 will serve as the basis for the design of an effective GRF antagonist.

ACKNOWLEDGMENTS. We are grateful to Fred Niemeyer for preparation of the hexapeptide analogues and to Clint Nerenberg for help with the RIAs. We thank Mrs. Mary Smith for her excellent assistance in the manuscript preparation. The support and encouragement of Drs. Anthony C. Allison and John G. Moffatt are gratefully acknowledged.

REFERENCES

Böhlen, P., Brazeau, P., Bloch, B., Ling, N., Gaillard, R., and Guillemin, R., 1983, Human hypothalamic growth hormone releasing factor (GRF): Evidence for two forms identical to tumor derived GRF-44-NH_2 and GRF-40, *Biochem. Biophys. Res. Commun.* **114:**930–936.

Brazeau, P., Böhlen, P., Esch, F., Ling, N., Wehrenberg, W. B., and Guillemin, R., 1984, Growth hormone-releasing factor from ovine and caprine hypothalamus: Isolation, sequence analysis, and total synthesis, *Biochem. Biophys. Res. Commun.* **125:**606–614.

Coy, D. H., Murphy, W. A., Sueiras-Diaz, J., Coy, E. J., and Lance, V. A., 1985, Structure–activity studies on the N-terminal region of growth hormone releasing factor, *J. Med. Chem.* **28:**181–185.

Epand, R. M., Epand, R. F., and Orlowski, R. C., 1985, Presence of an amphiphilic helical segment and its relationship to biological potency of calcitonin analogs, *Int. J. Peptide Protein Res.* **25:**105–111.

Esch, F., Böhlen, P., Ling, N. C., Brazeau, P. E., Wehrenberg, W. B., and Guillemin, R., 1983, Primary structures of three human pancreas peptides with growth hormone-releasing activity, *J. Biol. Chem.* **258:**1806–1812.

Gubler, U., Monahan, J. J., Lomedico, P. T., Bhatt, R. S., Collier, K. J., Hoffman, B. J., Böhlen, P., Esch, F., Ling, N., Zeytin, F., Brazeau, P., Poonian, M. S., and Gage, L. P., 1983, Cloning and sequence analysis of cDNA for the precursor of human growth hormone-releasing factor, somatocrinin, *Proc. Natl. Acad. Sci. U.S.A.* **80:**4311–4314.

Guillemin, R., Brazeau, P., Böhlen, P., Esch, F., Ling, N., and Wehrenberg, W. B., 1982, Growth hormone-releasing factor from a human pancreatic tumor that caused acromegaly, *Science* **218:**585–587.

Gysin, B., and Schwyzer, R., 1984, Hydrophobic and electrostatic interactions between adrenocorticotropin-(1–24)-tetracosapeptide and lipid vesicles. Amphiphilic primary structures, *Biochemistry* **23:**1811–1818.

Heiman, M. L., Nekola, M. V., Murphy, W. A., Lance, V. A., and Coy, D. H., 1985, An extremely sensitive *in vitro* model for elucidating structure–activity relationships of growth hormone-releasing factor analogs, *Endocrinology* **116:**410–415.

Laburthe, M., Amiranoff, B., Bioge, N., Rouyer-Fessard, C., Tatemoto, K., and Moroder, L., 1983, Interaction of GRF with VIP receptors and stimulation of adenylate cyclase in rat and human intestinal epithelial membranes: Comparison with PHI and secretin, *FEBS Lett.* **159:**89–92.

Lance, V. A., Murphy, W. A., Sueiras-Diaz, J., and Coy, D. H., 1984, Superactive analogs of growth hormone-releasing factor (1–29)-amide, *Biochem. Biophys. Res. Commun.* **119:**265–272.

Ling, N., Esch, F., Böhlen, P., Brazeau, P., Wehrenberg, W. B., and Guillemin, R., 1984, Isolation, primary structure, and synthesis of human hypothalamic somatocrinin: Growth hormone-releasing factor, *Proc. Natl. Acad. Sci. U.S.A.* **81:**4302–4306.

Mayo, K. E., Vale, W., Rivier, J., Rosenfeld, M. G., and Evans, R. M., 1983, Expression-cloning and sequence of a cDNA encoding human growth hormone-releasing factor, *Nature* **306:**86–88.

Momany, F. A., Bowers, C. Y., Reynolds, G. A., Chang, D., Hong, A., and Newlander, K., 1981, Design, synthesis, and biological activity of peptides which release growth hormone *in vitro*, *Endocrinology* **108:**31–39.

Nestor, J. J., Jr., Tahilramani, R., Ho, T. L., McRae, G. I., Vickery, B. H., and Bremner, W. J., 1983, New luteinizing hormone-releasing factor antagonists, in: *Peptides—Structure and Function* (V. J. Hruby and D. H. Rich, eds.), Pierce Chemical, Rockford, IL, pp. 861–864.

Nestor, J. J., Jr., Ho, T. L., Tahilramani, R., Horner, B. L., Simpson, R. A., Jones, G. H., McRae, G. I., and Vickery, B. H., 1984, in: *LHRH and Its Analogs—Contraceptive and Therapeutic Applications* (B. H. Vickery, J. J. Nestor, Jr., and E. S. E. Hafez, eds.), MTP Press, Lancaster, pp. 23–33.

Nestor, J. J., Jr., Newman, S. R., DeLustros, B. M., and Schreiber, A. G., 1985a, Antagonistic analogs of human transforming growth factor alpha, in: *Peptides—Structure and Function—Proceedings of the Ninth American Peptide Symposium* (C. M. Deber, V. J. Hruby, and K. D. Kopple, eds.), Pierce Chemical, Rockford, IL pp. 39–42.

Nestor, J. J., Jr., Tahilramani, R., Ho, T. L., McRae, G. I., and Vickery, B. H., 1985b, Potent LHRH agonists containing N^G,$N^{G'}$-dialkyl-D-homoarginines, in: *Peptides—Structure and Function—Proceedings of the Ninth American Peptide Symposium* (C. M. Deber, V. J. Hruby, and K. D. Kopple, eds.), Pierce Chemical, Rockford, IL pp. 557–560.

Pert, C. B., Bowie, D. L., Fong, B. T. W., and Chang, J.-K., 1976, Synthetic analogues of Met-enkephalin which resist enzymatic destruction, in: *Opiates and Endogenous Opioid Peptides* (H. W. Kosterlitz, ed.), North Holland, Amsterdam, New York, pp. 79–86.

Rivier, J., Spiess, J., Thorner, M., and Vale, W., 1982, Characterization of a growth hormone-releasing factor from a human pancreatic islet tumor, *Nature* **300:**276–278.

Schwyzer, R., Gremlich, H.-U., Gysin, B., and Sargent, D. F., 1983, Specific interactions between peptide hormones and artificial lipid membranes, in: *Peptides—Structure and Function* (V. J. Hruby and D. H. Rich, eds.), Pierce Chemical, Rockford, IL, pp. 657–664.

Seifert, H., Perrin, M., Rivier, J., and Vale, W., 1985, Binding sites for growth hormone releasing factor on rat anterior pituitary cells, *Nature* **313:**487–489.

Spiess, J., Rivier, J., and Vale, W., 1983, Characterization of rat hypothalamic growth hormone-releasing factor, *Nature* **303:**532–535.

Vale, W., Grant, G., Amoss, M., Blackwell, R., and Guillemin, R., 1972, Culture of enzymatically dispersed anterior pituitary cells: Functional validaton of a method, *Endocrinology* **91:**562–572.

Zeytin, F. N., Reyl-Desmares, F., and Rathbun, T., 1985, Rat hypothalamic GRF elicits its biologic action in GH3 cells by interaction with VIP preferring receptor sites, *Biochem. Biophys. Res. Commun.* **127:**992–998.

12

Circadian Variation of Immunoreactive CRF in Rat Hypothalamus

ABDUL-BADI ABOU-SAMRA,
MICHELLE FEVRE-MONTANGE, FRANCOISE BORSON,
HENRI DECHAUD, and JACQUES TOURNIAIRE

1. INTRODUCTION

The activity of the hypothalamic–pituitary–adrenal axis is characterized by episodic secretion and circadian variations. In normal human subjects, ACTH and cortisol secretion is pulsatile in nature; pulse amplitude and frequency dramatically increase during the second half of the night, giving high plasma concentrations of those hormones at 0800 hr, and then gradually decrease during the day, giving low plasma concentrations at 2400–0200 hr. This diurnal rhythmicity seems synchronized with the environment day–night cycles, which condition the sleep–activity cycles. In rats, whose normal activity occurs during the night, the ACTH rhythm does not resemble that in humans; instead, plasma ACTH and corticostrone concentrations are high in the evening (1600–2000 hr) and low in the morning (0800–1200 hr). Although hypothalamus-lesioned rats still have significant circadian variations in corticosterone production (Nicholson *et al.*, 1985), the hypothalamus seems to be the main regulator of the rhythmic activity of the hypothalamic–pituitary–adrenal axis.

Earlier studies had shown that the ACTH-releasing activity of hypothalamic extracts exhibits significant diurnal variations (David-Nelson and Brodish, 1969; Hiroshige and Sato, 1970; Takebe *et al.*, 1972; Kamstra *et al.*, 1983). However, the hypothalamic extract contains several factors that possess ACTH-releasing ac-

ABDUL-BADI ABOU-SAMRA, MICHELLE FEVRE-MONTANGE, FRANCOISE BORSON, HENRI DECHAUD, and JACQUES TOURNIAIRE • Clinique Endocrinologique Hôpital de l'Antiquaille, 69321 Lyon Cedex 5, France. *Present address of A.-B.A.-S.:* Endocrinology and Reproduction Research Branch, National Institute of Child Health and Human Development, National Institutues of Health, Bethesda, Maryland 20892.

tivity (Vale *et al.*, 1983), and it could not be determined which of these was the main factor regulating hormone rhythmicity. Recently, a 41-residue peptide isolated from ovine hypothalamus has been shown to have potent ACTH-releasing activity both *in vitro* and *in vivo* and was therefore labeled corticotropin-releasing factor (CRF) (Vale *et al.*, 1981). Following the characterization of CRF, it has been possible to determine hypothalamic CRF content using a specific radioimmunoassay. The present work was undertaken to explore the circadian variations of rat hypothalamic CRF content compared to those of plasma corticosterone and β-endorphin (β-end) concentrations. At the same time, the β-end content of the hypothalamus and the pituitary lobes was measured and analyzed for the presence of circadian cyclicity.

2. MATERIAL AND METHODS

2.1. Animals

Adult male Sprague–Dawley rats were kept for 2 weeks on a 12-hr light–dark schedule (light on at 0700 hr). Foods were given *ad libitum*. At each time point, the rats were taken individually to an adjacent room and decapitated. A red light was used when animals were sacrificed during the night. The brain was rapidly removed, and the median eminence (ME), anterior pituitary (AP), and neurointermediate lobe (NIL) were dissected and immediately frozen in dry ice. Three milliliters of trunk blood was collected into plastic tubes containing 1 mg/ml EDTA and 1 mg/ml bacitracin and rapdily centrifuged, and the plasma was frozen until assayed.

2.2. Tissue Extraction

After addition of 1 ml of 0.1 M hydrochloric acid, tissue fragments were sonicated at 0°C and centrifuged at 3000 r.p.m. at 4°C for 20 min, and the supernatant was kept frozen until assayed.

2.3. Hormone Assays

The CRF immunoreactivity was measured by a specific radioimmunoassay (Abou-Samra *et al.*, 1984a). The antibody was raised in the rabbit against oCRF coupled to bovine serum albumin and used at 1:10,000 final dilution; rCRF was used as tracer and standard. The ^{125}I-labeled rCRF was prepared by the transfer iodogen method (Salacinski *et al.*, 1981) and purified by chromatography on a 1- × 50-cm AcA 202 Ultrogel column. No cross reaction was detected with any of the following peptides used up to 1 μg/tube: GRF, GnRH, TRH, AVP, β-end, α-end, Met-enk, Leu-enk, h-β-LPH, h-α-LPH, β-MSH, and α-MSH. Serial dilutions of rat, bovine, and ovine hypothalamic extracts were parallel to the standard curve. The assay sensitivity was 30 pg/tube, and the concentration of standard that displaced

50% of the tracer was 150 pg/tube. Intra- and interassay variations were 8% and 13%, respectively.

β-Endorphin immunoreactivity was measured by a RIA that equally cross-reacted with h-β-LPH on a molar basis (Abou-Samra *et al.*, 1985b). The assay sensitivity was 1 pg/tube, and the intra- and interassay variations were less than 10%. Corticosterone was extracted from plasma with ethanol and assayed by radiocompetition. The assay sensitivity was 0.1 μg/dl, and the intra- and interassay variations were less than 3%.

2.4. Chromatographic Analysis of β-Endorphin and CRF Tissue Content

To study whether hormone processing undergoes circadian variations, the tissue extracts obtained at 1200 and 2400 hr were analyzed by chromatography (Abou-Samra *et al.*, 1984c). An Ultrogel AcA 202 column (1 × 50 cm) was equilibrated and eluted with 0.1 M acetic acid, and 1-ml fractions were collected, the pH was carried to 7 with 1 M NaOH, and the fractions were assayed for CRF and β-end immunoreactivities.

2.5. Statistics

The results are given as means ± standard error of the mean. A one-way variance analysis was used to determine the presence of significant time-dependent variations. Comparison between the means was carried using the least significant difference (LSD) calculated at 5% significance.

3. RESULTS

Significant circadian variations were observed in plasma β-end and corticosterone concentrations (Fig. 1), hypothalamic CRF (Fig. 2), and anterior pituitary β-end (Fig. 3) contents.

3.1. Circadian Variation of Plasma Concentrations of β-Endorphin and Corticosterone

Plasma β-end and corticosterone concentrations were low at 0800–1200 hr (0.41 ± 0.09 and 0.36 ± 0.08 ng/ml for β-end and 5.70 ± 1.51 and 4.40 ± 1.9 μg/dl for corticosterone at 0800 and 1200 hr, respectively); they started to increase significantly at 1600 hr (0.74 ± 0.05 ng/ml for β-end and 9.51 ± 0.91 μg/dl for corticosterone) to reach their peak values at 2000 hr (0.82 ± 0.11 ng/ml for β-end and 14.30 ± 1.91 μg/dl for corticosterone) and then progressively decreased to values of 0.46 ± 0.82 ng/ml for β-end and 5.60 ± 1.80 μg/dl for corticosterone at 0800 hr (Fig. 1).

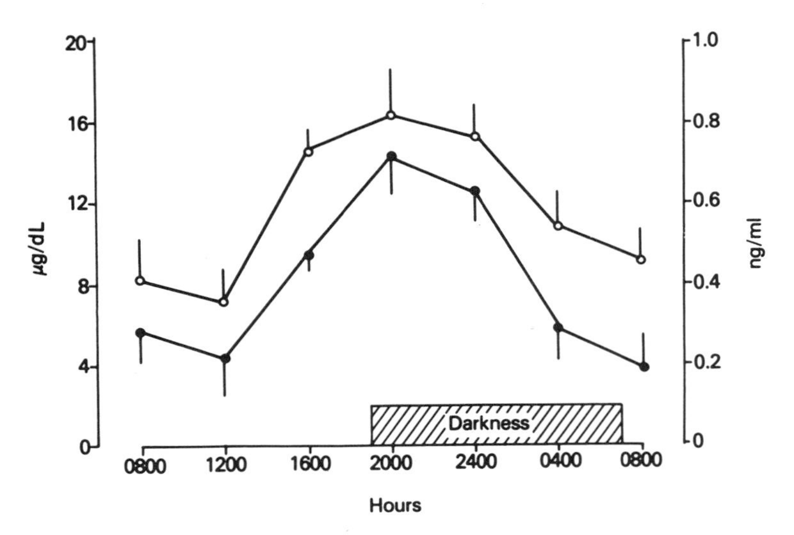

FIGURE 1. Circadian rhythm of plasma β-end, ○—○, and corticosterone, ●—●, concentration. Both rhythms are significant at $P < 0.001$.

3.2. Circadian Variation of Hypothalamic CRF Content

Hypothalamic CRF content at 0800 hr was 2.8 ± 0.2 ng/ME; it significantly decreased at 2000 hr to values of 1.71 ± 0.62 ng/ME, rapidly increased to a peak value of 3.55 ± 1.01 ng/ME at 2400 hr, and then progressively decreased to 2.65 ± 0.61 ng/ME at 0800 hr (Fig. 2).

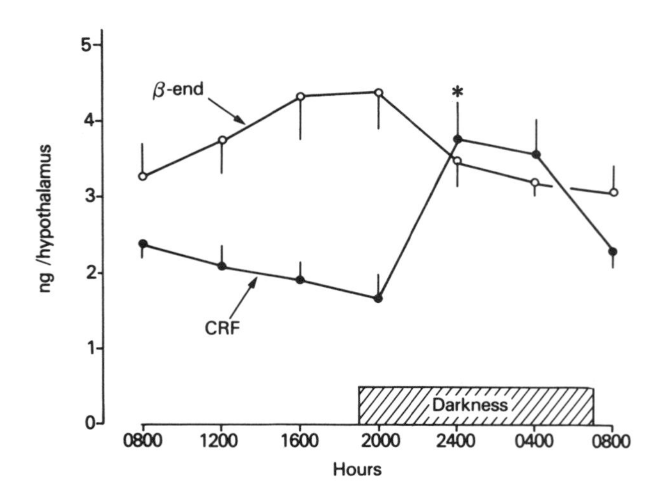

FIGURE 2. Circadian rhythm of CRF and β-end content of the hypothalamus. Only CRF rhythm was significant at $P < 0.01$.

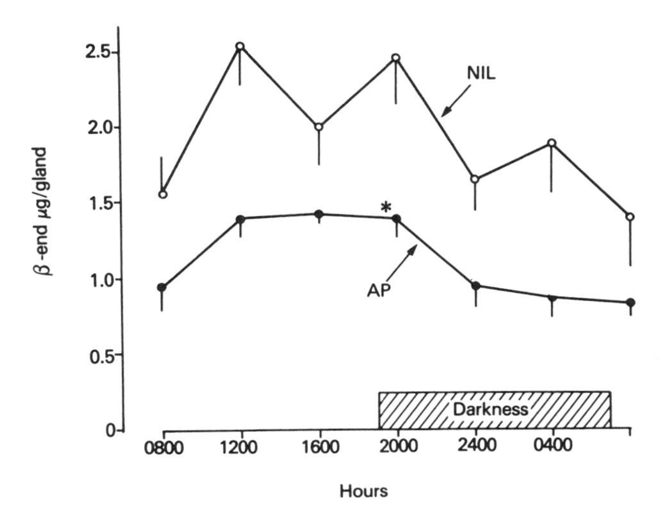

FIGURE 3. Circadian rhythm of β-end content of the anterior pituitary (AP) and the neurointermediate lobe (NIL). Only anterior pituitary β-end rhythm was significant at $P < 0.05$.

3.3. Circadian Variation of β-Endorphin Content of Anterior Pituitary, Neurointermediate Lobe, and Hypothalamus

Anterior pituitary β-end content increased from 0.95 ± 0.17 μg/AP at 0800 to a plateau at 1200–2000 hr (1.42 ± 0.15, 1.44 ± 0.07, and 1.40 ± 0.12 μg/AP at 1200, 1600, and 2000 hr, respectively), then significantly decreased at 2400 hr to 0.96 ± 0.15 μg/AP, and remained near this level at 0400 and 0800 hr (0.85 ± 0.14 and 0.82 ± 0.10 μg/AP, respectively).

No significant circadian variations were observed in the β-end content of the hypothalamus (Fig. 2) and the neurointermediate lobe (Fig. 3). No CRF immunoreactivity was detected in the pituitary lobe extracts or in the plasma.

3.4. Chromatographic Analysis of CRF Immunoreactivity

Figure 4 shows the chromatographic analysis of hypothalamic immunoreactive CRF; two distinct peaks were detected, one coeluting with rCRF (89% of the immunoreactivity) and the second peak in the void volume (11% of CRF immunoreactivity). The void-volume CRF peak may represent a high molecular weight of CRF. No diurnal variations were observed in the elution pattern of hypothalamic CRF immunoreactivity.

3.5. Chromatographic Analysis of β-Endorphin Immunoreactivity

The chromotographic patterns of immunoreactive β-end in hypothalamus (HT), anterior pituitary (AP), and neurointermediate lobe (NIL) are shown in Fig. 5. Three

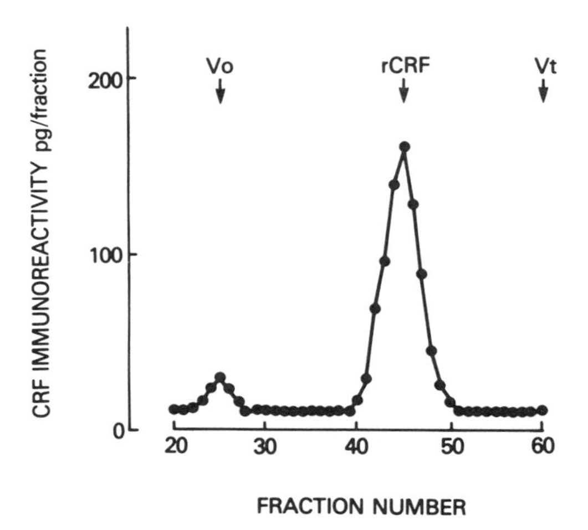

FIGURE 4. Chromatography of hypothalamic CRF immunoreactivity. The hypothalamic extract was applied on an AcA202 ultrogel column (1 × 50 cm) eluted with 0.1 M acetic acid. V_0 and V_t refer to the void volume and total volume, respectively.

immunoreactive peaks are present in the tissue extracts; two coeluted with β-end and β-LPH, and the third eluted in the void volume. No diurnal variations were observed in the chromatographic pattern of β-end immunoreactivity in the three tissue extracts, and the data were pooled. The distribution of β-end and β-LPH in the HT, AP, and NIL is shown in Fig. 6. Ten percent of β-end immunoreactivity

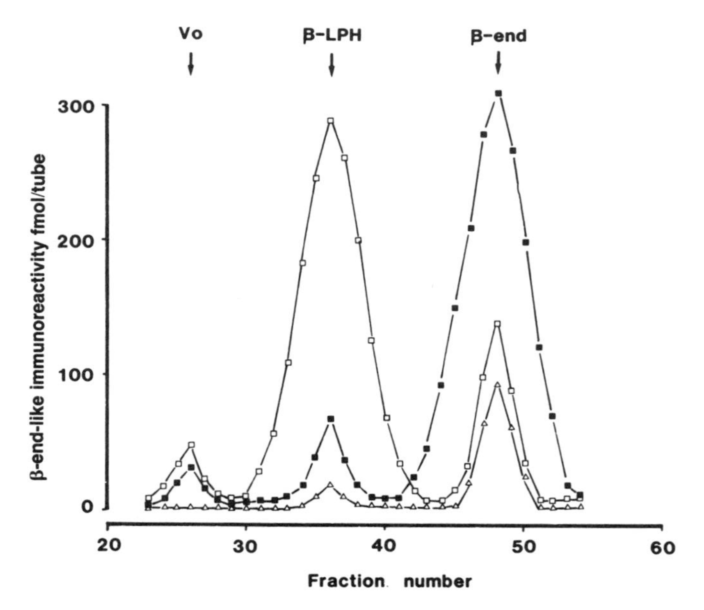

FIGURE 5. Chromatography of β-end immunoreactivity of the anterior pituitary, □—□; the neurointermediate lobe, ■—■; and the hypothalamus, △—△. The column was similar to that described in the legend of Fig. 4.

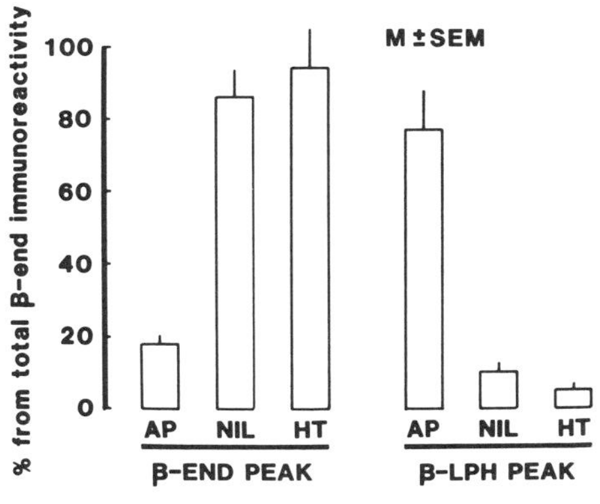

FIGURE 6. Distribution of β-end and β-LPH in the extracts of the anterior pituitary (AP), the neurointermediate lobe (NIL), and the hypothalamus (HT). The immunoreactivity eluted in the β-end on the β-LPH peaks is expressed as percentage of total β-end immunoreactivity.

of the AP coeluted with β-LPH. Conversely, 85% and 95% of β-end immunoreactivity of the NIL and HI coeluted wtih β-end.

4. DISCUSSION

The present work demonstrates significant circadian rhythms in the contents of hypothalamic CRF and anterior pituitary β-end as well as in the plasma concentrations of β-end and corticosterone. However, these rhythms were not parallel; plasma β-end and corticosterone levels peaked at 2000 hr, whereas hypothalamic CRF content peaked 4 hr later (at 2400 hr). Since tissue hormone content depends on hormone synthesis, release, and degradation, there is usually an inverse relationship between the magnitude of the store pools and secretion rate. For instance, plasma β-end concentrations at 0800–1200 hr were low, while at the same times pituitary β-end content started to increase, suggesting a low β-end release rate; conversely, at 2000–2400 hr, plasma β-end concentrations were high while pituitary β-end content started to decrease, suggesting a high β-end release rate (compare Figs. 1 and 3). Since CRF concentration in the hypothalamic–pituitary portal system was not measured, and that in the peripheral circulation was undetectable, there is no direct evidence of a significant diurnal rhythm in hypothalamic CRF release that might regulate the circadian rhythm of the pituitary–adrenal axis. However, the low hypothalamic CRF content at 1600 hr suggests that a high CRF release rate might be responsible for the significant increase in plasma β-end and corticosterone concentrations. The rapid increase in hypothalamic CRF content at 2400 hr may reflect high CRF synthesis, although the CRF release rate may still be high, maintaining high plasma β-end and corticosterone concentrations.

The chromatographic analysis of CRF and β-end immunoreactivities in tissue extracts did not show circadian changes. Accordingly, the circadian variations of hypothalamic CRF and anterior pituitary β-end are not caused by a circadian change in posttranslational hormone processing but are directly related to the balance between hormone synthesis and release.

Previous studies using biological methods to determine CRF activity in rat hypothalamus have shown significant diurnal rhythms in the content of hypothalamic CRF activity (David-Nelson and Brodish, 1969; Takebe *et al.*, 1972; Kamstra *et al.*, 1983). However, the patterns of the rhythms described were not consistent, showing either two peaks of activity with a trough at about 1 hr before lights off (David-Nelson and Brodish, 1969) or only one peak occurring approximately 1 hr before lights off (Takabe *et al.*, 1972; Kamstra *et al.*, 1983). Recently, Moldow and Fischman (1984) reported a significant circadian rhythm in rat hypothalamic CRF immunoreactivity with two peaks occurring at 1100 and 2000 hr and a trough at 1600 hr. Their study agrees with our data that hypothalamic CRF content is low at the time where the pituitary–adrenal axis begins its peak activity and that CRF content increases a few hours later. However, we did not observe the first CRF peak described by these authors. This discrepancy could not be explained by methodological variation and requires further investigations.

In conclusion, the data in this study demonstrate significant time-dependent variations in hypothalamic CRF content, which support the concept that changes in hypothalamic CRF may be responsible for the circadian rhythmicity of the activity of the pituitary–adrenal axis.

REFERENCES

Abou-Samra, A.-B., Durand, A., Loras, B., and Bertrand, J., 1984a, Development of a specific radioimmunoassay for the corticotropin releasing factor, in: *Sixth International Symposium on Radioimmunology*, Lyon, p. 96.

Abou-Samra, A.-B., Loras, B., Tourniaire, J., and Bertrand, J., 1984b, Demonstration of an antiglucocorticoid action of progesterone on the corticosterone inhibition of β-endorphin release by rat anterior pituitary in primary culture, *Endocrinology* **115:**1371–1375.

Abou-Samra, A.-B., Pugeat, M., Dechaud, H., Nachury, L., and Tourniaire, J., 1984c, Acute dopaminergic blockade by sulpiride stimulates β-endorphin secretion in pregnant women, *Clin. Endocrinol. (Oxford)* **21:**583–588.

David-Nelson, M. A., and Brodish, A., 1969, Evidence for a diurnal rhythm of corticotropin-releasing factor (CRF) in the hypothalamus, *Endocrinology* **85:**861–866.

Hiroshige, T., and Sato, T., 1970, Circadian rhythm and stress-induced changes in hypothalamic content of corticotropin-releasing activity during postnatal development in the rat, *Endocrinology* **86:**1184–1186.

Kamstra, G. S., Thomas, P., and Sadow, J., 1983, Evaluation of changes in the secretion of corticotropin releasing activity using the isolated rat hypothalamus incubated *in vitro*, *J. Endocrinol.* **97:**291–300.

Moldow, R. L., and Fischman, A. J., 1984, Circadian rhythm of corticotropin-releasing factor-like immunoreactivity in rat hypothalamus, *Peptides* **5:**1213–1215.

Nicholson, S., Lin, J. H., Mahmoud, S., Campbell, E., Gilham, B., and Jones, M., 1985, Diurnal variations in responsiveness of the hypothalamo–pituitary–adrenocortical axis of the rat, *Neuroendocrinology* **40:**217–224.

Salacinski, P. R. P., McLean, C., Sykes, J. E., Clement-Jones, V. V., and Lowry, P. J., 1981, Iodination of proteins, glycoproteins, and peptides using a solid-phase oxidizing agent, 1,3,4,6-tetrachloro-3β,6β-diphenyl glycoluril (Iodogen®b), *Ann. Biochem.* **117:**136–146.

Takebe, K., Sakakura, M., and Mashimo, K., 1972, Continuance of diurnal rhythmicity of CRF activity in hypophysectomized rats, *Endocrinology* **90:**1515–1520.

Vale, W., Spiess, J., Rivier, C., and Rivier, J., 1981, Characterization of a 41-residue ovine hypothalamic peptide that stimulates secretion of corticotropin and β-endorphin, *Science* **213:**1394–1397.

Vale, W., Rivier, C., Brown, M. R., Spiess, J., Koob, G., Swanson, L., Bilezikjian, L., Bloom, F., and Rivier, J., 1983, Chemical and biological characterization of corticotropin releasing factor, *Recent Prog. Horm. Res.* **39:**245–266.

13

Ultrastructural Studies of Peptide Coexistence in Corticotropin-Releasing Factor- and Arginine-Vasopressin-Containing Neurons

MARK H. WHITNALL and HAROLD GAINER

1. INTRODUCTION

It has become clear over the last several years that the coexistence of more than one transmitter in the same neuron is a widespread phenomenon in the nervous system, yet the physiological implications of this coexistence are only beginning to be understood (Hökfelt *et al.*, 1984). A complete description of the roles of coexisting neurotransmitters should include (1) knowledge of the locations, actions, and dose–response curves of the receptors for the multiple transmitters, (2) analysis of the quantity and kinetics of release of the transmitters, and (3) determination of whether the release of multiple transmitters in the same cell can be independently regulated. An important determinant of the regulation of release of transmitters is their intracellular packaging. If different transmitters are localized in different populations of secretory vesicles in the same nerve ending, then the possibility of separate regulation of their release exists. However, if different transmitters are packaged in the same secretory vesicles, then independent regulation of their release by the nerve ending is unlikely.

The most direct technique for studying the intracellular localization of multiple transmitters is ultrastructural immunocytochemistry. This technique has been used to determine that immunoreactive substance P and serotonin are present in the same dense-core vesicles in rat raphe nuclei and dorsal horn (Pelletier *et al.*, 1981) and

MARK H. WHITNALL and HAROLD GAINER • Laboratory of Neurochemistry and Neuroimmunology, National Institute of Child Health and Human Development, National Institutes of Health, Bethesda, Maryland 20892.

that substances related to enkephalin and somatostatin coexist in the same neurosecretory vesicles (NSVs) in the external zone (EZ) of the median eminence (ME) in guinea pig (Beauvillain *et al.*, 1984). We have used this technique in two model neurosecretory systems to study the intracellular packaging of coexisting neuropeptides: the hypothalamo–neurohypophyseal system (HNS) and the arginine vasopressin (AVP)/corticotropin-releasing factor (CRF) projection to the hypophyseal portal capillaries in the EZ of the ME.

The HNS is a classic system for studying the synthesis, packaging, axonal transport, processing, storage, and release of neuropeptides (Castel *et al.*, 1984). Arginine vasopressin and oxytocin (OT) are synthesized in separate cells in the paraventricular nucleus (PVN) and supraoptic nucleus (SON) of the hypothalamus. These peptides are synthesized as parts of precursor molecules containing the sequences of their "carrier" proteins, the neurophysins (NP-AVP and NP-OT), and are packaged into 160-nm neurosecretory vesicles (NSVs). The prohormones are posttranslationally processed during axonal transport to the posterior pituitary, where AVP, OT, and the NPs are released near capillaries. A number of other neuropeptides have been reported to coexist with AVP in magnocellular neurosecretory cells, including dynorphin (Watson *et al.*, 1983; Whitnall *et al.*, 1983, 1985a), enkephalin (Coulter *et al.*, 1981), leucine-enkephalin (Martin *et al.*, 1983), and substance P (Stoeckel *et al.*, 1982). Similarly, OT magnocellular neurons have been found to contain enkephalin (Coulter *et al.*, 1981), methionine- (and possibly leucine-) enkephalin (Martin *et al.*, 1983), cholecystokinin (Vanderhaeghen *et al.*, 1980; Martin *et al.*, 1983), and CRF (Burlet *et al.*, 1983; Dreyfuss *et al.*, 1984; Sawchenko and Swanson, 1985; Castel *et al.*, 1985).

On the ultrastructural level, CRF (Dreyfuss *et al.*, 1984; Castel *et al.*, 1985), cholecystokinin (Martin *et al.*, 1983), and methionine- (and possibly leucine-) enkephalin-related material (Coulter *et al.*, 1981; Martin *et al.*, 1983) have been found within OT NSVs, and dynorphin (Whitnall *et al.*, 1983, 1985a) and leucine-enkephalin-related material (Martin *et al.*, 1983) were found in AVP NSVs. Neurohypophyseal levels of dynorphin, leucine-enkephalin, and methionine-enkephalin fall in response to water deprivation, as do those of AVP and OT (Zamir *et al.*, 1985). These results indicate simultaneous release of opioid and other peptides from the AVP- and OT-containing terminals in the neurohypophysis. However, the functional significance of these examples of coexistence will not be known until the receptors for the coexisting peptides in the pituitary are more fully localized and characterized (see references in Whitnall *et al.*, 1983; Zamir *et al.*, 1985).

One of the advantages of the AVP-containing neurosecretory systems as models for studying peptidergic neurons is that a rat strain (Brattleboro) exists that does not synthesize AVP (Sokol and Valtin, 1982). Dynorphin is present in normal AVP-containing magnocellular neurons at a molar concentration of about 0.5–1% that of AVP (Cox *et al.*, 1980; Geis *et al.*, 1982; Molineaux *et al.*, 1982), and the homozygous Brattleboro (DI) rat contains 25–100% of the normal level of dynorphin in its HNS (Cox *et al.*, 1980; Geis *et al.*, 1982). The AVP-deficient cells in DI rats thus presented an opportunity to study the effect of a drastic reduction in total peptide synthesis on the formation and morphology of NSVs. In particular, we wished to determine whether the small (100-nm) dense-core vesicles that had been observed in the AVP-deficient magnocellular neurons (Morris, 1982) were dynorphin-containing NSVs. We used postembedding electron microscopic (EM) immunocytochemistry (ICC) to determine that dynorphin is contained exclusively within 100-nm NSVs in DI rats as compared to 160-nm NSVs in controls (Whitnall *et al.*, 1985a). The results indicate that NSV size is partially regulated by the total level of peptide synthesis of a neurosecretory cell.

The AVP/CRF projection to the hypophyseal portal system has long been a subject of interest with regard to the regulation of corticotropin (ACTH) release from the anterior pituitary (Zimmerman *et al.*, 1977). After CRF was isolated and characterized, and CRF antibodies became available (Vale *et al.*, 1981, 1983a), it was possible to determine immunocytochemically that CRF and AVP are coexisting peptides in the same neurons (Tramu *et al.*, 1983; Kiss *et al.*, 1984; Sawchenko *et al.*, 1984a). The CRF/AVP system is thus a good model for relating the physiology of release of coexisting transmitters to their effects on target cells, given the intense ongoing research devoted to characterizing the effects of AVP and CRF on the corticotrophs in the anterior pituitary (see below).

2. THE ROLE AND ORGANIZATION OF CRF- AND AVP-CONTAINING NERVE ENDINGS IN THE EXTERNAL ZONE OF THE MEDIAN EMINENCE

2.1. The Parvocellular Projection to the Hypophyseal Portal Capillaries

It was already suspected in the 1950s that AVP played a role in the regulation of ACTH secretion, but the pharmacological differences between AVP and unidentified corticotropin-releasing substances in hypothalamic extracts made it unlikely that AVP was the principal CRF (Zimmerman *et al.*, 1977). One finding that strengthened the hypothesis that AVP participates in the control of ACTH release was the immunocytochemical demonstration of AVP- or NP-containing nerve fibers terminating near the bed of hypophyseal portal capillaries in the EZ of the ME (Chateau *et al.*, 1974; Vandesande *et al.*, 1974; Watkins *et al.*, 1974; Silverman and Zimmerman, 1975; Dube *et al.*, 1976; Stillman *et al.*, 1977; Tramu and Pillez, 1982). Further support for the involvement of AVP came from the finding that adrenalectomy causes a striking increase in the amount of AVP or NP in the EZ of the ME (Chateau *et al.*, 1974; Vandesande *et al.*, 1974; Watkins *et al.*, 1974; Dube *et al.*, 1976; Stillman *et al.*, 1977; Robinson *et al.*, 1983) because of an increase in prohormone synthesis (Russell *et al.*, 1980), which can be prevented by administration of glucocorticoids (Stillman *et al.*, 1977). The NP-containing fibers in the EZ contain NP-AVP and AVP almost exclusively (as opposed to NP-OT and OT), as demonstrated by the facts that (1) OT fibers in the EZ are extremely scarce and are not affected by adrenalectomy and (2) AVP-deficient DI rats possess only a few NP fibers in the EZ, and the amount of stainable NP fibers in DI rats is not affected by adrenalectomy (Stillman *et al.*, 1977).

Corticotropin-releasing factor is also present in the EZ of the ME (Antoni *et al.*, 1983; Bloom *et al.*, 1982; Bugnon *et al.*, 1982, 1983; Burlet *et al.*, 1983; Leranth *et al.*, 1983; Liposits *et al.*, 1983; Merchenthaler *et al.*, 1982; Olschowka *et al.*, 1982; Pelletier *et al.*, 1982; Swanson *et al.*, 1983; Tramu and Pillez, 1982). After adrenalectomy, the EZ contents of CRF and AVP undergo a rapid depletion (within 12 hr) through release, followed by a period of accumulation (Bugnon *et al.*, 1983; Tramu and Pillez, 1982; Suda *et al.*, 1983; Schipper *et al.*, 1984). The initial depletion is prevented by glucocorticoid treatment (Bugnon *et al.*, 1983; Schipper *et al.*, 1984).

CRF-immunoreactive cell bodies are present in many areas of the rat brain, with the densest concentration found in the parvocellular regions of the PVN (Antoni *et al.*, 1983; Bloom *et al.*, 1982; Bugnon *et al.*, 1982; Burlet *et al.*, 1983; Hökfelt *et al.*, 1983; Leranth *et al.*, 1983; Liposits *et al.*, 1983; Merchenthaler *et al.*, 1982; Olschowka *et al.*, 1982; Pelletier *et al.*, 1982; Swanson *et al.*, 1983). Adrenalectomy causes the CRF-positive parvocellular cell bodies in the PVN to accumulate detectable levels of AVP (Tramu *et al.*, 1983; Kiss *et al.*, 1984; Sawchenko *et al.*, 1984a) and AVP messenger RNA (Wolfson *et al.*, 1985) between 5 and 60 days after surgery. Adrenalectomy also causes enhancement of CRF immunoreactivity in these cell bodies, and the increases in both peptides after adrenalectomy can be prevented by glucocorticoid treatment (Sawchenko and Swanson, 1985). A variety of ICC, tracing, and lesion studies have established the PVN as the major source of CRF- and AVP-containing fibers in the EZ of the ME (Antoni *et al.*, 1983; Armstrong *et al.*, 1980; Bloom *et al.*, 1982; Bruhn *et al.*, 1984; Ixart *et al.*, 1982; Lechan *et al.*, 1982; Makara, 1985; Swanson and Kuypers, 1980; Wiegand and Price, 1980; Zimmerman *et al.*, 1977).

2.2. Physiological Evidence for Synergistic Actions of AVP and CRF in Modulating ACTH Release

Although there are probably less potent corticotropin-releasing substances in addition to CRF and AVP (Vale *et al.*, 1983b; Gillies *et al.*, 1984), the present discussion focuses on these two peptides. Observations that stress results in the simultaneous release of AVP from the HNS and elevations of plasma ACTH led to the speculation in the 1950s that AVP might cause the release of ACTH (Martini, 1966). Correlations between deficits in the AVP content of the HNS and diminished adrenocortical responses to stress and the ability of large systemic doses of AVP to cause ACTH release strengthened this hypothesis. During the late 1950s and 1960s it became clear that a CRF could be extracted from hypothalamic tissue that was not AVP or any other known hormone but that AVP could cause release of ACTH *in vitro* and could potentiate the actions of the CRF-like hypothalamic extracts *in vivo* (Yates *et al.*, 1971; Yates and Maran, 1974). Physiologically significant levels of immunoreactive CRF (0.1 nM) and AVP (1 nM) have now been found in portal blood in a number of laboratories (Zimmerman *et al.*, 1977; Gibbs, 1985a; Plotsky, 1985). The levels of CRF and AVP rise and fall in portal blood in response to various types of stress (Gibbs, 1985a; Plotsky, 1985). Interestingly, there seem to be some types of stress in which the level of AVP in portal blood is modulated, but not that of CRF (Gibbs, 1985b; Plotsky, 1985).

Administration of CRF or AVP peripherally (Yates *et al.*, 1971; C. Rivier *et al.*, 1984) or CRF centrally (Rock *et al.*, 1984) causes a rise in plasma ACTH. After peripheral administration, the two peptides act in a synergistic manner; i.e., at certain doses of AVP and CRF, the combined effect is much greater than the sum of the effects of the two peptides alone (Yates *et al.*, 1971; C. Rivier *et al.*, 1984). The ACTH response to stress can be diminished by peripheral administration of antisera or antagonists to AVP (C. Rivier *et al.*, 1984; Linton *et al.*, 1985) or CRF (C. Rivier *et al.*, 1982; J. Rivier *et al.*, 1984; Linton *et al.*, 1985; Nakane *et al.*, 1985). Studies in AVP-deficient Brattleboro (DI) rats are consistent with a potentiating role for AVP in regulation of ACTH release (Sokol and Valtin, 1982, part VI). However, the *in vivo* effects of CRF and AVP are complicated by the various possible sites of action of these agents. For example, there is evidence that central AVP inhibits release of CRF, and vice versa (Plotsky *et al.*, 1984, 1985). However, evidence for direct synergistic effects of AVP and CRF on anterior pituitary corticotrophs comes from *in vitro* studies (see below).

Gillies and Lowry (1979) demonstrated that AVP and other hypothalamic factors potentiated each others' abilities to cause ACTH release from isolated anterior pituitary cells. It has now been shown that although it is a much weaker ACTH secretagogue than CRF, AVP strongly potentiates the action of CRF in stimulating ACTH release *in vitro* (Gillies *et al.*, 1982, 1984; Turkelson *et al.*, 1982; Vale *et al.*, 1983b). *In vitro* binding studies have revealed specific and distinct high-affinity ($K_d < 1$ nM) receptors for AVP (Antoni, 1984; Antoni *et al.*, 1985; Lutz-Bucher and Koch, 1983; Koch and Lutz-Bucher, 1985) and CRF (Holmes *et al.*, 1984; Wynn *et al.*, 1983, 1985) on anterior pituitary cells. Binding and ACTH release studies show that the receptor for AVP is not identical to the "V_1" pressor receptor or to the "V_2" antidiuretic receptor (Antoni, 1984; Antoni *et al.*, 1984). Both the AVP and the CRF receptors can be down-regulated by their homologous ligands but not by ligands for the other receptor (Antoni *et al.*, 1985; Holmes *et al.*, 1984; Koch and Lutz-Bucher, 1985). Cyclic AMP has been implicated in mediating ACTH release (Vale and Rivier, 1977). Although AVP alone does not affect the basal level of adenylate cyclase activity in anterior pituitary cells, it potentiates the CRF-induced stimulation of adenylate cyclase in whole cells (Giguere and Labrie, 1982) but not in membrane preparations (Holmes *et al.*, 1984).

Adrenalectomy results in decreases in AVP receptor density (Koch and Lutz-Bucher, 1985; Antoni *et al.*, 1985) and CRF receptor density (Wynn *et al.*, 1985) on anterior pituitary cells by 80%. Nevertheless, down-regulated cultured anterior pituitary cells from adrenalectomized rats exhibited a threefold increase in CRF-stimulated ACTH release despite a 40% reduction in CRF-induced cAMP accumulation (Wynn *et al.*, 1985). These results were interpreted to mean that the corticotrophs in adrenalectomized animals possess elevated ACTH levels and secretory capacity (consistent with the ICC findings of Westlund *et al.*, 1985) and that only a small number of CRF receptors need be occupied to elicit maximum ACTH release (Wynn *et al.*, 1985).

In spite of the increasing amount of information available on the actions of AVP and CRF on the anterior pituitary, their roles in regulating ACTH release in various physiological states will not be understood until it is known what controls their release into portal blood. The pools of AVP and CRF that can contribute to the portal system must be defined, and the physiology of their release must be characterized. We present below the results of an ultrastructural analysis that shed some light on the organization of pools of CRF and AVP in the EZ of the ME.

2.3. Colocalization of AVP and CRF in Neurosecretory Vesicles in the External Zone of the Median Eminence: Implications for Regulation of ACTH Release

Corticotropin-releasing factor (Pelletier *et al.*, 1982; Leranth *et al.*, 1983; Liposits *et al.*, 1983), AVP, and NP-AVP (Silverman and Zimmerman, 1975; Dube *et al.*, 1976) are contained within 100-nm NSVs in the EZ of the ME. However, it was not known whether these peptides are localized within the same NSVs or even the same axons, although light microscopic evidence indicated that NP may be located within some of the CRF-containing fibers (Tramu and Pillez, 1982). We felt it important to determine whether the two peptides are copackaged in the same NSVs in the EZ, since this would indicate that they are released concomitantly into the portal blood. We demonstrated, using postembedding EM ICC, that AVP and NP-AVP are present within a subpopulation of CRF-containing axons in the EZ of normal rats and that they are copackaged with CRF in the same 100-nm NSVs (Whitnall *et al.*, 1985b). In adrenalectomized rats, almost every CRF-positive axon in the EZ of the ME was immunoreactive for NP-AVP and AVP, again colocalized in CRF-containing NSVs (Whitnall *et al.*, 1985b).

Adult male rats were decapitated and their brains immersed in 4% glutaraldehyde, 0.2% picric acid, 0.1 M sodium cacodylate, pH 6. In some cases, bilateral adrenalectomy was performed under ether anesthesia 3 days before fixation. Horizontal ventral surface slabs containing the ME were then dissected out, and some blocks were postfixed in 1% osmium tetroxide. Tissues were embedded in LR white resin (Newman *et al.*, 1983), and serial sections were picked up on parlodion-coated nickel slot grids so that staining could be done on the back of one section and the front of the next section. Immunoperoxidase staining of the sections was performed as previously described (Whitnall *et al.*, 1985a). Sections of osmicated tissue were preincubated in sodium metaperiodate to restore antigenicity (Bendayan and Zollinger, 1983). No differences in staining patterns for CRF and NP-AVP were seen between the osmicated and unosmicated tissues (Table I), but the parvocellular axons did not stain for AVP after osmication.

TABLE I. Axonal Profiles in the EZ Staining for CRF and AVP or NP-AVP in Serial Sections from Normal and Adrenalectomized Rats[a]

Staining	Numbers (and percentages) of axonal profiles			
	Normal[f]		Adrenalectomized[g]	
	n	(%)	*n*	(%)
CRF +, NP or AVP +[b]	82	(43)	141	(89)
CRF +, NP or AVP ±[c]	48	(25)	9	(6)
CRF +, NP or AVP −[d]	57	(30)	7	(4)
CRF −, NP +[e]	3	(2)	1	(1)
Total	190		158	

[a] Profiles of parvocellular axons, axonal swellings, and terminals were pooled.
[b] Intensely stained for both CRF and NP or AVP (cf. Figs. 1–3).
[c] Intensely stained for CRF, faintly stained for NP or AVP (cf. Fig. 2).
[d] Intensely stained for CRF, unstained for NP or AVP (cf. Fig. 3).
[e] Unstained for CRF, intensely stained for NP.
[f] Data from osmicated tissue (110 profiles) and unosmicated tissue (80 profiles) were similar and were pooled; 45% of the profiles in osmicated tissue and 40% of the profiles in unosmicated tissue stained intensely for both CRF and NP. Data are from sections stained for NP only, not AVP.
[g] Three days after adrenalectomy. Data are from unosmicated tissue only. Results from sections stained for NP (115 profiles) and from sections stained for AVP (43 profiles) were pooled. The percentages of profiles in the four categories in the sections stained for NP were 86, 7, 6, and 1, respectively. The corresponding percentages from sections stained for AVP were 98, 2, 0, and 0.

The CRF antibody used in these studies was a rabbit antiserum directed against rat CRF, and its immunocytochemical specificity has been characterized elsewhere (Sawchenko *et al.*, 1984b). This CRF antiserum was used at 1:800 and stained 100-nm NSVs in the EZ of the ME (Fig. 1B). Magnocellular axons (containing 160-nm NSVs) in the internal zone of the ME did not stain for CRF under these conditions (Fig. 1B), and staining of the parvocellular axons was completely blocked by the addition of 5×10^{-5} M synthetic ovine CRF to the antiserum (Fig. 1C). An antiserum against AVP (obtained from G. Valiquette and E. A. Zimmerman) was used at 1 : 500. This antiserum is specific for AVP in radioimmunoassay (M. Altstein and G. Valiquette, personal communication) and does not stain the OT-containing axons in the internal zone of the ME as shown by serial section studies (not shown), and staining can be absorbed out with AVP-sepharose beads (not shown). We also used a monoclonal antibody (PS 45) that was highly specific for NP-AVP in liquid-phase assays (Ben-Barak *et al.*, 1985) to stain for NP-AVP (Fig. 1A). Since the EZ did not stain with a monoclonal antibody (PS 36) specific for NP-OT in ICC (Ben-Barak *et al.*, 1985; Whitnall *et al.*, 1985c), and OT is extremely scarce in the EZ (Zimmerman *et al.*, 1977), staining of the parvocellular axons with PS 45 indicates the presence of NP-AVP and AVP. Staining with PS 45 can be completely absorbed out with NP-AVP (not illustrated).

In these experiments, three types of CRF-positive parvocellular axons were observed in the EZ of the ME: (1) axons that stained intensely for both CRF and NP-AVP or AVP (Figs. 1–3); (2) axons that stained intensely for CRF but only faintly for NP-AVP or AVP (Fig. 2); and (3) axons that stained intensely for CRF

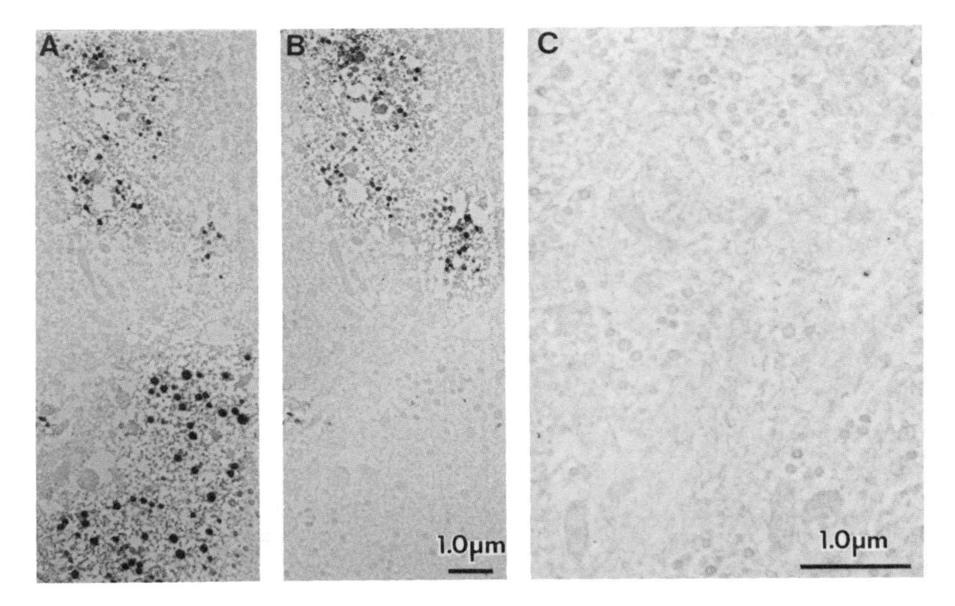

FIGURE 1. Colocalization of CRF and NP-AVP in the EZ of the ME of adrenalectomized rats. A and B are serial LR white sections of unosmicated tissue stained for (A) NP-AVP or (B) CRF. The parvocellular axons (top) contain 100-nm NSVs, which are stained for both NP-AVP and CRF. In contrast, the magnocellular axon near the edge of the EZ (bottom) contains 160-nm NSVs, which are stained for NP-AVP but not CRF. Absorption of the CRF antiserum with 5×10^{-5} M synthetic ovine CRF prevented staining of the parvocellular axons (C).

but did not contain immunocytochemically detectable NP-AVP or AVP (Fig. 3). In addition, a few axons were observed that contained NP-positive 100-nm NSVs but did not stain for CRF (Table I). When the distributions of the three main types of axons were compared in normal and adrenalectomized rats, a dramatic shift toward the first axon type (intensely stained for both peptides) was found to have occurred (Table I). The percentage of CRF-positive/NP-AVP- or AVP-positive axons increased from 43% to 89% 3 days after adrenalectomy (Table I). Whereas 55% of the CRF-positive axons in the EZ were unstained or faintly stained for NP-AVP in normal animals, only 10% were unstained or faintly stained for NP-AVP or AVP in adrenalectomized animals (Table I). This shift toward CRF-positive/NP-AVP- or AVP-positive axons is most likely responsible for the increase in NP or AVP in the EZ of the ME after adrenalectomy seen previously by light microscopic ICC (Chateau *et al.*, 1974; Vandesande *et al.*, 1974; Watkins *et al.*, 1974; Stillman *et al.*, 1977) and EM ICC (Dube *et al.*, 1976) and in the ME by biochemical techniques (Russell *et al.*, 1980; Robinson *et al.*, 1983).

The serial sections stained for CRF and NP-AVP revealed extensive colocalization of CRF and NP-AVP in the same NSVs, as shown by near-identical patterns of stained NSVs on each pair of serial sections (Figs. 2 and 3). Similar results were

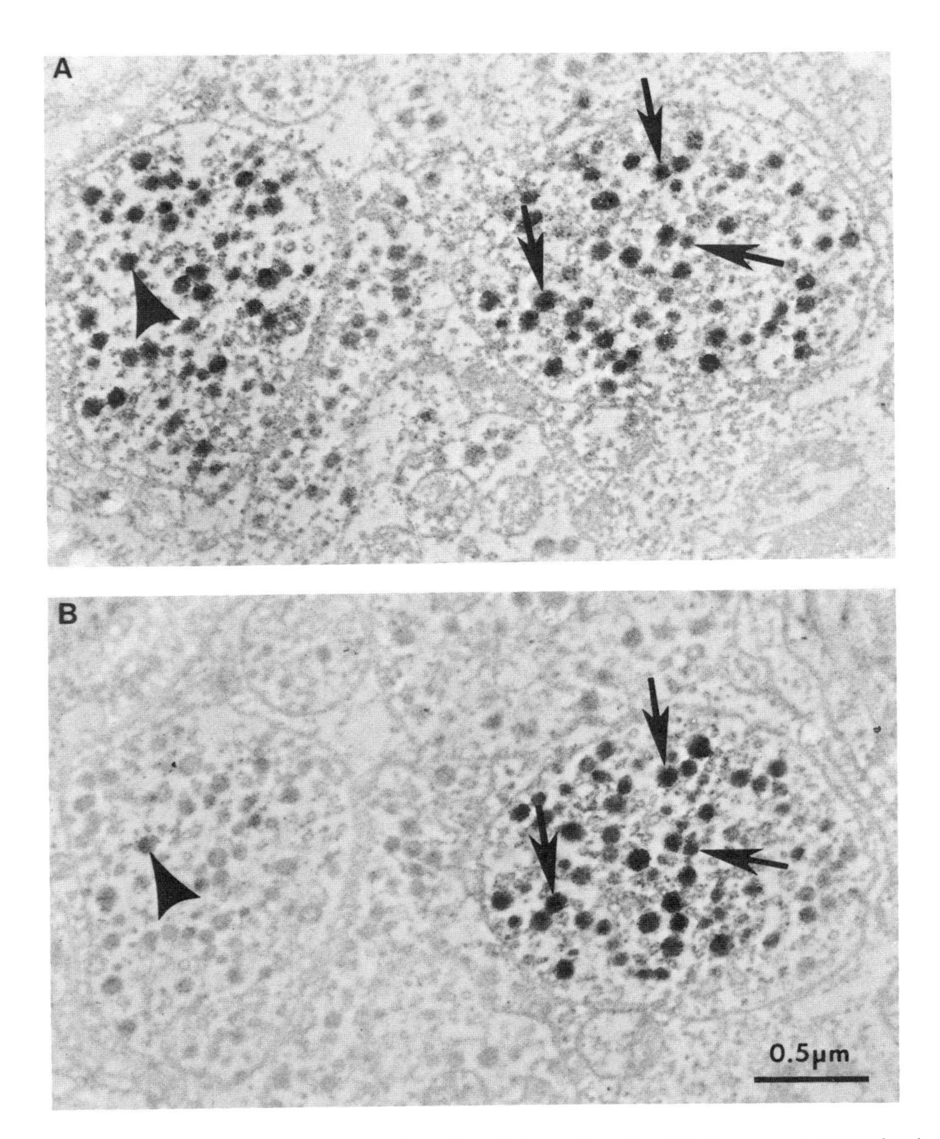

FIGURE 2. Serial LR white sections of osmicated tissue from the EZ of a normal rat, stained for (A) CRF or (B) NP-AVP. The axonal swelling on the right stained intensely for both CRF and NP-AVP. Examples of identical NSVs labeled on both sections are indicated by arrows. The swelling on the left stained intensely for CRF, but only a few NSVs are very faintly stained for NP-AVP. One NSV that stained intensely for CRF (A) and faintly for NP-AVP (B) is indicated by arrowheads.

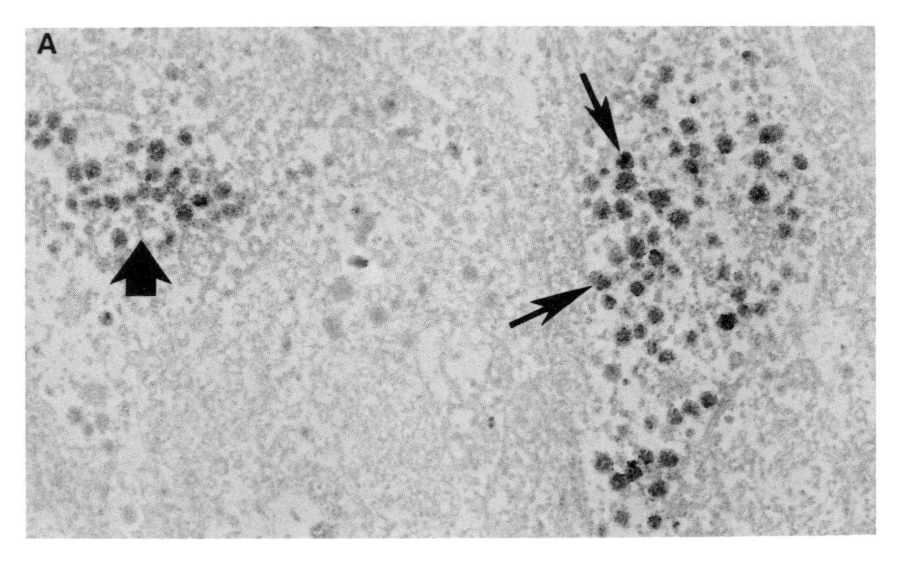

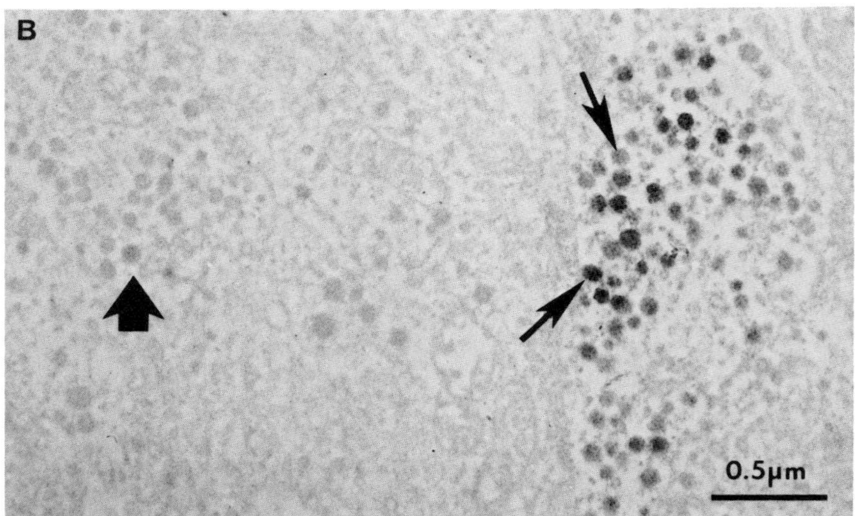

FIGURE 3. Example of an axonal profile in serial sections that stained intensely for CRF (A, large arrow) but did not stain for NP-AVP (B). The axonal swelling on the right stained intensely for both CRF (A) and NP-AVP (B). Examples of identical NSVs stained for both CRF and NP-AVP are indicated by small arrows. Serial LR white sections of osmicated tissue from the EZ of the ME of a normal rat.

obtained using the antiserum to AVP (not shown). Many of the axonal swellings also contained accumulations of small clear vesicles (Figs. 2 and 3), which are usually interpreted to indicate release sites in the posterior pituitary (Castel *et al.*, 1984).

The colocalization of CRF and AVP in the same 100-nm NSVs in axons in the EZ of the ME indicates that these peptides are released simultaneously to a large extent in normal rats and almost always in adrenalectomized rats. It may be significant that in normal rats, the CRF/AVP projection to the hypophyseal portal system can be divided into two populations of axons, i.e., axons that contain high amounts of both CRF and AVP and axons that contain abundant CRF but little or no AVP (Table I). If release from these two populations of axons is regulated independently, the ratio of CRF to AVP in portal blood could be modulated by differential activation of these two parvocellular populations. In adrenalectomized rats, on the other hand, the parvocellular PVN projection seems to be relatively homogeneous. Almost all the CRF-positive axons in the EZ of the ME contain stainable NP-AVP or AVP. If independent modulation of CRF and AVP release from the parvocellular system is of physiological significance in the normal rat, it is apparently overridden by the strong stimulus for maximum ACTH secretion after removal of circulating adrenal corticosteroids.

Especially intriguing in light of the present results are the findings that portal AVP levels can be modulated in certain stress conditions in normal animals while CRF levels remain steady (Gibbs, 1985b; Plotsky, 1985). Two independent mechanisms could account for such a situation. Given the two populations of CRF-containing axons discussed above for the normal rat, an increase in secretion rate by the AVP- and CRF-containing axons accompanied by an equivalent decrease in secretion rate by the CRF (only) axons could produce an increase in AVP but no change in CRF levels in portal blood. A reversal in this pattern of secretion rates could produce a decrease in AVP levels in portal blood with no change in CRF. Alternatively, the parvocellular axons in the EZ might maintain base-line levels of release, and another source of AVP could increase or decrease its contribution to portal blood.

This mechanism is related to various reports in the literature indicating a contribution of the magnocellular axons in the HNS to portal AVP. Oxytocin is present in portal plasma at a concentration similar to that of AVP (Gibbs, 1985a; Plotsky, 1985). Since OT is present in large amounts in the magnocellular system but absent in the parvocellular system, the HNS is a likely source. The levels of OT and AVP in portal plasma both rise in response to certain conditions that cause elevations in circulating ACTH (Gibbs, 1985a; Plotsky, 1985), and in hypothermia, although peripheral ACTH falls and portal CRF remains constant, both AVP and OT decrease in parallel in portal and peripheral blood (Gibbs, 1985b). Possible mechanisms by which AVP and OT might reach portal plasma from the HNS are by retrograde flow from the neurohypophysis or by secretion from the internal zone of the ME. The latter has been suggested by Holmes *et al.* (1986).

3. CONCLUSIONS

The physiological roles of coexisting transmitters must be considered in the context of their subcellular localization, since this factor limits the mechanisms by which they can be released. Therefore, some important goals in the study of coexisting transmitters are the identification of the populations of secretory vesicles that exist in neurons that contain these transmitters, the ways these transmitters are synthesized, processed, and stored in vesicles, and the specific physiological conditions that cause the release of each population of vesicles. Ultrastructural ICC, in combination with physiological and biochemical techniques, will undoubtedly prove to be a valuable tool in the study of these problems. Neurosecretory cells are attractive models in which to study the cell biology of secretory vesicles because of their high rates of synthesis of neuropeptides and their dramatic responses to physiological conditions (Castel *et al.,* 1984). In the examples discussed above, coexisting peptides have been found to be contained within the same NSVs, indicating simultaneous release. Other neuronal types exist, however, in which subpopulations of secretory vesicles containing different transmitters probably exist, e.g., neurons that contain a classical transmitter and one or more peptides (Hökfelt *et al.,* 1984). It will be important to correlate the physiological effects of various conditions of stimulation of these neurons on their target cells with ultrastructural studies of vesicle subpopulations and biochemical measurements of transmitter release.

ACKNOWLEDGMENTS. We thank Dr. Eva Mezey for assistance with the adrenalectomy experiments, Dr. W. Vale for donating the CRF antiserum, Drs. Guy Valiquette and Earl Zimmerman for donating the antiserum to AVP, Drs. Mona Castel and Ferenz Antoni for providing copies of manuscripts submitted for publication and in press, and Sharon Key for technical assistance.

REFERENCES

Antoni, F. A., 1984, Novel ligand specificity of pituitary vasopressin receptors in the rat, *Neuroendocrinology* **39:**186–188.

Antoni, F. A., Palkovits, M., Makara, G. B., Linton, E. A., Lowry, P. J., and Kiss, J. Z., 1983, Immunoreactive corticotropin-releasing hormone in the hypothalamoinfundibular tract, *Neuroendocrinology* **36:**415–423.

Antoni, F. A., Holmes, M. C., Makara, G. B., Karteszi, M., and Laszlo, F. A., 1984, Evidence that the effects of arginine-8-vasopressin (AVP) on pituitary corticotropin (ACTH) release are mediated by a novel type of receptor, *Peptides* **5:**519–522.

Antoni, F. A., Holmes, M. C., and Kiss, J. Z., 1985, Pituitary binding of vasopressin is altered by experimental manipulations of the hypothalamo–pituitary–adrenocortical axis in normal as well as homozygous (*di/di*) Brattleboro rats, *Endocrinology* **117:**1293–1299.

Armstrong, W. E., Warach, S., Hatton, G. I., and McNeill, T. H., 1980, Subnuclei in the rat hypothalamic paraventricular nucleus. A cytoarchitectural, horseradish perioxidase and immunocytochemical analysis, *Neuroscience* **5:**1931–1958.

Beauvillain, J.-C., Tramu, G., and Garaud, J.-C., 1984, Coexistence of substances related to enkephalin and somatostatin in granules of the guinea-pig median eminence: Demonstration by use of colloidal gold immunocytochemical methods, *Brain Res.* **301:**389–393.

Ben-Barak, Y., Russell, J. T., Whitnall, M. H., Ozato, K., and Gainer, H., 1985, Neurophysin in the hypothalamo–neurohypophysial system. I. Production and characterization of monoclonal antibodies, *J. Neurosci.* **5:**81–97.

Bendayan, M., and Zollinger, M., 1983, Ultrastructural localization of antigenic sites on osmium-fixed tissues applying the protein A–gold technique, *J. Histochem. Cytochem.* **31:**101–109.

Bloom, F. E., Battenberg, E. L. F., Rivier, J., and Vale, W., 1982, Corticotropin releasing factor (CRF) immunoreactive neurons and fibers in rat hypothalamus, *Regul. Peptides* **4:**43–48.

Bruhn, T. O., Sutton, R. E., Rivier, C. L., and Vale, W. W., 1984, Corticotropin-releasing factor regulates proopiomelanocortin messenger ribonucleic acid levels *in vivo, Neuroendocrinology* **39:**170–175.

Bugnon, C., Fellman, D., Gouget, A., and Cardot, J., 1982, Corticoliberin in rat brain: Immunocytochemical identification and localization of a novel neuropeptide system, *Neurosci. Lett.* **30:**25–30.

Bugnon, C., Fellmann, D., and Gouget, A., 1983, Changes in corticoliberin and vasopressin-like immunoreactivities in the zona externa of the median eminence in adrenalectomized rats. Immunocytochemical study, *Neurosci. Lett.* **37:**43–49.

Burlet, A., Tonon, M.-C., Tankosic, P., Coy, D., and Vaudry, H., 1983, Comparative immunocytochemical localization of corticotropin releasing factor (CRF-41) and neurohypophysial peptides in the brain of Brattleboro and Long Evans rats, *Neuroendocrinology* **37:**64–72.

Castel, M., Gainer, H., and Dellman, H.-D., 1984, Neuronal secretory systems, *Int. Rev. Cytol.* **88:**303–459.

Castel, M., Varndell, I. M., Aguilera, G., and Polak, J. M., 1985, Colocalization of corticotropin releasing factor (CRF) and the neurohypophysial hormones, *Neurosci. Lett.* (Suppl.) **22:**s118.

Chateau, M., Burlet, A., and Marchetti, J., 1974, La "vasopressine-like" du lobe anterieur de l'hypophyse: Isolement et identification par son activite biologique et immunologique, *J. Physiol. (Paris)* **68:**10B–11B.

Coulter, H. D., Elde, R. P., and Unvezagt, S. L., 1981, Co-localization of neurophysin- and enkephalin-like immunoreactivity in cat pituitary, *Peptides* **2**(Suppl. 1)**:**51–55.

Cox, B. M., Ghazarossian, V. E., and Goldstein, A., 1980, Levels of immunoreactive dynorphin in brain and pituitary of Brattleboro rats, *Neurosci. Lett.* **20:**85–88.

Dreyfuss, F., Burlet, A., Tonon, M. C., and Vaudry, H., 1984, Comparative immunoelectron microscopic localization of corticotroping-releasing factor (CRF-41) and oxytocin in the rat median eminence, *Neuroendocrinology* **39:**284–287.

Dube, D., Leclerc, R., and Pelletier, G., 1976, Electron microscopic immunohistochemical localization of vasopressin and neurophysin in the median eminence of normal and adrenalectomized rats, *Am. J. Anat.* **147:**103–108.

Geis, R., Weber, E., Martin, R., and Voigt, K. H., 1982, Hypothalamo–posterior pituitary system in Brattleboro rats: Immunoreactive levels of leucine-enkephalin, dynorphin (1–17), dynorphin (1–8) and α-neo-endorphin, *Life Sci.* **31:**1809–1812.

Gibbs, D. M., 1985a, Measurement of hypothalamic corticotropin-releasing factors in hypophyseal portal blood, *Fed. Proc.* **44:**203–206.

Gibbs, D. M., 1985b, Inhibition of corticotropin release during hypothermia: The role of corticotropin-releasing factor, vasopressin, and oxytocin, *Endocrinology* **116:**723–727.

Giguere, V., and Labrie, F., 1982, Vasopressin potentiates cyclic AMP accumulation and ACTH release induced by corticotropin releasing factor (CRF) in rat anterior pituitary cells in culture, *Endocrinology* **111:**1752–1754.

Gillies, G., and Lowry, P., 1979, Corticotropin releasing factor may be modulated by vasopressin, *Nature* **278:**463–464.

Gillies, G. E., Linton, E. A., and Lowry, P. J., 1982, Corticotropin releasing activity of the new CRF is potentiated several times by vasopressin, *Nature* **299:**355–357.

Gillies, G. E., Puri, A., Linton, E. A., and Lowry, P. J., 1984, Comparative chromatography of hypothalamic corticotropin-releasing factors, *Neuroendocrinology* **38:**17–24.

Hökfelt, T., Fahrenkrug, J., Tatemoto, K., Mutt, V., Werner, S., Hulting, A.-L., Terenius, L., and Chang, K. J., 1983, The PHI (PHI-27)/corticotropin-releasing factor/enkephalin immunoreactive hypothalamic neuron: Possible morphological basis for integrated control of prolactin, corticotropin, and growth hormone secretion, *Proc. Natl. Acad. Sci. U.S.A.* **80:**895–898.

Hökfelt, T., Johansson, O., and Goldstein, M., 1984, Chemical anatomy of the brain, *Science* **225:**1326–1334.

Holmes, M. C., Antoni, F. A., and Szentendrei, T., 1984, Pituitary receptors for corticotropin-releasing factor: No effect of vasopressin on binding or activation of adenylate cyclase, *Neuroendocrinology* **39:**162–169.

Holmes, M. C., Antoni, F. A., Aguilera, G., and Catt, K. J., 1986, Magnocellular axons in passage through the median eminence release vasopressin, *Nature* **319:**326–329.

Ixart, G., Alonso, G., Szafarzyk, A., Malaval, F., Nouguier-Soule, J., and Assenmacher, I., 1982, Adrenocorticotropic regulations after bilateral lesions of the paraventricular or supraoptic nuclei and in Brattleboro rats, *Neuroendocrinology* **35:**270–276.

Kiss, J. Z., Mezey, E., and Skirboll, L., 1984, Corticotropin-releasing factor-immunoreactive neurons of the paraventricular nucleus become vasopressin positive after adrenalectomy, *Proc. Natl. Acad. Sci. U.S.A.* **81:**1854–1858.

Koch, B., and Lutz-Bucher, B., 1985, Specific receptors for vasopressin in the pituitary gland: Evidence for down-regulation and desensitization to adrenocorticotropin-releasing factors, *Endocrinology* **116:**671–676.

Lechan, R. M., Nestler, J. L., and Jacobson, S., 1982, The tuberoinfundibular system of the rat as demonstrated by immunohistochemical localization of retrogradely transported wheat germ agglutinin (WGA) from the median eminence, *Brain Res.* **245:**1–15.

Leranth, C., Antoni, F. A., and Palkovits, M., 1983, Ultrastructural demonstration of ovine CRF-like immunoreactivity (oCRF-LI) in the rat, *Regul. Peptides* **6:**179–188.

Linton, E. A., Tilders, F. J. H., Hodgkinson, S., Berkenbosch, F., Vermes, I., and Lowry, P. J., 1985, Stress-induced secretion of adrenocorticotropin in rats is inhibited by administration of antisera to ovine corticotropin-releasing factor and vasopressin, *Endocrinology* **116:**966–970.

Liposits, Z., Gorcs, T., Setalo, G., Lenguari, I., Flerko, B., Vigh, S., and Schally, A. V., 1983, Ultrastructural characteristics of immunolabeled, corticotropin releasing factor (CRF)-synthesizing neurons in the rat brain, *Cell Tissue Res.* **229:**191–196.

Lutz-Bucher, B., and Koch, B., 1983, Characterization of specific receptors for vasopressin in the pituitary gland, *Biochem. Biophys. Res. Commun.* **115:**492–498.

Makara, G. B., 1985, Mechanisms by which stressful stimuli activate the pituitary adrenal system, *Fed. Proc.* **44:**149–153.

Martin, R., Geis, R., Holl, R., Schafer, M., and Voigt, K. H., 1983, Coexistence of unrelated peptides in oxytocin and vasopressin terminals of rat neurohypophyses: Immunoreactive methionine^5-enkephalin-, leucine^5-enkephalin- and cholecystokinin-like substances, *Neuroscience* **8:**213–227.

Martini, L., 1966, Neurohypophysis and anterior pituitary activity, in: *The Pituitary Gland,* Vol. 3 (G. W. Harris and B. T. Donovan, eds.), University of California Press, Berkeley, pp. 535–577.

Merchenthaler, I., Vigh, S., Petrusz, P., and Schally, A. V., 1982, Immunocytochemical localization of corticotropin-releasing factor (CRF) in rat brain, *Am. J. Anat.* **165:**385–396.

Molineaux, C. J., Feuerstein, G., Faden, A. L., and Cox, B. M., 1982, Distribution of immunoreactive dynorphin in discrete brain nuclei; comparison with vasopressin, *Neurosci. Lett.* **33:**179–184.

Morris, J. F., 1982, The Brattleboro magnocellular neurosecretory system: A model for the study of peptidergic neurons, in: *The Brattleboro Rat. Annals of the New York Academy of Sciences,* Vol. 394 (H. W. Sokol and H. Valtin, eds.), The New York Academy of Sciences, New York, pp. 54–71.

Nakane, T., Audhya, T., Kanie, N., and Hollander, C. S., 1985, Evidence for a role of endogenous corticotropin-releasing factor in cold, ether, immobilization, and traumatic stress, *Proc. Natl. Acad. Sci. U.S.A.* **82:**1247–1251.

Newman, G. R., Jasani, B., and Williams, E. D., 1983, A simple post-embedding system for the rapid demonstration of tissue antigens under the electron microscope, *Histochem. J.* **15:**543–555.

Olschowka, J. A., O'Donohue, T. L., Mueller, G. P., and Jacobowitz, D. M., 1982, The distribution of corticotropin releasing factor-like immunoreactive neurons in rat brain, *Peptides* **3:**995–1015.

Pelletier, G., Steinbusch, H. W. M., and Verhofstad, A. A. J., 1981, Immunoreactive substance P and serotonin present in the same dense vesicles, *Nature* **293:**71–72.

Pelletier, G., Desy, L., Cote, J., Lefevre, G., Vaudry, G., and Labrie, F., 1982, Immunoelectron microscopic localization of corticotropin-releasing factor in the rat hypothalamus, *Neuroendocrinology* **35:**402–404.

Plotsky, P. M., 1985, Hypophyseotropic regulation of adenohypophyseal adrenocorticotropin secretion, *Fed. Proc.* **44:**207–213.

Plotsky, P. M., Bruhn, T. O., and Vale, W., 1984, Central modulation of immunoreactive corticotropin-releasing factor secretion by arginine vasopressin, *Endocrinology* **115:**1639–1641.

Plotsky, P. M., Bruhn, T. O., and Otto, S., 1985, Central modulation of immunoreactive arginine vasopressin and oxytocin secretion into the hypophysial-portal circulation by corticotropin-releasing factor, *Endocrinology* **116:**1669–1671.

Rivier, C., Rivier, J., and Vale, W., 1982, Inhibition of adrenocorticotropic hormone secretion in the rat by immunoneutralization of corticotropin-releasing factor, *Science* **218:**377–379.

Rivier, C., Rivier, J., Mormede, P., and Vale, W., 1984, Studies of the nature of the interaction between vasopressin and corticotropin-releasing factor on adrenocorticotropin release in the rat, *Endocrinology* **115:**882–886.

Rivier, J., Rivier, C., and Vale, W., 1984, Synthetic competitive antagonists of corticotropin-releasing factor: Effect on ACTH secretion in the rat, *Science* **224:**889–891.

Robinson, A. G., Seif, S. M., Verbalis, J. G., and Brownstein, M. J., 1983, Quantitation of changes in the content of neurohypophyseal peptides in hypothalamic nuclei after adrenalectomy, *Neuroendocrinology* **36:**347–350.

Rock, J. P., Oldfield, F. M., Schulte, H. M., Gold, P. W., Kornbluth, P. L., Loriaux, L., and Chrousos, G. P., 1984, Corticotropin releasing factor administered into the ventricular CSF stimulates the pituitary–adrenal axis, *Brain Res.* **323:**365–368.

Russell, J. T., Brownstein, M. J., and Gainer, H., 1980, [^{35}S]Cysteine-labeled peptides transported to the neurohypophyses of adrenalectomized, lactating, and Brattleboro rats, *Brain Res.* **201:**227–234.

Sawchenko, P. E., and Swanson, L. W., 1985, Localization, colocalization, and plasticity of corticotropin-releasing factor immunoreactivity in rat brain, *Fed. Proc.* **44:**221–227.

Sawchenko, P. E., Swanson, L. W., and Vale, W., 1984a, Co-expression of corticotropin-releasing factor and vasopressin immunoreactivity in parvocellular neurosecretory neurons of the adrenalectomized rat, *Proc. Natl. Acad. Sci. U.S.A.* **81:**1883–1887.

Sawchenko, P. E., Swanson, L. W., and Vale, W. W., 1984b, Corticotropin-releasing factor: Co-expression within distinct subsets of oxytocin-, vasopressin-, and neurotensin-immunoreactive neurons in the hypothalamus of the male rat, *J. Neurosci.* **4:**1118–1129.

Schipper, J., Werkman, T. R., and Tilders, F. J. H., 1984, Quantitative immunocytochemistry of corticotropin-releasing factor (CRF). Studies on nonbiological models and on hypothalamic tissues of rats after hypophysectomy, adrenalectomy, and dexamethasone treatment, *Brain Res.* **293:**111–118.

Silverman, A., and Zimmerman, E. A., 1975, Ultrastructural immunocytochemical localization of neurophysin and vasopressin in the median eminence and posterior pituitary of the guinea pig, *Cell Tissue Res.* **159:**291–301.

Sokol, H. W., and Valtin, H. (eds.), 1982, *The Brattleboro Rat. Annals of the New York Academy of Sciences,* Vol. 394, The New York Academy of Sciences, New York.

Stillman, M. A., Recht, L. D., Rosario, S. L., and Zimmerman, E. A., 1977, The effects of adrenalectomy and glucocorticoid replacement on vasopressin and vasopressin-neurophysin in the zona externa of the median eminence of the rat, *Endocrinology* **101:**42–49.

Stoeckel, M. E., Porte, A., Klein, M. J., and Cuello, A. C., 1982, Immunocytochemical localization of substance P in the neurohypophysis and hypothalamus of the mouse compared with the distribution of other neuropeptides, *Cell Tissue Res.* **223:**533–544.

Suda, T., Tamori, N., Tozawa, F., Mouri, T., Demura, H., and Shizume, K., 1983, Effects of bilateral adrenalectomy on immunoreactive corticotropin-releasing factor in the rat median eminence and intermediate-posterior pituitary, *Endocrinology* **113:**1182–1184.

Swanson, L. W., and Kuypers, H. G. J. M., 1980, The paraventricular nucleus of the hypothalamus: Cytoarchitectonic subdivisions and organization of projections to the pituitary, dorsal vagal complex, and spinal cord as demonstrated by retrograde fluorescence double-labeling methods, *J. Comp. Neurol.* **194:**555–570.

Swanson, L. W., Sawchenko, P. E., Rivier, J., and Vale, W. W., 1983, Organization of ovine corticotropin-releasing factor immunoreactive cells and fibers in the rat brain: An immunocytochemical study, *Neuroendocrinology* **36:**165–186.

Tramu, G., and Pillez, A., 1982, Localisation immunohistochimique des terminaisons nerveuses à corticoliberine (CRF) dans l'eminence mediane du cobaye et du rat, *C.R. Acad. Sci. (Paris)* **294:**107–114.

Tramu, G., Croix, C., and Pillez, A., 1983, Ability of the CRF immunoreactive neurons of the paraventricular nucleus to produce a vasopressin-like material, *Neuroendocrinology* **37:**467–469.

Turkelson, C. M., Thomas, C. R., Arimura, A., Chang, D., Chang, J. K., and Shimizu, M., 1982, *In vitro* potentiation of the ability of synthetic ovine corticotropin-releasing factor by arginine vasopressin, *Peptides* **3:**111–113.

Vale, W., and Rivier, C., 1977, Substances modulating the secretion of ACTH by cultured anterior pituitary cells, *Fed. Proc.* **36:**2049–2099.

Vale, W., Spiess, J., Rivier, C., and Rivier, J., 1981, Characterization of a 41-residue ovine hypothalamic peptide that stimulates secretion of corticotropin and β-endorphin, *Science* **213:**1394–1397.

Vale, W. W., Rivier, C., Spiess, J., and Rivier, J., 1983a, Corticotropin releasing factor, in: *Brain Peptides* (D. Krieger, J. B. Martin, and M. J. Brownstein, eds.), John Wiley & Sons, New York, pp. 961–974.

Vale, W., Vaughan, J., Smith, M., Yamamota, G., Rivier, J., and Rivier, C., 1983b, Effects of synthetic ovine corticotropin-releasing factor, glucocorticoids, catecholamines, neurohypophysial peptides, and other substances on cultured corticotropic cells, *Endocrinology* **113:**1121–1131.

Vanderhaeghen, J. J., Lotstra, F., DeMey, J., and Gillies, C., 1980, Immunohistochemical localization of cholecystokinin- and gastrin-like peptides in the brain and hypophysis of the rat, *Proc. Natl. Acad. Sci. U.S.A.* **77:**1190–1194.

Vandesande, F., DeMey, J., and Dierickx, K., 1974, Identification of neurophysin producing cells. I. The origin of the neurophysin-like substance-containing nerve fibers of the external zone of the median eminence of the rat, *Cell Tissue Res.* **151:**187–200.

Watkins, W. B., Schwabedal, P., and Bock, R., 1974, Immunohistochemical demonstration of a CRF-associated neurophysin in the external zone of the rat median eminence, *Cell Tissue Res.* **152:**411–421.

Watson, S. J., Akil, H., Fischli, W., Goldstein, A., Zimmerman, E., Nilaver, G., and van Wimersma Greidanus, T. B., 1983, Dynorphin and vasopressin: Common localization in magnocellular neurons, *Science* **216:**85–87.

Westlund, K. N., Aguilera, G., and Childs, G. V., 1985, Quantification of morphological changes in pituitary corticotropes produced by *in vivo* corticotropin-releasing factor stimulation and adrenalectomy, *Endocrinology* **116:**439–445.

Whitnall, M. H., Gainer, H., Cox, B. M., and Molineaux, C. J., 1983, Dynorphin-A-(1–8) is contained within vasopressin neurosecretory vesicles in rat pituitary, *Science* **222:**1137–1139.

Whitnall, M. H., Castel, M., Key, S., and Gainer, H., 1985a, Immunocytochemical identification of dynorphin-containing vesicles in Brattleboro rats, *Peptides* **6:**241–247.

Whitnall, M. H., Mezey, E., and Gainer, H., 1985b, Co-localization of corticotropin-releasing factor and vasopressin in median eminence neurosecretory vesicles, *Nature* **317:**248–250.

Whitnall, M. H., Key, S., Ben-Barak, Y., Ozato, K., and Gainer, H., 1985c, Neurophysin in the hypothalamo–neurohypophysial system. II. Immunocytochemical studies of the ontogeny of oxytocinergic and vasopressinergic neurons, *J. Neurosci.* **5:**98–109.

Wiegand, S. J., and Price, J. L., 1980, The cells of origin of afferent fibers to the median eminence in the rat, *J. Comp. Neurol.* **192:**1–19.

Wolfson, B., Manning, R. W., Davis, L. G., Arentzen, R., and Baldino, F., Jr., 1985, Co-localization of corticotropin releasing factor and vasopressin mRNA in neurones after adrenalectomy, *Nature* **315:**59–61.

Wynn, P. C., Aguilera, G., Morell, J., and Catt, K. J., 1983, Properties and regulation of high-affinity pituitary receptors for corticotropin-releasing factor, *Biochem. Biophys. Res. Commun.* **110:**602–608.

Wynn, P. C., Harwood, J. P., Catt, K. J., and Aguilera, G., 1985, Regulation of corticotropin-releasing factor (CRF) receptors in the rat pituitary gland: Effects of adrenalectomy on CRF receptors and corticotroph responses, *Endocrinology* **116:**1653–1659.

Yates, F. E., and Maran, J. W., 1974, Stimulation and inhibition of adrenocorticotropin release, in: *Handbook of Physiology,* Section 7: *Endocrinology,* Vol. 4, Part 2 (E. Knobil and W. H. Sawyer, eds.), American Physiological Society, Washington, pp. 367–404.

Yates, F. E., Russell, S. M., Dallman, M. F., Hedge, G. A., McCann, S. M., and Dhariwal, A. P. S., 1971, Potentiation by vasopressin of corticotropin release induced by corticotropin-releasing factor, *Endocrinology* **88:**2–15.

Zamir, N., Zamir, D., Eiden, L. E., Palkovits, M., Brownstein, M. J., Eskay, R. L., Weber, E., Faden, A. I., and Feuerstein, G., 1985, Methionine and leucine enkephalin in rat neurohypophysis: Different responses to osmotic stimuli and T_2 toxin, *Science* **228:**606–608.

Zimmerman, E. A., Stillman, M. A., Recht, L. D., Antunes, J. L., Carmel, P. W., and Goldsmith, P. C., 1977, Vasopressin and corticotropin-releasing factor: An axonal pathway to portal capillaries in the zona externa of the median eminence containing vasopressin and its interaction with adrenal corticoids, *Ann. N.Y. Acad. Sci.* **297:**405–419.

14

Interconnections between Neurotransmitter- and Neuropeptide-Containing Neurons Involved in Gonadotrophin Release in the Rat

CSABA LERANTH, NEIL J. MacLUSKY, and FREDERICK NAFTOLIN

1. INTRODUCTION

A knowledge of interneuronal intrahypothalamic connections is essential for complete understanding of the neuroendocrine control of pituitary function. The Golgi studies of Szentagothai *et al.* (1972) and Millhouse (1973a,b) called attention to the abundant internal connections of hypothalamic nuclei. These early descriptions have been underscored by electrophysiological (Dyer, 1973; Harris and Sanghera, 1974; Makara and Hodacs, 1975; Renaud, 1976, 1977), autoradiographic (Conrad and Pfaff, 1976a,b; Swanson, 1976; Saper *et al.*, 1976; Krieger *et al.*, 1979), and electron microscopic degeneration (Koves and Rethelyi, 1976; Zaborszky and Makara, 1979) studies, which have clearly established the complexity of the anatomic interrelationships among different hypothalamic neuronal populations.

This laboratory has been interested for several years in the actions of gonadal steroids on the brain and, in particular, the effects of these steroids on the mechanisms regulating gonadotrophin release. It is well established that the hypothalamic control of gonadotrophin release is primarily mediated through the secretion of luteinizing hormone-releasing hormone (LHRH) into the pituitary portal circulation.

CSABA LERANTH and NEIL J. MacLUSKY • Department of Obstetrics and Gynecology and Section of Neuroanatomy, Yale University School of Medicine, New Haven, Connecticut 06510. FREDERICK NAFTOLIN • Department of Obstetrics and Gynecology, Yale University School of Medicine, New Haven, Connecticut 06510.

The available evidence, however, suggests that the steroidal feedback control of LHRH secretion is not exercised simply through direct effects of the steroids on the LHRH neurons. Combined autoradiographic and immunocytochemical studies have failed to demonstrate a measurable concentration of [^{3}H]estradiol in LHRH-immunoreactive perikarya in the rat brain, suggesting that these neurons do not themselves contain estrogen receptors (Shivers *et al.*, 1983). Moreover, a variety of experimental manipulations that give rise to failure of the capacity to support cyclic ovarian function do not appreciably affect the ultrastructure or distribution of the LHRH-immunoreactive neurons (King *et al.*, 1980; Leranth *et al.*, 1986). These observations are consistent with the hypothesis that the effects of gonadal steroids on LHRH release may be primarily mediated through inputs from other steroid-sensitive systems.

The nature of the mechanisms responsible for controlling the activity of the LHRH neurons remains poorly understood. A large number of different neurotransmitters and neuropeptides have been implicated in the control of gonadotrophin release, including the catecholamines, serotonin, γ-aminobutyric acid (GABA), and the endogenous opioid peptides (Coen *et al.*, 1980; Drouva *et al.*, 1981; Kalra and Simpkins, 1981; Rance *et al.*, 1981; Kalra and Crowley, 1982; Mansky *et al.*, 1982; Ferin *et al.*, 1984). The roles played by each of these systems and the manner in which they are integrated remain ill defined. We have approached this problem in two ways: first, by attempting to define better the anatomic interconnections between hypothalamic neuronal populations for which there is evidence of responsiveness to gonadal steroids, and second, by characterizing the synaptic input to the LHRH-containing neurons. This chapter summarizes the results of our ultrastructural immunocytochemical studies on two cell groups, the arcuate nucleus and the medial preoptic area (MPO), both of which have been shown to be important in the regulation of reproductive function (Halàsz, 1969; Gallo and Osland, 1976; Wilson *et al.*, 1984).

2. INTRINSIC CONNECTIONS BETWEEN TH- AND GAD-IMMUNOREACTIVE NEURONS IN THE ARCUATE NUCLEUS

The arcuate nucleus consists of a band of cells extending through the hypothalamus on either side of the third ventricle. In the rat it is approximately 1.2 mm long and contains a number of subdivisions that can be distinguished topographically and cytoarchitectonically (Bodoky and Rethelyi, 1977; Zaborszky, 1982). The neurotransmitter and neuropeptide content of the arcuate nucleus is extensive, including dopamine, GABA, acetylcholine, ACTH, and other proopiomelanocortin-derived peptides, enkephalin, neuropeptide Y, somatostatin, LHRH, growth hormone-releasing hormone, prolactin, and neurotensin (reviewed by Palkovits, 1978, 1980; Chronwall, 1985). As much as 40% of the dopaminergic innervation of the median eminence may be derived from perikarya in the arcuate nucleus (Kizer *et al.*, 1976; Zaborszky, 1982). Although many of the arcuate nucleus neurons project to the median eminence, a substantial proportion of them also project to other hypothalamic nuclei (Zaborszky, 1982). The proopiomelanocortin-derived peptide-containing neurons apparently fall into this latter category; this is discussed further, below.

Gonadal steroids produce dramatic changes in perikaryal morphology in the arcuate nucleus. Estrogen treatment increases the observed frequency of neurons containing large finely granulated cytoplasmic inclusions, termed nematosomes. In contrast, gonadectomy in either males or females results in the appearance of increased numbers of complex, lamellar cytoplasmic organelles, called "whorl" bodies (Brawer, 1971; King *et al.*, 1974). These latter structures seem to be induced by a variety of hormonal and pharmacological manipulations, including adrenalectomy, thyroidectomy, and morphine administration (Ford and Milks, 1978; Naftolin and Brawer, 1978).

The whorl-body-containing neurons observed in gonadectomized rats receive a particularly rich synaptic input (Naftolin *et al.*, 1985). Ultrastructural immunocytochemical studies in our laboratory have demonstrated that some of the synaptic contacts on whorl-body-containing cells are immunoreactive for the enzyme responsible for GABA synthesis, glutamic acid decarboxylase (GAD), and are therefore presumably GABAergic. After colchicine treatment, GAD immunoreactivity is also found in the perikarya of whorl-body- and nematosome-containing neurons (Leranth *et al.*, 1985a). Total deafferentation of the arcuate nucleus–median eminence, using the Halàsz–Pupp knife (Halàsz and Pupp, 1967), leaves the great majority of the GAD-reactive axon terminals intact (C. Leranth, unpublished observations).

These results suggest that GABAergic cells within the arcuate nucleus are morphologically responsive to gonadal steroids and receive synaptic input from intrinsic GABAergic terminals. This conclusion is consistent with previous studies showing that estrogens modulate GAD activity in the arcuate nucleus (Wallis and Luttge, 1980) and with evidence for GABA involvement in neuroendocrine feedback mechanisms, including the control of gonadotrophin and prolactin release (Vijayan and McCann, 1978; Mansky *et al.*, 1982). What we do not know at present is whether the arcuate nucleus GABAergic cells are directly responsive to estrogen or whether they are affected by gonadal hormones indirectly through changes in input from other, estrogen-receptor-containing cells. Light microscopic autoradiographic studies have shown that GAD-immunoreactive neurons in the preoptic area concentrate [^{3}H]estradiol or its metabolites (Sar *et al.*, 1983). It is not known, however, whether the same is true for the arcuate nucleus.

One possibility is that the effects of estrogen on the GABA neurons may at least partially be mediated through interactions with other estrogen target cells. One of the candidates for such an estrogen-sensitive modulatory input is the hypothalamic dopaminergic system. A subpopulation of the hypothalamic dopaminergic neurons (predominantly those in the arcuate nucleus and periventricular nuclei) have been shown to concentrate radioactivity after [^{3}H]estradiol administration (Sar, 1984). Tyrosine hydroxylase (TH)-immunoreactive neurons in the arcuate nucleus are also morphologically and biochemically responsive to estrogen (Luine *et al.*, 1977;

Sakamoto *et al.*, 1984). It therefore seemed possible that the actions of estrogens on the arcuate nucleus GAD-immunoreactive cells might at least in part reflect a change in dopaminergic input.

In view of reports indicating effects of GABA on dopamine turnover in the hypothalamus (Lamberts *et al.*, 1983), it also seemed possible that such connections, if they existed, might be of a reciprocal nature. To investigate this hypothesis, we attempted to define better the intranuclear connections of the arcuate nucleus dopaminergic and GABAergic cells. A study of the distribution and synaptic connections of TH-immunoreactive elements in the arcuate nucleus revealed a number of synaptic connections in which both the pre- and postsynaptic elements were immunopositive for TH as well as synapses in which only the postsynaptic element was immunopositive (Leranth *et al.*, 1985b). These synapses were essentially unaffected by transection of the afferents from more caudal dopaminergic cell groups. In contrast, when the lateral and medial parts of the arcuate nucleus were separated by a knife cut, degenerating axons terminating on TH-immunoreactive cell bodies in the lateral part of the arcuate nucleus were far more common (Leranth *et al.*, 1985b). In view of the fact that GABAergic neurons are predominantly found in the medial part of the arcuate nucleus (Tappaz *et al.*, 1983), these findings are clearly consistent with the hypothesis that connections exist between the arcuate TH- and GAD-immunoreactive cells.

To demonstrate this connection, it was necessary to develop a procedure for differential staining of the pre- and postsynaptic elements at the electron microscopic level. Since homologous GAD–GAD and TH–TH immunoreactive connections are both present in the arcuate nucleus, it was not possible to use single-label immunochemistry to test for the presence of connections between the GAD and TH systems. Several authors have reported methods for electron microscopic immunocytochemical double labeling based on labeling the antigens with colloidal gold particles of different sizes (Geuze *et al.*, 1981) or peroxidase combined with gold (Van den Pol, 1985). The method developed in our laboratory is conceptually similar but utilizes ferritin and peroxidase as the contrasting electron-dense markers (Leranth *et al.*, 1985c). A schematic outline of this procedure, as applied to the simultaneous staining of TH and LHRH, is illustrated in Fig. 1. Using this technique in the rat arcuate nucleus, we have observed GAD-immunoreactive synapses on TH-reactive cells (Leranth *et al.*, 1985c) as well as a somewhat more limited number of TH-immunopositive synapses on GAD cells (C. Leranth, unpublished observations).

3. CONNECTIONS AMONG GAD-, TH-, AND LHRH-IMMUNOREACTIVE ELEMENTS IN THE MEDIAL PREOPTIC AREA

The interconnections between the GAD- and TH-immunoreactive elements in the MPO appear to be at least superficially similar to those in the arcuate nucleus. Using electron microscopic double-label immunocytochemistry, we have identified

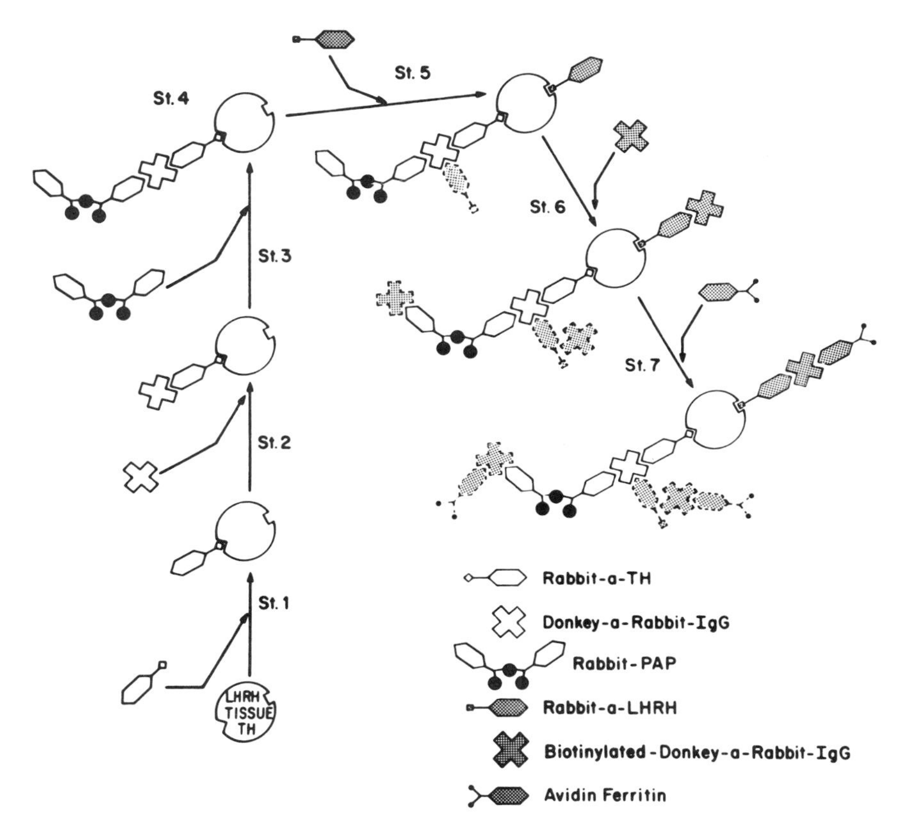

FIGURE 1. Schematic representation of the double-staining, double-label procedure as applied to the simultaneous electron microscopic identification of TH and LHRH in the preoptic area. Key: Rabbit-a-TH, rabbit anti-TH primary antibody (Pickel *et al.*, 1975); Rabbit-a-LHRH, rabbit anti-LHRH primary antibody (Merchentaler *et al.*, 1980); PAP, peroxidase–antiperoxidase complex.

synapses between GAD-immunoreactive elements as well as GAD-reactive synapses on TH-immunopositive profiles (unpublished observations).

Both the TH and GAD systems also appear to make direct contact with the LHRH neurons. Using peroxidase labeling for LHRH immunoreactivity and ferritin as a marker for GAD, we have reported the presence of symmetric (Gray II) synapses between GAD-labeled axons and LHRH-reactive dendrites and perikarya in the medial preoptic area (Leranth *et al.*, 1985d). More recently, we have obtained similar results after double-label staining for TH and LHRH (unpublished observations).

In addition to afferent input to the LHRH-immunoreactive neurons, there is also evidence that LHRH may itself act as a neurotransmitter or neuromodulator within the preoptic area. The existence of intrinsic terminations of LHRH axons in the preoptic area has been demonstrated by a number of authors (Kiss and Halàsz, 1985; Leranth *et al.*, 1985e). The question of whether there are connections between the LHRH cells, perhaps serving to synchronize their activity, has, however, remained controversial (Silverman, 1984; Pelletier, 1984). Our own studies have confirmed the existence of a direct LHRH-to-LHRH connection, demonstrating contacts between LHRH-immunopositive presynaptic boutons and LHRH-immunoreactive dendrites and perikarya in the MPO (Leranth *et al.*, 1985e). These contacts are symmetrical (Gray II) and appear to differ from LHRH-immunoreactive axons in the median eminence in that they contain fewer dense-core vesicles.

4. PROJECTIONS OF ARCUATE NUCLEUS PROOPIOMELANOCORTIN-CONTAINING NEURONS TO THE MPO

Studies of the connections between hypothalamic nuclei have provided electrophysiological and morphological evidence that a population of arcuate nucleus neurons projects directly to the MPO. Zaborszky and Makara (1979) observed degenerated terminals in the MPO following microlesions placed in the medial part of the arcuate nucleus. According to Renaud (1977), electrical stimulation in the MPO is associated with orthodromic responses from at least 11.5% of neurons in the arcuate nucleus.

One of the candidates for this projection was the axons of proopiomelanocortin (POMC)-derived peptide-containing neurons of the arcuate nucleus. Such a connection would be of considerable interest from a physiological standpoint in view of the evidence supporting a role for POMC-derived peptides in the control of gonadotrophin release (Kalra and Simpkins, 1981; Adler and Crowley, 1984; Ferin *et al.*, 1984). Recently, Mezey *et al.* (1985) reported that ACTH-, α-MSH-, and β-endorphin-immunoreactive axons originating from the arcuate nucleus form a dense network in the MPO. These studies were carried out at the light microscopic level, raising the question of whether the observed immunoreactive axons do in fact make synaptic contact in the MPO, or if they might simply represent fibers of passage.

To answer this question, we first examined the distribution of ACTH-immunoreactive elements in the preoptic area using conventional single-label immunocytochemistry, employing the antibody characterized by Watson and Akil (1982). The majority of the immunolabeled axons were found within the borders of the MPO, at particularly high densities in the medial and the ventrolateral regions of the nucleus (Fig. 2a). Under the electron microscope, the labeled axons had the characteristic appearance of ACTH terminals, containing small, clear spherical vesicles and large (120- to 130-nm) dense-core vesicles (Fig. 2b–d). In sections taken from the medial part of the MPO, ACTH-immunopositive axons were found in synaptic connection with dendritic shafts (Fig. 2b). In contrast, in the ventrolateral part of the MPO, the ACTH terminals synapsed primarily on dendritic spines (Fig.

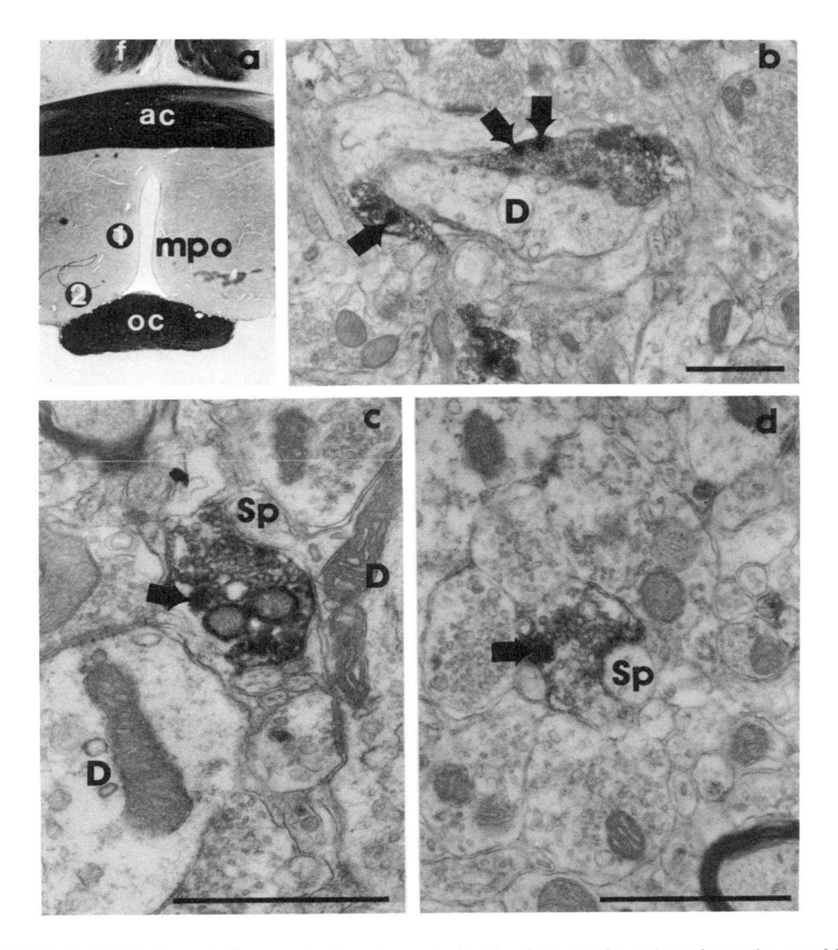

FIGURE 2. Light (panel a) and electron (panels b–d) micrographs taken from the rat MPO immunostained for ACTH. ACTH-immunoreactive axons (panel b) in the dorsomedial part of the MPO (1 in panel a) were mostly found in synaptic contact with dendritic shafts (D in panel b), whereas in the ventrolateral MPO, the ACTH-immunopositive axons synapse on spines (Sp in panels c and d). Arrowheads in panels b, c, and d indicate large dense-core vesicles. Bar scales, 1 μm.

2c,d). The type of the synaptic membrane specializations could not be precisely determined but appeared to represent a transitional form between symmetrical and asymmetrical connections (Fig. 2).

To determine whether some of these synaptic contacts might represent connections between the ACTH-immunoreactive neurons and LHRH cells in the preoptic area, further studies were performed using double-label immunostaining with peroxidase and avidin–ferritin. Electron microscopic analysis was performed on sections taken from both the medial and ventrolateral parts of the MPO of double-stained vibrotome sections labeled with peroxidase for LHRH and ferritin for ACTH. Peroxidase-labeled LHRH neurons and axons were examined only in the medial part of the MPO (Fig. 3a,b). In this area, synaptic connections between ferritin-labeled ACTH axons and peroxidase-labeled LHRH dendrites were frequently observed (Fig. 3e). Reversing the order in which the primary antibodies were applied to the sections resulted, as expected, in a reversal of the staining pattern, i.e., ferritin labeling of the LHRH dendrites with peroxidase labeling of the ACTH-immunoreactive synaptic contacts (Fig. 3d).

Finally, we examined the origin of the ACTH-immunoreactive fibers in the MPO using a combination of retrograde horseradish peroxidase (HRP) transport and single-label immunocytochemical techniques. A modification of the glucose oxidase method was used to visualize the transported HRP (Zaborszky and Leranth, 1985). Vibrotome sections from both the MPO and the arcuate nucleus were examined. Arcuate nucleus sections were taken for further examination only where the spread of HRP from the injection site was essentially limited to the MPO (Fig. 4a,b). In the arcuate nucleus sections from these animals, three types of labeled cells could be recognized. A small number of cells, mostly located in the ventromedial portion of the nucleus, were found to be doubly labeled; i.e., they contained very dark transported HRP granules in ACTH-immunostained cytoplasm (Fig. 4c–e). The immunoperoxidase reaction product was easily distinguished from transported HRP granules. A somewhat larger number of cells, homogeneously distributed throughout the arcuate nucleus, contained only the transported HRP granules (Fig. 4c,f). Synaptic contacts from labeled ACTH axons were frequently observed on these HRP-positive but ACTH-immunonegative neurons (Fig. 4f). The majority of labeled cells were found to be immunopositive for ACTH without recognizable HRP granules. These cells were observed mostly in the ventrolateral parts of the arcuate nucleus, conforming to earlier descriptions (Watson *et al.,* 1978) of the location of ACTH immunopositive neurons.

5. CONCLUSIONS

The Golgi method has been the conventional technique for studying short connections (Szentagothai, 1969; Szentagothai *et al.,* 1972). However, in Golgi preparations of hypothalamus, axons can be followed in the intricate neuropil for only a few tenths of micrometers (Szentagothai *et al.,* 1972). This imposes serious limitations on the utility of this method. Moreover, it is not possible, using single Golgi impregnation, to determine the neurotransmitter or neuropeptide contents of the pre- and postsynaptic elements. These considerations led us to devise an electron microscopic double-stain, double-label technique (Leranth *et al.,* 1985c) that provides the possibility of visualizing synaptic connections between different neuropeptide- or transmitter-containing neurons in the hypothalamus. By combining immunocytochemical staining with anterograde (Zaborszky *et al.,* 1984) or, as described

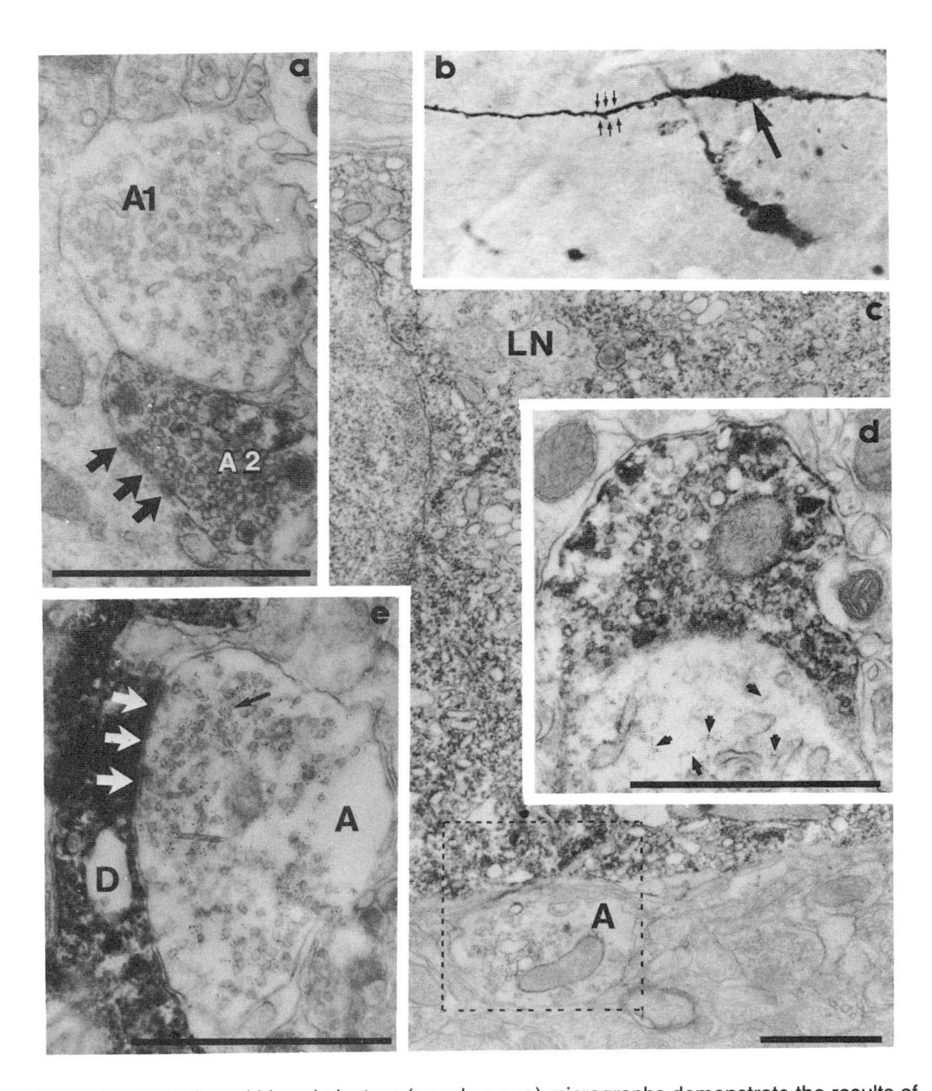

FIGURE 3. Light (panel b) and electron (panels a,c–e) micrographs demonstrate the results of the double-stain, double-label procedure for LHRH and ACTH in the rat MPO. In this case, the typical fusiform LHRH neurons (panel b) were labeled with peroxidase, and the ACTH terminals with avidin–ferritin. Panel a shows two labeled axons: the first (A1), labeled with ferritin, is immunopositive for ACTH, whereas the second, LHRH-immunopositive axon (A2) labeled with peroxidase synapses with an unlabeled dendrite. Panel c shows a ferritin-labeled ACTH axon (A) in the vicinity of a peroxidase-labeled LHRH cell body. Panel 3 demonstrates a synaptic connection (arrowheads) between a ferritin-labeled ACTH axon (A) and an immunoperoxidase-labeled LHRH dendrite (D). This high-power electron micrograph is taken from the indicated part (arrows) of the thick section (shown in the light micrograph in panel b), rotated 90° counterclockwise compared to the orientation of the light micrograph. Panel d is taken from another experiment in which the ACTH-immunoreactive axons in the MPO were labeled with immunoperoxidase and the LHRH neurons with avidinated ferritin. Original magnification of panel b, × 40. Bar scales on the electron micrographs, 1 μm.

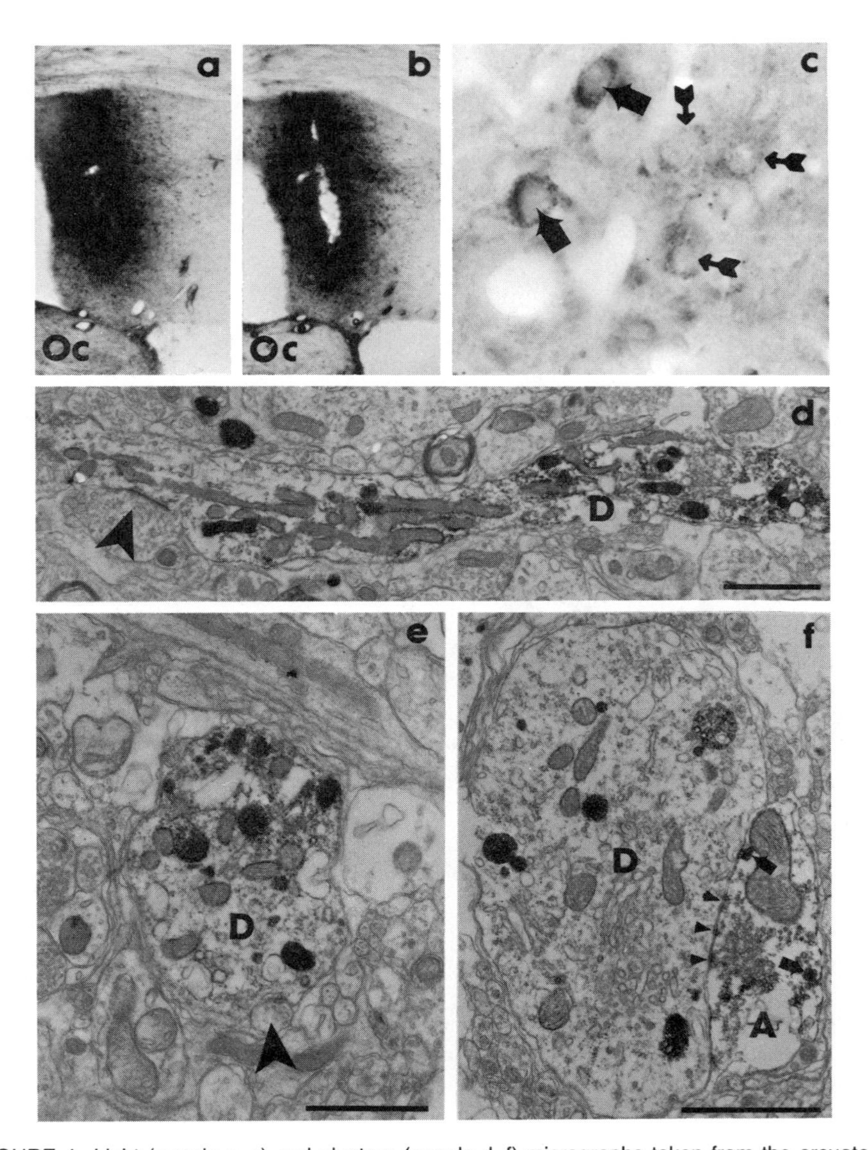

FIGURE 4. Light (panels a–c) and electron (panels d–f) micrographs taken from the arcuate nucleus (panels c–f) and MPO (panels a and b) of a rat following HRP injection into the MPO and immunostaining for ACTH. Panels a and b show two examples of the spread of HRP around the injection site. Panel c: Two types of arcuate nucleus neurons could be distinguished, labeled only with transported peroxidase (arrows) or containing both the transported HRP granules and immunoperoxidase reaction product for ACTH (arrowheads). Electron micrographs (panels d and e) taken from the arcuate nucleus show dendrites (D) immunopositive for ACTH containing transported HRP granules. The dendrites (D) receive synaptic input from unlabeled axons (arrowheads). Panel f shows a dendrite containing only the transported peroxidase granules. This dendrite establishes a synaptic connection (arrowheads) with an axon (A) immunoreactive for ACTH. Arrows in the axons indicate dense-core vesicles. Original magnifications: panels a and b, $\times 10$; panel c, $\times 40$. Bar scales on the electron micrographs, 1 μm.

here for ACTH-immunoreactive neurons, with retrograde HRP tracing, it is possible to describe not only intranuclear but also internuclear connections together with neurochemical characterization of the participating neurons.

The picture we currently have of the intranuclear connections of the arcuate nucleus and preoptic area is summarized in Fig. 5. Much of the evidence for this scheme is based on the double-label immunocytochemical studies reviewed in this chapter. In addition, the diagram includes information on the afferent connections of the 5-HT system based on the combined autoradiographic and ultrastructural immunocytochemical work of Kiss and collaborators (Kiss, 1984; Kiss *et al.*, 1984; Kiss and Halàsz, 1985). These 5-HT axons, terminating on proopiomelanocortin-derived peptide and TH-immunoreactive cells in the arcuate nucleus as well as the LHRH neurons in the preoptic area, may play a role in the serotonergic control of gonadotrophin secretion (Coen *et al.*, 1980).

In some cases, the organization of the intrinsic and extrinsic connections of the arcuate nucleus and preoptic area presented in Fig. 5 must be regarded as speculative. Although the GAD and TH connections to the LHRH neurons are shown as being derived from neurons within the preoptic area, there is no proof that this is the case. These connections may well originate from cell groups elsewhere in the brain, for example, the rostral periventricular dopaminergic cell group (A14) and/or catecholaminergic neurons in the brainstem in the case of the TH-immunoreactive axons. Additional studies on the effects on the synaptic input to the LHRH neurons of discrete lesions to structures known to project to the preoptic area, as well as deafferentation of the preoptic area, will be required to resolve this issue.

The results of the studies of the connections of ACTH-immunoreactive axons establish unequivocally that these axons make synaptic contacts in the medial preoptic area.ACTH-immunopositive axons synapsing with dendritic shafts were found in the medial part of the MPO. At least some of these contacts appear, from the results of the double-label experiments, to be on the dendrites of LHRH-immunoreactive neurons (Fig. 3). The other synapses formed by ACTH terminals were the axospine connections, which were recognized only in the ventrolateral part of the MPO. This is consistent with the Golgi studies of Szentagothai *et al.* (1972) and Millhouse (1973a,b) showing that spinous dendrites are found predominantly in the lateral preoptic area. Since, according to present knowledge, the perikarya of hypothalamic proopiomelanocortin-peptide-containing neurons are confined to the arcuate nucleus (Palkovits, 1978, 1980; Nilaver *et al.*, 1979), it seems reasonable to assume that the ACTH-immunoreactive boutons synapsing with LHRH neurons represent direct arcuate nucleus–MPO connections. This hypothesis is consistent with the results of the HRP experiments and with previous electrophysiological (Makara and Hodacs, 1975; Renaud, 1977) and morphological (Zaborszky and Makara, 1979) observations of the existence of a direct arcuate nucleus–MPO pathway.

In addition, the results suggest that there are at least two different populations of neurons projecting to the MPO, only some of which are ACTH immunopositive.

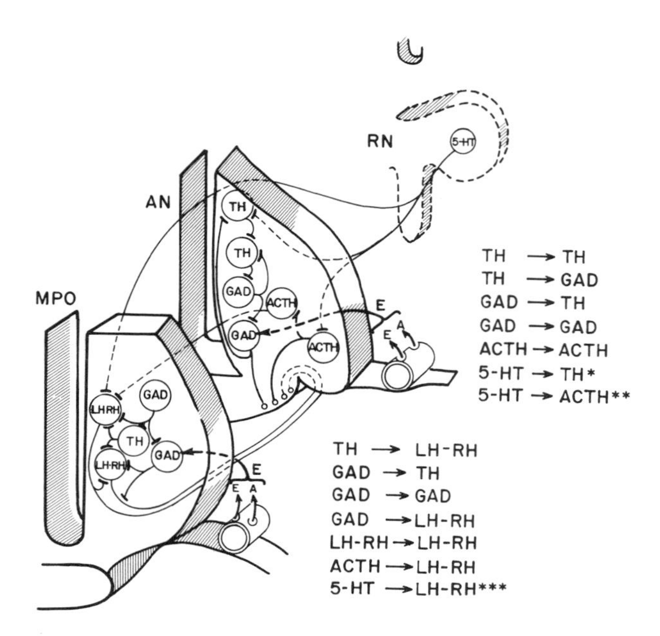

FIGURE 5. Schematic illustration summarizing the putative interconnections of medial preoptic area (MPO), arcuate nucleus (AN), and dorsal raphe (DR) neurons involved in the regulation of LHRH secretion in the rat. Pathways indicated with asterisks were described by other authors: *Kiss, 1984; **Kiss *et al.*, 1984; ***Kiss and Halàsz, 1985.

The ACTH-immunopositive HRP-labeled perikarya were found mostly close to the third ventricle. This is in good agreement with the results of the electron microscopic degeneration experiments of Zaborszky and Makara (1979) showing that lesions placed in the medial part of the arcuate nucleus cause the largest number of degenerated terminals in the MPO. However, since HRP can be retrogradely transported not only through axons terminating around the injection site but also by fibers of passage, we cannot be sure that all of the HRP-labeled cells in the arcuate nucleus connect to cells in the MPO.

The physiological role of these connections remains to be established. Although the studies reviewed in this chapter evolved from our interest in the mechanisms controlling gonadotrophin secretion, it is not possible definitively to assign such a role to any of the synaptic contacts that we have observed. In the case of the ACTH-immunoreactive (presumably POMC)–LHRH connections, it is tempting to speculate that this connection may be involved in the opioid-mediated regulation of gonadotrophin secretion. However, it is also possible that this regulation is mediated primarily through connections of POMC neurons to other regions of the brain, in particular the median eminence (O'Donohue *et al.*, 1979; Hisano *et al.*, 1982). Moreover, it must be acknowledged that the presence of synaptic contacts on LHRH-immunoreactive neurons cannot be taken to imply that these contacts are involved in the control of gonadotrophin secretion, since the role of LHRH in the brain is not confined to regulation of the pituitary gland (Pfaff, 1980; Sirinathsinghji, 1983). Similar considerations apply to the TH–LHRH and GAD–LHRH synaptic contacts observed in the preoptic area. Although studies of the actions of GABA and the GABAmimetic muscimol on gonadotrophin release suggest a role for GABAergic neurons in the preoptic area (Lamberts *et al.*, 1983), it would be premature to rule out the possibility that the effects of this neurotransmitter may be mediated indirectly through other neurotransmitter- or neuropeptide-containing neurons rather than a direct action on the LHRH-producing cells.

The results obtained in the arcuate nucleus suggest that intrinsic GAD- and TH-immunoreactive neurons may form a local reciprocally innervated network within this nucleus. The fact that the morphology of both components of this network is influenced by estrogen suggests that it is responsive to gonadal steroids. Thus, it may be involved in mediating feedback effects of circulating gonadal steroids on pituitary hormone secretion. These effects could include the feedback control of gonadotrophin and/or prolactin release (Neill, 1980; Lamberts *et al.*, 1983).

Thus, the development of a procedure for double-label electron microscopic immunostaining allows simultaneous characterization of the neurotransmitter or neuropeptide contents of the pre- and postsynaptic elements in synaptic contacts within the central nervous system. Using this technique in the rat arcuate nucleus and medial preoptic area, we have identified a number of synaptic connections that may be involved in the feedback control of the pituitary gland. The demonstration of GAD-, TH-, and ACTH-immunoreactive synapses on LHRH-immunoreactive profiles in the medial preoptic area suggests that GABAergic, catecholaminergic, and POMC-derived peptide-containing neurons may exert direct effects on the activity of the LHRH neurons.

6. SUMMARY

Pharmacological and physiological studies suggest that interactions between luteinizing hormone-releasing hormone- (LHRH) and GABA-containing neurons of the preoptic area and GABA-dopamine-, and proopiomelanocortin-derived peptide-producing neurons of the arcuate nucleus may be important for the control of gonadotrophin release. The studies summarized in this chapter were performed to investigate the possibility that there may be direct synaptic connections between these neurotransmitter- and neuropeptide-containing neurons in the rat.

Intrinsic synaptic connections in the arcuate nucleus and preoptic area were studied using a newly developed preembedding double-label electron microscopic immunostaining procedure in which peroxidase and avidin–ferritin serve as contrasting electron-dense markers (Leranth *et al.*, 1985c). Using this method, we have identified interneuronal connections in the arcuate nucleus among tyrosine hydroxylase (TH)-immunopositive pre- and postsynaptic elements, glutamic acid decarboxylase (GAD)-immunopositive pre- and postsynaptic elements, as well as heterologous synapses of the type GAD → TH and TH → GAD. In the medial preoptic area, LHRH → LHRH, GAD → GAD, and GAD → LHRH synapses were identified. To study interconnections between neurons in the arcuate nucleus and preoptic area, a combination of electron microscopic immunostaining and retrograde HRP tracing techniques was employed. Direct synaptic connections were observed between axons from ACTH-immunoreactive neurons in the ventromedial part of the arcuate nucleus and LHRH neurons in the medial preoptic area. These results suggest that there are local, reciprocal interconnections among TH-, GAD-,and LHRH-immunoreactive neurons in the arcuate nucleus and preoptic area. They also suggest that some of the reported pharmacological effects of neurotransmitters and neuropeptides on gonadotrophin release may reflect underlying physiological mechanisms involving direct synaptic connections between the neuronal systems containing these substances and the LHRH-producing cells.

ACKNOWLEDGMENTS. Supported by an Andrew Mellon Foundation fellowship in the Reproductive Sciences to C. L. and by grant No. HD13587 from NIH, NICHHD.

REFERENCES

Adler, B. A., and Crowley, W. R., 1984, Modulation of luteinizing hormone release and catecholamine activity by opiates in the female rat, *Neuroendocrinology* **38:**248–253.

Bodoky, M., and Rethelyi, M., 1977, Dendritic arborization and axon trajectory of neurons in the hypothalamic arcuate nucleus of the rat, *Exp. Brain Res.* **28:**543–555.

Brawer, J. R., 1971, The role of the arcuate nucleus in the brain–pituitary–gonad axis, *J. Comp. Neurol.* **143:**411–46.

Chronwall, B. M., 1985, Anatomy and physiology of the neuroendocrine arcuate nucleus, *Peptides* **6:**1–12.

Coen, C. W., Franklin, M., Laynes, R. W., and MacKinnon, P. C. B., 1980, Effects of manipulating serotonin on the incidence of ovulation in the rat, *J. Endocrinol.* **87:**195–201.

Conrad, L. C. A., and Pfaff, D. W., 1976a, Efferents from medial basal forebrain and hypothalamus in the rat I. An autoradiographic study of the medial preoptic area, *J. Comp. Neurol.* **169:**185–220.

Conrad, L. C. A., and Pfaff, D. W., 1976b, Efferents from medial basal forebrain and hypothalamus in the rat II. An autoradiographic study of the anterior hypothalamus, *J. Comp. Neurol.* **169:**221–261.

Drouva, S. V., Epelbaum, J., Tapia-Arancibia, L., Laplante, E., and Kordon, C., 1981, Opiate receptors modulate LHRH and SRIF release from the mediobasal hypothalamus, *Neuroendocrinology* **32:**163–167.

Dyer, R. G., 1973, An electrophysiological dissection of the hypothalamic regions which regulate the pre-ovulatory secretion of the luteinizing hormone in the rat, *J. Physiol. (Lond.)* **234:**421–442.

Ferin, M., Van Vugt, D., and Wardlaw, S., 1984, The hypothalamic control of the menstrual cycle and the role of endogenous opioid peptides, *Recent Prog. Horm. Res.* **40:**441–485.

Ford, D. H., and Milks, L. C., 1978, Smooth endoplasmic reticular whorls in neurons of the arcuate nucleus in male rats following adrenalectomy, *Psychoneuroendocrinology* **3:**65–83.

Gallo, R. V., and Osland, R. B., 1976, Electrical stimulation of the arcuate nucleus in ovariectomized rats inhibits episodic luteinizing hormone (LH) release but excites LH release after estrogen priming, *Endocrinology* **99:**659–668.

Geuze, H. J., Slot, J. W., Van Der Ley, P. A., and Scheffer, R. C., 1981, Use of colloidal gold particles in double-labelling immunoelectron microscopy of ultrathin frozen tissue sections, *J. Cell Biol.* **89:**653–665.

Halàsz, B., 1969, The endocrine effects of isolation of the hypothalamus from the rest of the brain, in: *Frontiers in Neuroendocrinology,* Vol. 1 (W. F. Ganong and L. Martini, eds.), Oxford University Press, London, pp. 307–342.

Halàsz, B., and Pupp, L., 1967, Hormone secretion of the anterior pituitary gland after physical interruption of all nervous pathways to the hypophysiotrophic area, *Endocrinology* **77:**553–562.

Harris, C., and Sanghera, M., 1974, Projection of medial basal hypothalamic neurones to the preoptic anterior hypothalamic areas and paraventricular nucleus in the rat, *Brain Res.* **91:**404–412.

Hisano, S., Kawano, H., Nishiyama, T., and Daikoku, S., 1982, Immunoreactive ACTH/β-endorphin neurons in the tuberoinfundibular hypothalamus of rats, *Cell Tissue Res.* **224:**303–314.

Kalra, S. P., and Crowley, W. R., 1982, Epinephrine synthesis inhibitors block naloxone-induced LH release, *Endocrinology* **111:**1403–1405.

Kalra, S. P., and Simpkins, J. W., 1981, Evidence for noradrenergic mediation of opioid effects on luteinizing hormone secretion, *Endocrinology* **109:**776–782.

King, J. C., Williams, T. H., and Gerall, A. A., 1974, Transformation of hypothalamic arcuate neurons. I. Changes associated with stages of the estrous cycle, *Cell Tissue Res.* **153:**497–515.

King, J. C., Tobet, S. A., Snavely, F. L., and Arimura, A. A., 1980, The LHRH system in normal and neonatally androgenized female rats, *Peptides* **1:**85–100.

Kiss, J., 1984, Synaptic connections between serotonergic nerve terminals and LH-RH or catecholamine containing neurons in the rat hypothalamus, *Neurosci. Lett. Suppl.* **18:**S186.

Kiss, J., and Halàsz, B., 1985, Demonstration of serotonergic axons terminating on luteinizing hormone-releasing hormone neurons in the preoptic area of the rat using combination of immunocytochemistry and high resolution autoradiography, *Neuroscience* **14:**69–78.

Kiss, J., Leranth, C., and Halàsz, B., 1984, Serotonergic endings of VIP-neurons in the suprachiasmatic nucleus and on ACTH-neurons in the arcuate nucleus of the rat hypothalamus. A combination of high resolution autoradiography and electron microscopic immunocytochemistry, *Neurosci. Lett.* **44:**119–124.

Kizer, J. S., Palkovits, M., and Brownstein, M. J., 1976, The projections of the A8, A9 and A10 dopaminergic cell bodies: Evidence for a nigral–hypothalamic median eminence dopaminergic pathway, *Brain Res.* **108:**363–370.

Koves, R., and Rethelyi, M., 1976, Direct neuronal connections from the medial preoptic area to the hypothalamic arcuate nucleus of the rat, *Exp. Brain Res.* **25:**529–539.

Krieger, M. S., Conrad, L. C., and Pfaff, D. W., 1979, Autoradiographic study of the efferent connections of the ventromedial nucleus of the hypothalamus, *J. Comp. Neurol.* **183:**785–816.

Lamberts, R., Vijayan, E., Graf, M., Mansky, T., and Wuttke, W., 1983, Involvement of preoptic–anterior hypothalamic GABA neurons in the regulation of pituitary LH and prolactin release, *Exp. Brain Res.* **52:**356–362.

Leranth, C., Sakamoto, H., MacLusky, N. J., Shanabrough, M., and Naftolin, F., 1985a, Estrogen responsive cells in the arcuate nucleus of the rat contain glutamic acid decarboxylase (GAD): An electron microscopic immunocytochemical study, *Brain Res.* **331:**376–381.

Leranth, C., Sakamoto, H., MacLusky, N. J., Shanabrough, M., and Naftolin, F., 1985b, Intrinsic tyrosine hydroxylase (TH) immunoreactive neurons synapse with TH immunopositive neurons in the rat arcuate nucleus, *Brain Res.* **331:**371–375.

Leranth, C., Sakamoto, H., MacLusky, N. J., Shanabrough, M., and Naftolin, F., 1985c, Application of avidin–ferritin and peroxidase as contrasting electron-dense markers for simultaneous electron microscopic immunocytochemical labeling of glutamic acid decarboxylase and tyrosine hydroxylase in the rat arcuate nucleus, *Histochemistry* **82:**165–168.

Leranth, C., MacLusky, N. J., Sakamoto, H., Shanabrough, M., and Naftolin, F., 1985d, Glutamic acid decarboxylase-containing neurons synapse on LHRH neurons in the rat medial preoptic area, *Neuroendocrinology* **40:**536–539.

Leranth, C., Segura, L. M. G., Palkovits, M., MacLusky, N. J., Shanabrough, M., and Naftolin, F., 1985e, The LHRH-containing network in the preoptic area of the rat: Demonstration of LHRH-containing nerve terminals in synaptic contact with LHRH neurons, *Brain Res.* **345:** 332–336.

Leranth, C., Palkovits, M., MacLusky, N. J., Shanabrough, M., and Naftolin, F., 1986, Normal numbers and morphology of LHRH immunoreactive axons in the medial basal hypothalami of rats with reproductive failure induced by estrogen injection, constant light exposure or neonatal testosterone administration, *Neuroendocrinology* (in press).

Luine, V. N., McEwen, B. S., and Black, I. B., 1977, Effect of 17β-estradiol on hypothalamic tyrosine hydroxylase activity, *Brain Res.* **120:**188–192.

Makara, G. B., and Hodacs, L., 1975, Rostral projections from the hypothalamic arcuate nucleus, *Brain Res.* **84:**23–29.

Mansky, T., Mestres-Ventura, P., and Wuttke, W., 1982, Involvement of GABA in the feedback action of estradiol on gonadotropin and prolactin release: Hypothalamic GABA and catecholamine turnover rates, *Brain Res.* **231:**353–364.

Merchenthaler, I., Kovacs, G., Lovasz, G., and Setalo, G., 1980, The preoption–infundibular LH-RH tract of the rat, *Brain Res.* **198:**63–74.

Mezey, E., Kiss, J., O'Donohue, T., Eskay, R., Palkovits, M., 1985, Distribution of pro-opiomelanocortin derived peptides (ACTH, α-MSH, β-endorphin) in the rat hypothalamus, *Brain Res.* **328:**341–348.

Millhouse, O. E., 1973a, The organization of the ventromedial hypothalamic nucleus, *Brain Res.* **55:**71–87.

Millhouse, O. E., 1973b, Ventromedial hypothalamic afferents, *Brain Res.* **55:**89–105.

Naftolin, F., and Brawer, J. R., 1978, The effect of estrogens on hypothalamic structure and function, *Am. J. Obstet. Gynecol.* **132:**758–765.

Naftolin, F., Bruhlman-Papayzan, M., Baetens, D., and Garcia-Segura, L. M., 1985, Neurons with whorl bodies have increased numbers of synapses, *Brain Res.* **329:**289–293.

Neill, J. D., 1980, Neuroendocrine regulation of prolactin secretion, in: *Frontiers in Neuroendocrinology,* Vol. 6 (W. F. Ganong and L. Martini, eds.), Oxford University Press, London, pp. 129–156.

Nilaver, G., Zimmerman, E. A., Defendi, R., Liotta, A., Krieger, D. T., and Brownstein, M. J., 1979, Adrenocorticotropin and beta-lipotropin in the hypothalamus. Localization in the same arcuate neurons by sequential immunocytochemical procedures, *J. Cell Biol.* **81:**50–58.

O'Donohue, T. L., Miller, R. L., and Jacobowitz, D. M., 1979, Identification, characterization and stereotaxic mapping of intraneural α-melanocyte-stimulating hormone like immunoreactive peptides in discrete regions of the rat brain, *Brain Res.* **176:**101–123.

Palkovits, M., 1978, Topography of chemically identified neurons in the central nervous system: A review, *Acta Morphol. Acad. Sci. Hung.* **26:**211–290.

Palkovits, M., 1980, Topography of chemically identified neurons in the central nervous system: Progress in 1977–1979, *Med. Biol.* **58:**188–227.

Palkovits, M., Fekete, M., Makara, G. B., and Herman, J. P., 1977, Total and partial hypothalamic deafferentations for topographical identification of catecholaminergic innervations of certain preoptic and hypothalamic nuclei, *Brain Res.* **127:**127–136.

Pelletier, G., 1984, Localization and interactions of RF neurons in the central nervous system, in: *Proceedings of the 7th International Congress of Endocrinology, Quebec, Canada,* Excerpta Medica International Congress Series 652, Amsterdam, p. 39.

Pfaff, D. W., 1980, *Estrogens and Brain Function,* Springer-Verlag, New York.

Pickel, V. M., Joh, T. H., and Reis, D. J., 1975, Ultrastructural localization of tyrosine hydroxylase in noradrenergic neurons in brain, *Proc. Natl. Acad. Sci. U.S.A.* **72:**659–663.

Rance, N., Wise, P. M., Selmanoff, M. K., and Barraclough, C. A., 1981, Catecholamine turnover rates in discrete hypothalamic areas and associated changes in median eminence luteinizing hormone-releasing hormone and serum gonadatropins on proestrus and diestrous day 1, *Endocrinology* **108:**1795–1802.

Renaud, L. P., 1976, Tuberoinfundibular neurons in the basomedial hypothalamus of the rat: Electrophysiological evidence for axon collaterals to hypothalamic and extrahypothalamic areas, *Brain Res.* **105:**59–72.

Renaud, L. P., 1977, Influence of medial preoptic anterior hypothalamic area stimulation on the excitability of mediobasal hypothalamic neurons in the rat brain, *J. Physiol. (Lond.)* **264:**541–564.

Sakamoto, H., Leranth, C., MacLusky, N. J., Hurlburt, C., and Naftolin, F., 1984, Estrogen (E) inhibition of the rat brain dopamine system is area specific and reversible, *Soc. Neurosci. Abstr.* **14(7)** p. 478.

Saper, C. B., Swanson, L. B., and Cowan, W. M., 1976, The afferent connections of the ventromedial nucleus of the hypothalamus of the rat, *J. Comp. Neurol.* **169:**409–442.

Sar, M., 1984, Estradiol is concentrated in tyrosine hydroxylase-containing neurons of the hypothalamus, *Science* **223:**938–940.

Sar, M., Stumpf, W. E., and Tappaz, M. L., 1983, Localization of ^{3}H estradiol in preoptic GABAergic neurons, *Fed. Proc.* **42:**495.

Shivers, B. D., Harlan, R. E., Morrell, J. I., and Pfaff, D. W., 1983, Absence of oestradiol concentration in cell nuclei of LHRH-immunoreactive neurones, *Nature* **304:**345–347.

Silverman, A.-J., 1984, Luteinizing hormone releasing hormone containing synapses in the diagonal band and preoptic area of the guinea pig, *J. Comp. Neurol.* **227:**452–458.

Sirinathsinghji, D. J. S., 1983, β-Endorphin regulates lordosis in female rats by modulating LHRH release, *Nature* **301:**62–64.

Swanson, L. W., 1976, Autoradiographic study of the efferent connections of the preoptic region in the rat, *J. Comp. Neurol.* **167:**227–256.

Szentagothai, J., 1969, The synaptic architecture of the hypothalamo–hypophyseal neuron system, *Acta Neurol. (Belg.)* **69:**453–468.

Szentagothai, J., Flerko, B., Mess, B., and Halàsz, B., 1972, *Hypothalamic Control of the Anterior Pituitary,* Akademiai Kiado, Budapest, pp. 40–68.

Tappaz, M. L., Wassef, M., Oertel, W. H., Paut, L., and Pujol, J. F., 1983, Light and electron microscopic immunochemistry of glutamic acid decarboxylase (GAD) in the basal hypothalamus: Morphologic evidence for neuroendocrine gamma-aminobutyrate (GABA), *Neuroscience* **9:**271–287.

Van den Pol, A. N., 1985, Silver-intensified gold and peroxidase as dual ultrastructural immunolabels for pre- and postsynaptic neurotransmitters, *Science* **228:**332–335.

Vijayan, E., and McCann, S. M., 1978, The effect of intraventricular injection of gamma-aminobutyric acid (GABA) on prolactin and gonadotropin release in conscious female rats, *Brain Res.* **155:**35–42.

Wallis, C., and Luttge, W. G., 1980, Influence of estrogen and progesterone on glutamic acid decarboxylase activity in discrete regions of rat brain, *J. Neurochem.* **34:**609–613.

Watson, G. J., and Akil, H., 1982, Opioid peptides and related substances: Immunocytochemistry, in: *Neurosecretion and Brain Peptides* (J. B. Martin, J. Reichlin, and K. L. Bictes, eds.), Raven Press, New York, pp. 77–78.

Watson, S. J., Richard, C. W., and Barchas, J. D., 1978, Adrenocorticotropin in rat brain: Immunocytochemical localization in cells and axons, *Science* **200:**1180–1182.

Wilson, R. C., Kesner, J. S., Kaufman, J. M., Uemura, T., Akema, T., and Knobil, E., 1984, Central electrophysiological correlates of pulsatile luteinizing hormone secretion in the rhesus monkey, *Neuroendocrinology* **39:**256–560.

Zaborzsky, L., 1982, Afferent connections of the medial basal hypothalamus, in: *Advances in Anatomy, Embryology and Cell Biology,* Vol. 69 (W. Hild, J. van Limborgh, R. Ortmann, J. E. Pauly, and T. H. Schiebler, eds.), Springer-Verlag, New York.

Zaborszky, L., and Leranth, C., 1985, Simultaneous ultrastructural demonstration of retrogradely transported horseradish peroxidase and choline acetyltransferase immunoreactivity, *Histochemistry* **82:**529–537.

Zaborszky, L., and Makara, G. B., 1979, Intrahypothalamic connections: An electron microscopic study of the rat, *Exp. Brain Res.* **34:**201–215.

Zaborszky, L., Leranth, C., and Heimer, L., 1984, Ultrastructural evidence of amygdalofugal axons terminating on cholinergic cells of the rostral forebrain, *Neurosci. Lett.* **521:**219–225.

15

Corticotrophin-Releasing Hormone

Effects on Stress-Related Behavior in Rats

JACK E. SHERMAN and NED H. KALIN

1. INTRODUCTION

1.1. Background Information on Corticotrophin-Releasing Hormone

For over 25 years, investigators searched for the elusive hypothalamic hormone corticotrophin-releasing factor, hypothesized to stimulate the pituitary release of ACTH. Finally, in 1981, Vale *et al.* (1981) identified a 41-amino-acid peptide that potently and specifically promoted the release of ACTH from the pituitary corticotroph. Further work demonstrated that this peptide met the criteria established for a releasing hormone. Like other hypothalamic releasing hormones, corticotrophin-releasing hormone (CRH) and its receptors were found to have extrahypothalamic brain distribution (Olschowka *et al.*, 1982; Schipper *et al.*, 1983; Swanson *et al.*, 1983). From results of initial studies, it was hypothesized that CRH systems play an integrative role in an organism's response to stress.

Evidence linking CRH to the stress response comes from a number of observations. First, CRH and its receptors are found in brain regions associated with behavioral arousal and anxiety (De Souza *et al.*, 1984; Olschowka *et al.*, 1982; Swanson *et al.*, 1983). Second, CRH administered intracerebroventricularly (ICV) increases neuronal activity in these areas (Valentino *et al.*, 1983) and produces electroencephalographic changes suggestive of increased arousal (Ehlers *et al.*, 1983). Third, ICV CRH affects the sympathetic nervous system, resulting in increased arterial pressure, heart rate (Brown *et al.*, 1982, 1983; Fisher *et al.*, 1982), oxygen consumption (Brown *et al.*, 1982), and plasma concentrations of both ACTH

JACK E. SHERMAN and NED H. KALIN • Department of Psychiatry, University of Wisconsin–Madison, and Psychiatry Service, William S. Middleton Memorial Veterans Hospital, Madison, Wisconsin 53705.

and cortisol (Donald *et al.*, 1983; Insel *et al.*, 1984; Kalin *et al.*, 1983; Veldhuis and DeWied, 1984). Lastly, ICV CRH induces behaviors suggestive of enhanced emotional reactivity (Britton *et al.*, 1982, 1985; Kalin *et al.*, 1983; Morley and Levine, 1982; Sutton *et al.*, 1982; Veldhuis and DeWied, 1984), some of which have been demonstrated to be independent of the peptide's effects on the pituitary–adrenal axis (Morley and Levine, 1982; Veldhuis and DeWied, 1984).

1.2. Goals of Chapter

This chapter presents data from our recent work with rats exploring the hypothesized integrative function of CRH in the stress response and discusses issues concerning the interpretation of these and other relevant data. In this study we measured a constellation of possible stress-related behaviors, including analgesia, after either ICV or peripheral administration of CRH. Our interest in assessing possible antinociceptive effects of ICV CRH in the rat was motivated by the large literature demonstrating a wide range of stressors that activate endogenous mechanisms of pain inhibition (Amir *et al.*, 1980; Chance, 1980; Watkins *et al.*, 1982; Terman *et al.*, 1984). Because CRH plays a significant role in pituitary–adrenal activation and has been shown to elicit sympathetic and behavioral responses similar to those activated by stress, it was of interest to explore the possibility that CRH also would elicit analgesia.

2. EXPERIMENTAL METHODS

2.1. Subjects

Naive male rats obtained from Sasco–King Laboratories, Oregon, Wisconsin, were used for all three experiments described. The rats were individually housed on pine chip bedding in clear polypropylene cages measuring 30.2 by 26.2 by 13.5 cm. The cages were fitted with a wire cover with food and bottle holder. Food and water were available *ad libitum* in the colony. Procedures were conducted no sooner than 1 week after the rats arrived at our colony and only during the light phase of the 12-hr light–dark cycle (0600 to 1800).

2.2. Apparatus and Drugs

Assessment of pain sensitivity was conducted with a hot-plate apparatus consisting of a Haake water bath and pump that heated water to 51.3°C and circulated it under the surface of an aluminum plate during testing (for further details, see Sherman, 1979). Constant white noise was maintained in the test environment, and food (but not water) was available during testing. The CRH solutions were prepared as previously described (Kalin *et al.*, 1983), using synthetic rat CRH (Bachem Co., Torrance, CA). The vehicle was 0.9% sterile saline.

2.3. Intracerebroventricular Procedures

For delivery of ICV CRH, rats under pentobarbital anesthesia were stereotactically implanted with a stainless steel guide cannula (Plastic Products, Roanoke, VA) placed dorsal to the left lateral ventricle. With the head horizontal, and using bregma as a reference point, the cannula tip was placed 1.0 mm posterior, 1.8 mm lateral, and −3.3 mm vertical to the skull surface. When drug was administered, a stainless steel inner cannula was inserted into the guide cannula, extending 1.0 mm beyond the guide, which was placed within the lateral ventricle. An infusion pump administered 5 μl of CRH solution or vehicle over a period of 90 sec. At all other times, a stylet was maintained in the guide. After the experiment, verification of cannula placement was made by visually inspecting the dispersion of 5 μl of methylene blue dye throughout the ventricular system.

2.4. Dependent Measures and Testing Procedures

Twenty minutes after drug administration, each rat was placed on the hot-plate surface for 30 sec, and the latency until a paw lick or jump was recorded. Four subsequent hot-plate tests were conducted at 20-min intervals. Two minutes before each hot-plate test, each rat was observed and behaviorally rated for sleeping, rearing, walking, grooming, self-gnawing, pica (chewing of pine chips or fecal material), burrowing in the pine chips, and pellet chewing. The single behavior in which the rat was engaging was rated every 10 sec, so that a total of 12 ratings could be made in each 2-min interval. All testing was conducted between 1100 and 1500 hr, during the light phase of the cycle, by an observer who was unaware of the treatment conditions.

2.5. Experimental Design

All rats were exposed to the test environment on three occasions before surgery. This included exposure to the unheated hot-plate apparatus. In addition, rats were exposed to the test environment once after surgery and again after the first test. For testing, the rats, in their plastic home cages, were returned to the test environment 5 to 8 days after surgery. A within-subject counterbalanced design was employed to compare the effects of CRH and vehicle. Rats were infused with either CRH solution or vehicle, and observational and hot-plate testing were conducted. Five to 7 days later, the other drug was administered, and testing was repeated. Two doses of CRH were tested. The low dose (experiment 1A, $n = 13$) was 0.3 μg, and the high dose (experiment 1B, $n = 14$) was 3.0 μg.

In experiment 2 ($n = 28$), CRH (0.3 or 3.0 μg) and vehicle were administered subcutaneously in a volume of 0.3 ml to rats that had not undergone surgery. Ten rats received CRH first, and ten received vehicle first. The design was completely balanced with respect to CRH dose, order of testing, and number of subjects in each treatment group.

2.6. Statistical Analyses

Initially, a multivariate analysis of variance (MANOVA) was conducted to assess the overall effect of drug (CRH versus vehicle), order of testing (CRH or vehicle first), and the interaction of these two independent variables for all nine dependent variables taken together. The experimental factors in the MANOVA that yielded an overall significant effect were then evaluated by the analysis of variance (ANOVA) for each behavior considered separately. The criterion for statistical significance was $P < 0.05$. Statistical comparisons across experiments 1A and 1B were not made because the two experiments were done at different times and with different shipments of animals. Thus, no inferences can be made regarding the effects of different doses on behavior in the ICV CRH experiments.

3. RESULTS AND DISCUSSION

3.1. Overall Effects of ICV CRH

Figure 1 presents results for experiments 1A and 1B, showing the mean total frequency of each of the eight observational measures and the average response

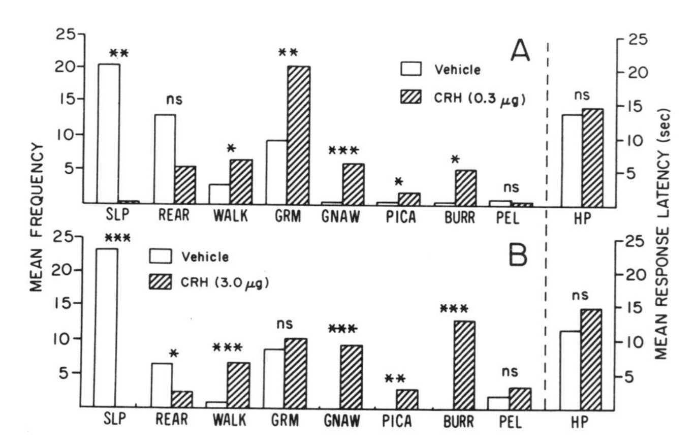

FIGURE 1. Effect of CRH and vehicle on the mean frequency of sleeping (SLP), rearing (REAR), walking (WALK), grooming (GRM), self-gnawing (GNAW), chewing on inedible objects (PICA), burrowing in pine-chip bedding (BURR), and chewing food pellets (PEL), and on mean latency to respond in the hot-plate test (HP). (A) Experiment 1A, low dose. (B) Experiment 1B, high dose. *$P < 0.05$; **$P < 0.01$; *** $P < 0.001$; ns, $P > 0.05$. [Reproduced with permission of Pergamon Press, Ltd. (Sherman and Kalin, 1986).]

latencies on the hot plate during the entire test session; CRH had a marked effect on behavior. The MANOVAs for each experiment revealed statistically significant effects for CRH versus vehicle [$F(9,3) = 105.85$ and $F(9,4) = 30.77$, respectively] but no other effects. The absence of a significant effect of either drug order or interaction of drug order and drug treatment suggests that use of a within-subject design for assessing the effects of CRH is valid and does not pose interpretive problems.

Subsequent ANOVAs revealed that 0.3 μg of CRH (Fig. 1A) significantly increased the mean frequency of walking, grooming, self-gnawing, burrowing, and pica relative to vehicle and significantly decreased the mean frequency of sleeping. The decremental effect of CRH on rearing just missed statistical significance ($P = 0.052$). No statistically significant effect on pellet chewing was observed, and there was no overall effect on latency to respond on the hot-plate test. The 3.0-μg dose of CRH (Fig. 1B) significantly increased the overall mean frequency of walking, gnawing, pica, and burrowing relative to vehicle and significantly decreased the mean frequency of sleeping and rearing. No statistically significant overall effect on grooming, pellet chewing, or latency to respond on the hot-plate test was obtained with the higher dose.

It is notable that whereas burrowing, pica, and self-gnawing rarely occurred in vehicle-treated rats, they were frequently observed in CRH-treated rats. These behaviors appear to be highly stereotyped, especially at the 3.0-μg dose. Because CRH at higher doses has been shown to induce electroencephalographic changes indicative of psychomotor seizures (Ehlers *et al.*, 1983), one possible mechanism underlying these stereotypic behaviors might be seizure activity. Although the effects of CRH on burrowing, pica, and self-gnawing have not been previously reported, the observed effects of CRH on sleeping, grooming, walking, and rearing are consistent with previously reported effects of ICV CRH in rats (Britton *et al.*, 1982; Morley and Levine, 1982; Sutton *et al.*, 1982; Veldhuis and DeWied, 1984).

3.2. Time Course Analyses of ICV CRH Effects

Because CRH blocked sleep, it was interesting to consider the extent to which the greater frequency of sleeping behavior in vehicle-treated rats accounts for some of the behavioral effects attributed to CRH. In other words, because sleeping animals do not engage in active behavior, simply the maintenance of wakefulness by CRH might account for some of the apparent effects of the drug. In at least one other behavioral study, it was noted that control rats slept, whereas CRH-treated rats did not (Koob *et al.*, 1982). The occurrence of sleep—or, at least, decreased arousal—in control rats may be a more frequent phenomenon than has been reported, because many of the reported behavioral studies in the rat have been conducted during the rat's sleep cycle.

Inspection of the sleep pattern of the vehicle-treated rats in our study suggested that almost all of these animals were awake at the time of the first 20-min test. Statistical analysis (ANOVA) of the mean differences between CRH- and vehicle-

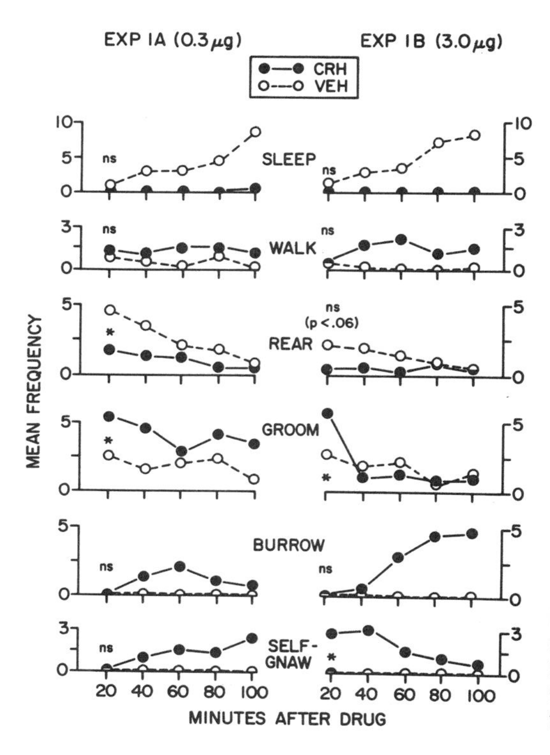

FIGURE 2. Effect of CRH and vehicle over time on the mean frequency of rated behaviors for each test interval. Significance tests were conducted only for the first test interval. *$P < 0.05$.

treated rats at each test interval indicated that the two groups were not different at 20 min but were afterwards. This provided an opportunity to assess the effects of CRH under conditions in which controls were displaying comparable levels of wakefulness, as indexed by the frequency of sleep. Figure 2 shows the time course results for a subset of all the behaviorally rated dependent measures in experiments 1A and 1B.* Variables in which ANOVA revealed a significant difference at the 20-min test are indicated by an asterisk. In experiment 1A, only the mean frequencies of grooming and rearing differed significantly for the CRH and vehicle treatments. In experiment 1B, the mean frequencies of grooming and self-gnawing were significantly different, and rearing was marginally significant. Thus, under conditions of comparable arousal as defined here, CRH-treated rats displayed specific effects that are unlikely to be attributable to differences in arousal alone.

The behavioral effects of ICV CRH observed at both doses may be characterized as increased grooming and decreased rearing, two behaviors that have been

* The observational variables selected for presentation were those showing a statistically significant main effect of drug or of drug × test interval interaction. Only pellet consumption failed to meet these criteria. Although the measure of pica did meet these criteria in both experiments 1A and 1B, the average frequency of occurrence for this behavior was so low at each test inverval (always less than 1.0) that a time course analysis was not presented. The hot-plate results are discussed.

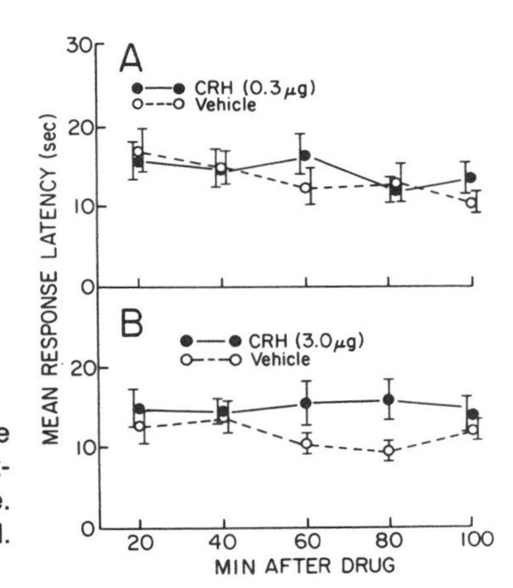

FIGURE 3. Effect of CRH and vehicle over time on the mean latency to respond for each hot-plate test. (A) Experiment 1A, low dose. (B) Experiment 1B, high dose. [Reproduced with permission of Pergamon Press, Ltd. (Sherman and Kalin, 1986).]

shown to be elicited by novelty stress (Britton and Britton, 1981) and enhanced by ICV CRH (Britton *et al.*, 1982). It is interesting that these behaviors were evidenced in a familiar environment, suggesting that environmental novelty may not be necessary for their elicitation by CRH. The higher dose of CRH yielded more frequent self-gnawing than was exhibited by controls. It is not clear whether this behavior reflects a component of normal stress-related behavior.

Figure 3 presents results of hot-plate testing for each of the test intervals in experiments 1A and 1B. CRH did not yield statistically significant effects on hot-plate latencies averaged across all test intervals, nor was there a statistically significant interaction of CRH and test interval. These data suggest that CRH does not activate antinociceptive processes, a finding that is consistent with findings of Britton *et al.* (1985). They found that a 1.0-μg dose of ICV CRH did not cause analgesia. In a recent experiment employing a between-subjects design and both the hot-plate and tail-flick tests of analgesia, we (Sherman and Kalin, 1986) again found that 3.0 μg of ICV CRH did not produce significantly longer latencies. In contrast, 10 μg of ICV morphine yielded potent evidence of analgesia with both measures, suggesting the adequacy of our procedures to detect analgesia. The absence of an antinociceptive effect of ICV CRH is of interest in view of evidence showing that activation of antinociceptive processes is a common response to stress (Amir *et al.*, 1980; Terman *et al.*, 1984) and ICV CRH produces what otherwise appears to be a generalized physiological response to stress. Thus, the results of the analgesic assessments suggest some limitations of the hypothesized integrative role of CRH brain systems in the stress response.

3.3. Effects of Peripheral CRH

In contrast to analyses of experiments 1A and 1B, an overall MANOVA failed to show a statistically significant effect of peripherally administered CRH

[$F(9,8) = 0.53$]. To confirm this apparent absence of a drug effect, univariate analyses were conducted. In no instance was a statistically significant effect of CRH found.

Thus, the peripheral administration of CRH did not produce the behavioral effects observed after central administration of the peptide. These results are consistent with other studies in rats showing that peripheral CRH is without effect at doses that yield potent effects when administered centrally (Sutton *et al.*, 1982). The results of this experiment suggest that none of the CRH effects observed in experiments 1 and 2 were likely to be mediated by peripheral mechanisms alone.

4. SUMMARY

Because CRH has been hypothesized to serve an integrative function in the stress response, and stress has been shown to activate endogenous mechanisms of pain inhibition, a focus of the present study was on the possible antinociceptive effects of CRH. The effect of CRH on the latency to respond on the hot-plate test was studied along with other possible stress-related behaviors. The effects of low-dose (0.3 μg) and high-dose (3.0 μg) ICV CRH were compared to those of vehicle, employing a within-subject design. At both doses, ICV CRH elicited changes in behavior reported by other investigators to be stress related (viz., increased grooming and walking and decreased sleeping and rearing) as well as a set of new behaviors rarely seen in controls (viz., burrowing, pica, and self-gnawing). Twenty minutes after drug administration, at a time when CRH- and vehicle-treated rats displayed comparable levels of behavioral arousal, CRH-treated rats displayed a greater frequency of grooming and a decreased frequency of rearing at both doses studied. The higher dose yielded a greater frequency of self-gnawing, as well. The results of the hot-plate test suggested that CRH does not activate antinociceptive processes. In a second experiment, peripherally administered CRH was without effect on this measure. The results obtained here generally are consistent with the hypothesis that central CRH plays a role in the integration of the stress response; the exception thus far is the absence of a role in the activation of antinociception.

The task of assessing whether CRH is involved in the mediation of stress-related behavior is a difficult one. Largely, the difficulty is caused by the lack of specificity in the stress construct, especially with regard to behavior. However, addressing this task will undoubtedly be fruitful. Not only will we learn about the role of CRH in brain function but of necessity we will be forced to deal with the larger issue of specifying the meaning of stress.

ACKNOWLEDGMENT. This work was supported in part by the Veterans Administration.

REFERENCES

Amir, S., Brown, Z. W., and Amit, Z., 1980, The role of endorphins in stress: Evidence and speculations, *Neurosci. Biobehav. Rev.* **4:**77–86.

Britton, D. R., and Britton, K. T., 1981, A sensitive open field measure of anxiolytic drug activity. *Pharmacol. Biochem. Behav.* **15:**577–582.

Britton, D. R., Koob, G. F., Rivier, J., and Vale, W., 1982, Intraventricular corticotropin-releasing factor enhances behavioral effects of novelty, *Life Sci.* **31:**363–367.

Britton, K. T., Morgan, J., Rivier, J., Vale, W., and Koob, G., 1985, Chlordiazepoxide attenuates response suppression induced by corticotropin-releasing factor in the conflict test, *Psychopharmacology* **86:**170–174.

Brown, M. R., and Fisher, L. A., 1983, Central nervous system effects of corticotropin-releasing factor in the dog, *Brain Res.* **280:**75–79.

Brown, M. R., Fisher, L A., Rivier, J., Spiess, J., Rivier, C., and Vale, W., 1982, Corticotropin-releasing factor: Effects on the sympathetic nervous system and oxygen consumption, *Life Sci.* **30:**207–210.

Chance, W. T., 1980, Autoanalgesia: Opiate and non-opiate mechanisms, *Neurosci. Biobehav.* **4:** 55–67.

De Souza, E. B., Perrin, M. H., Insel, T. R., Rivier, F., Vale, M., and Kuhar, M., 1984, Corticotropin-releasing factor receptors in rat forebrain autoradiographic identification, *Science* **224:** 1449–1451.

Donald, R. A., Redekopp, C., Cameron, V., Nicholls, M. G., Solton, J., Livesey, J., Espina, E. A., Rivier, J., and Vale, W., 1983, The hormonal actions of corticotropin-releasing factor in sheep: Effect of intravenous and intracerebroventricular injection, *Endocrinology* **113:**866–870.

Ehlers, C. L., Henrikson, S. J., Wang, M., Rivier, J., Vale, W., and Bloom, F. E., 1983, Corticotropin-releasing factor produces increases in brain excitability and convulsive seizures in rats, *Brain Res.* **278:**332–336.

Fisher, L. A., Rivier, J., Rivier, C., Spiess, J., Vale, W., and Brown, M. R., 1982, Corticotropin-releasing factor (CRF): Central effects on mean arterial pressure and heart rate in rats, *Endocrinology* 110:2222–2224.

Insel, T. R., Aloi, J. A., Goldstein, D., Wood, J. W., and Jimerson, D. C., 1984, Plasma cortisol and catecholamine response to intracerebroventricular administration of CRF to rhesus monkeys, *Life Sci.* **34:**1873–1878.

Kalin, N. H., 1985, Behavioral effects of ovine corticotropin-releasing factor administered to rhesus monkeys, *Fed. Proc.* **44:**249–253.

Kalin, N. H., Shelton, S. E., Kraemer, G. W., and McKinney, W. T., 1983, Corticotropin-releasing factor administered intraventricularly to rhesus monkeys, *Peptides* **4:**217–220.

Koob, G. F., Le Moal, M., Bloom, F. E., Sutton, R. E., Rivier, J., and Vale, W., 1982, Corticotropin releasing factor produces behavioral activation and improves learning in rats, *Neurosci. Abstr.*

Morley, J. E., and Levine, A. S., 1982, Corticotropin-releasing factor, grooming and ingestive behavior, *Life Sci.* **31:**1459–1464.

Olschowka, J. A., O'Donohue, T. L., Mueller, G. P., and Jacobowitz, D. M., 1982, Hypothalamic and extra-hypothalamic distribution of CRF-like immunoreactive neurons in the rat brain, *Neuroendocrinology* **35:**305–308.

Schipper, J., Steinbusch, H. W. M., Vermes, I., and Tilders, F. J. H., 1983, Mapping of CRF-immunoreactive nerve fibers in the medulla oblongata and spinal cord of the rat, *Brain Res.* **267:**145–150.

Sherman, J. E., 1979, The effects of conditioning and novelty on the rat's analgesic and pyretic responses to morphine, *Learn. Motivat.* **10:**383–418.

Sherman, J. E., and Kalin, N. H., 1986, ICU-CRH potently affects behavior without altering antinociceptive responding, *Life Sci.* **39:**433–441.

Sutton, R. E., Koob, G. F., LeMoal, M., Rivier, J., and Vale, W., 1982, Corticotropin-releasing factor produces behavioral activation in rats, *Nature* **297:**331–333.

Swanson, L. W., Sawchenko, P. E., Rivier, J., and Vale, W., 1983, Organization of ovine corticotropin-releasing factor immunoreactive cells and fibers in the rat brain: An immunohistochemical study, *Neuroendocrinology* **36:**165–186.

Terman, G. W., Shavit, Y., Lewis, J. W., Cannon, J. T., and Liebeskind, J. C., 1984, Intrinsic mechanisms of pain inhibition: Activation by stress, *Science* **226:**1270–1277.

Vale, W., Spiess, J., Rivier, C., and Rivier, J., 1981, Characterization of a 41-residue ovine hypothalamic peptide that stimulates secretion of corticotropin and β-endorphin, *Science* **213:**1394–1397.

Valentino, R. J., Foote, S. L., and Aston-Jones, G., 1983, Corticotropin-releasing factor activates noradrenergic neurons of the locus coeruleus, *Brain Res.* **270:**363–367.

Veldhuis, H. D., and DeWied, D., 1984, Differential behavioral actions of corticotropin-releasing factor (CRF), *Pharmacol. Biochem. Behav.* **21:**707–713.

Watkins, L. R., and Mayer, D. J., 1982, The neural organization of endogenous opiate and nonopiate pain control systems, *Science* **216:**1185–1192.

III

Brain Peptides and Receptors

16

Cholecystokinin

ROSALYN S. YALOW and JOHN ENG

1. INTRODUCTION

In his 1904 Nobel lecture, Ivan Pavlov described the chronic esophageal fistula preparation with which he had made many fundamental discoveries concerning canine digestion. In a particularly intriguing experiment, Pavlov observed that fasted dogs ate voraciously and continually while ingested food was removed through the fistula, becoming satiated only when gastric chyme was placed in the duodenum. This observation led to the concept of postgastric satiety mechanisms and, after the discovery of intestinal hormones, provided the basis for studying intestinal hormones as satiety-inducing agents. Many years later, Schally *et al.* (1967) demonstrated that enterogastrone, a gut extract undoubtedly rich in cholecystokinin (CCK), inhibited feeding by fasted mice. Since that time there have been numerous reports concerning the possible role of CCK-related peptides peripherally administered as a putative satiety factor with a variety of functions other than its well-known classical roles, i.e., its effect on gallbladder contractions and release of pancreatic enzymes.

2. PERIPHERAL ADMINISTRATION OF CHOLECYSTOKININ PEPTIDES

The question first to be addressed is whether the effects reported to be observed following peripheral administration of CCK-like peptides are a consequence of circulating levels that are in the physiological or pharmacological range. What are the fasting and food-stimulated circulating levels of CCK? The initial reports by Young *et al.* (1969) had suggested that stimulated plasma levels of CCK were as high as 25–50 fmol/ml. More recent studies by Walsh *et al.* (1982) and Calam *et*

ROSALYN S. YALOW and JOHN ENG • Solomon A. Berson Laboratory, Veterans Administration Medical Center, Bronx, New York 10468.

al. (1982) were in agreement in demonstrating that the primary form of CCK in plasma is the C-terminal octapeptide (CCK8), with little or no larger CCK peptides present. They also reported that basal plasma levels of CCK8 are about 1 fmol/ml or less and that peak concentrations following oral administration of a mixed meal or fat are 4–6 fmol/ml. Walsh *et al.* (1982) also observed that following CCK8 administration at the rates of 16, 43, and 98 pmol/kg per hr for 45-, 30-, and 30-min periods, respectively, over a background dose of intravenous secretin administered at 41 pmol/kg per hr, the increases in plasma CCK during the three dose levels were 5, 10, and 17 fmol/ml, respectively. The highest dose was the maximal that could be given without side effects. Furthermore, one-half of the maximal observed amylase response was obtained with an increase in CCK8 immunoreactivity of only 7–8 fmol/ml plasma. It is necessary to keep these infusion rates and concentrations in mind when considering other reports concerning biological effects in primates following exogenous administration of CCK.

There have been several reports concerning administration of CCK to man and its role in food intake. Sturdevant and Goetz (1976) reported that continuous infusion in man of 300 pmol CCK33/kg per hr resulted in an increase, not a decrease, in food intake over a 20-min period. This group also reported that acute administration of 50 pmol CCK33/kg resulted in decreased food intake. Obviously plasma levels achieved in these experiments were greatly in excess of the physiological range. Gibbs *et al.* (1976) reported that intravenous administration of CCK33 or CCK8 to rhesus monkeys at dose rates 10 to 40 times those employed by Sturdevant and Goetz (1976) in human subjects resulted in suppression of food intake, a 70% decrease at the highest dose administered. Since physiological levels of CCK are not likely to be that different in man and monkey, it would seem that the levels achieved were well into the pharmacological range. More recently, Kissileff *et al.* (1981) have reported that continuous administration of CCK8 at an infusion rate of 240 pmol/kg per hr resulted in decreased food intake through reduction in meal duration in nine of 12 subjects studied. This is more than twice the peak doses used by Walsh *et al.* (1982), but none of the subjects was reported to have complained of side effects such as diarrhea. From this sampling of studies, it appears that if there were CCK-induced inhibition of food intake, the associated plasma levels would be considerably in excess of those achieved in response to normal feeding. Thus, peripheral CCK *per se* does not appear to be the physiological satiety factor.

3. CHOLECYSTOKININ PEPTIDES IN BRAIN

The discovery by Vanderhaeghen *et al.* (1975) of a brain peptide cross reacting with antigastrin antibodies was followed by numerous reports of the localization, concentration, and molecular forms of brain CCK. There is general agreement from many laboratories that CCK is widely distributed throughout the brain except in the cerebellum and that the concentrations are highest in the cortex (Table I).

TABLE I. Cholecystokinin-like Immunoreactivity (pmol/g) in Tissues of Adult Pigs[a]

		Methanol extractant		O.1 N HCl extractant	
N	Location	G1[b]	R71[b]	Gl[c]	R71[c]
	Brain				
3	Cortex	<3	315 ± 16	356 ± 18	80 ± 9
3	Olfactory bulb	<3	99 ± 29	131 ± 37	45 ± 3
3	Hypothalamus	<3	25 ± 5	38 ± 9	12 ± 4
3	Thalamus	<3	14 ± 1	26 ± 2	6 ± 1
3	Midbrain	<3	25 ± 4	37 ± 5	10 ± 2
2	Pons	<3	6 ± 1	17 ± 5	5 ± 2
3	Medulla	<3	6 ± 1	16 ± 3	5 ± 2
	Gut				
3	Duodenal mucosa	<3	318 ± 141	983 ± 271	295 ± 13
1	Duodenum	<3	118 ± 21	442 ± 31	240 ± 13
1	Jejunum	<3	63 ± 2	347 ± 31	161 ± 1
1	Ileum	<3	5 ± 1	45 ± 4	23 ± 5

[a] Reproduced from Eng *et al.* (1982).
[b] CCK-8 standard, mean ± SEM.
[c] CCK-33 standard, mean ± SEM.

The finding that CCK concentrations in brain are comparable to those in gut suggests synthesis of this peptide in neuronal as well as in gut mucosal tissues. Our immunohistochemical studies of rabbit cortical tissues employing a C-terminal CCK antiserum revealed deep staining of many neurons throughout the cortical gray matter and more diffuse staining in the subcortical white matter (Straus *et al.*, 1977) (Fig. 1). This finding has been confirmed and extended by several groups, and positive-staining cell bodies and fibers have been observed in hippocampal areas, in the olfactory bulb, and in hypothalamic, preoptic, and amygdaloid nuclei as well as in the dorsal horn of the spinal cord (Hökfelt *et al.*, 1978; Larsson and Rehfeld, 1979; Vanderhaeghen *et al.*, 1978; Loren *et al.*, 1979; Innis *et al.*, 1979). There have been no reports of similar studies with an N-terminal antiserum.

The finding that the abundant CCK immunoreactivity in cortical gray matter was restricted to neurons stimulated our studies (Pinget *et al.*, 1978) of the distribution of CCK peptides in subcellular fractions of rat cerebral cortex prepared according to the method of Whittaker (1964). These studies demonstrated the presence of immunoreactive CCK in the pellet identified by electron microscopy to contain a high proportion of synaptic vesicles. The recovery in this pellet of 40% of the toal immunoreactivity in the initial cortical extract (Fig. 2) is quite comparable to that of other peptides such as vasoactive intestinal peptide (VIP) (Giachetti *et al.*, 1977) and somatostatin (Epelbaum *et al.*, 1977), which have also been reported to localize in the synaptosomes. Immunoreactive CCK was released from synaptosomes incubated in solutions containing K^+ and Ca^{2+} according to the paradigms

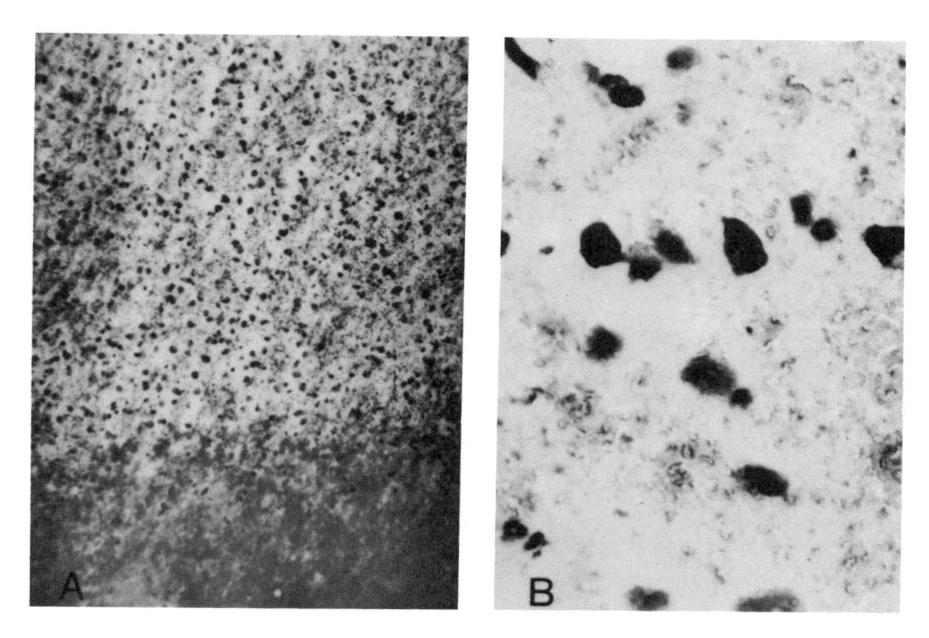

FIGURE 1. (A) Low-power photomicrograph of rabbit cerebral cortex (frontal lobe). The tissue was stained by the immunoperoxidase technique using rabbit B antiserum in a 1:10 dilution. Staining of individual cell bodies can be seen in all layers of cortical gray matter, and diffuse staining can be seen at the bottom in subcortical white matter (×33). (B) High-power photomicrograph showing staining of cell bodies in cortical gray matter (×208). (Reproduced from Straus *et al.*, 1977.)

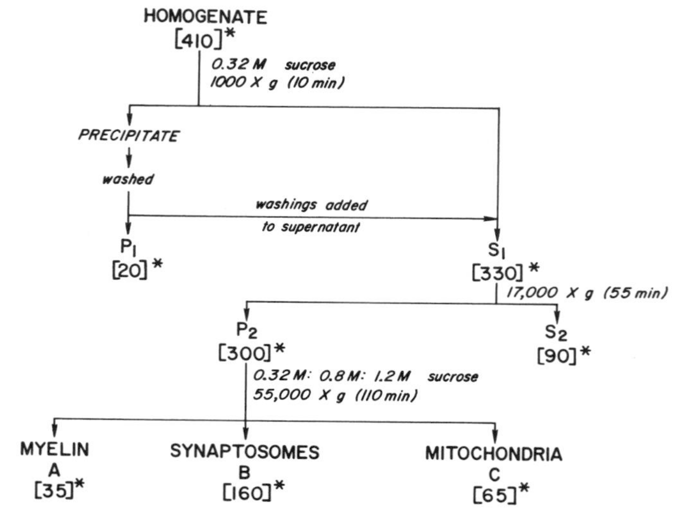

FIGURE 2. Flow diagram and recovery of CCK-like immunoreactivity for subcellular fractionation of a rat cerebral cortical extract. [ ]* = ng CCK-8 equivalent/g wet weight tissue. (Reproduced from Pinget *et al.*, 1978.)

usually employed to evaluate neuronal chemicals purported to have a role in synaptic physiology (Pinget *et al.*, 1979). Subsequently, our findings were confirmed by Emson *et al.* (1980), who extended their studies to synaptosomal fractions of the hypothalamus, and Dodd *et al.* (1980), who studied synaptosomes from brainstem and corpus striatum as well as cortex and hypothalamus. More recently, Klaff *et al.* (1981) performed similar studies and extended them to rat striatum and thalamus and used not only K^+ but also veratrine as depolarizing agents.

4. MOLECULAR FORMS OF CHOLECYSTOKININ

These studies taken together suggest a role for CCK as a neuroregulator. Before we discuss what it might be regulating, it would be of interest to compare the concentrations and molecular forms of brain and gut CCK. In the adult pig, the concentrations of CCK peptides in the brain cortex and duodenum as measured with a C-terminal antiserum are quite comparable, averaging about 0.4 nmol/g in both organs (Ichihara *et al.*, 1984a). In the brain of the adult rat, the concentration averages about 0.3 nmol/g, which is about three times the duodenal concentration (Ichihara *et al.*, 1983). Thus, in these two species, the concentrations of CCK peptides in the brain are quite comparable to those in the duodenum. These relative concentrations are quite different from those of another brain–gut peptide, vasoactive intestinal peptide, (VIP). The concentration of VIP in the rat duodenum (~400 pmol/g) is about ten times that in the brain (~40 pmol/g) (Ichihara *et al.* 1983); in the pig duodenum the concentration of VIP (~200 pmol/g) is almost seven times that in the brain (~30 pmol/g) (Ichihara *et al.*, 1984a).

The molecular forms of CCK in brain and duodenum differ (Ichihara *et al.*, 1984a,b). The predominant molecular form in brain, as determined with a C-terminal antiserum, is CCK8 (Fig. 3). There is a relatively minor component corresponding to a precursor molecule, CCK58. When assayed with an N-terminal antiserum, peptides corresponding to CCK33, -39, and -58 desoctapeptides have been purified and sequenced from pig brain (Eng *et. al.*, 1983) (Fig. 4). This suggests that the cleavage of CCK8 occurs very early in the processing of the CCK precursor in the brain. The molecular forms of CCK in the pig duodenum are quite different from those found in brain (Ichihara *et. al.*, 1984a). Although in both organs CCK8 predominates in methanol extracts, the acid extracts of the duodenum contain CCK33, -39, and -58 in about equal concentrations as the corresponding desoctapeptides (Fig. 4). Processing in the rat duodenum is similar to that in the pig duodenum in that the intact larger molecular forms are as prominent as CCK8 (Fig. 5) (Ichihara *et al.*, 1984b). It is of interest that in methanol extracts of the rat duodenum there is a larger C-terminal fragment that is equally as prominent as CCK8. This fragment has been purified and identified as a 22-amino-acid CCK peptide (Eng *et al.*, 1984).

The observation that there are differences in the CCK peptides in brain and gut raises the question of whether the differences could be a consequence of differences in posttranslational processing or represent expression of different mRNAs coding for CCK in the two tissues. Through molecular cloning it has been dem-

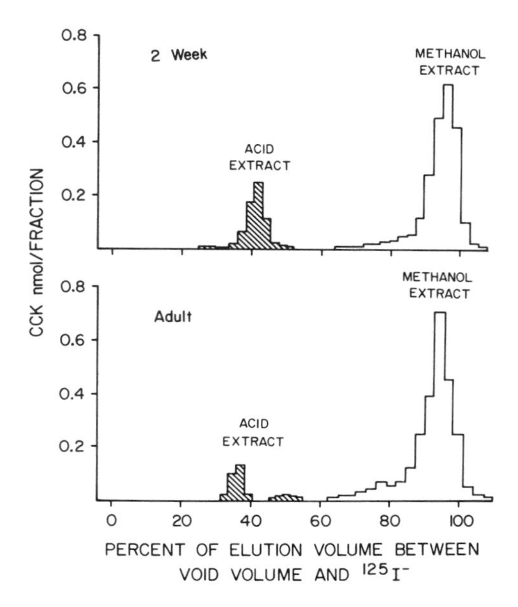

FIGURE 3. Sephadex G50 superfine gel filtration of extracts of rat brain. The methanol extracts were dried in a vacuum centrifuge, reconstituted in 0.02 M barbital buffer, and applied to the Sephadex columns, which were equilibrated and eluted with barbital buffer. The peak in the methanol extract is CCK8. Acid extracts were applied to columns equilibrated and eluted with 0.1 M HCl. The peak in the acid extract corresponds to a form with a molecular weight larger than CCK33 or CCK39, and in other species it has been identified as CCK58. The column eluates were assayed with a C-terminal antiserum. Elution volumes for authentic CCK8, -12, and -33 in columns eluted with 0.02 M barbital are 96, 80, and 50%, respectively. The elution volume for authentic CCK33 in columns eluted with 0.1 M HCl is 58%. (Reproduced from Ichihara *et al.*, 1984b.)

onstrated that there is a 114-amino-acid preprocholecystokinin that is identical in both tissues (Gubler *et al.*, 1984). What accounts for the differences in posttranslational processing of CCK in neuronal and mucosal tissues has not yet been well studied, although a nontrypsin converting enzyme capable of cleaving smaller CCK peptides has been identified in brain extracts (Ryder *et al.*, 1980).

5. ONTOGENY OF CHOLECYSTOKININ PEPTIDES

Not only does posttranslational processing of preprocholecystokinin differ in brain and gut but so too does the ontogeny of CCK peptides. The CCK concentrations in methanol and acid extracts of rat cortex, duodenum, and jejunum are shown in

Fig. 6 (Ichihara *et al.*, 1983). The hormonal forms in these extracts have been described earlier (Figs. 3 and 5). The CCK concentration in the brain cortex of the 3-day neonate is less than 10% of the adult level, which is reached by day 28. In contrast, the concentrations of duodenal and jejunal CCK in the 3-day neonate equal adult levels with a significant peaking at 14 days. Similar studies of the ontogeny of VIP and secretin in the rat suggest that the concentrations of neuronal peptides (VIP or CCK in brain) are low at birth, reflecting relative immaturity of these tissues, whereas the concentrations of peptides in endocrine tissues (gut CCK and secretin among others) peak at or shortly after birth (Ichihara, 1983). However, in the pig, which is more mature at birth than is the rat, both CCK (Fig. 7) and VIP in the cortex at birth average about 50% of adult levels (Ichihara *et al.*, 1984a).

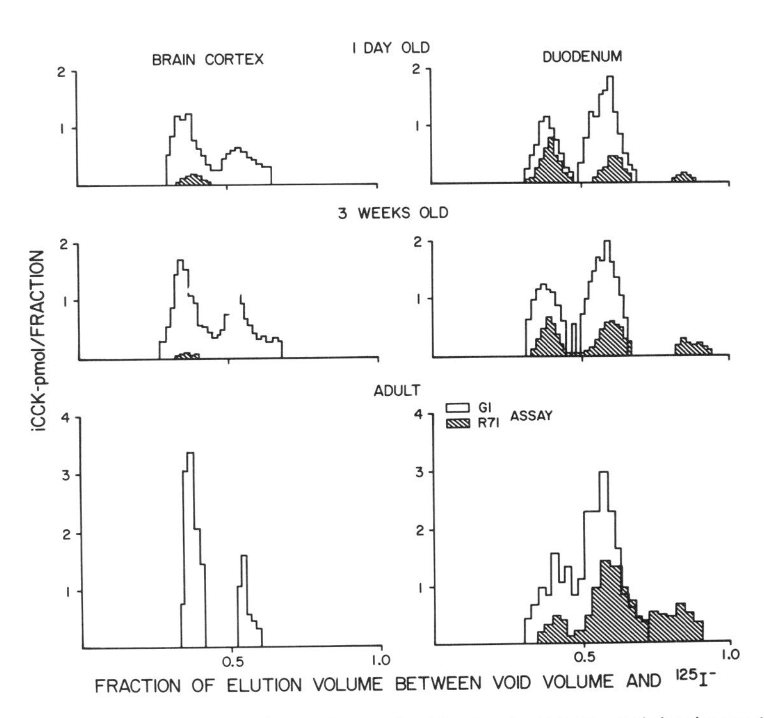

FIGURE 4. Sephadex G50 gel filtration patterns of acid extracts of brain and duodenum taken from the 1-day, 3-week, and adult pig. The conditions for equilibration and elution of columns are given in the legend to Fig. 3. In the brain, CCK desoctapeptides predominate in the acid extracts, since very little C-terminal immunoreactivity is measurable. In the duodenum, the intact peptides are found in concentrations similar to those of the desoctapeptides. G1 is an N-terminal antiserum, and R71 is a C-terminal antiserum. (Reproduced from Ichihara *et al.*, 1984a.)

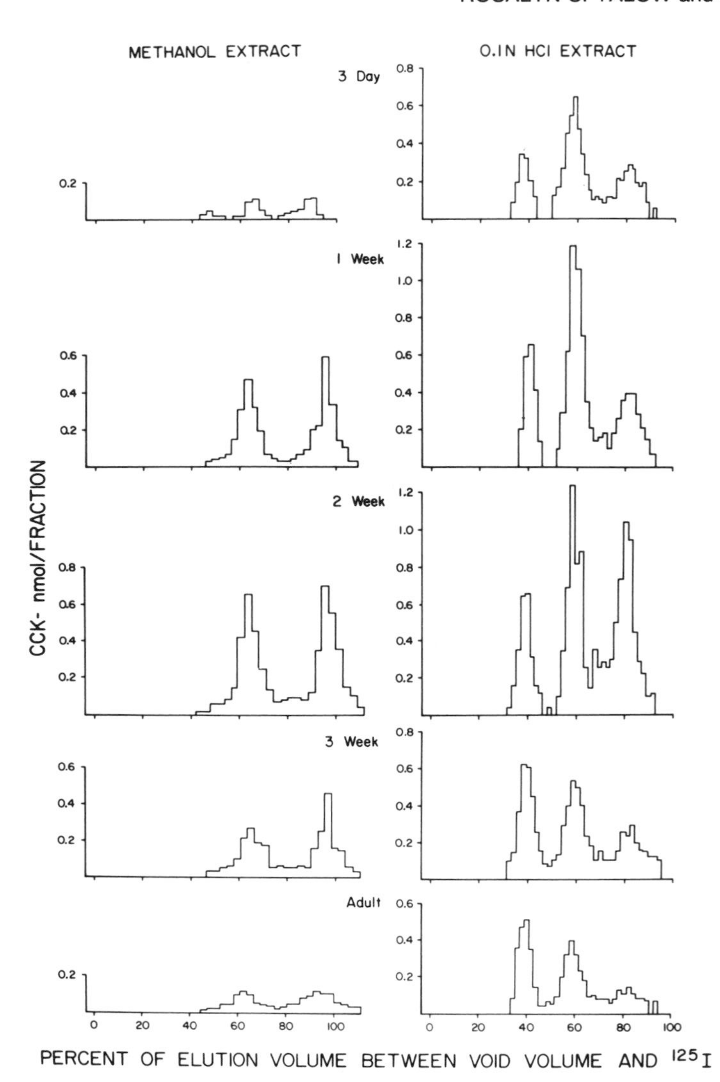

FIGURE 5. Sephadex G50 gel filtration of rat duodenal extracts. The conditions for equilibration and elution of columns are given in the legend to Fig. 3. (Reproduced from Ichihara *et al.*, 1984b.)

6. ROLE OF CHOLECYSTOKININ PEPTIDES IN BRAIN

The observations that both posttranslational processing and ontogeny of brain and gut CCK differ strikingly are consistent with very different roles for these peptides. Does CCK serve the same or different functions in different portions of

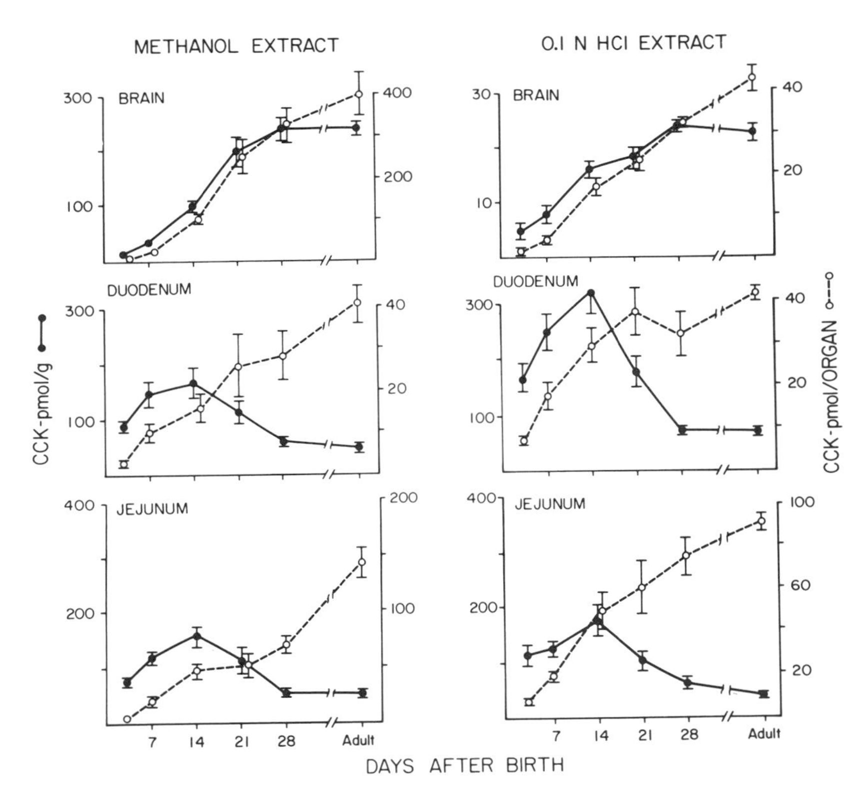

FIGURE 6. C-terminal immunoreactive cholecystokinin (iCCK) concentration in methanol (left) and 0.1 N HCl (right) extracts of rat brain (top), duodenum (middle), and jejunum (bottom) as a function of time after birth. Shown also are the total organ contents for the various tissues. In this and the subsequent figure, the vertical bars represent the standard error of the mean. (Reproduced from Ichihara *et al.*, 1983.)

the brain? The question must first be addressed as to whether CCK peptides exert their effect by release into the cerebrospinal fluid (CSF) followed by transport to sites of action or by transmission along neuronal pathways or both. Rehfeld and Kruse-Larsen (1978) reported that concentrations of CCK immunoreactivity in human CSF averaged 14 fmol/ml. Of particular interest is the observation by Della-Fera *et al.* (1981) that food intake increased 100% during a continuous lateral ventricular infusion into sheep of antibody to CCK. They interpreted their data as suggesting that binding to administered antibody prevented CCK peptides released into the CSF from binding to receptors involved in satiety. This study appears to be consistent with the concept that release of CCK peptides into CSF accounts for at least part of its mode of action.

Other experiments suggesting that CCK action is through CSF are those of a

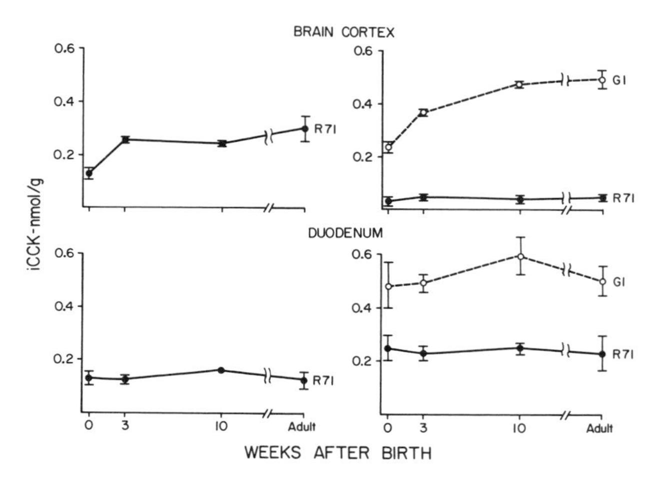

FIGURE 7. Immunoreactive CCK as a function of age in methanol (left) and acid (right) extracts of pig brain cortex (top) and pig duodenum (bottom) as determined with N-terminal (G1) and C-terminal (R71) radioimmunoassays. (Reproduced from Ichihara *et al.,* 1984a.)

Hungarian group who demonstrated that intraventricular administration of CCK8 increased the dopamine and norepinephrine content of the brain in a time- and dose-dependent manner (Fekete *et al.*, 1981a) and that administration of CCK antiserum in the same region decreased norepinephrine in the hypothalamus but not in the cortex (Fekete *et al.*, 1981b). The question has been considered whether the action of CCK on feeding is direct or through a noradrenergic pathway. McCaleb and Myers (1980) have suggested it is the latter, since they have observed that administration of norepinephrine into the hypothalamus of satiated rats provoked feeding that could be attenuated or blocked by prior hypothalamic administration of CCK.

It has also been reported that CCK8 can alter pituitary hormone release via hypothalamic action (Vijayan *et al.*, 1979). Thus, injection of as little as 4 pmol into the third ventricle of conscious ovariectomized rats appears to result in a threefold decrease in LH and a threefold increase in growth hormone in jugular vein samples. A tenfold higher dose appeared to decrease TSH in the same samples by a factor of two. Zetler (1981) has described other central effects of CCK including analgesia, ptosis, and inhibition of picrotoxin-induced convulsions and methylphenidate-induced gnawing.

7. SUMMARY

Over the past decade an enormous body of data has been accumulated about the nature and role of CCK peptides. With so many effects reported, it is difficult at present to be certain which are real and which are the physiologically important functions. It can be expected that another decade will be required to sort out and define fully the multiple roles of the CCK peptides.

ACKNOWLEDGMENT. This work was supported by the Medical Research Program of the Veterans Administration.

REFERENCES

Calam, J., Ellis, A., and Dockray, G. J., 1982, Identification and measurement of molecular variants of cholecystokinin in duodenal mucosa and plasma, *J. Clin. Invest.* **69:**218–225.

Della-Fera, M. A., Baile, C. A., Schneider, B. S., Grinkler, J. A., 1981, Cholecystokinin antibody injected in cerebral ventricles stimulates feeding in sheep, *Science* **212:**687–689.

Dodd, P. R., Edwardson, J. A., and Dockray, G. J., 1980, The depolarization-induced release of cholecystokinin C-terminal octapeptide (CCK-8) from rat synaptosomes and brain slices, *Regul. Peptides* **1:**17–29.

Emson, P. C., Lee, C. M., and Rehfeld, J. F., 1980, Cholecystokinin octapeptide: Vesicular localization and calcium dependent release from rat brain *in vitro, Life Sci.* **26:**2157–2163.

Eng, J., Shiina, Y., Straus, E., and Yalow, R. S., 1982, Post-translational processing of cholecystokinin in pig brain and gut, *Proc. Natl. Acad. Sci. U.S.A.* **79:**6060–6064.

Eng, J., Shiina, Y., Pan, Y.-C. E., Blacher, R., Chang, M., Stein, S., and Yalow, R. S., 1983, Pig brain contains CCK-octapeptide and several CCK-desoctapetides, *Proc. Natl. Acad. Sci. U.S.A.* **80:**6381–6385.

Eng. J., Du, B.-H., Pan, Y.-C. E., Chang, M., Hulmes, J. D., and Yalow, R. S., 1984, Purification and sequencing of a rat intestinal 22 amino acid C-terminal CCK fragment, *Peptides* **5:**1203–1206.

Epelbaum, J., Brazeau, P., Tsang, D., Brawer, J., and Martin, J. B., 1977, Subcellular distribution of radioimmunoassayable somatostatin in rat, *Brain Res.* **126:**309–323.

Fekete, M., Várszegi, M., Kádár, T., Penke, B., Kovács, K., and Telegdy, G., 1981a, Influence of cholecystokinin octapeptide sulfate ester on brain monoamine metabolism in rats, *J. Neural Transm.* **50:**80–88.

Fekete, M., Kádár, T., and Telegdy, G., 1981b, Effect of cholecystokinin antiserum on the brain monoamine content in rats, *Acta Physiol. Acad. Sci. Hung.* **57**(2)**:**177–183.

Giachetti, A., Said, S.-I., Reynolds, R. C., and Koniges, F. C., 1977, Vasoactive intestinal polypeptide in brain: Localization in and release from isolated nerve terminals, *Proc. Natl. Acad. Sci. U.S.A.* **74:**3424–3428.

Gibbs, J., Falasco, J. D., and McHugh, P. R., 1976, Cholecystokinin-decreased food intake in rhesus monkeys, *Am. J. Physiol.* **230:**15–18.

Gubler, U., Chua, A. O., Hoffman, B. J., Collier, K. J., and Eng. J., 1984, Cloned cDNA to cholecystokinin mRNA predicts an identical preprocholecystokinin in pig brain and gut. *Proc. Natl. Acad. Sci. U.S.A.* **81:**4307–4310.

Hökfelt, T., Elde, R., Fuxe, K., Johansson, O., Ljungdahl, A., Goldstein, M., Luft, R., Efendic, S., Nilsson, G., Terenius, L., Ganten, D., Jeffcoate, S. L., Rehfeld, J. F., Said, S., Perez de la Mora, M., Possani, L., Tapia, R., Teran, L., and Palacios, R., 1978, Aminergic and peptidergic pathways in the nervous system with special reference to the hypothalamus, in: *The Hypothalamus* (S. Reichlin, R. J. Baldessarini, and J. B. Martin, eds.), Raven Press, New York, pp. 69–135.

Ichihara, K., Eng, J., and Yalow, R. S., 1983, Ontogeny of immunoreactive CCK, VIP and secretin in rat brain and gut, *Biochem. Biophys. Res. Commun.* **112:**891–898.

Ichihara, K., Eng, J., Pond, W. G., Yen, J. T., Straus, E., and Yalow, R. S., 1984a, Ontogeny of immunoreactive CCK and VIP in pig brain and gut, *Peptides* **5:**623–626.

Ichihara, K., Eng, J., and Yalow, R. S., 1984b, Ontogeny of molecular forms of CCK-peptides in rat brain and gut, *Life Sci.* **34:**93–98.

Innis, R. B., Correa, F. M. A., Uhl, G., Schneider, R., and Snyder, S. H., 1979, Cholecystokinin octapeptide-like immunoreactivity: Histochemical localization in rat brain, *Proc. Natl. Acad. Sci. U.S.A.* **76:**521–525.

Kissileff, H. R., Pi-Sunyer, X. F., Thornton, J., and Smith, G. P., 1981, C-terminal octapeptide of cholecystokinin decreases food intake in man, *Am. J. Clin. Nutr.* **34:**154–160.

Klaff, L. J., Hudson, A., Sheppard, M., and Tyler, M., 1981, *In vitro* release of cholecystokinin octapeptide-like immunoreactivity from rat brain synaptosomes, *S. Afr. Med. J.* **59:**158–160.

Larsson, L. I., and Rehfeld, J. F., 1979, A peptide resembling COOH-terminal tetrapeptide amide of gastrin from a new gastrointestinal endocrine cell type, *Nature* **277:**575–577.

Loren, I., Alumets, J., Hakanson, R., and Sundler, F., 1979, Distribution of gastrin and CCK-like peptides in rat brain. An immunocytochemical study, *Histochemistry* **59:**249–257.

McCaleb, M. L., and Myers, R. D., 1980, Cholecystokinin acts on the hypothalamic "noradrenergic system" involved in feeding. *Peptides* **1:**47–49.

Pinget, M., Straus, E., and Yalow, R. S., 1978, Localization of cholecystokinin-like immunoreactivity in isolated nerve terminals, *Proc. Natl. Acad. Sci. U.S.A.* **75:**6324–6326.

Pinget, M., Straus, E., and Yalow, R. S., 1979, Release of cholecystokinin peptides from a synaptosome-enriched fraction of rat cerebral cortex, *Life Sci.* **25:**339–342.

Rehfeld, J. F., and Kruse-Larsen, C., 1978, Gastrin and cholecystokinin in human cerebrospinal fluid. Immunochemical determination of concentrations and molecular heterogeneity, *Brain Res.* **155:**19–26.

Ryder, S., Straus, E., and Yalow, R. S., 1980, Further characterization of brain cholecystokinin converting enzymes, *Proc. Natl. Acad. Sci. U.S.A.* **77:**3669–3671.

Schally, A. V., Redding, T. W., Lucien, H. W., and Meyer, J., 1967, Enterogastrone inhibits eating by fasted mice, *Science* **157:**210–211.

Straus, E., Muller, J. E., Choi, H.-S., Paronetto, F., and Yalow, R. S., 1977, Immunohistochemical localization in rabbit brain of a peptide resembling the C-terminal cholecystokinin octapeptide, *Proc. Natl. Acad. Sci. U.S.A.* **74:**3033–3034.

Sturdevant, R. A. L., and Goetz, H., 1976, Cholecystokinin both stimulates and inhibits human food intake, *Nature* **261:**713–715.

Vanderhaeghen, J. J., Signeau, J. C., and Gepts, W., 1975, New peptide in the vertebrate CNS reacting with antigastrin antibodies, *Nature* **257:**604–605.

Vanderhaeghen, J. J., DeMey, J., Lotstra, F., and Giles, C., 1978, Localization of gastrin–cholecystokinin like peptides in the brain and hypophysis of the rat. Symposium on neural peptides and proteins, *Acta Neurol. (Belg.)* **79:**62–63.

Vijayan, E., Samson, W. K., and McCann, S. M., 1979, *In vivo* and *in vitro* effects of cholecystokinin on gonadotropin, prolactin, growth hormone and thyrotropin release in the rat, *Brain Res.* **172:**295–302.

Walsh, J. H., Lamers, C. B., and Valenzuela, J. E., 1982, Cholecystokinin octapeptide like immunoreactivity in human plasma, *Gastroenterology* **82:**438–444.

Whittaker, V. P., Michaelson, J. A., and Kirkland, R. J. A., 1964, The separation of synaptic vesicles from nerve-ending particles (synaptosomes), *Biochem. J.* **90:**293–303.

Young, J. D., Lazarus, L., Chisholm, D. J., and Atkinson, F. F. V., 1969, Radioimmunoassay of pancreozymin–cholecystokinin in human serum, *J. Nucl. Med.* **10:**743.

Zetler, G., 1981, Central effects of ceruletide analogues, *Peptides* **2:**65–69.

17

The Neurobiology of Neurotensin

PETER J. ELLIOTT and CHARLES B. NEMEROFF

1. INTRODUCTION

This chapter summarizes the available data supporting a neuroregulatory role for neurotensin (NT) in the mammalian central nervous system (CNS). The major focus is on effects of centrally administered NT, and a secondary focus is the possible importance of this peptide in the pathophysiology of schizophrenia and in the mechanism of action of antipsychotic drugs.

The tridecapeptide NT was originally isolated (Carraway and Leeman, 1973) and later sequenced by Carraway and Leeman (1975, 1976) from bovine hypothalamus. Subsequently, others have shown NT to be present in all of the vertebrate classes, including mammals (Carraway and Leeman, 1976; Kitabgi *et al.*, 1976; Kobayashi *et al.*, 1977; Uhl *et al.*, 1977a; Dupont *et al.*, 1979; Kataoka *et al.*, 1979; Cooper *et al.*, 1981; Emson *et al.*, 1982c; Grant *et al.*, 1982; Jennes *et al.*, 1982; Manberg *et al.*, 1982a,b), birds, amphibians, and reptiles (Reinecke *et al.*, 1980a,b; Bhatnagar and Carraway, 1981; Carraway *et al.*, 1982; Eldred and Karten, 1983; Goedert *et al.*, 1984a), fish (Langer *et al.*, 1979), as well as in invertebrates (Reinecke *et al.*, 1980b; Grimmelikhuijzen *et al.*, 1981; Price *et al.*, 1982) and even bacteria (Bhatnagar and Carraway, 1981). However, the structure of chicken NT is distinct from that of mammalian NT.

The present chapter deals solely with NT in mammals, where the peptide is distributed throughout the gastrointestinal tract (Sundler *et al.*, 1977; Schultzberg *et al.*, 1980; Reinecke *et al.*, 1983), cardiovascular system (Mashford *et al.*, 1978; Blackburn and Bloom, 1979), CNS (Uhl and Snyder, 1977a,b; Uhl *et al.*, 1979; Bissette *et al.*, 1984; Ghatei *et al.*, 1984), CSF (Fernstrom *et al.*, 1980), adrenal

PETER J. ELLIOTT • Department of Psychiatry, Duke University Medical Center, Durham, North Carolina 27710. CHARLES B. NEMEROFF • Departments of Psychiatry and Pharmacology and Center for Aging and Human Development, Duke University Medical Center, Durham, North Carolina 27710. *Present address of P.J.E.*: Department of Psychiatry, Yale University School of Medicine, New Haven, Connecticut 06508.

medulla (Lee *et al.*, 1981; Rokaeus *et al.*, 1984), and parts of the autonomic nervous system.

2. DISTRIBUTION IN THE CNS

With radioimmunoassay (RIA) and immunohistochemical techniques, NT has been shown to have a heterogeneous CNS distribution in a variety of mammals including man. High concentrations of the peptide are found in the bed nucleus of the stria terminalis, central amygdaloid nucleus, habenula, pituitary, and in certain hypothalamic nuclei; low concentrations are found in the cortex, hippocampus, and cerebellum (Uhl and Snyder, 1977a,b; Uhl *et al.*, 1979; Kahn *et al.*, 1980; Jennes *et al.*, 1982; Roberts *et al.*, 1982; Goedert *et al.*, 1983; Inagaki *et al.*, 1983a; Reinecke *et al.*, 1983; Ibata *et al.*, 1984a; Kalivas, 1984; Kohler and Eriksson, 1984; Milner *et al.*, 1984). Table I shows the concentration of NT in micropunched brain regions of the rat (C. D. Kilts, C. B. Nemeroff, and G. Bissette, unpublished observations).

The locations of NT-containing cell bodies, nerve terminals, and fibers have been identified by immunohistochemistry, and recently a few NT pathways have been identified. These include one from the central amygdaloid nucleus passing to (and through) the bed nucleus of the stria terminalis (Uhl and Snyder, 1979; Leranth *et al.*, 1981). Another pathway is comprised of neuronal cell bodies in the hippocampus that send fibers into the anterior cingulate cortex (Roberts *et al.*, 1981; Bloom and Polak, 1982), although this circuit has not been observed by others. A NT pathway from the endopiriform nucleus and adjacent prepiriform cortex to the anterior olfactory nucleus and nucleus of the diagonal band of Broca has been identified (Inagaki *et al.*, 1983b). A pathway from cells located in the periaqueductal gray/nucleus solitary tract to the nucleus raphe magnus has also been reported (Beitz, 1982). Finally, a VTA–nucleus accumbens NT-containing pathway has been proposed (Kalivas and Miller, 1984) based on studies of NT immunohistochemistry and retrograde transport of HRP. Measurement of NT immunoreactivity by immunocytochemistry and radioimmunoassay depends ultimately on the specificity of the antisera used in such studies. Carraway (1982) and Emson *et al.* (1982a) have written detailed reviews of problems associated with RIAs in neurotensin research.

3. COLOCALIZATION WITH OTHER NEUROTRANSMITTERS

Neurotensin has been localized by immunohistochemistry in a small percentage of dopamine (DA) neurons in the mesencephalon (Hokfelt *et al.*, 1984; Kalivas and Miller, 1984) and in the periventricular and arcuate nuclei of the hypothalamus (Ibata *et al.*, 1983, 1984a; Hokfelt *et al.*, 1984). However, destruction of DA neurons with 6-hydroxydopamine (6OHDA) does not significantly alter NT concentrations (Bissette *et al.*, 1983; Emson *et al.*, 1985a; G. Bissette and C. B. Nemeroff, unpublished observations). In postmortem brain tissue of patients with

TABLE I. Distribution of Neurotensin in Microdissected Brain Nuclei

Nuclei	Neurotensin concentration[a] (pg/mg protein, mean ± S.E.M.)
Nucleus accumbens	395 ± 24
Septal	
Lateral	827 ± 80
Medial	681 ± 42
Caudate nucleus	
Anterior	143 ± 25
Posterior (with putamen)	374 ± 48
Amygdaloid	
Central	4078 ± 339
Lateral	560 ± 113
Basal	1051 ± 84
Medial	573 ± 33
Cortical	544 ± 19
Posterior	218 ± 35
Medial posterior	314 ± 56
Basal posterior	477 ± 31
Hippocampus	
Dorsal	102 ± 19
Ventral	108 ± 15
Locus coeruleus	1259 ± 198
Stria terminalis	
interstitial (bed) nucleus	1704 ± 220
Substantia nigra	
Pars reticulata	340 ± 19
Pars lateralis	815 ± 117
Ventral tegmental area	1466 ± 158
Raphe	
Dorsal	2362 ± 269
Medial	1048 ± 74
Preoptic area	1782 ± 184
Anterior hypothalamic nucleus	1124 ± 131
Cortex	
Prefrontal	146 ± 5
Cingulate	78 ± 5
Entorhinal	209 ± 33
Piriform	317 ± 36

[a] Brain nuclei of interest were micropunch-dissected from 300-μm-thick coronal brain slices. The neurotensin content was determined by radioimmunoassay using an antibody generated against the midportion of the amino acid sequence of neurotensin. Values represent the mean ± S.E.M. of six determinations.

Parkinson's disease in which the nigrostriatal and mesolimbic DA pathways have degenerated, NT concentrations are not altered (Uhl *et al.*, 1984; Bissette *et al.*, 1985; Sadoul *et al.*, 1984c) except in the hippocampus. It seems difficult to explain these discrepant findings; the histochemical studies reveal the presence of neurons that contain both DA and NT, whereas the neurochemical studies do not. However, it is possible that those neurons containing DA and NT are unaffected in Parkinson's disease or by 6OHDA.

4. SYNTHESIS AND STORAGE

No studies have been published concerning NT biosynthesis in the CNS, though it is probably synthesized, like most biologically active peptides, first as a prohormone, followed by cleavage to the active form. Neurotensin is known to be preferentially localized in the synaptosomal fraction of brain following density gradient centrifugation (Kataoka *et al.*, 1979). Moreover, NT-rich synaptic vesicles have been identified in synaptic terminals by electron microscopy (Uhl and Snyder, 1977a; Uhl *et al.*, 1977a).

5. RELEASE AND DEGRADATION

In vitro release of NT can be stimulated by depolarizing agents such as potassium (Iversen *et al.*, 1978) and by a variety of other agents including dibutyryl cAMP (Maeda and Frohman, 1981). The process has been also shown to be calcium dependent (Iversen *et al.*, 1978).

Once released from nerve terminals, NT is degraded by peptidases in brain (Dupont and Merand, 1978; Martins *et al.*, 1980; Emson *et al.*, 1982b,c; McDermott *et al.*, 1982). The major site of enzymatic inactivation is the Arg^8-Arg^9 peptide bond (Checler *et al.*, 1982; McDermott and Kidd, 1984), although the Pro^7-Arg^8 and Pro^{10}-Tyr^{11} bonds can also be broken (Camargo *et al.*, 1983). These peptidases release, originally, NT_{1-8} NT_{9-13}, and NT_{1-10}. The N-terminal fragments are apparently resistent to further catabolism, whereas NT_{9-13} is further degraded. One of the peptidases that inactivates NT is a metalloendopeptidase because chelating agents such as EDTA, 10-phenanthroline, and dithiothreitol all inhibit degradation (Checler *et al.*, 1982) of the peptide. Captopril (Ondetti *et al.*, 1977) and thiorphan (Roques *et al.*, 1980), substances that respectively inhibit angiotensin-converting enzyme and "enkephalinase," only partially inhibit NT degradation (Checler *et al.*, 1982). To date, no specific peptidases have been discovered, nor have specific inhibitors of NT degradation been identified. Figure 1 shows the major degradation products of NT.

It should be noted here that degradation of a peptide such as NT does not necessarily correlate with biological inactivation. Certain NT fragments have been shown to possess biological activity (see Section 8).

Finally, it has recently been reported that different vertebrates inactivate NT differently. Fish and reptiles produce NT_{1-8}, NT_{1-10}, and NT_{1-7} as NT degradation products as well as NT_{1-11} and NT_{1-12}; birds and rats produce NT_{1-8}, NT_{9-13}, NT_{1-10}, and NT_{1-12} (Griffiths *et al.*, 1984).

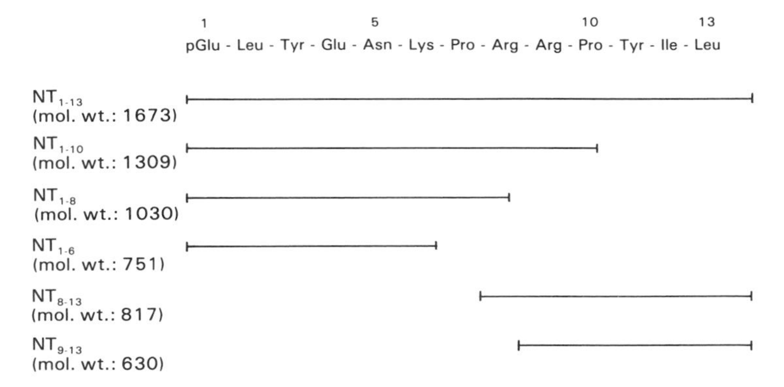

FIGURE 1. Neurotensin and its fragments.

6. RECEPTORS

High-affinity binding sites for NT are heterogeneously distributed throughout the CNS in a pattern similar to that of the endogenous peptide (Young and Kuhar, 1979, 1981; Ninkovic *et al.*, 1981; Palacios and Kuhas, 1981; Quirion *et al.*, 1982b–d). However, a few disparities do exist; e.g., in cortical areas where the concentration of NT is relatively low, the number of binding sites is very high (Lazarus *et al.*, 1977a; Uhl *et al.*, 1977b). An inverse relationship between NT receptor density and NT-like immunoreactivity has also been reported in the rat striatum (Goedert *et al.*, 1984a).

The NT binding sites identified show properties similar to those of other neurotransmitter receptors. They are specific, saturable, and have reversible high-affinity binding sites (Uhl and Snyder, 1977b). Characterization of NT receptors has been achieved with both [^{3}H]NT and [^{125}I]NT (Kitabgi *et al.*, 1977; Lazarus *et al.*, 1977a–d; Uhl *et al.*, 1977b; Kitabgi *et al.*, 1980; Quirion *et al.*, 1982a,c; Vincent *et al.*, 1982; Mazella *et al.*, 1983; Goedert *et al.*, 1984a; Nakagawa *et al.*, 1984; Poustis *et al.*, 1984; Sadoul *et al.*, 1984a,b), with NT fragments, and with structural analogues of NT. In general, the ability of these NT analogues to displace radiolabeled NT binding is similar in both CNS and peripheral preparations. However, in contrast, D-Arg^{8}-NT has 0.5% and 640% of the binding potency (relative to NT) in CNS and peripheral (mast cell) preparations, respectively (Kitabgi *et al.*, 1977, 1980; Lazarus *et al.*, 1977b). These findings as well as other reports on very-high-affinity binding sites for NT have led to the hypothesis that multiple classes of NT receptors (Fox *et al.*, 1982; Jolicoeur and Barbeau, 1982; Jolicoeur *et al.*, 1985) exist in the CNS.

There is evidence that NT binding sites are present on DA perikarya in the substantia nigra and ventral tegmental area; 6-hydroxydopamine (6OHDA), which destroys these neurons, produces a decrease in NT binding sites (Palacios and Kuhar, 1981; Quirion *et al.*, 1982d; Goedert *et al.*, 1984b). Moreover, the number of NT binding sites in the DA terminal region of the nigroneostriatal system is decreased following intranigral 6OHDA administration, indicating that NT receptors are also found on presynaptic DA terminals (Quirion *et al.*, 1982d; Goedert *et al.*, 1984b). This effect is not observed in the mesolimbic DA pathway, indicating that these DA terminals do not possess NT receptors (Quirion *et al.*, 1982d; Goedert *et al.*, 1984b).

The exact nature of the interaction of NT with its receptor is not known, but structure–activity studies have shown that the C-terminal region of the peptide is necessary for binding (Rivier *et al.*, 1977; Quirion *et al.*, 1981; Rioux *et al.*, 1980) and that NT probably folds into a B-type chain conformation for binding, with breaks or "corners" at Tyr^3, Asn^5, Lys^6, Arg^8, and Tyr^{11} (Finn *et al.*, 1984). However, it should be stressed that binding potency does not always correlate with biological activity.

Evidence exists that is consistent with the view that cAMP is involved in the stimulatory action of NT on mast cells (Rossie and Miller, 1982). The tridecapeptide appears to have a stimulatory effect on cGMP synthesis in cultured neuroblastoma (Gilbert and Richelson, 1984; Gilbert *et al.*, 1984) cells and also in rat cerebellar slices (Gilbert *et al.*, 1984). The significance of these findings with cyclic nucleotides is as yet not fully understood.

7. ELECTROPHYSIOLOGY

Inhibitory (Young *et al.*, 1978; McCarthy *et al.*, 1979; Marwaha *et al.*, 1980; Shiraishi *et al.*, 1980) and excitatory (Nicoll, 1978; Zieglgansberger *et al.*, 1978; Phillis and Kirkpatrick, 1979, 1980; Sawada and Yamamoto, 1980; Sawada *et al.*, 1980; Andrade and Aghajanian, 1981; Suzue *et al.*, 1981) actions as well as mixed (Miletic and Randic, 1979; Henry, 1982) and null effects (Guyenet and Aghajanian, 1977) have been reported after iontophoretic or pressure-ejected application of NT. In most studies, however, NT produces an excitatory action.

8. STRUCTURE–ACTIVITY RELATIONSHIPS

As described above, degradation of the native peptide resulted in the production of various NT fragments. These and other synthetic NT fragments have been studied in a variety of biological systems (Jolicoeur *et al.*, 1984). Generally, the N-terminal peptides NT_{1-6}, NT_{1-8}, and NT_{1-10} are inactive, whereas the C-terminal peptides NT_{8-13} and NT_{9-13} are biologically active. Figure 2 contains a list of available congeners of NT.

Both NT_{1-8} and NT_{8-13} have been shown to inhibit TRH-induced wet dog shakes after intracerebral administration (Griffiths *et al.*, 1982; Widdowson *et al.*,

Amino acid substitutions of $NT_{1\text{-}13}$	D-Amino Acid substitutions of $NT_{1\text{-}13}$	Amino acid substitutions of NT fragments	Acetylated amino acid substitutions of NT fragments
$(Gln^4)\ NT_{1-13}$	$(D\text{-}Tyr^3)\ NT_{1\text{-}13}$	$(Ala^{12})\ NT_{8\text{-}13}$	$AC - NT_{8\text{-}13}$
$(Ala^8)\ NT_{1\text{-}13}$	$(D\text{-}Arg^8)\ NT_{1\text{-}13}$	$(Ala^{13})\ NT_{8\text{-}13}$	$(AC - Cit^8)\ NT_{8\text{-}13}$
$(Ala^9)\ NT_{1\text{-}13}$	$(D\text{-}Arg^9)\ NT_{1\text{-}13}$		$(AC - Cit^9)\ NT_{8\text{-}13}$
$(Phe^{11})\ NT_{1\text{-}13}$	$(D\text{-}Arg^{8,9})\ NT_{1\text{-}13}$		$(AC - Phe^{11})\ NT_{8\text{-}13}$
$(Trp^{11})\ NT_{1\text{-}13}$	$(D\text{-}Tyr^{11})\ NT_{1\text{-}13}$		$(AC - Phe^{12})\ NT_{8\text{-}13}$
	$(D\text{-}Trp^{11})\ NT_{1\text{-}13}$		$(AC - Me\ Ile^{12})\ NT_{8\text{-}13}$
	$(D\text{-}Leu^{11})\ NT_{1\text{-}13}$		
	$(D\text{-}Leu^{13})\ NT_{1\text{-}13}$		

FIGURE 2. Some synthetic analogues of neurotensin.

1983). Like NT, NT_{1-8} and NT_{1-10} have been reported to possess a cytoprotective action against cold-restraint stress-induced gastric ulcers after intracerebroventricular injection (Hernandez *et al.*, 1984b); NT_{8-13} and NT_{1-6} were ineffective (Hernandez *et al.*, 1984b).

Like NT, the C-terminal NT fragments NT_{8-13} and NT_{9-13} both decrease locomotor activity and body temperature after central infusions (Jolicoeur *et al.*, 1981a, 1982a,b; Kitabgi *et al.*, 1980; Quirion *et al.*, 1982b); NT_{10-13} is inactive (Jolicoeur *et al.*, 1981a, 1982a,b; Kitabgi *et al.*, 1980; Quirion *et al.*, 1982b). After intracisternal (IC) or intraventricular (ICV) administration, NT produces antinociception (see Section 11) in mice and rats (Luttinger *et al.*, 1983a). In the mouse tail pressure test (Furuta *et al.*, 1984), NT_{8-13}, NT_{8-10}, NT_{8-11}, NT_{9-13}, and NT_{9-11} all exhibit antinociceptive activity. Many analogues of NT have been synthesized by substitution of the naturally occurring L-amino acids with their D isomers. Other congeners of NT have been synthesized by replacing original amino acid moieties with other amino acids or cyclizing NT fragments.

In general, modification of positions 2 to 7 has little effect on biological activity (Nagai and Frohman, 1976; Lazarus *et al.*, 1977a,d; Loosen *et al.*, 1978). However, substitution at position 11 results in dramatic effects (Nagai and Frohman, 1976; Lazarus *et al.*, 1977a,d; Loosen *et al.*, 1978). Usually, replacing an L-amino acid with its D enantiomer produces a congener with opposite actions, although this has not always been found to be true. $DOPA^{11}$-neurotensin increases locomotor activity, whereas D-Leu^{11}-NT has no effect (Jolicoeur *et al.*, 1985). Differences in activity and potency of peptide analogues can be explained, in part, by their relative ability to diffuse through CNS neuropil and their metabolic stability. Peptide analogues that are enzymatically resistant usually exhibit a prolonged duration of action as well as increased potency (Checler *et al.*, 1983).

There are other naturally occurring peptides that have amino acid sequences homologous to mammalian NT. Such peptides have been isolated from chicken small intestine (Checler *et al.*, 1983), from extracts of *Xenopus* (xenopsin) (Araki *et al.*, 1973), and from porcine spinal cord (neuromedin N) (Minamino *et al.*, 1984). The structures of these peptides, which are naturally occurring homologues of NT, are shown in Fig. 3.

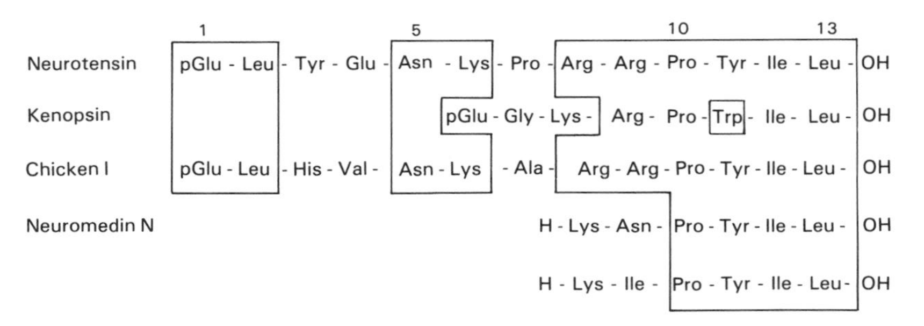

FIGURE 3. Neurotensin and related peptides.

9. INTERACTIONS WITH BARBITURATES AND ETHANOL

The sedative and hypothermic actions of barbiturates (Nemeroff *et al.*, 1977; Bissette *et al.*, 1978) and ethanol (Luttinger *et al.*, 1981a) are potentiated by IC administration of NT in mice. This effect is associated with a decline in the rate of barbiturate metabolism, but IC NT, in contrast, does not alter blood or brain ethanol concentrations (Luttinger *et al.*, 1981a, 1983b). Finally, the motor impairment induced by ethanol is potentiated by IC NT (Frye *et al.*, 1981).

10. THERMOREGULATORY EFFECTS

After IC or ICV administration, NT produces hypothermia (Mason *et al.*, 1980; Kalivas *et al.*, 1982d); in fact, it is one of the most potent hypothermic agents known. The precise mechanisms by which it produces a reduction in body temperature are unknown. Structure–activity studies have shown that the C-terminal portion of the NT molecule is essential for its hypothermic action (Rivier *et al.*, 1977; Loosen *et al.*, 1978).

In mice (Bissette *et al.*, 1976; Brown *et al.*, 1977; Clineschmidt and McGuffin, 1977; Lipton *et al.*, 1977; Nemeroff *et al.*, 1977; Brown and Vale, 1980; Jolicoeur *et al.*, 1981a; Hernandez *et al.*, 1984a), guinea pigs, gerbils, rats, hamsters, and monkeys (Nemeroff *et al.*, 1980a,b), NT produces hypothermia, although the finding in monkeys has been questioned (Mora *et al.*, 1984). In rabbits (Nemeroff *et al.*, 1980a; Metcalf *et al.*, 1980), ground squirrels, woodchucks (Nemeroff *et al.*, 1980a), frogs, lizards, and bluegills (Prange *et al.*, 1979), NT does not produce hypothermia. Thus, the tridecapeptide produces no change in body temperature in lower vertebrates including reptiles and amphibians. The hypothermic effect in rodents is both dose dependent and temperature dependent. At colder ambient temperatures, NT-induced hypothermia is more robust (Bissette *et al.*, 1976; Chandra *et al.*, 1981). The peptide also produces hypothermia in rats as young as 10 days of age (Hernandez *et al.*, 1982). Moreover, prior adaptation to a cold environment significantly antagonizes the hypothermia induced by IC NT in mice (Merritt *et al.*, 1986). This effect on NT-induced hypothermia is attenuated by indomethacin but not by acetylsalicylic acid (Merritt *et al.*, 1986), indicating that CNS prostaglandins are essential for this response. This thermoregulatory effect of NT is believed to be independent of serotonin, acetylcholine, norepinephrine, and opiate systems, but the importance of dopaminergic pathways seems likely (Nemeroff *et al.*, 1980b), although this has also been disputed (Lee *et al.*, 1983). Depletion of CNS dopamine (DA) potentiates NT-induced hypothermia in rats, as does pretreatment with the DA receptor antagonist haloperidol. Furthermore, three indirect DA agonists, *d*-amphetamine, cocaine, and methylphenidate, significantly attenuate NT-induced hypothermia in mice (Bissette *et al.*, 1981).

Neurotensin-induced hypothermia can be enhanced by a reduction in the availability of CNS prostaglandins (Clark, 1979; Mason *et al.*, 1982; Kalivas *et al.*, 1983a) as well as by IC administration of TRH (Nemeroff *et al.*, 1980b; Morley *et al.*, 1982). A neuroanatomic map of the CNS sites in which microinjection of NT elicits hypothermia has been produced (Martin *et al.*, 1981b; Kalivas *et al.*, 1982d); these include the preoptic area of the hypothalamus, the periaqueductal gray, and the ventral tegmental area. All of these areas contain substantial quantities of NT-like immunoreactivity (see Table I for NT distribution).

11. ANTINOCICEPTIVE EFFECTS

Because of the high concentration of NT in the substantia gelatinosa of the spinal cord and trigeminal nerve nucleus, areas known to subserve nociceptive functions, it seemed plausible to evaluate the effects of exogenously administered NT on pain processes. Although some contradictory data concerning the effects of centrally administered NT on nociception have been obtained, these differences might be explained, in part, by the vast differences in methods used to test for analgesic activity as well as by differences in dose and route of administration. Furthermore, NT might exert different antinociceptive actions in mice and rats (Cowan and Gmerek, 1981, 1982).

In the acetic acid writhing test, a model of visceral pain, both intrathecal infusions of NT in rats (Yaksh *et al.*, 1982) and intracisternal administration of NT in mice (Clineschmidt *et al.*, 1979; Kalivas *et al.*, 1981; Osbahr *et al.*, 1981) have been shown to produce marked antinociception.

In the mouse tail-flick test, a model of acute pain, intrathecal NT did not inhibit the nociceptive spinal reflex. In contrast, NT was found to produce dose-related antinociception after subcutaneous hypertonic saline (Hylden and Wilcox, 1983).

In mice, the antinociceptive action of intrathecal NT has been reported to be blocked by naloxone but not by TRH (Nemeroff *et al.*, 1979). In contrast, our group reported (Osbahr *et al.*, 1981) that NT-induced antinociception after IC administration in the hot-plate, tail-immersion, and acetic writhing tests is blocked by pretreatment with TRH but not naloxone. In rats, NT has no significant effect in the tail-flick test (Clineschmidt *et al.*, 1979) and relatively weak antinociceptive activity in the hot-plate test (Kalivas *et al.*, 1982c).

In an electric footshock regimen in rats, ICV infusion of NT has been shown to reduce responsivness to shock. This effect is observed even after repeated NT administration (Rinkel *et al.*, 1983).

As noted above, IC injection of NT produces analgesia in a hot-plate test in mice (Hernandez *et al.*, 1984a) and rats (Martin *et al.*, 1981a; Martin and Naruse, 1982). In cats, intrathecal NT administration produces antinociception in the hot-plate test but not in the tail-flick paradigm (Yaksh *et al.*, 1982). This effect of NT in cats is naloxone reversible (Yaksh *et al.*, 1982).

The loci where NT acts within the CNS to produce antinociception have been studied in detail in the rat. Positive sites include the central nucleus of the amygdala, the medial preoptic area of the hypothalamus, the ventral thalamus, and the medial pontine reticular formation, as demonstrated by increases in hot-plate response latencies following bilateral NT administration (Clineschmidt *et al.*, 1982; Kalivas *et al.*, 1982d). The neuronal pathways through which NT might be operating are unknown, but evidence exists to suggest serotonergic (Long *et al.*, 1984) and/or opioid involvement (Clineschmidt *et al.*, 1979; Osbahr *et al.*, 1981; Hylden and Wilcox, 1983). Although the exact importance of opioid systems is controversial (Luttinger *et al.*, 1981c; Yaksh *et al.*, 1982), in mice chronically treated with morphine, a significant attenuation of NT-induced antinociception, but not NT-induced hypothermia (Luttinger *et al.*, 1983a), was found. Furthermore, IC NT inhibits naloxone-precipitated jumping in morphine-dependent mice (Luttinger *et al.*, 1983a).

12. CARDIOVASCULAR EFFECTS

Intracerebroventricular administration of NT produces a dose-dependent decrease in arterial blood pressure of pentobarbital-anesthetized rats (Rioux *et al.*, 1981). However, this effect is considerably less pronounced in conscious rats, probably indicative of a NT–pentobarbital interaction (Nemeroff *et al.*, 1977), as we had described nearly a decade ago.

Structure–activity studies (Quirion *et al.*, 1981) on the cardiovascular effects of NT have provided data consistent with the view that the C terminus of the peptide is essential for its hypotensive action.

13. EFFECTS ON MUSCLE TONE

Intracisternal (IC) injection of NT produces muscle relaxation in mice (Osbahr *et al.*, 1979) as assessed in the Jolou–Courvoisier test, a paradigm used to screen compounds with neurolepticlike activity (Courvoisier, 1956). Furthermore, in another traction test, IC injections of NT were shown to decrease muscle tone in mice without affecting their grasping response (Jolicoeur *et al.*, 1981b). These indirect measures of muscle tone cannot be used to identify neuronal pathways or mechanisms involved in muscular events but are useful paradigms to implicate CNS neurotransmitters involved in control of muscular activity.

Finally, like some antipsychotic drugs, ICV NT has been shown to produce a dose-dependent catalepsy in mice (Snijders *et al.*, 1982) but not rats (Jolicoeur *et al.*, 1981a).

14. ENDOCRINE EFFECTS

Intraventricular (ICV) infusion of NT decreases plasma concentrations of various pituitary hormones including thyroid-stimulating hormone (TSH), prolactin (PRL), and growth hormone (GH) in rats (Maeda and Frohman, 1978; Vijayan and McCann, 1979; Nemeroff *et al.*, 1980b). In contrast, plasma arginine vasopressin concentrations in cats (Skowsky *et al.*, 1978) are increased after ICV NT. The reduction in plasma TSH after ICV NT has also been observed in thyroidectomized animals (Nemeroff *et al.*, 1980b). Furthermore, ICV injections of NT can block the rise in plasma PRL and GH induced by acute cold stress in rats (Tache *et al.*, 1979) as well as TRH-induced PRL release (Maeda and Frohman, 1978; Vijayan and McCann, 1979; Nemeroff *et al.*, 1980b).

15. EFFECTS ON LEARNING AND MEMORY

Very little is known about the effects of NT on learning and memory. However, NT has been reported to both decrease (Luttinger *et al.*, 1982b) and increase (van Wimersma Greidanus *et al.*, 1982) performance in conditioned avoidance paradigms. Moreover, in the conditioned place preference test, NT administered bilaterally into the ventral tegmental area produces effects consistent with the peptide possessing rewarding properties (Glimcher *et al.*, 1982; 1984; Hoebel *et al.*, 1982).

16. EFFECTS ON FEEDING AND DRINKING

When administered ICV to rats, NT decreases the food consumption of food-deprived animals (Luttinger *et al.*, 1982a; Hoebel, 1984). This effect has also been observed when NT was infused into the paraventricular nucleus of the hypothalamus (Stanley *et al.*, 1982), a brain area rich in NT receptors and NT-like immunoreactivity (see Table I of NT distribution) and previously associated with eating behavior (Leibowitz, 1978). However, NT does not alter water intake in water-deprived animals (Stanley *et al.*, 1982). Finally, ICV administration of NT does not alter stress-induced eating in rats (Morley and Levine, 1981), although the peptide reduces dynorphin- (Levine *et al.*, 1983) and norepinephrine- (Stanley *et al.*, 1982) but not insulin-induced (Bueno *et al.*, 1983) feeding when infused into the paraventricular area of the hypothalamus.

17. EFFECTS ON LOCOMOTOR ACTIVITY

The dopamine-containing perikarya in the ventral tegmental area (A10 or VTA) project to the nucleus accumbens (NAS) (Dahlstrom and Fuxe, 1964; Moore and Bloom, 1978) and other forebrain areas; NT receptors have been localized on these cells. Bilateral administration of NT into the VTA elicits an increase in locomotion, rearing, and sniffing that is significant 20 min after the infusion and persist for a further 60–70 min (Kalivas *et al.*, 1981, 1982b, 1983b; Kalivas, 1984; P. J. Elliott and C. B. Nemeroff, unpublished results). However, injection of NT into the NAS inhibits dopamine-induced locomotion and rearing (Kalivas *et al.*, 1984) as well as psychostimulant-induced hyperactivity (Ervin *et al.*, 1981; Luttinger *et al.*, 1981b). The complex interactions of NT and DA systems are described in more detail in Section 21.

In the open field, ICV NT produces a decrease in locomotor activity (van Wimersma Greidanus *et al.*, 1982). This effect is, however, reduced after repeated infusions, probably through a reduction in NT receptor number and/or affinity. These effects of NT on locomotor activity are unrelated to its hypothermic action (van Wimersma Greidanus *et al.*, 1984).

18. INTERACTIONS WITH THYROTROPIN-RELEASING HORMONE

Many of the central effects of NT can be blocked or attenuated by prior or concomitant administration of TRH, and these data have been reviewed (Burgess *et al.*, 1983). In mice, the hypothermic effects (Burgess *et al.*, 1983), analgesic properties (Osbahr *et al.*, 1981), and muscle relaxation (Osbahr *et al.*, 1979) produced by NT are all antagonized, to some extent, by TRH. However, intra-VTA infusion of TRH in rats does not block the locomotor activity induced by intra-VTA injections of NT (Burgess *et al.*, 1983).

19. INTERACTIONS WITH SPECIFIC NEUROTRANSMITTER SYSTEMS

Increases in serotonin turnover have been found in a few CNS areas following ICV administration of high doses of NT (Garcia-Sevilla *et al.*, 1978). Furthermore, ICV injection of NT was reported to increase serotonergic activity in the diagonal band of Broca while decreasing serotonergic activity in the medial forebrain bundle and periaqueductal gray (Long *et al.*, 1981).

Intraventricular administration of NT has been reported to slightly but significantly increase the rate of turnover of norepinephrine (NE) in grossly dissected rat brain areas (Garcia-Sevilla *et al.*, 1978). An NT–NE interaction is further supported by a report that indicated that NT increases both spontaneous and potassium-induced release of NE in hypothalamic slice preparations (Okuma and Osumi, 1982). Furthermore, 6OHDA-induced lesions of ascending noradrenergic pathways produce a decrease in the hypothalamic concentration of both NT and NE, further suggesting a relationship between these two neurotransmitters (Ibata *et al.*, 1984b).

As discussed above and in other places (Nemeroff and Cain, 1985; Nemeroff, 1986), NT is closely associated with CNS DA systems (Haubrich *et al.*, 1982; Nemeroff *et al.*, 1983a; Quirion, 1983; Reches *et al.*, 1983; Goedert, 1984; Kilts *et al.*, 1986). Both neurotransmitter candidates have CNS distributions that significantly overlap (Anderson *et al.*, 1984), and NT and DA have even been reported to coexist in certain brain areas such as the arcuate nucleus of the hypothalamus (Hökfelt *et al.*, 1984; Ibata *et al.*, 1983; 1984a; Kalivas and Miller, 1984).

As noted earlier, NT binding sites are present on both DA cell bodies (Palacios and Kuhar, 1981; Quirion *et al.*, 1982d) and terminals (Quirion *et al.*, 1982d; Goedert *et al.*, 1984b) as demonstrated in autoradiographic and biochemical studies, and electrophysiological studies have shown that iontophoretically applied NT can excite midbrain DA neurons (Andrade and Aghajanian, 1981).

In vitro experiments have shown that NT releases DA from slices of rat striatum, hypothalamus, and nucleus accumbens (Starr, 1982; De Quidt and Emson, 1983; Okuma *et al.*, 1983; Myers and Lee, 1984a,b). Furthermore, NT administered ICV stimulates tyrosine hydroxylase activity in the same areas (Reches *et al.*, 1982). Moreover, IC and ICV infusions of NT produce a dose-dependent increase in DA turnover in several brain areas (Garcia-Sevilla *et al.*, 1978; Reches *et al.*, 1982, 1983; Widerlov *et al.*, 1982a) but especially in mesolimbic DA terminal areas, as assessed by measurement of DA metabolites as well as by measurement of DOPA accumulation after inhibition of DOPA decarboxylase with NSD 1015.

These NT–DA interactions render plausible the study of NT-like immunoreactivity and NT receptor numbers in neuropsychiatric disorders where DA is believed to be important (see Section 21) such as Parkinson's disease and schizophrenia.

20. INTERACTIONS WITH DOPAMINERGIC AGENTS

Chronic administration of antipsychotic drugs (DA receptor antagonists) increase the concentration of NT-like immunoreactivity in certain CNS areas (Govoni *et al.*, 1980), particularly in terminal areas of the mesolimbic DA system such as the nucleus accumbens. These results have recently been replicated in our laboratory, and, furthermore, a distinction between types of neuroleptic drugs has been found. Haloperidol, but not clozapine, an atypical neuroleptic (Simpson and Varga, 1974), increased NT concentrations in the striatum, whereas both DA antagonists increased NT concentrations in the nucleus accumbens. This result is consistent with their clinical effects; clozapine has no propensity to induce extrapyramidal side effects or to produce tardive dyskinesia, whereas haloperidol does (Gerlach *et al.*, 1975).

More than 5 years ago, our group proposed (Nemeroff, 1980; Nemeroff *et al.*, 1982, 1984) that NT may function as an endogenous neuroleptic agent because it shares many pharmacological properties with these DA receptor antagonists. Although NT does not alter the binding of [^{3}H]spiroperidol (a neuroleptic) in the rat CNS (Nemeroff *et al.*, 1982, 1983a; Reches *et al.*, 1982), it has been reported to decrease the affinity but not the number of [^{3}H]N-propylnorapomorphine (a DA agonist) binding sites in limbic forebrain *in vitro* (Agnati *et al.*, 1983). This ligand has been reported to label preferentially D_1 (Seeman, 1980) rather than D_2 receptor sites to which spiroperidol binds (Fuxe *et al.*, 1983). Chronic administration of neuroleptics in rats has been reported to produce an increase in the number of NT receptors in the substantia nigra (Uhl and Kuhar, 1984) and, as noted above, an increase in the concentration of NT-like immunoreactivity in the nucleus accumbens (Govoni *et al.*, 1980; Goedert *et al.*, 1985), caudate, and prefrontal cortex (Goedert *et al.*, 1985). Furthermore, haloperidol decreases the *in vitro* release of NT from hypothalamic tissue (Maeda and Frohman, 1981). The opposite effect is seen with infusions of dopamine (Maeda and Frohman, 1981).

As noted above, not all antipsychotic drugs have the same characteristics, and this has led to classification of one particular group of these agents as "atypical neuroleptics" (Costall and Naylor, 1976; Bentall and Herberg, 1980). Neurotensin shares many common properties with this subgroup of DA antagonists in that it does not block *d*-amphetamine-induced stereotypy (Jolicoeur *et al.*, 1983). Finally, NT differs from dopamine antagonists in several ways; the peptide has antinociceptive activity and does not alter DA-stimulated adenylate cyclase activity (Nemeroff *et al.*, 1982).

21. CLINICAL STUDIES

Neurotensinlike immunoreactivity (NT-LI) has been found to be increased in postmortem samples of the nucleus caudatus and globus pallidus (Emson *et al.*, 1985b) of Huntington's chorea patients. This is most likely merely evidence that NT neuronal elements are spared in this disease, in contrast to other neurons. In Parkinson's disease patients, the NT receptor number in the substantia nigra (SN) is reportedly decreased (Sadoul *et al.*, 1984a,c; Uhl *et al.*, 1984). However, in Parkinson's disease, NT-LI is unchanged in the SN and VTA and a variety of other brain regions but decreased in the hippocampus (Bissette *et al.*, 1985).

In schizophrenic patients, NT-LI has been found to be increased in an area of frontal cortex (BA32) (Nemeroff *et al.*, 1983b). However, a report by Kleinman *et al.* (1983) found no change in NT-LI in postmortem brain tissue of schizophrenic patients, though cortical areas were not assayed. In our first study (Widerlov *et al.*, 1982b) of the concentration of NT-LI in cerebrospinal fluid (CSF) of (drug-free) patients with schizophrenia, a subgroup of patients exhibited low levels of NT. This finding has now been replicated in a second study. In Alzheimer's disease, a significant reduction of NT-LI was found in the amygdala but not in other brain areas (Nemeroff *et al.*, 1983c).

22. CONCLUSION

It is evident from the data presented in this chapter that NT produces a variety of effects after direct CNS administration. These effects are observed after microinjection into a relatively few brain areas, rendering specificity to these findings. Whether many or all of these effects are representative of physiological roles of NT or merely are pharmacological actions of the peptide is uncertain. At present, it is plausible to postulate a role for NT in thermoregulation, nociception, and regulation of dopaminergic activity, with its consequent effects on behavior. It might be an important neuromodulator in the pathogenesis of psychiatric disorders such as schizophrenia. We have learned much about this peptide since its discovery a little more than a decade ago, and in the next decade we should witness further progress.

ACKNOWLEDGMENTS. We are grateful to Mary Lassiter for typing this manuscript. Supported by NIMH-39415, MH-40524, MH-40159, and the Schizophrenia Research Foundation (Society of the Scottish Rite). Charles B. Nemeroff is the recipient of a Nanaline H. Duke Fellowship from Duke University Medical Center. Peter J. Elliott was the recipient of a Gorrell Family Psychiatric Research Fellowship from Duke University Medical Center.

REFERENCES

Agnati, L. F., Fuxe, K., Benfenati, F., and Battistini, N., 1983, Neurotensin *in vitro* markedly reduces the affinity in subcortical limbic ^{3}H-N-propylnorapomorphine binding sites, *Acta Physiol. Scand.* **119**:459–461.

Anderson, C. M., Bissette, G., Nemeroff, C. B., and Kilts, C. D., 1984, Microtopographic distribution of neurotensin and catecholamine concentrations at the level of individual rat brain nuclei, *Soc. Neurosci. Abstr.* **10**:437.

Andrade, R., and Aghajanian, G. K., 1981, Neurotensin selectively activates dopamine neurons in the substantia nigra, *Soc. Neurosci. Abstr.* **7**:573.

Araki, K., Tachibana, S., Uchiyama, M., Nakajima, T., and Yasuhara, T., 1973, Isolation and structure of a new active peptide "xenopsin" on the smooth muscle, especially on a strip of fundus from a rat stomach, from the skin of *Xenopus laevis*, *Chem. Pharm. Bull.* **21**:2801–2804.

Beitz, A. J., 1982, The sites of origin of brain stem neurotensin and serotonin projections to the rodent nucleus raphe magnus, *J. Neurosci.* **2**:829–842.

Bentall, A. C. C., and Herberg, L. J., 1980, Blockade of amphetamine-induced locomotor activity and stereotypy in rats by spiroperidol but not by an atypical neuroleptic thioridazine, *Neuropharmacology* **19:**699–703.

Bhatnagar, Y. M., and Carraway, R., 1981, Bacterial peptides with C-terminal similarities to bovine neurotensin, *Peptides* **2:**51–59.

Bissette, G., Nemeroff, C. B., Loosen, P. T., Prange, A. J., Jr., and Lipton, M. A., 1976, Hypothermia and intolerance to cold induced by intracisternal administration of the hypothalamic peptide neurotensin, *Nature* **262:**607–609.

Bissette, G., Nemeroff, C. B., Loosen, P. T., Breese, G. R., Burnett, G. B., Lipton, M. A., and Prange, A. J., Jr., 1978, Modification of pentobarbital-induced sedation by natural and synthetic peptides, *Neuropharmacology* **17:**229–237.

Bissette, G., Luttinger, D., Nemeroff, C. B., and Prange, A. J., Jr., 1981, The effects of *d*-amphetamine, methylphenidate, and apomorphine on neurotensin-induced hypothermia in mice, *Soc. Neurosci. Abstr.* **7:**32.

Bissette, G., Jennes, L., Prange, A. J., Jr., Breese, G. R., and Nemeroff, C. B., 1983, Neurotensin and dopamine are not co-localized in rat brain, *Soc. Neurosci. Abstr.* **9:**290.

Bissette, G., Richardson, C., Kizer, J. S., and Nemeroff, C. B., 1984, Ontogeny of brain neurotensin in the rat: A radioimmunoassay study, *J. Neurochem.* **43:**283–287.

Bissette, G., Nemeroff, C. B., Decker, M. W., Kizer, J. S., Agid, Y., and Javoy-Agid, F., 1985, Alterations in regional brain concentrations of neurotensin and bombesin in Parkinson's disease, *Ann. Neurol.* **17:**324–329.

Blackburn, A. M., and Bloom, S. R., 1979, A radioimmunoassay for neurotensin in human plasma, *J. Endocrinol.* **83:**175–181.

Bloom, S. R., and Polak, J. M., 1982, Aspects of neurotensin physiology and pathology, *Ann. N.Y. Acad. Sci.* **400:**105–116.

Brown, M. R., and Vale, W., 1980, Peptides and thermoregulation, in: *Thermoregulatory Mechanisms and Their Therapeutic Implication* (B. Cox, P. Lomax, A. S. Milton, and E. Schonbaum, eds.), pp. 186–194, S. Karger, Basel.

Brown, M., Rivier, J., and Vale, W., 1977, Bombesin: Potent effects on thermoregulation in the rat, *Science* **196:**998–1000.

Bueno, L., Ferre, J. P., Fioramonti, J., and Honde, C., 1983, Effects of intracerebroventricular administration of neurotensin, substance P and calcitonin on gastrointestinal motility in normal and vagotomized rats, *Regul. Peptides* **6:**197–205.

Burgess, S. K., Luttinger, D. E., Hernandez, D., Kalivas, P. W., Nemeroff, C. B., and Prange, A. J., Jr., 1983, Antogonism of the central nervous system effects of neurotensin by thyrotropin-releasing hormone, in: *Thyrotropin-Releasing Hormone* (E. C. Griffiths and G. W. Bennett, eds.), pp. 291–302, Raven Press, New York.

Camargo, A. C. M., Caldo, H., and Emson, P. C., 1983, Degradation of neurotensin by rabbit brain endo-oligopeptidase A and endo-oligopeptidase B, *Biochem. Biophys. Res. Commun.* **116:**1151–1159.

Carraway, R. E., 1982, A critical analysis of three approaches to radioimmunoassay of peptides: Applications to the study of the neurotensin family, *Ann. N.Y. Acad. Sci.* **400:**17–36.

Carraway, R., and Leeman, S. E., 1973, The isolation of a new hypotensive peptide, neurotensin, from bovine hypothalami, *J. Biol. Chem.* **250:**1907–1911.

Carraway, R. E., and Leeman, S. E., 1975, The amino acid sequence of a hypothalamic peptide, neurotensin, *J. Biol. Chem.* **250:**1907–1911.

Carraway, R., and Leeman, S. E., 1976, Characterization of radioimmunoassayable neurotensin in the rat: Its differential distribution in the central nervous system, small intestine, and stomach, *J. Biol. Chem.* **251:**7045–7052.

Carraway, R., Ruane, S. E., and Kim, H. R., 1982, Distribution and immunochemical character of neurotensin-like material in representative vertebrates and invertebrates: Apparent conservation of the COOH-terminal region during evolution, *Peptides* **3:**115–123.

Chandra, A., Chov, H. C., Chang, C., and Lin, M. T., 1981, Effects if intraventricular administration and somatostatin on thermoregulation in the rat, *Neuopharmacology* **20:**715–718.

Checler, F., Kitabgi, P., and Vincent, J. P., 1982, Degradation of neurotensin by brain synaptic membranes, *Ann. N.Y. Acad. Sci.* **400:**413–414.

Checler, F., Vincent, J. P., and Kitabgi, P., 1983, Neurotensin analogs [D-Tyr^{11}] and [D-Phe^{11}] neurotensin resist degradation by brain peptidases *in vitro* and *in vivo, J. Pharmacol. Exp. Ther.* **227:**743–748.

Clark, W. G., 1979, Changes in body temperature after administration of amino acids, peptides, dopamine, neuroleptics and related agents, *Neurosci. Biobehav. Res.* **3:**179–231.

Clineschmidt, B. V., and McGuffin, J. C., 1977, Neurotensin administered intracisternally inhibits responsiveness of mice to noxious stimuli, *Eur. J. Pharmacol.* **46:**395–396.

Clineschmit, B.V., McGuffin, J.C., and Bunting, P. B., 1979, Neurotensin: Antinocisponsive action in rodents, *Eur. J. Pharmacol.* **54:**129–139.

Clineschmidt, B. V., Martin, G. E., and Veber, D. F., 1982, Antinocisponsive effects of neurotensin and neurotensin-related peptides, *Ann. N.Y. Acad. Sci.* **400:**283–306.

Cooper, P. E., Fernstrom, M. H., Rorstad, O. P., Leeman, S. E., and Martin, J. B., 1981, The regional distribution of somatostain, substance P and neurotensin in human brain, *Brain Res.* **218:**219–232.

Costall, B., and Naylor, R. J., 1976, A comparison of the abilities of typical neuroleptic agents and of thioridazine, clozapine, sulpiride, and metaclopramide to antagonize the hyperactivity induced by dopamine applied intracerebrally to areas of the extrapyramidal and mesolimbic systems, *Eur. J. Pharmacol.* **40:**9–19.

Courvoisier, S., 1956, Pharmacodynamic basis for the use of chlorpromazine in psychiatry, *J. Clin. Exp. Psychopathol.* **17:**25–37.

Cowan, A., and Gmerek, D. E., 1981, Studies on the antinocisponsive action of neurotensin, *Fed. Proc.* **40:**273.

Cowan, A., and Gmerek, D. E., 1982, Defining the antinociceptive actions of neurotensin, *Ann. N.Y. Acad. Sci.* **400:**438–439.

Dahlstrom, A., and Fuxe, K., 1964, Evidence for the existence of monoamine containing neurons in the central nervous system. I. Demonstration of monoamines in the cell bodies of brain stem neurons, *Acta Physiol. Scand.* **62:**1–55.

DeQuidt, M., and Emson, P. C., 1983, Neurotensin facilitates dopamine release *in vitro* from rat striatal slices, *Brain Res.* **274:**376–380.

Dupont, A., and Merand, Y., 1978, Enzymic inactivation of neurotensin by hypothalamic and brain extracts of the rat, *Life Sci.* **22:**1623–1630.

Dupont, A., Largelier, P., Merand, Y., Cote, J., and Barden, N., 1979, Radioimmunoassay studies of the regional distribution of neurotensin, substance P, somatostatin, thyrotropin, corticotropin, Lys-vasopressin, and opiate peptides in bovine brain, in: *Proceedings of the Endocrine Society 61st Annual Meeting,* p. 127, Endocrine Society, Washington, D.C.

Eldred, W. D., and Karten, H. J., 1983, Characterization and quantification of peptidergic amacrine cells in the turtle retina: Enkephalin, neurotensin and glucagon, *J. Comp. Neurol.* **221:**371–381.

Emson, P. C., Goedert, M., Williams, B., Ninkovic, N., and Hunt, S. P., 1982a, Neurotensin: Regional distribution, characterization and inactivation, *Ann. N.Y. Acad. Sci.* **400:**198–215.

Emson, P. C., Goedert, M., Benton, H., St. Pierre, S., and Rioux, F., 1982b, The regional distribution and chromatographic characterization of neurotensin-like immunoreactivity in the rat, *Adv. Biochem. Psychopharmacol.* **33:**477–487.

Emson, P. C., Goedert, M., Horsfield, P., Rioux, F., and St. Pierre, S., 1982c, The regional distribution and chromatographic characterization of neurotensin-like immunoreactivity in the rat central nervous system, *J. Neurochem.* **38:**992–999.

Emson, P. C., Goedert, M., and Mantyh, P. W., 1985a, Neurotensin-containing neurons, in: *Handbook of Chemical Neuroanatomy,* Vol. 4 (T. Hökfelt and A. Bjorklund, eds.), pp. 355–405, Elsevier, Amsterdam.

Emson, P. C., Horsfield, P. M., Goedert, M., Rossor, M. N., and Hawkes, C. H., 1985b, Neurotensin in human brain: Regional distribution and effects of neurological illness, *Brain Res.* **347:**239–244.

Ervin, G. N., Birkemo, L. S., Nemeroff, C. B., and Prange, A. J., Jr., 1981, Neurotensin blocks certain amphetamine behaviors, *Nature* **291:**73–76.

Fernstrom, M. H., Carraway, R. E., and Leeman, S. E., 1980, Neurotensin, in: *Frontiers in Neuroendocrinology,* Vol. 6 (L. Martini and W. F. Ganong, eds.), pp. 103–127, Raven Press, New York.

Finn, P. W., Robson, B., and Griffiths, E. C., 1984, Conformational study of neurotensin, *Int. J. Peptide Protein Res.* **24:**407–413.

Fox, J. E. T., Sakai, Y., Jury, J., McLean, J., and Daniel, E. E., 1982, Neurotensin, evidence for multiple receptors for gastrointestinal motility action in dogs, *Ann. N.Y. Acad. Sci.* **400:**398–399.

Frye, G. D., Luttinger, D., Nemeroff, C. B., Vogel, R. A., Prange, A. J., Jr., and Breese, G. R., 1981, Modification of the actions of ethanol by centrally active peptides, *Peptides* **2**(Suppl. 1):99–106.

Furuta, S., Kisara, K., Sakurada, S., Sakurada, T., Sasaki, V., and Suzuki, K., 1984, Structure–antinociceptive activity studies with neurotensin, *Br. J. Pharmacol.* **83:**43–48.

Fuxe, K., Agnati, L. F., Benfenati, F., Celani, M., Zini, I., Zoli, M., and Mutt, V., 1983, Evidence for the existence of receptor–receptor interactions in the central nervous system. Studies on the regulation of monoamine receptors by neuropeptides, *J. Neural Transm. (*Suppl.*)* **19:**165–179.

Garcia-Sevilla, J. A., Magnusson, T., Carlsson, A., Leban, J., and Folkers, K., 1978, Neurotensin and its amide analogue [Gln^4]-neurotensin: Effects on brain monoamine turnover, *Naunyn Schmiedebergs Arch. Pharmacol.* **305:**213–218.

Gerlach, J., Thorse, K., and Fog, R., 1975, Extrapyramidal reactions and amine metabolites in cerebrospinal fluid during haloperidol and clozapine treatment of schizophrenic patients, *Psychopharmacologia* **40:**341–350.

Ghatei, M. A., Bloom, S. R., Langerin, H., McGregor, G. P., Lee, Y. C., Adrian, T. E., O'Shaughessy, D. J., Blank, M. A., and Uttenthal, L. O., 1984, Regional distribution of bombesin and seven other regulatory peptides in the human brain, *Brain Res.* **293:**101–109.

Gilbert, J. A., and Richelson, E., 1984, Neurotensin stimulates formation of cyclic GMP in murine neuroblastoma clone N1E-115, *Eur. J. Pharmacol.* **99:**245–246.

Gilbert, J. A., McKinney, M., and Richelson, E., 1984, Neurotensin stimulates cyclic GMP formation in neuroblastoma clone N1E-115 and rat cerebellum, *Soc. Neurosci. Abstr.* **10:**378.

Glimcher, P., Margolin, D., and Hoebel, B. G., 1982, Rewarding effects of neurotensin in the brain, *Ann. N.Y. Acad. Sci.* **400:**422–424.

Glimcher, P. W., Margolin, D. H., Giovino, A. A., and Hoebel, B. G., 1984, Neurotensin: A new "reward peptide," *Brain Res.* **291:**119–124.

Gmerek, D. E., and Cowan, A., 1981, Tolerance and cross-tolerance studies on the grooming and shaking effects of ACTH-(1–24), bombesin and RX 336-M, *Fed. Proc.* **223:**274.

Goedert, M., 1984, Neurotensin—a status report, *Trends Neurosci.* **7:**3–5.

Goedert, M., Mantyh, P. W., Hunt, S. P., and Emson, P. C., 1983, Mosaic distribution of neurotensin-like immunoreactivity in the cat striatum, *Brain Res.* **274:**176–179.

Goedert, M., Mantyh, P. W., Emson, P. C., and Hunt, S. P., 1984a, Inverse relationship between neurotensin receptors and neurotensin-like immunoreactivity in cat striatum, *Nature* **307:**543–546.

Goedert, M., Pittaway, K., and Emson, P. C., 1984b, Neurotensin receptors in the rat striatum: Lesion studies, *Brain Res.* **299:**164–168.

Goedert, M., Pittaway, K., Williams, B. J., and Emson, P. C., 1984c, Specific binding of tritiated neurotensin to rat brain membranes: Characterization and regional distribution, *Brain Res.* **304:**71–81.

Goedert, M., Sturmey, N., Williams, B., and Emson, P. C., 1984d, The comparative distribution of xenopsin- and neurotensin-like immunoreactivity in *Xenopus laevis* and rat tissues, *Brain Res.* **308:**273–280.

Goedert, M., Iversen, S. D., and Emson, P. C., 1985, The effects of chronic neuroleptic treatment on neurotensin-like immunoreactivity in the rat central nervous system, *Brain Res.* **335:**334–336.

Govoni, S., Hong, J. S., Yang, H.-Y. T., and Costa, E., 1980, Increase of neurotensin content elicited by neuroleptics in nucleus accumbens, *J. Pharmacol. Exp. Ther.* **215:**413–417.

Grant, L. D., Nemeroff, C. B., and Bissette, G., 1982, Neuroanatomical distribution of neuroactive peptides in mammalian species, in: *Peptides, Hormones and Behavior* (C. B. Nemeroff and A. J. Dunn, eds.), pp. 37–98, Spectrum Press, New York.

Griffiths, E. C., Widdowson, P., and Slater, P., 1982, Antagonism of TRH-induced wet-dog shaking in rats by neurotensin and a neurotensin fragment, *Neurosci. Lett.* **31:**17.

Griffiths, E. C., McDermott, J. R., and Smith, A. I., 1984, A comparative study of neurotensin inactivation by brain peptidases from different vertebrate species, *Comp. Biochem. Physiol.* **77C:**363–366.

Grimmelikhuijzen, C. J. P., Carraway, R. E., Rokaeus, A., and Sundler, F., 1981, Neurotensin-like immunoreactivity in the nervous system of hydra, *Histochemistry* **72:**199–209.

Guyenet, P. G., and Aghajanian, G. K., 1977, Excitation of neurons in the nucleus locus coeruleus by substance P and related peptides, *Brain Res.* **136:**178–184.

Haubrich, D. R., Martin, G. E., Pflueger, B., and Williams, M., 1982, Neurotensin effects on brain dopaminergic systems, *Brain Res.* **231:**216–221.

Henry, J. L., 1982, Electrophysiological studies on the neuroactive properties of neurotensin, *Ann. N.Y. Acad. Sci.* **400:**216–227.

Hernandez, D. E., Nemeroff, C. B., and Prange, A. J., Jr., 1982, Ontogeny of the hypothermic response to centrally administered neurotensin in rats, *Dev. Brain Res.* **3:**497–501.

Hernandez, D., Nemeroff, C. B., Valderrama, M. M., and Prange, A. J., Jr., 1984a, Neurotensin-induced antinociception and hypothermia in mice: Antagonism by TRH and structural analogs of TRH, *Regul. Peptides* **8:**41–49.

Hernandez, D. E., Richardson, C. M., Nemeroff, C. B., Orlando, R. C., St.-Pierre, S., Rioux, F., and Prange, A. J., 1984b, Evidence for biological activity of two N-terminal fragments of neurotensin, neurotensin$_{1-8}$ and neurotensin$_{1-10}$, *Brain Res.* **301:**153–156.

Hoebel, B. G., 1984, Neurotransmitters in the control of feeding and its rewards: Monoamines, opiates and brain–gut peptides, in: *Eating and Its Disorders* (A. J. Stunkard and E. Stellar, eds.), pp. 15–38, Raven Press, New York.

Hoebel, B. G., Hernandez, L., McLean, S., Stanley, B. G., Aulissi, E. F., Elimacher, P., and Margolin, D., 1982, Catecholamines, enkephalin, and neurotensin in feeding and reward, in: *The Neural Basis of Feeding and Reward* (B. G. Hoebel and D. Novin, eds.), pp. 465–478, Haer Institute, Sebasco, ME.

Hokfelt, T., Everitt, B. J., Theodorsson-Norheim, E., and Goldstein, M., 1984, Occurrence of neurotensin-like immunoreactivity in subpopulations of hypothalamic, mesencephalic, and medullary catecholamine neurons, *J. Comp. Neurol.* **222:**543–559.

Hylden, J. L. K., and Wilcox, G. L., 1983, Antinociceptive action of intrathecal neurotensin in mice, *Peptides* **4:**517–520.

Ibata, Y., Okamura, F. H., Kawakami, T., Tanaka, M., Obata, H. L., Tsuto, O. T., Terubayashi, H., Yanaihara, C., and Yanaihara, N., 1983, Coexistence of dopamine and neurotensin in hypothalamic arcuate and periventricular neurons, *Brain Res.* **269:**177–179.

Ibata, Y., Kawakami, F., Fukui, K., Obata-Tsuto, H. L., Tanaka, M., Kubo, T., Okamura, H., Morimoto, N., Yanaihara, C., and Yanaihara, N., 1984a, Light and electron microscopic immunocytochemistry of neurotensin-like immunoreactive neurons in the rat hypothalamus, *Brain Res.* **302:**221–230.

Ibata, Y., Kawakami, G., Fukui, K., Okamura, H., Obata-Tsuto, H. L., Tsuto, T., and Terubayashi, H., 1984b, Morphological survey of neurotensin-like immunoreactive neurons in the hypothalamus, *Peptides* **5:**109–120.

Inagaki, S., Yamano, M., Shiosaka, S., Takagi, H., and Tohyama, M., 1983a, Distribution and origins of neurotensin-containing fibers in the nucleus ventromedialis hypothalamai of the rat: An experimental immunohistochemical study, *Brain Res.* **273:**229–235.

Inagaki, S., Shinoda, K., Kubota, Y., Shiosaka, S., Marsuzaki, T., and Tohyama, M., 1983b, Evidence for the existence of a neurotensin-containing pathway from the endopiriform nucleus and the adjacent prepiriform cortex to the anterior olfactory nucleus and nucleus of diagonal band (Broca) of the rat, *Neuroscience* **8:**487–493.

Iversen, L. L., Iversen, S. D., Bloom, F., Douglas, C., Brown, M., and Vale, W., 1978, Calcium-dependent release of somatostatin and neurotensin from rat brain *in vitro, Nature* **273:**161.

Jennes, L., Stumpf, W. E., and Kalivas, P. W., 1982, Neurotensin: Topographical distribution in rat brain by immunohistochemistry, *J. Comp. Neurol.* **210:**211–224.

Jolicoeur, F. B., and Barbeau, A., 1982, Pharmacological evidence for a heterogeneity of receptors underlying various central and peripheral effects of neurotensin, *Ann. N.Y. Acad. Sci.* **400:**440–441.

Jolicoeur, F. B., Barbeau, A., Rioux, F., Quirion, R., and St. Pierre, S., 1981a, Differential neurobehavioral effects of neurotensin and structural analogues, *Peptides* **2:**171–175.

Jolicoeur, F. B., Rondeau, D., St. Pierre, S., Rioux, F., and Barbeau, A., 1981b, Peptides and the basal ganglia, in: *Clinical Pharmacology of Apomorphine and Other Dopamimetrics* (J. L. Gessa and G. U. Corsini, eds.), pp. 19–25, Raven Press, New York.

Jolicoeur, F. B., Barbeau, A., Quirion, R., Rioux, F., and St. Pierre, S., 1982a, Pharmacological evidence for a heterogeneity of receptors in developing various central and peripheral effects of neurotensin, *Ann. N.Y. Acad. Sci.* **400:**440–441.

Jolicoeur, F. B., Rioux, F., St.-Pierre, S., and Barbeau, A., 1982b, Structure–activity studies of neurotensin's neurobehavioral effects, in: *Brain Neurotransmitters and Hormones* (R. Collu, A. Barbeau, J. R. Ducharme, and G. Tollis, eds.), Raven Press, New York, pp. 171–183.

Jolicoeur, F. B., De Michele, G., and Barbeau, A., 1983, Neurotensin affects hyperactivity but not stereotypy induced by pre- and post-synaptic dopaminergic stimulation, *Neurosci. Biobehav. Rev.* **7:**385–390.

Jolicoeur, F. B., St. Pierre, S., Aube, C., Riuest, R., and Grogne, M. A., 1984, Relationships between structure and duration of neurotensin's central action: Emergence of long-lasting analogs, *Neuropeptides* **4:**467–476.

Jolicoeur, F. B., Rioux, F., and St. Pierre, S., 1985, Neurotensin, in: *Handbook of Neurochemistry,* Vol. 8 (A. Lajtha, ed.), pp. 94–114, Plenum Press, New York.

Kahn, D., Abrams, G. M., Zimmerman, E. A., Carraway, R., and Leeman, S. E., 1980, Neurotensin neurons in the rat hypothalamus: An immunocytochemical study, *Endocrinology* **107:**47–54.

Kalivas, P. W., 1984, Neurotensin in the ventromedial mesencephalon of the rat: Anatomical and functional considerations, *J. Comp. Neurol.* **226:**495–507.

Kalivas, P. W., and Miller, J. S., 1984, Neurotensin neurons in the ventral tegmental area project to the medial nucleus accumbens, *Brain Res.* **300:**157–160.

Kalivas, P. W., Nemeroff, C. B., and Prange, A. J., Jr., 1981, Increase in spontaneous motor activity following infusion of neurotensin into the ventral tegmental area, *Brain Res.* **229:**525–529.

Kalivas, P. W., Nemeroff, C. B., and Prange, A. J., Jr., 1982b, Neuroanatomical sites of action of neurotensin, *Ann. N.Y. Acad. Sci.* **400:**307–318.

Kalivas, P. W., Gau, B., Nemeroff, C. B., and Prange, A. J., Jr., 1982c, Antinociception after microinjection of neurotensin into the central amygdaloid nucleus of the rat, *Brain Res.* **243:**279–286.

Kalivas, P. W., Jennes, L., Nemeroff, C. B., and Prange, A. J., Jr., 1982d, Neurotensin: Topographic distribution of brain sites involved in hypothermia and antinociception, *J. Comp. Neurol.* **210:**225–238.

Kalivas, P. W., Hernandez, D. E., Bissette, G., Mason, G. A., Nemeroff, C. B., and Prange, A. J., Jr., 1983a, Neurotensin-induced hypothermia: Neuroanatomical and neurochemical mechanisms, in: *Proceedings Pharmacology Thermoregulation 5th International Meeting* (P. Lomax and E. Schonbaum, eds.), pp. 108–112, S. Karger, Basel.

Kalivas, P. W., Burgess, S. K., Nemeroff, C. B., and Prange, A. J., Jr., 1983b, Behavioral and neurochemical effects of neurotensin microinjection into the ventral tegmental area of the rat, *Neuroscience* **8:**495–505.

Kalivas, P. W., Nemeroff, C. B., and Prange, A. J., Jr., 1984, Neurotensin microinjection into the nucleus accumbens antagonizes dopamine-induced increase in locomotion and rearing, *Neuroscience* **11:**919–930.

Kataoka, K., Mizumo, N., and Frohman, L. A., 1979, Regional distribution of immunoreactive neurotensin in monkey brain, *Brain Res. Bull.* **4:**57–60.

Kilts, C. D., Bissette, G., Cain, S. T., Skoog, K. M., and Nemeroff, C. B., 1986, Neuropeptide-dopamine interactions: Focus on neurotensin, in: *Dopamine '84* (G. N. Woodruff and P. J. Roberts, eds.), Macmillan, London (in press).

Kitabgi, P., Carraway, R., and Leeman, S. E., 1976, Isolation of a tridecapeptide from bovine intestinal tissue and its partial characterization as neurotensin, *J. Biol. Chem.* **251:**7053–7058.

Kitabgi, P., Carraway, R., Van Rietschoten, J., Granier, C., Morgat, J. L., Mendez, A., Leeman, S., and Freychet, P., 1977, Neurotensin: Specific binding to synaptic membranes from rat brain, *Proc. Natl. Acad. Sci. U.S.A.* **74:**1846–1850.

Kitabgi, P., Poustis, C., Granier, C., Van Rietschoten, J., Rivier, J., Morgat, J. L., and Freychet, P., 1980, Neurotensin binding to extraneural and neural receptors: Comparison with biological activity and structure–activity relationships, *Mol. Pharmacol.* **18:**11–19.

Kleinman, J. E., Iadarola, M., Govoni, S., Hong, J., Gillin, J. C., and Wyatt, R. J., 1983, Postmortem measurements of neuropeptides in human brain, *Psychopharmacol Bull.* **19:**375–377.

Kobayashi, R. M., Brown, M. R., and Vale, W., 1977, Regional distribution of neurotensin and somatostatin in rat brain, *Brain Res.* **126:**584–588.

Kohler, C., and Eriksson, L. G., 1984, An immunohistochemical study of somatostatin and neurotensin positive neurons in the septal nuclei of the rat brain, *Anat. Embryol.* **170:**1–10.

Langer, M., Van Noorden, S., Polak, J. M., and Pearse, A. G. E., 1979, Peptide hormone-like immunoreactivity in the gastrointestinal tract and endocrine pancreas of eleven teleost species, *Cell Tissue Res.* **199:**493–508.

Lazarus, L. H., Brown, M. R., and Perrin, H. H., 1977a, Distribution, localization, and characterization of neurotensin binding sites in the rat brain, *Neuropharmacology* **16:**625–629.

Lazarus, L. H., Perrin, M. H., Brown, M. R., and Rivier, J. E., 1977b, Mast cell binding of neurotensin. II. Molecular conformation of neurotensin involved in the stereospecific binding of mast cell receptor sites, *J. Biol. Chem.* **252:**7180–7183.

Lazarus, L. H., Brown, M., Perrin, M. H., and Rivier, J. E., 1977d, Neurotensin: Mast cell receptor binding with comparison to biological activity, *Fed. Proc.* **36:**1015.

Lee, Y. C., Terenghi, G., Polak, J. M., and Bloom, S. R., 1981, Neurotensin in adrenal medulla of the rat, *Acta Endocrinol.* (*Kbh.*) **97**(Suppl. 243)**:**84.

Lee, T. F., Hepler, J. R., and Myers, R. D., 1983, Evaluation of neurotensin's thermolytic action by ICV infusion with receptor antagonists and a Ca^{++} chelator, *Pharmacol. Biochem. Behav.* **19:** 477–481.

Leibowitz, S. F., 1978, Paraventricular nucleus: A primary site mediating adrenergic stimulation of feeding and drinking, *Pharmacol. Biochem. Behav.* **8:**163–175.

Leranth, C. S., Jew, J. Y., Williams, T. H., and Palkovits, M., 1981, Stria terminalis axons ending on substance P and neurotensin-containing cells of rat central amygdaloid nucleus: An electron microscopic immunocytochemical study, *Neuropeptides* **1:**261–272.

Levine, A. S., Kneip, J., Grace, M., and Morley, J. E., 1983, Effect of centrally administered neurotensin on multiple feeding paradigms, *Pharmacol. Biochem. Behav.* **18:**19–23.

Lipton, M. A., Bissette, G., Nemeroff, C. B., Loosen, P. T., and Prange, A. J., Jr., 1977, Neurotensin: A possible mediator of thermoregulation in the mouse, in: *Drugs, Biogenic Amines, and Body Temperature—Third Symposium on the Pharmacology of Thermoregulation* (P. Lomax, E. Schombaum, and B. Cox, eds.), pp. 54–57, S. Karger, Basel.

Long, J. B., Kalivas, P. W., Youngblood, W. W., Prange, A. J., Jr., and Kizer, J. S., 1981, Selective effects of centrally administered neurotensin on *in vivo* tryptophan hydroxylase activity (THA) and serotonin (5HT) levels in discrete rat brain areas, *Pharmacologist* **23:**133.

Long, J. B., Kalivas, P. W., Youngblood, W. W., Prange, A. J., and Kizer, S., 1984, Possible involvement of serotonergic neurotransmission in neurotensin but not morphine analgesia, *Brain Res.* **310:**35–42.

Loosen, P. T., Nemeroff, C. B., Bissette, G., Burnett, G., Prange, A. J., Jr., and Lipton, M. A., 1978, Neurotensin-induced hypothermia in the rat: Structure–activity studies, *Neuropharmacology* **17:**108–113.

Luttinger, D., Nemeroff, C. B., Mason, G. A., Frye, G. D., Breese, G. R., and Prange, A. J., Jr., 1981a, Enhancement of ethanol-induced sedation and hypothermia by centrally administered neurotensin, β-endorphin, and bombesin, *Neuropharmacology* **20:**305–309.

Luttinger, D., Nemeroff, C. B., King, R. A., Ervin, G. N., and Prange, A. J., Jr., 1981b, The effect of injection of neurotensin into the nucleus accumbens on behaviors mediated by the mesolimbic dopamine system in the rat, in: *The Neurobiology of the Nucleus Accumbens* (R. B. Chronister and J. F. DeFrance, eds.), pp. 322–332, Haer Institute for Electrophysiological Research, Sebasco, ME.

Luttinger, D., Nemeroff, C. B., and Prange, A. J., Jr., 1981c, Inhibition of neurotensin-induced antinociception in morphine-tolerant mice, *Pharmacologist* **23:**120.

Luttinger, D., King, R. A., Sheppard, D., Strupp, J., Nemeroff, C. B., and Prange, A. J., Jr., 1982a, The effect of neurotensin on food consumption in the rat, *Eur. J. Pharmacol.* **81:**499–504.

Luttinger, D., Nemeroff, C. B., and Prange, A. J., Jr., 1982b, The effects of neuropeptides on discrete-trial conditioned avoidance responding, *Brain Res.* **237:**183–192.

Luttinger, D. E., Burgess, S. K., Nemeroff, C. B., and Prange, A. J., Jr., 1983a, The effect of chronic morphine treatment on neurotensin-induced antinociception, *Psychopharmacology* **81:**10–13.

Luttinger, D., Frye, G. D., Nemeroff, C. B., and Prange, A. J., Jr., 1983b, The effects of neurotensin, β-endorphin and bombesin on ethanol-induced behaviors in mice, *Psychopharmacology* **79:**357–363.

Maeda, K., and Frohman, L. A., 1978, Dissociation of systemic and central effects of neurotensin on the secretion of growth hormone, prolactin, and thyrotropin, *Endocrinology* **103:**1903–1909.

Maeda, K., and Frohman, L. A., 1981, Neurotensin release by rat hypothalamic fragments *in vitro*, *Brain Res.* **210:**261–269.

Manberg, P. J., Youngblood, W. W., Nemeroff, C. B., Rossor, M., Iversen, L. L., Prange, A. J., Jr., and Kizer, J. S., 1982a, Regional distribution of neurotensin in human brain, *J. Neurochem.* **38:**1777–1780.

Manberg, P. J., Nemeroff, C. B., Iversen, L. L., Rosser, M. N., Kizer, J. S., and Prange, A. J., Jr., 1982b, Human brain distribution of neurotensin in normals, schizophrenics and Huntington's choreics, *Ann. N.Y. Acad. Sci.* **400:**354–367.

Martin, G. E., and Naruse, S. T., 1982, Differences in the pharmacological actions of intrathecally administered neurotensin and morphine, *Regul. Peptides* **3:**97–103.

Martin, G. E., Naruse, T., and Papp, N. L., 1981a, Antinociceptive and hypothermic actions of neurotensin administered centrally in the rat, *Neuropeptides* **1:**447–454.

Martin, G. E., Bacino, C. B., and Papp, N. L., 1981b, Hypothermia elicited by intracerebral microinjection of neurotensin, *Peptides* **1:**333–339.

Martins, A. R., Caldo, H., Coelho, H. L., Moreira, A. C., Antines-Rodriguez, J., Greene, L. J., and de Camarco, A. C. M., 1980, Screening for rabbit brain neuropeptide-metabolizing peptidases. Inhibition of endopeptidase B by bradykinin-potentiating peptide 9a (SQ 20881), *J. Neurochem.* **34:**100–107.

Marwaha, J., Hoffer, B., and Breedman, R., 1980, Electrophysiological actions of neurotensin in rat cerebellum, *Regul. Peptides* **1:**115–125.

Mashford, M. L., Nilsson, G., Rokaeus, A., and Rosell, S., 1978, The effect of food ingestion on circulating neurotensin-like immunoreactivity (NTLI) in the human, *Acta Physiol. Scand.* **104:** 244–246.

Mason, G. A., Nemeroff, C. B., Luttinger, D., Hatley, O. L., and Prange, A. J., Jr., 1980, Neurotensin and bombesin: Differential effects on body temperature of mice after intracisternal administration, *Regul. Peptides* **1:**53–60.

Mazella, J., Poustis, C., Labbe, C., Checler, F., Kitabgi, P., Granier, C., van Rietschoten, J., and Vincent, J. P., 1983, Monoiodo-[Trp^{11}]-neurotensin, a highly radioactive ligand of neurotensin receptors, *J. Biol. Chem.* **258:**3476–3481.

McCarthy, P. S., Walker, R. J., Yajima, H., Kitagawa, K., and Woodruff, G. N., 1979, The action of neurotensin on neurones in the nucleus accumbens and cerebellum of the rat, *Gen. Pharmacol.* **10:**331–333.

McDermott, J. R., and Kidd, A. N., 1984, Ion-exchange gel-filtration and reversed-phase high-performance liquid chromatography in the isolation of neurotensin-degrading enzymes from rat brain, *J. Chromatogr.* **290:**231–239.

McDermott, J. R., Smith, A. I., Edwardson, J. A., and Griffiths, E. C., 1982, Mechanism of neurotensin degradation by rat brain peptidases, *Regul. Peptides* **3:**397–404.

Merritt, W. D., Bissett, G., Luttinger, D., Prange, A. J., Jr., and Nemeroff, C. B., 1986, Adaptation to cold antagonizes neurotensin-induced hypothermia in mice, *Brain Res.* **369:**136–142.

Metcalf, G., Dettmar, P., and Watson, T., 1980, The role of neuropeptides in thermoregulation, in: *Thermoregulatory Mechanisms and Their Therapeutic Implications* (B. Cox, P. Lomax, A. S. Milton, and E. Schonbaum, eds.), pp. 175–179, S. Karger, Basel.

Miletic, V., and Randic, M., 1979, Neurotensin excites cat spinal neurones in laminae 1–111, *Brain Res.* **169:**600–604.

Milner, T. A., Joh, T. H., Miller, R. J., and Pickel, V. M., 1984, Substance P, neurotensin, enkephalin, and catecholamine-synthesizing enzymes: Light microscopic localizations compared with autoradiographic label in solitary efferents to the rat parabrachial region, *J. Comp. Neurol.* **220:**434–447.

Minamino, N., Kargawa, K., and Matsuo, H., 1984, Neuromedin N: A novel neurotensin-like peptide identified in porcine spinal cord, *Biochem. Biophys. Res. Commun.* **122:**542–549.

Moore, R. Y., and Bloom, F. E., 1978, Central catecholamine neurone systems: Anatomy and physiology of the dopamine system, *Annu. Rev. Neurosci.* **1:**129–169.

Mora, F., Lee, T. F., and Myers, R. D., 1984, Is neurotensin in the brain involved in thermoregulation of the monkey? *Peptides* **5:**125–128.

Morley, J. E., and Levine, A. S., 1981, Bombesin inhibits stress-induced eating, *Pharmacol. Biochem. Behav.* **14:**149–151.

Morley, J. E., Levine, A. S., Oken, M. M., Grace, M., and Kneip, J., 1982, Neuropeptides and thermoregulation: The interactions of bombesin, neurotensin, TRH, somatostatin, naloxone and prostaglandins, *Peptides* **3:**1–6.

Myers, R. D., and Lee, T. F., 1984a, Neurotensin perfusion of rat hypothalamus: Dissociation of dopamine release from body temperature change, *Neuroscience* **12:**241–253.

Myers, R. D., and Lee, T. F., 1984b, *In vivo* release of dopamine during perfusion of neurotensin in the substantia nigra of the unrestrained rat, *Peptides* **4:**955–961.

Nagai, K., and Frohman, L. A., 1976, Hyperglycemia and hyperglucagonemia following neurotensin administration, *Life Sci.* **19:**273–280.

Nakagawa, Y., Higashida, H., and Miki, N., 1984, A single class of neurotensin receptors with high affinity in neuroblastoma and glioma NG108-15 hybrid cells that mediate facilitation of synaptic transmission, *J. Neurosci.* **4:**1053–1601.

Nemeroff, C. B., 1980, Neurotensin: Perchance an endogenous neuroleptic? *Biol. Psychiatry* **15:**283–302.

Nemeroff, C. B., 1986, The interaction of neurotensin with dopaminergic pathways in the central nervous system: Basic neurobiology and implications for the pathogenesis and treatment of schizophrenia, *Psychoneuroendocrinology* **11:**15–37.

Nemeroff, C. B., and Cain, S. T., 1985, Neurotensin–dopamine interactions in the central nervous stem, *Trends Pharmacol. Sci* **6:**201–205.

Nemeroff, C. B., Bissette, G., Prange, A. J., Jr., Loosen, P. T., Barlow, T. S., and Lipton, M. A., 1977, Neurotensin: Central nervous system effects of a hypothalamic peptide, *Brain Res.* **128:**485–496.

Nemeroff, C. B., Osbahr, A. J. III, Manberg, P. J., Ervin, G. N., and Prange, A. J., Jr., 1979, Alterations in nociception and body temperature after intracisternally administered neurotensin, β-endorphin, other endogenous peptides, and morphine, *Proc. Natl. Acad. Sci. U.S.A.* **76:**5368–5371.

Nemeroff, C. B., Bissette, G., Manberg, P. J., Moore, S. I., Ervin, G. N., Osbahr, A. J. III, and Prange, A. J., Jr., 1980a, Hypothermic responses to neurotensin in vertebrates, in: *Thermoregulatory Mechanisms and Their Therapeutic Implications* (B. Cox, P. Lomax, A. S. Milton, and E. Schonbaum, eds.), pp. 180–185, S. Karger, Basel.

Nemeroff, C. B., Bissette, G., Manberg, P. J., Osbahr, A. J. III, Breese, G. R., and Prange, A. J., Jr., 1980b, Neurotensin-induced hypothermia: Evidence for an interaction with dopaminergic systems and the hypothalamic–pituitary–thyroid axis, *Brain Res.* **195:**69–84.

Nemeroff, C. B., Hernandez, D. E., Luttinger, D., Kalivas, P., and Prange, A. J., Jr., 1982, Interactions of neurotensin with brain dopamine systems, *Ann. N.Y. Acad. Sci.* **400:**330–344.

Nemeroff, C. B., Luttinger, D., Hernandez, D. E., Mailman, R. B., Mason, G. A., Davis, S. D., Widerlov, E., Frye, G. D., Kilts, C. D., Beaumont, K., Breese, G. R., and Prange, A. J., Jr., 1983a, Interactions of neurotensin with brain dopamine systems: Biochemical and behavioral studies, *J. Pharmacol. Exp. Ther.* **225:**337–345.

Nemeroff, C. B., Youngblood, W., Manberg, P. J., Prange, A. J., Jr., and Kizer, J. S., 1983b, Regional brain concentrations of neuropeptides in Huntington's chorea and schizophrenia, *Science* **221:**972–975.

Nemeroff, C. B., Bissette, G., Busby, W. H., Youngblood, W. W., Rossor, M., Roth, M., and Kizer, J. S., 1983c, Regional concentrations of neurotensin, thyrotropin-releasing hormone, and somatostatin in Alzheimer's disease, *Soc. Neurosci. Abstr.* **9:**1052.

Nemeroff, C. B., Kalivas, P. W., and Prange, A. J., Jr., 1984, Interaction of neurotensin and dopamine in limbic structures, in: *Catecholamines: Neuropharmacology and Central Nervous System—Theoretical Aspects* (E. Usdin, A. Carlsson, A. Dahlstrom, and J. Engel, eds.), pp. 199–205, Alan R. Liss, New York.

Nicoll, R. A., 1978, The action of thyrotropin-releasing hormone, substance P, and related peptides on frog spinal motoneurons, *J. Pharmacol. Exp. Ther.* **207:**817–824.

Ninkovic, M., Hunt, S. P., and Kelly, J. S., 1981, Effect of dorsal rhizotomy on the autoradiographic distribution of opiate and neurotensin receptors and neurotensin-like immunoreactivity within the rat spinal cord, *Brain Res.* **230:**111–119.

Okuma, Y., and Osumi, Y., 1982, Neurotensin-induced release of endogenous noradrenaline from rat hypothalamic slices, *Life Sci.* **30:**77–84.

Okuma, Y., Fukuda, Y., and Osumi, Y., 1983, Neurotensin potentiates the potassium-induced release of endogenous dopamine from rat striatal slices, *Eur. J. Pharmacol.* **93:**27–33.

Ondetti, M. A., Rubin, B., and Cushman, D. W., 1977, Design of specific inhibitors of angiotensin-converting enzyme: New class of orally active antihypertensive agents, *Science* **196:**441–444.

Osbahr, A. J. III, Nemeroff, C. B., Manberg, P. H., and Prange, A. J., Jr., 1979, Centrally administered neurotensin: Activity in the Julou–Courvosier muscle relaxation test in mice, *Eur. J. Pharmacol.* **54:**299–302.

Osbahr, A. J. III, Nemeroff, C. B., Luttinger, D., Mason, G. A., and Prange, A. J., Jr., 1981, Neurotensin-induced analgesia in mice: Antagonism by thyrotropin-releasing hormone, *J. Pharmacol. Exp. Ther.* **217:**645–651.

Palacios, J. M., and Kuhar, M. J., 1981, Neurotensin receptors are located on dopamine-containing neurons in rat midbrain, *Nature* **294:**587–589.

Phillis, J. W., and Kirkpatrick, J. R., 1979, Action of various gastrointestinal peptides on the isolated amphibian spinal cord, *Can. J. Physiol. Pharmacol.* **57:**887–889.

Phillis, J. W., and Kirkpatrick, J. R., 1980, The actions of motilin, leuteinizing-hormone-releasing hormone, cholecystokinin, somatostatin, vasoactive intestinal peptide, and other peptides on rat cerebral neurons, *Can. J. Physiol. Pharmacol.* **58:**612–623.

Poustis, C., Mazella, J., Kitabgi, P., and Vincent, S. P., 1984, High affinity neurotensin binding sites in differentiated neuroblastoma N1E 115 cells, *J. Neurochem.* **42:**1094–1100.

Prange, A. J., Jr., Nemeroff, C. B., Bissette, G., Manberg, P. J., Osbahr, A. J. III, Burnett, G. B., Loosen, P. T., and Kraemer, G. W., 1979, Neurotensin: Distribution of hypothermic response in mammalian and submammalian vertebrates, *Pharmacol. Behav.* **11:**473–477.

Price, C. H., Ruane, S. E., and Carraway, R. E., 1982, Biochemistry and physiology of neurotensin-like peptides in the brain and gut of the mollusc *Aplysia, Ann. N.Y. Acad. Sci.* **400:**409–410.

Quirion, R., 1983, Interactions between neurotensin and dopamine in the brain: An overview, *Peptides* **4:**609–615.

Quirion, R., Rioux, F., St. Pierre, S., Belanger, F., Jolicoeur, F. B., and Barbeau, A., 1981, Hypotensive effects of centrally and peripherally administered neurotensin and neurotensin derivatives in rats, *Neuropeptides* **1:**253–259.

Quirion, R., Gaudreau, P., St. Pierre, S., and Rioux, F., 1982a, Localization of neurotensin binding sites in rat kidney, *Peptides* **3:**765–769.

Quirion, R., Gaudreau, P., St. Pierre, S., Rioux, F., and Pert, C., 1982b, Autoradiographic distribution of [^{3}H]neurotensin receptors in rat brain, *Peptides* **3:**757–763.

Quirion, R., Larsen, T. A., Caine, D., Chase, T., Rioux, F., St. Pierre, S., Evarist, H., Pert, A., and Pert, C. B., 1982c, Analysis of [^{3}H]neurotensin receptors by computerized densitometry: Visualization of central and peripheral neurotensin receptors, *Ann. N.Y. Acad. Sci.* **400:** 415–417.

Quirion, R., Everist, H. D., and Pert, A., 1982d, Nigrostriatal dopamine terminals bear neurotensin receptors but the mesolimbic do not, *Soc. Neurosci. Abstr.* **8:**582.

Reches, A., Burke, R. E., Jiang, D., Wagner, H. R., and Fahn, S., 1982, The effect of neurotensin on dopaminergic neurons in rat brain, *Ann. N.Y. Acad. Sci.* **400:**420–421.

Reches, A., Burke, R. E., Jiant, D. H., Wagner, H. R., and Fahn, S., 1983, Neurotensin interactions with dopaminergic neurons in rat brain, *Peptides* **4:**43–48.

Reinecke, M., Almasan, K., Carraway, R., Helmstaedter, V., and Forssman, W. G., 1980a, Distribution patterns of neurotensin-like immunoreactive cells in the gastrointestinal tract of vertebrates, *Cell Tissue Res.* **205:**383–395.

Reinecke, M., Carraway, R. E., Falkemer, S., Feurley, G. E., and Forssman, W. G., 1980b, Occurrence of neurotensin-immunoreactive cells in the digestive tract of lower vertebrates and deuterostomian invertebrates, *Cell Tissue Res.* **212:**173–183.

Reinecke, M., Forssmann, W. G., Thiekotter, G., and Triepel, J., 1983, Localization of neurotensin-immunoreactivity in the spinal cord and peripheral nervous system of the guinea pig, *Neurosci. Lett.* **37:**37–42.

Rinkel, G. J. E., Hoeke, E. C., and van Wimersma-Greidanus, T.-J. B., 1983, Elective tolerance to behavioral effects of neurotensin, *Pharmacol. Behav.* **31:**467–470.

Rioux, F., Quirion, R., Regoli, D., Leblanc, M.-A., and St. Pierre, S., 1980, Pharmacological characterization of neurotensin receptors in the rat isolated portal vein using analogues and fragments of neurotensin, *Eur. J. Pharmacol.* **66:**273–279.

Rioux, F., Quirion, R., St. Pierre, S., Regoli, D., Jolicoeur, F., Belanger, F., and Barbeau, A., 1981, The hypotensive effect of centrally administered neurotensin in rats, *Eur. J. Pharmacol.* **69:**241–247.

Rivier, J. E., Lazarus, L. H., Perrin, M. H., and Brown, M. R., 1977, Neurotensin analogues. Structure–activity relationships, *J. Med. Chem.* **20:**1409–1411.

Roberts, G. W., Crow, T. J., and Polak, J. M., 1981, Neurotensin: First report of a cortical pathway, *Peptides* **2**(Suppl. 1)**:**37–43.

Roberts, G. W., Woodhams, P. L., Polak, J. M., and Crow, T. J., 1982, Distribution of neuropeptides in the limbic system of the rat: The amygdala complex, *Neuroscience* **7:**99–131.

Rokaeus, A., Fried, G., and Lundberg, J. M., 1984, Occurrence, storage and release of neurotensin-like immunoreactivity from the adrenal gland, *Acta Physiol. Scand.* **120:**373–380.

Roques, B. P., Fournie-Zaluski, M. C., Soroca, E., Lecomte, J. H., Malfroy, B., Llorens, C., and Schwartz, J. C., 1980, The enkephalinase inhibitor thiorphan shows antinociceptive activity in mice, *Nature* **288:**286–288.

Rossie, S. S., and Miller, R. J., 1982, Regulation of mast cell histamine release by neurotensin, *Ann. N.Y. Acad. Sci.* **400:**394–395.

Sadoul, J.-L, Kitabgi, P., Rostene, W., Javoy-Agid, F., Agid, Y., and Vincent, J.-P., 1984a, Characterization and visualization of neurotensin binding to receptor sites in human brain, *Biochem. Biophys. Res. Commun.* **120:**206–213.

Sadoul, J.-L., Mazella, J., Amar, S., Kitabgi, P., and Vincent, J.-D., 1984b, Preparation of neurotensin selectivity iodinated on the tyrosine 3 residue. Biological activity and binding properties on mammalian neurotensin receptors, *Biochem. Biophys. Res. Commun.* **120:**812–819.

Sadoul, J.-L., Checler, F., Kitabgi, P., Rostrene, W., Javoy-Agid, F., and Vincent, J.-P., 1984c, Loss of high affinity neurotensin receptors in substantia nigra from parkinsonian subjects, *Biochem. Biophys. Res. Commun.* **125:**395–404.

Sawada, S., and Yamamoto, C., 1980, Effects of neurotensin, enkephalin and dopamine on electrical activity of the interstitial nucleus of the stria terminalis, *Neurosci. Lett.* **17**(Suppl. 4)**:**58.

Sawada, S., Takada, S., and Yamamoto, C., 1980, Electrical activity recorded from thin sections of the bed nucleus of the stria terminalis, and the effects of neurotensin, *Brain Res.* **188:**578–581.

Schultzberg, M., Hokfelt, T., Nilsson, G., Terenius, L., Rehfield, J. F., Brown, M., Elde, R., Goldstein, M., and Said, S., 1980, Distribution of peptide- and catecholamine-containing neurons in the gastrointestinal tract of rat and guinea pig: Immunohistochemical studies with antisera to substance P, vasoactive intestinal polypeptide, enkephalins, somatostatin, gastrin/cholecystokinin, neurotensin, and dopamine β-hydroxylase, *Neuroscience* **5:**689–744.

Seeman, P., 1980, Dopamine receptors, *Pharmacol. Rev.* **3:**229–313.

Shiraishi, T., Inoue, A., and Yanaihara, N., 1980, Neurotensin and bombesin effects on LHA-gastrosecretory relations, *Brain Res. Bull.* **4**(suppl. 5)**:**133–142.

Simpson, W. M., and Varga, E., 1974, Clozapine—a new antipsychotic agent, *Curr. Ther. Res.* **16:**679–686.

Skowsky, R., Smith, P., and Swan, L., 1978, The effects of enkephalins, substance P and neurotensin on arginine vasopressin (AVP) release in the unanesthetized cat, *Clin. Res.* **26:**108A.

Snidjers, R., Kramarcy, N. R., Hurd, R. W., Nemeroff, C. B., and Dunn, A. J., 1982, Neurotensin induces catalepsy in mice, *Neuropharmacology* **21:**465–468.

Stanley, B. G., Eppel, N., and Noebel, B. G., 1982, Neurotensin injected into the paraventricular hypothalamus suppresses feeding in rats, *Ann. N.Y. Acad. Sci.* **400:**425–427.

Starr, M. S., 1982, Influence of peptides on ^{3}H-dopamine release from superfused rat striatal slices, *Neurochem. Int.* **4:**233–240.

Sundler, F., Hakanson, R., Hammer, R. A., Alumets, J., Carraway, R., Leeman, S. E., and Zimmerman, E. A., 1977, Immunohistochemical localization of neurotensin in endocrine cells of the gut, *Cell Tissue Res.* **178:**313–321.

Suzue, T., Yanaihara, N., and Otsuka, M., 1981, Actions of vasopressin, gastrin releasing peptide and other peptides on neurons of newborn rat spinal cord *in vitro, Neurosci. Lett.* **26:**137–142.

Tache, Y., Brown, M. R., and Collu, R., 1979, Effects of neuropeptides on adenohypophyseal hormone response to acute stress in male rats, *Endocrinol.* **105:**220–224.

Uhl, G. R., and Kuhar, M. J., 1984, Chronic neuroleptic treatment enhances neurotensin receptor binding in human and rat substantia nigra, *Nature* **309:**350–352.

Uhl, G. R., and Snyder, S. H., 1977a, Regional and subcellular distribution of brain neurotensin, *Life Sci.* **19:**1827–1832.

Uhl, G. R., and Snyder, S. H., 1977b, Neurotensin receptor binding: Regional and subcellular distributions favor transmitter role, *Rue. J. Pharmacol.* **41:**89–91.

Uhl, G. R., and Snyder, S. H., 1979, Neurotensin: A neuronal pathway projecting from amygdala through stria terminalis, *Brain Res.* **161:**522–526.

Uhl, G. R., and Snyder, S. H., 1981, Neurotensin, in: *Neurosecretion and Brain Peptides* (J. B. Martin, S. Reichlin, and K. L. Bick, eds.), pp. 87–106, Raven Press, New York.

Uhl, G. R., Kuhar, M. J., and Snyder, S. H., 1977a, Neurotensin: Immunohistochemical localization in rat central nervous system, *Proc. Natl. Acad. Sci. U.S.A.* **74:**4059–4063.

Uhl, G. R., Bennett, J. P., and Snyder, S. H., 1977b, Neurotensin: A central nervous system peptide; apparent receptor binding in brain membranes, *Brain Res.* **130:**299–313.

Uhl, G. R., Goodman, R. R., and Snyder, S. H., 1979, Neurotensin-containing cell bodies, fibers, and nerve terminals in the brainstem of the rat: Immunohistochemical mapping, *Brain Res.* **167:**77–91.

Uhl, G. R., Whitehouse, P. J., Price, D. L., Tourtelofte, W. W., and Kuhar, M. J., 1984, Parkinson's disease: Depletion of substantia nigra neurotensin receptors, *Brain Res.* **308:**186–190.

van Wimersma Greidanus, T. B., van Praag, M. C. G., Kalmann, R., Rinkel, G. J. E., Croiset, G., Hoeke, E. C., van Egmond, M. A. H., and Fekete, M., 1982, Behavioral effects of neurotensin, *Ann. N.Y. Acad. Sci.* **400:**319–329.

van Wimersma Greidanus, T. J. B., Schijff, J. A., Woteboom, J. L., Spit, M. C., Bruins, L., van Zumberen, M., and Rinkel, J. R., 1984, Neurotensin and bombesin, a relationship between their effects on body temperature and locomotor activity? *Pharmacol. Biochem. Behav.* **21:**197–202.

Vijayan, E., and McCann, S. M., 1979, *In vivo* and *in vitro* effects of substance P and neurotensin on gonadotropin and prolactin release, *Endocrinology* **105:**64–68.

Vincent, J. P., Mazella, J., Poustis, C., Checker, F., Kitabgi, P., Labbe, L., Granier, C., and von Rietschoten, J., 1982, Monoido-Trp[11]-neurotensin, a new ligand to study the interaction of neurotensin with its receptor, *Ann. N.Y. Acad. Sci.* **400:**436–437.

Widdowson, P. S., Griffiths, E. C., Slater, P., and Yajima, H., 1983, Effects of mammalian and avian neurotensins and neurotensin fragments on wet-dog shaking and body temperature in the rat, *Reg. Peptides* **7:**357–365.

Widerlov, E., Kilts, C. D., Mailman, R. B., Nemeroff, C. B., McCown, T. J., Prange, A. J., Jr., and Breese, G. R., 1982a, Increase in dopamine metabolites in rat brain by neurotensin, *J. Pharmacol. Exp. Ther.* **222:**1–6.

Widerlov, E., Lindstrom, L. H., Besev, G., Manberg, P. J., Nemeroff, C. B., Breese, G. R., Kizer, J. S., and Prange, A. J., Jr., 1982b, Subnormal CSF levels of neurotensin in a subgroup of schizophrenic patients: Normalization after neuroleptic treatment, *Am. J. Psychiatry* **139:**1122–1126.

Yaksh, T. L., Schmouss, C., Micevych, P. E., Abay, E. O., and Go, V. L. M., 1982, Pharmacological studies on the application, disposition and release of neurotensin in the spinal cord, *Ann. N.Y. Acad. Sci.* **400:**228–243.

Young, W. S. III, and Kuhar, M. J., 1979, Neurotensin receptors: Autoradiographic localization in rat CNS, *Eur. J. Pharmacol.* **59:**161–163.

Young, W. S. III, and Kuhar, M. J., 1981, Neurotensin receptor localization by light microscopic autoradigraphy in rat brain, *Brain Res.* **220:**273–285.

Young, W. S., Uhl, G. R., and Kuhar, M. J., 1978, Iontophoresis of neurotensin in the area of the locus coeruleus, *Brain Res.* **150:**431–435.

Zieglgansberger, W., Siggins, G., Brown, M., Vale, W., and Bloom, F., 1978, Actions of neurotensin upon single neurone activity in different regions of the rat brain, in: *Seventh International Congress on Pharmacology,* p. 126, London.

18

Calcitonin Gene-Related Peptide in the Central Nervous System

Neuronal and Receptor Localization, Biochemical Characterization, and Functional Studies

DAVID M. JACOBOWITZ and GERHARD SKOFITSCH

1. INTRODUCTION

Calcitonin gene-related peptide (CGRP) is a 37-amino-acid peptide that has been identified in rat brain and peripheral tissue by immunocytochemistry (Amara *et al.*, 1982; Rosenfeld *et al.*, 1983; Skofitsch and Jacobowitz, 1985c) and in the human brain by radioimmunoassay (Morris *et al.*, 1984; Tschopp *et al.*, 1984, 1985). It was isolated from human medullary thyroid carcinomas and sequenced (Morris *et al.*, 1984). The initial immunocytochemical description of the localization of CGRP revealed a unique neuronal distribution in the CNS and periphery (Rosenfeld *et al.*, 1984). A widespread distribution of immunoreactive cells and fibers was observed at all brain levels and in numerous peripheral organs.

It is becoming apparent that because of the ubiquitous distribution of most brain peptides, rational approaches to the study of the function of these neurochemicals will require a knowledge of their discrete localization within the CNS. Therefore detailed atlases of the distribution of neuronal systems are necessary adjuncts for stereotactic manipulation of sites of localization of peptide-containing neurites and receptor binding sites.

In this work we present detailed stereotactic maps of the distribution of CGRP-immunoreactive neurons and receptor binding sites in the rat brain using immunohistochemistry and an *in vitro* labeling autoradiographic technique. Furthermore, the concentration of CGRP in discrete brain regions was determined by radioim-

DAVID M. JACOBOWITZ • Clinical Neuroscience Branch, National Institute of Mental Health, National Institutes of Health, Bethesda, Maryland 20205. GERHARD SKOFITSCH • Laboratory of Clinical Science, National Institute of Mental Health, National Institutes of Health, Bethesda, Maryland 20205.

munoassay. Localization of the peptide and its binding sites, coupled to quantitative correlation in small brain regions, has set the stage for a study of the influence of CGRP on cardiovascular dynamics following injection into a discrete brain nucleus.

2. STEREOTACTIC ATLAS OF DISTRIBUTION OF CGRP

Preparation of complete stereotactic maps of the discrete localization of CGRP-immunoreactive neurons in the rat CNS was accomplished by standard procedures previously described in other communications from this laboratory (Jacobowitz and O'Donohue, 1978; Jacobowitz *et al.,* 1981, 1983; Olschowka *et al.,* 1981, 1982; Skofitsch *et al.,* 1985; Skofitsch and Jacobowitz, 1985a,b).

Normal and colchicine-treated rats were perfused with 10% formalin. Cryostat sections were processed for immunofluorescence evaluation. Specific antisera obtained commercially (Peninsula, Belmont, CA) were raised in rabbits against human CGRP coupled to BSA by carbodiimide using standard immunization procedures. Tissue sections were incubated in the antiserum (diluted 1:2000) for 3 days followed by incubation in FITC-conjugated goat antirabbit IgG (diluted 1:400). Appropriate immunohistochemical controls were preabsorbed with 10^{-6} M synthetic human or rat CGRP, which resulted in essentially complete absence of immunofluorescence. What was observed histochemically as "CGRP immunoreactivity" or "CGRP-positive nerves" will be referred to simply as "CGRP." A mapping of the discrete localization of CGRP fibers (left) and cell bodies (right) is shown in Figs. 1–14. CGRP cells were seen only in colchicine-treated rats; nerve fibers were observed in colchicine-treated animals as well as in untreated ones. A summary of areas with cell bodies and/or nerve processes is given in Table I. Cells and fibers were "semiquantitatively" rated as ± (sparse), + (low), + + (moderate), + + + (dense), or + + + + (very dense).

A widespread distribution of CGRP-positive cells and fibers were observed throughout the central nervous system. Comparatively large numbers of CGRP-positive cell bodies were noted in the preoptic periventricular area (Fig. 4), the perifornical area (Fig. 5), in the peripeduncular area (Figs. 9,16A), the nuclei of the third, fourth, fifth, sixth, seventh, and twelfth nerves (Figs. 5,9,11–13,17), the dorsal and ventral parabrachial nuclei (Figs. 11–16B), the superior olive (Figs. 11D–12A), the ambiguus nucleus (Figs. 12,13,17), and in the ventral horn of the spinal cord (Figs. 14,17). A more modest number of perikarya were observed in the anterior hypothalamic nucleus (Fig. 5A), medial forebrain bundle (Figs. 6,7), premamillary nucleus (Figs. 7C,D), amygdala medialis (Figs. 7C,D), hippocampus and dentate gyrus (Fig. 8C), central gray (Fig. 9A), ventromedial nucleus of the thalamus (Fig. 7), and ventral tegmental nucleus (Figs. 10D,11A). Dense accumulations of fibers were noted in the most caudal part of the caudate–putamen (Figs. 6,7), the central amygdaloid nucleis (Figs. 6,7), and the sensory trigeminal and spinal cord areas (Figs. 11–14,24). Other areas not listed in Table I are (1) the organum vasculosum lamina terminalis (OVLT), which was densely innervated;

(2) the plexiform and glomerulos laminae of the olfactory bulb contained a moderate number of CGRP fibers; (3) no nerves were observed in the supraoptic nucleus, and (4) some nerves were observed in the locus coeruleus (Fig. 11).

Nerves containing CGRP were seen in a few cortical areas. A small number of fibers were observed in the medial frontal cortex (Fig. 1). More moderate numbers of fine CGRP nerve fibers were observed in the insular cortex (Figs. 4–8,15). The temporal cortex just dorsal to the insular cortex contained a small number of fibers (Fig. 15). A few cell bodies were seen only at one level in the entorhinal cortex (Fig. 15). The insular cortex is thought to play a role in gustatory function. Benjamin and Akert (1959) showed that destruction of the rat cortical taste area produces retrograde degeneration of thalamic neurons confined to the ventral posteromedial nucleus of the thalamus. It is therefore of interest that we have noted CGRP cell bodies in the ventromedial thalamic nucleus (Figs. 7B–D). Whether or not these CGRP thalamic cells project to the insular cortex remains for further study.

The most densely innervated regions of the CNS were the central amygdaloid nucleus (Figs. 6,7), caudal part of the caudate putamen (Figs. 6D–7D), substantia gelatinosa (Fig. 13E), lateral part of the nucleus tractus spinalis of the trigeminal nerve (Fig. 13), and the dorsal horn of the spinal cord (Fig. 14). The most prominent density of CGRP perikarya was observed in the peripeduncular area (Figs. 9A,B) (which includes the peripeduncular nucleus ventral to the medial geniculate body) and the dorsal and ventral parabrachial nuclei (Figs. 10D–11D). A unique feature of the CGRP neuronal distribution is its presence in motor neurons of the brain (III, IV, V, VI, VII, XII), the nucleus ambiguus of the vagal complex, and the ventral horn cells of the spinal cord. This is the first peptide found to coexist with cholinergic motor cells of the brain and spinal cord.

3. QUANTITATION AND IDENTIFICATION OF CGRP BY RADIOIMMUNOASSAY

3.1. Quantitation

Using an antiserum directed against human calcitonin gene-related peptide (hCGRP), which fully cross reacts with rat CGRP, we developed a sensitive radioimmunoassay (RIA). The quantitative distribution of CGRP immunoreactivity was measured by RIA. Rats were decapitated, and their brains were frozen, cut in a cryostat (300 μm), and microdissected by the method of Palkovits (1973). The approximate size and coordinates of the microdissected regions have been reported previously (Skofitsch and Jacobowitz, 1985d). Micropunches were delivered into 250 μl of 0.1 N HCl in 1.5-ml plastic conical test tubes on ice. Samples were boiled for 10 min and then homogenized by sonication. A 10 to 20-μl aliquot was removed for protein determination. Samples were centrifuged at 8000 × *g* for 5 min, and 200-μl aliquots of supernatant were removed and evaporated to dryness in a vacuum centrifuge.

FIGURES 1–14. Schematic drawings of the rat brain and spinal cord according to Jacobowitz and Palkovits (1974) and Palkovits and Jacobowitz (1974). Consult these atlases for identification of unlabeled structures. Coordinates are given in micrometers anterior or posterior to the intraaural plane according to König and Klippel (1963). The density and distribution of CGRP-like immunoreactive cell bodies (filled circles on the right half of the panels) and fibers (small dots on the left side of the panels) are indicated.

a	Nucleus accumbens
aa	Area amygdaloidea anterior
ab	Nucleus amygdaloideus basalis
abd	Nucleus amygdaloideus basalis posterior
ac	Nucleus amygdaloideus centralis
aco	Nucleus amygdaloideus corticalis
al	Nucleus amygdaloideus lateralis
am	Nucleus amygdaloideus medialis
amb	Nucleus ambiguus
ap	Area postrema
apm	Area pretectalis medialis
apo	Nucleus amygdaloideus posterior
BCI	Brachium colliculi inferioris
c	Nucleus caudatus
CA	Commissura anterior
CAI	Capsula interna
CC	Crus cerebri
CCA	Corpus callosum
ccgm	Nucleus centralis corporis geniculati medialis
cd	Cornu dorsale
CFV	Commissura fornicis ventralis
ci	Colliculus inferior
cl	Claustrum
CO	Chiasma opticum
cod	Nucleus cochlearis dorsalis
cov	Nucleus cochlearis ventralis
cp	Nucleus caudatus putamen
CSDV	Commissura supraoptica dorsalis, pars ventralis (Meynert)
CT	Corpus trapezoideum
ct	Nucleus corporis trapezoidei
cul	Nucleus cuneatus lateralis
cv	Cornu ventrale
dcgl	Nucleus dorsalis corporis geniculati lateralis
DP	Decussatio pyramidis
DPCS	Decussatio pedunculi cerebellarium superiorum
F	Fornix
FC	Fasciculus cuneatus
FD	Funiculus dorsalis
FH	Fimbria hippocampi
FL	Funiculus lateralis
FLM	Fasciculus longitudinalis medialis
FMI	Forceps minor
FMT	Fasciculus mamillothalamicus
FMTG	Fasciculus mamillotegmentalis
FR	Fasciculus retroflexus
FS	Fornix superior
FV	Funiculus ventralis
g	Nucleus gelatinosus thalami
GD	Gyrus dentatus
gp	Globus pallidus
HI	Hippocampus
HIA	Hippocampus anterior
hl	Nucleus habenulae lateralis
hm	Nucleus habenulae medialis
ic	Nucleus interstitialis (Cajal)
io	Nucleus olivaris inferior
ip	Nucleus interpeduncularis
lc	Locus coeruleus
LGA	Lamina glomerulosa bulbi olfactorii accessorii
LGR	Lamina granularis bulbi olfactorii
LL	Lemniscus lateralis
lld	Nucleus lemnisci lateralis dorsalis
llr	Nucleus lemnisci lateralis rostralis
llv	Nucleus lemnisci lateralis ventralis
LM	Lemniscus medialis
LMA	Lamina cellularum mitralium bulbi olfactorii accessorii
LP	Lamina plexiformis externa bulbi olfactorii
mcgm	Nucleus marginalis corporis geniculati medialis
ME	Eminentia mediana
MFB	Fasciculus medialis prosencephali
na	Nucleus arcuatus
nco	Nucleus commissuralis
ncs	Nucleus centralis superior
ncu	Nucleus cuneiformis
ndm	Nucleus dorsomedialis hypothalami
nha	Nucleus hypothalamicus anterior
nhp	Nucleus hypothalamicus posterior
nic	Nucleus intercalatus
nist	Nucleus interstitialis striae medullaris
nistd	Nucleus interstitialis striae terminalis pars dorsalis
nistv	Nucleus interstitialis striae terminalis pars ventralis
nlo	Nucleus linearis oralis
nml	Nucleus mamillaris lateralis
nmm	Nucleus mamillaris medialis
np	Nucleus parabrachialis ventralis
npd	Nucleus parabrachialis dorsalis
npe	Nucleus periventricularis hypothalami
npf	Nucleus perifornicalis
npmv	Nucleus premamillaris ventralis
npv	Nucleus paraventricularis hypothalami
npV	Nucleus sensorius principalis nervi trigemini
nrd	Nucleus reticularis medullae oblongatae, pars dorsalis
nrp	Nucleus reticularis paramedianus
nrv	Nucleus reticularis medullae oblongatae, pars ventralis
nsc	Nucleus suprachiasmaticus
nso	Nucleus supraopticus
ntd	Nucleus tegmenti dorsalis (Gudden)

ntdl Nucleus tegmenti dorsalis lateralis
nts Nucleus tractus solitarii
ntv Nucleus tegmenti ventralis (Gudden)
ntV Nucleus tractus spinalis nervi trigemini
ntVd Nucleus tractus spinalis nervi trigemini, pars dorsomedialis
nvm Nucleus ventromedialis hypothalami
nIII Nucleus originis nervi oculomotorii
nIV Nucleus originis nervi trochlearis
nV Nucleus originis nervi trigemini
nVI Nucleus originis nervi abducentis
nVII Nucleus originis nervi facialis
nX Nucleus originis dorsalis nervi vagi
nXII Nucleus originis nervi hypoglossi
oad Nucleus olfactorius anterior, pars dorsalis
oal Nucleus olfactorius anterior, pars lateralis
oam Nucleus olfactorius anterior, pars medialis
oap Nucleus olfactorius anterior, pars posterior
ol Nucleus tractus olfactorii lateralis
ope Nucleus preolivaris externus
os Nucleus olivaris superior
P Tractus corticospinalis
p Nucleus pretectalis
PCMA Pedunculus corporis mamillaris
PCS Pedunculus cerebellaris superior
pf Nucleus parafascicularis
ph Nucleus prepositus hypoglossi
pi Cortex piriformis
po Nucleus pontis
pol Nucleus preopticus lateralis
pom Nucleus preopticus medialis
pop Nucleus preopticus periventricularis
pos Nucleus preopticus suprachiasmaticus
ps Nucleus parasolitarius
pt Nucleus paratenialis
pv Nucleus periventricularis thalami
r Nucleus ruber
rc Area retrochiasmatica
rd Nucleus raphe dorsalis
re Nucleus reuniens
rgi Nucleus reticularis gigantocellularis
rh Nucleus rhomboideus
rl Nucleus reticularis lateralis
rm Nucleus raphe magnus
ro Nucleus raphe obscurus
rpe Nucleus reticularis parvocellularis
rpo Nucleus raphe pontis
rpoc Nucleus reticularis pontis caudalis
rpoo Nucleus reticularis pontis oralis
rtp Nucleus reticularis tegmenti pontis
SAM Stratum album mediale colliculi superioris
sd Nucleus dorsalis septi
sdi Nucleus dorsalis septi, pars intermedia
sdl Nucleus dorsalis septi, pars lateralis
sf Nucleus fimbrialis septi
sfo Organum subfornicale
sg Nucleus suprageniculatus facialis
SGC Substantia grisea centralis
SGCd Substantia grisea centralis dorsalis
SGCv Subtantia grisea centralis ventralis
sgm Stratum griseum mediale colliculi superioris
sgp Stratum grieseum profundum colliculi superioris
sgs Stratum griseum superficiale colliculi superioris
sl Nucleus lateralis septi
SM Stria medullaris thalami
sm Nucleus medialis septi
snc Substantia nigra zona compacta
snl Substantia nigra zona lateralis
snr Substantia nigra zona reticularis
SO Stratum opticum colliculi superioris
spf Nucleus subparafascicularis
ST Stria terminalis
st Nucleus triangularis septi
SUM Decussatio supramamillaris
sut Nucleus subthalamicus
tam Nucleus anterior medialis thalami
tav Nucleus anterior ventralis thalami
td Nucleus tractus diagonalis Broca
tl Nucleus lateralis thalami
tlp Nucleus lateralis posterior thalami
tm Nucleus medialis thalami
tml Nucleus medialis thalami, pars lateralis
tmm Nucleus medialis thalami, pars medialis
TO Tractus opticus
TOI Tractus olfactorius intermedius
TOL Tractus olfactorius lateralis
tol Nucleus tractus optici, pars lateralis
tpm Nucleus posteromedianus thalami
tpo Nucleus posterior thalami
tr Nucleus reticularis thalami
TSHT Tractus septohypothalamicus
TSTH Tractus striohypothalamicus
TSV Tractus spinalis nervi trigemini
TT Tractus tectospinalis
tu Tuberculum olfactorium
tv Nucleus ventralis thalami
tvd Nucleus ventralis thalami, pars dorsalis
tvm Nucleus ventralis medialis thalami pars magnocellularis
vcgl Nucleus ventralis corporis geniculati lateralis
VII Nervus facialis
vl Nucleus vestibularis lateralis
vm Nucleus vestibularis medialis
vs Nucleus vestibularis superior
vsp Nucleus vestibularis spinalis
X Nervus vagus
zi Zona incerta

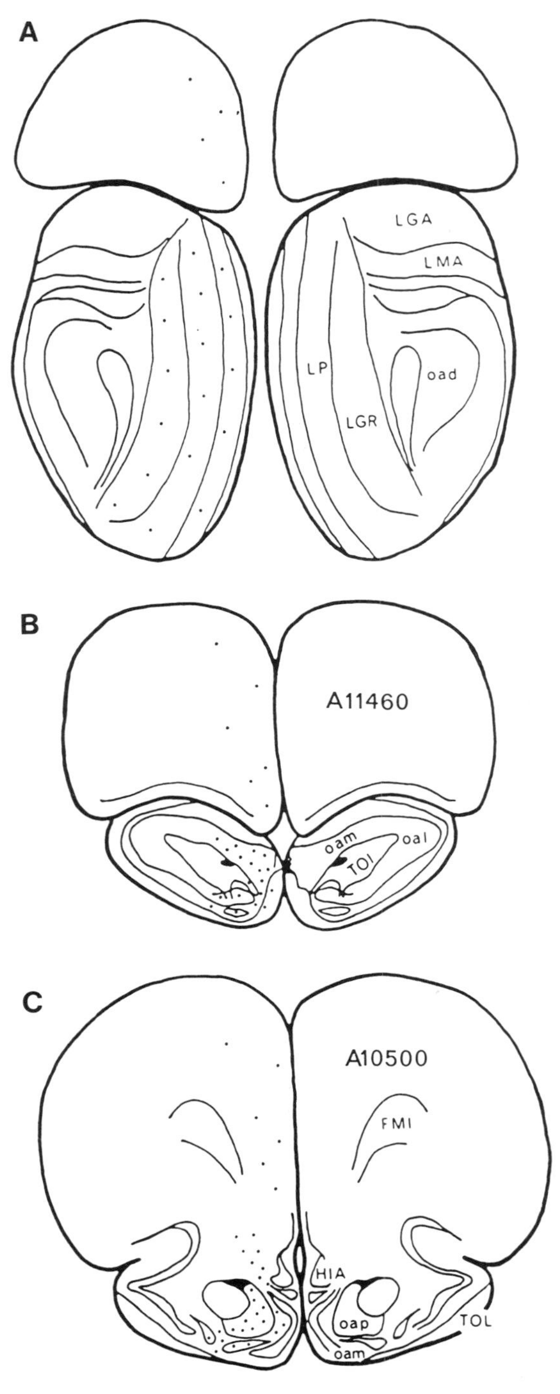
A
LGA
LMA
LP
oad
LGR
B
A11460
oam
oal
TOl
C
A10500
FMI
HIA
oap
TOL
oam

FIGURE 1

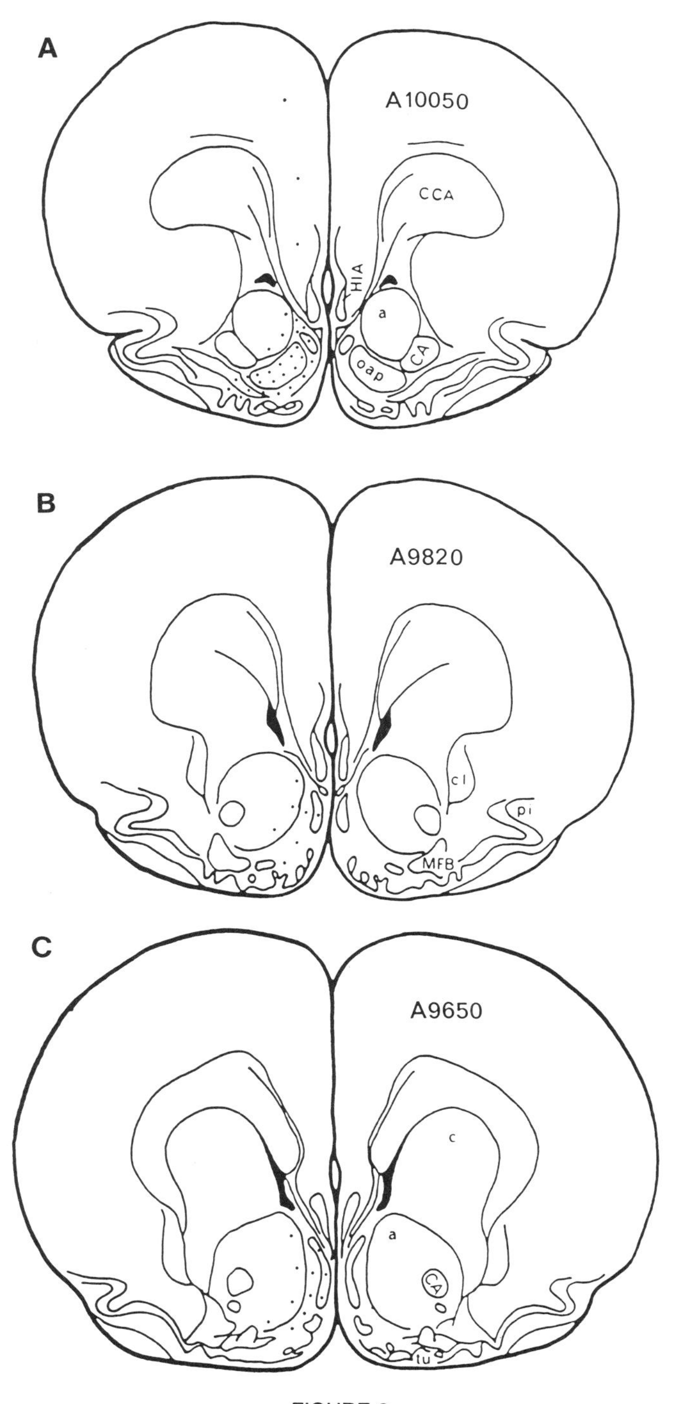

FIGURE 2

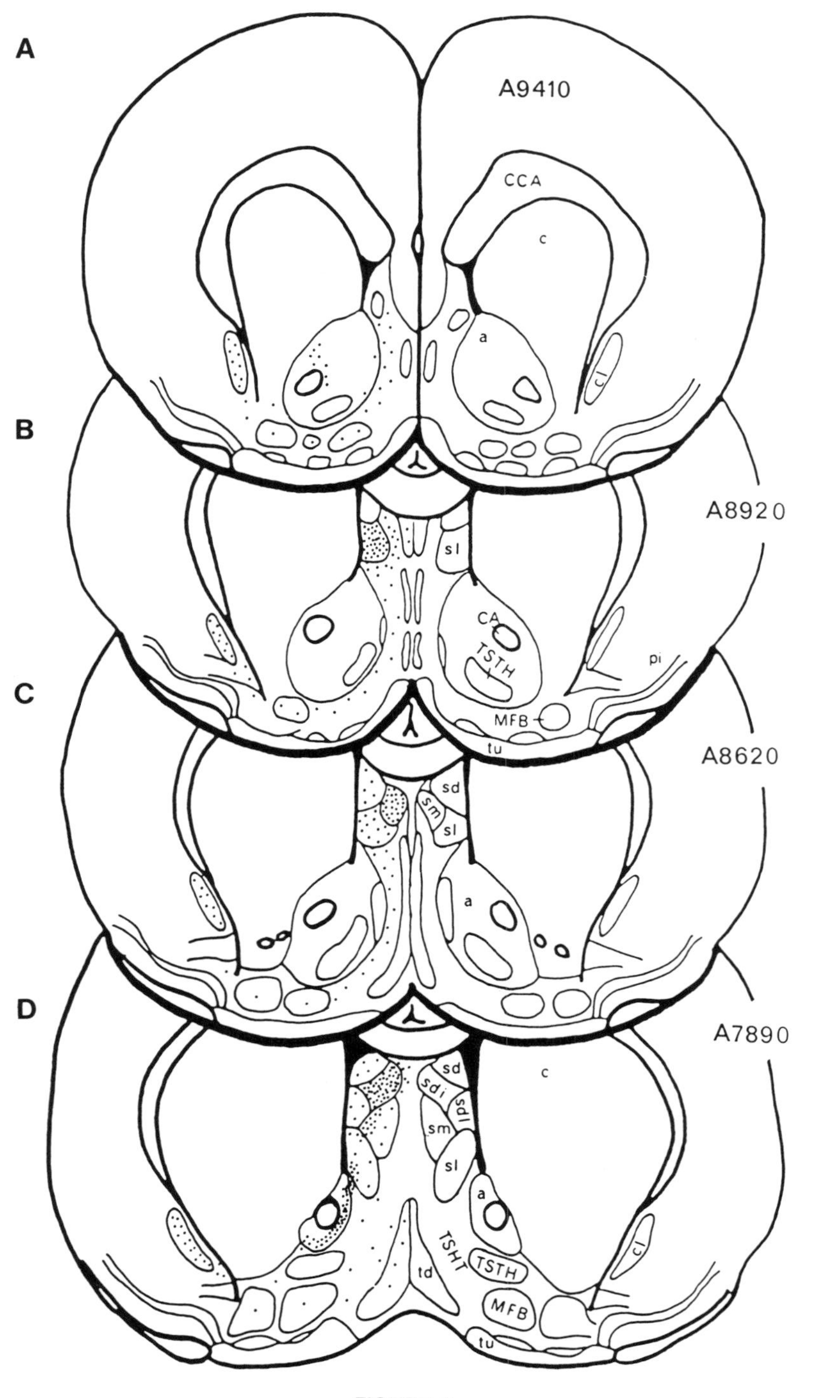
A
A9410
CCA
c
a
cl
B
A8920
sl
CA
TSTH
pi
MFB
tu
C
A8620
sd
sm
sl
a
D
A7890
c
sd
sdi
sdl
sm
sl
a
TSHT
td
TSTH
MFB
tu
cl

FIGURE 3

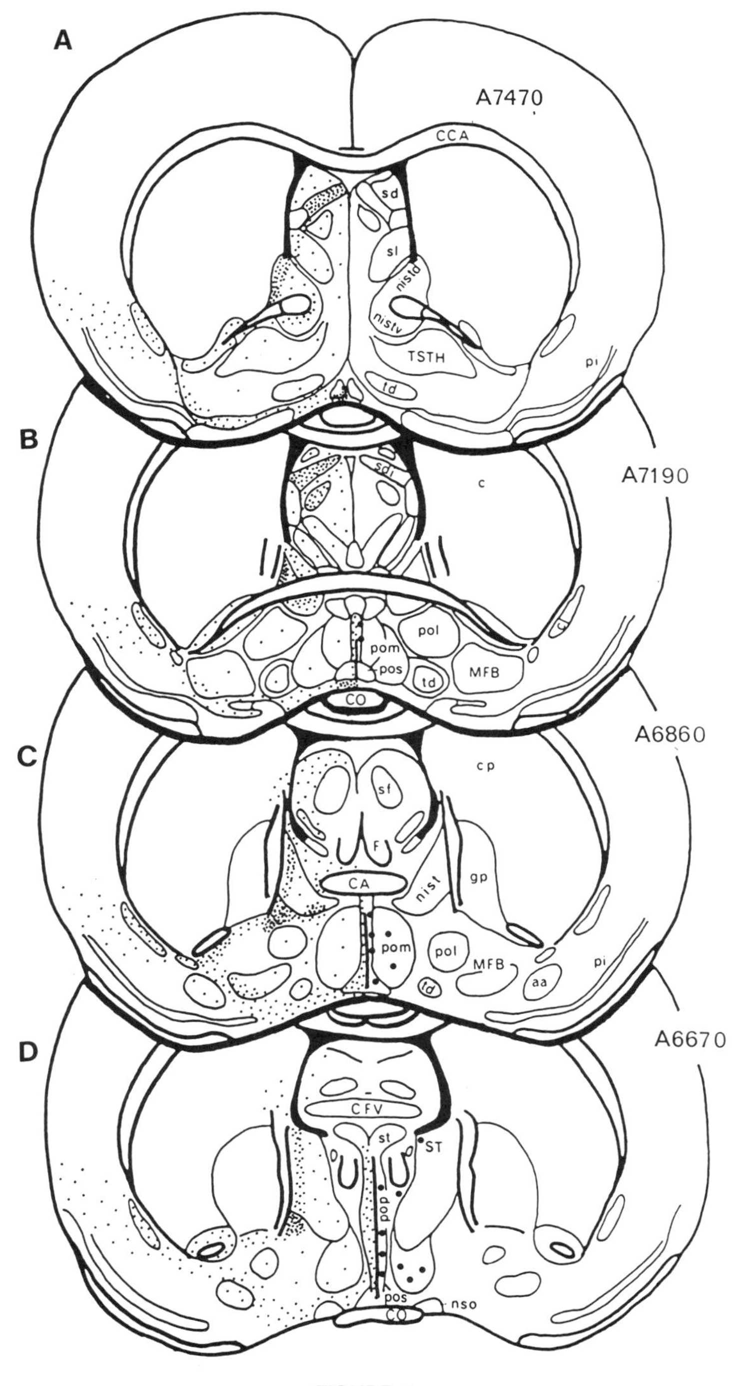
A
A7470
CCA
sd
sl
nistd
nistv
TSTH
pi
td
B
A7190
sdi
c
pol
cl
pom
pos
MFB
td
CO
C
A6860
cp
sf
F
gp
nist
CA
pom
pol
MFB
pi
td
aa
D
A6670
CFV
st
ST
dpd
pos
nso
CO

FIGURE 4

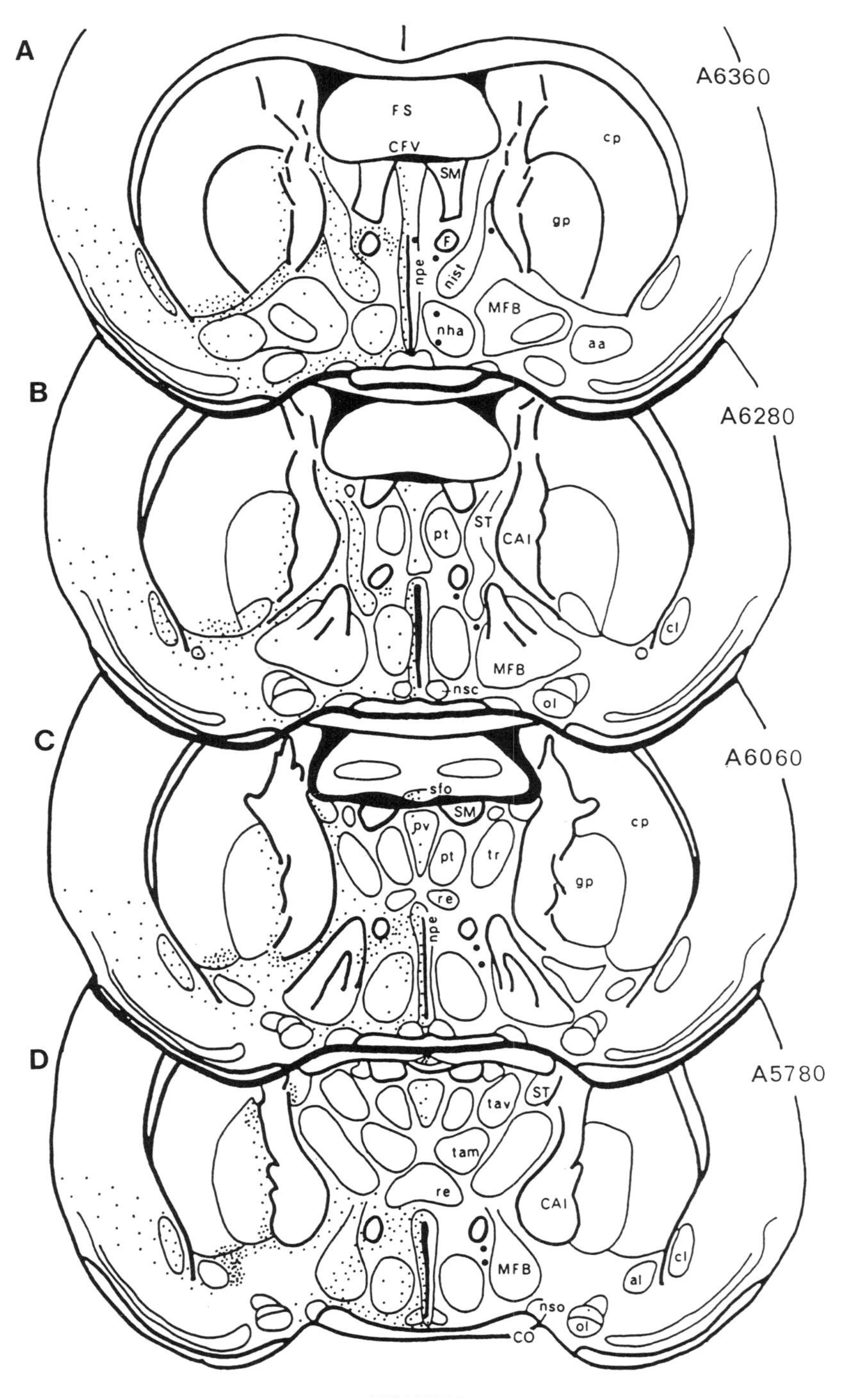
A
A6360
FS
CFV
SM
cp
gp
F
npe
nist
MFB
nha
aa
B
A6280
pt
ST
CAI
cl
MFB
nsc
ol
C
A6060
sfo
SM
pv
pt
tr
cp
gp
re
npe
D
A5780
ST
tav
tam
re
CAI
cl
MFB
al
nso
ol
CO

FIGURE 5

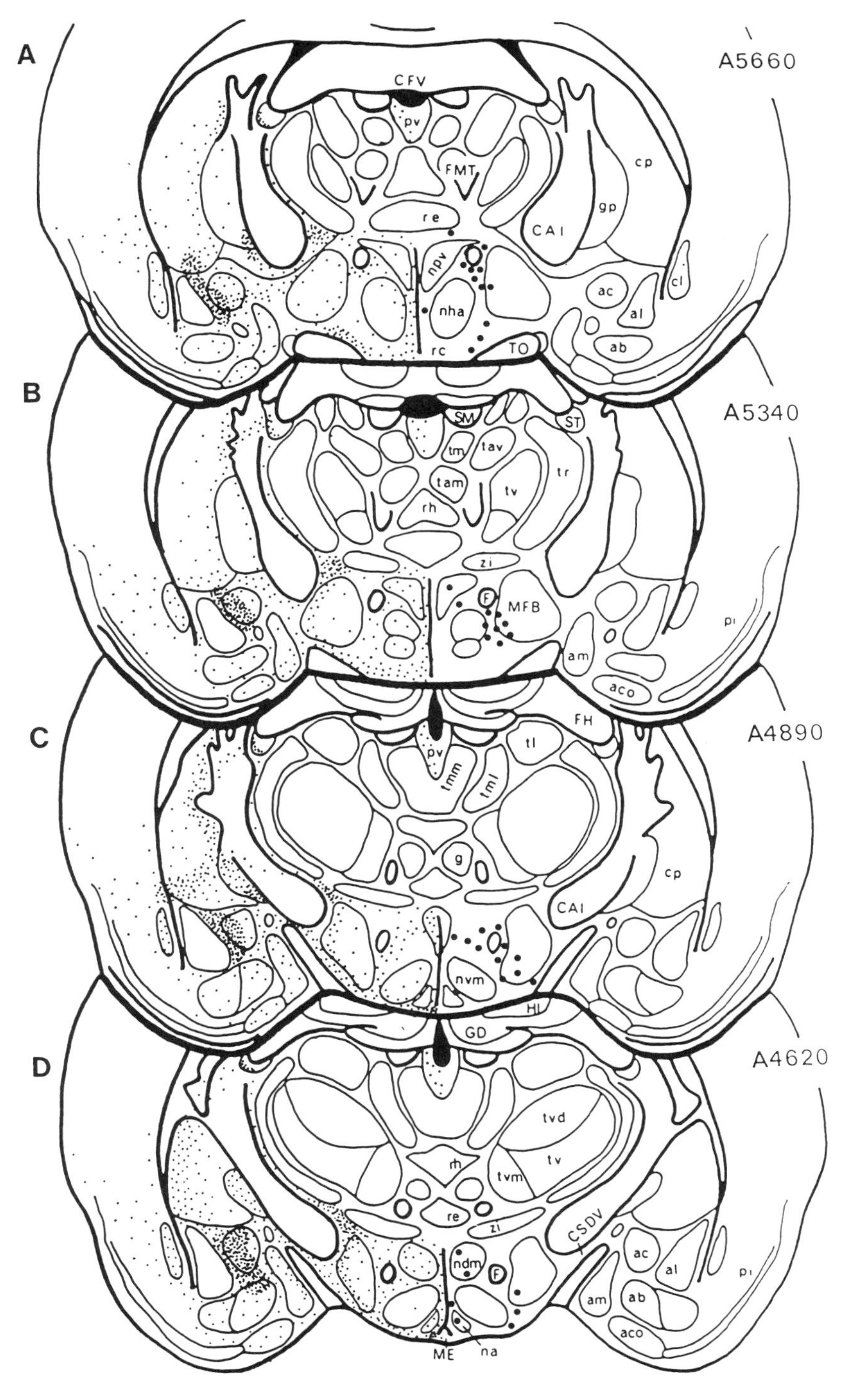
A
A5660
CFV
pv
FMT
cp
gp
re
CAI
npv
nha
cl
ac
al
ab
rc
TO
B
A5340
SM
ST
tm
tav
tam
tr
tv
rh
zi
F
MFB
pi
am
aco
C
A4890
FH
pv
tl
tmm
tml
g
cp
CAI
nvm
HI
GD
D
A4620
tvd
rh
tv
tvm
re
zi
CSDV
ndm
F
ac
al
pi
am
ab
aco
ME
na

FIGURE 6

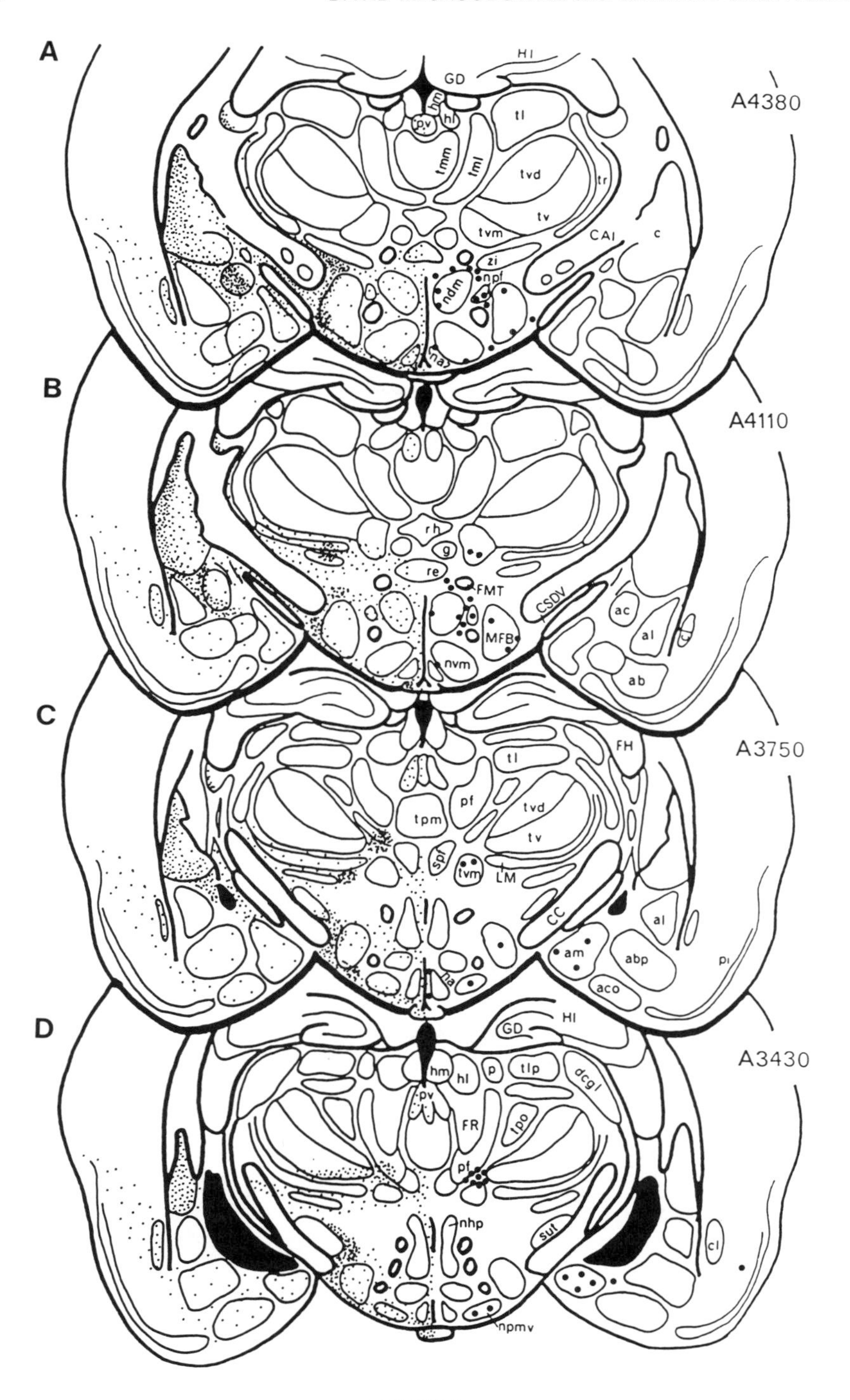
A
HI
GD
hm
pv
hl
tl
A4380
tmm
tml
tvd
tr
tv
tvm
CAI
c
zi
npf
ndm
na
B
A4110
rh
g
re
FMT
CSDV
ac
al
MFB
nvm
ab
C
FH
A3750
tl
pf
tvd
tpm
tv
spf
tvm
LM
CC
al
am
abp
aco
na
D
GD
HI
A3430
hm
hl
p
tlp
dcgl
pv
FR
tpo
pf
nhp
sul
cl
npmv

FIGURE 7

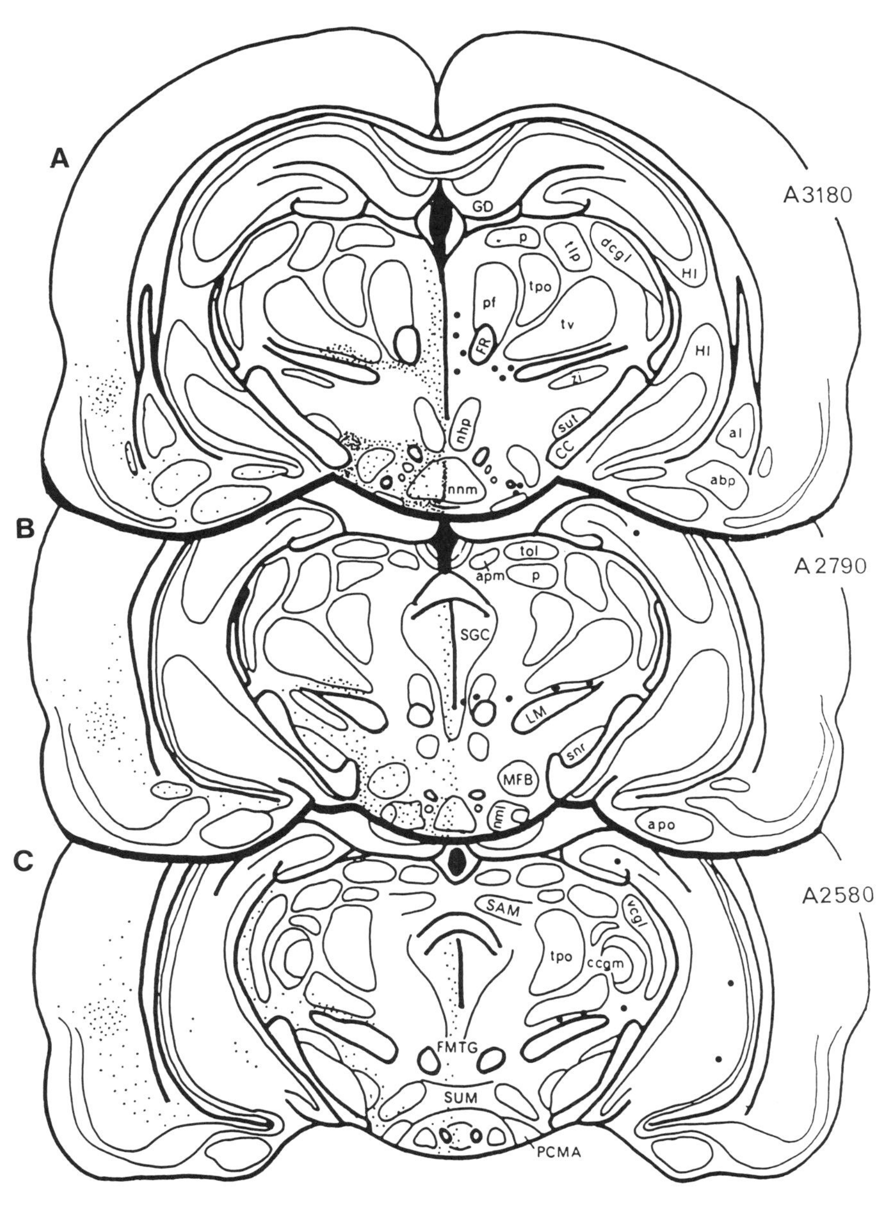
A
A3180
GD
p
tlp
dcgl
HI
pf
tpo
tv
FR
HI
zi
nhp
sut
CC
al
abp
nnm
B
A2790
tol
apm
p
SGC
LM
snr
MFB
nml
apo
C
A2580
SAM
vcgl
tpo
ccgm
FMTG
SUM
PCMA

FIGURE 8

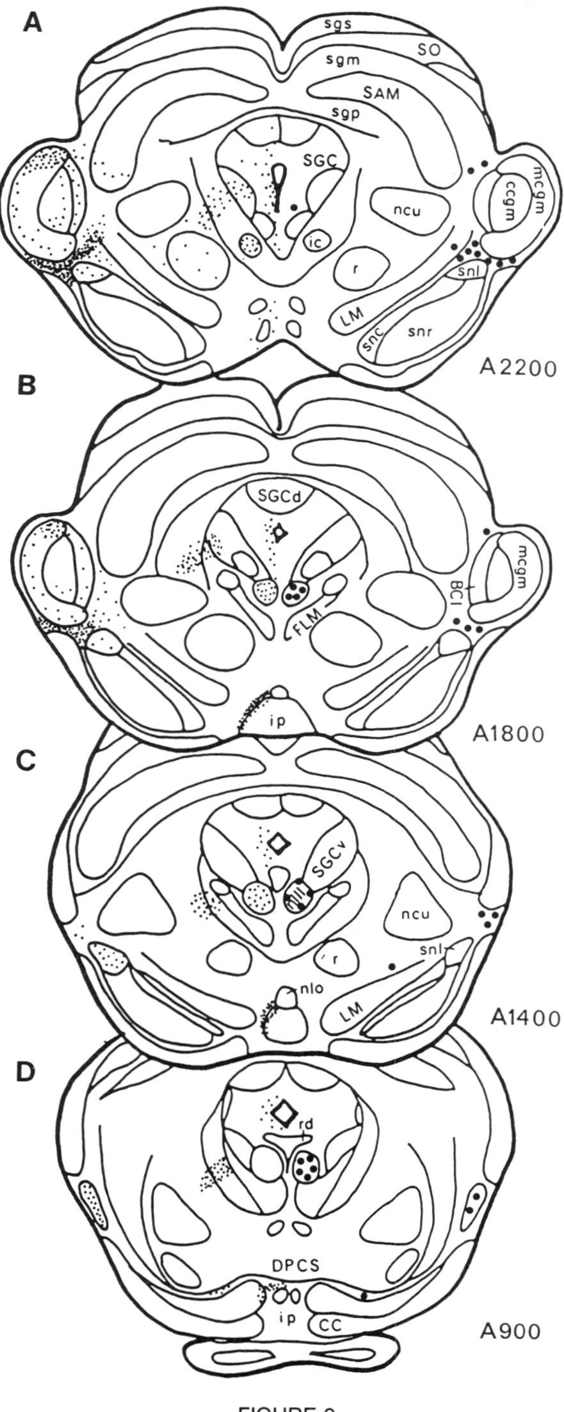
A
sgs
SO
sgm
SAM
sgp
SGC
mcgm
ccgm
ncu
ic
r
snl
LM
snc
snr
A2200
B
SGCd
mcgm
BCI
FLM
ip
A1800
C
SGCv
ncu
snl
r
nlo
LM
A1400
D
rd
DPCS
ip
CC
A900

FIGURE 9

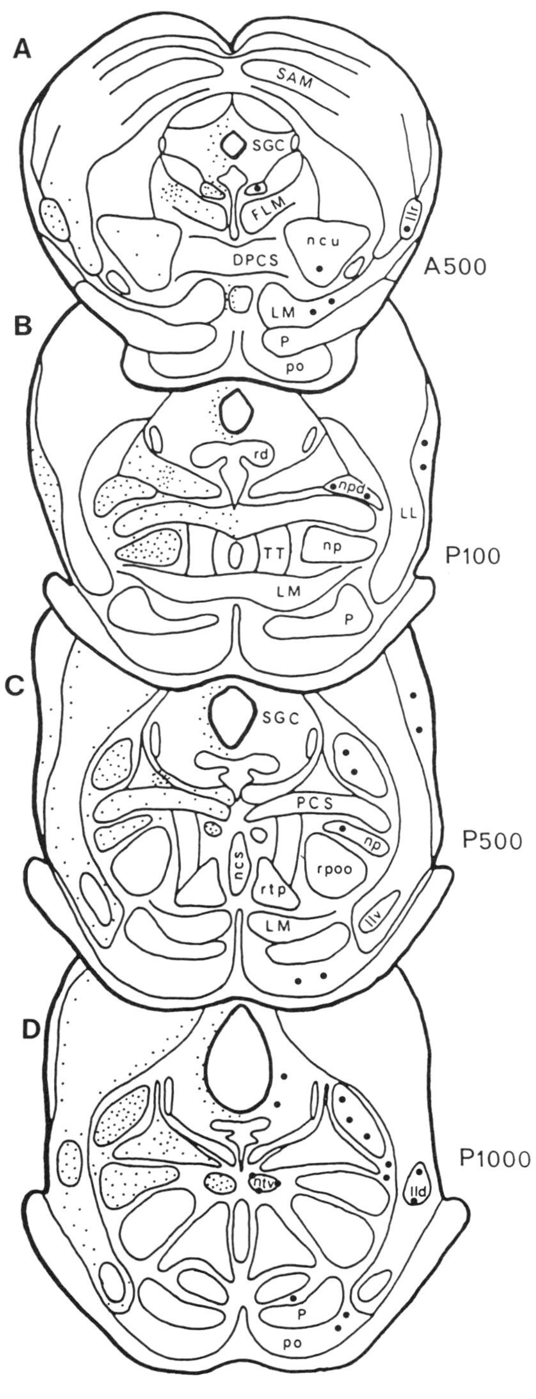
A
SAM
SGC
FLM
llr
ncu
DPCS
A500
LM
P
po
B
rd
npd
LL
TT
np
P100
LM
P
C
SGC
PCS
np
P500
ncs
rpoo
rtp
LM
llv
D
P1000
ntv
lld
P
po

FIGURE 10

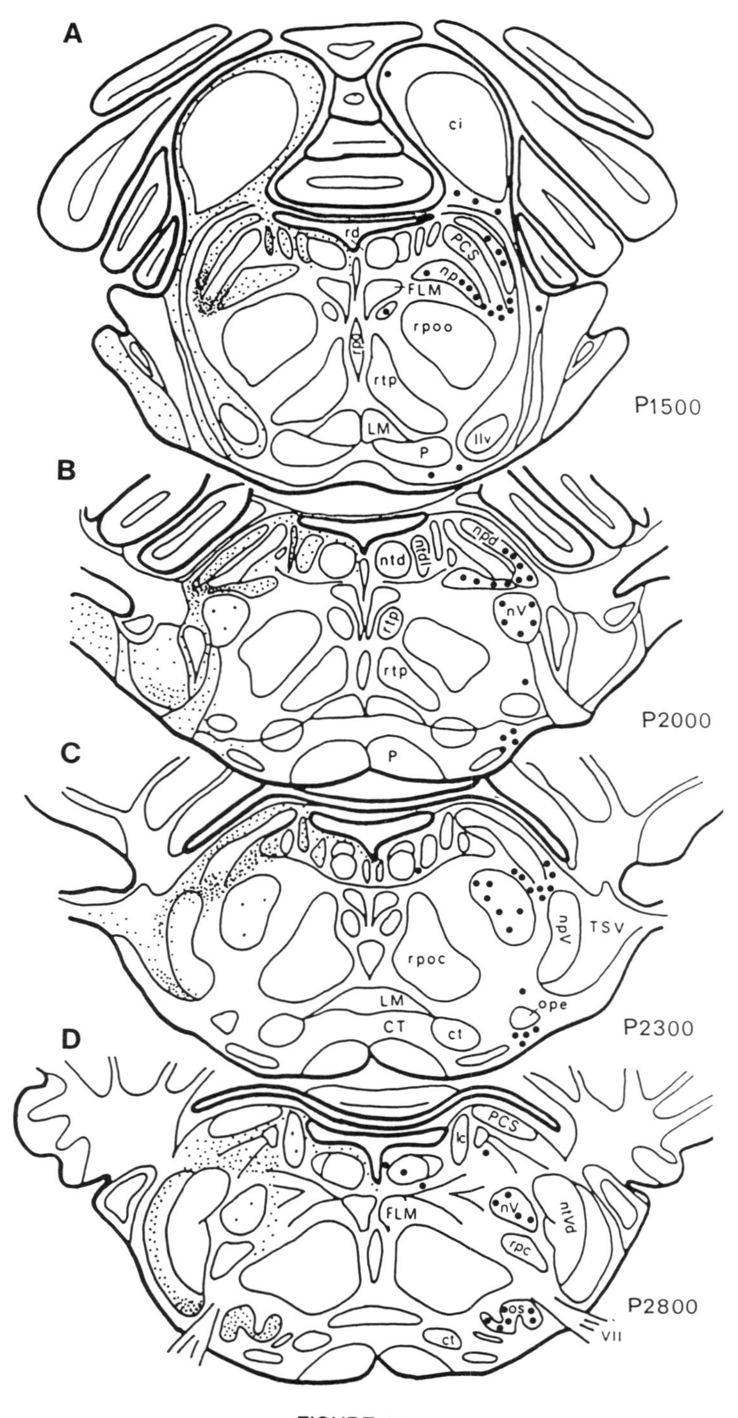
A
ci
rd
PCS
np
FLM
rpoo
rpo
rtp
P1500
LM
P
llv
B
npd
ntd
ntdl
nV
rtp
rtp
P2000
P
C
npV
TSV
rpoc
LM
ope
CT
ct
P2300
D
PCS
kc
FLM
nV
ntVd
rpc
os
ct
P2800
VII

FIGURE 11

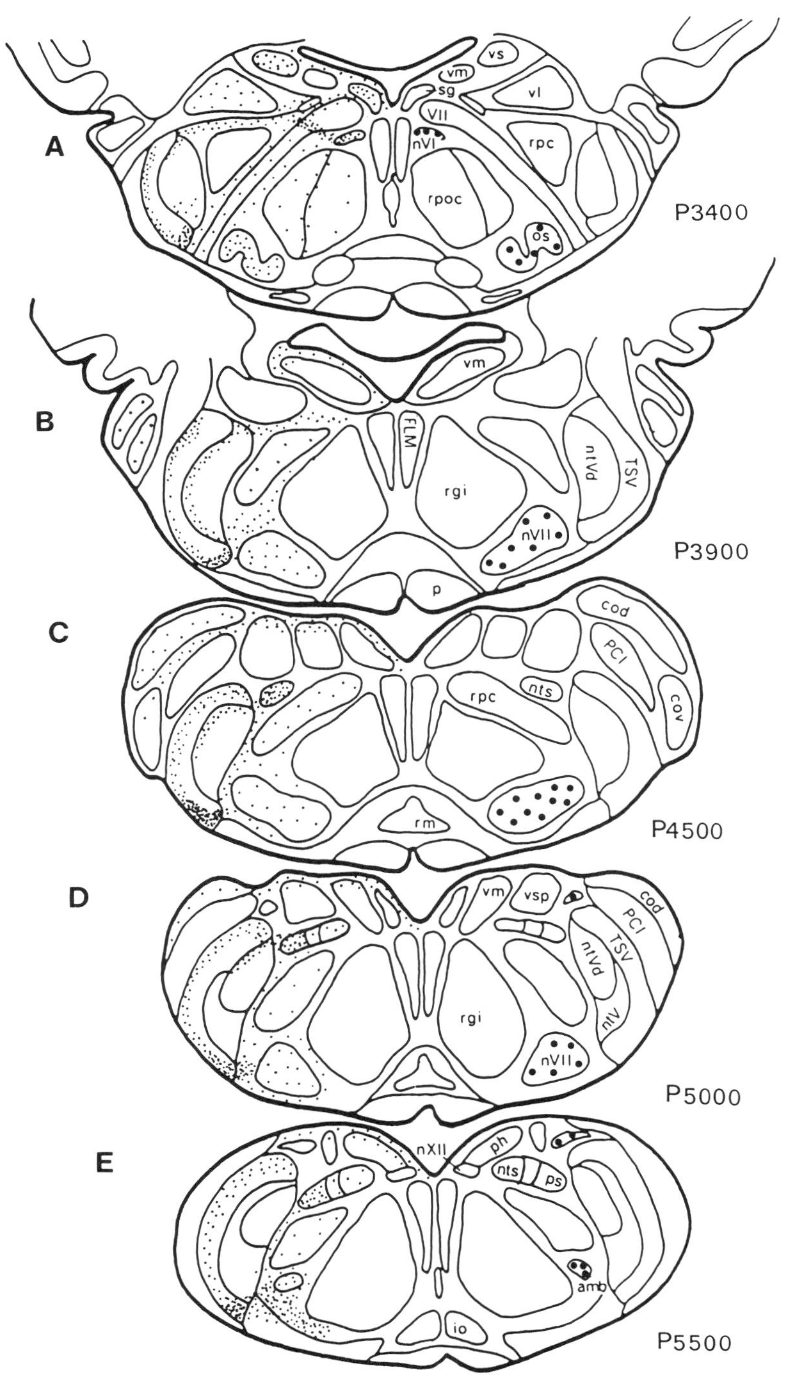
A
vs
vm
sg
vl
VII
nVI
rpc
rpoc
os
P3400
B
vm
FLM
ntVd
TSV
rgi
nVII
p
P3900
C
cod
PCl
nts
rpc
cov
rm
P4500
D
vm
vsp
cod
PCl
TSV
ntVd
rgi
nV
nVII
P5000
E
nXII
ph
nts
ps
amb
io
P5500

FIGURE 12

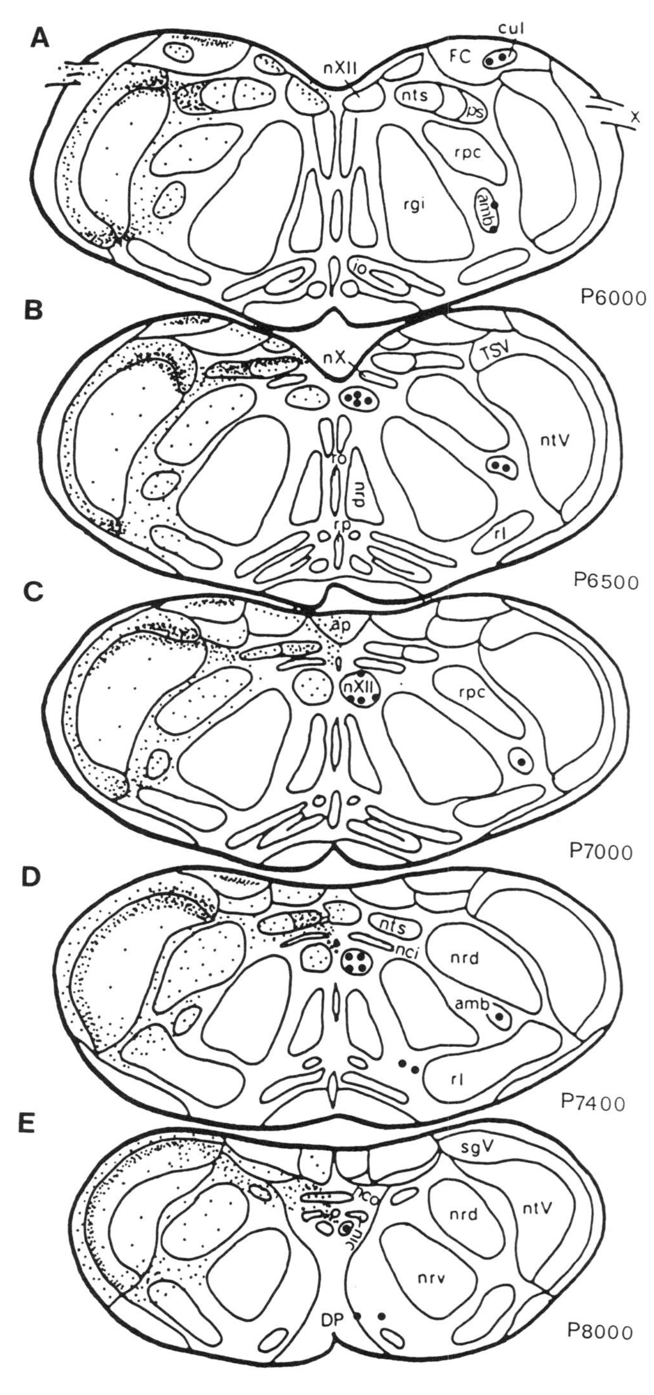
A
cul
FC
nXII
nts
ps
X
rpc
rgi
amb
io
P6000
B
nX
TSV
ntV
ro
nrp
rp
rl
P6500
C
ap
nXII
rpc
P7000
D
nts
nci
nrd
amb
rl
P7400
E
sgV
nco
nrd
ntV
nic
nrv
DP
P8000

FIGURE 13

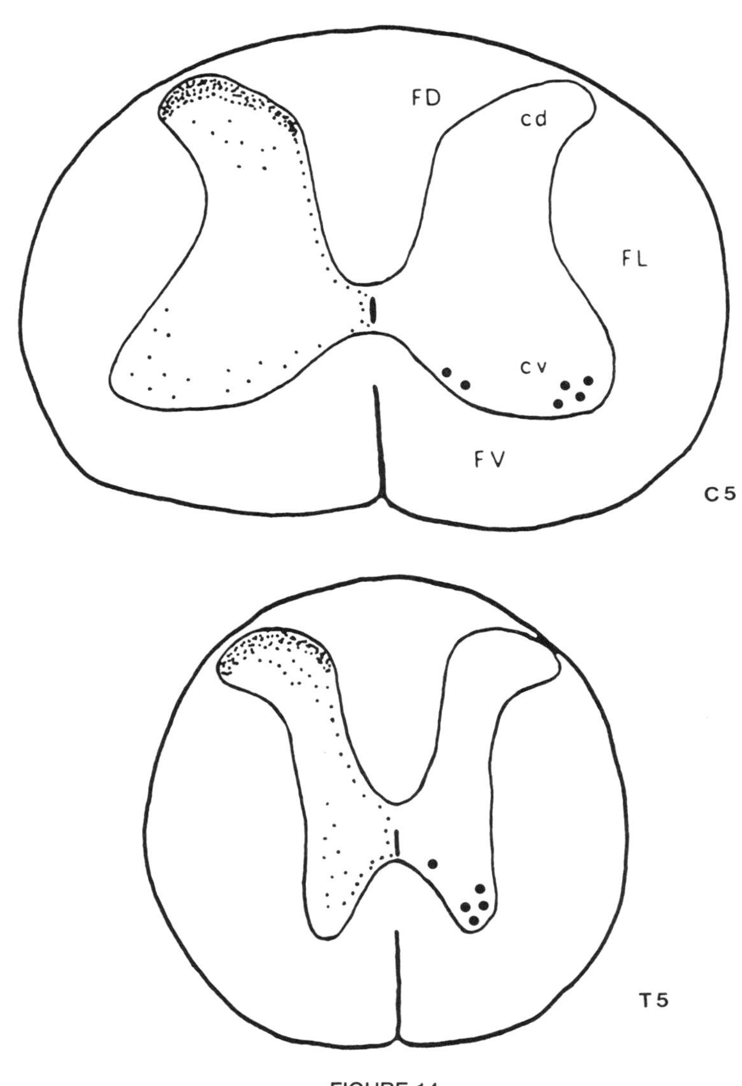

FIGURE 14

Radioimmunoassay

Disequilibrium RIA was performed in a 50 mM sodium phosphate buffer at pH 7.6 containing 0.1% Triton X-100, 0.1% gelatin, 0.1% bovine serum albumen, and 0.01% merthiolate. To 12 × 75 mm polystyrene tubes were added 100 μl standard (rat CGRP) in RIA buffer or 100 μl buffer to rehydrate the samples and 100 μl antiserum diluted in RIA buffer. The tubes were vortexed and kept at 4°C for 48 hr. Approximately 10,000 cpm of [^{125}I-Tyr°]rat CGRP (Peninsula, Belmont, CA; 1500 Ci/mmol) were added in a volume of 100 μl in RIA buffer to each tube

TABLE I. Distribution of CGRP in Various Regions of Rat Brain[a]

Region	Map figure numbers	Rating: Cell bodies	Rating: Nerve fibers	CGRP (fmol/mg protein ± S.E.M.)
Frontal cortex	1–2	–	+	40.2 ± 8.0
Nucleus accumbens	2–3	–	±	73.4 ± 9.6
Nucleus caudatus	2–4	–	–	81.2 ± 25.5
Septum medialis	3B–4A	–	+ +–+ + +	190.6 ± 60.7
Nucleus preopticus	4C	+	±	71.1 ± 19.8
Nucleus preopticus periventricularis	4C–D	+ +	+ +	181.0 ± 34.7
Nucleus interstitialis striae terminalis ventralis	4	±	+ +–+ + +	248.8 ± 108.4
Nucleus interstitialis striae terminalis dorsalis	4	–	+ +–+ + +	432.6 ± 79.2
Insular cortex	4–8	–	+ +	101.6 ± 23.9
Subfornical organ	5C	–	±	372.5 ± 177.9
Globus pallidus	5	–	+ +	52.1 ± 13.0
Nucleus hypothalamicus anterior	5	±	+	174.8 ± 60.1
Medial forebrain bundle	5–6	+–+ +	+–+ +	81.8 ± 19.7
Nucleus periventricularis thalami	5–7	–	+ +	159.5 ± 27.0
Nucleus amygdaloideus centralis	6–7	–	+ + + +	2457.6 ± 443.5
Nucleus amygdaloideus lateralis	6–7	–	±	158.5 ± 40.1
Nucleus paraventricularis	6C	±	±	415.8 ± 59.3
"Dopamine bundle"	6–8	–	+ +	329.4 ± 98.3
Nucleus caudatus putamen (caudal)	6	–	+ + + +	1159.4 ± 203.5
Nucleus arcuatus	6–7	±	±	304.8 ± 76.1
Median eminence	6–7	–	+ +–+ + +	296.5 ± 32.5
Nucleus ventromedialis	6–7	±	±	87.7 ± 22.2
Temporal cortex	7–8	–	±	172.0 ± 18.7
Nucleus dorsomedialis	7A–B	+	+ +	245.6 ± 75.4
Hippocampus gyrus dentatus	8	±	+	80.4 ± 14.9
Medial geniculate body	9	–	+–+ + +	64.9 ± 20.1
Peripeduncular area	9A–C	+ +–+ + +	+ +–+ + +	75.5 ± 19.5
Colliculus superior	9–10	–	–	115.7 ± 21.5
"Dorsal bundle"	9	–	+–+ +	101.7 ± 14.6
Nucleus of the third nerve	9	+ +	+ +	88.2 ± 32.4
Colliculus inferior	10B–11A	±	±–+	65.8 ± 16.9
Nucleus parabrachialis dorsalis	10B–D	+ + +	+ + +	760.2 ± 146.3
Nucleus parabrachialis ventralis	10B–D	+ + +	+ + +	136.5 ± 3.0
Nucleus interpeduncularis	10	–	+	92.5 ± 14.3
Nucleus of the fifth nerve	11B–D	+ +	±	166.3 ± 47.6
Nucleus olivaris superior	11D–12A	+ +	+ +–+ + +	58.7 ± 19.3
Cerebellum	11–12	–	–	32.4 ± 4.7
Nucleus of the seventh nerve	12B–D	+ +	+	43.3 ± 3.0
Nucleus tractus solitarii	12C–13	–	+ +	618.0 ± 212.2
Tractus spinalis nervi trigemini	12–13	–	+ +–+ + +	4682.5 ± 526.6
Nucleus ambiguus	12D–13	+ +	+ +	370.9 ± 104.8
Nucleus of the twelfth nerve	13	+ +	+	182.5 ± 21.3
Substantia gelatinosa	13E	–	+ + + +	1579.4 ± 230.2

TABLE I. (*Continued*)

Region	Map figure numbers	Rating: Cell bodies	Rating: Nerve fibers	CGRP (fmol/mg protein ± S.E.M.)
Nucleus tractus spinalis nervi trigemini (lateral part)	13	−	+ + +	3977.4 ± 299.1
Dorsal horn	14	−	+ + + +	3289.3 ± 241.2
Ventral horn	14	+ +	+ − + +	74.0 ± 14.7
Pituitary anterior lobe	−	−	+ − + +	179.5 ± 38.8
Pituitary neurointermediate lobe	−	−	−	47.1 ± 7.9

[a] The CGRP concentrations in discrete regions of the rat brain were obtained by RIA using an antiserum directed against human CGRP coupled to bovine thyroglobulin. Synthetic rat CGRP is used as a standard. Each value represents the mean ± S.E.M. of four to six samples. See atlases of Jacobowitz and Palkovits (1974) and Palkovits and Jacobowitz (1974) for coordinates. Semiquantitative ratings of densities of cell bodies and nerve fibers are also tabulated.

and incubated at 4°C for an additional 24 hr. Thereafter, 100 μl normal rabbit serum (10% in RIA buffer) and 100 μl sheep antirabbit serum (35% in RIA buffer) were added. Precipitates were allowed to form for 12 hr at 4°C. Precipitates were collected by centrifugation at 5000 × g for 30 min. The supernatants were carefully removed by aspiration, and the pellets counted for 2 min each in a γ counter.

In RIA the antiserum to human CGRP showed 100% cross reactivity to rat CGRP, about 5% cross reactivity towards bombesin and kassinin, and 2% towards galanin. Substance P, substance K, the neuromedins K, B, and C, eledoisin, physalaemin, and vasoactive intestinal polypeptide showed less than 1% cross reactivity.

The quantitative distribution of CGRP in discrete areas of the brain is presented in Table I. Perusal of these values reveals that CGRP was unevenly distributed in

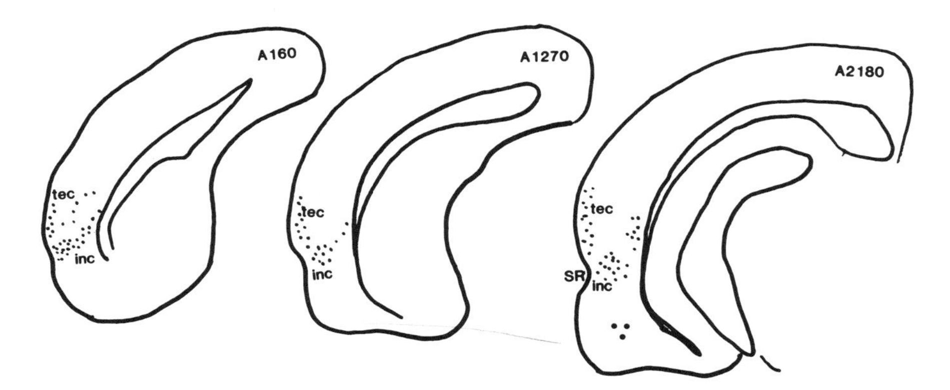

FIGURE 15. Schematic drawing showing the localization of CGRP-immunoreactive fibers (fine dots) in the insular cortex (inc) and the temporal cortex (tec). Cell bodies (filled circles) are localized in the entorhinal cortex. Stereotactic coordinates are indicated above.

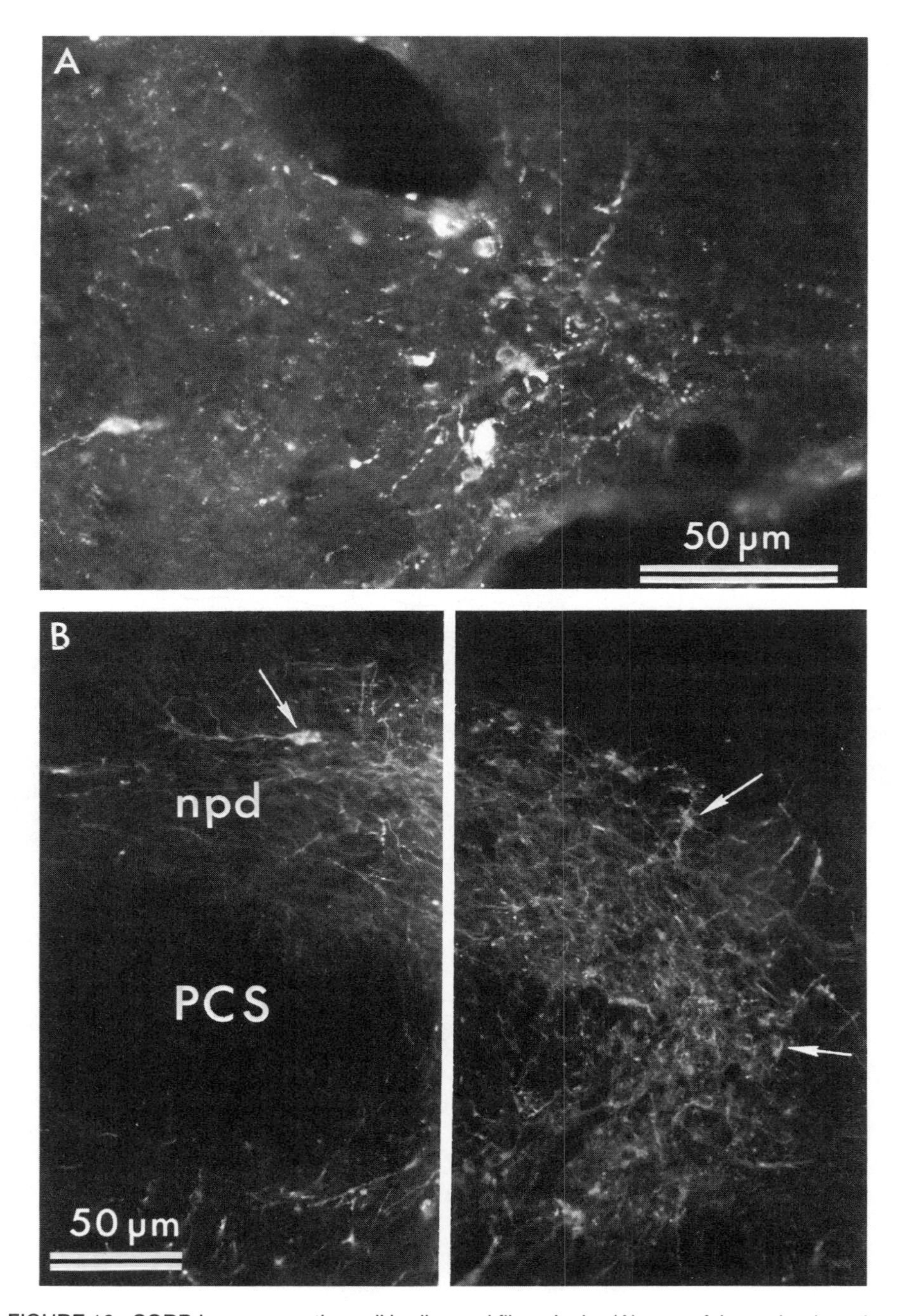

FIGURE 16. CGRP-immunoreactive cell bodies and fibers in the (A) area of the peripeduncular nucleus ventral to the medial geniculate body of a colchicine-treated rat and the (B) nucleus parabrachialis dorsalis (npd) dorsal to the superior cerebellar peduncle (PCS). Arrows indicate cell bodies.

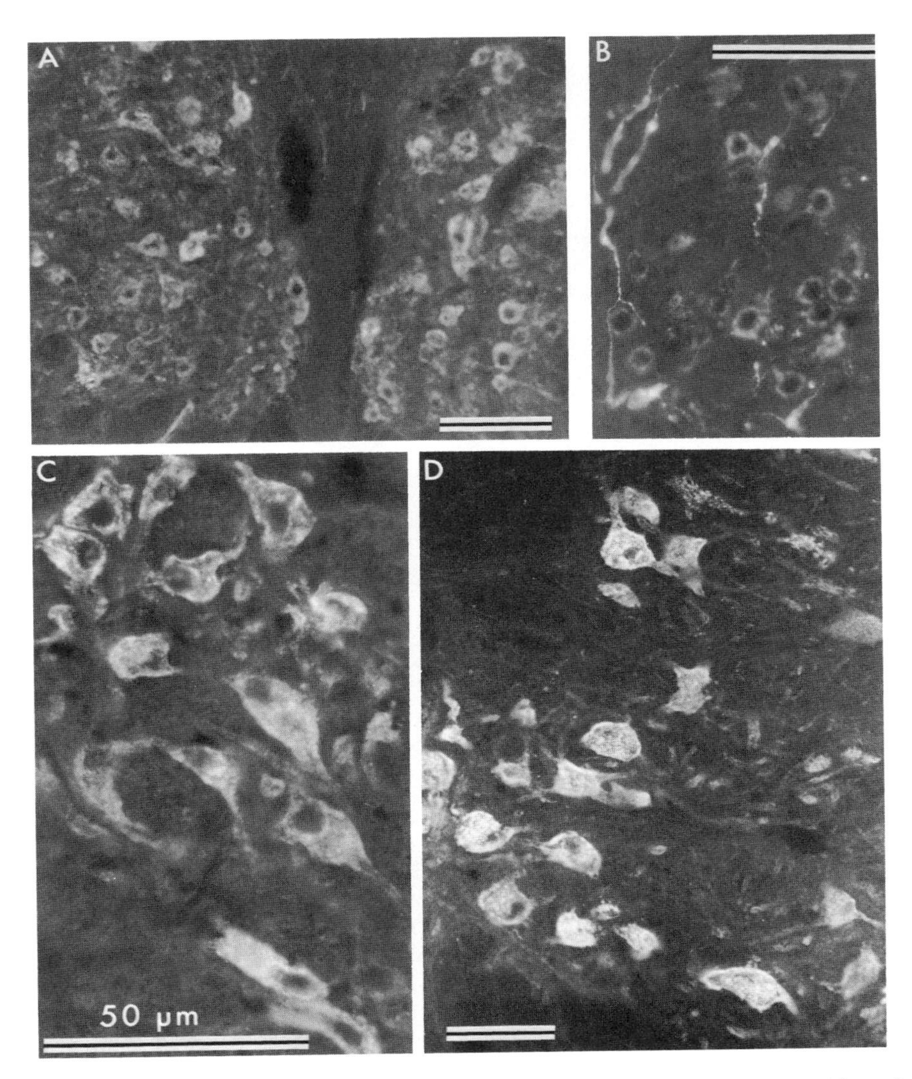

FIGURE 17. CGRP-immunoreactive cell bodies of colchicine-treated rats. (A) nIII, bilateral, level A1800; (B) nucleus ambiguus, level P5500; (C) nIV, level A300; (D) spinal cord, motor horn cell bodies, cervical level.

individual nuclei and regions throughout the CNS. The highest concentrations were observed in the amygdala centralis, caudal portion of the caudate–putamen, tractus spinalis of the trigeminal nerve, nucleus of the tractus spinalis of the trigeminal nerve (lateral portion), substantia gelantinosa, and dorsal horn of the spinal cord. More moderate concentrations were observed in areas such as the subfornical organ, bed nucleus of the stria terminalis (especially the dorsal part), paraventricular nucleus, arcuate nucleus, median eminence, nucleus tractus solitarius, nucleus parabrachialis dorsalis, and nucleus ambiguus. The remaining areas contained low to trace quantities of CGRP. A 100-fold range exists between the lowest and highest values in the brain.

A correlation of these quantitative values with "semiquantitative" estimations of densities of nerve fibers (Table I) reveals an excellent correlation in those areas cited as having the highest concentrations of CGRP. Overall, about 80% of the areas cited in the table show a good correlation between neuronal density and RIA values. The remaining 20%, where moderate densities of nerves were observed and low values of CGRP were measured or vice versa, may be explained in terms of uniformity of fibers within a dissected ("punched") area. A dense cluster of fibers contained in an otherwise sparsely populated region may partly explain some of the discrepant correlations. A high RIA value in an area with a sparse number of nerves is more difficult to explain. It is possible that in areas such as the subfornical organ and arcuate nucleus there are lower concentrations of protein, which will yield comparatively higher values of peptides. The CGRP-containing cell bodies in an area cannot be correlated with RIA values since only colchicine-treated rats reveal immunoreactive peptides in perikarya.

3.2. Identification by High-Performance Liquid Chromatography

Micropunches of the dorsal horn of the spinal cord, the borderline area where the nucleus and tract of the spinal trigeminal nerve meet, and of the amygdala centralis of seven animals were pooled in 1 ml 0.1 N HCl on ice, boiled for 10 min, and homogenized by sonication. A 20-μl aliquot was taken for protein assay. Samples were centrifuged at 8000 × *g* for 30 min. A 500-μl aliquot of the supernatant was applied to a high-performance liquid chromatography (HPLC) system. The remainder of the supernatant was diluted in consecutive 1:1 steps, evaporated to dryness, and used for parallel displacement studies. The HPLC system used was described previously (Skofitsch and Jacobowitz, 1985a). As a reference standard rat CGRP (2.5 μg in 0.1 N HCl) was used.

With HPLC and RIA several peaks were identified in tissue extracts of the dorsal horn of the spinal cord, the nucleus and tract area of the spinal trigeminal nerve, and the amygdala centralis (Fig. 18). The first peak eluted with the void volume and might result from unspecific reaction with salts and high acidity. A second small peak was seen in the amygdala centralis extract only and was coeluted with substance K at 7.5 min. The third peak identified in the spinal trigeminal area extracts coeluted with kassinin at 9.2 min. The fourth peak (eluted at 16.5 min together with eledoisin and galanin) and the fifth peak (eluted at 19.5 min earlier than neuromedin K) were also present in the standard and thus thought to represent oxidized or breakdown products of the CGRP. The main peak of rat CGRP eluted at 23 min from the column and was at least ten times the amount of the previous

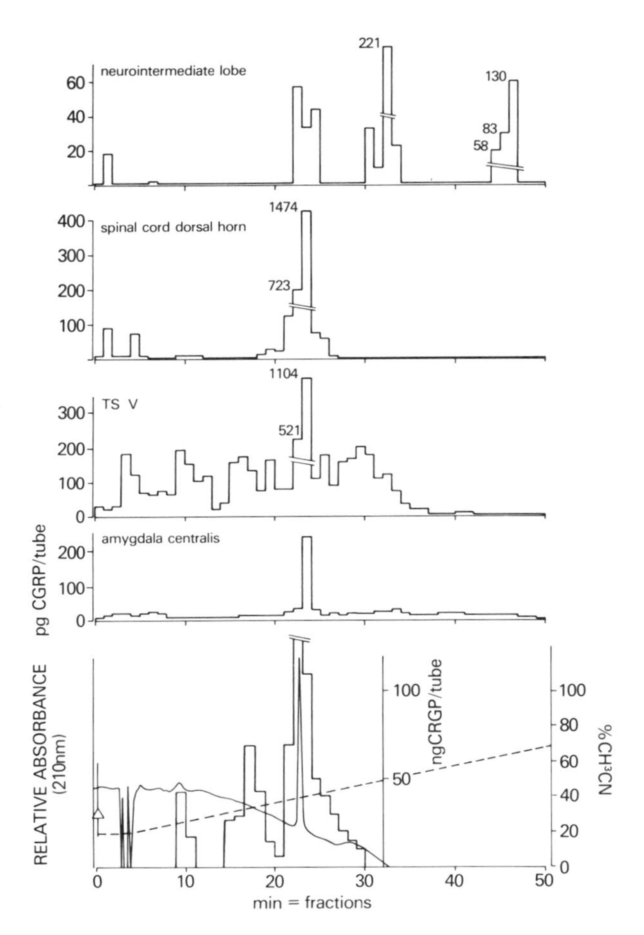

FIGURE 18. High-performance liquid chromatography and radioimunoassay of tissue extracts of the neurointermediate lobe of the pituitary (1800 μg protein), the dorsal horn of the spinal cord (1121 μg protein), the nucleus and tract area of the spinal trigeminal nerve (670 μg protein), and the central amygdaloid nucleus (291 μg protein). The bottom panel shows the original tracing of the rat CGRP at 210 nm relative absorbance in the column effluent and superimposes the RIA analysis showing two small front peaks as well as the linear gradient by which the concentration of acetonitrile was increased in the mobile phase.

peaks. An unidentified seventh peak was observed at 34 min in spinal trigeminal area extracts. The neurointermediate lobes of the pituitary showed two major peaks following the CGRP peak.

Increasing concentrations of tissue extracts of the dorsal horn of the spinal cord, the spinal trigeminal area, the amygdala centralis, and the medial preoptic area displaced the radioactive ligand in parallel to increasing concentrations of synthetic CGRP from the antiserum. Parallel displacement and HPLC results indicate that the antiserum and RIA are highly specific for CGRP with negligible cross reactivity towards other substances in most brain areas.

The similarity of HPLC elution patterns of synthetic CGRP with one immunoreactive peak is suggestive of chemical identity. The additional immunoreactive peaks suggest the possibility that several immunoreactive forms exist in some parts of the brain and pituitary. The nature of such heterogeneous forms, whether precursor molecules or metabolic products, cannot be assessed here.

4. RECEPTOR AUTORADIOGRAPHY

An attractive adjunct to the histochemical localization of brain neuroactive substances is the utilization of iodinated peptides for the autoradiographic demonstration of receptor binding sites in discrete areas of the brain. It is becoming apparent that by no means does the presence of neuronal fibers imply the close proximity of receptors to neuroactive substances contained within these nerves (Leroux and Pelletier, 1984; Shults *et al.*, 1984; Wynn *et al.*, 1984). Localization of receptor binding sites by autoradiography, therefore, presents an important dimension that allows us to focus attention on potential sites of action of peptides in the brain.

Rats were perfused via the ascending aorta with 150 ml of ice-cold 50 mM phosphate-buffered saline (PBS) containing 0.5% sodium nitrite and 10% sucrose. The brain and the cervical spinal cord were removed quickly, cut into 8-mm slices, frozen on dry ice, and cut into serial 20-μm sections in a cryostat. The sections were thaw mounted on chrom-alum-coated slides, dried under a stream of cold air, and frozen at −20°C until they were used for autoradiography.

Sections that contained cerebellum were used for time course and saturation curve studies. To displace endogenous ligands, slides were preincubated twice for 15 min at room temperature in 50 mM tris-HCl buffer (pH 7.7) containing 5 mM $MgCl_2$, 2 mM EGTA. Thereafter, sections were incubated in the same solvent containing 1% (w/v) bovine serum albumin, 0.05% leupeptin, 0.001% pepstatin A, and various concentrations of $[^{125}I\text{-Tyr}^{\circ}]$rat CGRP with a specific activity of 1400 Ci/mmol. For time course experiments, 0.69 nM iodinated ligand was used. For saturation curve and Scatchard plot studies, concentrations of 0.05 to 2.9 nM of iodinated ligand were used. The autoradiographic localization of receptors was carried out at a concentration of 0.69 nM iodinated ligand. To determine the amount of unspecific binding, cold synthetic rat CGRP was added to controls in a concentration of 1 μM.

For binding experiments all slides were incubated with the iodinated ligand for 90 min at room temperature, rinsed quickly in 50 mM phosphate-buffered saline containing 1% (w/v) bovine serum albumin, and further washed in this solution on ice twice for 10 min each. After a final rinse in distilled water, the sections were rapidly dried under a stream of warm dry air. Autoradiograms were developed using Kodak X-Omat XAR-5 film after 4–7 days of exposure at 4°C. Tissue sections were stained with 0.1% thionin. Binding for saturation curves and time course was determined by scraping off the dried sections and counting in a γ counter. Thereafter, the material was rehydrated with 0.1 N HCl (1 ml) and homogenized by sonication, and the amount of protein was analyzed.

Only representative areas of the brain were taken for autoradiography. Photomicrographs of most of the areas investigated are shown in Figs. 19–21. Receptors (binding sites) were discretely localized in the brain and spinal cord. Densely and moderately labeled areas of the rat brain are listed in Table II.

Of particular interest is the striking density of binding sites contained in the nucleus accumbens (Figs. 19C,D), the nucleus amygdaloideus centralis (Figs. 20A–D), and the caudal part of the caudate–putamen (Fig. 20H). Furthermore, there seems to be a direct continuity between the nucleus accumbens and the ventrolateral part of the caudate known as the fundus striatum (Figs., 19E–H,20A–C), which joins the ventral pallidum (Fig. 19F) medially. This continues with the caudal part of the caudate–putamen (Fig. 20D), which blends with the amygdala centralis. This is aligned most caudally with the lateral amygdaloid nucleus (Figs. 20C–E).

Only discrete portions of the prefrontal (Fig. 19A), insular (Figs. 19–20D), and temporal (Figs. 20E–H) cortices contain binding sites. Midline thalamic and hypothalamic regions were also densely labeled (Figs. 19C–H,20A–E). In the midbrain and hindbrain, dense labeling was observed in the superior and inferior colliculi (Figs. 20F–21A), the medial geniculate body (Fig. 20F), raphe dorsalis (Fig. 20G), cerebellum (Fig. 21), dorsal tegmental nuclei (Figs. 20G,H), trigeminal nuclei (Figs. 20H–21D), inferior olive (Fig. 21B), and dorsal horn of the spinal cord (Fig. 21D).

A comparison of the distribution of CGRP binding sites to the localization of CGRF fibers and perikarya revealed that most of the areas that contained CGRP-positive nerve fibers also contained appreciable binding sites with the exception of the tractus of the spinal trigeminal nerve, which contained a low density of binding sites. This is understandable in view of the fact that axons would not necessarily be expected to have receptors to the peptide contained within these tracts. Two prominent regions that showed a striking lack of concordance between CGRP nerves and peptide binding were the cerebellum and superior colliculus, which contained no visible nerve fibers but very dense receptor binding. Likewise, the dorsal portion of the hippocampus–dentate gyrus contained only rare fibers, whereas binding was observed in the pyramidal layer of Ammons horn and the granular layer of the dentate gyrus. However, very small amounts of CGRP could be measured by RIA. Regions such as the nucleus accumbens, preoptic medial nucleus, and amygdala lateralis, which contain only trace amounts of immunoreactive fibers and very low concentrations of CGRP (RIA), had very dense receptor binding sites. Generally, with few exceptions, there was a good relationship between the distribution of CGRP fibers (terminal fields) and CGRP receptor binding sites.

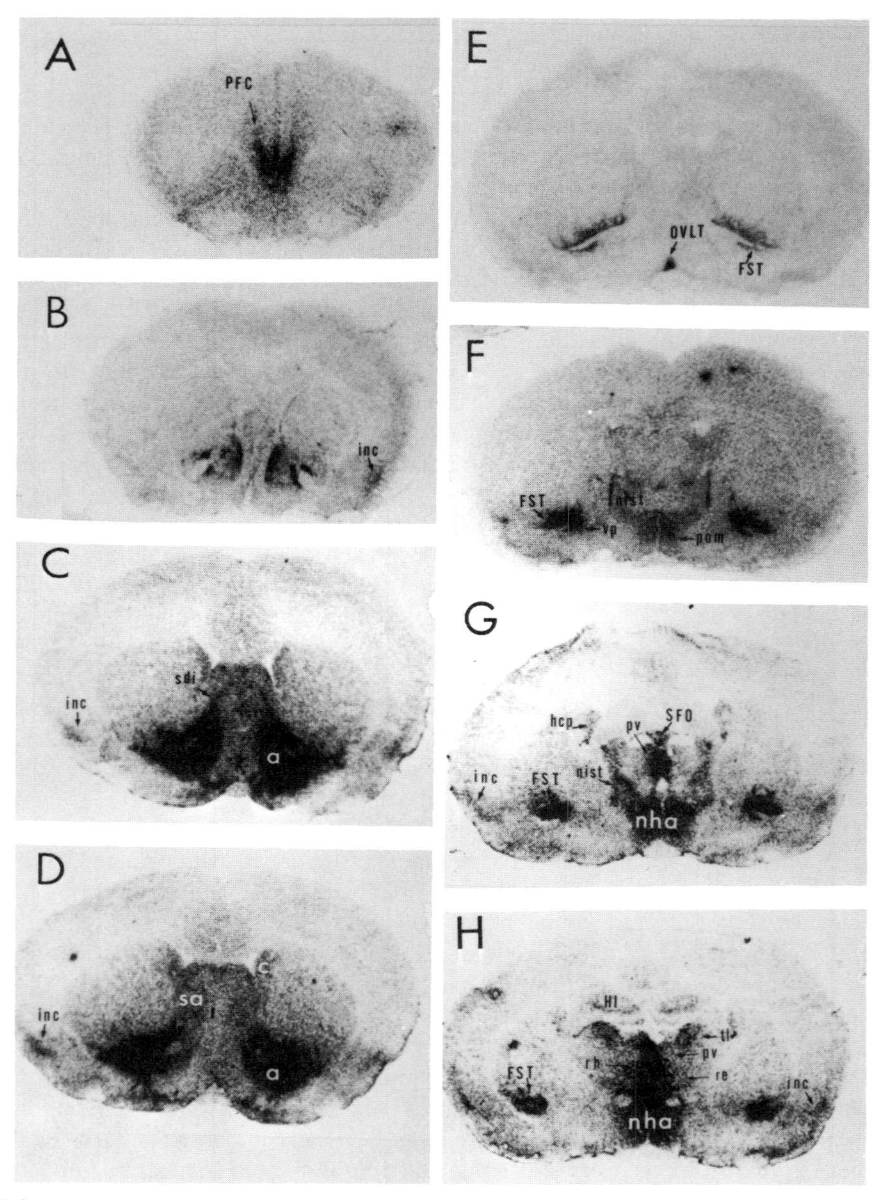

FIGURES 19–21. Distribution of CGRP receptor binding sites in the rat brain. Approximate coordinates according to the atlases of König and Klippel (1963) and Palkovits and Jacobowitz (1974). These brain atlases should be consulted for identification of unlabeled structures. See Table II for figure abbreviations.

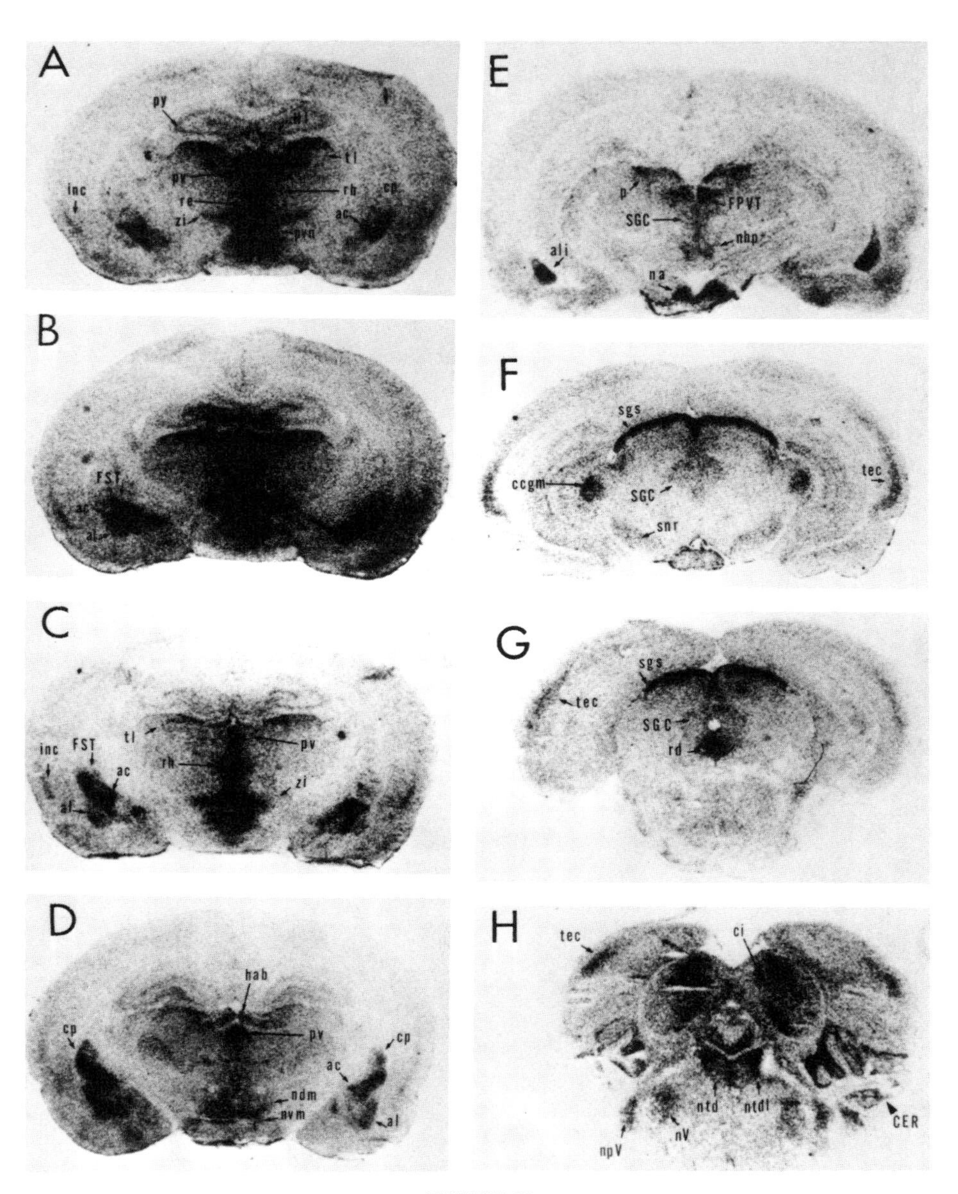
A
py
tl
pv
re
zi
inc
cp
ac
B
FST
ac
al
C
inc
FST
tl
ac
al
pv
rh
zi
D
hab
pv
cp
ac
ndm
nvm
al
E
p
FPVT
SGC
ali
na
F
sgs
ccgm
SGC
tec
snr
G
sgs
tec
SGC
rd
H
tec
ci
ntd
ntdl
CER
nV
npV

FIGURE 20

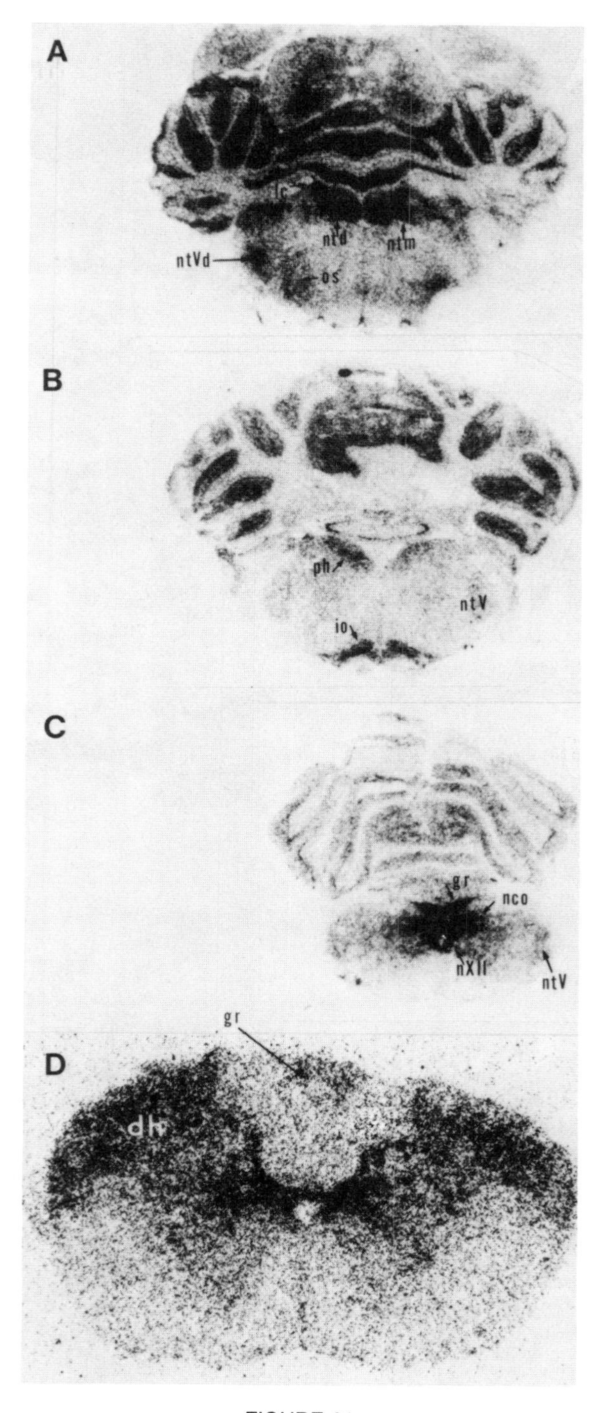

FIGURE 21

TABLE II. Distribution of CGRP Receptor-Labeled Areas of the Brain

Area (abbreviation)	Figure
Dense areas	
Prefrontal cortex (PFC)	19A
Nucleus accumbens (a)	19B–D
Organum vasculosum lamina terminalis (OVLT)	19E
Fundus striatum (FST)	19E–20C
Ventral pallidum (vp)	19F
Subfornical organ (SFO)	19G
Nucleus interstitialis stria terminalis (NIST)	19F,G
Periventricular nucleus of the thalamus (pv)	19G–20D
Nucleus amygdaloideus centralis (ac)	20A–D
Caudate–putamen (caudal area) (cp)	20A–D
Nucleus lateralis thalami (tl)	20A–C
Nucleus rhomboideus (rh)	20A–C
Nucleus reuniens (re)	19H–20A
Nucleus paraventricularis (pvn)	20A,B
Zona incerta (ZI)	20A–C
Habenula (hab)	20D
Nucleus dorsomedialis (ndm)	20D
Nucleus pretectalis (p)	20E
Fibrae periventriculares thalamus (FPVT)	20E
Nucleus arcuatus (na)	20E
Nucleus amygdaloideus lateralis inferior (ali)	20E
Nucleus centralis corporis geniculati medialis (ccgm)	20F
Stratum grieseum superficiale colliculi superioris (sgs)	20F,G
Temporal cortex (tec)	20F–H
Nucleus raphe dorsalis (rd)	20G
Colliculus inferior (ci)	20H–21A
Cerebellum (molecular layer) (CER)	20H–21C
Nucleus tegmenti dorsalis lateralis (ntdl)	20H
Nucleus principalis nervi trigemini (npV)	20H
Nucleus originis nervi trigemini (nV)	20H
Locus coeruleus (lc)	21A
Nucleus tegmenti dorsalis (ntd)	20H–21A
Nucleus tractus spinalis nervi trigemini pars dorsomedialis (ntVd)	21A
Nucleus olivaris superior (os)	21A
Nucleus olivaris inferior (io)	21B
Nucleus prepositus hypoglossi (ph)	21B
Nucleus originis nervi hypoglossi (nXII)	21C
Nucleus commissuralis (nco)	21C
Nucleus gracilis (gr)	21C
Dorsal horn spinal cord (dh)	21D
Moderate areas	
Insular cortex (inc)	19C–20D
Nucleus dorsalis septi pars intermedia (sdi)	19C
Septal area (sa)	19D

(Continued)

TABLE II. (*Continued*)

Area (abbreviation)	Figure
Nucleus preopticus medialis (pom)	19F
Hippocampus (HI) and dentate gyrus	19H–20F
Substantia grisea centralis (SGC)	20E–G
Nucleus hypothalamicus posterior (nhp)	20E
Substantia nigra zona reticularis (snr)	20F
Nucleus tractus spinalis nervi trigemini (ntV)	21B,C
Nucleus parabrachialis dorsalis	20H
Reticular formation	20H–21C
Nucleus cuneatus	21C
Nucleus lateral lemniscus	20G
Nucleus vestibular medialis	—
Nucleus tractus solitarius	—

From a functional standpoint it is noteworthy that receptor binding sites were observed in several sensory systems of the brain and spinal cord. In the auditory system, binding was noted in the superior olive, inferior colliculus, medial geniculate body, nucleus of the lateral lemniscus, and a small portion of the temporal lobe cortex (auditory cortex). The somesthetic system of the body (somatosensory information to the cortex) contained binding sites in the dorsal horn of the spinal cord and the gracilis and cuneate nuclei. Although binding in the dorsal root ganglia was not studied, perikarya in these ganglia do contain CGRP (Fig. 22B). This system of nerves transmits pain, temperature, touch, and pressure to the somatosensory cortex. Efferent pathways from the cortex (area 3,1,2) transmit signals through the trigeminal sensory nucleus, nucleus tractus solitarius, and reticular formation, all of which contained CGRP binding sites. In addition, the occipital cortex (area 17) projects fibers to the superior colliculus, which contained a large density of binding sites. The cerebellar afferent pathway, i.e., the inferior olive and the cerebellum, contained abundant receptor binding sites. The taste pathway (gustatory pathway) is a multisynaptic pathway that projects sensory fibers (from the tongue, epiglottis, and larynx) to the rostral part of the nucleus tractus solitarius, nucleus parabrachialis "pontine taste area" (Norgren and Leonard, 1973), central amygdaloid nucleus (Norgren, 1976), medioventral posterior nucleus of the thalamus, and the insular cortex, all of which contained binding sites.

The significance of CGRP receptors in multisynaptic pathways is not clear. The binding technology as used in this study does not reveal receptor localization, i.e., cell body and/or neuronal (or glial) fibers. Therefore, information as to actual pathways of the receptor-containing neurons is not available. Lesion studies of pathways will begin to unravel these mysteries.

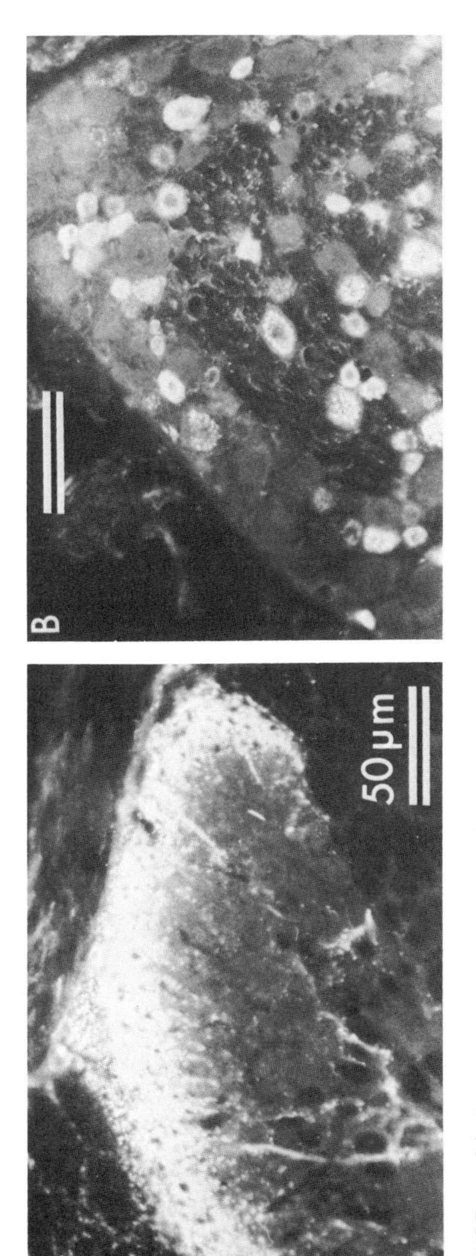

FIGURE 22. CGRP immunoreactivity in the (A) fibers of the cervical dorsal horn of the spinal cord and (B) cell bodies in the dorsal root ganglion of a colchicine-treated rat.

5. COEXISTENCE OF CGRP WITH SUBSTANCE P IN SENSORY NEURONS

The above immunocytochemical, autoradiographic, and RIA data revealed that CGRP nerves and binding sites are present in regions known to contain primary sensory neurons (e.g., spinal trigeminal area, spinal cord). Furthermore, spinal sensory ganglia and trigeminal ganglia contained CGRP (Lee *et al.*, 1985; Rosenfeld *et al.*, 1983; Skofitsch and Jacobowitz, 1985e). It was therefore of interest to study the discrete localization of CGRP immunoreactivity in primary sensory neurons and the influence of treatment of neonatal rats with capsaicin on the CGRP immunoreactivity in the medulla oblongata, spinal cord, and sensory ganglia. Capsaicin is a neurotoxic agent known to cause selective degeneration of a certain population of primary sensory neurons of the C-fiber type and of type B cells of sensory ganglia (Jancso *et al.*, 1977; Jancso and Kiraly, 1980; Scadding, 1980). Since substance-P-containing neurons are both present in sensory ganglia and affected by capsaicin, we studied the possible coexistence of substance P (SP) and CGRP in the dorsal root and trigeminal ganglion cells.

Male rats were treated systemically with capsaicin (50 mg/kg, s.c.) on the second day of life according to Jancso *et al.* (1977). Control animals received solvent only (10% ethanol, 10% Tween 80®, 80% saline). After 3 months the animals were tested for impaired chemosensitivity of the eye (Szolcsanyi *et al.*, 1975) by instillation of 50 μl of a capsaicin solution (100 μg/ml saline) into the eye and counting the protective wiping–scratching movements of the forepaw. Controls responded with 27–34 wipes (31.4 ± 2.3 wipes, $n = 7$); capsaicin treatment was regarded to be sufficient when the animals did not or only rarely responded to the instillation of the capsaicin solution into the eye (0.6 ± 0.4 wipes, $n = 7$). Three animals of each, the control and the capsaicin-treated groups, received an intracerebroventricular injection of 100 μg colchicine and were processed for indirect immunohistochemistry as described above. For coexistence studies of CGRP and SP, adjacent sections of spinal dorsal root ganglia and the trigeminal ganglion were cut at 6 μm and incubated either in CGRP or SP antiserum (1:1000) for 3 days.

For RIA, male rats treated with capsaicin or solvent only on the second day of life were killed by decapitation. The brain and the cervical spinal cord were removed, frozen on dry ice, and cut in coronal 300-μ sections in a cryostat. Microdissected tissues (Palkovits, 1973) were delivered in 250 μl of 0.1 N HCl, boiled in a water bath for 10 min, and homogenized by sonication. A 20-μl aliquot was taken for protein determination, and samples were centrifuged at 5000 × *g* for 20 min; the supernatants were divided into four aliquots of 50 μl and evaporated to dryness in a vacuum centrifuge. Aliquots of the same samples were used for CGRP and SP determination: CGRP and SP RIA was performed as described above for CGRP.

The dorsal root ganglia of colchicine-treated animals revealed a moderate number of cell bodies (15–25 μm) with dense-staining cytoplasm. Similarly, the trigeminal ganglion (Fig. 23) contained a moderate number of positive CGRP-immunoreactive perikarya. No fluorescent cells were observed in non-colchicine-treated rat ganglia. In rats treated with capsaicin as neonates, CGRP-like immunoreactivity was found to be markedly reduced at the border between the spinal trigeminal tract and the nucleus of the spinal trigeminal tract (Fig. 24), the substantia gelatinosa nervi trigemini, and the nucleus commissuralis but not the nucleus tractus

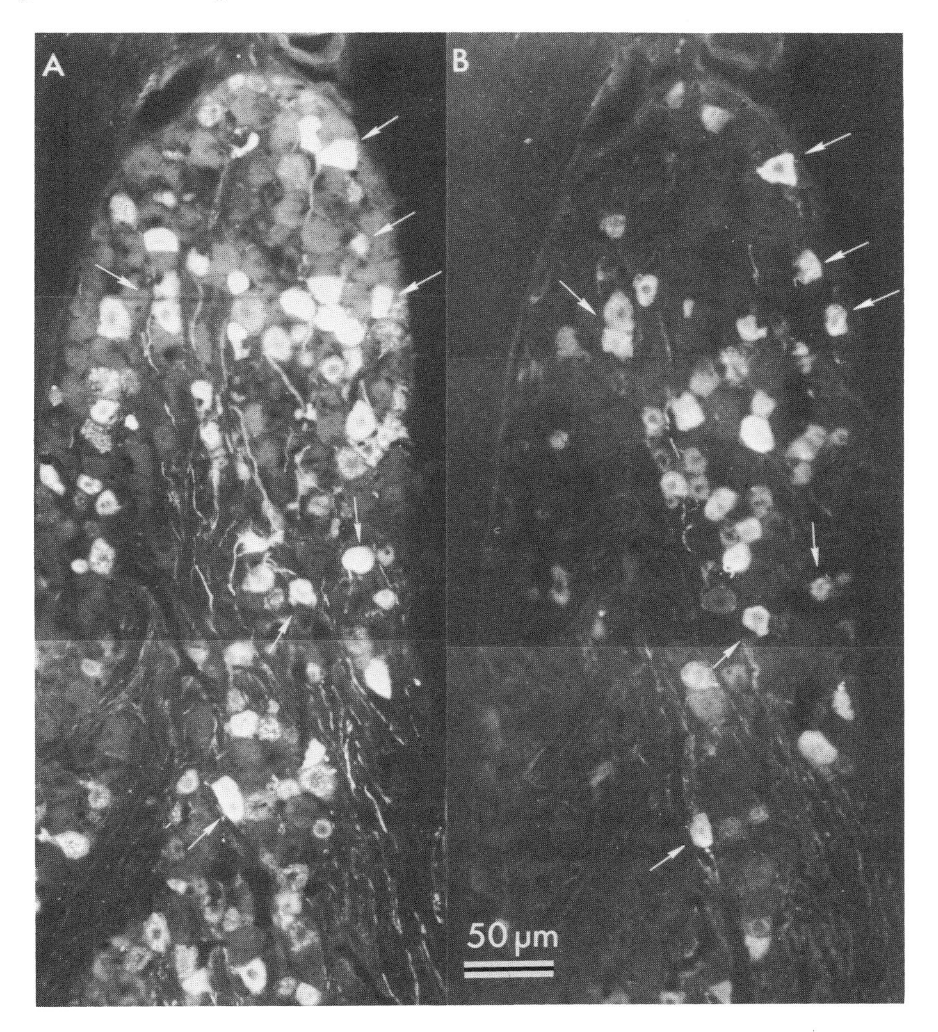

FIGURE 23. Adjacent 6-μm sections of the trigeminal ganglion of a colchicine-treated rat: (A) CGRP and (B) SP immunoreactivity. Arrows indicate corresponding cell bodies, thereby demonstrating coexistence of CGRP and SP.

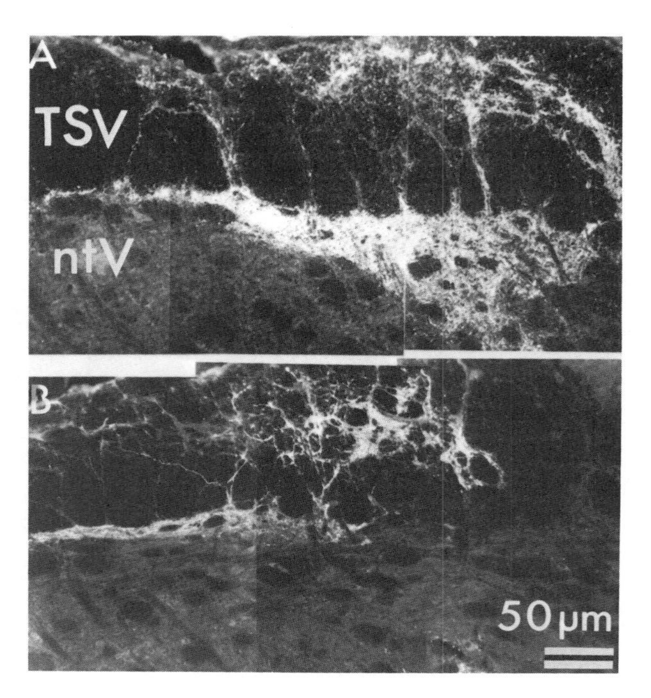

FIGURE 24. CGRP immunoreactivity in the posterior dorsomedial spinal trigeminal tract (TSV) and nucleus (ntV). (A) Control rat; (B) rat treated with capsaicin on day 2 postnatally. Note reduction in the density of fibers.

solitarii and the dorsal horn of the spinal cord. Cell bodies of the trigeminal and dorsal root ganglia of colchicine-treated rats appeared to be markedly decreased in number.

Adjacent sections revealed that CGRP-like immunoreactivity was colocalized with SP-like immunoreactivity in capsaicin-sensitive neurons of the spinal trigeminal tract and nucleus, the substantia gelatinosa, the dorsal horn of the spinal cord (laminae I and II), and the nucleus commissuralis. Adjacent thin sections of the dorsal root and the trigeminal ganglia of colchicine-treated animals revealed that a certain number of perikarya, which appeared to be identical, stained with both CGRP and SP antisera (Fig. 23), thereby demonstrating coexistence of both peptides in the same neuron. The relative percentage of CGRP-positive perikarya in the sensory ganglia is about twice that of SP-positive perikarya.

Radioimmunoassay values of selected brain areas are summarized in Table III. Capsaicin treatment reduced the CGRP-like immunoreactivity in the area of the spinal trigeminal tract and nucleus (at the level of the area postrema) by 84%, in the substantia gelatinosa nervi trigemini by 65%, and in the dorsal horn of the cervical spinal cord by 68%. The ventral spinal cord levels and those of higher

TABLE III. Quantitation of CGRP-like and SP-like Immunoreactive Material in Microdissected Areas of the Central Nervous System of Vehicle-Treated Controls and Rats Treated with Capsaicin on the Second Day of Life[a]

Area	*n*	CGRP		Substance P	
		Control	Capsaicin	Control	Capsaicin
Nucleus preopticus medialis	5	188 ± 29	227 ± 45	3029 ± 351	4038 ± 719
Nucleus interstitialis striae terminalis	5	190 ± 25	268 ± 37	1548 ± 338	1718 ± 425
Nucleus caudatus putamen	5	97 ± 11	115 ± 31	488 ± 72	429 ± 69
Globus pallidus	5	134 ± 25	99 ± 11	729 ± 251	809 ± 123
Nucleus hypothalamicus anterior	5	89 ± 41	121 ± 57	1724 ± 222	1235 ± 236
Nucleus dorsomedialis	5	444 ± 108	313 ± 48	1496 ± 459	1414 ± 418
Substantia nigra reticularis	5	88 ± 22	120 ± 20	5086 ± 393	4392 ± 406
Nucleus tractus solitarii	5	245 ± 64	160 ± 36	1126 ± 91	1347 ± 259
Spinal trigeminal area					
Nucleus and tract	8	1171 ± 106	185 ± 36*	7749 ± 697	829 ± 189*
Substantia gelatinosa	8	521 ± 80	184 ± 29*	3809 ± 347	1932 ± 345*
Spinal cord					
Dorsal horn	8	800 ± 136	255 ± 24*	5123 ± 384	1716 ± 376*
Ventral horn	8	40 ± 16	105 ± 31	469 ± 120	496 ± 98*

[a] Values are in fmol/mg protein ± S.E.M.; $*P < 0.005$.

brain areas of capsaicin-treated animals were not significantly different from controls. The RIA results therefore confirmed the immunohistochemical observations.

Using aliquots of the same samples for SP-RIA, we found similar results (Table III). Following capsaicin treatment, SP-like immunoreactivity was markedly decreased in the borderline area of the spinal trigeminal tract and nucleus (89%), the substanti gelatinosa nervi trigemini (49%), and the dorsal spinal cord (67%). The ventral spinal cord levels and those of higher brain areas were not significantly different from controls.

The observation that CGRP coexists with substance P (Gibson *et al.*, 1984; Wiesenfeld-Hallin *et al.*, 1984; Lee *et al.*, 1985; Skofitsch and Jacobowitz, 1985e) in capsaicin-sensitive neurons suggests that an interaction might take place at the site of peptide release. Since substance P and CGRP are both potent vasodilators in the periphery (Fisher *et al.*, 1983; Brain *et al.*, 1985), we would suggest that both peptides have modulatory effects on nociception and peripheral vasodilatory processes.

6. CARDIOVASCULAR ACTIONS OF CGRP

We have recently demonstrated that microinjections of peptides into discrete nuclei of the hypothalamus and preoptic area of anesthetized rats result in cardiovascular effects in a variety of different nuclei (Diz *et al.*, 1984; Diz and Jacobowitz, 1983, 1984a–d). The present study has focused attention on the presence of a dense

CGRP neuronal innervation and receptor binding in the nucleus amygdaloideus centralis. It was therefore of interest to pursue functional studies that might reveal CGRP actions in this area of the brain. In order to determine if injections of CGRP into the central amygdaloid nucleus result in cardiovascular changes, we monitored blood pressure and heart rate during microinjections of the peptide in halothane-anesthetized rats.

Experiments were performed on male rats anesthetized with 0.9% halothane in 95% oxygen–5% carbon dioxide with a cannuala inserted into the femoral artery for blood pressure measurements. Heart rate and blood pressure recordings and stereotactic procedures were performed as described previously (Diz and Jacobowitz, 1983). Glass micropipettes (20–70 μm tip) were filled with either CGRP in phosphate-buffered saline (PBS) or PBS only. The pipette was lowered into the brain, ±4.5 mm lateral to midline, to a depth of −6.8 mm below dura and +6.0 mm from bregma. Injection volumes were 200 nl, and injections were given on either the right or left side of the brain.

Injection of CGRP (100 pmol) into the central amygdaloid nucleus of the halothane-anesthetized rat resulted in a 21% increase in heart rate that was accompanied by a moderate increase in blood pressure (15%) (Fig. 25). The response was characterized by a gradual onset, with a peak increase at 31 min and a duration of 45–120 min. Tachyphylaxis to the response was apparent following initial exposure to the peptide.

This work departs from the commonly utilized procedure of injecting neuroactive substances into indwelling cannulae placed in the lateral ventricle of the rat. The direct injection of peptides into discrete nuclei has the advantage of revealing potential behavioral responses that can be elicited from a single brain nucleus. Prior studies have revealed that electrical stimulation and ablation of the amygdaloid nuclear complex have produced behavioral, visceral, somatic, and endocrine changes (Kaada, 1951; MacLean and Delgado, 1953). The cardiovascular influence of CGRP on the central amygdaloid nucleus does not imply that this is the sole function of

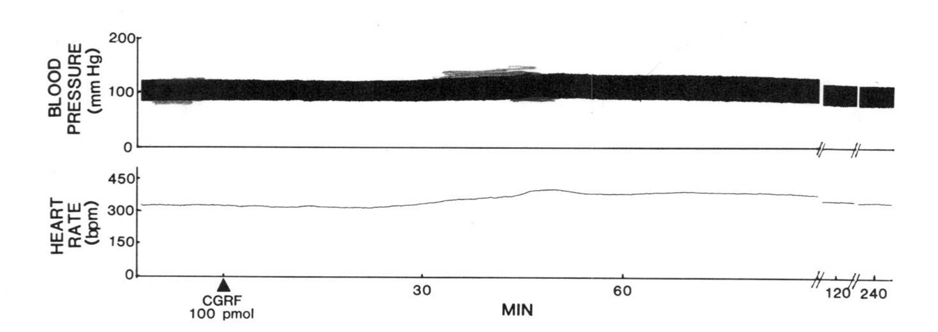

FIGURE 25. Effects of CGRP (100 pmol) on blood pressure and heart rate following injection (200 nl) into the central amygdaloid nucleus.

this nucleus. It is more likely that CGRP stimulates a population of cells in this anatomically heterogeneous nucleus that affects autonomic pathways (adrenergic, cholinergic) to cardiovascular structures by multisynaptic projections. Other cells in this nucleus that project to a variety of brain regions (Krettek and Price, 1978), in all likelihood, are responsible for other behavioral changes measured following electrical stimulation and ablation studies. Therefore, release of CGRP in this nucleus, which contains dense CGRP receptor binding sites, may simultaneously stimulate and/or inhibit a variety of behavioral, visceral, somatic, and endocrine changes.

7. CONCLUSION

This work describes an extensive system of neurons containg CGRP immunoreactivity in the rat brain, spinal cord, and sensory ganglia. The immunocytochemical and radioimmunochemical data demonstrate a regional distribution of immunoreactive CGRP throughout the CNS.

The results obtained by RIA generally correlate with the immunocytochemical distribution of nerve fibers and terminal fields in the rat brain. The dense accumulation of CGRP-like immunoreactivity in the sensory hindbrain and spinal cord areas (e.g., the nucleus and tract area of the spinal trigeminal nerve, the substantia gelatinosa nervi trigemini) indicates a possible involvement of the peptide in the perception or modulation of peripheral nociceptive information. The presence of binding sites in regions containing sensory structures is interesting from a functional standpoint. The auditory, somesthetic, and gustatory systems, the cerebellar afferent pathway, and the efferent pathways from the cortex all contain receptor binding sites and therefore implicate CGRP in modulation of the above functions.

A motor function of CGRP is also implicated in skeletal muscle of the periphery, since this peptide is present in motor neurons of the brain (III, IV, V, VI, VII, XII) and in the ventral horn cells of the spinal cord. Thus, CGRP is the first peptide found to coexist with acetylcholine-containing motor cells of the brain and spinal cord. The existence of CGRP has recently been reported in motor end plates of the striated muscle of the esophagus (Rodrigo *et al.*, 1985). The possibility that a CGRP-like peptide is involved in the action of acetylcholine at the motor end plates of striated muscle is of great physiological and clinical significance.

The demonstration that CGRP injected into a discrete nucleus can alter cardiovascular dynamics is a first approach to understanding the function of this very unusual peptide in the brain.

ACKNOWLEDGMENTS. We thank Mr. Khanh Q. Nguyen and Ms. Reingard Resch for expert technical assistance and Mrs. Lois Brown for typing the manuscript. G. S. is an International Research Fellow of the Fogarty International Center, Fellowship 3 FO5 TWO 3293-01S2 BI-5.

REFERENCES

Amara, S. G., Jonas, V., Rosenfeld, M. G., Ong, E. S., and Evans, R. M., 1982, Alternative RNA processing in calcitonin gene expression generates mRNAs encoding different polypeptide products, *Nature* **298:**240–244.

Benjamin, R. M., and Akert, K., 1959, Cortical and thalamic areas involved in taste discrimination in the albino rat, *J. Comp. Neurol.* **111:**231–260.

Brain, S. D., Williams, T. J., Tippins, H. R., and MacIntyre, I., 1985, Calcitonin gene-related peptide is a potent vasodilator, *Nature* **323:**54–56.

Diz, D. I., and Jacobowitz, D. M., 1983, Cardiovascular effects of intrahypothalamic injection of α-melanocyte stimulating hormone, *Brain Res.* **270:**265–272.

Diz, D. I., and Jacobowitz, D. M., 1984a, Effects of adrenalectomy, propanolol and atropine on the increase in heart rate induced by injection of dermorphin in the rat anterior hypothalamic nucleus, *Brain Res.* **293:**196–199.

Diz, D. I., and Jacobowitz, D. M., 1984b, Cardiovascular effects of discrete intrahypothalamic and preoptic injections of bradykinin, *Brain Res. Bull.* **12:**409–417.

Diz, D. I., and Jacobowitz, D. M., 1984c, Cardiovascular effects produced by injection of thyrotropin-releasing hormone in specific preoptic and hypothalamic nuclei in the rat, *Peptides* **5:**801–808.

Diz, D. I., and Jacobowitz, D. M., 1984d, Cardiovascular actions of four neuropeptides in the rat hypothalamus, *Clin. Exp. Hypertension* **A6:**2085–2090.

Diz, D. I., Vitale, J. A., and Jacobowitz, D. M., 1984, Increases in heart rate and blood pressure produced by injections of dermorphin into discrete hypothalamic sites, *Brain Res.* **294:**47–57.

Fisher, L. A., Kikkawa, D. O., Rivier, J. E., Amara, S. G., Evans, R. M., Rosenfeld, M. G., Vale, W. W., and Brown, M. R., 1983, Stimulation of noradrenergic sympathetic outflow by calcitonin gene-related peptide, *Nature* **305:**534–536.

Gibson, S. J., Polak, J. M., Bloom, S. R., Sabate, I. M., Mulderry, P. M., Ghatel, M. A., McGregor, G. P., Morrison, J. F. B., Kelley, J. S., Evans, R. M., and Rosenfeld, M. G., 1984, Calcitonin gene-related peptide immunoreactivity in the spinal cord of man and of eight other species, *J. Neurosci.* **4:**3101–3111.

Jacobowitz, D. M., and O'Donohue, T. L., 1978, α-MSH: Immunohistochemical identification and mapping in neurons of the rat brain, *Proc. Natl. Acad. Sci. U.S.A.* **75:**6300–6304.

Jacobowitz, D. M., and Palkovits, M., 1974, Topographic atlas of catecholamine and acetylcholinesterase-containing neurons in the brain. I. Forebrain (telencephalon, diencephalon), *J. Comp. Neurol.* **157:**29–41.

Jacobowitz, D. M., O'Donohue, T. L., Chey, W. Y., and Chang, T.-M., 1981, Mapping of motilin-immunoreactive neurons of the rat brain, *Peptides* **2:**479–487.

Jacobowitz, D. M., Schulte, H., Chrousos, G. P., and Loriaux, D. M., 1983, Localization of CRF-like immunoreactive neurons in the rat brain, *Peptides* **4:**521–524.

Jansco, G., and Kiraly, E., 1980, Distribution of chemosensitive primary sensory afferents in the central nervous system of the rat, *J. Comp. Neurol.* **190:**781–792.

Jancso, G., Kiraly, E., and Jansco-Gabor, A., 1977, Pharmacologically induced selective degeneration of chemosensitive primary sensory neurons, *Nature* **270:**741–743.

Kaada, B. R., 1951, Somatomotor, autonomic and electrographic responses to electrical stimulation of "rhinencephalic" and other structures in primates, cat and dog, *Acta Physiol. Scand.* (Suppl.) **83:** 285.

König, J. F. R., and Klippel, R. A., 1963, *The Rat Brain. A Stereotaxic Atlas,* Krieger Publishing, New York.

Krettek, J. E., and Price, J. L., 1978, A description of the amygdaloid complex in the rat and cat with observations on intra-amygdaloid axonal connections, *J. Comp. Neurol.* **178:**255–279.

Lee, Y., Kawai, Y., Shiosaka, S., Takami, K., Kiyama, H., Hillyard, C. J., Girgis, S., MacIntyre, I., Emson, P. C., and Tohyama, M., 1985, Coexistence of calcitonin gene-related peptide and substance P-like peptide in single cells of the trigeminal ganglion of the rat: Immunohistochemical analysis, *Brain Res.* **330:**194–196.

Leroux, P., and Pelletier, G., 1984, Radioautographic localization of somatostatin-14 and somatostatin-28 binding sites in the rat brain, *Peptides 5*:503–506.

MacLean, P. D., and Delgado, J. M. R., 1953, Electrical and chemical stimulation of fronto-temporal portion of limbic system in the waking animal, *Electroencephalogr. Clin. Neurophysiol.* **5**:91–100.

Morris, H. R., Panico, M., Etienne, T., Tippins, J., Girgis, S. I., and MacIntyre, I., 1984, Isolation and characterization of human calcitonin gene-related peptide, *Nature* **308**:746–748.

Norgren, R., 1976, Taste pathways to hypothalamus and amygdala, *J. Comp. Neurol.* **166**:17–30.

Norgren, R., and Leonard, C. M., 1973, Ascending central gustatory pathways, *J. Comp. Neurol.* **150**:217–238.

Olschowka, J. A., O'Donohue, T. L., and Jacobowitz, D. M., 1981, The distribution of bovine pancreatic polypeptide-like immunoreactive neurons in rat brain, *Peptides* **2**:309–331.

Olschowka, J. A., O'Donohue, T. L., Mueller, G. P., and Jacobowitz, D. M., 1982, The distribution of corticotropin releasing factor-like immunoreactive neurons in rat brain, *Peptides* **3**: 995–1015.

Palkovits, M., 1973, Isolated removal of hypothalamic or other brain nuclei of the rat, *Brain Res.* **59**:449–450.

Palkovits, M., and Jacobowitz, D. M., 1974, Topographic atlas of catecholamine and acetylcholinesterase-containing neurons in the brain. II. Hindbrain (mesencephalon, rhombencephalon), *J. Comp. Neurol.* **157**:29–41.

Rodrigo, J., Polak, J. M., Fernandez, L., Ghatei, M. A., Mulderry, P., and Bloom, S. R., 1985, Calcitonin gene-related peptide immunoreactive sensory and motor nerves of the rat, cat, and monkey esophagus, *Gastroenterology* **88**:444–451.

Rosenfeld, M. G., Mermod, J.-J., Amara, S. G., Swanson, L. W., Sawchenko, P. E., Rivier, J., Vale, W. W., and Evans, R. M., 1983, Production of a novel neuropeptide encoded by the calcitonin gene via tissue-specific RNA processing, *Nature* **304**:129–135.

Rosenfeld, M. G., Amara, S. F., and Evans, R. M., 1984, Alternative RNA processing: Determining neuronal phenotype, *Science* **225**:1315–1320.

Scadding, J. W., 1980, The permanent anatomical effects of neonatal capsaicin on somatosensory nerves, *J. Anat.* **131**:473–484.

Shults, C. W., Quirion, R., Chronwall, B., Chase, T. N., and O'Donohue, T. L., 1984, A comparison of the anatomical distribution of substance P and substance P receptors in the rat central nervous system, *Peptides* **5**:1097–1128.

Skofitsch, G., and Jacobowitz, D. M., 1985a, Distribution of corticotropin releasing factor-like immunoreactivity in the rat brain by immunohistochemistry and radioimmunoassay: Comparison and characterization of ovine and rat/human CRF antisera, *Peptides* **6**:319–336.

Skofitsch, G., and Jacobowitz, D. M., 1985b, Immunohistochemical mapping of galanin-like neurons in the rat central nervous system, *Peptides* **6**:509–546.

Skofitsch, G., and Jacobowitz, D. M., 1985c, Calcitonin gene-related peptide: Detailed immunohistochemical distribution in the rat central nervous system, *Peptides* **6**:721–746.

Skofitsch, G., and Jacobowitz, D. M., 1985d, Quantitative distribution of calcitonin gene-related peptide in the rat central nervous system. *Peptides* **6**:1069–1073.

Skofitsch, G., and Jacobowitz, D. M., 1985e, Calcitonin gene-related peptide coexists with substance P in capsaicin sensitive neurons and ganglia of the rat, *Peptides* **6**:747–754.

Skofitsch, G., Jacobowitz, D. M., Eskay, R. L., and Zamir, N., 1985, Distribution of atrial natriuretic factor-like immunoreactive neurons in the rat brain, *Neuroscience* **16**:917–948.

Szolcsanyi, J., Jancso-Gabor, A., and Joo, F., 1975, Functional and fine structural characteristics of the sensory neuron blocking effect of capsaicin, *Naunyn Schmiedebergs Arch. Pharmacol.* **287**:157–163.

Tschopp, F. A., Tobler, P. H., and Fischer, J. A., 1984, Calcitonin gene-related peptide in the human thyroid, pituitary and brain, *Mol. Cell. Endocrinol.* **36**:53–57.

Tschopp, F. A., Henke, H., Petermann, J. B., Tobler, P. H., Janzer, R., Hökfelt, T., Lundberg, J. M., Cuello, C., and Fischer, J. A., 1985, Calcitonin gene-related peptide and its binding sites in the human central nervous system and pituitary, *Proc. Natl. Acad. Sci. U.S.A.* **82**:248–252.

Wisenfeld-Hallin, Z., Hökfelt, T., Lundberg, J. M., Forssmann, W. G., Reinecke, M., Tschopp, F. A., and Fischer, J. A., 1984, Immunoreactive calcitonin gene-related peptides and substance P coexist in sensory neurons to the spinal cord and interact in spinal behavioral responses of the rat, *Neurosci. Lett.* **52:**199–204.

Wynn, P. C., Hauger, R. L., Holmes, M. C., Millan, M. A., Catt, K. J., and Auguilera, G., 1984, Brain and pituitary receptors for corticotropin releasing factor: Localization and differential regulation after adrenalectomy, *Peptides* **5:**1077–1084.

19

Insulin Receptors in the Brain

DEREK LeROITH, WILLIAM LOWE,
ROBERT J. WALDBILLIG, CELESTE HART,
JEAN SIMON, JOSHUA SHEMER, JUAN C. PENHOS,
and MAXINE A. LESNIAK

1. INTRODUCTION

This review briefly outlines the studies demonstrating specific insulin receptors in brain, describes the recent studies on the structure and function of these receptors, and concludes with suggestions for future studies for exploring the function of insulin on nervous tissue.

Insulin is one of the major hormones regulating intermediary metabolism. Despite the enormous effort spent studying the mechanisms of insulin action at the target cell, clear understanding of the molecular events is still lacking. Binding of insulin to specific receptors is considered the essential first stage in this process. Furthermore, the insulin receptor itself contains the full program for initiation of the hormone-sensitive pathway of the cell.

In traditional terms, the central nervous system is not considered to be insulin sensitive. This is especially pertinent *in vivo* because of the lower permeability of the blood–brain barrier to insulin. However, the circumventricular regions, which lack a blood–brain barrier, provide access to plasma insulin. Also, the CSF, which contains insulin, provides indirect access. More recently several lines of evidence suggest the possible presence of a specific brain insulin–effector system. Insulin-specific receptors have been demonstrated in many regions of the brain, and insulin-related material has been isolated from brain and other neural elements (Havrankova *et al.*, 1978a,b, 1979; LeRoith *et al.*, 1983; Raizada *et al.*, 1982). These as well

DEREK LeROITH, WILLIAM LOWE, ROBERT J. WALDBILLIG, CELESTE HART, JEAN SIMON, JOSHUA SHEMER, JUAN C. PENHOS, and MAXINE A. LESNIAK • Diabetes Branch, National Institute of Arthritis, Diabetes, Digestive and Kidney Diseases, National Institutes of Health, Bethesda, Maryland 20892.

as other data have led a number of investigators to postulate various physiological roles for insulin on the nervous system including neurotransmission, neuromodulation, and even a major role in growth and development of nervous tissue (Rhoads *et al.*, 1984).

2. STUDIES ON IDENTIFICATION AND CHARACTERIZATION OF INSULIN RECEPTORS IN BRAIN

2.1. Binding Studies

Following the demonstration of specific binding of [^{125}I]insulin to membrane preparations from whole brain of rat, monkey, pigeon, and pig (Posner *et al.*, 1974; Landau *et al.*, 1983), specific insulin receptors were demonstrated on membrane preparations from various regions of the rat brain, with the highest concentrations of receptors in the olfactory bulb (Fig. 1, Table I). Furthermore, studies of insulin binding in rat brain tissue gave results indistinguishable from insulin receptors on classical target tissues (liver, fat, and muscle), i.e., pH, time, and temperature dependence as well as the specificity based on binding of insulin-related peptides and analogues (Fig. 2) (Havrankova *et al.*, 1978a,b, 1979). Similar to peripheral (nonneural) insulin receptors, studies of insulin binding to CNS receptors revealed curvilinear Scatchard plots (Van Schravendijk *et al.*, 1984; Zahniser *et al.*, 1984). In contrast, however, dissociation of labeled hormone from receptors was not accelerated when receptors were diluted with buffer containing excess unlabeled hormone. This was interpreted as representing two classes of insulin binding sites and not as "negative cooperativity," which has been described for classic peripheral nonnneural insulin receptors (Gammeltoft *et al.*, 1984a,b).

2.2. Cellular Localization Studies

Several groups have demonstrated specific insulin receptors in subcellular fractions enriched in synaptosomes or membrane vesicles isolated from different regions of rat brain (Table I). With *in vitro* and *in vivo* autoradiographic techniques using [^{125}I]insulin, specific binding has been more localized on predominantly neural elements of brain (Table I). More recently, binding has been demonstrated on primary cell cultures of glial cells from rat brain (Clarke *et al.*, 1984). Insulin receptors have also been found on brain microvessels (Van Houten and Posner, 1979; Frank and Pardridge, 1981). These studies corroborate the initial suggestions of wide distribution of insulin receptors on cellular elements of the CNS.

2.3. Regulation of CNS Insulin Receptors

Concentrations of insulin receptors on classic target tissues (liver, fat, and muscle) often are affected by the ambient insulin concentration. Down-regulation of receptors occurs in the presence of elevated insulin levels, and an increase in

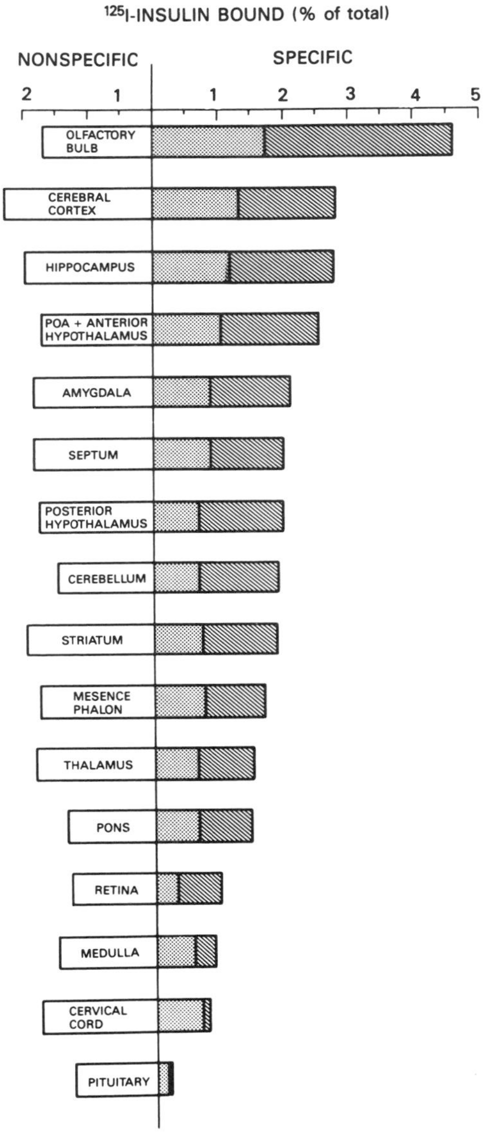

FIGURE 1. Regional distribution of the insulin receptor in the CNS of the rat. Membranes were exposed to porcine [^{125}I]insulin (130–180 μCi/μg) in the absence or presence of unlabeled insulin, 20 ng/ml and 100 μg/ml. The nonspecific binding (in the presence of 100 μg/ml of unlabeled insulin), indicated by empty bars to the left of the base line, was subtracted from total binding (shown as percentage of total radioactivity) to obtain specific binding, indicated by height of bars to the right of the base line. The stippled part represents the specific binding in the presence of 20 ng/ml of unlabeled insulin (indicating roughly the affinity of the receptors). The total, specific, and nonspecific binding were adjusted for each region to 500 μg of protein per milliliter. (From Havarankova, 1979, with permission.)

receptor number occurs when insulin levels are decreased. For example, hyperinsulinemia in the obese mouse caused a 60% decrease in liver insulin receptors, whereas hypoinsulinemic diabetic rats (induced with streptozotocin) had increased liver insulin receptors (Havrankova *et al.*, 1979; Pacold and Blackard, 1979). The number of brain insulin receptors did not, however, change despite these marked differences in plasma insulin concentrations, which may reflect isolation of most

TABLE I. Insulin Binding to Brain

Types of study	Major areas of insulin binding	References
Autoradiography		
In vivo		
Monkey	Median eminence, infundibular nucleus, microvessels	Landau *et al.* (1983)
Rat	Organum vasculosum, lamina terminalis, subfornical organ, median eminence, hypothalamus, arcuate nucleus, ventromedial nucleus, area postrema, paravagal region (binding specifically to dendrites)	van Houten and Posner (1981), van Houten *et al.* (1979)
	Microvessels in region of hypothalamus, hippocampus, neocortex	van Houten and Posner (1979)
In vitro		
Rat	Widespread throughout brain, maximum on external plexiform layer of olfactory bulb	Young *et al.* (1980)
Membrane/synaptosome preparations		
Rat	Olfactory bulb > cerebral cortex = hippocampus > hypothalamus > cerebellum > pituitary	Gammeltoft *et al.* (1984b), Havrankova *et al.* (1978a), Kappy *et al.* (1984), Pacold and Blackard (1979), Zahniser *et al.* (1984)
	Glial cells	Albrecht *et al.* (1982)
Pig and monkey	Hypothalamus > cortex = thalamus	Landau *et al.* (1983
Microvessel preparations		
Cow	Retina, cortex	Frank and Partridge (1981), Haskell *et al.* (1984), Pillion *et al.* (1982)
Primary cell cultures		
Mouse	Fetal cortical cells	Van Schravendijk *et al.* (1984)
Rat	Fetal and neonatal neurons and glial cells	Clarke *et al.* (1984), Raizada *et al.* (1982), Weyhenmeyer (1984)

regions of the brain by the blood–brain barrier or, alternatively, a different form of insulin receptor.

To circumvent the problem of the blood–brain barrier, studies were performed with direct exposure of neuron-enriched primary cultures of neonatal and adult rat brain as well as slices of rat brain to high insulin concentrations; insulin receptors failed to down-regulate (Boyd and Raizada, 1983; Zahniser *et al.*, 1984). On the other hand, studies have shown different results when studying young animals. Thus, intracerebral injection of insulin in newborn rabbit pups caused a 50% de-

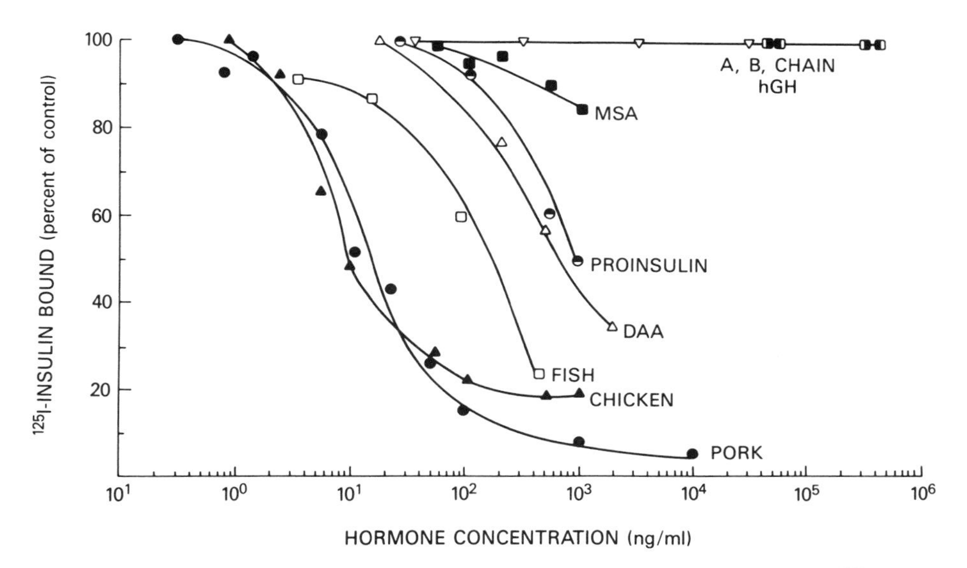

FIGURE 2. [^{125}I]Insulin binding to cerebral cortex membranes of the rat. Porcine [^{125}I]insulin (130–180 μCi/μg) and unlabeled peptides were used to compete for [^{125}I]insulin binding to membranes. Porcine insulin (●), chicken insulin (▲), fish insulin (□), desalanine-desasparagine insulin (DAA) (△), proinsulin (◐), multiplication-stimulating activity (MSA) (■), hGH (▽), A chain (◧), and B chain (◨) of insulin. The specific binding (total binding minus nonspecific binding) is expressed as a percentage of [^{125}I]insulin specifically bound in the presence of increasing concentrations of unlabeled peptide. (From Havrankova, 1979, with permission.)

crease in brain insulin receptor number (Devaskar and Holekamp, 1984), whereas exposure of dispersed murine fetal cortical cells to insulin resulted in an apparent increase in number of insulin receptors (Van Schravendijk *et al.*, 1984). In addition, fluctuations in insulin receptor concentrations in fetal and neonatal rat brain have been observed (Young *et al.*, 1980; Kappy *et al.*, 1984).

In summary, in contrast to extraneural receptors, the major regulators of insulin receptors in brain are unknown. Alterations in plasma insulin have no effect, but developmental events may be important.

3. STUDIES ON INSULIN RECEPTOR STRUCTURE–FUNCTION

3.1. General Structure of the Insulin Receptor

Recently, a uniform picture of the insulin receptor of classic tissues has emerged (Fig. 3). The insulin receptor is composed of two glycoprotein subunits: one subunit (α) with apparent molecular weight of 135,000 appears to be the binding site of insulin; a second smaller subunit (β) with an apparent molecular weight of 95,000 possesses a tyrosine-specific protein kinase, which can cause autophosphorylation

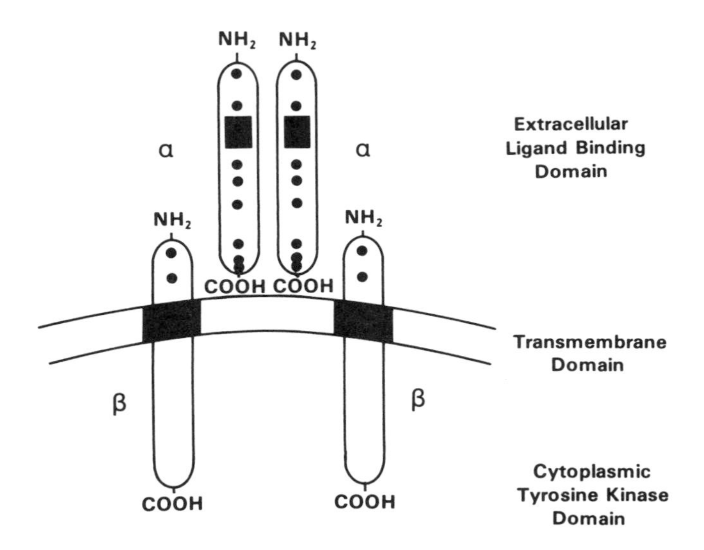

FIGURE 3. Schematic representation of the insulin receptor. The 135,000-molecular-weight (α) subunit binds insulin on the outer surface of the plasma membrane. The 95,000-molecular-weight (β) subunit is attached to the subunit by disulfide bonds and has tyrosine kinase activity on its cytoplasmic aspect. (From Perrotti *et al.,* 1985, with permission.)

and phosphorylation of other protein and peptide substrates. The two subunits are bound to one another by interchain disulfide bonds as well as noncovalent interactions to form a heterodimer. Binding of insulin to the α subunit activates the protein kinase activity of the β subunit (Kasuga *et al.,* 1982a,b; Van Obberghen, 1984; Perroti *et al.,* 1985).

3.2. Brain Insulin Receptors

3.2.1. α Subunit Structure

The α subunit of brain insulin receptor has been identified by either photolabeling it with light-sensitive insulin derivatives or covalently cross linking it with [^{125}I]insulin (Table II) (Heidenreich *et al.,* 1983; Hendricks *et al.,* 1984; Gammeltoft *et al.,* 1984a,b; Yip *et al.,* 1980). When examined under reducing conditions on SDS gel electrophoresis, the α subunit of the brain insulin receptors migrates with an apparent molecular weight significantly lower (~10,000) than the corresponding α subunits of nonneural (peripheral) insulin receptors derived from rat liver and fat as well as human IM-9 lymphocytes (Table II) (Heidenreich *et al.,* 1983; Hendricks *et al.,* 1984; Gammeltoft *et al.,* 1984a,b; Yip *et al.,* 1980). Some antireceptor antibodies are apparently capable of distinguishing between brain insulin receptors and liver insulin receptors (Yip *et al.,* 1980; Heidenreich *et al.,* 1983).

TABLE II. Comparison of Insulin Receptors from Brain and Nonneural Tissues

	Brain	Nonneural
Molecular weight		
Subunit α	≈125K	≈135K
Subunit β	≈85K	≈ 90K
Binding to lectin (wheat germ agglutinin)	±[a]	Yes
Neuraminidase alters apparent MW	No	Yes
Phosphorylation		
Of receptor	Yes	Yes
Of exogenous substrates	Yes	Yes
Recognition by anti-insulin-receptor antibodies	±[a]	Yes

[a] ±, conflicting results by two independent laboratories.

Other studies suggest that differences between peripheral (nonneural) and brain insulin receptors may reside in the glycosylation of the receptor. Neuraminidase, a glycosidase capable of cleaving terminal sialic acid residues, resulted in an apparent decrease in molecular weight of insulin receptors from liver and fat but had no effect on brain insulin receptors (Heidenreich *et al.*, 1983; Hendricks *et al.*, 1984). Tunicamycin, an antibiotic capable of inhibiting normal glycosylation of the receptor during biosynthesis, failed to affect insulin binding to primary rat neuronal cultures (Boyd and Raizada, 1983). Binding of brain insulin receptors to lectin (wheat germ agglutinin) was decreased in one study (Heidenreich *et al.*, 1983), though another group showed normal binding (Hendricks *et al.*, 1984). These preliminary studies suggest that the α subunit of the brain insulin receptor seems to differ from that found in peripheral (nonneural) tissues, possibly because of a difference in glycosylation.

3.2.2. β Subunit

Some studies of the β subunit of brain insulin receptors have demonstrated that its apparent molecular weight on SDS gel electrophoresis may be slightly lower than that of nonneural peripheral insulin receptors (Table II). Despite this apparent difference in molecular weight, insulin is capable of stimulating autophosphorylation of the β subunit as well as phosphorylation of exogenous substrates by brain insulin receptors (Gammeltoft *et al.*, 1984a,b; Rees-Jones *et al.*, 1984). The phosphorylation of exogenous substrates reveals predominant incorporation of ^{32}P on tyrosine residues (Rees-Jones *et al.*, 1984).

These studies suggest that the α and β subunits of brain insulin receptors are functionally coupled, as was seen in other tissues studied thus far; insulin binding to the α subunit results in stimulation of the receptor-associated tyrosine kinase on the β subunit. Furthermore, the apparent structural differences in the α subunits of the brain insulin receptors do not apparently affect coupling between α and β subunits.

4. CONCLUSIONS AND FURTHER STUDIES

As outlined in this review, specific insulin receptors are present on a variety of cellular elements in the brain. Structural and functional studies as well as attempts to regulate the receptors demonstrate differences between brain insulin receptors and insulin receptors on peripheral (nonneural) target tissues.

To date, however, the biological function of insulin in the brain has not been elucidated. Some have speculated that insulin may be involved in brain growth and development. Insulin is capable of stimulating ornithine decarboxylase activities in developing rat brains (Roger and Fellows, 1980) as well as thymidine and uridine incorporation in cells from primary cultures of fetal rat brain (Raizada *et al.*, 1980). In addition, insulin affects neuronal maturation and differentiation (Kessler *et al.*, 1984; Puro and Agardh, 1984).

The first step in the biological action of insulin is binding to the receptor. Whether the insulin-stimulated phosphorylation of the β subunit is an early integral part of the cascade of cellular action by insulin is not yet known. In an attempt to determine the function of insulin on brain, studies of the classic effects of insulin, such as glucose transport, as well as studies of the effect of insulin on neural elements, such as membrane ion flux, should be undertaken.

ACKNOWLEDGMENTS. Thanks to Violet Katz for excellent secretarial assistance and to Jesse Roth for reviewing the manuscript.

REFERENCES

Albrecht, J., Wroblewska, B., and Mossakowski, M. J., 1982, The binding of insulin to cerebral capillaries and astrocytes of the rat, *Neurochem. Res.* **7**:489–494.

Boyd, F. T., Jr., and Raizada, M. K., 1983, Effects of insulin and tunicamycin on neuronal insulin receptors in culture, *Am. J. Physiol.* **245**:C283–C287.

Clarke, D. W., Boyd, F. T., Jr., Kappy, M. S., and Raizada, M. K., 1984, Insulin binds to specific receptors and stimulates 2-deoxy-*d*-glucose uptake in cultured glial cells from rat brain, *J. Biol. Chem.* **259**:11672–11675.

Devaskar, S. U., and Holekamp, N., 1984, Insulin down-regulates neonatal brain insulin receptors, *Biochem. Biophys. Res. Commun.* **120**:359–367.

Frank, H. J. L., and Pardridge, W., 1984, A direct *in vitro* demonstration of insulin binding to isolated brain microvessels, *Diabetes* **30**:757–761.

Gammeltoft, S., Kowalski, A., Fehlmann, M., and van Obberghen, E., 1984a, Insulin receptors in rat brain: Insulin stimulates phosphorylation of its receptor β-subunit, *FEBS Lett.* **172**:87–90.

Gammeltoft, S., Staun-Olsen, P., Ottesen, B., and Fahrenkrug, J., 1984b, Insulin receptors in rat brain cortex. Kinetic evidence for a receptor subtype in the central nervous system, *Peptides* **5**:937–944.

Haskell, J. F., Meezan, E., and Pillion, D. J., 1984, Identification and characterization of the insulin receptor of bovine retinal microvessels, *Endocrinology* **115**:698–704.

Havrankova, J., Roth, J., and Brownstein, M., 1978a, Insulin receptors are widely distributed in the central nervous system of the rat, *Nature* **272**:827–829.

Havrankova, J., Schmechel, D., Roth, J., and Brownstein, M., 1978b, Identification of insulin in rat brain, *Proc. Natl. Acad. Sci. U.S.A.* **75**:5737–5741.

Havrankova, J., Roth, J., and Brownstein, M. J., 1979, Concentrations of insulin and of insulin receptors in the brain are independent of peripheral insulin levels, *J. Clin. Invest.* **64:**636–642.

Heidenreich, K. A., Zahniser, N. R., Berhanu, P., Brandenburg, D., and Olefsky, J. M., 1983, Structural differences between insulin receptors in the brain and peripheral target tissues, *J. Biol. Chem.* **258:**8527–8530.

Hendricks, S. A., Agardh, C.-D., Taylor, S. I., and Roth, J., 1984, Unique features of the insulin receptor in rat brain, *J. Neurochem.* **43:**1302–1309.

Kappy, M., Sellinger, S., and Raizada, M., 1984, Insulin binding in four regions of the developing rat brain, *J. Neurochem.* **42:**198–203.

Kasuga, M., Karlsson, F. A., and Kahn, C. R., 1982a, Insulin stimulates the phosphorylation of the 95,000 dalton subunit of its own receptor, *Science* **215:**185–187.

Kasuga, M., Hedo, J. A., Yamada, K. C., and Kahn, C. R., 1982b, The structure of insulin receptor and its subunits, *J. Biol. Chem.* **257:**10392–10399.

Kessler, J. A., Spray, D. C., Saez, J. C., and Bennett, M. V. L., 1984, Determination of synaptic phenotype: Insulin and cAMP independently initiate development of electronic coupling between cultured sympathetic neurons, *Proc. Natl. Acad. Sci. U.S.A.* **81:**6235–6239.

Landau, B. R., Takaoka, Y., Abrams, M. A., Genuth, S. M., van Houten, M., Posner, B. I., White, R. J., Ohgaku, S., Horvat, A., and Hemmelgarn, E., 1983, Binding of insulin by monkey and pig hypothalamus, *Diabetes* **32:**284–292.

LeRoith, D., Hendricks, S. A., Lesniak, M. A., Rishi, S., Becker, K. L., Havrankova, J., Rosenzweig, J. L., Brownstein, M. J., and Roth, J., 1983, Insulin in brain and other extrapancreatic tissues of vertebrates and non-vertebrates, *Adv. Metab. Dis.* **10:**303–340.

Pacold, S. T., and Blackard, W. G., 1979, Central nervous system insulin receptors in normal and diabetic rats, *Endocrinology* **105:**1452–1457.

Perrotti, N., Rees-Jones, R., and Taylor, S. I., 1985, The role of the receptor-associated protein kinase in the mechanism of insulin action and mechanisms of insulin resistance, in: *Diabetes, Obesity, and Hyperlipidemias* (G. Crepaldi, ed.), Elsevier, Amsterdam, pp. 443–450.

Pillion, D. J., Haskell, J. F., and Meezan, E., 1982, Cerebral cortical microvessels: An insulin-sensitive tissue, *Biochem. Biophys. Res. Commun.* **104:**686–692.

Posner, B. I., Kelly, P. A., Shiu, R. P. C., and Friesen, H. G., 1974, Studies of insulin, growth hormone and prolactin binding: Tissue distribution, species variation and characterization, *Endocrinology* **95:**521–531.

Puro, D. G., and Agardh, E., 1984, Insulin-mediated regulation of neuronal maturation, *Science* **225:**1170–1172.

Raizada, M. K., Yang, J. W., and Fellows, R. E., 1980, Binding of [^{125}I]insulin to specific receptors and stimulation of nucleotide incorporation in cells cultured from rat brain, *Brain Res.* **200:**389–400.

Raizada, M. K. Stamler, J. F. Quinlan, J. T., Landas, S., and Phillips M. I., 1982, Indentification of insulin receptor-containing cells in primary cultures of rat brain, *Cell. Mol. Neurobiol.* **2:**47-52.

Rees-Jones, R. W., Hendricks, S. A., Quarum, M., and Roth, J., 1984, The insulin receptor of rat brain is coupled to tyrosine kinase activity, *J. Biol. Chem.* **259:**3470–3474.

Rhoads, D. E., DiRocco, R. J., Osburn, L. D., Peterson, N. A., and Raghupathy, E., 1984, Stimulation of synaptosomal uptake of neurotransmitter amino acids by insulin: Possible role of insulin as a neuromodulator, *Biochem. Biophys. Res. Commun.* **119:**1198–1204.

Roger, I. J., and Fellows, R. E., 1980, Stimulation of ornithine decarboxylase activity by insulin in developing rat brain, *Endocrinology* **106:**619–625.

van Houten, M., and Posner, B. I., 1979, Insulin binds to brain blood vessels *in vivo, Nature* **282:**623–625.

van Houten, M., and Posner, B. I., 1981, Specific binding and internalization of blood-borne [^{125}I]-iodoinsulin by neurons of the rat area postrema, *Endocrinology* **109:**853–859.

van Houten, M., Posner, B. I., Kopriwa, B. M., and Brawer, J. R., 1979, Insulin-binding sites in the rat brain: *In vivo* localization to the circumventricular organs by quantitative radioautography, *Endocrinology* **105:**666–673.

Van Obberghen, E., 1984, The insulin receptor: Its structure and function, *Biochem. Pharmacol.* **33:**889–896.

Van Schravendijk, F. H., Hooghe-Peters, E. L., DeMeyts, P., and Pipeleers, D. G., 1984, Identification and characterization of insulin receptors on foetal-mouse brain-cortical cells, *Biochem. J.* **220:**165–172.

Weyhenmeyer, J. A., 1984, Localization of insulin receptors in primary cultures of fetal rat brain by immunocytochemistry, *Monogr. Neural Sci.* **10:**140–157.

Yip, C. C., Moule, M. L., and Yeung, C. W. T., 1980, Characterization of insulin receptor subunits in brain and other tissues by photoaffinity labeling, *Biochem. Biophys. Res. Commun.* **96:**1671–1678.

Young, W. S., Kuhar, M. J., Roth, J., and Brownstein, M. J., 1980, Radiohistochemical localization of insulin receptors in the adult and developing rat brain, *Neuropeptides* **1:**15–22.

Zahniser, N. R., Goens, M. B., Hanaway, P. J., and Vinych, J. V., 1984, Characterization and regulation of insulin receptors in rat brain, *J. Neurochem.* **42:**1354–1362.

20

Atrial Natriuretic Factors

The Brain–Heart Peptides

RÉMI QUIRION

1. INTRODUCTION

Various groups have recently reported on the existence of atrial natriuretic peptides (ANP) in secretorylike granules present in the myocytes of mammalian atria. These newly isolated peptides include the atrial natriuretic factors (ANF) (Grammer *et al.*, 1983; Thibault *et al.*, 1983a,b, 1984; Lazure *et al.*, 1984; Napier *et al.*, 1984a; Seidah *et al.*, 1984), atrial natriuretic polypeptides (Atlas *et al.*, 1984; Kangawa and Matsuo, 1984), auriculins (Atlas *et al.*, 1984), atriopeptins (Currie *et al.*, 1983, 1984; Geller *et al.*, 1984), cardionatrins (deBold and Flynn, 1983; deBold and Salerno, 1983; Flynn *et al.*, 1983), and cardiodilatin (Frossman *et al.*, 1983; Lazure *et al.*, 1984).

All these peptides are likely to be derived from a single multihormone precursor. Recent studies have shown that the cardiodilatin/ANP gene sequence contains a putative signal peptide at the N terminal followed by cardiodilatin, "spacers" flanked by arginine residues, and, finally, all the ANP/ANF-derived peptides at its C terminus (Greenberg *et al.*, 1984; Maki *et al.*, 1984; Nakayama *et al.*, 1984; Nemer *et al.*, 1984; Oikawa *et al.*, 1984; Seidman *et al.*, 1984; Yamanaka *et al.*, 1984; Zivin *et al.*, 1984). Very little is currently known on the processing of the ANP/ANF prohormone. Moreover, it is not clear if all or only a few of the isolated ANP are physiologically relevant.

2. DISTRIBUTION IN PERIPHERAL TISSUES

Various ANP-like peptides have been fully isolated from mammalian atria (Table I). Thus far, ANP-like peptides have been shown to be present in rat (deBold

RÉMI QUIRION • Douglas Hospital Research Centre and Department of Psychiatry, Faculty of Medicine, McGill University, Verdun, Quebec H4H 1R3, Canada. The literature has been reviewed as of May 1985.

TABLE I. Some of the ANP/ANF-like Peptides That Have Been Isolated

Name	Length (number of amino acids)
β-ANP$_{14-48}$	35
β-ANP$_{16-48}$	33
β-ANP$_{17-48}$	32
β-ANP$_{18-48}$	31
α-ANP or ANF	28
ANF$_{8-33}$	26
Auriculin A	24
Auriculin B	25
Atriopeptin I	21
Atriopeptin II	23
Atriopeptin III	24
Cardiodilatin	16

et al., 1981; Garcia *et al.*, 1982; Grammer *et al.*, 1983; Seidah *et al.*, 1984; Atlas *et al.*, 1984), mouse (Seidman *et al.*, 1984), rabbit (Trippodo *et al.*, 1984), frog (deBold and Salerno, 1983), monkey (Nakayama *et al.*, 1984), and human (Kangawa and Matsuo, 1984; Trippodo *et al.*, 1983; Nakayama *et al.*, 1984) atria. All these peptides are contained in the secretorylike granules of the atrial myocytes. Rat and human ANP-like peptides are strikingly similar (different in only one amino acid), demonstrating the preservation of structure during mammalian evolution.

Only trace amounts of ANP/ANF-like peptides are present in other peripheral tissues (deBold and Salerno, 1983). It has been shown that the kidney and the liver are capable of metabolizing ANP-like peptides (Tang et al., 1984). In the plasma, ANP/ANF-like peptide concentrations range between 50 and 400 pM (Gutkowska *et al.*, 1984; Tanaka *et al.*, 1984; Lang *et al.*, 1985) and markedly increase following volume loading (Lang *et al.*, 1985).

3. BIOLOGICAL EFFECTS IN PERIPHERAL TISSUES

3.1. Effects on the Kidney

DeBold and colleagues (1981) were among the first to demonstrate that materials present in atrial extracts induce marked natriuresis and diuresis in rats. Subsequently, various groups have reported similar effects: ANP/ANF-like peptides increase the excretion of water, sodium, chloride, and potassium in a dose-dependent manner (for review see Palluk *et al.*, 1985). Increases in renal blood flow (Borenstein *et al.*, 1983) and glomerular filtration rate have also been reported (Borenstein *et al.*, 1983; Atlas *et al.*, 1984; Yukimura *et al.*, 1984; Burnett *et al.*, 1984; Huang *et al.*, 1985).

The mechanism of action of ANF/ANP-like peptides in the kidney is not fully understood. However, it is clear that these renal hemodynamic effects are not mediated by catecholaminergic, cholinergic, dopaminergic, or histaminergic systems (Oshima *et al.*, 1984). It is possible that these effects could be associated with a modulation of calcium influx and/or cyclic nucleotide synthesis, especially cGMP (Grammer *et al.*, 1983; Hamet *et al.*, 1984; Tremblay *et al.*, 1985). Structure–activity studies have shown that α-human ANP > atriopeptin III > atriopeptin II > atriopeptin I in stimulating natriuresis (Tang *et al.*, 1984).

3.2. Effects on the Adrenal Gland

The ANP/ANF-like peptides inhibit the basal production of aldosterone by zona glomerulosa cells *in vitro* (Chartier *et al.*, 1984; Goodfriend *et al.*, 1984; DeLean *et al.*, 1984a; Kudo and Baird, 1984). These peptides also antagonize the stimulating effects of ACTH, angiotensin II, prostaglandins, and potassium on aldosterone secretion (Chartier *et al.*, 1984; DeLean *et al.*, 1984a; Kudo and Baird, 1984). It has been suggested that the inhibition of adenylate cyclase could mediate the inhibiting effect of ANP/ANF peptides on steroidogenesis (Anand-Srivastava *et al.*, 1985).

3.3. Effects on Blood Vessels and Blood Pressure

The ANP/ANF-like peptides exert powerful smooth muscle relaxant effects on various vascular and nonvascular tissues. The ANP-induced relaxation of precontracted vascular smooth muscle is dose dependent and appears to be the result of a direct action of these peptides on the tissues (Currie *et al.*, 1983; Grammer *et al.*, 1983; Atlas *et al.*, 1984; Atarashi *et al.*, 1984; Winquist *et al.*, 1984; Kleinert *et al.*, 1984; Garcia *et al.*, 1984). Moreover, ANP/ANF effects on blood vessels are likely to be via an endothelium-independent mechanism (Winquist *et al.*, 1984), possibly related to the stimulation of adenylate cyclase (Anand-Srivastava *et al.*, 1984; Winquist *et al.*, 1984).

The ANP/ANF-like peptides also decrease blood pressure in a dose-dependent manner (deBold *et al.*, 1981; Kangawa and Matsuo, 1984; Atlas *et al.*, 1984; Tang *et al.*, 1984). Structure–activity studies have shown that the C-terminal portion of the ANF/ANP molecule is essential for the induction of cardiovascular effects (Currie *et al.*, 1984; Sugiyama *et al.*, 1984; Tang *et al.*, 1984; Cantin *et al.*, 1985).

4. RECEPTORS IN PERIPHERAL TISSUES

Very little is currently known on ANP/ANF receptor binding sites in peripheral tissues. DeLean *et al.* (1984b) have recently demonstrated the presence of high-affinity $[^{125}I]ANF_{8-33}$ binding sites in bovine adrenal zona glomerulosa. These sites have very high affinity with an apparent K_d in the picomolar range. Various ions and nucleotides affected $[^{125}I]ANF_{8-33}$ binding in a dose-dependent manner. Unrelated peptides and neurotransmitters do not interact with $[^{125}I]ANF_{8-33}$ binding. Moreover, these sites are almost exclusively present in the zona glomerulosa with very low densities in other regions of the adrenal gland.

Napier *et al.* (1984b) have recently shown the existence of specific receptors for ANF in renal and vascular tissues. These sites appear to possess similar properties to those found in the bovine adrenal gland. Finally, Hirata *et al.* (1984) have reported the presence of specific $[^{125}I]\alpha$-human ANP binding sites in cultured vascular smooth muscle cells. The apparent dissociation constant was found to be in the low-nanomolar range, and various unrelated substances such as angiotensin II, bradykinin, substance P, neurotensin, glucagon, secretin, histamine, and serotonin were unable to compete for these sites, even at micromolar concentrations (Hirata *et al.*, 1984). These data clearly demonstrate the existence of specific receptor binding sites for ANP/ANF related peptides in various peripheral tissues.

5. RECEPTORS IN THE BRAIN

It is very well known that various peptides are simultaneously found in peripheral tissues (e.g., gut) and in the central nervous system. Moreover, it is possible that certain effects of peripherally released peptides could be centrally mediated, mostly through an action on circumventricular organs. For these two reasons, we decided to investigate the possible existence of specific receptors for ANF/ANP-like peptides in brain and related structures.

As we have recently reported, specific $[^{125}I]ANF_{8-33}$ binding sites are found in various regions of the rat brain (Quirion *et al.*, 1984). Using an autoradiographic technique, we have shown that high densities of sites are present in the external plexiform layer of the olfactory bulb, subfornical organ (Fig. 1), area postrema, and tractus solitarius. Much lower densities of sites are seen in remaining brain regions. White matter areas such as the choroid plexus, corpus callosum, and linings of the ventricles contain moderate densities of $[^{125}I]ANF_{8-33}$ binding sites (Quirion *et al.*, 1984).

The high densities of sites in areas relevant for the central regulation of water intake and blood pressure suggest that certain effects of ANP/ANF-like peptides could be centrally mediated. Since the subfornical organ and the area postrema are located outside the blood–brain barrier, plasma ANP-like peptides could act on these receptors and "centrally" modulate cardiovascular parameters. However, the presence of $[^{125}I]ANF_{8-33}$ binding sites in the olfactory bulb suggests the existence of ANP/ANF-like peptides in the brain (see below).

The presence of $[^{125}I]ANF_{8-33}$ binding sites in the choroid plexus and linings of ventricles could be related to a role of ANP/ANF-related peptides in the control of ion fluxes across the brain cerebrospinal fluid barrier. This is of interest especially in regard to the known effects of ANF on ion fluxes in the kidney (Keller, 1982; Oshima *et al.*, 1984). The eye is also enriched in $[^{125}I]ANF_{8-33}$ binding sites, principally in the ciliary body and its processes (Quirion *et al.*, 1984). These

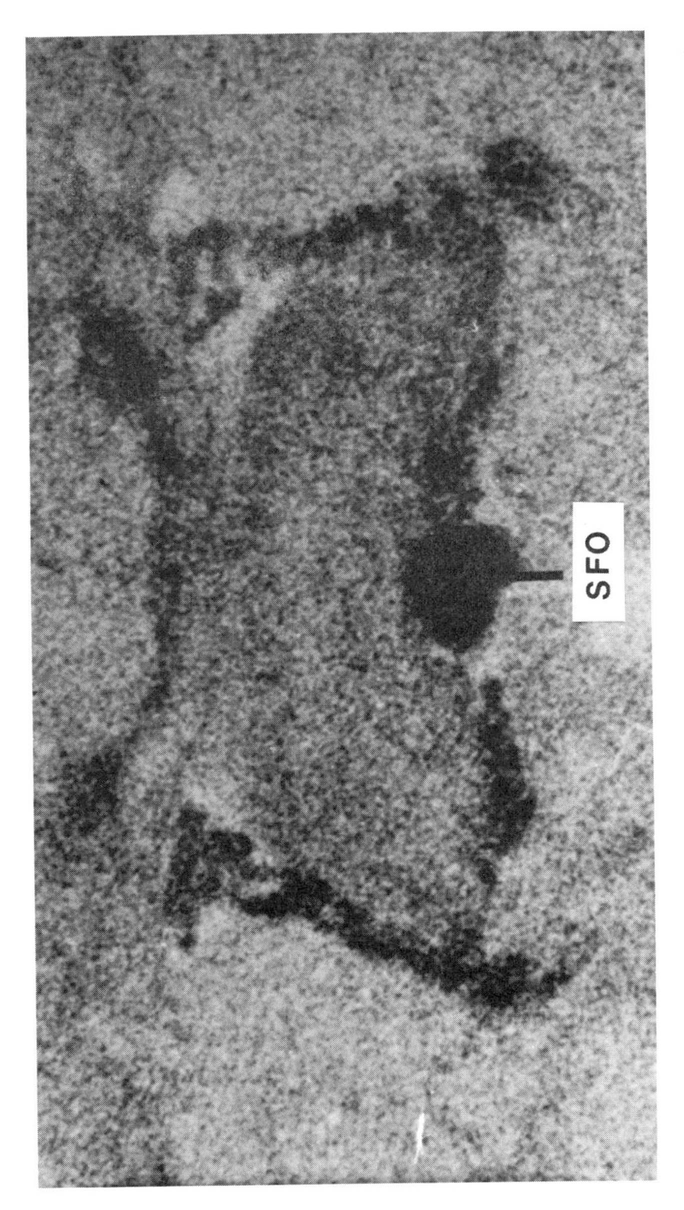

FIGURE 1. Autoradiographic localization of [^{125}I]ANF$_{8-33}$ (0.05 nM) binding sites in rat brain coronal section at the level of the subfornical organ (SFO). Note the high density of sites in the SFO. For experimental details, see Quirion *et al.* (1984).

structures are responsible for the production of fluids, and it has been shown that ANF-like peptides are important regulators of body fluid volume (deBold *et al.*, 1981; Garcia *et al.*, 1982; Geller *et al.*, 1984). Thus, ANP/ANF-like peptides could be important regulators of fluid volumes and ion fluxes in peripheral as well as brain tissues.

More recently, we used rat and human [^{125}I]ANP to characterize mammalian brain ANP receptors using a membrane binding assay and autoradiography (Quirion *et al.*, 1985). Both ligands bind with high affinity to an apparent single class of saturable sites in guinea pig brain membrane preparations. The respective affinities of the two ligands for these binding sites in the cerebellum and the thalamus/hypothalamus area are similar and in the picomolar range. The number of sites (B_{max}) in both brain regions is much lower than the one reported for the bovine adrenal zona glomerulosa (DeLean *et al.*, 1984b).

The relative potency of ANP/ANF-related peptides in displacing [^{125}I]ANP binding is shown in Table II: ANF_{8-33} is very potent, and atriopeptin III, atriopeptin II, and atriopeptin I are much less active. Thus, it is likely that the tyrosine residue at the C-terminal portion of the molecule is important for the binding of ANP/ANF-like peptides. Similar results have been obtained in various bioassays (Tang *et al.*, 1984; Sugiyama *et al.*, 1984; Cantin *et al.*, 1985). This suggests that the structural requirements of central and peripheral ANP/ANF receptors are fairly similar.

The distribution of [^{125}I]ANP binding sites in guinea pig brain is unique. High densities of sites are found in the olfactory bulb, subfornical organ (Fig. 2), paraventricular, paratenial, paracentral, and centrolateral nuclei of the thalamus (Figs. 2 and 3), medial geniculate nucleus (Fig. 4), and cerebellum (Fig. 5). Moderate to high densities of sites are present in the rostral portion of the central nucleus of the amygdala (Fig. 2), dorsal portion of the nucleus of the lateral olfactory tract (Fig. 2), granule cell layer of the dentate gyrus (Figs. 3 and 4), pyramidal cell layer of the hippocampus (Fig. 4), and area postrema. Most remaining areas such as the striatum, cortex, and spinal cord contain very low densities of sites (Figs. 2–5). White matter areas are practically devoid of [^{125}I]ANP binding sites. Thus, ANP/ANF binding sites are differentially distributed in rat and guinea pig brain

TABLE II. Relative Potency of Various ANP/ANF-Related Peptides in Guinea Pig Brain Membrane Binding Assays

Peptide	Relative potency[a]	
	Thalamus/hypothalamus	Cerebellum
Rat ANF_{8-33}	100	100
Human α-ANP	26	3.9
Rat atriopeptin III	3.9	2.7
Rat atriopeptin II	0.4	0.3
Rat atriopeptin I	<0.01	<0.01

[a] Relative potency of various peptides in displacing 0.05 nM [^{125}I]human α-ANP.

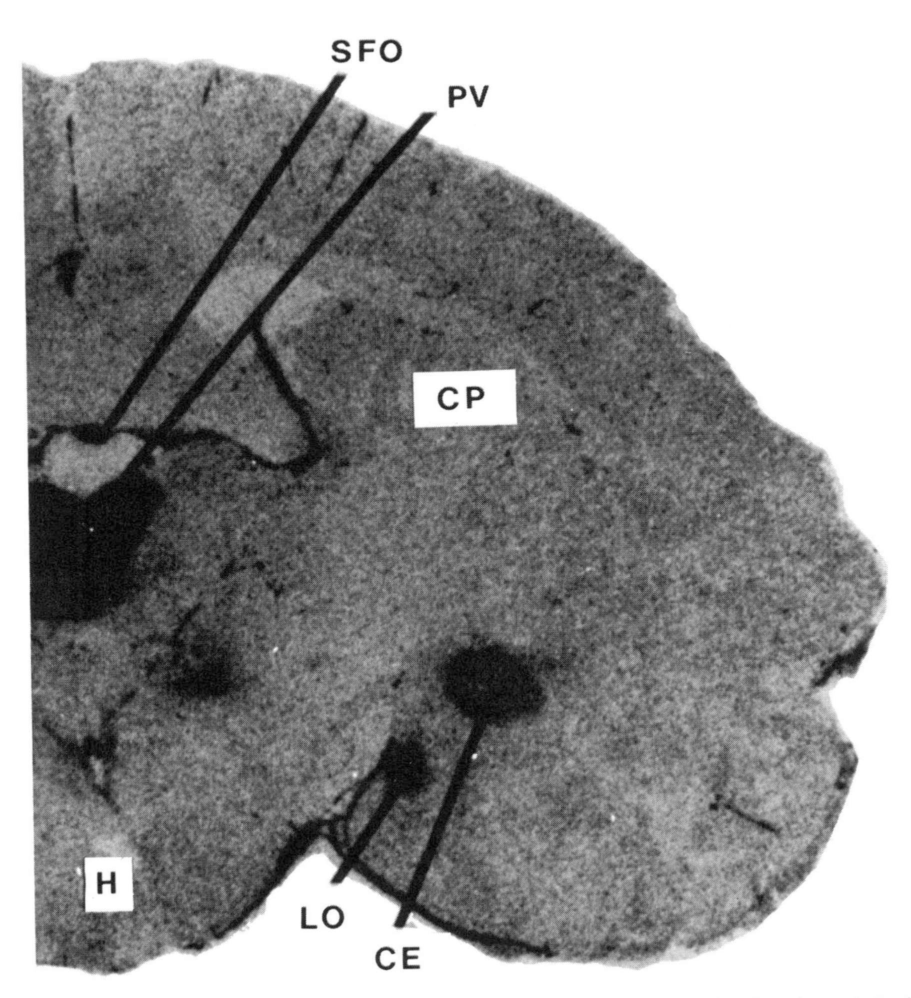

FIGURE 2. Autoradiographic localization of [^{125}I]ANP (0.05 nM) binding sites in guinea pig brain coronal section at the level of the subfornical organ (SFO). High densities of sites are present in the SFO and paraventricular nucleus of the thalamus (PV). Moderate to high densities are found in the rostral portion of the central nucleus of the amygdala (CE) and dorsal portion of the nucleus of the lateral olfactory tract (LO). Much lower densities are seen in the caudate–putamen (CP) and the hypothalamus (H). For experimental details, see Quirion *et al.* (1986).

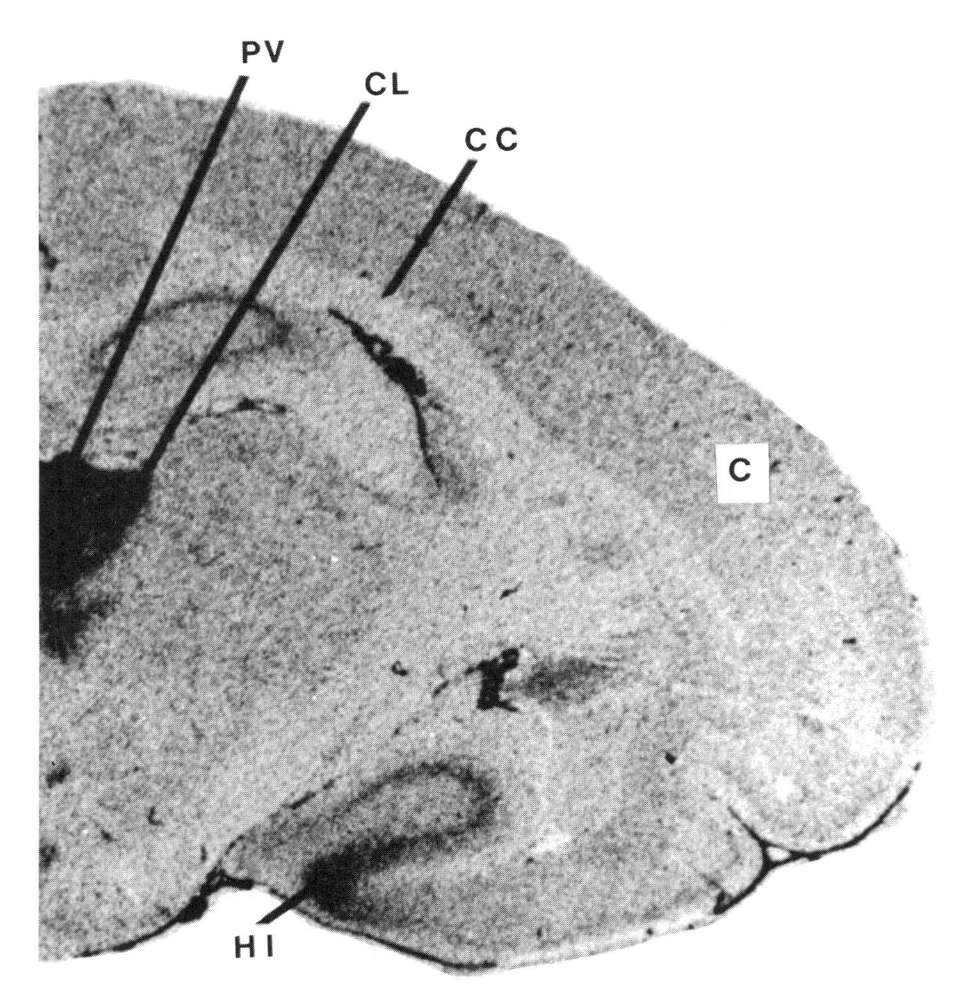

FIGURE 3. Autoradiographic localization of [^{125}I]ANP (0.05 nM) binding sites in guinea pig brain coronal section. High densities of sites are present in the paraventricular (PV) and centrolateral (CL) nuclei of the thalamus and in the hippocampus (HI). Much lower densities are found in the cortex (C). Very low densities are seen in white matter areas such as the corpus callosum (CC). For experimental details, see Quirion *et al.* (1986).

except in a few regions such as the olfactory bulb, subfornical organ, and area postrema.

The possible physiological relevance of ANP receptor binding sites in limbic structures such as the amygdala and in thalamus and cerebellum is totally unknown. However, studies on the possible effects of ANP/ANF-related peptides in the co-

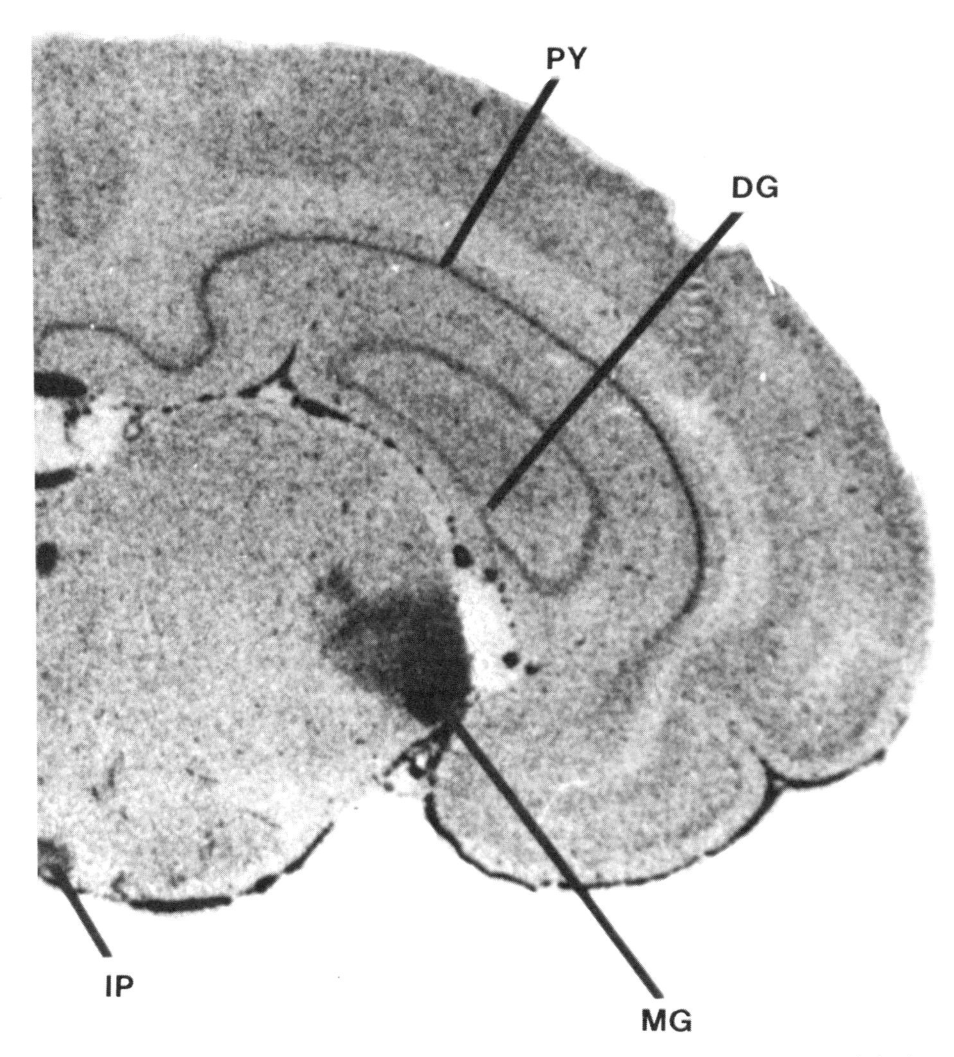

FIGURE 4. Autoradiographic localization of [^{125}I]ANP (0.05 nM) binding sites in guinea pig brain coronal section. Moderate to high densities of sites are present in the medial geniculate nucleus (MG), granule cell layer of the dentate gyrus (DG), and pyramidal cell layer of the hippocampus (PY). Moderate densities of sites are also present in the surrounding of the interpeduncular nucleus (IP). For experimental details, see Quirion *et al.* (1986).

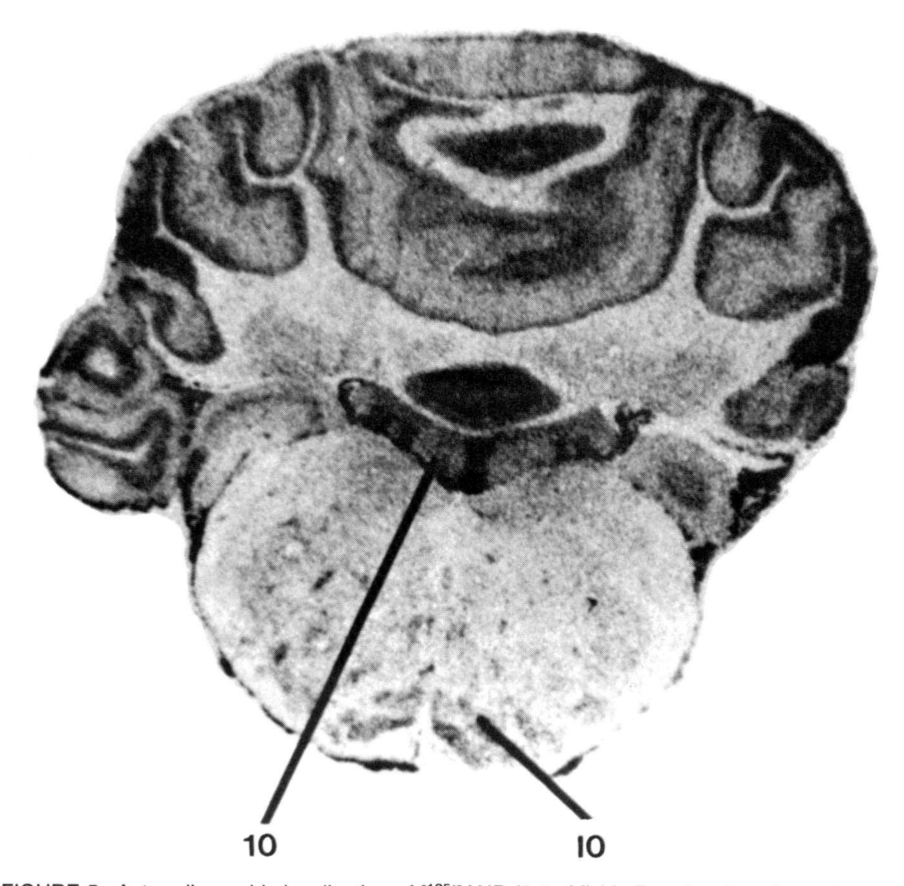

FIGURE 5. Autoradiographic localization of [^{125}I]ANP (0.5 nM) binding sites in guinea pig cerebellar sections. High densities of sites are present in the cerebellum, especially in lobules 9 and 10 (10). Low densities of sites are found in the inferior olive (IO). For experimental details, see Quirion *et al.* (1986).

ordination of movements, postural control, and certain sensory pathways are certainly warranted.

6. ATRIAL NATRIURETICLIKE FACTORS IN THE BRAIN

The presence of ANP/ANF receptor binding sites in various brain regions located inside the blood–brain barrier strongly suggests the existence of ANP-like peptides in the central nervous system. Already, two groups have reported on the presence of ANP/ANF-like immunoreactivity in rat brain (Jacobowitz *et al.*, 1985;

Saper *et al.*, 1985). Moreover, Gardner *et al.* (1985) have just demonstrated the presence of ANF mRNA in the rat hypothalamus. Thus, brain ANP/ANF-related peptides will be isolated and fully characterized in the near future.

7. CONCLUSION

In summary, highly specific ANP/ANF receptors are present in various regions of the mammalian brain, suggesting that certain of the biological effects of these peptides could be centrally mediated. Moreover, the presence of brain ANP/ANF receptors strongly suggests the existence of a new family of brain–heart peptides. The possible physiological relevance of these peptides and their receptors in the central nervous system should be an area of great interest for the next few years.

ACKNOWLEDGMENTS. This research was supported by the Medical Research Council of Canada and the Fonds de la Recherche en Santé du Québec (FRSQ). Rémi Quirion is a *Chercheur Boursier* of the FRSQ. We wish to thank Dr. K. G. McFarthing (Amersham International) for providing rat and human [^{125}I]ANP. The expert secretarial assistance of Mrs. Prema Jayaraman is acknowledged.

REFERENCES

Anand-Srivastava, M. B., Franks, D. J., Cantin, M., and Genest, J., 1984, Atrial natriuretic factor inhibits adenylate cyclase activity, *Biochem. Biophys. Res. Commun.* **121:**855–862.

Anand-Srivastava, M. B., Genest, J., and Cantin, M., 1985, Inhibitory effect of atrial natriuretic factor on adenylate cyclase activity in adrenal cortical membranes, *FEBS Lett.* **181:**199–202.

Atarashi, K., Mulrow, P. J., Franco-Saez, R., Snajdar, R., and Rapp, J., 1984, Inhibition of aldosterone production by an atrial extract, *Science* **224:**992–994.

Atlas, S. A., Kleinert, H. D., Camargo, M. J., Januszewicz, A., Sealey, J. E., Laragh, J. H., Schilling, J. W., Lewicki, J. A., Johnson, L. K., and Maack, T., 1984, Purification sequencing and syntheses of natriuretic and vasoactive rat atrial peptide, *Nature* **309:**717–719.

Borenstein, H. B., Cupples, W. A., Sonnenberg, H., and Veress, A. T., 1983, The effect of a natriuretic atrial extract on renal haemodynamics and urinary excretion in anesthetized rats, *J. Physiol. (Lond.)* **334:**130–140.

Burnett, J. C., Granger, J. P., and Opgenorth, T. J., 1984, Effects of synthetic atrial natriuretic factor on renal function and renin release, *Am. J. Physiol.* **247:**F863–866.

Cantin, M., Thibault, G., Garcia, R., Gutkowska, J., De Lean, A., Schiffrin, E., Seidah, N. G., Lazure, C., and Chretien, M., 1985, Structure–activity relationships of atrial natriuretic factor (ANF), *Clin. Res.* **33:**607A.

Chartier, L., Schiffrin, E., and Thibault, G., 1984, Effect of atrial natriuretic factor (ANF)-related peptides on aldosterone secretion by adrenal glomerulosa cells: Critical role of the intramolecular disulphide bond, *Biochem. Biophys. Res. Commun.* **122:**171–174.

Currie, M. G., Geller, D. M., Cole, B. R., Boylan, J. G., Wu, Y. S., Holmberg, S. W., and Needleman, P., 1983, Bioactive cardiac substance: Potent vasorelaxant activity in mammalian atria, *Science* **221:**71–73.

Currie, M. G., Geller, D. M., Cole, B. R., Siegel, N. R., Fok, F. K., Adams, S. P., Eubanks, S. R., Galluppi, G. R., and Needleman, P., 1984, Purification and sequence analysis of bioactive atrial peptides (atriopeptins), *Science* **223:**67–69.

deBold, A. J., and Flynn, T. G., 1983, Cardionatrin I—a novel heart peptide with potent diuretic and natriuretic properties, *Life Sci.* **33:**297–302.

deBold, A. J., and Salerno, T. A., 1983, Natriuretic activity of extracts obtained from hearts of different species and from various rat tissues, *Can. J. Physiol. Pharmacol.* **61:**127–130.

deBold, A. J., Borenstein, H. B., Veress, A. T., and Sonnenberg, H. A., 1981, Rapid and potent natriuretic response to intravenous injection of atrial myocardial extract in rats, *Life Sci.* **28:**89–94.

DeLean, A., Racz, K., Gutkowska, J., Nguyen, T. T., Cantin, M., and Genest, J., 1984a, Specific receptor-mediated inhibition by synthetic atrial natriuretic factor of hormone-stimulated steroidogenesis in cultural bovine adrenal cells, *Endocrinology* **115:**1636–1638.

DeLean, A., Gutkowska, J., McNicoll, N., Schiller, P. W., Cantin, M., and Genest, J., 1984b, Characterization of specific receptors for atrial natriuretic factor in bovine adrenal zona glomerulosa, *Life Sci.* **35:**2311–2318.

Flynn, T. G., deBold, M. L., and deBold, A. J., 1983, The amino acid sequence of an atrial peptide with potent diuretic and natriuretic properties, *Biochem. Biophys. Res. Commun.* **117:**859–865.

Frossman, W. G., Hock, D., Lottspeich, F., Henschem, A., Kreye, V., Christmann, M., Reinecke, M., Metz, J., Carlquist, M., and Mutt, V., 1983, The right auricle of the heart is an endocrine organ. Cardiodilatin as a peptide hormone candidate, *Anat. Embryol.* **168:**307–313.

Garcia, R., Cantin, M., Thibault, G., Ong, H., and Genest, J., 1982, Relationship of specific granules to the natriuretic and diuretic activity of rat atria, *Experientia* **38:**1071–1073.

Garcia, R., Thibault, G., Cantin, M., and Genest, J., 1984, Effect of a purified atrial natriuretic factor on rat and rabbit vascular strips and vascular beds, *Am. J. Physiol.* **247:**R34–R39.

Gardner, D. G., Lewecki, J. A., Fiddes, J. C., Metzler, C. H., Ramsay, D. J., Trachewsky, D., Hane, S., and Baxter, J. D., 1985, Preauriculin gene expression: Detection in extra-atrial tissues and regulation of cardiac and extracardiac mRNA levels by glucocorticoids and mineralocorticoids, *Clin. Res.* **33:**553A.

Geller, D. M., Currie, M. G., Wakitani, K., Cole, B. R., Adams, S. P., Fok, K. F., Seigal, N. R., Eubanks, S. R., Calluppi, G. R., and Needleman, P., 1984, Atriopeptins: A family of potent biologically active peptides derived from mammalian atria, *Biochem. Biophys. Res. Commun.* **120:**333–338.

Goodfriend, T. L., Elliott, M. E., and Atlas, S. A., 1984, Actions of synthetic atrial natriuretic factor on bovine adrenal glomerulosa, *Life Sci.* **35:**1675–1682.

Grammer, R. T., Fukumi, H., Inagami, T., and Misono, K. S., 1983, Rat atrial natriuretic factor. Purification and vasorelaxant activity, *Biochem. Biophys. Res. Commun.* **116:**696–703.

Greenberg, B. D., Bencen, G. H., Seilhamer, J. J., Lewicki, J. A., and Fiddes, J. C., 1984, Nucleotide sequence of the gene encoding human atrial natriuretic factor precursor, *Nature* **312:**656–658.

Gutkowska, J., Thibault, G., Januszewicz, P., Cantin, M., and Genest, J., 1984, Direct radioimmunoassay of atrial natriuretic factor, *Biochem. Biophys. Res. Commun.* **122:**593–599.

Hamet, P., Tremblay, J., Pang, S. C., Garcia, R., Thibault, G., Gutkowska, J., Cantin, M., and Genest, J., 1984, Effect of native and synthetic atrial natriuretic factor on cyclic GMP, *Biochem. Biophys. Res. Commun.* **123:**515–527.

Hirata, Y., Tomita, M., Yoshimi, H., and Ikeda, M., 1984, Specific receptors for atrial natriuretic factor (ANF) in cultured vascular smooth muscle cells of rat aorta, *Biochem. Biophys. Res. Commun.* **125:**562–568.

Huang, C. L., Lewicki, J., Johnson, L. K., and Cogan, M. G., 1985, Renal mechanism of action of rat atrial natriuretic factor, *J. Clin. Invest.* **75:**769–773.

Jacobowitz, D. M., Skofitsch, G., Keiser, H. R., Eskay, R. L., and Zamir, N., 1985, Evidence for the existence of atrial natriuretic factor-containing neurons in the rat brain, *Neuroendocrinology* **40:**92–94.

Kangawa, K., and Matsuo, H., 1984, Purification and complete amino acid sequence of α-human atrial natriuretic polypeptide (α-hANP), *Biochem. Biophys. Res. Commun.* **118:**131–139.

Keller, R., 1982, Atrial natriuretic factor has a direct, prostaglandin-independent action on kidneys, *Can. J. Physiol. Pharmacol.* **60:**1078–1082.

Kleinert, H. D., Maack, T., Atlas, S. A., Januszewicz, A., Sealey, J. E., and Laragh, J. H., 1984, Atrial natriuretic factor inhibits angiotensin-, norepinephrine-, and potassium-induced vascular contractility, *Hypertension* **6**(Suppl. I)**:**I143–I147.

Kudo, T., and Baird, A., 1984, Inhibition of aldosterone production in the adrenal glomerulosa by atrial natriuretic factor, *Nature* **312:**756–757.

Lang, R. E., Thölken, H., Ganten, D., Luft, F. C., Ruskoaho, H., and Unger, T., 1985, Atrial natriuretic factor, a circulating hormone stimulated by volume loading, *Nature* **314:**264–266.

Lazure, C., Seidah, N. G., Chrétien, M., Thibault, G., Garcia, R., Cantin, M., and Genest, J., 1984, Atrial pronatriodilatin: A precursor for natriuretic factor and cardiodilatin, *FEBS Lett.* **122:**80–86.

Maki, M., Takayanagi, R., Misono, K. S., Pandez, K. N., Tibbetts, C., and Inagami, T., 1984, Structure of rat atrial natriuretic factor precursor deduced from cDNA sequence, *Nature* **309:**722–724.

Nakayama, K., Ohkubo, H., Hirose, T., Inayama, S., and Nakanishi, S., 1984, mRNA sequence for human cardiodilatin–atrial natriuretic factor precursor and regulation of precursor mRNA in rat atria, *Nature* **310:**699–701.

Napier, M. A., Dewey, R. S., Albers-Schonberg, G., Bennett, C. D., Rodkey, J. A., Marsh, E. A., Whinnery, M., Seymor, A. A., and Blaine, E. H., 1984a, Isolation and sequence determination of peptide components of atrial natriuretic factor, *Biochem. Biophys. Res. Commun.* **120:**981–988.

Napier, M. A., Vandlen, R. L., Albers-Schonberg, G., Nutt, R. F., Brady, S., Lyle, T., Winquist, R., Faison, E. P., Heinel, L. A., and Blaine, E. H., 1984b, Specific membrane receptors for atrial natriuretic factor in renal and vascular tissues, *Proc. Natl. Acad. Sci. U.S.A.* **81:**5946–5950.

Nemer, M., Chamberland, M., Sirois, D., Argentin, S., Drouin, J., Dixon, R. A. F., Zivin, R. A., and Condra, J. H., 1984, Gene structure of human cardiac hormone precursor, pronatriodilatin, *Nature* **312:**654–656.

Oikawa, S., Imai, M., Uena, A., Tanaka, S., Noguchi, T., Nakazato, W., Kangawa, K., Fukuda, A., and Matsuo, H., 1984, Cloning and sequence analysis of cDNA encoding a precursor for human atrial natriuretic polypeptide, *Nature* **309:**724–726.

Oshima, T., Currie, M. G., Geller, D. M., and Needleman, P., 1984, An atrial peptide is a potent renal vasodilator substance, *Circ. Res.* **54:**612–616.

Palluck, R., Gaida, W., and Hoefke, W., 1985, Atrial natriuretic factor, *Life Sci.* **36:**1415–1425.

Quirion, R., Dalpé, M., De Lean, A., Gutkowska, J., Cantin, M., and Genest, J., 1984, Atrial natriuretic factor (ANF) binding sites in brain and related structures, *Peptides* **5:**1167–1172.

Quirion, R., Dalpé, M., and Dam, T. V., 1986, Characterization and distribution of receptors for the atrial natriuretic peptides in mammalian brain, *Proc. Natl. Acad. Sci. U.S.A.* **83:**174–178.

Saper, C. B., Standert, D. G., Currie, M. G., Schwartz, D., Geller, D. M., and Needleman, P., 1985, Atriopeptin-immunoreactive neurons in the brain: Presence in cardiovascular regulatory areas, *Science* **227:**1047–1049.

Seidah, N. G., Lazure, C., Chrétien, M., Thibault, G., Garcia, R., Cantin, M., Genest, J., Nutt, R. F., Brady, S. F., Lyle, T. A., Paleveda, W. J., Colton, C. D., Ciccarone, T. M., and Veber, D. F., 1984, Amino acid sequence of homologous rat atrial peptides: Natriuretic activity of native and synthetic forms, *Proc. Natl. Acad. Sci. U.S.A.* **81:**2640–2644.

Seidman, C. E., Duby, A. D., Choi, E., Graham, R. M., Haber, E., Homcy, C., Smith, J. A., and Seidman, J. G., 1984, The structure of rat preproatrial natriuretic factor as defined by a complementary DNA clone, *Science* **225:**324–326.

Sugiyama, M., Fukumi, H., Grammer, R. T., Misono, K. S., Yabe, Y., Morisawa, Y., and Inagami, T., 1984, Synthesis of atrial natriuretic peptides and studies on structural factors and tissue specificity, *Biochem. Biophys. Res. Commun.* **123:**338–344.

Tanaka, I., Misono, K. S., and Inagami, T., 1984, Atrial natriuretic factor in rat hypothalamus, atria and plasma: Determination by specific radioimmunoassay, *Biochem. Biophys. Res. Commun.* **124:**663–668.

Tang, J., Webber, R. J., Chang, D., Chang, J. K., Kiang, J., and Wei, E. T., 1984, Depressor and natriuretic activities of several atrial peptides, *Regul. Peptides* **9:**53–59.

Thibault, G., Garcia, R., Cantin, M., and Genest, J., 1983a, Atrial natriuretic factor. Characterization and partial purification, *Hypertension* **5**(Suppl. 1):175–180.

Thibault, G., Garcia, R., Seidah, N. G., Lazure, C., Cantin, M., Chrétien, M., and Genest, J., 1983b, Purification of three rat atrial natriuretic factors and their amino acid composition, *FEBS Lett.* **164**:286–290.

Thibault, G., Garcia, R., Cantin, M., Genest, J., Lazure, C., Seidah, N. G., and Chrétien, M., 1984, Primary structure of a high M_r form of rat atrial natriuretic factor, *FEBS Lett.* **167**:352–356.

Tremblay, J., Gerzer, R., Vinay, P., Pang, S. C., Beliveau, R., and Hamet, P., 1985, The increase in cGMP by atrial natriuretic factor correlates with the distribution of particulate guanylate cyclase, *FEBS Lett.* **181**:17–22.

Trippodo, N. C., MacPhee, A. A., and Cole, F. A., 1983, Partially purified human and rat atrial natriuretic factor, *Hypertension* **5**(Suppl. 1):181–188.

Trippodo, N. C., Ghai, R. D., MacPhee, A. A., and Cole, F. E., 1984, Atrial natriuretic factor: Atrial conversion of high to low molecular weight forms, *Biochem. Biophys. Res. Commun.* **119**:282–288.

Winquist, R. J., Faison, E. P., Waldman, S. C., Schwartz, K., Murad, F., and Rapoport, R. M., 1984, Atrial natriuretic factor elicits an endothelium-independent relaxation and activates particulate guanylate cyclase in vascular smooth muscle, *Proc. Natl. Acad. Sci. U.S.A.* **81**:7661–7664.

Yamanaka, M., Greenberg, B., Johnson, L., Seilhamer, J., Brewer, M., Freedemann, T., Miller, J., Atlas, S., Laragh, J., Lewicki, J., and Fiddes, J., 1984, Cloning and sequence analysis of the cDNA for the rat atrial natriuretic factor precursor, *Nature* **309**:719–722.

Yukimura, T., Ito, K., Takenaga, T., Yamamoto, K., Kangawa, K., and Matsuo, H., 1984, Renal effects of a synthetic α-human atrial natriuretic polypeptide (α-hANP) in anesthetized dogs, *Eur. J. Pharmacol.* **103**:363–366.

Zivin, R. A., Condra, J. H., Dixon, R. A. F., Seidah, N. G., Chrétien, M., Nemer, M., Chamberland, M., and Drouin, J., 1984, Molecular cloning and characterization of DNA sequences encoding rat and human atrial natriuretic factors, *Proc. Natl. Acad. Sci. U.S.A.* **81**:6325–6329.

21

Effects of Cations and Nucleotides on Neurotensin Binding to Rat Brain Synaptic Membranes

PATRICK KITABGI and JEAN-PIERRE VINCENT

1. INTRODUCTION

The tridecapeptide neurotensin (NT) (pGlu-Leu-Tyr-Glu-Asn-Lys-Pro-Arg-Arg-Pro-Tyr-Ile-Leu-OH) is thought to act as a neurotransmitter/modulator in the central nervous system (Nemeroff *et al.*, 1982). The existence of functional specific NT receptors that exist in both high- and low-affinity states in brain tissues has been well documented (Kitabgi *et al.*, 1985). In some cases, it has been possible to identify the neurons that bear NT receptors. For instance, there is strong evidence for the presence of somatodendritic NT receptors on dopaminergic neurons from the substantia nigra and the ventral tegmental area (Quirion, 1983). These receptors presumably mediate, at least in part, the regulatory effects that NT is thought to exert on the nigrostriatal and mesolimbic dopaminergic systems (Nemeroff and Cain, 1985).

If a great deal of information has accumulated regarding the pharmacology, biochemistry, and anatomic distribution of NT receptors in the central nervous system (Kitabgi *et al.*, 1985), little is known about the early molecular events that are generated on NT binding to its nerve cell receptors. Generally, the binding of a regulatory substance to its cell surface receptors regulates a second messenger system, which, in turn, triggers a series of intracellular biochemical reactions. There is now a large body of evidence that coupling between receptor occupancy and second messenger regulation occurs through a membrane component that belongs to a family of guanyl nucleotide binding proteins (N or G). Among the best known

PATRICK KITABGI and JEAN-PIERRE VINCENT • Centre de Biochimie du CNRS, Faculté des Sciences, Parc Valrose, 06034 Nice Cedex, France.

of these proteins are N_s and N_i, which respectively mediate effector-induced stimulation and inhibition of adenylate cyclase (Rodbell, 1980).

A characteristic of effector–receptor complexes that interact with N proteins is their sensitivity to cations and guanyl nucleotides (Rodbell, 1980). In the present study, we have investigated the effects of mono- and divalent cations and of nucleotides on the binding of NT to its receptors in rat brain synaptic membranes. These results show that cations and guanine nucleotides selectively modulate NT binding to its high-affinity receptors, thus suggesting an interaction of these receptors with a membrane-bound nucleotide binding protein.

2. MATERIALS AND METHODS

The radiolabeled ligand used here was monoiodo[^{125}I-Tyr^3]NT ([^{125}I]NT). It was prepared at specific radioactivities of 100 and 2000 Ci/mmol as previously described (Sadoul *et al.*, 1984). Binding assays were carried out at 25°C for 20 min in 250 μl of binding assay buffer (50 mM tris-HCl, pH 7.5, 0.2% bovine serum albumin) that contained appropriate concentrations of [^{125}I]NT and, when needed, cations, nucleotides, and unlabeled NT. The binding reaction was started by adding 50–100 μg of protein per assay tube of rat brain synaptic membranes prepared according to Jones and Matus (1974). Total, specific, and nonspecific binding were determined as previously reported (Mazella *et al.*, 1983; Sadoul *et al.*, 1984). Rat brain synaptic membranes have been shown to contain two classes of NT receptor sites (Mazella *et al.*, 1983; Sadoul *et al.*, 1984). The dissociation constant, K_d, and maximal binding capacity, B_m, for each class of sites were derived from computerized Scatchard analysis of data obtained from either saturation experiments using increasing concentrations of [^{125}I]NT or competition experiments in which varying concentrations of unlabeled NT were added to a fixed concentration (0.05–0.1 nM) of [^{125}I]NT. All cations were used as their chloride salt except Mg^{2+}, which was used as its sulfate salt.

3. RESULTS

3.1. Effects of Monovalent Cations

The binding of 0.1 nM [^{125}I]NT to rat brain synaptic membranes was inhibited in a concentration-dependent manner by Na^+ and to a lesser extent by K^+ and Li^+ (Fig. 1). Half-maximally effective concentrations (EC_{50}) were 45 mM for Na^+, 113 mM for K^+, and >150 mM for Li^+. Scatchard plots of NT binding in the absence of Na^+ or in the presence of 33 mM and 100 mM Na^+ are shown in Fig. 2. Consistent with previous data (Mazella *et al.*, 1983; Sadoul *et al.*, 1984), Scatchard plots were curvilinear, and this was interpreted as reflecting the existence of two classes of NT binding sites in rat brain synaptic membranes. Binding parameters

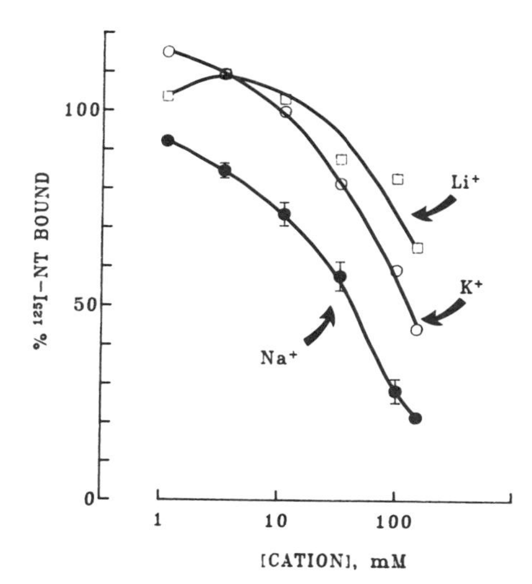

FIGURE 1. Effect of monovalent cations on the binding of 0.1 nM [^{125}I]NT to rat brain synaptic membranes. Data are expressed as percentage of [^{125}I]NT binding in the absence of cation. Points are mean ± S.E.M. from three experiments for Na^+ and mean from two experiments for Li^+ and K^+, with duplicate determinations within each experiment.

derived from Scatchard analysis are shown in Table I. At 33 mM, Na^+ induced a threefold increase in the dissociation constant K_d(h) for the high-affinity NT binding sites without affecting other binding parameters. At 100 mM, Na^+ further increased K_d(h) up to sixfold compared to control and also increased by threefold the dissociation constant K_d(l) for the low-affinity NT binding sites. There was no significant change in B_m values for either class of binding sites in the presence of 100 mM Na^+.

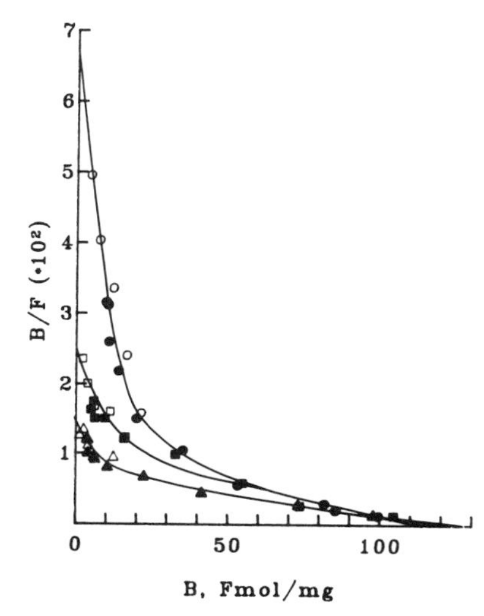

FIGURE 2. Scatchard analysis of [^{125}I]NT binding to rat brain synaptic membranes in the absence (○, ●) or presence of 33 mM (□, ■) and 100 mM (△, ▲) Na^+. Open symbols represent data points obtained by using increasing concentrations of [^{125}I]NT, and closed symbols are data points obtained in competition experiments with unlabeled NT (see Section 2). Data are from a typical experiment.

TABLE I. Effect of Na^+, Mg^{2+}, and GTP on NT Binding Parameters in Rat Brain Synaptic Membranes[a]

Binding parameter	Control	Na^+ 33 mM	Na^+ 100 mM	Mg^{2+} (5 mM)	GTP (0.1 mM)
$K_d(h)$[b] (nM)	0.11 ± 0.02	0.35 ± 0.11	0.60 ± 0.06	0.07	0.44 ± 0.03
$B_m(h)$ (fmol/mg)	14.7 ± 1.3	15.4 ± 3.1	14.8 ± 2.4	15.0	22.9 ± 1.4
$K_d(l)$[c] (nM)	7.2 ± 1.3	9.8 ± 3.1	21.7 ± 3.8	7.1	7.0 ± 1.4
$B_m(l)$ (fmol/mg)	111 ± 5	126 ± 16	152 ± 30	112	108 ± 8

[a] Values were derived from computer analysis of Scatchard plots such as those shown in Fig. 2. They represent mean ± S.E.M. from six control experiments and three experiments with Na^+ and GTP, and mean from two experiments with Mg^{2+}.
[b] (h) refers to high-affinity NT binding sites.
[c] (l) refers to low-affinity NT binding sites.

3.2. Effects of Divalent Cations

At low concentrations of divalent cations (0.2–5 mM), the binding of 0.1 nM [^{125}I]NT to synaptic membranes was increased by Mg^{2+} and Mn^{2+}, virtually unaffected by Ca^{2+}, and strongly inhibited by Co^{2+} (Fig. 3). High concentrations of divalent cations (>25 mM) resulted in a general, nonspecific inhibition of NT binding. The enhancing effect of Mg^{2+} on NT binding was further investigated. Scatchard analysis of binding experiments in the presence of 5 mM Mg^{2+} revealed that the cation decreased K_d(h) by about 40% without affecting other binding parameters (Table I).

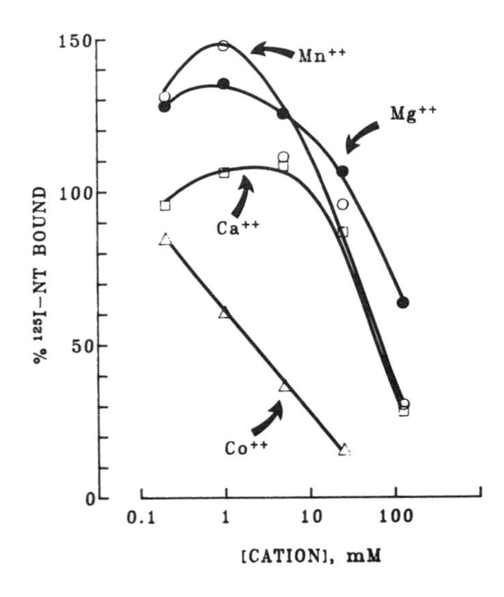

FIGURE 3. Effect of divalent cations on the binding of 0.1 nM [^{125}I]NT to rat brain synaptic membranes. Data are expressed as percentage of [^{125}I]NT binding in the absence of cation. Points are mean from two experiments with duplicate determinations within each experiment.

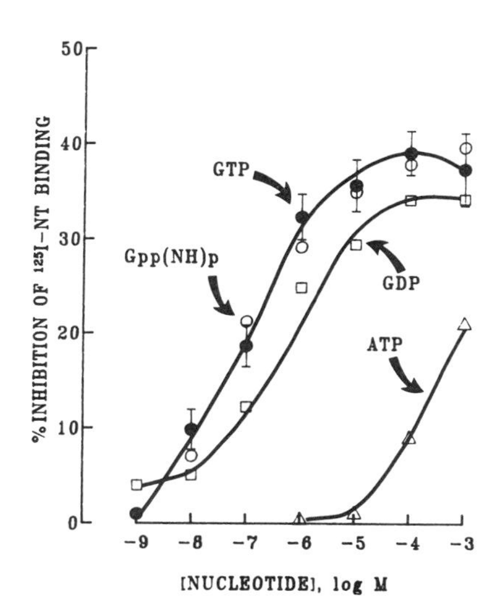

FIGURE 4. Effects of nucleotides on the binding of 0.1 nM [^{125}I]NT to rat brain synaptic membranes. Data are expressed as percentage inhibition of [^{125}I]NT binding in the absence of nucleotide. Points are mean ± S.E.M. from three experiments for GDP, Gpp(NH)p, and ATP with duplicate determinations within each experiment.

3.3. Effects of Nucleotides

GTP, GDP, and the nonhydrolyzable GTP analogue guanyl-5′-ylimidodiphosphate [Gpp(NH)p] inhibited in a concentration-dependent manner the binding of 0.1 nM [^{125}I]NT to synaptic membranes (Fig. 4). The EC_{50} values were 0.1 μM for GTP and Gpp (NH)p and 0.32 μM for GDP. For all three nucleotides, maximal inhibition was observed at 0.1 mM and reached 35–40%. ATP was about 10,000 times less efficient than GTP in inhibiting NT binding (Fig. 4). Other nucleotides, GMP, ADP, and App(NH)p, had little or no inhibitory effect. Scatchard analysis of NT binding to synaptic membranes in the presence of 0.1 mM GTP showed that the nucleotide induced a fourfold increase in K_d(h) and a 50% increase in B_m(h) without affecting the binding parameters of the low-affinity NT binding sites (Table I).

4. DISCUSSION

The data presented here clearly show that NT binding to rat brain synaptic membranes is inhibited by Na^+ and GTP and increased by Mg^{2+}. Furthermore, they show that these agents essentially modulate the affinity of NT for its high-affinity NT receptor and, in contrast, have little effect on the low-affinity NT receptor sites. Thus, low-affinity NT receptors are insensitive to Mg^{2+} and GTP and are sensitive only to high Na^+ concentrations. This may explain why no effect of GTP and Mg^{2+} could be observed in a study of [^{3}H]NT binding to rat brain membranes at ligand concentrations that saturated high-affinity NT receptors (Goedert *et al.*, 1984).

The specificity and properties of NT binding regulation by cations and nucleotides, as described here, strongly suggest that high-affinity NT receptors in brain interact with a synaptic membrane component analogous to members of a family of guanyl nucleotide binding proteins generally thought to regulate second messenger systems. In the introduction, we mentioned the existence of the now well characterized N_s and N_i involved in the regulation of adenylate cyclase activity (Rodbell, 1980; Gilman, 1984). Recent functional evidence suggests the possible existence of a GTP binding protein, putatively designated N_p, that would regulate the breakdown of phosphatidylinositol-4,5-bisphosphate (PIP_2) into inositol triphosphate and diacylglycerol, two cellular second messengers involved in Ca^{2+} mobilization and protein phosphorylation (Cockroft and Gomperts, 1985). Also, a GTP binding protein distinct from N_s and N_i has recently been isolated from bovine brain membranes (Sternweiss and Robishaw, 1984). The function of this protein, designated G_o, remains to be elucidated.

From the above discussion, the question arises as to whether high-affinity NT receptors in brain may interact with N_s or N_i, thereby modulating adenylate cyclase activity. Previous attempts to demonstrate a stimulatory or inhibitory effect of NT on brain adenylate cyclase have failed (P. Kitabgi, unpublished observations; Goedert *et al.*, 1984b). However, brain tissue is highly heterogeneous, and NT receptors are likely to represent only a minor proportion of all brain adenylate-cyclase-coupled receptors. It may therefore be inherently difficult to detect an effect of NT on brain adenylate cyclase. Such an effect, if there is any, would better be observed in a homogeneous NT-receptor-bearing population of cells. In this context, we have recently found that in the neuroblastoma N1E115 cell line (taken as a nerve cell model), which possesses high-affinity NT receptors (Poustis *et al.*, 1984), NT–receptor interaction leads to an inhibition of adenylate cyclase through coupling with N_i (Bozou *et al.*, 1986).

Alternatively, could high-affinity NT receptors interact with N (or G) proteins other than N_s and N_i such as N_p and G_o? An interaction with N_p should logically lead to breakdown of PIP_2. In support of this hypothesis, NT has been found to induce PIP_2 hydrolysis in rat brain slices (Goedert *et al.*, 1984b) and in N1E115 cells (our unpublished observations). Finally, the possibility exists that high-affinity NT receptors interact with a GTP binding protein of as yet unknown regulatory function such as G_o.

Obviously, much remains to be learned about the molecular events generated on binding of NT to its receptors. The present data show that high-affinity NT receptors likely interact with a GTP-binding regulatory protein present in brain synaptic membranes. Identification of this protein should provide essential information about the mechanisms by which NT transmits its message to nerve cells.

ACKNOWLEDGMENTS. We wish to thank G. Clénet for expert secretarial assistance. This work was supported by the Centre National de la Recherche Scientifique (ATP1057), the Institut National de la Santé et de la Recherche Médicale (CRE846021), and the Fondation pour la Recherche Médicale.

REFERENCES

Bozou, J. C., Amar, S., Vincent, J.-P., and Kitabgi, P., 1986, Neurotensin-mediated inhibition of cyclic AMP formation in neuroblastoma N1E115 cells: Involvement of the inhibitory GTP-binding component of adenylate cyclase, *Mol. Pharmacol.* (in press).

Cockroft, S., and Gomperts, B. D., 1985, Role of guanine nucleotide binding protein in the activation of polyphosphoinositide phosphodiesterase, *Nature* **314:**534–536.

Gilman, A. G., 1984, Guanine nucleotide-binding regulatory proteins and dual control of adenylate cyclase, *J. Clin. Invest.* **73:**1–4.

Goedert, M., Pittaway, K., Williams, B. J., and Emson, P. C., 1984a, Specific binding of tritiated neurotensin to rat brain membranes: Characterization and regional distribution, *Brain Res.* **304:**71–81.

Goedert, M., Pinnock, R. D., Downes, C. P., Mantyh, P. W., and Emson, P. C., 1984b, Neurotensin stimulates inositol phospholipid hydrolysis in rat brain slices, *Brain Res.* **323:**193–197.

Jones, D. H., and Matus, A. I., 1974, Isolation of synaptic plasma membrane from brain by combined flotation–sedimentation density gradient centrifugation, *Biochim. Biophys. Acta* **356:**276–287.

Kitabgi, P., Checler, F., Mazella, J., and Vincent, J.-P., 1986, Pharmacology and biochemistry of neurotensin receptors, *Rev. Clin. Basic Pharmacol.* (in press).

Mazella, J., Poustis, C., Labbé, C., Checler, F., Kitabgi, P., Granier, C., Van Rietschoten, J., and Vincent, J.-P., 1983, Monoiodo-Trp^{11} neurotensin, a highly radioactive ligand of neurotensin receptors. Preparation, biological activity and binding properties to rat brain synaptic membranes, *J. Biol. Chem.* **258:**3476–3481.

Nemeroff, C. B., and Cain, S. T., 1985, Neurotensin–dopamine interactions in the central nervous system, *Trends Pharmacol. Sci.* **6:**201–205.

Nemeroff, C. B., Luttinger, D., and Prange, A. J., 1982, Neurotensin and bombesin, *Handbook Psychopharmacol.* **16:**363–467.

Poustis, C., Mazella, J., Kitabgi, P., and Vincent, J.-P., 1984, High affinity neurotensin binding sites in differentiated neuroblastoma N1E115 cells, *J. Neurochem.* **42:**1094–1100.

Quirion, R., 1983, Interactions between neurotensin and dopamine in the brain: An overview, *Peptides* **4:**609–614.

Rodbell, M., 1980, The role of hormone receptors and GTP-regulatory proteins in membrane transduction, *Nature* **284:**17–22.

Sadoul, J. L., Mazella, J., Amar, S., Kitabgi, P., and Vincent, J.-P., 1984, Preparation of neurotensin selectively iodinated on the tyrosine 3 residue. Biological activity and binding properties on mammalian neurotensin receptors, *Biochem. Biophys. Res. Commun.* **120:**812–819.

Sternweiss, P. C., and Robishaw, J. D., 1984, Isolation of two proteins with high affinity for guanine nucleotides from membranes of bovine brain, *J. Biol. Chem.* **259:**13806–13813.

22

Neuropeptides Modulate Carbachol-Stimulated "Boxing" Behavior in the Rat Medial Frontal Cortex

JACQUELINE N. CRAWLEY, JILL A. STIVERS, and DAVID M. JACOBOWITZ

1. INTRODUCTION

The discovery of neuropeptides in mammalian brain was followed by the iconoclastic revelation that peptides coexist in the same neuron with classical neurotransmitters (Hökfelt *et al.*, 1984). In 1983, coexistence of substance P (SP) and acetylcholinesterase (AchE) or choline acetyltransferase (ChAt) was described in neurons of the nucleus tegmentalis dorsalis lateralis (NTDL) in rat brain (Vincent *et al.*, 1983; Olschowka *et al.*, 1983). Preliminary data suggested that some of the NTDL cells also contained corticotropin-releasing factor (CRF) (Olschowka *et al.*, 1983). In addition, immunohistochemical studies have shown that SP- and CRF-containing cells in the NTDL project to the medial frontal cortex, lateral septum, and thalamus (Crawley *et al.*, 1985a).

We have been interested in the functional significance of coexisting peptides and transmitters in the central nervous system. In particular, behavioral techniques in rats have been applied to the investigation of peptide–transmitter coexistences in discrete neuroanatomic sites (Crawley *et al.*, 1985b). Our approach involves cannulating the postsynaptic nucleus containing the nerve terminals of pathways in which neuropeptides and neurotransmitters coexist. Behavioral actions of the transmitter, the peptide(s), and combinations of transmitter and peptide(s), injected directly into the postsynaptic site, are then evaluated to test for potential interactions between the behavioral effects of the transmitter and the peptide(s).

JACQUELINE N. CRAWLEY and JILL A. STIVERS • Clinical Neuroscience Branch, National Institute of Mental Health, National Institutes of Health, Bethesda, Maryland 20205 DAVID M. JACOBOWITZ • Laboratory of Clinical Science, National Institute of Mental Health, National Institutes of Health, Bethesda, Maryland 20205.

2. METHODS AND RESULTS

2.1. Immunocytochemistry

2.1.1. Coexistence of SP and CRF

The triple coexistence of SP, CRF, and AchE was revealed by a combination of imunofluorescence techniques. Male Sprague–Dawley rats (290–340 g body weight) were treated with colchicine (100 μg in 25 μl i.c.) 48 hr before perfusion through the ascending aorta with 75 ml of ice-cold phosphate-buffered saline (PBS) containing 0.5% sodium nitrite followed by 200–300 ml of ice-cold 10% formalin in PBS. The brain was rapidly removed, cut into 3-mm slices, and postfixed 30 min in the fixative. The tissue slices were rinsed for 48 hr in PBS with 20% sucrose (w/v), frozen on dry ice, cut into 16-mm coronal sections in a cryostat, mounted on chrom–alum-coated slides, and processed for indirect immunofluorescence. The antibody to CRF was raised in rabbits against synthetic ovine CRF coupled to bovine thyroglobulin as described previously (Olschowka *et al.*, 1982). The antibody to SP was purchased from ImmunoNuclear, Inc. and Peninsula Labs. All antisera have been demonstrated to be specific for their respective antigens following incubation of the antisera with 1 μM of the two peptides. Cryostat sections were first stained for CRF immunofluorescence and photographed. The first antiserum was eluted, and the sections were then restrained for SP immunofluorescence by the method of Tramu *et al.* (1978). Figure 1 illustrates cell bodies of the NTDL containing both SP and CRF.

2.1.2. Coexistence of AchE and Peptides (SP, CRF)

The coexistence of AchE and SP and of AchE and CRF were demonstrated by serial staining for peptide followed by acetylcholinesterase (AchE) (Jacobowitz and Creed, 1983; modification of Koelle, 1955). Following peptide staining, the NTDL region was photographed, landmarks recorded, coverslip carefully removed, and slides washed in PBS (2 min). The sections were then processed for AchE histochemistry and rephotographed. The NTDL was found to contain a dense cluster of AchE-positive cells, of which 10–25% contained SP (Fig. 2). There appeared to be somewhat fewer AchE cells that also contained CRF (Fig. 3).

2.1.3. Retrograde Tracing Studies

The projections of the NTDL neurons were determined by combining the use of the retrograde fluorescence tracers true blue or fast blue with indirect immunofluorescence for SP and CRF (Olschowka *et al.*, 1983; Crawley *et al.*, 1985a). Following dye injections into the medial frontal cortex, septum, or thalamus, moderate numbers of cells within the ipsilateral NTDL were labeled, of which 20–50%

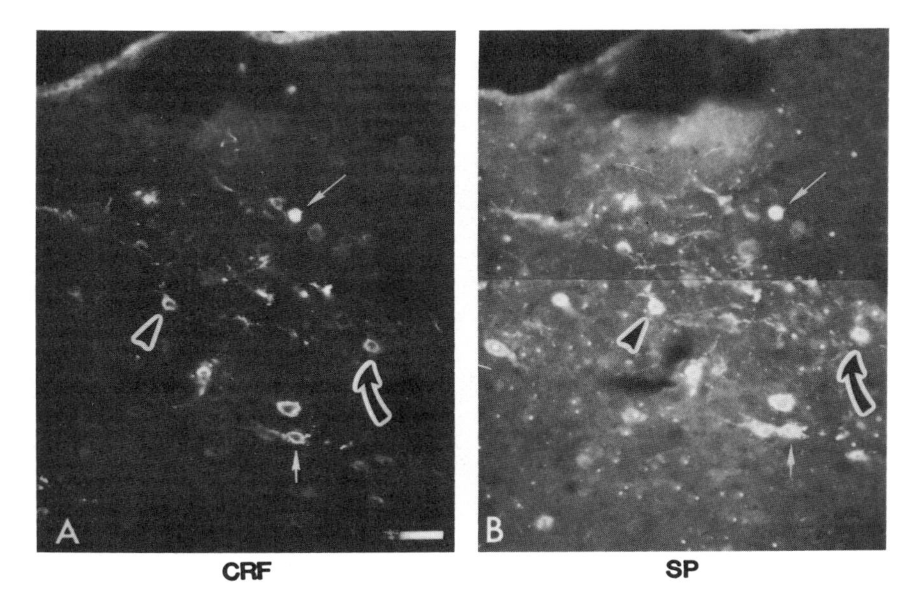

FIGURE 1. Immunofluorescence photomicrographs of the same cryostat section of the NTDL of the rat. (A) FITC marker for corticotropin-releasing factor (CRF)-positive neurons. (B) Rhodamine marker for substance P (SP)-positive neurons. The section was first stained with CRF antibody and photographed; the CRF antiserum was then eluted off, and the section was then stained with SP antibody and rephotographed. Note that the CRF-positive neurons also contain SP immunoreactivity (see corresponding arrows). Bar: 50 μm.

were observed to contain SP or CRF immunoreactivity. Dye injections into the cingulate or parietal cortices, areas with no SP innervation, failed to label NTDL neurons (Olschowka *et al.*, 1983; Crawley *et al.*, 1985a). Taken together, these studies provided evidence that SP, CRF, and AchE coexist within a subpopulation of NTDL neurons and that these cells project to a number of forebrain regions.

2.2. Behavioral Studies

2.2.1. Carbachol, SP, and CRF in Rat Medial Frontal Cortex

To investigate possible behavioral interactions among SP, CRF, and Ach in terminal regions of the NTDL, indwelling cannulae were implanted bilaterally in one of the terminal sites, the medial frontal cortex, of male Sprague–Dawley rats. Figure 4 illustrates the placement of 24-gauge hypodermic stainless steel cannulae closed by 31-gauge stylets (Small Parts, Inc., Miami, FL) at stereotactic coordinates of AP + 3.8, LAT ± 0.08, VERT −3.5 mm from bregma (Paxinos and Watson,

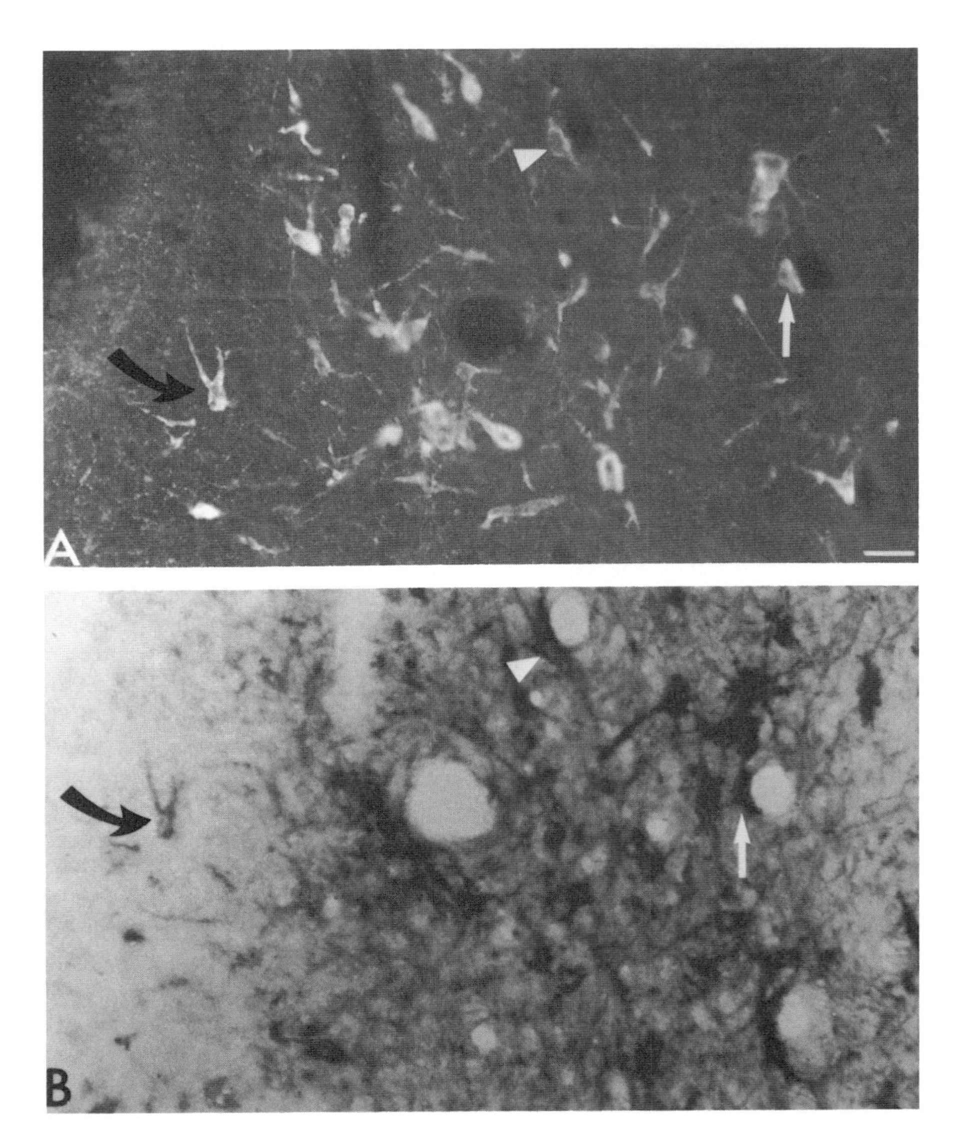

FIGURE 2. Consecutive staining of SP (A) and AchE (B) in the same section of the NTDL. Identical SP-immunoreactive and AchE-containing cell bodies are indicated by corresponding arrows. Size bar: 50 μm.

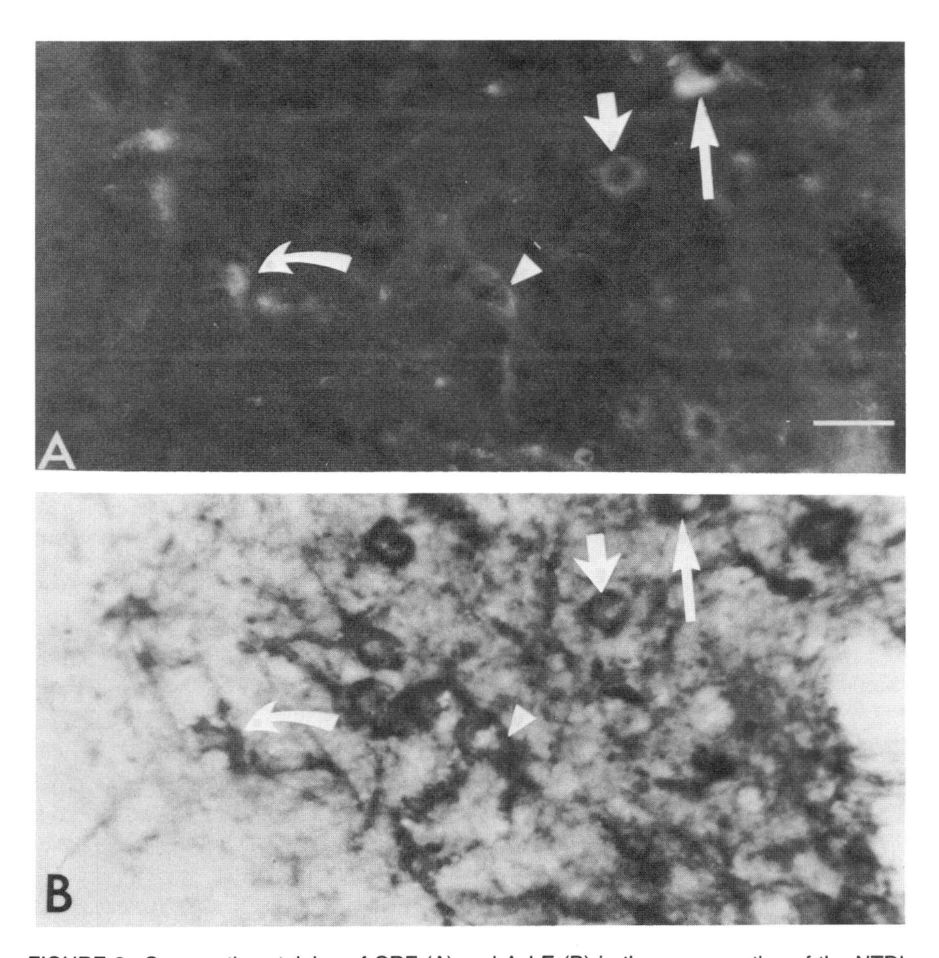

FIGURE 3. Consecutive staining of CRF (A) and AchE (B) in the same section of the NTDL. Identical CRF-immunoreactive and AchE-containing cell bodies are indicated by corresponding arrows. Size bar: 50 μm.

1982) under sodium pentobarbital anesthesia. One week after surgery, awake, gently restrained rats were injected through the implanted cannulae using a 31-gauge injection tube inserted 1 mm ventral to the ventral tip of the implanted cannula with a volume of 0.2 ml over a 1-min period using a Sage microinfusion pump (Orion Research, Inc., Cambridge, MA). Each rat was used only once. At the end of each experiment, fast green dye was identically injected, and brains were histologically processed to evaluate the location of the guide cannula and site of the injection.

In the first series of experiments, SP or CRF (Peninsula Laboratories, Belmont,

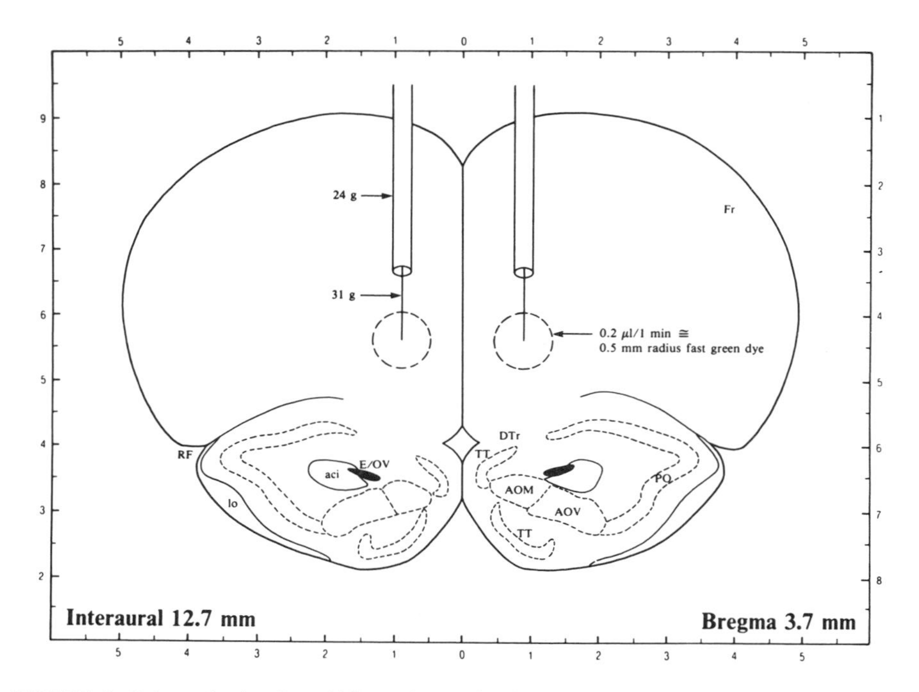

FIGURE 4. Schematic drawing of bilateral cannula placements in the medial anterior cortex of the rat. Stereotactic coordinates, from Paxinos and Watson (1982), were AP + 3.2, LAT ± 0.8, VERT − 3.5 mm from bregma. Injections were administered through a 31-gauge injection tube inserted through the 24-gauge implanted outer barrel in a volume of 0.2 μl over a 1-min period.

CA), saline, or the long-lasting Ach analogue carbachol (Sigma Chemical Co., St. Louis, MO) was bilaterally microinjected into the medial frontal cortex over a wide dose range. Beginning immediately after injection and ending 2 hr after injection, a wide range of behaviors were scored, including grooming, sniffing, locomotion, exploration, vertical movements, ptosis, resting bouts, resting postures, limb orientation, tail orientation, and forepaw movements. No spontaneous behaviors were induced by either SP or CRF over a dose range of 1 ng to 5 μg injected bilaterally into the medial frontal cortex. A dramatic behavioral syndrome was observed after administration of carbachol into the medial frontal cortex. Rats reared up on their hindlegs to a vertical posture and rapidly and repetitively moved their forepaws in the air for bouts of 5–30 sec. The vertical posture and repetitive forepaw treading were nicknamed "boxing." Figure 5 shows that doses of 1 μg to 5 μg of carbachol per bilateral microinjection elicited dose-related amounts of "boxing," scored as cumulative number of seconds by an observer uninformed of the treatment condition, for up to 90 min after injection. No "boxing" was ever observed after injection of saline, SP, or CRF into the medial frontal cortex at any dose.

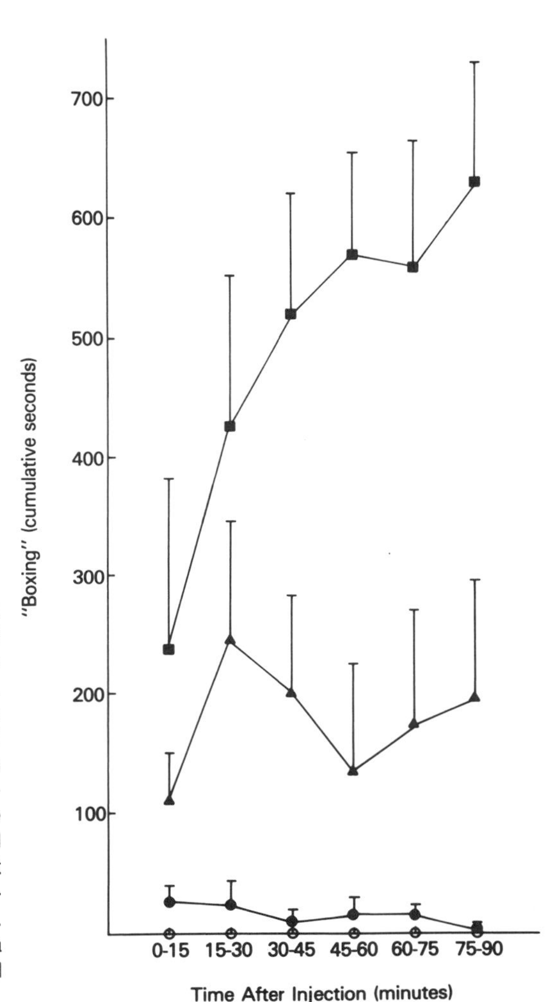

FIGURE 5. "Boxing" behavior elicited by carbachol, 1 μg (●), 2 μg (▲), or 5 μg (■), microinjected bilaterally into the medial anterior cortex (○, saline). "Boxing" consisted of bouts in which the rat reared up on his hindlegs and repetitively treaded the air with his forepaws. Cumulative number of seconds per 15-min interval spent in the "boxing" mode was dose related, and "boxing" persisted for over 90 min at high doses. No "boxing" was observed after microinjection of saline. $N = 4$ for each treatment group. Data are expressed as mean $\pm$ standard error of the mean in Figs 5–9.

2.2.2. Cholinergic Mediation of Boxinglike Seizures in Rat Medial Frontal Cortex

The cholinergic receptor mediating carbachol-induced "boxing" was investigated by pretreating rats with the muscarinic antagonist atropine, 5 mg/kg i.p., or the nicotinic antagonist mecamylamine, 3 mg/kg s.c., 15 min before microinjection of carbachol, 2 μg, into the medial frontal cortex. Table I shows that atropine blocked carbachol-induced "boxing" whereas mecamylamine did not, providing evidence that the muscarinic cholinergic receptor mediates carbachol-induced "boxing."

TABLE I. Cholinergic Antagonists

Treatment	"Boxing" (cumulative sec/15 min)
Saline + carbachol i.p. (2 μg)	263 ± 112
Atropine + carbachol 5 mg/kg i.p. (2 μg)	6 ± 4 ($P < 0.01$)
Mecamylamine + carbachol 3 mg/kg s.c. (2 μg)	210 ± 85

A comprehensive literature search of rodent behaviors mediated by the medial frontal cortex failed to uncover any reports of syndromes or fixed action motor patterns resembling "boxing." To test the possibility that "boxing" is a component of aggressive behavior (Eichelman and Thoa, 1973), two carbachol-treated rats were placed together in a large polycarbonate cage for 30 min. No episodes of fighting were observed; each rat continued "boxing" independently at opposite ends of the cage. To test the possibility that "boxing" is a component of seizures, which have been described by Racine (1975) after electrical stimulation of the frontal–cingulate cortex in rats, EEG electrodes were implanted into the bifrontal and frontal occipital cortex as previously described (Skolnick *et al.*, 1983) at the same time as bilateral cannulae were implanted into the medial frontal cortex. Table II shows that high-amplitude spiking (>150 μV amplitude, >2.5 sec duration) was recorded for up to 2 hr after microinjection of carbachol. Observation of the rats during this period found that the periods of "boxing" were coincident with periods of high-amplitude spiking (Crawley *et al.*, 1986).

To test further the interpretation that "boxing" represents a component of seizures, rats were pretreated with standard anticonvulsant drugs at doses that inhibit seizures elicited by convulsants such as pentylenetetrazole and strychnine before carbachol injection. Figure 6 shows that intraperitoneal administration of three anticonvulsants, diazepam, clonazepam, and pentobarbital, attenuated "boxing" elicited by carbachol microinjected into the medial frontal cortex (Crawley *et al.*, 1986). Another clinically effective anticonvulsant, diphenylhydantoin, had no effect. "Boxing," therefore, appears to be a stereotyped motor component of a form of seizure elicited by carbachol administered into the medial frontal cortex of rats. This motor pattern does not resemble the tonic–clonic motor activity of rats and mice during seizures elicited by agents such as pentylenetetrazole or strychnine.

TABLE II. Electroencephalographic Analysis

Treatment	High-amplitude spiking (sec/75 min ± S.E.M.)
Saline	19 ± 10
Carbachol (1 μg)	1234 ± 441
Carbachol (4 μg)	2004 ± 1137

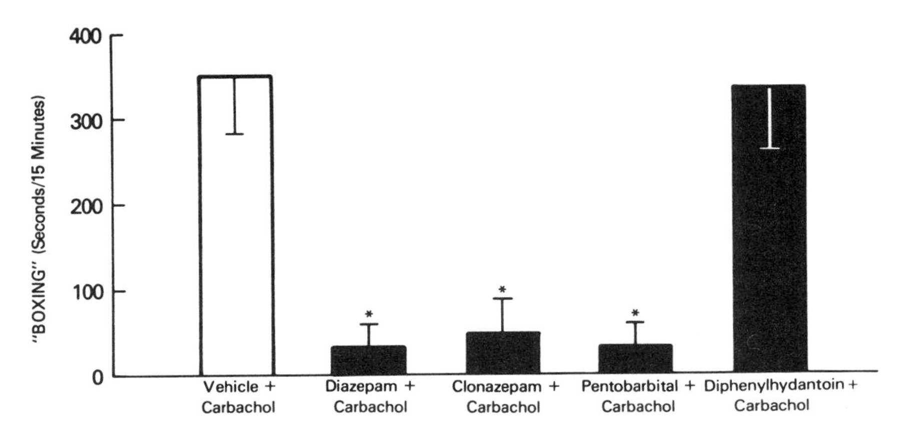

FIGURE 6. Antagonism of carbachol-induced "boxing" by standard anticonvulsant drugs. Diazepam, 2 mg/kg i.p., clonazepam, 0.5 mg/kg i.p., and pentobarbital, 10 mg/kg i.p., administered 30 min before the beginning of behavioral scoring, significantly attenuated "boxing" elicited by microinjection of carbachol, 4 μg, into the medial anterior cortex, whereas diphenylhydantoin, 16 mg/kg i.p., had no effect on carbachol-elicited "boxing" (ANOVA $F_{4,14} = 10.66$, $^*P < 0.01$ by Newman–Keuls analysis of individual means). $N = 4$ for each treatment group.

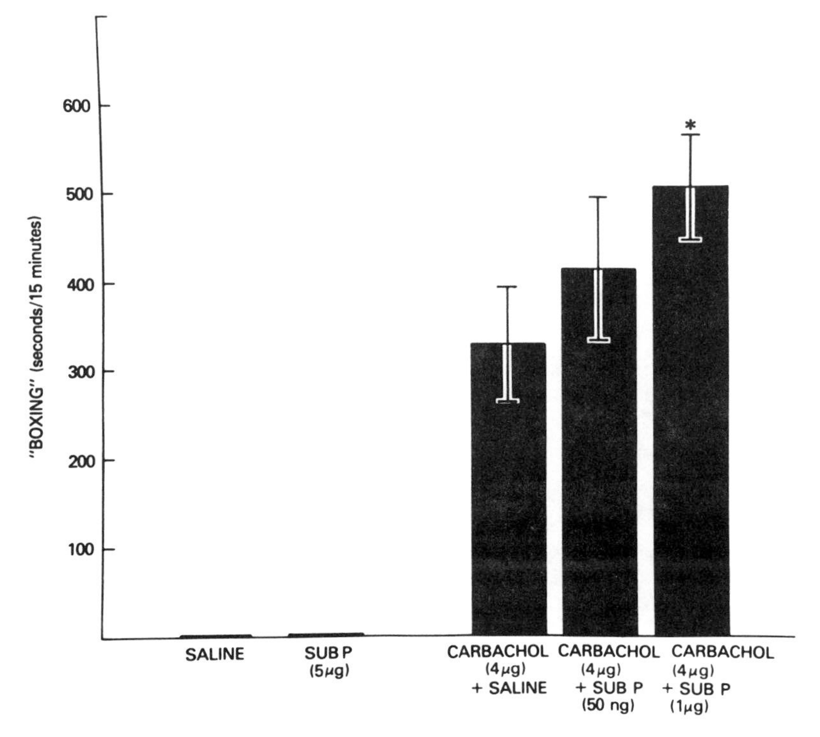

FIGURE 7. Substance P (SUB P, 1 μg) significantly potentiated carbachol-induced "boxing" when coinjected with carbachol (4 μg) into the medial frontal cortex of the rat [ANOVA $F_{3,19} = 7.36$, $P < 0.01$; $^*P < 0.05$ SUB P (1 μg) + carbachol as compared to carbachol alone]. Substance P alone did not include "boxing" at doses up to 5 μg. $N = 6$ for each treatment group.

2.2.3. Modulation of Carbachol-Induced "Boxing" by Peptides (SP, CRF)

To test the hypothesis that the peptides, SP and CRF, that coexist with AchE affect the behavioral response to the cholinergic agonist carbachol, several doses of each peptide were administered in combination with carbachol (4 μg) bilaterally into the medial frontal cortex. "Boxing" was scored by an observer, uninformed of the treatment condition, beginning 15 min after microinjection of the peptide–carbachol mixture for a period of 15 min. Data were analyzed for statistical significance by analysis of variance, followed by Newman–Keuls test for significance of individual means. Scores from animals with cannula placements or injection sites outside of the medial frontal cortex were eliminated from data analysis. Figure 7 illustrates the finding that SP significantly potentiated carbachol-induced "boxing" at a dose of 1 μg of SP. Figure 8 illustrates the finding that CRF significantly inhibited carbachol-induced "boxing" at doses of 1 ng and 10 ng. Substance P and CRF, therefore, appear to act as modulators of carbachol in this behavioral paradigm, having no effect alone but, respectively, increasing and decreasing the behavioral response to carbachol.

Another peptide recently localized in the medial frontal cortex (Skofitsch and Jacobowitz, 1985) but not found in the projections from the NTDL is calcitonin gene-related peptide (CGRP) (see also Rosenfeld *et al.*, 1983). In a protocol identical

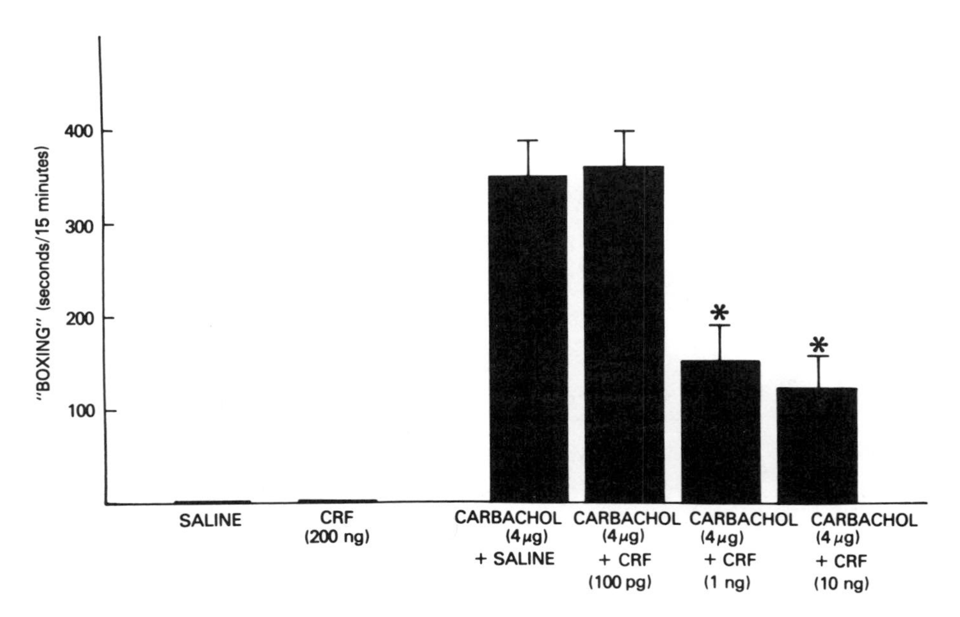

FIGURE 8. Corticotropin-releasing factor (CRF, 1 ng and 10 ng) significantly inhibited carbachol-induced "boxing" when coinjected with carbachol (4 μg) into the medial frontal cortex of the rat (ANOVA $F_{4,24} = 14.49$, $P < 0.01$, $^*P < 0.05$ for CRF + carbachol compared to carbachol alone). CRF alone did not induce "boxing" at doses up to 200 ng. $N = 6$ for each treatment group.

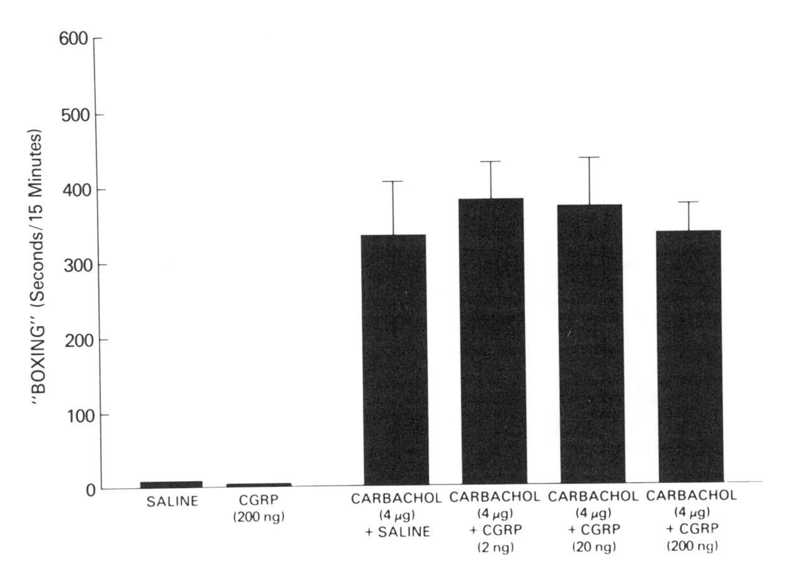

FIGURE 9. Calcitonin gene-related peptide (CGRP, 2 ng, 20 ng, and 200 ng) had no effect on carbachol-induced "boxing" when coinjected with carbachol (4 μg) into the medial frontal cortex of the rat (ANOVA $F_{3,20} = 0.17$, nonsignificant). CGRP alone did not induce "boxing" at doses up to 200 ng. $N = 6$ for each treatment group.

to that described for SP and CRF, CGRP was bilaterally microinjected alone or in combination with carbachol (4 μg) into the medial frontal cortex. Figure 9 illustrates the lack of effect of CGRP over the dose range of 2 ng to 200 ng on carbachol-induced "boxing."

3. CONCLUSIONS

In summary, a triple coexistence of AchE, SP, and CRF has been described in NTDL neurons projecting to the medial frontal cortex, lateral septum, and thalamus. Microinjection of the cholinergic agonist carbachol into the medial frontal cortex elicited a stereotyped motor pattern resembling "boxing," which appears to be mediated by the muscarinic cholinergic receptor and to represent a form of seizure. Although neither SP or CRF microinjected into the medial frontal cortex elicited "boxing" or any other detectable spontaneous motor behavior pattern, SP potentiated carbachol-induced "boxing," and CRF inhibited carbachol-induced "boxing." Further testing of other peptides found in the medial frontal cortex and of anatomic sites adjacent to and distant to the NTDL terminals in medial frontal cortex is necessary to characterize the peptidergic modulation of carbachol-induced "boxing." It is interesting to speculate that the functional significance of a triple coexistence involves peptidergic modulation of the primary transmitter, with one peptide increasing and the second peptide decreasing the postsynaptic actions of acetylcholine.

REFERENCES

Crawley, J. N., Olschowka, J. A., Diz, D. I., and Jacobowitz, D. M., 1985a, Behavioral significance of the coexistence of substance P, cortocotropin-releasing factor, and acetylcholinesterase in lateral dorsal tegmental neurons projecting to the medial frontal cortex of the rat, *Peptides* **6:**891–901.

Crawley, J. N., Stivers, J. A., Blumstein, L. K., and Paul, S. M., 1985b, Cholecystokinin potentiates dopamine-mediated behaviors: Evidence for modulation specific to a site of coexistence, *J. Neurosci.* **5:**1972–1983.

Crawley, J. N., Stivers, J. A., Martin, J. V., and Mendelson, W. B., 1986, Cholinergic induction of seizures in the rat prefrontal cortex, *Life Sci.* **38:**2347–2354.

Eichelman, B. S., and Thoa, N. B., 1973, The aggressive monoamines, *Biol. Psychiatry* **6:**143–164.

Hökfelt, T., Johansson, O., and Goldstein, M., 1984, Chemical anatomy of the brain, *Science* **225:**1326–1334.

Jacobowitz, D. M., and Creed, G. J., 1983, Cholinergic projection sites of the nucleus of tractus diagonalis, *Brain Res. Bull.* **10:**365–371.

Koelle, G. B., 1955, The histochemical identification of acetylcholinesterase in cholinergic, adrenergic and sensory neurons, *J. Pharmacol. Exp. Ther.* **114:**167–184.

Olschowka, J. A., O'Donohue, T. L., Mueller, G. P., and Jacobowitz, D. M., 1982, The distribution of corticotropin releasing factor-like immunoreactive neurons in rat brain, *Peptides* **3:**995–1015.

Olschowka, J. A., Diz, D. I., and Jacobowitz, D. M., 1983, Coexistence of substance P, CRF, and AchE in neurons of the nucleus tegmenti dorsalis lateralis, *Soc. Neurosci. Abstr.* **167:**5.

Paxinos, G., and Watson, C., 1982, *The Rat Brain in Stereotaxic Coordinates,* Academic Press, New York.

Racine, R. J., 1975, Modification of seizure activity by electrical stimulation: Cortical areas, *Electroencephalogr. Clin. Neurophysiol.* **38:**1–21.

Rosenfeld, M. G., Mermod, J.-J., Amara, S. G., Swanson, L. W., Sawchenko, P. E., Rivier, J., Vale, W. W., and Evans, R. M., 1983, Production of a novel neuropeptide encoded by the calcitonin gene via tissue-specific RNA processing, *Nature* **304:**129–135.

Skofitsch, G., and Jacobowitz, D. M., 1985, Calcitonin gene-related peptide: Detailed immunohistochemical distribution in the rat central nervous system, *Peptides* **6:**721–745.

Skolnick, P., Schweri, M. M., Paul, S. M., Martin, J. V., Wagner, R. L., and Mendelson, W. B., 1983, 3-Carboethoxyl-β-carboline (β-CCE) elicits electroencephalographic seizures in rats: Reversal by the benzodiazepine antagonist CGS 8216, *Life Sci.* **32:**2439–2445.

Tramu, G., Pillez, A., and Leonardelli, J., 1978, An efficient method of antibody elution for the successive or simultaneous localization of two antigens by immunocytochemistry, *J. Histochem. Cytochem.* **26:**322–324.

Vincent, S. R., Satoh, K., Armstrong, D. M., and Fibiger, H. C., 1983, Substance P in the ascending cholinergic reticular system, *Nature* **306:**688–691.

23

Brain Peptides and the Control of Eating Behavior

SARAH FRYER LEIBOWITZ and B. GLENN STANLEY

1. INTRODUCTION

Our understanding of brain function has been dramatically enhanced by the discovery of endogenous brain peptides and by subsequent evidence for their role as neurotransmitters, neuromodulators, and neurohormones (Krieger, 1983; Palkovits and Brownstein, 1985). This review focuses on an area of research that has benefited greatly from these discoveries, namely, the neurochemistry of eating behavior. Research conducted in this field during the past two decades has clearly demonstrated that food intake is controlled via multiple neurotransmitters in the brain, which are responsive to a complex array of metabolic, hormonal, and neural signals. Initially, the focus of this research was on the brain monoamines, which are now believed to have a critical and diverse role in the normal control of food ingestion (Leibowitz, 1980, 1985, 1986a). More recently, however, significant advances have been made with the brain peptides, which also have an important function in feeding, in part through their interaction with the endogenous monoaminergic system (Morley *et al.*, 1983, 1985; Leibowitz, 1985, 1986a,b).

The initial proposal that the peptide cholecystokinin may be involved in producing satiety (Gibbs *et al.*, 1973) has provided key inspiration for this area of research. Over the past several years, numerous peptides have been examined for their potential impact on food intake, and so far, it has been established that most are effective in inhibiting eating behavior, whereas only a few cause a potentiation of eating. Two families of peptides have received the most attention with respect to their role as stimulators of feeding. These are the opioid peptides and, more recently, the pancreatic polypeptides. The evidence obtained with these peptides,

SARAH FRYER LEIBOWITZ and B. GLENN STANLEY • The Rockefeller University, New York, New York 10021.

as well as the results from a recent study on growth hormone-releasing factor, is described in this review. In contrast to these feeding-stimulating peptides, however, there are a wide variety of other peptides, more than 20 in number, that are now found to cause a reduction in food ingestion. This review summarizes the results obtained with a few of these peptides, those that have the strongest evidence for a potent and central site of action, for pharmacological and behavioral specificity, and for a specific physiological role in the mediation of satiety.

With regard to questions of central versus peripheral sites of action, it appears that, in general, systemically administered peptides act predominantly within the periphery, only indirectly involving central neural mechanisms for control of satiety (Morley, 1982; Stanley *et al.*, 1983; Gibbs and Smith, 1984, 1986; Morley *et al.*, 1985). This seems to be the case for circulating cholecystokinin as well as for bombesin, pancreatic glucagon, and somatostatin. With the exception of bombesin, each of these peptides is believed to act through the abdominal vagus nerve, which carries appropriate visceral information to central eating control mechanisms. In the case of insulin, it has been proposed (Woods and Porte, 1983) that this peptide in the peripheral circulation, released in proportion to body adiposity, may actually pass into the CSF and then interact with central insulin receptors controlling satiety. Whereas there is potential for an inhibitory role of peripheral calcitonin and thyrotropin-releasing hormone in eating, studies with peripheral injection of other peptides, namely, gastrin, neurotensin, secretin, gastric insulinotropic peptide, and neuropeptide Y, have failed to reveal any consistent effect, at least in the rat.

The strength of these studies on peripheral peptide systems lies in the extensive effort put forth to determine the chemical and behavioral specificity of the effects observed. Although in a few cases research with centrally injected peptides has advanced along similar lines, as described below, a great deal of work is still lacking in this area in terms of having adequate control experiments and a proper focus on issues of neurobiological significance. In particular, important questions have arisen from studies of peptides that cause a decrease in eating behavior. Do these drug effects merely result from such factors as malaise, general debilitation, or behavioral competition, or do they reflect the physiological action of the endogenous neuropeptides in specific satiety systems of the brain? In light of the rapidly growing list of centrally injected peptides that are found to suppress eating behavior, it has become even more urgent that we answer questions of this nature as well as evaluate the significance of this research in our ultimate goal to understand the central neurochemical substrates essential to normal behavior.

This long list of centrally injected peptides that inhibit food intake now includes cholecystokinin, bombesin, neurotensin, adrenocorticotropin, anorexigenic peptide, corticotropin-releasing factor, vasopressin, insulin, calcitonin, somatostatin, thyrotropin-releasing hormone, substance P, satietin, and glucagon (for reviews, see Baile and Della-Fera, 1984; Morley *et al.*, 1985; Leibowitz, 1986a). In many studies, these peptides have been injected intraventricularly and tested at high doses (1–50 μg), which, in the absence of additional behavioral analyses, may argue for a pharmacological rather than a physiological effect. It is important to bear in mind, however, that with the central injection technique, particularly with substances administered into the ventricles, a great deal of the peptide is possibly lost, either by very rapid degradation or up the inside or outside of the cannula tract. Thus, it is difficult to evaluate the amount of peptide that is actually reaching its mediating receptors. Based on these intraventricular studies, it is also difficult to draw any conclusions regarding site(s) and thus mechanism of action, particularly in light of the potential spread of ventricularly injected peptides into the peripheral circulation (Passaro *et al.*, 1982). Additional questions pertain to the behavioral specificity of these peptide effects, which in some cases are associated with additional nonspecific aversive properties and in most cases are accompanied by a suppression of drinking as well as eating behavior. A number of the peptides also cause a dramatic rise in grooming behavior, indicating that the reduction in food intake may not reflect a primary action on systems controlling eating and satiation but rather may result secondarily from the enhancement of a competing behavior. Thus, to evaluate the brain peptides in terms of their potential physiological role in food ingestion, we need to consider each of these questions, of dose, chemical specificity, behavioral specificity, and site of action, before significant progress can be made on the question of whether, where, and in what manner these peptides act in the brain.

Since it is believed that the brain peptides function, in part, either through or in conjunction with central catecholamine systems that control food intake, a brief summary of what is currently known about these aminergic systems will be helpful in providing a more complete understanding of the peptides' mechanism of action (for reviews, see Leibowitz, 1980, 1985, 1986a; Leibowitz and Shor-Posner, 1986). Through extensive mapping studies with centrally injected catecholamines, it has been established that the medial hypothalamus, in particular the paraventricular nucleus (PVN), is most responsive to the stimulatory action of these neurotransmitters on feeding, in contrast to the lateral hypothalamus, in particular the perifornical area (PLH), which is most responsive to their inhibitory action. Through pharmacological studies, the receptors mediating these effects have been identified as α_2-noradrenergic in nature for the feeding-stimulatory effects of norepinephrine (NE) in the PVN and β-adrenergic and dopaminergic in nature for the feeding-suppressive effects of epinephrine and dopamine in the PLH. The effects of these catecholamines are behaviorally specific and can occur at doses near physiological levels. In particular, NE injected into the PVN at a dose as low as 4 ng potentiates food intake through a specific increase in the size, as opposed to the frequency, of meals consumed. It also enhances eating via a selective increase in preference for carbohydrate as opposed to protein or fat. These results contrast, once again, with those obtained with the catecholamines in the PLH, where they reduce food intake, apparently by delaying meal onset and preferentially decreasing protein ingestion.

A critical and revealing feature of the α_2-noradrenergic system in the PVN is that its activity is dependent on and closely correlated with circulating levels of corticosterone and possibly insulin. Specifically, the feeding response elicited by PVN noradrenergic stimulation is abolished by adrenalectomy and hypophysectomy and is selectively restored by corticosterone replacement in a dose-dependent fashion. Further, NE-induced feeding is attenuated by surgical dissection of the vagus nerve, in particular the celiac branch, and is potentiated by systemic administration of insulin. Both corticosterone and insulin have been known to be involved in the control of food intake and now appear to work in close association with the catecholamine neurotransmitters in the hypothalamus. As described in this review, similar neurotransmitter–endocrine relationships have been identified for specific peptides, thereby strengthening the idea that the amines and peptides may function similarly and in close association as they exert their control over feeding behavior.

With regard to the specific physiological role of hypothalamic catecholamines in feeding behavior, the available biochemical and pharmacological evidence has generated the hypothesis that these classical neurotransmitters, as part of their overall effort to rapidly replenish body energy stores, become specifically activated under conditions involving energy expenditure, e.g., during food deprivation, stress, or at the start of the active period of the diurnal cycle. At these times, NE in the PVN and catecholamines in the PLH are believed to have a critical function in coordinating neural, hormonal, and metabolic signals of energy depletion and translating these signals into a very specific and optimal pattern of macronutrient selection (Leibowitz, 1986a; Leibowitz and Shor-Posner, 1986).

2. PANCREATIC POLYPEPTIDES

Recently, evidence has accumulated suggesting that the pancreatic polypeptides, a family of structurally related 36-amino-acid peptides, play an important role in the control of feeding behavior. This family of peptides, consisting of neuropeptide Y (NPY), peptide YY (PYY), and pancreatic polypeptide, have all been shown to elicit feeding behavior in satiated rats (Clark *et al.*, 1984; Levine and Morley, 1984; Morley *et al.*, 1985; Stanley and Leibowitz, 1984a, 1985a). Clark *et al.* (1984) first demonstrated that injections of NPY, and to a lesser extent pancreatic polypeptide, into the third ventricle elicit a feeding response in satiated rats. Levine and Morley (1984) observed similar effects in response to lateral ventricular injection of NPY.

Recently, we have shown that administration of NPY directly into the PVN elicits feeding behavior in satiated rats and that this response is larger, and occurs at lower doses, than that induced by ventricular injection (Stanley and Leibowitz, 1984a, 1985b). These findings, suggesting a hypothalamic site of action for NPY, are supported by studies that demonstrate that NPY-like immunoreactivity and receptor binding sites are most abundant in this structure (Allen *et al.*, 1983; Unden *et al.*, 1983). They are also consistent with the results of a cannula mapping study that show that NPY injections into several hypothalamic areas, namely, the PVN, lateral hypothalamus, and ventromedial hypothalamus, produce large increases in food intake, whereas injections into sites outside the hypothalamus, either anterior, posterior, lateral, or dorsal to this structure, are ineffective (Stanley *et al.*, 1985a). Although the exact site within the hypothalamus where NPY is most effective remains to be determined, these results suggest that the hypothalamus may contain critical receptors that mediate the powerful eating response elicited by this peptide.

We have recently shown that NPY is the most powerful chemical stimulant of feeding behavior and is several times more powerful than any other putative neurotransmitter tested to date. Injection of NPY into the PVN of satiated rats causes a dose-dependent stimulation of feeding behavior, with a low dose of 24 pmol producing a significant increase and a 235-pmol dose causing a peak effect. The subjects begin to eat after a relatively short latency, within an average of about 10 min after the injection and in some animals consistently within 1 or 2 min. At the lower end of the dose range, the feeding is completed within 1 hr. Higher doses, however, not only produce greater intakes initially (15 g in 1 hr) but also cause a continued eating response such that food intake 4 hr postinjection is equivalent to normal total daily food intake. There is also a significant increase in total daily food intake after a single PVN injection of NPY (Stanley and Leibowitz, 1985b). These results, indicating that hypothalamic stimulation by NPY can override mechanisms of satiety, may implicate NPY not only in the control of food intake but also in the regulation of body weight.

A recent study employing chronic PVN injections of NPY supports this possibility (Stanley *et al.*, 1985b). This study demonstrates that animals injected with NPY three times a day can double their daily food consumption and gain weight at a rate of more than 10 g/day (over a 10-day period), which is comparable to that of rats with electrolytic lesions in the ventromedial hypothalamus (Hetherington and Ranson, 1940). These results, demonstrating that chronic injection of NPY into the PVN can produce hyperphagia and obesity, presumably by overriding long- and short-term signals of satiety and control mechanisms for body weight, suggest the possibility that elevated release of endogenous NPY may contribute to some types of natural or experimentally induced obesity. In this regard, it may be noted that the PVN has one of the most abundant supplies of NPY-containing presynaptic terminals in the brain (Everitt *et al.*, 1984; Olschowka, 1984) and that, to date, NPY is the most abundant peptide in the brain of rats and humans, with a similar distribution in the brains of both species (Adrian *et al.*, 1983; Allen *et al.*, 1983). Whether some cases of bulimia or obesity in humans may be mediated by NPY remains to be determined.

Another issue concerns the behavioral specificity of NPY injection in the PVN. To address this, we have administered into the PVN a dose of NPY that elicits somewhat greater feeding than occurs in a normal meal and then observed its impact on the animals' behavioral pattern (Stanley and Leibowitz, 1984a). In response to the peptide stimulation, the rats are found to eat a single large meal, which lasts approximately 5 min. At the termination of the meal, they exhibit the sequence of behaviors, including grooming and sleeping, that is typical of rat behavior (Antin *et al.*, 1975). Except for a small increase in water intake, no alterations in any other behaviors (e.g., levels of activity, grooming, resting, sleeping) are observed in response to NPY. These findings demonstrate that NPY's effect is specific to ingestive behavior, producing a strong orientation toward the ingestion of food without other behavioral effects. This specificity supports a role for NPY in the control of feeding behavior.

In addition to the feeding-stimulatory effect of NPY, it has been shown that a structurally related peptide, PYY, produces feeding when injected into the PVN (Stanley and Leibowitz, 1985a; Stanley *et al.*, 1985d). To the extent that PYY and NPY have been compared, they appear to exhibit almost identical effects, except that PYY is approximately three times more potent than NPY. Thus, PYY, although not more powerful than NPY in terms of its maximal response, does appear to be more potent, eliciting feeding at doses at least as low as 8 pmol. Since the effects of these two structurally related peptides are almost identical, it has been suggested that NPY and PYY stimulate feeding by acting on the same receptor subtype.

Recently, in a macronutrient self-selection study, we have examined the stimulatory effects of NPY and PYY on consumption of pure macronutrients (Stanley and Leibowitz, 1985a; Stanley *et al.*, 1985d). Both NPY and PYY are found to produce a dramatic and selective increase in consumption of carbohydrate, with no change in ingestion of either protein or fat. This pattern occurs whether the carbohydrate diet is placed with just one or both of the other two macronutrients. In the absence of the carbohydrate diet, NPY and PYY are found to potentiate protein and fat intake; however, these responses are significantly smaller than the potentiating effect of these peptides on carbohydrate ingestion.

What factors may account for the carbohydrate preference as well as the shift to protein and fat in the absence of this diet? One possibility is that carbohydrate may be strongly preferred because this diet most rapidly provides usable energy (Steffens, 1969). In the absence of this preferred diet, however, the animals may then turn to the protein or fat diets as alternative, albeit less rapid, sources of usable energy. This interpretation, which focuses on metabolic rather than sensory features of these diets, leads to the suggestion that endogenous NPY may be involved in mediating signals of negative energy balance, which then translate into the elicitation of compensatory behavior. A similar function has been proposed for the PVN α_2-noradrenergic feeding system (Leibowitz, 1985, 1986a), and given the similar effects of NE and NPY, it is possible that NPY may perform this function in concert with this amine.

The idea for a functional interaction between NPY and NE in relation to feeding is suggested by recent studies demonstrating a close functional relationship between these two neurotransmitters in other systems and by the remarkable similarities in their effects on feeding behavior. Neuropeptide Y and NE coexist in central neurons, including those innervating the PVN (Sawchenko *et al.*, 1985), and in nerves of the sympathetic nervous system (Everitt *et al.*, 1984). They are released from sympathetic nerves apparently simultaneously, interacting to produce the same complex pattern of responses as produced by nerve stimulation (Lundberg *et al.*, 1984). Neuropeptide Y has also been shown specifically to up-regulate α_2-noradrenergic receptors (Agnati *et al.*, 1983). With regard to feeding behavior, both NE and NPY elicit a strong feeding response and a small drinking response when injected into the medial hypothalamus (Leibowitz, 1980; Stanley and Leibowitz,

1985b). Both produce virtually identical effects on macronutrient consumption, and the feeding response produced by either is attenuated by adrenalectomy (Stanley and Leibowitz, 1984b; Leibowitz *et al.*, 1984b, 1985).

Although these results suggest that NPY and NE may interact in some manner to control feeding, it is clear that the effects are not mediated through activation of identical neural substrates. This is demonstrated by the findings that feeding elicited by NPY, unlike NE, is unattentuated by PVN injection of an α-adrenergic receptor blocker. It is also unaffected by midbrain knife cuts, which strongly attentuated feeding elicited by NE (Stanley and Leibowitz, 1984b, 1985b; Weiss and Leibowitz, 1985). Thus, although it is possible that NPY and NE may interact to control feeding, the precise nature of this potential interaction remains to be elucidated.

3. OPIOID PEPTIDES

Evidence is accumulating in support of the idea that endogenous opioid peptides, particularly in unpredictable environmental conditions, have important functions as mediators of various physiological processes involved in maintenance of bodily homeostasis (for reviews, see Margules, 1979; Sanger, 1981; Henry, 1982; Morley *et al.*, 1983; Olson *et al.*, 1984; Leibowitz, 1986b). These opioid functions appear to manifest themselves, at least in part, through their interaction with the various monoamine neurotransmitters as well as through other neuroendocrine systems, in particular adrenal glucocorticoids.

This is found to be the case with brain opioid control of eating behavior. Opiate drugs, peripherally or centrally injected, are generally found to stimulate eating behavior in various species. Based on this and a variety of other biochemical and pharmacological evidence, it is proposed that brain opioids, sometimes in conjunction with hypothalamic catecholamines, participate in the stimulation of food ingestion and become activated, in particular, under stressful conditions perhaps involving food deprivation or painful stimulation. The opiates are well known for their analgesic properties, and it is also known that food deprivation decreases nociception and potentiates hypothalamic opiate activity in addition to enhancing appetite for food (McGivern *et al.*, 1979, Gambert *et al.*, 1980; Morley *et al.*, 1983; Olson *et al.*, 1984). Different forms of stress elicit food ingestion, and it has been proposed that this response is mediated through brain or perhaps hypothalamic opiates.

Recent mapping studies with centrally injected opioids have found the hypothalamus, in contrast to forebrain and hindbrain structures, to be most responsive to these peptides (Gosnell *et al.*, 1984; Scott *et al.*, 1984; Stanley *et al.*, 1984; Lanthier *et al.*, 1985; Woods and Leibowitz, 1985). The opioids tested are β-endorphin, Met-enkephalin, and dynorphin, as well as various opiate receptor agonists and antagonists. All compounds are effective in stimulating eating in satiated animals, indicating that different receptor subtypes, specifically, μ, δ, and κ, are involved. Although the opiates are effective in different medial and lateral hypo-

hypothalamic sites, two particularly sensitive sites appear to be the PVN and the PLH. This eating-stimulatory effect of the opiate agonists in these hypothalamic sites is sensitive to blockade by peripheral and locally administered opiate antagonists, which themselves, at somewhat higher concentrations, can suppress the natural eating response caused by food deprivation (Grandison and Guidotti, 1977; Tepperman *et al.,* 1981a; Leibowitz and Hor, 1982; Gosnell *et al.,* 1984; Woods and Leibowitz, 1985). Anatomic studies have identified in these hypothalamic areas dense concentrations of opiate terminals and receptor sites (see Leibowitz, 1986b, for review), and discrete lesions within the PVN or the dorsomedial nucleus, in contrast to the ventromedial nucleus, significantly disrupt the impact of peripheral morphine and naloxone on food ingestion (King *et al.,* 1979; Bellinger *et al.,* 1983; Shor-Posner *et al.,* 1985).

In a number of studies, the eating effect of medial hypothalamic opiate activation has been shown to be reversed by α-adrenergic receptor antagonists as well as by opiate antagonists (Tepperman *et al.,* 1981b; Leibowitz and Hor, 1982). This raises the possibility that the opioid peptides and NE, in particular within the PVN, may interact in some fashion in their effort to produce eating. There is a variety of evidence linking opioid and α_2-noradrenergic systems in the brain (Olson *et al.,* 1984; Unnerstall *et al.,* 1984). Electrophysiological experiments have revealed inhibitory effects of morphine and enkephalin, like NE, on PVN neuronal firing (Moss *et al.,* 1972; Pittman *et al.,* 1980). The opioids may in part operate through the release of NE in the PVN, since peripheral morphine injection at doses that enhance eating significantly decreases NE concentration and increases NE's turnover in the PVN, and the eating-stimulatory effect of morphine, like that of NE and clonidine, is attenuated by adrenalectomy and restored by corticosterone (Bhakthavatsalam and Leibowitz, 1986a; Jhanwar-Uniyal *et al.,* 1984; Leibowitz *et al.,* 1984b; Leibowitz, 1986b). The opiates and NE are positively correlated in their effectiveness within the PVN, and in the rat, they both exhibit a similar diurnal pattern of responsiveness, peaking in the early phase of the dark cycle when circulating corticosterone, hypothalamic opioids, and α_2-noradrenergic receptors, nociceptive thresholds, and spontaneous eating are also at their highest (Krieger and Hauser, 1978; McGivern and Berntson, 1980; Olson *et al.,* 1984; Leibowitz *et al.,* 1984a; Leibowitz, 1986a,b; Bhakthavatsalam and Leibowitz, 1986a,b). Further, under conditions of chronic morphine administration, a potentiation of PVN α_2-noradrenergic receptor function can be seen (Leibowitz, 1986b).

This association between the opiate and α_2-noradrenergic systems does not occur in all brain areas and under all conditions. The eating-stimulatory effects of δ and κ agonists, in contrast to that of a μ agonist, are unaffected by adrenalectomy (McLean and Hoebel, 1982; Levine and Morley, 1983). In addition, PVN lesions only attenuate but do not abolish morphine-induced eating (Shor-Posner *et al.,* 1985), consistent with the results of cannula mapping studies (see above) showing sensitivity to opiate stimulation in other hypothalamic areas besides the PVN. Whereas NE selectively stimulates preference for carbohydrate (Leibowitz *et al.,* 1985), morphine preferentially enhances protein intake under satiated conditions and fat ingestion under food-deprived conditions while actually reducing relative preference for carbohydrate (Marks-Kaufman, 1982; Leibowitz, 1986a; Shor-Posner *et al.*, 1985; Leibowitz and Shor-Posner, 1986). Thus, one must consider multiple sites and perhaps multiple modes of action for the brain opioid peptides in contrast to the relatively localized system for NE (Leibowitz, 1978). In light of the findings that catecholamines in the PLH are involved in regulating protein ingestion (Leibowitz, 1986a; Leibowitz and Shor-Posner, 1986), it is possible that the opioids may interact with this lateral hypothalamic system to induce its stimulatory effect on protein ingestion. Feeding elicited by electrical stimulation of this lateral hypothalamic area can be antagonized by naloxone administration (Carr and Simon, 1983).

The possibility exists that through their action on taste mechanisms, brain opioids may have impact on the rewarding (e.g., taste) qualities of food and hence ingestion. Opiate agonists, which have naloxone-sensitive electrophysiological effects on parabrachial taste units (Hermann and Novin, 1980), enhance the palatability of certain foods and broaden the range of acceptable sweet solutions (e.g., LeMagnen *et al.*, 1980; Apfelbaum and Mendenoff, 1981; Cooper, 1983; Dum *et al.*, 1983; Siviy and Reid, 1983). The opiate antagonist naloxone, in contrast, evokes the opposite response. In fact, ingestion of a highly palatable diet apparently releases β-endorphin specifically within the hypothalamus, and long-term hyperphagia generated by such diets can be attenuated by an opiate antagonist at a dose that has no effect on ingestion of a less palatable, standard diet. Stress, which releases hypothalamic opioid peptides and increases nociceptive threshold (Olson *et al.*, 1984), also potentiates appetite specifically for sucrose, an effect antagonized by naloxone (Bertierre *et al.*, 1984). In each of these effects, the opioid peptides in the PVN may be involved; this nucleus has direct projections to the parabrachial nucleus (Swanson and Sawchenko, 1983), is able to sustain self-stimulation reward (Atrens and von Vietinghoff-Riesch, 1972), and has direct impact on ingestion, particularly of carbohydrate (Leibowitz, 1986a; Leibowitz and Shor-Posner, 1986).

4. GROWTH HORMONE-RELEASING FACTOR

A recent study has demonstrated that within 30 min of injection into the lateral ventricles, growth hormone-releasing factor (GRF) potentiates food intake in rats maintained on a 23-hr food-deprivation schedule (Vaccarino *et al.*, 1985). This peptide is effective at very low doses, at least as low as 0.2 pmol. In contrast to the effectiveness of the intact peptide, fragments of GRF have no significant impact on food intake, suggesting that the effect is pharmacologicially specific. The stimulation of feeding is observed in the absence of changes in locomotor activity, indicating that the effect is behaviorally specific. Further, the increase in feeding appears to be independent of GRF's effect on growth hormone as well as its peripheral effects, since peripheral injection of this peptide, which stimulates growth hormone release, has no influence on food ingestion. Taken together, these results point to GRF as a third potential candidate for endogenous peptide-induced stimulation of eating behavior.

5. CHOLECYSTOKININ

Although studies originally focused on a role for circulating cholecystokinin (CCK) in satiety (Gibbs *et al.*, 1973; Gibbs and Smith, 1984, 1986), more recently a role for central CCK has been suggested. Experiments conducted with CCK in ruminants are particularly convincing, showing a suppressive effect on feeding with ventricular CCK injection in low, perhaps near physiological, doses (Dell-Fera and Baile, 1979). Continuous infusions are considerably more effective than bolus injections, and the doses required to reduce meal size in food-deprived sheep increase with the length of the deprivation, revealing an adaptive interaction between brain CCK and energy depletion (Della-Fera and Baile, 1980a,b). These investigators have also demonstrated that only biologically active forms of CCK reduce eating in sheep (Della-Fera and Baile, 1981). Conversly, ventriclar infusion of CCK antibodies has been shown to have the opposite effect, that of increasing food intake (Della-Fera *et al.*, 1981). These findings, suggesting that endogenous CCK may act to inhibit feeding, also imply that this peptide may function in a primary and behaviorally specific manner. Since CCK is actively taken up from cerebrospinal fluid (Della-Fera *et al.*, 1982), these investigators suggested that during eating, this peptide may be released into the ventricular system and then transported to the brain site(s) of action. In contrast to the consistent effects observed in sheep, studies conducted with ventricular CCK injection in the rat have been less successful in revealing a reliable effect on feeding.

Although extensive cannula mapping studies have not been conducted, evidence to date points to the hypothalamus, perhaps in addition to other brain structures, as a site mediating central CCK's effect on feeding behavior. It has been shown that CCK injected directly into either the PVN or ventromedial hypothalamus of the rat causes a significant decrease in eating and, further, that as little as 5 ng of CCK in the PVN is effective (Faris *et al.*, 1983b; Ritter and Ladenhein, 1984; Willis *et al.*, 1984). Conversely, PVN injection of proglumide, a CCK antagonist, potentiates feeding (Hoebel, 1985). It is of interest that this hypothalamic nucleus may also be involved in relaying information, via ascending vagal afferents, from peripheral CCK receptors to central satiety systems (Crawley and Knas, 1984). These findings support the PVN as a brain site for CCK-induced satiety and suggest that endogenous PVN CCK receptors may have a function in satiety.

Further evidence for hypothalamic CCK receptors and endogenous CCK in satiety is provided by investigations examining the impact of eating and fasting on endogenous CCK levels and receptors. For example, it has been shown that fasting decreases hypothalamic levels of CCK while causing an increase in CCK receptor number. Feeding, in contrast, increases CCK levels exclusively in the hypothalamus (Saito *et al.*, 1981; McLaughlin *et al.*, 1985; Scallet *et al.*, 1984).

Studies have suggested that hypothalamic CCK may exert its inhibitory effect on feeding by altering the effect of catecholamines in both the medial and lateral hypothalamus. Specifically, the stimulation of feeding produced by PVN injection of NE is strongly inhibited by CCK (75 ng) administered into the PVN but not other hypothalamic areas (Myers and McCaleb, 1981). Further, it has been shown that CCK infused into the PVN suppresses the release of NE in the PVN. Of particular interest is the observation that CCK suppresses PVN release of NE in food-deprived but not satiated animals (Myers, 1985). This suggests that only PVN terminals that release NE, presumably in response to low energy stores (Leibowitz 1986a), are affected by activation of CCK receptors. This pattern of results contrasts with that observed in the PLH, where CCK increases the release of NE in deprived but not satiated animals. A consequence of this CA release in the PLH would be expected to be a reduction of food intake (Leibowitz, 1980, 1986a) similar to that seen with central CCK injection. Since CCK has also been suggested to act as an opioid antagonist (Faris *et al.*, 1983a), an additional mode of action may be via its influence on opioid peptide systems that potentiate eating (see pp. 339–341).

6. NEUROTENSIN

Evidence is accumulating to suggest that neurotensin (NT), a tridecapeptide found in the gut and brain, may induce satiety when injected centrally in the rat. It has been shown that intraventricular injection of NT causes a dose-related reduction in deprivation-induced and nocturnal feeding (Luttinger *et al.*, 1982; Levine *et al.*, 1983). We have demonstrated that injection of NT directly into the PVN is effective in suppressing feeding in deprived rats and that this effect may be mediated by an action on the PVN noradrenergic system for feeding (Stanley *et al.*, 1982, 1983, 1985c).

A convergence of evidence suggests that this effect of NT is behaviorally specific. Specifically, PVN injection of NT suppresses consumption of solid and liquid food, but it does not affect drinking induced by water deprivation (Stanley *et al.*, 1983). Also, doses of NT that cause a large suppression of feeding do not produce a conditioned taste aversion, suggesting that NT does not induce gastrointestinal malaise (Luttinger *et al.*, 1982). We have shown that PVN injections of NT that produce large decreases in feeding behavior do not affect other behaviors, namely, drinking, grooming, resting, rearing, sleeping, or levels of locomotor activity (Stanley *et al.*, 1983). Furthermore, NT appears to act within the general vicinity of the PVN injection site, presumably on the NE feeding-stimulatory system; this is reflected by the finding that unilaterally administered NT is significantly more effective in suppressing PVN NE-induced feeding when administered ipsilateral, as opposed to contralateral, to the NE injection (Stanley *et al.*, 1985c). These results arguing for behavioral specificity are supported and extended by the finding that ventricular injection of NT suppresses feeding elicited by central injection of NE or dynorphin but not feeding elicited by central muscimol or peripheral insulin injection (Levine *et al.*, 1983).

There is evidence to suggest that exogenous NT produces its satiating effect through activation of endogenous NT receptors. This is indicated by a study which employed NT fragments to evaluate the amino acid structure required for this peptide's suppressive action on feeding (Stanley *et al.*, 1985c). It has been found that the structural requirements for NT to suppress feeding closely parallel the structure required to activate NT receptors in other biological systems. Specifically, the C-terminal hexapeptide is critical; the absence of a single C-terminal amino acid abolishes NT's suppression of feeding. This evidence argues for pharmacological specificity despite the relatively high dose range (0.25 to 1.25 nmol) required to produce a reliable response with NT. Since this peptide has an extremely short half-life, less than 45 sec (Checler *et al.*, 1983), it is logical to conclude that these high doses are needed to overcome the counteracting effects of the brain's degradative enzymes.

Evidence also points to a central site of action for NT. Intravenous injections of NT at doses comparable to those that were effective centrally have no effect on feeding. Only at higher doses does peripheral NT suppress feeding, and, in contrast to the behaviorally specific effects of central NT, this effect of peripheral NT is characterized by a general nonspecific suppression of behavior (Stanley *et al.*, 1983). This conclusion that NT is acting via central receptors is strengthened by the finding (see above) that NT is anatomically site specific in its suppressive effect on feeding elicited by PVN NE injection. Although these results suggest that NT may act in the medial hypothalamus, preliminary cannula mapping studies indicate that NT, in addition to its effects in the PVN, also acts in other hypothalamic as well as extrahypothalamic brain sites to suppress feeding behavior (B. G. Stanley and S. F. Leibowitz, unpublished results).

The finding that NT has an inhibitory effect on feeding induced by exogenous NE is consistent with other evidence indicating that NT may act, in part, via inhibition of endogenous NE. Specifically, it has recently been shown that PVN infusion of NT, similar to CCK, inhibits the release of NE from the PVN of food deprived but not satiated animals (Myers, 1985). Although this may be only one mode of action for NT, these biochemical and pharmacological results, together, present a strong case for a satiety function of medial hypothalamic NT receptors. Perhaps in concert with CCK, these peptides may induce satiety, in part, through antagonism of the well-documented stimulatory influence of NE on food intake and meal size (Leibowitz, 1986a).

7. CALCITONIN

The potent suppressive effect of calcitonin (CT) on feeding was first established by Freed and his colleagues (1979; Perlow *et al.*, 1980). This and subsequent work have shown ventricular injection of CT to be remarkably potent and long acting as a satiety agent, effective at doses of 75–300 ng in food-deprived rats and of 1 ng or less in rats induced to eat by tail pinch (Levine and Morely, 1981; Twery *et al.*, 1982; Morley *et al.*, 1985). The latency of this suppressive effect is relatively short, while at the same time the effect is very long-lasting, at least 24 hr after a single injection of CT. Despite its potency and long duration of action, chronic ventricular administration of CT fails to have a sustaining effect on daily food intake, with tolerance or adaptation developing by the fourth day of the injection sequence.

With injection directly into the brain parenchyma, it has been established that CT reliably suppresses food intake at doses considerably lower than peripherally effective doses (Levine and Morley, 1981; Twery *et al.*, 1982). Thus, it is clear that this peptide is acting centrally to alter behavior. Although its specific central site of action has yet to be established via mapping studies, the available evidence argues for a hypothalamic site, possibly involving the PVN (DeBeaurepaire and Freed, 1983). This nucleus causes a reliable satiety effect in response to a low dose of only 15 ng. This result, of course, does not preclude the possibility, or even likelihood in light of recent peptide mapping studies (see above), that other hypothalamic and extrahypothalamic sites may also be involved.

Questions concerning the behavioral specificity of these effects of CT have yet to be adequately explored. In addition to food intake, CT is known to suppress drinking behavior (Twery *et al.*, 1982). Also, the calcitonin gene-related peptide, which similarly decreases food intake, produces a taste aversion, thus calling into question the specificity of this peptide's action (Morley *et al.*, 1985). Further work of this nature will need to be conducted with CT in order to establish its physiological role in satiety mechanisms. One possible mode of action within the brain may be via CT's potent effects on calcium metbolism and/or on the calcium/sodium ratio in the hypothalamus. Hypothalamically administered calcium has long been known to enhance food intake (Myers *et al.*, 1972), and thus CT, which reduces calcium flux by neuronal tissue, may reduce feeding through this cellular mechanism. In support of this hypothesis, CT is effective in reversing calcium-induced eating and in reducing CA^{2+} uptake in hypothalamic tissue (Levine and Morley, 1981).

8. OTHER FEEDING-INHIBITORY PEPTIDES

There is a recent study that yields promising data for a potential role of the peptide glucagon in lateral hypothalamic control of eating (Inokuchi *et al.*, 1984a,b). Ventricular injection of pancreatic glucagon, at low doses of 5–25 ng, reliably suppresses food intake in rats, and a similar effect can be seen with lateral hypothalamic glucagon injection, which also suppresses the firing of local glucosensitive neurons (Oomura *et al.*, 1982). It is interesting that in the medial hypothalamus, in contrast to the lateral hypothalamus, the pancreatic peptide insulin acts similarly to reduce feeding (Hatfield *et al.*, 1974), possibly via its impact on the neural firing of glucose-responsive neurons in this area. A potential physiological role for insulin as a central satiety agent was more strongly indicated by the finding that insulin antibodies injected into the medial hypothalamus potentiate food intake (Strubbe and Mein, 1977). Chronic ventricular infusion of insulin is effective in decreasing both daily food intake and body weight gain (Brief and Davis, 1984; Woods *et al.*, 1979). A resistance to this effect has been detected in obese Zucker rats in contrast to their lean littermates (Ikeda *et al.*, 1983), suggesting the possibility that the insulin satiety system of the brain may be disturbed in these obese animals.

Bombesin, another peptide that acts peripherally to control food intake, is similarly found to suppress eating when administered into the lateral hypothalamus of the rat (Stuckey and Gibbs, 1982), apparently by decreasing meal size rather than meal frequency (Grinker *et al.*, 1984). However, extensive mapping studies indicate that this effect of bombesin can be observed throughout all hypothalamic areas as well as in many forebrain and hindbrain sites and that the magnitude of the eating suppression is strongly positively correlated with a potent enhancement in grooming behavior (Kyrkouli *et al.*, 1986).

The peptide corticotropin-releasing factor (CRF) seems to be burdened with similar complications, effective only at relatively high doses and inhibiting eating in association with enhanced grooming and signs of behavioral aversion (Gosnell *et al.*, 1983a). Although only limited mapping studies have been conducted to date, the evidence obtained points to the PVN as a site mediating CRF's anorexic action (Krahn *et al.*, 1984). It is of interest that this peptide functions specifically within the PVN to control corticosterone release (Ixart *et al.*, 1982; Swanson and Sawchenko, 1983) and that corticosterone itself acts within this nucleus to influence natural eating as well as α_2-noradrenergic control of eating (Leibowitz *et al.*, 1984b; Roland *et al.*, 1985; Leibowitz, 1986a). Although CRF is effective in inhibiting NE-induced feeding (Morley *et al.*, 1985), a similar interaction has been described for several other feeding-inhibitory peptides (see above), which argues against a unique role for CRF in relation to NE in the PVN. Furthermore, although the action of NE via α_2 receptors is dependent on adrenal release of corticosterone as well as on vagal control of the pancreas (Sawchenko *et al.*, 1981; Leibowitz *et al.*, 1984b), CRF is found to function independently of both of these endocrine–autonomic systems (Gosnell *et al.*, 1983b). Further mapping studies are required to determine whether CRF, like most other peptides, act via other brain sites in addition to the PVN.

9. CONCLUSION

It is clear from this review on peptides and eating behavior that these substances are highly active in altering food ingestion when injected directly into the brain as well as peripherally. This evidence provides an initial and important step in our quest to understand the neurochemical substrates of feeding behavior. As emphasized throughout this review, it is imperative that we concern ourselves with such critical questions as the behavioral and pharmacological specificity of these peptide effects. At the same time, however, we must advance beyond these injection studies to focus our attention on the endogenous peptide systems and to determine whether they are, in fact, active in controlling the normal feeding process. Needless to say, this task will be a formidable one in view of the apparent involvement of multiple brain areas, the complex interactions between different neurotransmitter systems, and the various technical limitations inherent in the peptide field. We are encouraged, however, by the significant advances made to date and expect very rapid progress over the next few years that should help us to establish the critical features of peptide control of meal patterns and macronutrient selection.

ACKNOWLEDGMENTS. The preparation of this manuscript and this laboratory's research described herein was supported by Grant MH 22879 and by a grant from the Whitehall Foundation.

REFERENCES

Adrian, T. E., Allen, J. M., Bloom, S. R., Ghatei, M. A., Rossor, M. N., Crow, T. J., Tatemoto, K., and Polak, J. M., 1983, Neuropeptide Y distribution in human brain, *Nature* **306:**584–586.

Agnati, L. F., Fuxe, K., Benfenati, F., Battistini, N., Harfstrand, K., Tatemoto, K., Hokfelt, T., and Mutt, V., 1983, Neuropeptide Y *in vitro* selectively increases the number of α_2-adrenergic binding sites in membranes of the medulla oblongata of the rat, *Acta Physiol. Scand.* **118:**293–295.

Allen, Y. S., Adrian, T. E., Allen, J. M., Tatemoto, K., Crow, T. J., Bloom, S. R., and Polak, J. M., 1983, Neuropeptide Y distribution in the rat brain, *Science* **221:**877–879.

Antin, J., Gibbs, J., Holt, J., Young, R. C., and Smith, G. P., 1975, Cholecystokinin elicits the complete behavioral sequence of satiety in rats, *J. Comp. Physiol. Psychol.* **87:**784–790.

Apfelbaum, M., and Mandenoff, A., 1981, Naltrexone suppresses hyperphagia induced in the rat by a highly palatable diet, *Pharmacol. Biochem. Behav.* **15:**89–91.

Atrens, D. M., and von Vietinghoff-Riesch, F., 1972, The motivational properties of electrical stimulation of the medial and paraventricular hypothalamic nuclei, *Physiol. Behav.* **9:**229–235.

Baile, D. A., and Della-Fera, M. A., Peptidergic control of food intake in food-producing animals, 1984, *Fed. Proc.* **43:**2898–2902.

Bellinger, L. L., Bernardis, L. L., and Williams, F. E., 1983, Naloxone suppression of food and water intake and cholecystokin reduction of feeding is attenuated in weanling rats with dorsomedial hypothalamic lesions, *Physiol. Behav.* **31:**839–846.

Bertierre, M. C., Mame Sy, T., Baigts, F., Mandenoff, A., and Apfelbaum, M., 1984, Stress and sucrose hyperphagia: Role of endogenous opiates, *Pharmacol. Biochem. Behav.* **20:**675–679.

Bhakthavatsalam, P., and Leibowitz, S. F., 1986a, Morphine-elicited feeding: Diurnal rhythm, circulation corticosterone and macronutrient selection, *Pharmacol. Biochem. Behav.* **24:**911–917.

Bhakthavatsalam, P., and Leibowitz, S. F., 1986b, α_2-Noradrenergic feeding rhythm in paraventricular nucleus: Relation to corticosterone, *Am. J. Physiol.* **250:**R83–R88.

Brief, D. J., and Davis, J. D., 1984, Reduction of food intake and body weight by chronic intraventricular insulin infusion, *Brain Res. Bull.* **12:**571–575.

Carr, K. D., and Simon, E. J., 1983, Effects of naloxone and its quarternary analogues on stimulation-induced feeding, *Neuropharmacology* **22:**127–130.

Checler, F., Vincent, J.-P., and Kitabgi, P., 1983, Neurotensin analogs [D-Tyr II] and [D-Phe II] neurotensin resist degradation by brain peptidases *in vitro* and *in vivo, J. Pharmacol. Exp. Ther.* **227:**743–748.

Clark, J. T., Kalra, P. S., Crowley, W. R., and Kalra, S. P., 1984, Neuropeptide Y and human pancreatic polypeptide stimulate feeding behavior in rats, *Endocrinology* **115:**427–429.

Cooper, S. J., 1983, Benzodiazepine–opiate antagonist interactions in relation to feeding and drinking behavior, *Life Sci.* **32:**1043–1051.

Crawley, J. N., and Knas, J. Z., 1984, Tracing the sensory pathway from gut to brain regions mediating the actions of cholecystokinin on feeding and exploration, *Soc. Neurosci. Abstr.* **10:**533.

DeBeaurepaire, R., and Freed, W. J., 1983, Anorectic effect of calcitonin: Localization to the paraventricular neucleus of the hypothalamus, *Soc. Neurosci. Abstr.* **9:**188.

Della-Fera, M. A., and Baile, C. A., 1979, Cholecystokinin octapeptide: Continuous picomole injections into the cerebral ventricles of sheep suppress feeding, *Science* **206:**471–473.

Della-Fera, M. A., and Baile, C. A., 1980a, CCK-octapeptide injected in CSF decreases meal size and daily food intake in sheep, *Peptides* **1:**51–54.

Della-Fera, M. A., and Baile, C. A., 1980b, Cerebral ventricular injection of CCK-octapeptide and food intake: The importance of continuous injection, *Physiol. Behav.* **24:**1133–1138.

Della-Fera, M. A., and Baile, C. A., 1981, Peptides with CCK-like activity administered intracranially elicit satiety in sheep, *Physiol. Behav.* **26:**979–983.

Della-Fera, M. A., Baile, C. A., Schneider, B. S., and Grinker, J. A., 1981, Cholecystokinin antibody injected in cerebral ventricles stimulates feeding in sheep, *Science* **212:**687–689.

Della-Fera, M. A., Baile, C. A., and Beinfeld, M. C., 1982, Cerebral ventricular transport and uptake: Importance of CCK-mediated satiety, *Peptides* **3:**963–968.

Dum, J., Gramsch, C. H., and Herz, A., 1983, Activation of hypothalamic β-endorphin pools by reward induced by highly palatable food, *Pharmacol. Biochem. Behav.* **18:**443–447.

Everitt, B. J., Hokfelt, T., Terenius, L., Tatemoto, K., Mutt, V., and Goldstein, M., 1984, Differential coexistence of neuropeptide Y (NPY)-like immunoreactivity with catecholamine in the central nervous system of the rat, *Neuroscience* **11:**443–462.

Faris, P. L., Komisaruk, B. R., Watkins, L. R., and Mayer, D. J., 1983a, Evidence for the neuropeptide cholecystokinin as an antagonist of opiate analgesia, *Science* **219:**310–312.

Faris, P. L., Scallet, A. C., Olney, J. W., Della-Fera, M. A., and Baile, C. A., 1983b, Behavioral and immunohistochemical analysis of the function of cholecystokinin in the hypothalamic paraventricular nucleus, *Soc. Neurosci. Abstr.* **9:**184.

Freed, W. J., Perlow, M. J., and Wyatt, R. D., 1979, Calcitonin: Inhibitory effect on eating in rats, *Science* **206:**850–852.

Gambert, S. R., Garthwaite, T. L., Pontzer, C. H., and Hagen, T. C., 1980, Fasting associated with decrease in hypothalamic β-endorphin, *Science* **210:**1271–1272.

Gibbs, J., and Smith, G. P., 1984, The neuroendocrinology of postprandial satiety, in: *Frontiers in Neuroendocrinology,* Vol. 8 (L. Martin and W. F. Ganong, eds.), Raven Press, New York, pp. 223–246.

Gibbs, J., and Smith, G. P., 1986, Satiety: The roles of peptides from stomach and intestine, *Fed. Proc.* **45:**1391–1395.

Gibbs, J., Young, R. C., and Smith, G. P., 1973, Cholecystokinin elicits satiety in rats with open gastric fistulas, *Nature* **245:**323–325.

Gosnell, B. A., Morley, J. E., and Levine, A. S., 1983a, A comparison of the effects of corticotropin releasing factor and sauvagine on food intake, *Pharmacol. Biochem. Behav.* **19:**771–775.

Gosnell, B. A., Morley, J. E., and Levine, A. S., 1983b, Adrenal modulation of the inhibitory effects of corticotropin releasing factor on feeding, *Peptides* **4:**807–812.

Gosnell, B. G., Morley, J. E., and Levine, A. S., 1984, Localization of naloxone-sensitive brain areas in relation to food intake, *Soc. Neurosci. Abstr.* **10:**306.

Grandison, L., and Guidotti, A., 1977, Stimulation of food intake by muscimol and β-endorphin, *Neuropharmacology* **16:**533–536.

Grinker, J. A., Schneider, B. S., Leibowitz, S. F., Cohen, A., and Gruen, R., 1984, Bombesin and CCK: Effects of central injections on feeding patterns and metabolism, in: *Proceedings of the Satellite Symposium of the 1984 Society for Neuroscience Meeting, on The Neural and Metabolic Bases of Feeding,* University of California, Davis, p. 24.

Hatfield, J. S., Millard, J., and Smith, C. J. V., 1974, Short-term influence of intra-ventromedial hypothalamic administration of insulin on feeding in normal and diabetic rats, *Pharmacol. Biochem. Behav.* **2:**223–226.

Henry, J. L., 1982, Circulating opioids: Possible physiological roles in cerebral nervous function, *Neurosci. Biobehav. Rev.* **6:**229–245.

Hermann, G., and Novin, D., 1980, Morphine-inhibition of parabrachial taste units reversed by anoxone, *Brain Res. Bull.* **5**(Suppl. 4)**:**169–173.

Hetherington, A. W., and Ranson, S. W., 1940, Hypothalamic lesions and adiposity in the rat, *Anat. Rec.* **78:**149–179.

Hoebel, B. G., 1985, Integrative Peptides, *Brain Res. Bull.* **141:**525–528.

Ikeda, H., West, D. B., Pustek, J. J., and Woods, S. C., 1983, Insulin infused intraventricularly reduces food intake and body weight of lean but not obese *(fa/fa)* Zucker rats, *Diabetes* **32**(Suppl. 1)**:**61A.

Inokuchi, A., Oomura, Y., and Nishimura, H., 1984, Effect of intracerebroventricularly infused glucagon on feeding behavior, *Physiol. Behav.* **33:**397–400.

Inokuchi, A., Nishimura, H., Zierger, U., Oomura, Y., and Shimizu, N., 1984b, The effects of glucagon on feeding behavior, *Neurosci. Lett.* **17**(Suppl.)**:**S99.

Ixart, G., Alonso, G., Szafarczyk, A., Malaval, F., Nouguier-Soulé, J., and Assenmacher, I., 1982, Adrenocorticotropic regulations after bilateral lesions of the paraventricular or supraoptic nuclei and in Brattleboro rats, *Neuroendocrinology* **35:**270–276.

Jhanwar-Uniyal, M., Woods, J. S., Levin, B. E., and Leibowitz, S. F., 1984, Cannula mapping and biochemical studies of the hypothalamic opiate and noradrenergic systems in relation to eating behavior, *Proc. East. Psychol. Assoc.* **55:**106.

King, B. M., Castellanos, F. X., Kastin, A. J., Berzas, M. C., Mauk, M. D., Olson, G. A., and Olson, R. D., 1979, Naloxone-induced suppression of food intake in normal and hypothalamic obese rats, *Pharmacol. Biochem. Behav.* **11:**729–732.

Krahn, D. D., Gosnell, B. A., Levine, A. S., Morley, J. E., 1984, Localization of the effects of corticotropin releasing factor on feeding, *Soc. Neurosci. Abstr.* **10:**302.

Krieger, D. T., 1983, Brain peptides: What, where and why? *Science* **222:**975–985.

Krieger, D. T., and Hauser, H., 1978, Comparison of synchronization of circadian corticosteroid rhythms by photoperiod and food, *Proc. Natl. Acad. Sci. U.S.A.* **75:**1577–1581.

Kyrkouli, S. E., Stanley, B. G., and Leibowitz, S. F., 1986, Bombesin induced anorexia: Sites of action in the rat brain, *Peptides* (in press).

Lanthier, D., Stanley, B. G., and Leibowitz, S. F., 1985, Feeding elicited by central morphine injection: Sites of action in the brain, *Proc. East. Psychol. Assoc.* **56:**30.

Leibowitz, S. F., 1978, Paraventricular nucleus: A primary site mediating adrenergic stimulation of feeding and drinking, *Pharmacol. Biochem. Behav.* **8:**163–175.

Leibowitz, S. F., 1980, Neurochemical systems of the hypothalamus. Control of feeding and drinking behavior and water-electrolyte excretion, in: *Handbook of the Hypothalamus,* Vol. 3A (P. J. Morgane and J. Panksepp, eds.), Marcel Dekker, New York, pp. 297–437.

Leibowitz, S. F., 1985, Brain neurotransmitters and appetite regulation, *Psychopharmacol. Bull.* **21:**412–418.

Leibowitz, S. F., 1986a, Brain monoamines and peptides: Role in the control of eating behavior, *Fed. Proc.* **45:**1396–1403.

Leibowitz, S. F., 1986b, Opiate, α_2-noradrenergic and adrenocorticotropin systems of hypothalamic parventricular nucleus, in: *Perspectives on Behavioral Medicine,* Vol. 5 (A. Baum, ed.), Academic Press, New York (in press).

Leibowitz, S. F., and Hor, L., 1982, Endorphinergic and α-noradrenergic systems in the paraventricular nucleus: Effects on eating behavior, *Peptides* **3:**421–428.

Leibowitz, S. F., and Shor-Posner, G., 1986, Hypothalamic monoamine systems for control of food intake: Analysis of meal patterns and macronutrient selection, in: *Psychopharmacology of Eating Disorders: Theoretical and Clinical Development* (M. O. Carruba and J. E. Blundell, eds.), Raven Press, New York, pp. 29–49.

Leibowitz, S. F., Jhanwar-Uniyal, M., and Roland, C. R. 1984a, Circadian rhythms of circulating corticosterone and α_2-noradrenergic receptors in discrete hypothalamic and extra-hypothalamic areas of the rat brain, *Soc. Neurosci. Abstr.* **10:**294.

Leibowitz, S. F., Roland, C. R., Hor, L., and Squillari, V., 1984b, Noradrenergic feeding elicited via the paraventricular nucleus is dependent upon circulating corticosterone, *Physiol. Behav.* **32:**857–864.

Leibowitz, S. F., Weiss, G. F., Yee, F., and Tretter, J. R., 1985, Noradrenergic innervation of the paraventricular nucleus: Specific role in control of carbohydrate ingestion, *Brain Res. Bull.* **14:**561–567.

LeMagnen, J., Marfaing-Jallat, P., Miceli, D., and Devos, M., 1980, Pain modulating and reward systems: A single brain mechanism? *Pharmacol. Biochem. Behav.* **12:**729–733.

Levine, A. S., and Morley, J. E., 1981, Reduction of feeding in rats by calcitonin, *Brain Res.* **222:**187–191.

Levine, A. S., and Morley, J. E., 1983, Adrenal modulation of opiate induced feeding, *Pharmacol. Biochem. Behav.* **19:**403–406.

Levine, A. S., and Morley, J. E., 1984, Neuropeptide Y: A potent inducer of consummatory behavior in rats, *Peptides* **5:**1025–1029.

Levine, A. S., Kneip, J., Grace, M., and Morley, J. E., 1983, Effect of centrally administered neurotensin on multiple feeding paradigms, *Pharmacol. Biochem. Behav.* **18:**19–23.

Lundberg, J. M., Terenius, L., Hökfelt, T., and Tatemoto, K., 1984, Catecholamines, neuropeptide Y (NPY), and the pancreatic polypeptide family: Coexistence and interaction in the sympathetic response, in: *Catecholamines: Neuropharmacology and Central Nervous System—Theoretical Aspects* (E. Usdin, A. Carlsson, A. Dahlstrom, and J. Engel, eds.), Alan R. Liss, New York, pp. 179–189.

Luttinger, D., King, R. A., Sheppard, D., Struff, O., Nemeroff, C. B., and Prange, A. J., Jr., 1982, The effect of neurotensin on food consumption in the rat, *Eur. J. Pharmacol.* **81:**499–503.

Margules, D. L., 1979, Beta-endorphin and endoloxone: Hormones for the autonomic nervous system for the conservation or expenditure of bodily resources and energy in anticipation of famine or feast, *Neurosci. Biobehav. Rev.* **3:**155–162.

Marks-Kaufman, R., 1982, Increased fat consumption induced by morphine administration in rats, *Pharmacol. Biochem. Behav.* **16:**949–955.

McGivern, R. F., and Berntson, G. G., 1980, Mediation of diurnal fluctuations in pain sensitivity in the rat by food intake patterns: Reversal by naloxone, *Science* **210:**210–211.

McGivern, R. F., Berka, C., Berntson, G. B., Walker, J. M., and Snadman, C. A., 1979, Effect of naloxone on analgesia induced by food deprivation, *Life Sci.* **25:**885–888.

McLaughlin, C. L., Baile, C. A., Della-Fera, M. A., and Kasser, T. G., 1985, Meal-stimulated increased hypothalamic concentrations of CCK in the hypothalamus of Zucker obese and lean rats, *Physiol. Behav.* **35:**215–220.

McLean, S., and Hoebel, B. G., 1982, Opiate and norepinephrine-induced feeding from the paraventricular nucleus of the hypothalamus are dissociable, *Life Sci.* **31:**2379–2382.

Morley, J. E., 1982, The ascent of cholecystokinin (CCK)—from gut to brain, 1982, *Life Sci.* **30:**479–493.

Morley, J. E., Levine, A. S., Yim, G. K., and Lowy, M. T., 1983, Opioid modulation of appetite, *Neurosci. Biobehav. Rev.* **7:**281–305.

Morley, J. E., Levine, A. S., Gosnell, B. A., and Krahn, D. D., 1985, Peptides as central regulators of feeding, *Brain Res. Bull.* **14:**511–519.

Morley, J. E., Levine, A. S., Grace, M. and Kneip, J., 1985, Peptide YY(PYY), a potent oxerigenic agent, *Brain Res.* **341:**200–203.

Moss, R. L., Urban, I., and Cross, B. A., 1972, Microelectrophoresis of cholinergic and aminergic drugs on paraventricular neurons, *Am. J. Physiol.* **223:**310–318.

Myers, R. D., 1985, Peptide-catecholamine interactions: Feeding and satiety, *Psychopharmacol. Bull.* **21:**406–411.

Myers, R. D., and McCaleb, M. L., 1981, Peripheral and intrahypothalamic cholecystokinin acts on the noradrenergic "feeding circuit" in the rat's diencephalon, *Neuroscience* **6:**645–655.

Myers, R. D., Bender, S. A., Krstic, M., and Brophy, P. D., 1972, Feeding produced in the satiated rat by elevating the concentration of calcium in the brain, *Science* **176:**1124–1125.

Olschowka, J. A., 1984, Neuropeptide Y innervation of the rat paraventricular and supraoptic nuclei, *Soc. Neurosci. Abstr.* **10:**437.

Olson, G. A., Olson, R. D., and Kastin, A. J., 1984, Endogenous opiates: 1983, *Peptides* **5:**975–992.

Oomura, Y., Shimizu, N., Miyahara, S., and Hattori, K., 1982, Chemosensitive neurons in the hypothalamus—do they relate to feeding behavior? in: *The Neural Basis of Feeding and Reward* (B.

G. Hoebel and D. Novin, eds.), Haer Institute for Electrophysiological Research, Brunswick, Maine, pp. 551–566.

Palkovits, M., and Brownstein, M. J., 1985, Distribution of neuropeptides in the central nervous system using biochemical micromethods, in: *GABA and Neuropeptides in the CNS*, Vol. 4, *Handbook of Chemical Neuroanatomy* (A. Bjorklund and T. Hökfelt, eds.) Elsevier, Amsterdam, pp. 1–71.

Passaro, E., Debas, H., Oldendorf, W., and Yamada, T., 1982, Rapid appearance of intraventricularly administered neuropeptide in the peripheral circulation, *Brain Res.* **241**:335–340.

Perlow, M. J., Freed, W. J., Carman, J. S., and Wyatt, R. D., 1980, Calcitonin reduces feeding in man, monkey and rat, *Pharmacol. Biochem. Behav.* **12**:609–612.

Pittman, Q. J., Hatton, J. D., and Bloom, F. E., 1980, Morphine and opioid peptides reduce paraventricular neuronal activity: Studies on the rat hypothalamic slice preparation, *Proc. Natl. Acad. Sci. U.S.A.* **77**:5527–5531.

Ritter, R. C., and Ladenhein, E. E., 1984, Fourth ventricular cholecystokinin suppresses feeding in rats, *Soc. Neurosci. Abstr.* **10**:652.

Roland, C. R., Bhakthavatsalam, P., and Leibowitz, S. F., 1985, Interaction between corticosterone and α_2-noradrenergic system of the paraventricular nucleus in relation to feeding behavior, *Neuroendocrinology* **42**:296–305.

Saito, A., Williams, J. A., and Goldfine, I. D., 1981, Alterations in brain cholecystokinin receptors after fasting, *Nature* **289**:599–600.

Sanger, D. J., 1981, Endorphinergic mechanisms in the control of food and water intake, *Appetite* **2**:193–208.

Sawchenko, P. E., Gold, R. M., and Leibowitz, S. F., 1981, Evidence for vagal involvement in the eating elicited by adrenergic stimulation of the paraventricular nucleus, *Brain Res.* **225**: 249–269.

Sawchenko, P. E., Swanson, L.W., Grzanna, R., Howe, P.R.C., Bloom, S. R., and Polak, J.M., 1985, Co-localization of Neuropeptide Y-immuno-reactivity in brainstem catecholaminergic neurons that project to the paraventricular nucleus of the hypothalamus. *J. Comp. Neurol.* **241**:138–153.

Scallet, A. C., Della-Fera, M. A., and Baile, C. A., 1984, Changes in cholecystokinin (CCK) content of specific hypothalamic areas of sheep with feeding and fasting, *Soc. Neurosci. Abstr.* **10**:652.

Scott, P., Jawaharlal, K., and Hoebel, B. G., 1984, Feeding induced with kappa and mu agonists injected in the region of the paraventricular nucleus (PVN) in rats, *Proc. East. Psychol. Assoc.* **55**:106.

Shor-Posner, G., Azar, A., Filart, R., Tempel, D., and Leibowitz, S. F., 1985, Morphine-stimulated feeding: Analysis of macronutrient selection and paraventricular nucleus lesions, *Pharmacol. Biochem. Behav.* **248**:931–939.

Siviy, S. M., and Reid, L. D., 1983, Endorphinergic modulation of acceptability of putative reinforcers, *Appetite* **4**:249–257.

Stanley, B. G., 1982, Neurotensin: Evidence for a role in satiety, Ph.D. thesis, Princeton University.

Stanley, B. G., and Leibowitz, S. F., 1984a, Neuropeptide Y: Stimulation of feeding and drinking by injection into the paraventricular nucleus, *Life Sci.* **35**:2635–2642.

Stanley, B. G., and Leibowitz, S. F., 1984b, Neuropeptide Y injected into the hypothalamus elicits feeding behavior in the rat, in: *Proceedings of the Satellite Symposium of the 1984 Society for Neuroscience Meeting, on the Neural and Metabolic Bases of Feeding*, University of California, Davis, p. 7.

Stanley, B. G., and Leibowitz, S. F., 1985a, Regulation of feeding behavior by neuropeptide Y and peptide YY, in: *Proceedings of the Neural and Endocrine Peptides and Receptors Symposium*, Washington, D.C., p. 67.

Stanley, B. G., and Leibowitz, S. F., 1985b, Neuropeptide Y injected into the paraventricular hypothalamus: A powerful stimulant of feeding behavior, *Proc. Natl. Acad. Sci. U.S.A.* **82**: 3940–3943.

Stanley, B. G., Hoebel, B. G., and Leibowitz, S. F., 1983, Neurotensin: Effects of hypothalamic and intravenous injections on feeding and drinking in rats, *Peptides* **4**:493–500.

Stanley, B. G., Lanthier, D., and Leibowitz, S. F., 1984, Feeding elicited by the opiate peptide D-Ala-2-Met-enkephalinamide: Sites of action in the brain, *Neurosci. Abstr.* **10:**1103.

Stanley, B. G., Chin, A. S., and Leibowitz, S. F., 1985a, Feeding and drinking elicited by central injection of neuropeptide Y: Evidence for a hypothalamic site(s) of action, *Brain Res. Bull.* **14:**521–524.

Stanley, B. G., Kyrkouli, S. E., Lampert, S., and Leibowitz, S. F., 1985b, Hyperphagia and obsesity induced by neuropeptide Y injected chronically into the paraventricular hypothalamus of the rat, *Soc. Neurosci. Abstr.* **11:**36.

Stanley, B. G., Leibowitz, S. F., Eppel, N., St. Pierre, S., and Hoebel, B. G., 1985c, Suppression of norepinephrine-elicited feeding by neurotension: Evidence for behavioral, anatomical and pharmacological specificity, *Brain Res.* **343:**297–304.

Stanley, B. G., Daniel, D. R., Chin, A. S., and Leibowitz, S. F., 1985d, Paraventricular nucleus injections of peptide YY and neuropeptide Y selectively enhance carbohydrate ingestion, *Peptides* **6:**1205–1211.

Steffens, A. B., 1969, Rapid absorption of glucose in the intestinal tract of the rat after ingestion of a meal, *Physiol. Behav.* **4:**829–832.

Strubbe, J. H., and Mein, C. G., 1977, Increased feeding in response to bilateral injections of insulin antibodies in the VMH, *Physiol. Behav.* **19:**309–314.

Stuckey, J. A., and Gibbs, S., 1982, Lateral hypothalamic injection of bombesin decreases food intake in rats, *Brain Res. Bull.* **8:**617–621.

Swanson, L. W., and Sawchenko, P. E., 1983, Hypothalamic integration: Organization of the paraventricular and supraoptic nuclei, *Annu. Rev. Neurosci.* **6:**269–324.

Tepperman, F. S., Hirst, M., and Gowdey, C. W., 1981a, Hypothalamic injection of morphine: Feeding and temperature responses, *Life Sci.* **28:**2459–2467.

Tepperman, F. S., Hirst, M., and Gowdey, W., 1981b, A probable role for norepinephrine in feeding after hypothalamic injection of morphine, *Pharmacol. Biochem. Behav.* **15:**555–558.

Twery, M. J., Obie, J. F., and Cooper, C. W., 1982, Ability of calcitonin to alter food and water consumption in the rat, *Peptides* **3:**749–755.

Unden, A., Tatemoto, K., and Bartfai, T., 1983, Receptors for neuropeptide Y in rat brain, *Soc. Neurosci. Abstr.* **9:**170.

Unnerstall, J. R., Kopajtic, T. A., and Kuhar, M. J., 1984, Distribution of α_2-agonist binding sites in the rat and human central nervous system: Analysis of some functional, anatomic correlates of the pharmacologic effcts of clonidine and related adrenergic agents, *Brain Res. Rev.* **7:**69–101.

Vaccarino, F. J., Bloom, F. E., Rivier, J., Vale, W., and Koob, G. F., 1985, Stimulation of food intake in rats by centrally administered hypothalamic growth hormone-releasing factor, *Nature* **314:**167–168.

Weiss, G. F., and Leibowitz, S. F., 1985, Efferent projections from the paraventricular nucleus mediating α_2-noradrenergic feeding, *Brain Res.* **347:**225–238.

Willis, G. L., Hansky, J., and Smith, G. C., 1984, Ventricular, paraventricular and circumventricular structures involved in peptide-induced satiety, *Regul. Peptides* **9:**87–99.

Woods, S. C., Lotter, E. C., McKay, L. D., and Porte, Jr., D., 1979, Chronic intracerebroventricular infusion of insulin reduces food intake and body weight of baboons, *Nature* **282:**503–505.

Woods, J. S., and Leibowitz, S. F., 1985, Hypothalamic sites sensitive to morphine and naloxone: Effects on feeding behavior, *Pharmacol. Biochem. Behav.* **23:**431–438.

Woods, J. S., and Porte, D., 1983, The role of insulin as a satiety factor in the central nervous system, in: *Advances in Metabolic Disorders,* Vol. 10 (A. J. Szabo, ed.), Academic Press, New York, p. 457.

24

Neuropeptide Y

An Integrator of Reproductive and Appetitive Functions

SATYA P. KALRA, L. G. ALLEN, J. T. CLARK, W. R. CROWLEY, and P. S. KALRA

1. INTRODUCTION

Information processing in the brain largely involves communication among neurons through the release of neurotransmitters at synapses. In addition to the amino acids and biogenic amines, it now appears that neuropeptides may also mediate interneuronal communications to coordinate a variety of neuroendocrine and behavior functions. The list of neuropeptides as putative modulators or mediators in the brain is growing steadily along with the diversity of physiological functions ascribed to them at various peripheral and central sites. In addition, recent neurochemical and morphological studies indicate that single neurons might contain and release more than one transmitter (Hökfelt *et al.*, 1984; Chan-Palay and Palay, 1984). Several neuropeptides have been shown to coexist intraneuronally with the classical neurotransmitters—monoamines and acetylcholine. Thus, the emerging possibility that more than one neurotransmitter and neuromodulator may be coreleased and coact has prompted a reexamination of neuronal communications controlling the basic functions of reproductive and appetitive behaviors.

Neuropeptide Y (NPY), recently isolated and chemically characterized from porcine brain by Tatemoto (1982), is structurally related to the pancreatic polypeptides (Lin, 1980). It is a 36-amino-acid peptide and is reported to be one of the

SATYA P. KALRA, L. G. ALLEN, J. T. CLARK, and P. S. KALRA • Department of Obstetrics and Gynecology, University of Florida College of Medicine, Gainesville, Florida 32610. W. R. CROWLEY • Department of Pharmacology, University of Tennessee College of Medicine, Memphis, Tennessee 38163.

most widespread neuropeptides (Tatemoto, 1982; Tatemoto *et al.*, 1982). Mapping studies utilizing immunocytochemistry and radioimmunoassay have revealed a wide distribution of NPY-containing pathways in the rat brain (Allen *et al.*, 1983; Guy *et al.*, 1983; Everitt *et al.*, 1984; Chronwall *et al.*, 1985). Within the hypothalamus, perikarya containing NPY immunoreactivity were seen in the arcuate nucleus, periventricular nucleus, and in the septum. In addition, intensely NPY-positive nerve terminals were seen in the lateral septum, stria terminalis, and suprachiasmatic, paraventricular, and dorsomedial nuclei. Further, several investigations have affirmed the coexistence of NPY with catecholamines in peripheral and central neurons (Lundberg *et al.*, 1982a,b; Everitt *et al.*, 1984; Varndell *et al.*, 1984; Emson and DeQuidt, 1984). The coexistence of NPY with several norepinephrine (NE)- and epinephrine (E)-containing cells in the brainstem is of particular interest. Neuropeptide Y immunoreactivity was detected in a majority of NE-containing perikarya of the A1/C1 cell groups in the ventral lateral medulla oblongata and the A6 cell group in the locus coeruleus. Similarly, a large number of E-containing cell bodies of the C2 cell group displayed NPY-immunoreactivity (Everitt *et al.*, 1984).

Although it is premature to ascribe a neurotransmitter function to NPY, the widespread distribution pattern both centrally and peripherally and its coexistence with adrenergic transmitters are suggestive of a wide range of physiological roles of NPY. In our early studies, we focused on defining the significance of NPY coexistence with adrenergic transmitters in the control of gonadotropin secretion (Kalra and Crowley, 1984a,b). The subsequent serendipitous observation of stimulation of feeding behavior by human pancreatic peptide (hPP) and NPY led us to assess systematically the role of NPY as an integrator of several basic brain functions. What follows are our findings to date towards that goal.

2. NEUROPEPTIDE Y AND GONADOTROPIN SECRETION

It is well established that luteinizing hormone-releasing hormone (LHRH) is the primary neurochemical signal from the hypothalamus that regulates pituitary gonadotropin secretion (Kalra and Kalra, 1983; Kalra, 1986). Within the hypothalamus, adrenergic systems may mediate the discharge of LHRH into the hypophyseal portal system for transport to the pituitary (Kalra and Kalra, 1985; Kalra, 1985, 1986; Crowley *et al.*, 1982). Anatomic and physiological studies indicate that functional links between the adrenergic and LHRH neuronal networks may occur in the preoptic–tuberal pathway of the hypothalamus (Kalra and McCann, 1973; Simpkins and Kalra, 1979; Simpkins *et al.*, 1980). However, the precise mechanisms modulating the release of the adrenergic transmitters in the vicinity of LHRH neurons are poorly understood (Kalra and Kalra, 1984; Kalra, 1985, 1986). It is likely that a host of peptidergic neural circuits may control the output of adrenergic afferents locally within the hypothalamus (Kalra, 1983, 1985). For example, opioid neurons are believed to restrain LHRH secretion by inhibiting the influx of adrenergic transmitters in the proximity of LHRH inervations (Kalra and Simpkins, 1981; Kalra, 1981; Leadem *et al.,* 1985). Is there a complementary peptidergic system(s) that promotes LHRH secretion by augmenting either the adrenergic transmitter discharge or the response of LHRH neurons to the adrenergic transmitter?

Neuropeptide Y has been shown to coexist in several NE- and E-containing neurons in the brainstem, the afferents of which project extensively into the preoptic–tuberal pathway (Everitt *et al.,* 1984; Chronwall *et al.,* 1985; Emson and DeQuidt, 1984). Further, there is some evidence that NPY may modulate NE release peripherally, and it may even enhance adrenergic function in some instances (Edvinsson, 1985; Lundberg and Stjarne, 1984; Emson and DeQuidt, 1984; Füxe *et al.,* 1983; Agnati *et al.,* 1983). These observations led us to examine the role of the NPY neural system in the control of LHRH secretion. In the following section our recent findings on the effects of NPY on LH release and the interaction of NPY with adrenergic neural systems in excitation of LH release in the rat are summarized.

2.1. Effects of NPY on LH Release in Ovariectomized Rats

We observed that NPY evoked LH responses in ovariectomized rats in a manner previously documented for NE (Kalra and Crowley, 1984b). Injection of NPY into the third ventricle of the brain suppressed LH release for over 30 min in long-term ovariectomized rats (Fig. 1). Interestingly, LH suppression occurred in an "all-or-none" fashion, since the low dose of 0.1 μg was completely ineffective, and the attenuation of LH secretion produced by 0.5 and 2.0 μg was quantitatively similar. These observations have been confirmed by the recent report of McDonald *et al.* (1985a).

Recently, we have noted that the magnitude of LH response can be amplified and prolonged by continuous NPY treatment (S. P. Kalra, unpublished data). Within 30–60 min of intraventricular NPY infusion, LH secretion decreased to the basal

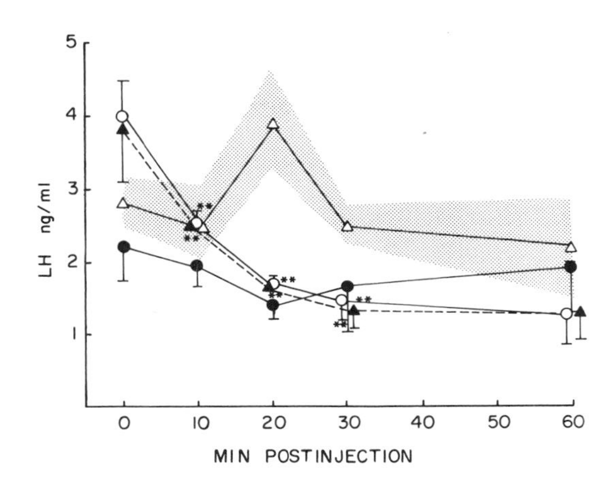

FIGURE 1. Effects of NPY on LH release in ovariectomized rats. $^{**}P < 0.01$ vs. 0 min. (With permission from Kalra and Crowley, 1984b.) △—△, saline (6); ●—●, NPY 0.1 μg (5); ○—○, NPY 0.5 μg (6); ▲---▲, NPY 2.0 μg (6).

range normally seen in gonad-intact rats, and the profound suppression, caused by a drastic reduction in the frequency and amplitude of LH pulses, persisted for the 2.5-hr duration of the experiment. The rapid suppression of LH secretion suggests that NPY may be more effective than gonadal steroids, which require several hours of continuous action to return LH secretion to the low basal range (Kalra *et al.*, 1973; Negro-Vilar, 1973; Kalra, 1983).

Intriguingly, the members of the pancreatic polypeptide family (Lin, 1980; Tatemoto, 1982; Emson and Dequidt, 1984), despite considerable structural differences from NPY, also suppress LH release. In ovariectomized rats, hPP decreased LH secretion in a similar fashion (Kalra and Crowley, 1984a), and McDonald *et al.* (1985b) reported that bovine and avian pancreatic polypeptides inhibited LH release. We suspect that a central action leading to inhibition of LHRH secretion may primarily be responsible for the decrease in LH secretion that follows NPY administration. In accord with this assumption are the findings that NPY, if anything, stimulated LH release from dispersed pituitary cells of ovariectomized rats (McDonald *et al.*, 1985a). It is possible that diminution in LHRH secretion may result from NPY-induced decrease in the influx of excitatory adrenergic transmitters or increase in the influx of inhibitory neurotransmitters such as dopamine or γ-aminobutyric acid (Kalra and Kalra, 1985; Colmers *et al.*, 1985; Hendry *et al.*, 1984). Alternatively, NPY may augment the release of a hypothalamic inhibitory factor(s), which may, in turn, counteract the LHRH-induced LH response of the gonadotrophs (Kalra, 1976; Leadem and Kalra, 1984).

2.2. Effects of NPY on LH Release in Steroid-Primed Ovariectomized and Intact Male Rats

In contrast to the inhibitory effects in ovariectomized rats, NPY was found to stimulate LH release in ovariectomized rats pretreated with estrogen and progesterone (Kalra and Crowley, 1984b; Fig. 2). Significant elevations in LH levels were

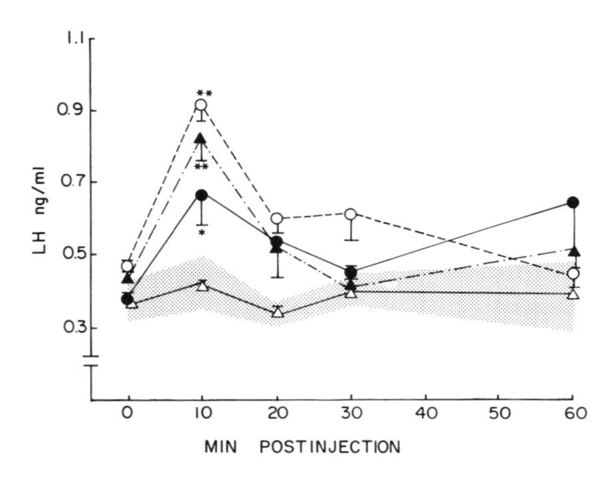

FIGURE 2. Effects of NPY on LH release in estrogen-, progesterone-primed ovariectomized rats. (With permission from Kalra and Crowley, 1984b.) *$P < 0.05$ vs. 0 min. **$P < 0.01$ vs. 0 min. △—△, saline (6); ●—●, NPY 0.5 μg (5); ○—○, NPY 2.0 μg (5); ▲---▲, NPY 10.0 μg (5).

seen within 10 min in a dose-related fashion after the administration of 0.5 or 2 μg NPY. These excitatory effects of NPY on LH release were also seen in gonad-intact male rats; however, a dose-related stimulation of LH release was not apparent (Allen *et al.*, 1985a). The relatively weaker LH response to NPY in male rats may be caused by circulating androgens, which may counteract the effects of LHRH at the level of pituitary gonadotrophs (Kalra and Kalra, 1983).

The excitation of LH release in steroid-primed ovariectomized rats and in intact male rats implies that NPY may play a physiological role in the control of LH secretion. Normally, NPY may stimulate the release of LHRH, and/or it may act at the pituitary level to potentiate the action of LHRH. Alternatively, NPY may interact with the excitatory hypothalamic adrenergic systems to stimulate LH release. We have used pharmacological probes to assess the role of the adrenergic systems in the NPY-induced LH release (Allen *et al.*, 1985b). Prior blockade of α-adrenergic receptors with phenoxybenzamine or α_1-adrenergic receptors with prazosin failed to block the NPY-induced LH release in the ovarian-steroid-primed rats. However, diethyldithiocarbamate, which suppresses NE and E brain levels (Estes *et al.*, 1982), and to a lesser extent LY 78335 (2,3-dicholoromethyl-α-benzylamine), which selectively decreases E levels (Kalra and Crowley, 1983; Adler *et al.*, 1983), attenuated the NPY-induced LH response. Since the disruption of central adrenergic transmission consistently failed to antagonize the stimulatory LH response, and the selective β-adrenergic and dopamine receptor antagonists were completely ineffective, it is likely that NPY may, in part, act independently of adrenergic systems to stimulate LH release.

In an attempt to ascertain further the physiological role of hypothalamic NPY neurons, we examined the temporal changes in NPY levels in several regions in the preoptic–tuberal pathway following the administration of gonadal steroids (Crowley *et al.*, 1985). Estrogen treatment decreased LH release in ovariectomized rats, and, interestingly, NPY concentrations in the median eminence–arcuate nucleus region were also significantly decreased. Furthermore, in agreement with our previous studies (Simpkins and Kalra, 1979; Simpkins *et al.*, 1980; Adler *et al.*, 1983), administration of progesterone to these estrogen-primed rats evoked a sequential rise and fall in LHRH levels in the median eminence in association with the LH surge. Surprisingly, the NPY concentrations in the median eminence also displayed sequential changes with a time course similar to that of the LHRH response. In addition, NPY concentrations in the medial preoptic nucleus, a site known to mediate the positive feedback effects of progesterone on LH release (Kalra and Kalra, 1985; Kalra, 1986), also decreased after progesterone treatment. It seems that ovarian steroid treatment regimens capable of influencing LH secretion can modify NPY levels in the regions innervated by LHRH neurons. Whether the concurrency of NPY and LHRH concentrations in the median eminence reflected a cause-and-effect relationship cannot be firmly stated, but our findings strongly support the proposal that the hypothalamic NPY circuit may participate in the inhibitory and stimulatory feedback effects of gonadal steroids on LH release.

3. NEUROPEPTIDE Y AND FEEDING BEHAVIOR

3.1. Effects of NPY on Feeding Behavior

While assessing the effects of hPP on LH release in 1983, we noted that the experimental rats displayed increased ingestive behavior after the intraventricular administration of hPP (Kalra and Crowley, 1984a). In a careful follow-up study, we not only reaffirmed these findings but further found that NPY also stimulated feeding in steroid-primed ovariectomized rats (Clark *et al.*, 1984). Successive intraventricular injections of NPY stimulated feeding by increasing the local eating rate. The onset of feeding was generally evident within 10–15 min after NPY administration. More recently, we have observed a dose-related stimulation of feeding in satiated intact male rats (Table I), and, surprisingly, NPY differentially influenced the parameters of feeding behavior (Allen *et al.*, 1985a; Clark *et al.*, 1985). For example, whereas 0.47 and 2.3 nM NPY stimulated a similar magnitude of cumulative food intake during the first hour, the increased intake after 0.47 nM was caused by an increase in local eating rate, whereas the increase after 2.3 nM was caused by increased time spent eating (Clark *et al.*, 1985). Neuropeptide Y was equally effective in satiated intact females, and, as in the case of males, the optimal response after 0.47 nM NPY occurred as a result of enhanced local eating rate (Clark and S. P. Kalra, 1985). These studies unequivocally show that NPY activates feeding in male and female rats during the lights-on period, when rats normally do not eat.

Extending these observations further, we have now observed that NPY is equally effective during the lights-off period (Clark *et al.*, 1985). Central injections of NPY during the nighttime stimulated food intake above the normal eating rate.

TABLE I. Food Intake in the Hour Following Injection of NE, E, or NPY into the Third Ventricle of the Rat Brain[a]

Exp.	Drug	Dose (nM)	No. eating/ no. rats	Latency to feeding (min)	Food intake (g/hr)
1	CSF	—	4/6	—	0.47 ± 0.29^{1}
	E	9.61	5/6	4.80 ± 0.79^{1}	$0.83 \pm 0.29^{1,2}$
	E	28.94	7/7	2.57 ± 0.37^{1}	1.50 ± 0.32^{2}
	E	86.97	5/5	3.40 ± 0.68^{1}	$0.71 \pm 0.29^{1,2}$
	NE	31.31	7/7	2.28 ± 0.18	1.54 ± 0.32^{2}
2	Saline	—	0/8	—	0^{1}
	NPY	0.12	3/7	19.00 ± 5.3	0.26 ± 0.13^{1}
	NPY	0.47	11/12	15.27 ± 2.3^{1}	1.42 ± 0.32^{2}
	NPY	2.35	12/12	12.25 ± 1.57^{1}	2.73 ± 0.36^{3}
	NPY	5.88	7/7	16.14 ± 4.10^{1}	$1.68 \pm 0.25^{2,3}$

[a] Data are presented as mean ± S.E.M. Within each experiment, similar superscript numbers indicate groups that were not significantly different ($P > 0.05$) as determined by the Mann–Whitney *U*-test. Control rats received either artificial cerebrospinal fluid (CSF) or saline.

In fact, cumulative food intake for the 4-hr test was increased in a dose-related fashion. Increase in feeding behavior during the nighttime occurred immediately after NPY administration and was primarily seen in the first 60-min period. The apparent decrease in the latency of response in conjunction with the ongoing feeding behavior is intriguing when compared with the relatively long latency of 10–15 min during the daytime in satiated rats (Clark *et al.*, 1984, 1985). Evidently, once the neural activation of feeding has begun, the central circuits become extremely sensitive to further stimulation with NPY. This additional versatility of NPY in enhancing food intake further supports the proposal that NPY may normally be released at crucial brain sites to activate feeding behavior.

3.2. Comparative Effects on Feeding of NPY and Members of the Pancreatic Polypeptides

Porcine NPY displays considerable amino acid sequence homologies with the members of the pancreatic polypeptide (PP) family (Tatemoto, 1982; Lin, 1980; Emson and DeQuidt, 1984; Kimmel *et al.*, 1984). Since NPY stimulated feeding in male and female rats, it was of interest to examine the effects on feeding of the members of the PP family and human NPY (hNPY, Cordor *et al.*, 1984). All the PP tested stimulated feeding in satiated rats; however, there were subtle differences in the parameters of appetitive behavior. Thus, hPP, which has identical residues in 20 positions, was not as effective as NPY, and the increase in food intake primarily reflected an increase in the time spent eating (Clark *et al.*, 1984). Rat pancreatic polypeptide (rPP), which shares only ten identical amino acid residues with NPY (Kimmel *et al.*, 1984), evoked a different pattern of feeding in male rats (Clark *et al.*, 1985; Clark and S. P. Kalra, 1985). It was considerably less effective than NPY or hPP, since only a high dose (0.45 nM) elicited a significant feeding response during the first hour after a latency of 12–15 min.

Interestingly, the gut peptide YY (PYY), which has identical residues in 25 positions (Tatemoto *et al.*, 1982; Lundberg *et al.*, 1982; Emson and DeQuidt, 1984), was only about half as effective as NPY in evoking food intake in satiated rats. Human NPY, isolated recently from human adrenal medullary pheochromocytoma tissue, differs from porcine NPY by substitution of methione in place of leucine at the 17 position (Corder *et al.*, 1984); it also stimulated feeding in satiated male rats (Clark and S. P. Kalra, 1985). The pattern and magnitude of cumulative feeding response after hNPY was quite similar to that evoked by other members of the PP family. Although preliminary in nature, these studies provide important new information on the structure–function relationships among NPY and members of the PP family. It seems that the PP, found either in the endocrine cells or in neurons, contain an amino acid sequence that reliably enhances feeding in satiated rats. Furthermore, substitutions in porcine NPY of either a single amino acid as in hNPY or multiple amino acids as in rPP and PYY profoundly alter the cumulative food intake response and differentially affect various parameters of feeding behavior.

3.3. Mode of Action of NPY in Stimulation of Feeding Behavior

Based on these findings, one can speculate that after injection into the third ventricle of the brain, NPY reaches receptive sites to stimulate feeding. These receptive sites may be different from those previously implicated in the control of feeding (Leibowitz, 1980). It is likely that the site(s) of NPY action may overlap or coincide with the sites where adrenergic transmitters have been shown to evoke a potent feeding response. Indeed, intracerebral injection of NPY into the paraventricular nucleus, a site of NE action, stimulated feeding in male rats (Stanley and Leibowitz, 1984).

In view of the reports that NPY coexists in many NE- and E-containing cells in the brainstem (Hökfelt *et al.*, 1984) and the overlapping distribution patterns of the afferents of these neurons with the innervations of hypothalamic NPY neurons (Allen *et al.*, 1983; Guy *et al.*, 1983; Everitt *et al.*, 1984; Chronwall *et al.*, 1985) in the diencephalic regions implicated in the control of feeding (Leibowitz, 1980), we have compared the effects of NPY and adrenergic transmitters on feeding in male rats (Allen *et al.*, 1985a). As shown in Table I, NPY stimulated feeding in a dose-related fashion, and an equivalent feeding response was evoked by 0.47 nM NPY, 31.31 nM NE, or 28.94 nM E. Thus, NPY is evidently more potent than either of the two catecholamines in eliciting the feeding response in satiated rats. However, there was a significant difference in the latency to feeding evoked by NPY and the catecholamines. Although NE or E activated feeding within 2–4 min, the NPY-induced response occurred after 10–15 min.

These findings, together with the fact that a long latency to feeding is seen after either intraventricular injection (Clark *et al.*, 1984, 1985) or injection directly into the paraventricular nucleus (Stanley and Leibowitz, 1984), raise the possibility that NPY may act at brain sites remote from injection sites. Additionally, these observations indicate that some other neurochemical signals may participate in NPY action. In this context, the reports that a number of NPY-immunoreactive perikarya, particularly those localized within the hypothalamus, have not yet been shown to colocalize NE or E are important (Allen *et al.*, 1983; Everitt *et al.*, 1984; Chronwall *et al.*, 1985). Perhaps NPY discharged from these neurons may provoke feeding at a hypothalamic site(s) independent of adrenergic involvement. Whatever may be the mode of NPY action, the discovery of the stimulatory effect of NPY on feeding, a response more potent than that seen after adrenergic transmitters (Table I) or that reported for opioid peptides (Morley and Levine, 1983), opens a new level of inquiry to further explain the complexities of the neural circuitry controlling feeding behavior in rats.

4. NEUROPEPTIDE Y AND SEX BEHAVIOR

4.1. Effects of NPY on Male Copulatory Behavior

The effects of NPY on copulatory behavior in sexually active male rats are shown in Fig. 3 (Clark *et al.*, 1985). Intraventricular NPY administration suppressed

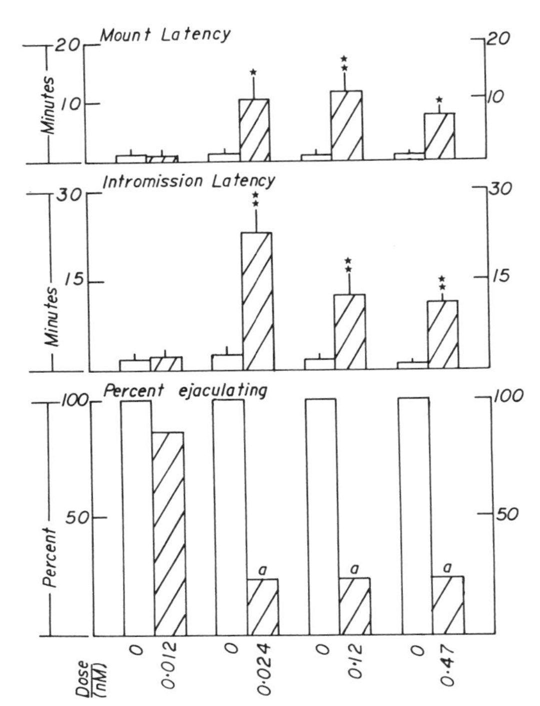

FIGURE 3. Effects of NPY or saline on mount and intromission latencies (min) and on the number of males achieving ejaculation in tests initiated 10 min after injection. 0, Test (open bars) after saline injection; others (hatched bars) after various doses of NPY; *saline values significantly lower than NPY values, $P < 0.05$; **saline values significantly lower than NPY values, $P < 0.01$; *a*, significantly fewer males ejaculated after NPY than after saline treatment, $P < 0.01$; $n = 7–9$ rats/treatment. (With permission from Clark *et al.*, 1985.)

various components of copulatory behavior. A drastic suppression of ejaculation was seen after administration of NPY in doses ranging from 0.02 to 0.47 nM, and the response displayed an "all-or-none" pattern reminiscent of the effects of NPY on LH release in unprimed ovariectomized rats (Fig. 1). In addition, NPY significantly increased the mount and intromission latencies. It is unlikely that this increased latency to initiate copulation played any major role in the dramatic suppression of ejaculatory behavior, since most NPY-treated rats displayed mounting and intromissive behavior but failed to ejaculate. Furthermore, failure to copulate was not caused by an imposed deficit in general sexual ability, since NPY produced no discernible adverse effect on the penile reflex activity. Interestingly, although rPP shares considerable amino acid homologies with NPY (Tatemoto, 1982; Kimmel *et al.*, 1984) and stimulates feeding (Clark *et al.*, 1985), it produced no untoward effects on any parameter of rat copulatory behavior.

4.2. Effects on NPY on Female Sexual Behavior

Encouraged by the findings of suppression of male copulatory behavior, we assessed the effects of NPY on female sexual behavior. Ovariectomized rats bearing permanent cannulae in the third ventricle were treated with a sequential regimen of estradiol and progesterone (Clark *et al.*, 1985; Clark and P. S. Kalra, 1985). Only those rats that displayed a good level of lordosis quotient (>70) were selected

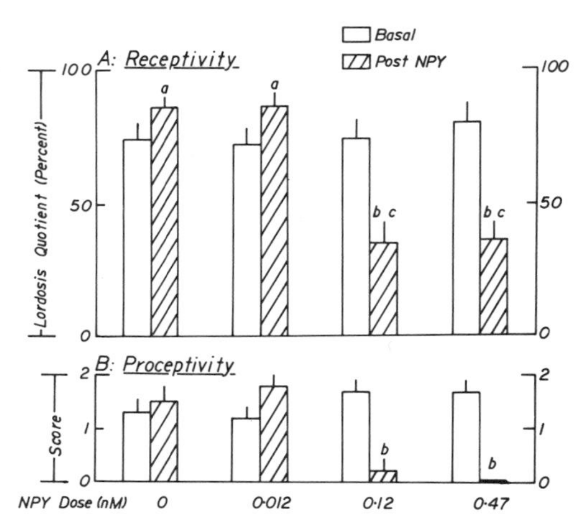

FIGURE 4. Effects of NPY on estrogen-plus progesterone-induced sexual behavior in ovariectomized rats. (A) Receptivity indicated by lordosis quotient; (B) proceptivity. Basal tests were conducted 4 hr after progesterone, and post-NPY tests were conducted 10 min after NPY (and 5 hr after progesterone). *a*, Basal test value $<$ post-NPY test values, $P < 0.01$; *b*, basal test values $>$ post-NPY test value, $P < 0.01$; *c*, 0.12 and 0.47 nM test values $<$ 0 or 0.012 nM test values; $n = 6$–8 rats/treatment. (With permission from Clark *et al.*, 1985.)

for these studies. As in the case of male rats, NPY promptly decreased female sexual behavior. Within 10 min of NPY administration in doses as small as 0.12 nM, there was a decrease in lordosis behavior and a virtual elimination of proceptive behavior (Fig. 4).

5. DUAL BEHAVIORAL EFFECTS OF NPY: A HYPOTHESIS

It is obvious from these findings that NPY rapidly suppessed sexual motivation while it concurrently stimulated feeding in the rat. What is the significance of these findings? How does NPY act to modulate the two behavioral responses? In an attempt to propose a unifying hypothesis on the role of NPY in integrating the two behavioral responses, the following salient features of our studies have been considered. First, NPY is apparently the most potent neuropeptide known to stimulate feeding in male and female rats: NPY stimulated feeding in satiated rats and also augmented the ongoing nocturnal feeding. The fact that other members of the PP family were less effective than NPY and that NPY has been localized in discrete central sites implicated in the regulation of feeding behavior strongly suggest that either NPY or a NPY-like peptide discharged from the neural network in these critical brain sites may normally stimulate feeding in the rat. Second, we have shown that NPY can drastically suppress sexual behaviors in male and female rats and that these suppressive effects manifest rapidly. Accordingly, it is possible that NPY-containing neural circuits may be involved in terminating sexual behavior in male and female rats.

Third, a comparison of the latencies of onset of these two behaviors after NPY administration clearly indicated that the suppression of sexual behavior was rapid and reached near-maximal levels at or near the time of onset of feeding. Thus, it is plausible that NPY may sequentially turn off the neural circuitry for sexual motivation and turn on the neural circuit that regulates appetite for food. Fourth, an examination of the NPY doses affecting feeding and sexual behaviors disclosed different thresholds of responsiveness for the inhibitory and excitatory behavioral responses. Lower concentrations of NPY were needed to suppress sexual motivation than were needed to stimulate feeding, thus implying that suppression of sexual motivation may manifest fully and precede the feeding response. Consequently, if one assumes that the sites for stimulation of feeding and inhibition of sexual behavior are coextensive, then it is logical to infer that stimulation of these sites with NPY may produce decrements in sexual motivation in parallel with or prior to the stimulation of ingestive behavior. The reports that feeding in rats is seen almost exclusively during periods when sexual arousal is decreased as after ejaculation (Sachs and Marsden, 1972; Brown and McFarland, 1979) and that transections of discrete neural pathways in the hypothalamus produce hyperphagia accompanied by markedly impaired copulatory behavior (Paxinos and Bindra, 1972, 1973) are in agreement with this inference. Therefore, we propose that NPY may be an important common neurochemical signal that suppresses sexual motivation in parallel with activation of a sustained feeding response.

6. CONCLUDING REMARKS

In conclusion, it appears that the discovery of NPY-containing pathways in the brain has had a major impact on further understanding of the control of reproduction and appetitive behaviors by the brain. Further, the manifold modulatory effects of NPY are reminiscent of a similar diversity displayed by the adrenergic systems. Although the overlapping anatomic disposition of these two systems in the brain may provide a partial explanation, the precise nature of interaction, if any, between them remains to be elucidated. Nevertheless, the concurent effects of NPY on food intake, sex behavior, and gonadotropin secretion have major clinical implications. They allow a critical reassessment, at neurochemical levels, of not only the pathophysiology of obesity but also of an implicit connection between the well-known feeding disorders and attendant sexual dysfunction in man.

ACKNOWLEDGMENTS. The research embodied in this review was supported by grants from the National Institutes of Health, HD 08634 and 14006 (S.P.K.); HD 11362 and AM 37273 (P.S.K.); HD 13703 (W.R.C.); and postdoctoral fellowship, HD 06660 (J.T.C.).

REFERENCES

Adler, B. A., Johnson, M. D., Lynch, C. O., and Crowley, W. R., 1983, Evidence that norepinephrine and epinephrine systems mediate the stimulatory effects of ovarian hormones on luteinizing hormone and luteinizing hormone releasing hormone, *Endocrinology* **113:**1431–1438.

Agnati, L., Füxe, K, Benfenati, F., Battistine, N., Harfstrand, A., Tatemoto, K., Hökfelt, T., and Mutt, V., 1983, Neuropeptide Y *in vitro* selectively increases the number of α_2-adrenergic binding sites in the membranes of the medulla oblongata of the rat, *Acta Physiol. Scand.* **118:**293–295.

Allen, L. G., Kalra, P. S., Crowley, W. R., and Kalra, S. P., 1985a, Comparison on the effects of neuropeptide Y and adrenergic transmitters on LH release and food intake in male rats, *Life Sci.* **37:**617–623.

Allen, L. G., Crowley, W. R., and Kalra, S. P., 1985b, Role of catecholamines and opiates in stimulation of LH release by neuropeptide Y (NPY), *Soc. Neurosci. Abstr.* **11:**1296.

Allen, Y. S., Adrian, T. E., Allen, J. M., Tatemoto, K., Crow, T. J., Bloom, S. R., and Polak, J. M., 1983, Neuropeptide Y distribution in the rat brain, *Science* **221:**877–879.

Brown, R. E., and McFarland, D. J., 1979, Interaction of hunger and sexual motivation in the male rat: A time-sharing approach, *Anim. Behav.* **27:**887–896.

Chan-Palay, V., and Palay, S. L., 1984, *Coexistence of Neuroactive Substances in Neurons,* John Wiley & Sons, New York.

Chronwall, B. M., DiMaggio, D. A., Massari, V. J., Vickel, V. M., Ruggiero, D. A., and O'Donohue, T. L., 1985, The anatomy of neuropeptide Y-containing neurons in the rat brain, *Neuroscience* **15:**1159–1166.

Clark, J. T., and Kalra, P. S., 1985, Suppression of female rat sexual behavior by intraventricular administration of neuropeptide Y, in: *67th Annual Meeting of the Endocrine Society,* Williams & Wilkins, Baltimore, MD. p. 275.

Clark, J. T., and Kalra, S. P., 1985, Neuropeptide Y (NPY)-induced feeding: Comparison with rat pancreatic polypeptide (rPP), human NPY, and peptide YY (PYY) in male and female rats, *Soc. Neurosci. Abstr.* **11:**619.

Clark, J. T., Kalra, P. S., Crowley, W. R., and Kalra, S. P., 1984, Neuropeptide Y and human pancreatic polypeptide stimulate feeding behavior in rats, *Endocrinology* **115:**427–429.

Clark, J. T., Kalra, P. S., and Kalra, S. P., 1985, Neuropeptide Y stimulates feeding and suppresses male sexual behavior, *Endocrinology* **117:**2435–2442.

Colmers, W., Lukowiak, K., and Pittman, Q. J., 1985, Neuropeptide Y reduces orthodromically-evoked population spike in rat hippocampal CA1 by a possibly presynaptic mechanism, *Brain Res.* **346:**404–408.

Cordor, R., Emson, P. C., and Lowry, P. J., 1984, Purification and characterization of human neuropeptide Y from adrenal-medullary phaechromocytoma tissue, *Biochem. J.* **219:**699–706.

Crowley, W. R., Terry, L. C., and Johnson, M. D., 1982, Evidence for the involvement of central epinephrine systems in the regulation of luteinizing hormone, prolactin, and growth hormone release in female rats, *Endocrinology* **110:**1102–1107.

Crowley, W. R., Tessel, R. E., O'Donohue, T. L., Adler, B. A., and Kalra, S. P., 1985, Effects of ovarian hormones on the concentrations of immunoreactive neuropeptide Y in discrete brain regions of the female rat: Correlation with serum LH and median eminence LHRH, *Endocrinology* **117:**1151–1155.

Edvinsson, L., 1985, Functional role of perivascular peptides in the control of cerebral circulation, *Trends Neurosci.* **8:**126–131.

Emson, P. C., and DeQuidt, M. E., 1984, NPY—a new member of the pancreatic polypeptide family, *Trends Neurosci.* **7:**31–36.

Estes, K. S., Simpkins, J. W., and Kalra, S. P., 1982, Resumption with clonidine of pulsatile LH release following acute norepinephrine depletion in ovariectomized rats, *Neuroendocrinology* **35:**56–62.

Everitt, B. J., Hökfelt, T., Terenius, L., Tatemoto, K., Mutt, V., and Goldstein, M., 1984, Differential co-existence of neuropeptide Y (NPY)-like immunoreactivity with catecholamines in the central nervous system of the rat, *Neuroscience* **11:**443–462.

Füxe, K., Agnati, L., Harsfstrand, A., Zini, I., Tatemoto, K., Pich, E., Hökfelt, T., Mutt, V., and Terenius, L., 1983, Central administration of neuropeptide Y induces hypotension bradypnea and EEG syndronization in the rat, *Acta Physiol. Scand.* **118:**189–192.

Guy, J., Allen, Y. S., Polak, J. M., and Pelletier, G., 1983, Immunocytochemical localization of neuropeptide Y (NPY) in the rat brain, *Soc. Neurosci. Abstr.* **13:**291.

Hendry, S. H. C., Jones, E. G., DeFelipe, J., Schechel, D., Brandon, C., and Emson, P. C., 1984, Neuropeptide-containing neurons of the cerebral cortex are also GABAergic, *Proc. Natl. Acad. Sci. U.S.A.* **81:**6526–6530.

Hökfelt, T., Johansson, O., and Goldstein, M., 1984, Chemical anatomy of the brain, *Science* **225:**1326–1334.

Kalra, P. S., and Kalra, S. P., 1985, Control of gonadotropin secretion, in: *The Pituitary Gland* (H. Imura, ed.), Raven Press, New York, pp. 189–220.

Kalra, P. S., Fawcett, C. P., Krulich, L., and McCann, S. M., 1973, The effects of gonadal steroids on plasma gonadotropins and prolactin in the rat, *Endocrinology* **92:**1256–1268.

Kalra, S. P., 1976, Ovarian steroids differentially augment pituitary FSH release in deafferented rats, *Brain Res.* **144:**541–544.

Kalra, S. P., 1981, Neural loci involved in naloxone-induced luteinizing hormone release: Effects of a norepinephrine synthesis inhibitor, *Endocrinology* **109:**1805–1810.

Kalra, S. P., 1983, Opioid peptides—Inhibitory neuronal systems in regulation of gonadotropin secretion, in: *Role of Peptides and Proteins in Control of Reproduction* (S. M. McCann and D. S. Dhindsa, eds.), Elsevier Biomedical, New York, pp. 63–87.

Kalra, S. P., 1985, Catecholamine involvement in preovulatory LH release: Reassessment of the role of epinephrine, *Neuroendocrinology* **40:**139–144.

Kalra, S. P., 1986, Neural circuitry involved in the control of LHRH secretion: A model for the preovulatory LH release, in: *Frontiers in Neuroendocrinology,* Vol. 9 (W. F. Ganong, and L. Martini, eds.), Raven Press, New York, pp. 31–75.

Kalra, S. P., and Crowley, W. R., 1983, Epinephrine synthesis inhibitors block naloxone-induced LH release, *Endocrinology* **82:**1403–1405.

Kalra, S. P., and Crowley, W. R., 1984a, Differential effects of pancreatic polypeptide on luteinizing hormone release in female rats, *Neuroendocrinology* **38:**511–513.

Kalra, S. P., and Crowley, W. R., 1984b, Norepinephrine-like effects of neuropeptide Y on LH release in the rat, *Life Sci.* **35:**1173–1176.

Kalra, S. P., and Kalra, P. S., 1983, Neural regulation of luteinizing hormone secretion in the rat, *Endocrinol. Rev.* **4:**311–351.

Kalra, S. P., and Kalra, P. S., 1984, Opioid–adrenergic–steroid connection in regulation of LH secretion in the rat, *Neuroendocrinology* **39:**45–48.

Kalra, S. P., and McCann, S. M., 1973, Effects of drugs modifying catecholamine synthesis on LH release induced by preoptic stimulation in the rat, *Endocrinology* **93:**356–362.

Kalra, S. P., and Simpkins, J. W., 1981, Evidence for noradrenergic mediation of opioid effects on luteinizing hormone secretion, *Endocrinology* **109:**776–782.

Kimmel, J. R., Pollock, H. G., Chance, R. E., Johnson, M. G., Reeve, J. R., Jr., Taylor, I. L., Miller, C., and Shively, J. E., 1984, Pancreatic polypeptide from rat pancreas, *Endocrinology* **114:**1725–1731.

Leadem, C. A., and Kalra, S. P., 1984, Stimulation with estrogen and progesterone of LHRH release from perifused adult female rat hypothalami: Correlation with the LH surge, *Endocrinology* **114:**51–56.

Leadem, C. A., Crowley, W. R., Simpkins, J. W., and Kalra, S. P., 1985, Effects of naloxone on catecholamine and LHRH release from the perifused hypothalamus of the steroid-primed rat, *Neuroendocrinology* **42:**497–500.

Leibowitz, S. F., 1980, Neurochemical systems of the hypothalamus. Control of feeding and drinking behavior and water-electrolyte excretion, in: *Handbook of the Hypothalamus,* Vol. 3, Part A (P. Morgane and J. Panksepp, eds.), Marcel Dekker, New York, pp. 299–437.

Lin, T. M., 1980, Pancreatic polypeptides: Isolation, chemistry and biological functions, in: *Gastrointestinal Hormones* (G. B. J. Glass, ed.), Raven Press, New York, pp. 275–306.

Lundberg, J., and Stjarne, L., 1984, Neuropeptide Y (NPY) depresses the secretion of H^3-noradrenaline and the contractile response evoked by field stimulation in rat vas deferens, *Acta Physiol. Scand.* **120**:477–479.

Lundberg, J., Tatemoto, K., Terenius, L., Hellstrom, P. M., Mutt, V., Hökfelt, T., and Hamberger, B., 1982a, Localization of peptide YY (PYY) in gastrointestinal endocrine cells and effects on intestinal flow and motility, *Proc. Natl. Acad. Sci. U.S.A.* **79**:4424–4427.

Lundberg, J., Terenius, L., Hökfelt, T., Martling, C. M., Tatemoto, M., Mutt, V., Polak, J., Bloom, S., and Goldstein, M., 1982b, Neuropeptide Y (NPY)-like immunoreactivity in peripheral noradrenergic neurons and effects of NPY on sympathetic function, *Acta Physiol. Scand.* **116**:477–480.

McDonald, J., Lumpkin, M. D., Samson, W., and McCann, S. M., 1985a, Neuropeptide Y affects secretion of luteinizing hormone and growth hormone in ovariectomized rats, *Proc. Natl. Acad. Sci. U.S.A.* **82**:561–564.

McDonald, J., Lumpkin, M., Samson, W. K., and McCann, S. M., 1985b, Pancreatic polypeptides affect luteinizing and growth hormone secretion in rats, *Peptides* **6**:79–84.

Morley, J. E., and Levine, A. S., 1983, The central control of appetite, *Lancet* **1**:398–401.

Negro-Vilar, A., Orias, R., and McCann, S. M., 1973, Evidence for a pituitary site of action for the acute inhibition of LH release by estrogen in the rat, *Endocrinology* **92**:1680–1684.

Paxinos, G., and Bindra, D., 1972, Hypothalamic knife cuts: Effects on eating, drinking, irritability, aggression, and copulation in the male rat, *J. Comp. Physiol. Psychol.* **79**:219–229.

Paxinos, G., and Bindra, D., 1973, Hypothalamic and midbrain neural pathways involved in eating, drinking, irritability, aggression, and copulation in rats, *J. Comp. Physiol. Psychol.* **82**:1–14.

Sachs, B. D., and Marsden, E., 1972, Male rats prefer sex to food after 6 days of food deprivation, *Psychonom. Sci.* **28**:47–49.

Simpkins, J. W., and Kalra, S. P., 1979, Blockade of progesterone-induced increase in hypothalamic luteinizing hormone-releasing hormone levels and serum gonadotropins by intrahypothalamic implantation of 6-hydroxydopamine, *Brain Res.* **170**:475–484.

Simpkins, J. W., Kalra, P. S., and Kalra, S. P., 1980, Temporal alterations in luteinizing hormone-releasing hormone concentrations in several discrete brain regions: Effects of estrogen–progesterone and norepinephrine synthesis inhibition, *Endocrinology* **107**:573–577.

Stanley, B. G., and Leibowitz, S. F., 1984, Neuropeptide Y; stimulation of feeding and drinking by injection into the paraventricular nucleus, *Life Sci.* **35**:2635–2642.

Tatemoto, K., 1982, Neuropeptide Y: Complete amino acid sequence of the brain peptide, *Proc. Natl. Acad. Sci. U.S.A.* **79**:5485–5489.

Tatemoto, K., Carlquist, M., and Mutt, V., 1982, Neuropeptide Y—a novel brain peptide with structural similarities to peptide YY and pancreatic polypeptide, *Nature* **296**:659–660.

Varndell, I. M., Polak, J. M., Allen, J. M., Terenghi, G., and Bloom, S. R., 1984, Neuropeptide Y tyrosine (NPY) immunoreactivity in norepinephrine-containing cells and nerves of the mammalian adrenal gland, *Endocrinology* **114**:1460–1463.

25

Kinins as Neuropeptides

DAVID C. PERRY

1. INTRODUCTION

The kinins, including the nonapeptide bradykinin (BK) as well as several other peptide analogues of BK, have been known since their initial discovery (Rocha e Silva, 1949) to be highly active in a number of peripheral organ systems. The recent explosion of research on brain peptides has renewed interest in the possibility that kinins may also play roles as neuropeptides. In this chapter I review evidence that BK and other kinins may be acting as neuroactive peptides, especially in the central nervous system.

1.1. Biochemistry and Physiology

Bradykinin is thought to play a role in several pathophysiological conditions, including inflammation, shock, hypertension, and pain (Kellermeyer and Graham, 1968; Erdos, 1970; Pisano and Austen, 1974; Regoli and Barabe, 1980; Douglas, 1980). Bradykinin is released from inactive protein precursors called kininogens by the action of serine proteinases (kininogenases), most notably the family of enzymes called kallikreins; the active peptide is rapidly destroyed by the action of various kininases (including kininase II, which is identical to angiotensin-converting enzyme). Each of the components of this system has multiple forms, and in fact it is thought that (at least) two separate kinin systems exist: one preferentially producing BK from high-molecular-weight kininogen (the "plasma system") and one preferentially producing kallidin from low-molecular-weight kininogen (the "tissue system"; Douglas, 1980). Kallidin is Lys-BK; it can be readily converted to BK by aminopeptidase action. Bradykinin, kallidin, and several other analogues, known collectively as kinins, share a spectrum of activity: they are potent vasodilators,

DAVID C. PERRY • Department of Pharmacology, George Washington University Medical Center, Washington, D. C. 20037.

causing lowered blood pressure; they increase capillary permeability, causing edema; they are potent algesic agents when injected intraarterially or applied to a blister base; they cause contraction of various smooth muscles; and they cause diuresis, natriuresis, and intestinal chloride secretion (Lim, 1970; Armstrong, 1970; Juan and Lembeck, 1974; Johnson 1979; Regoli and Barabe, 1980; Carretero and Scicli, 1980; Douglas, 1980; Manning *et al.*, 1982).

1.2. Receptors and Second Messengers

Binding studies using [^{3}H]BK or [^{125}I]analogue of BK have identified high-affinity receptors for kinins in several tissues, including guinea pig ileum, duodenum, and colon (Innis *et al.*, 1981; Manning *et al.*, 1982), heart (Innis *et al.*, 1981), kidney, ureter, and bladder (Innis *et al.*, 1981; Manning and Snyder, 1986), and spinal cord (Manning and Snyder, 1983); bovine myometrium (Odya *et al.*, 1980; Frederick *et al.*, 1984); rat uterus (Odya *et al.*, 1980; Innis *et al.*, 1981; Yasujima *et al.*, 1981); isolated rabbit nephron (Tomita and Pisano, 1984); and human fibroblasts (Roscher *et al.*, 1984). Regoli and co-workers have identified a kinin receptor in rabbit aorta that shows marked preference for [desArg^{9}]BK, a fragment that has little or no activity in other systems (Regoli and Barabe, 1980). They refer to this aorta receptor as "B1" and all others as "B2" (Regoli and Barabe, 1980).

Bradykinin has been shown to activate second messenger systems in various tissues. Reports demonstrate a stimulation of both cGMP (Stoner *et al.*, 1973; Clyman *et al.*, 1975; Frucht *et al.*, 1984) and cAMP (Stoner *et al.*, 1973; Bareis *et al.*, 1983; Roscher *et al.*, 1984; Zenser *et al.*, 1984). There is considerable evidence that at least some kinin effects may be mediated by phospholipase A_2 products, including prostaglandins and leukotrienes (Johnson, 1979; Bareis *et al.*, 1983; McGiff *et al.*, 1976; Roscher *et al.*, 1984).

2. NEUROPHARMACOLOGY OF KININS

2.1. Peripheral Nervous System

Kinins have long been known to be potent algesic agents, and in fact i.p. administration of BK to mice is a common pain stimulus used for testing analgesic agents (Lim, 1970). Bradykinin or similar substances have been detected in blister fluid and at injury sites; trauma is known to activate kallikreins; and the response to exogenously administered BK resembles that seen in an actual injury (Lim, 1970; Armstrong, 1970). For these reasons BK is believed to be one of the endogenous compounds mediating the effects of injury and inflammation, including the wheal (edema resulting from increased capillary permeability) and local reddening (because of vasodilation). The flare (diffuse reddening) and hyperalgesic state associated with inflammation are known to be neurogenic in origin via the axon reflex. Chapman described a BK-like peptide substance termed "neurokinin" that was produced following antidromic stimulation of the dorsal root; neurokinin caused flare and pain (Chapman *et al.*, 1961). Although neurokinin differed in some respects from either BK or substance P, it is conceivable that a mixture of these two peptides could account for the features of Chapman's material. This suggests the possibility that BK may be contained in free afferent nerves, being released on injury and acting as some kind of neurotransmitter or neurohormone. This contrasts with the standard concept that BK acting at sites of injury arises from plasma or perhaps vessel walls. Although other studies have shown that peripheral neural activity leads to release of BK-like material (e.g., Inoki *et al.*, 1978), so far proof of the direct neural origin of BK in the inflammation response is still lacking.

Whatever the source of BK, it is well accepted that it causes stimulation of afferent pain fibers, causing both immediate pain and a long-lasting hyperalgesic state. Prostaglandins play an important role in these functions; at a nonnoxious dose, PGE_2 can greatly sensitize pain fibers to certain stimuli, including BK; furthermore, BK is known to stimulate production of prostaglandins, presumably through stimulation of phospholipase A_2 (Ferreira, 1972; Juan and Lembeck, 1974; Lembeck *et al.*, 1976; Chahl and Iggo, 1977). Although it seems reasonable to presume that BK stimulates free afferents by direct interaction with receptors on the nerve endings, there is so far no direct proof of this. Numerous other endogenous compounds, including substance P, serotonin, histamine, acetylcholine, ATP, K^+, and H^+ may play roles in pain and inflammation, and it remains to be demonstrated which agents(s) is/are the proximate stimulus of free nerve endings.

The interactions of BK with the peripheral nervous system have been studied from several perspectives. Barasi and Wright (1979) showed that intraarterial BK and noxious mechanical stimulation both caused firing of putative substance P neurons in rat CNS. Belcher (1979) studied effects of intraarterial BK on the firing of cat dorsal horn interneurons; BK potently stimulated those neurons identified as nociceptive, whereas larger amounts of BK caused stimulation of both nociceptive and nonnociceptive cells. Bradykinin has also been shown to stimulate afferent vagal C fibers from the heart and lungs of dogs (Kaufman *et al.*, 1980a,b). Bradykinin causes pupillary constriction and inflammation; the mechanism appears to involve the stimulation of sensory nerves to release substance P (Bynke *et al.*, 1983; Ueda *et al.*, 1984; Wahlestedt *et al.*, 1985).

The smooth muscle effects of kinins may also involve interaction with intrinsic nerves. For instance, Goldstein *et al.* (1983) showed that BK and analogues potently reverse the opiate-induced inhibition of electrically stimulated constriction of the guinea pig ileum/myenteric plexus preparation; this effect was apparently a result of the stimulation of the release of acetylcholine by BK. These authors also report the release of BK from this preparation (Goldstein *et al.*, 1983). Bradykinin has been shown to excite autonomic ganglia directly (Lewis and Reit, 1965). Finally, sensory neurons from dorsal root and trigeminal ganglia of rats, when grown in culture, are excited by bradykinin and sensitized by PGE_2 (Baccaglini and Hogan, 1983), supporting the hypothesis of the direct action of BK on primary afferent nerve endings.

Recent evidence for the direct action of BK was reported by Manning and Snyder (1983), who described autoradiographic localization of binding sites for [^{3}H]BK in the guinea pig spinal cord, concentrated in laminae I and II. This region is the primary terminal field for nociceptive fibers, supporting the direct role of BK in pain generation. Assuming that these receptors are present on nociceptive fiber terminals, it remains to be seen if their presence in the spinal cord is functional or is an artifact caused by bidirectional transport of receptor molecules from the cell body to both ends of the nerve. These researchers also reported evidence of [^{3}H]BK binding sites in the guinea pig dorsal root ganglion and trigeminal ganglion as well as the stellate ganglion of the dog (Manning and Snyder, 1983).

2.2. Central Nervous System

Because kinins are present throughout the circulation, their putative roles as peripheral modulators of pain and inflammation are widely accepted. The possibility that kinins may also have functions within the brain, as neurotransmitters or neuromodulators, has been more difficult to demonstrate. An important part of this demonstration has been to show that exogenous kinins, when applied directly into the central nervous system (either intracerebroventricularly, i.c.v., or directly into brain tissue) elicit specific pharmacological effects.

2.2.1. Cardiovascular Effects

Numerous studies have examined the effects of central administration of BK, either i.c.v. or directly into various brain regions. The best-documented effects have been cardiovascular changes: an increase in blood pressure and a change (usually an increase) in heart rate. Peripherally, BK is generally hypotensive because of its potent vasodilator effects. This reversal of peripheral and central effects on blood pressure has been seen with other peptides as well, including VIP, substance P, and neurotensin (Phillips *et al.*, 1982).

Most studies employ i.c.v. injection of BK in doses generally of 1–5 μg, although significant increases in mean arterial blood pressure (MAP) have been detected with i.c.v. BK doses as small as 0.1 μg (Pearson *et al.*, 1969) and 0.05 μg (Unger *et al.*, 1980). Although all studies report a rise in MAP, usually within 1 min or so following i.c.v. injection, biphasic responses have also been reported. Graeff *et al.* (1969) reported an initial fall in MAP of unanesthetized rabbits at 30–60 sec postinjection, followed rapidly by an increase lasting 1 min. A similar but more prolonged pattern was seen by Pearson *et al.* (1969) in anesthetized cats and rats; however, this same group found only a pressor response in awake rats (Lambert and Lang, 1970). The initial depression phase could be abolished by pretreatment with α-adrenergic blockers (Pearson *et al.*, 1969). Others describe a different biphasic pattern: an initial sharp rise in MAP at 1–2 min with a second, broader peak at about 4 min (Kariya and Yamauchi, 1981; Kariya *et al.*, 1982a).

The anatomic substrate for these effects is unclear. Correa and Graeff (1975, 1976), using different injections of 0.5 μg BK into different brain regions of rats as well as various lesions, concluded that the pars ventralis of the lateral septum was the site of BK's hypertensive effects. Diz and Jacobowitz (1984) employed direct injection of 5 μg BK into hypothalamic regions of rats and found increases in MAP following injection into the dorsal medial, rostral anterior, posterior, and ventromedial hypothalamic nuclei but not with six other hypothalamic regions. Lewis and Phillips (1984) employed ventricular plugs and concluded that i.c.v. BK was acting on sites adjacent to the third ventricle to increase blood pressure. Hoffman and Schmid (1978) showed that median eminence lesions did not change the effect of i.c.v. BK on rat MAP.

Several studies investigated the interaction of other agents with the blood pressure response to central BK, with some conflicting conclusions. As noted, some investigators detected an initial fall in MAP, which was blocked by α-adrenergic blockers (Pearson *et al.*, 1969; Takahashi and Bunag, 1981). Lang and co-workers (Pearson *et al.*, 1969) reported that in anesthetized rats, the initial depressor phase was blocked by phentolamine, and the subsequent pressor phase was blocked by propranolol. However, in awake rats, these researchers reported that only a pressor response to i.c.v. BK was seen, which was blocked by phentolamine (i.p.) but not by propranolol. Similarly, Correa and Graeff (1974, 1976) reported that phentolamine administered centrally blocked the pressor response to i.c.v. and intracerebral injections of BK, but no effect was obtained from various other drugs, including propranolol, pyrilamine, hexamethonium, methysergide, capsaicin (all i.c.v.), or atropine (i.a.). Thus, the pressor response to i.c.v. BK appears to involve release of norepinephrine, apparently both centrally and peripherally, which is acting via an α-adrenergic mechanism.

Experiments to test potential cholinergic involvement yield a somewhat less clear story. Graeff *et al.* (1969) found that i.v. atropine enhanced the pressor effects of i.c.v. BK in unanesthetized rabbits. However, Pearson *et al.* (1969), using anesthetized rats, and Correa and Graeff (1974), using conscious rats, found no effect on the BK pressor response with atropine (i.v. and i.a., respectively); and Diz and Jacobowitz (1984) found that i.p. methylatropine blocked the pressor effects of BK injected into the dorsomedial and posterior hypothalamic nuclei. Buccafusco and Serra, (1985) found that a 1-hr pretreatment of rats with hemicholinium-3 i.c.v. to deplete brain stores of acetylcholine blocked the pressor effect; coinjection of choline chloride prevented the blockade. Thus, there appears to be a central cholinergic component to the pressor effect of i.c.v. BK even though studies with peripheral muscarinic blockade yield conflicting results. Finally, hexamethonium centrally did not affect BK responses (Correa and Graeff, 1974), but peripherally it abolished BK responses (Pearson *et al.*, 1969); this is probably because of ganglionic blockade of sympathetic outflow.

The involvement of other neurotransmitters has also been examined. Graeff *et al.* (1969) found that i.v. morphine gave a slight enhancement to the pressor effect of i.c.v. BK in unanesthetized rabbits. In contrast, s.c. morphine in unanesthetized rats decreased BK's pressor response (Correa and Graeff, 1974; Lambert and Lang, 1970). Although this result suggests that the pressor response to central BK may be secondary to stimulation of a pain response, other reports (Ribeiro *et al.*, 1971; Ribeiro and Rocha e Silva, 1973) show that i.c.v. BK causes analgesia. Histamine involvement has also been suggested. Diphenhydramine (i.a.) and pyrilamine (i.c.v.) caused a reduction in the pressor response to i.c.v. BK (Correa and Graeff, 1974), whereas the pressor response to intraseptal injection of BK was not blocked by central pyrilamine (Correa and Graeff, 1976).

Because kinin effects are linked with arachidonic acid metabolites in the periphery, the same might be true for central BK effects. Kondo *et al.* (1979) showed that i.c.v. BK and PGE_2 produce similar effects on blood pressure (but not heart rate) in conscious rats; pretreatment with i.c.v. indomethacin blocked the BK pressor response (it actually became depressor) but had no effect on the PGE_2 response. Very similar results with i.c.v. indomethacin were reported by Takahashi and Bunag (1981), Kariya *et al.* (1982a), and Lewis *et al.* (1983). Thus, there is good evidence that the pressor response to i.c.v. BK is mediated by production of prostaglandins.

The relationship of central BK to other centrally active peptides has also been explored. Pearson *et al.* (1969) and Lambert and Lang (1970) showed that i.c.v. eledoisin produces similar effects to BK, but the two peptides can be distinguished by their different responses to various drugs. Hoffman and Schmid (1978) showed that whereas central BK has antidiuretic effects via release of ADH, this is not the source of the accompanying pressor effect. Lewis *et al.* (1983) showed that the angiotensin partial agonist analogue saralasin, at a dose that inhibited i.c.v. angiotensin II pressor responses, enhanced the pressor and heart rate responses to 5 μg BK i.c.v., although angiotensin II itself did not alter BK responses. Thus, there appears to be some as yet unexplained relationship between these two peptides, at least for central cardiovascular control. Iwata *et al.* (1984) reported that i.c.v. injection of BK, angiotensin II, Leu-enkephalin, and neurotensin all caused pressor responses; however, BK caused no effect on plasma renin activity, whereas the others either suppressed (angiotensin II, neurotensin) or increased (Leu-enkephalin) this activity.

Several investigators also reported effects on heart rate following central BK; however, such changes appeared to be somewhat variable and inconsistent. Graeff *et al.* (1969) reported an initial bradycardia (30–60 sec) followed by about 1 min of tachycardia after 2.5–5.0 μg i.c.v. BK in awake rabbits; in contrast, Hoffman and Schmid (1978) reported the opposite trend following 5 μg i.c.v. BK in awake rats: a transient increase in rate followed by a long-term decrease. Correa and Graeff (1975) found no effect on heart rate with 1 μg BK in anesthetized rats, and Takahashi and Bunag reported little change following 1–10 μg i.c.v. in awake rats. Several groups have reported significant increases in heart rate following 1–10 μg doses i.c.v. in awake rats (Lewis *et al.*, 1983; Iwata *et al.*, 1984; Buccafusco and Serra, 1985). Finally Diz and Jacobowitz (1983) examined heart rate and blood pressure changes in anesthetized rats following direct injection of 5 μg BK into ten different hypothalamic sites: six sites showed tachycardia, two showed bradycardia, and two showed no significant change. The four injection sites that yielded blood pressure increases (see above) were all sites that also gave increases in heart rate.

2.2.2. Other Central Effects

Numerous other effects of centrally administered BK have been reported. A behavioral syndrome consisting of rapid and short-lived excitation followed by a more prolonged sedation has been described in rabbits (Graeff *et al.*, 1969; Melo and Graeff, 1975), mice (Iwata *et al.*, 1970; Capek *et al.*, 1969), and rats (Lambert and Lang, 1970; Kariya and Yamauchi, 1981; Kariya *et al.*, 1982a). The initial excitation was characterized by rapid movement, head shaking, struggling movements, and sometimes vocalization in rats, with similar responses in mice and rabbits. This phase began within 10–60 sec after i.c.v. injection and lasted only 1 or 2 min; it was noted to correlate with the time course of the pressor response in several studies (Graeff *et al.*, 1969; Lambert and Lang, 1970; Kariya *et al.*, 1982a). The subsequent sedation phase lasted longer, sometimes up to 40 min, and was characterized in rabbits by ptosis, a lack of spontaneous activity, dropped ears, and a tendency towards catalepsy (Graeff *et al.*, 1969) and catatonia (DaSilva and Rocha e Silva, 1971). The latter report notes that the excitation and sedation phases were distinct: excitation was enhanced, whereas sedation was greatly reduced, with cisterna magna injection compared to i.c.v. injection; tachyphylaxis occurred only to the sedation phase; and catatonia occurred much later than other symptoms, 20–30 min after injection. Also, small doses of i.c.v. morphine, which had no discernable behavioral effects, enhanced the catatonic effects of 2.5 μg BK (DaSilva and Rocha e Silva, 1971). The sedation phase in rats was characterized by piloerection, blepharoptosis, a crouching posture, and a lack of motor activity (Kariya and Yamauchi, 1981; Kariya *et al.*, 1982a). In mice, lower doses produced immediate sedation with rapid respiration and catatonia, whereas higher doses (≥750 μg/kg) caused the biphasic response, 1–2 min of excitation followed by 10–20 min of sedation (Iwata *et al.*, 1970).

Kariya et al. (1981) speculate that the excitation phase is a function of BK, whereas the subsequent sedation phase may be caused by one or more degradative fragments of BK. In support of this hypothesis, Okada *et al.* (1977) demonstrated that several such fragments injected centrally, as well as BK, prolong the pentobarbital-induced sleeping time in mice, and Iwata *et al.* (1970) showed similar effects using BK pretreated with kininase activity isolated from brain.

Electroencephalographic changes have also been reported. Graeff *et al.* (1969) report a desynchronization in rabbit EEG waves coinciding with the excitation phase, followed after 3–5 min by waves of high amplitude and an increased number of spindles. In rats, i.c.v. BK also produced an initial EEG desynchronization as well as an increased EMG voltage and rhythmic slow activity from the hippocampus, changing to convulsive spiking (Pearson *et al.*, 1969; Iwata *et al.*, 1970; Kariya and Yamauchi, 1981). At 1.5–2.0 min, the behavior abruptly changed to sedation phase; however, the EEG activation continued for approximately 5 min, changing to δ waves and a low EMG characteristic of deep sedation (Kariya and Yamauchi, 1981).

Central BK in rabbits has antinociceptive effects: the threshold for response to electrical stimulation of dental pulp increased in a dose-dependent manner, and the order of potency of BK and three analogues at raising this threshold was the inverse of their order of potency at the peripheral effects of BK on vascular permeability and blood pressure (Ribeiro *et al.*, 1971; Ribeiro and Rocha e Silva, 1973). In experiments with rabbit operant behavior, Melo and Graeff (1975) reported that BK caused dose-dependent decreases in variable-interval responding, and Graeff and Arisawa (1978) found that i.c.v. doses of 30–56 ng BK increased fixed-interval response rates without affecting fixed-ratio rates. Other effects of i.c.v. BK observed in rabbits were hyperthermia (Pela *et al.*, 1975; Almeida de Silva and Pela, 1978), hyperglycemia (Ribeiro *et al.*, 1970), and a long-lasting miosis (Graeff *et al.*, 1969; Melo and Graeff, 1975).

As mentioned above, BK has peripheral antidiuretic activity. Central administration to rats also causes antidiuretic effects, which are blocked by median eminence lesions, suggesting that this effect is mediated by stimulation of the release of ADH (Hoffman and Schmid, 1978). However, Lewis *et al.* (1983) reported that i.c.v. BK attenuates the drinking response to i.c.v. angiotensin II. Several investigators have searched for effects of i.c.v. BK on brain levels of amine neurotransmitters; the most consistent effects were seen with norepinephrine (NE). Graeff *et al.* (1969) found that NE was significantly decreased in rabbit brainstem, with slight or no effects on serotonin and dopamine. Capek (1969) reported in mice that 1 μg BK caused a 60% decrease in whole-brain levels of NE and suggested that this might be caused by an inhibition of NE reuptake by BK. Moniuszko-Jakoniuk *et al.* (1976) found significant decreases in both NE and its metabolites in rat striatum, midbrain, cerebellum, and hippocampus, with a small increase of serotonin. These decreases in brain NE seem to correlate with findings that at least some of the cardiovascular effects of i.c.v. BK are mediated by α-adrenergic neurons (see above). Finally, Phillis and Limacher (1974) showed BK acted to excite cortical neurons on direct iontophoresis.

3. ENDOGENOUS BRAIN KININS

In order to demonstrate that kinins serve as neuroactive brain peptides, it is necessary to demonstrate their presence in brain as well as their synthesis and degradation there.

3.1. Early Studies on Brain Kinins

Several early studies examined brain tissue for the presence of kininlike activity. In all cases, tissue extracts were treated to release active peptide from inactive precursors, so that total kininlike material was being measured. The brain was not perfused to remove possible contamination from blood. All of these reports relied on bioassays such as guinea pig ileum or rat uterus contraction for quantitation; also, few utilized extensive purification schemes. Thus, the possibility for false positive results must be considered.

Inouye *et al.* (1961) reported that ethanolic extracts of bovine spinal cord (dorsal section) and guinea pig brain contained a peptidelike material activated by trypsin treatment that had a spectrum of activity in several *in vitro* smooth muscle tests similar to BK; this material was shown not to be substance P. Hori (1968) used acidic extract of pooled brain tissue from cattle and bullfrogs and used a scheme for partial purification including anion exchange and gel filtration. He found that the substance behaved identically to BK in several smooth muscle preparations as well as chromatographically and with chemical and enzyme susceptibility (Hori, 1968). Shikimi *et al.* (1973) examined rat brain extracts utilizing the estrous rat uterus bioassay following treatment with trypsin; the resulting activity was assumed to be caused by BK released from kininogen. They found approximately 45 pmol/g acetone powder extract in cerebellum and brainstem and approximately 15 pmol/g in cortex (Shikimi *et al.*, 1973). Two other early studies report the presence of kininlike material (total kininogen) in mammalian tissue. Werle and Zach (1970) extracted a variety of animal tissues and used the estrous rat uterus bioassay; they report a value of 530 pmol/g kininogen in rat brain (this was the lowest value of any organ measured in this study). Pela *et al.* (1975) reported the presence of kininlike material in rabbit brain, located almost exclusively in the hypothalamus.

3.2. Kinin-Related Enzymes in Brain

An important criterion for the determination of a neuroactive agent is the identification of specific systems in the brain for formation and destruction of the agent. A number of studies have demonstrated the presence of kininase activity in brain extracts (Hori, 1968; Iwata *et al.*, 1969, 1970; Damas, 1972; Camargo *et al.*, 1972; Shikimi *et al.*, 1973; Oliveira *et al.*, 1976; Wilk and Orlowski, 1979; Kariya *et al,* 1981, 1982a). Kariya *et al.* (1982b) inferred the presence of kininase in conscious animals by following the disappearance of exogenous BK given i.c.v.; they found that the *in vivo* half-life was prolonged by injection of kininase inhibitors. The presence of kininase activity in brain can also be inferred from reports that pharmacological effects of exogenous i.c.v. BK are enhanced by coadministration of kininase inhibitors (Kondo *et al.*, 1979; Unger *et al.*, 1980). Enzymatic activities in extracts have been characterized to varying degrees as to the precise nature of substrate specificity; however, it is difficult to know for sure whether this activity has any physiological relationship to the presence of brain kinins. These enzymes may be relatively nonspecific peptidases or may, in fact, act specifically but not on BK. One well-known kininase, angiotensin-converting enzyme, is known to exist in the brain, with high concentrations in choroid plexus and caudate putamen; its role relative to either common substrate, i.e., angiotensin I or BK, has not been clarified (Yang and Neff, 1972).

Several studies have also demonstrated the presence in brain extracts of a kinin-generating activity, that is, a trypsinlike or kallikreinlike activity (Hori, 1968; Damas, 1972; Shikimi *et al.*, 1973). Chao *et al.* (1983) used monoclonal antibodies to tissue kallikrein as well as enzymatic assays to demonstrate the presence of kallikrein in rat brain and in the cell-free translation products of rat brain mRNA. This same group later reported the immunohistochemical localization of kallikrein in rat brain slices (Simson *et al.*, 1985).

3.3. Immunohistochemistry

Perhaps the most important single tool in the elucidation of brain neuropeptides has been immunohistochemistry, whereby specific neuronal systems containing immunologically similar material can be precisely localized. Correa *et al.* (1979) used antiserum to BK raised in rabbits to search for BK-like material in rat brain sections. The antiserum was highly specific, recognizing only BK, LysBK (kallidin), and MetLysBK. The only cell bodies staining for BK were found in the hypothalamus (Fig. 1); especially dense clusters were found in the paraventricular and dorsomedial hypothalamic nuclei (Correa *et al.*, 1979). These cells varied in shape; staining was seen only in the cytoplasm (see Fig. 1). Fibers staining for BK-like activity were seen in many brain regions, including the hypothalamus, the perirhinal and cingulate cortex, the periaqueductal gray, the ventral caudate–putamen, the globus pallidus, and the lateral septum (Correa *et al.*, 1979).

3.4. Recent Evidence

In order to identify definitively and to quantitate the BK-like material in the brain, Perry and Snyder (1984) examined acidic extracts of rat and guinea pig brain. Animals were throughly perfused to eliminate the possibility of contamination from residual blood, and peptidase inhibitors were employed. Extracts were purified using sequential chromatographic steps of gel filtration (Sephadex G-25) and two different reverse-phase HPLC gradient systems. Extracts were spiked with known quantities of [^{3}H]bradykinin, permitting assessment of recovery and of BK elution position (Perry and Snyder, 1984). The primary detection system was radioimmunoassay (RIA), using a high-titer rabbit antiserum raised against BK, which showed high specificity for the BK sequence. Confirmation was obtained by parallel use of a sensitive and specific radioreceptor assay (using guinea pig ileum tissue and [^{3}H]BK) and an estrous rat uterus bioassay as well as HPLC systems that separated the various BK analogues and fragments.

Low levels (approx. 0.6 pmol/g) of BK were detected in rat brain; activity was seen in all regions, but much higher levels were detected in the hypothalamus (Table I), corresponding to previous immunohistochemical findings (Correa *et al.*, 1979). This material coeluted with [^{3}H]BK in three different chromatographic systems; no evidence for any other kinin was seen. The material behaved qualitatively and quantitatively as authentic BK in the estrous rat uterus bioassay (Fig. 2). Extract

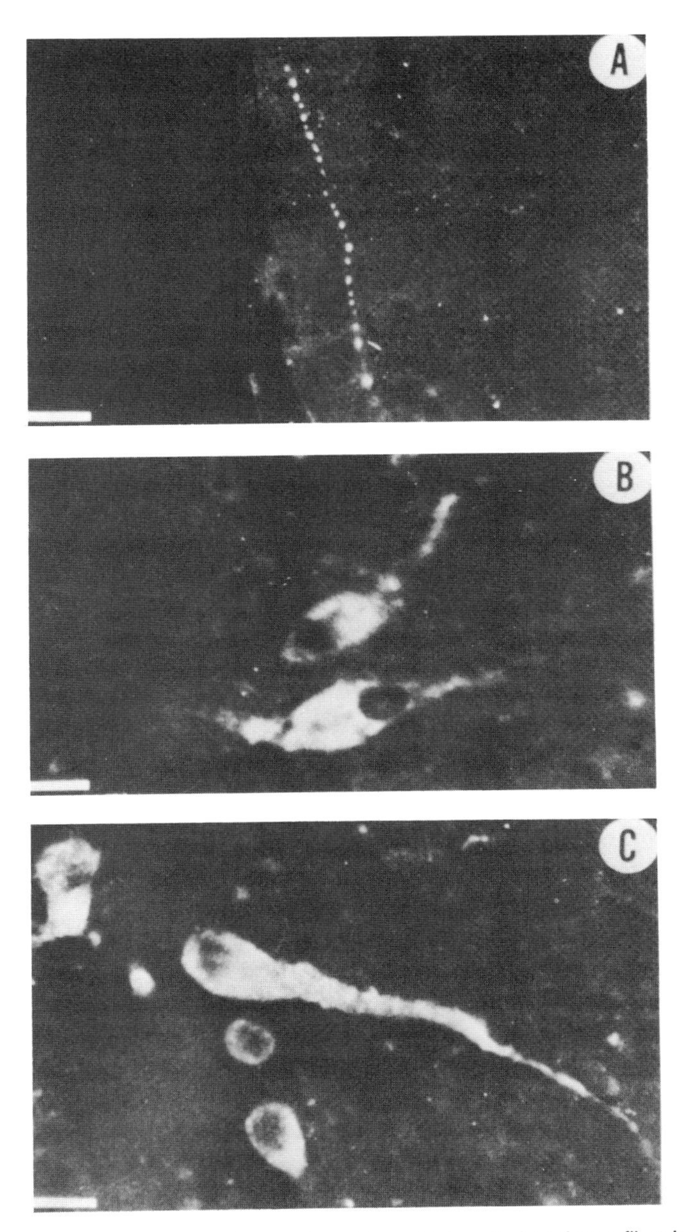

FIGURE 1. Immunofluorescence micrographs of bradykinin. (A) A varicose fiber in layer I of perirhinal cortex. The varicosities are large and prominent with short intervaricose segments. (B) Two positively stained perikarya from lateral hypothalamus. (C) A positively stained neuron in the lateral hypothalamus with fluorescence extending into a wide and long process. Bars, 20 μm. (Reprinted with permission from Correa *et al.*, 1979.)

TABLE I. Distribution of BK Immunoreactivity in Rat Brain[a]

Brain region	BK (pmol/g)
Hypothalamus	4.54
Cerebral cortex	0.43
Cerebellum	0.50
Midbrain	0.44
Pons and medulla	0.70
Corpus striatum	0.65
Hippocampus	0.60
Spinal cord	0.70

[a] Data are the means of two separate experiments, each one with tissue pooled from six animals.

of rat and guinea pig whole brain and hypothalamus were subjected to trypsin treatment; levels of BK were 7–23 times higher than before trypsinization (Table II), indicating that BK in the brain exists primarily in precursor form. Treatment of the brain extract with chymotrypsin or angiotensin-converting enzyme, both of which destroy BK, reduced levels 80–85% (Perry and Snyder, 1984). This work clearly establishes the presence of authentic BK in rat brain, in both precursor and active forms.

Immunoassay techniques have been employed by other investigators to measure kinins in the CNS. Kariya *et al.* (1982b) used an RIA to examine the disappearance of exogenous BK given i.c.v. to rats. The background level obtained was 5 ± 1 pmol/g, which presumably represented endogenous material, although perfusion was not performed, nor were any chromatographic steps employed (Kariya *et al.*, 1982b). Chao *et al.*. (1985) reported levels of 0.85 ng kinin equivalents/mg protein

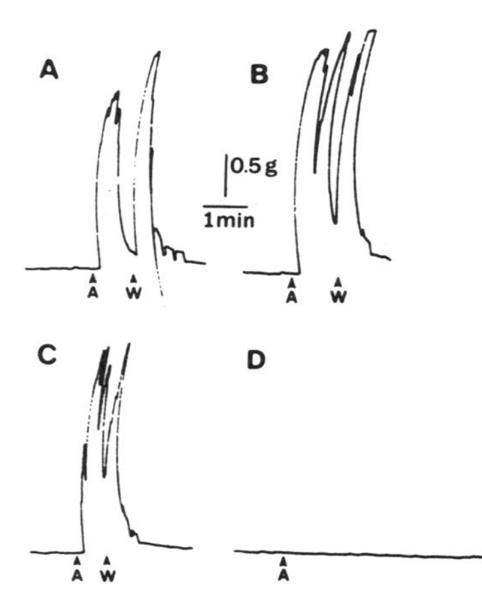

FIGURE 2. Pooled and concentrated HPLC fractions or BK standards were tested in the estrous rat uterus bioassay. Arrow A is addition; arrow W is wash. (A) BK, 2 pmol in 100 μl and 0.1 M acetic acid. (B) HPLC fractions corresponding to elution of immunoreactive BK (equivalent to 2.58 pmol BK immunoreactivity). (C) Same as A. (D) Control; pooled HPLC fractions from region with no immunoactivity. (Reprinted with permission from Perry and Snyder, 1984.)

TABLE II. Bradykinin Immunoreactivity in Rodent Brain Extracts Before and After Incubation with Trypsin[a]

Species	Tissue	BK (pmol/g)	
		Before trypsin	After trypsin
Rat	Whole brain	0.61 ± 0.05	13.79 ± 2.99
Rat	Hypothalamus	4.99 ± 0.47	49.95
Guinea pig	Whole brain	0.21	1.50
Guinea pig	Hypothalamus	2.08	ND

[a] Rat whole-brain data are the mean (± S.E.M.) from five separate determinations, each obtained from pooled tissue (from five to 28 animals per pool). Data for rat hypothalamus before trypsin are the mean (± S.E.M.) from three experiments, each with pooled tissue (6–16 rats); data after trypsin represent one determination (tissue pooled from six rats). Guinea pig data represent single determinations in individual animals. Data are expressed as pmol/g wet weight tissue. ND, not determined. (From Perry and Snyder, 1984, reprinted with permission.)

in rat brain extracts using a monoclonal antibody RIA. Thomas *et al.* (1984) used a BK RIA on dog CSF, which was found to contain 13 ± 13 pg/ml of immunoreactive kinin material. When melittin, an activator of membrane-bound kallikrein, was administered to the CSF *in situ,* levels of immunoreactive material rose over tenfold, indicating the presence of substantial amounts of kinin precursor in CSF or in paraventricular tissue (Thomas *et al.*, 1984). The increase in immunoreactive kinins was accompanied by a prolonged increase in MAP and a transient tachycardia (Thomas *et al.*, 1984). Thus, endogenous kinins have been directly linked to effects previously demonstrated to occur with administration of exogenous BK.

4. NEURAL CELL LINES

Neuroblastomas and related transformed cell lines have become valuable tools for studying neuroactive agents. These cells serve as model systems for neurons and allow ready demonstration of receptor binding and the consequent biochemical and electrophysiological changes as well as biosynthesis of neuropeptides.

4.1. NG108-15

This neuroblastoma × glioma hybrid is widely used in studies of neurotransmitters. Reiser and Hamprecht (1982) demonstrated that BK induced a 10- to 30-sec hyperpolarization in NG108-15 cells and in the rat glioma line C6-4-2. This response was apparently the result of increased K^+ conductance. The response showed desensitization to repeated application of BK and was not obtained with a variety of other peptides (Reiser and Hamprecht, 1982). This lab also demonstrated that BK stimulated cGMP production in NG108-15 cells (Reiser *et al.*, 1984). Half-maximal stimulation occurred at 0.1 nM; BK at 0.1 μM produced an eightfold increase in cGMP over basal levels. Stimulation was maximal at 30 sec, declining thereafter; desensitization was seen for approximately 20 min. Similar effects were seen with LysBK and MetLysBK, but only very slight stimulation was obtained using various BK fragments. A small BK effect on cGMP was also seen in rat glioma C6-4-2 cells (Reiser *et al.*, 1984).

Yano *et al.* (1984) also utilized NG108-15 cells to study electrophysiological and biochemical effects of BK. They demonstrated that BK produced a transient hyperpolarization followed by a longer-lasting depolarization; this ultimately led to the release of acetylcholine. They also found that BK stimulated the rapid breakdown of radiolabeled phosphatidylinositol-4,5-bisphosphate independently of a rise in Ca^{2+} concentration (Yano *et al.*, 1984). This group subsequently demonstrated the release of labeled inositol phosphates by BK (Yano *et al.*, 1985). Inositol triphosphate was maximum at 15 sec, and the others at slightly later times; 10^{-6} M BK gave a threefold increase in total inositol phosphates over base line (Yano *et al.*, 1985).

Receptor binding with [^{3}H]BK using NG108-15 cell membranes was done by Braas *et al.* (1985). Table III lists the results of these experiments, demonstrating the relative potencies of BK analogues in these cells. This represents the first direct demonstration of BK receptors in this cell line, which should allow further studies to correlate receptor binding and activation.

A recent abstract reported that NG108-15 cells contain both kallikrein and kininogen (Chao *et al.*, 1985). Thus, these cells may also serve as a model for neural biosynthesis of BK and its regulation.

4.2. N1E-115

The stimulation of cGMP in the neuroblastoma N1E-115 has been extensively studied by Richelson and colleagues; they utilized these cells in an elegant correlation of binding and biochemical effect for BK (Snider and Richelson, 1984). Binding and cGMP stimulation were done under the same conditions, affording direct correlation. Computer analysis of saturation experiments with [^{3}H]BK yielded three apparent binding sites with affinities of 0.83 pM, 1.0 nM, and 4.9 nM

TABLE III. Competition for [^{3}H]Bradykinin Binding[a]

	IC_{50} (nM)	
Analogue	Neuroblastoma	Guinea pig ileum
BK	0.35	0.23
DPhe^{7}BK	20	20
DPhe^{7}Thi^{5,8}BK	48	45
Thi^{5}BK	2.5	—
LysBK	1.3	—
MetLysBK	2.2	—

[a] Membranes were prepared from guinea pig ileum and pellets of NG108-15 cells, and binding was performed using 0.2 nM [^{3}H]BK as previously described (Manning *et al.*, 1982). IC_{50} values were obtained from plots of six concentrations run in triplicate. Thi, thienylalanine.

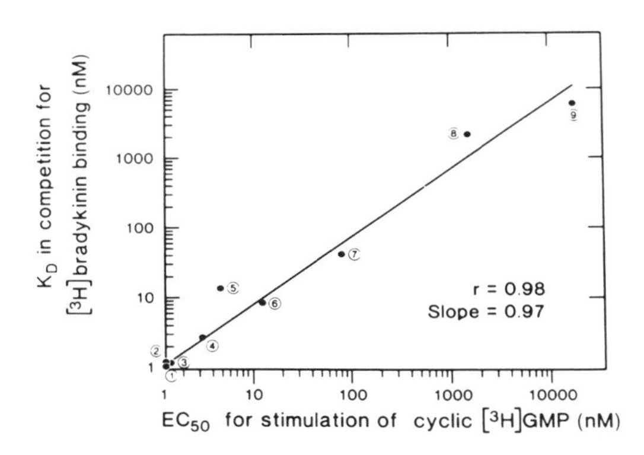

FIGURE 3. Correlation between EC_{50} for stimulation of cGMP and K_D for bradykinin and its analogues. 1, BK; 2, Hyp^{3}BK; 3, LysBK; 4, Thi^{5,8}BK; 5, MetLysBK; 6, Tyr^{8}BK; 7, AIB^{3}BK; 8, desArg^{9}BK; 9, GlyPheSerPro. Hyp, hydroxyproline; Thi, β-(2-thienyl)-alanine; AIB, α-aminoisobutyric acid. (Reprinted with permission from Snider and Richelson, 1984.)

(respective B_{max} values 12, 160, and 250 pmol/10^6 cells). However, computer analysis of kinetic and competition studies yielded only one site with an affinity approximately 1.0 nM; this was hypothesized to be the site responsible for cGMP stimulation (Snider and Richelson, 1984).

Cyclic GMP accumulation was very rapid, with a maximum at 45–60 sec; it required the presence of Ca^{2+}. Maximum levels were five- to 13-fold over basal levels. A series of ten BK analogues was tested for receptor affinity and potency at stimulating cGMP; an excellent correlation was obtained (see Fig. 3). The rank order of binding was almost identical to that reported by Innis *et al.* (1981) for [^{3}H]BK binding to guinea pig ileum membranes.

Braas *et al.* (1985) measured stimulation of labeled inositol phosphates in N1E-115 cells and correlated this to binding of [^{3}H]BK to a membrane preparation from these cells. Binding was similar to that described by Snider and Richelson (1984), although affinities were distinctly higher, since conditions were optimized for binding. Inositol metabolism was measured by release of total labeled inositol phosphates from the membrane; stimulation was rapid, peaking at 1 min (Fig. 4). Half-maximal

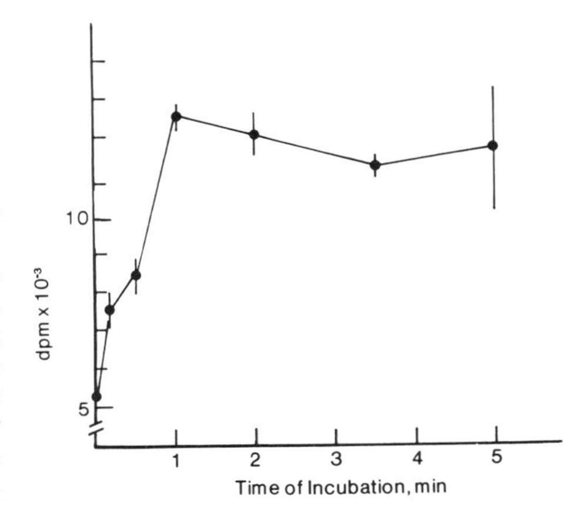

FIGURE 4. N1E-115 cells were prelabeled for 24 hr with [^{3}H]inositol; incubations with 1 μM BK were performed with triplicate aliquots of 180,000 cells each in the presence of 10 mM LiCl. Reaction was quenched at the time indicated and extracted with chloroform : methanol; the aqueous phase was applied to a Dowex X1-8 column, and after rinsing the total [^{3}H]inositol phosphates were eluted with 0.5 M HCl and counted by LSC.

stimulation was seen at 19.7 nM; peak values were two- to threefold higher than basal levels. The analogue $Thi^{5,8}$DPhe^{7}BK behaved as a competitive receptor antagonist in this system, inhibiting BK stimulation of inositol phosphate release with an AD_{50} of 1120 nM. The ratio of antagonist potency of the analogue to BK's potency was 57, which was quite similar to the relative binding potencies of those two peptides (Table III). The analogue did exhibit some slight agonist potency at high concentrations, reaching a plateau of 14% maximal stimulation at concentration of 10–30 μM; half-maximal stimulation (half of the plateau) was seen at approximately 4.6 μM.

5. SUMMARY

Considerable evidence has been presented to support the hypothesis that kinins including BK are present in nervous tissue and play a neuroactive role there. The role of BK in stimulating peripheral afferents is well accepted. Its presence in peripheral nerves and release from these nerves have been proposed, although evidence supporting this role is scanty. In the CNS, a great number of studies have shown that BK has pharmacological activity, especially in such autonomic functions as blood pressure control. There is now convincing evidence, both immunohistochemical and biochemical, for the presence of BK and precursors in mammalian brain tissue. The hypothalamus is particularly rich in BK. Work with neural cell lines shows that BK binds with high affinity and potently activates second messenger systems.

The case for kinins as neurotransmitters can be strengthened in several important ways. Perhaps the most striking omission has been the difficulty in measuring receptors for BK in the brain. The presence of active receptors on neuroblastoma cells is encouraging, certainly. Innis *et al.* (1981) reported the presence of specific binding in rat brain homogenates but were unable to characterize this binding further because of technical difficulties. Manning and Snyder (1983) used [^{3}H]BK autoradiography to demonstrate the presence of BK binding sites in the dorsal horn of the spinal cord as well as several ganglia. These tantalizing results suggest that this powerful technique may represent the best approach for demonstrating BK receptors in the rest of the brain.

Recently, great strides in the understanding of neuropeptides have been made with the techniques of modern molecular biology. Nakanishi and colleagues (Nawa *et al.*, 1983) reported the successful cloning of the bovine kininogen gene, which opens the way for these techniques to be applied to the kinin system in brain. For instance, *in situ* hybridization studies with labeled RNA or cDNA could be employed to determine the location of kininogen synthesis in brain tissue.

The lack of a true competitive antagonist has been the bane of kinin researchers for over three decades. However, Stewart and Vavreck recently reported the synthesis of a number of BK analogues that act as competitive antagonists in *in vitro* and *in vivo* assays (Stewart and Vavreck, 1985) as well as in neuroblastoma cells

(Braas *et al.*, 1985). These compounds will be of inestimable benefit in the effort to determine whether BK is an endogenous brain neurotransmitter. By administering antagonist i.c.v. and examining for resultant changes in biochemistry, physiology, and behavior, the role of endogenous brain BK can be inferred. This classical tool may prove to be the most important one of all, now that it is at last available to kinin researchers.

REFERENCES

Almeida de Silva, T. C., and Pela, I. R., 1978, Changes in rectal temperature of the rabbit by intracerebroventricular injection of bradykinin and related peptides, *Agents Actions* **8:**102–107.

Armstrong, D., 1970, Pain, in: *Bradykinin, Kallidin and Kallikrein. Handbook of Experimental Pharmacology,* Vol. 25 (E. G. Erdos, ed.), Springer-Verlag, Berlin, pp. 434–481.

Baccaglini, P. I., and Hogan, P. G., 1983, Some rat sensory neurons in culture express characteristics of differentiated pain sensory cells, *Proc. Natl. Acad. Sci. U.S.A.* **80:**594–598.

Barasi, S., and Wright, D. M., 1979, Effects of bradykinin on central neurons, in: *Advances in Pain Research and Therapy,* Vol. 3 (J. J. Bonica, ed.), Raven Press, New York, pp. 883–888.

Bareis, D. L., Manganiello, V. C., Hirata, F., Vaughan, M., and Axelrod, J., 1983, Bradykinin stimulates phospholipid methylation, calcium influx, prostaglandin formation, and cAMP accumulation in human fibroblasts, *Proc. Natl. Acad. Sci. U.S.A.* **80:**2514–2518.

Belcher, G., 1979, The effects of intra-arterial bradykinin, histamine, acetylcholine and prostaglandin E_1 on nociceptive and non-nociceptive dorsal horn neurones of the cat, *Eur. J. Pharmacol.* **56:**385–395.

Braas, K. M., Manning, D. C., Wilson, V. S., Perry, D. C., Stewart, J. M., Vavreck, R. J., and Snyder, S. H., 1985, Characterization of bradykinin antagonists in a cultured neuronal cell line, *Neuroscience Abstr.* **11:**414.

Buccafusco, J. J., and Serra, M., 1985, Role of cholinergic neurons in the cardiovascular responses evoked by central injection of bradykinin or angiotensin II in conscious rats, *Eur. J. Pharmacol.* **113:**43–51.

Bynke, G., Hakanson, R., Horig, J., and Leander, S., 1983, Bradykinin contracts the pupillary sphincter and evokes ocular inflammation through release of neuronal substance P, *Eur. J. Pharmacol.* **91:**469–475.

Camargo, A. C. M., Ramalho-Pinto, F. J., and Greene, L. J., 1972, Brain peptidases: Conversion and inactivation of kinin hormones, *J. Neurochem.* **19:**37–49.

Capek, R., Masek, K., Sramka, M., Krsiak, M., and Svec, P., 1969, The similarities of the angiotensin and bradykinin action on the central nervous system, *Pharmacology* **2:**161–170.

Carretero, O. A., and Scicli, A. G., 1980, The renal kallikrein–kinin system, *Am. J. Physiol.* **238:**F247–F255.

Chahl, L. A., and Iggo, A., 1977, The effects of bradykinin and prostaglandin E_1 on rat cutaneous afferent nerve activity, *Br. J. Pharmacol.* **59:**343–347.

Chao, J., Woodley, C., Chao, L., and Margolius, H. S., 1983, Identification of tissue kallikrein in rat brain and in the cell-free translation product encoded by rat brain mRNA, *J. Biol. Chem.* **258:**15173–15178.

Chao, J., Chao, L., Ando, T., and Margolius, H., 1985, Kallikrein, kallikrein binding protein and kininogen in rat brain and in neuroblastoma × glioma hybrid cells, *Clin. Res.* **33:**532A.

Chapman, L. F., Ramos, A. O., Goodell, H., and Wolff, H. G., 1961, Neurohumoral features of afferent fibers in man, *Arch. Neurol.* **4:**49–82.

Clark, W. G., 1979, Kinins and the peripheral and central nervous system, in: *Bradykinin, Kallidin and Kallikrein, Handbook of Experimental Pharmacology,* Vol. 25, Suppl. (E. G. Erdos, ed.), Springer-Verlag, Berlin, pp. 311–356.

Clyman, R. I., Blacksin, A. S., Manganiello, V. C., and Vaughan, M., 1975, Oxygen and cyclic nucleotides in the human umbilical artery, *Proc. Natl. Acad. Sci. U.S.A.* **72:**3883–3887.

Correa, F. M. A., and Graeff, F. G., 1974, Central mechanisms of the hypertensive action of intraventricular bradykinin in the unanesthetized rat, *Neuropharmacology* **13:**65–75.

Correa, F. M. A., and Graeff, F. G., 1975, Central site of the hypertensive action of bradykinin, *J. Pharmacol. Exp. Ther.* **192:**670–676.

Correa, F. M. A., and Graeff, F. G., 1976, On the mechanisms of the hypertensive action of intraseptal bradykinin in the rat, *Neuropharmacology* **15:**713–717.

Correa, F. M. A., Innis, R. B., Uhl, G. R., and Snyder, S. H., 1979, Bradykinin-like immunoreactive neuronal systems localized histochemically in rat brain, *Proc. Natl. Acad. Sci. U.S.A.* **76:**1489–1493.

Damas, J., 1972, La demi-vie de la bradykinine dans les espaces sous-arachnoides du rat, *C.R. Soc. Biol. (Paris),* **166:**740–744.

DaSilva, G. R., and Rocha e Silva, M., 1971, Catatonia induced in the rabbit by intracerebral injection of bradykinin and morphine, *Eur. J. Pharmacol.* **15:**180–186.

Diz, D. I., and Jacobowitz, D. M., 1984, Cardiovascular effects of discrete intrahypothalamic and preoptic injections of bradykinin, *Brain Res. Bull.* **12:**409–417.

Douglas, W. W., 1980, Polypeptides—angiotensin, plasmakinins, and others, in: *The Pharmacological Basis of Therapeutics,* 6th ed. (A. Goodman Gilman, L. S. Goodman, and A. Gilman, eds.), Macmillan, New York, pp. 647–667.

Erdos, E. G. (ed.), 1970, *Bradykinin, Kallidin and Kallikrein. Handbook of Experimental Pharmacology,* Vol. 25, Springer-Verlag, Berlin, Heidelberg, New York.

Ferreira, S. H., 1972, Prostaglandins, aspirin-like drugs and analgesia, *Nature* **240:**200–203.

Frederick, M. J., Vavreck, R. J., Stewart, J. M., and Odya, C. E., 1984, Further studies of myometrial bradykinin receptor-like binding, *Biochem. Pharmacol.* **18:**2887–2892.

Frucht, H., Lilling, G., and Beitner, R., 1984, Influence of bradykinin on glucose 1,6-bisphosphate and cyclic GMP levels and on the activities of glucose 1,6-bisphosphatase, phosphofructokinase and phosphoglucomutase in muscle, *Int. J. Biochem.* **16:**397–402.

Goldstein, D. J., Ropchak, T. G., Keiser, H. R., Atta, G. J., Argiolas, A., and Pisano, J. J., 1983, Bradykinin reverses the effect of opiates in the gut by enhancing acetylcholine release, *J. Biol. Chem.* **258:**12122–12124.

Graeff, F. G., and Arisawa, E. A. L., 1978, Effect of intracerebroventricular bradykinin, angiotensin II and substance P on multiple fixed-interval fixed-ratio responding in rabbits, *Psychopharmacology* **57:**89–95.

Graeff, F. G., Pela, I. R., and Rocha e Silva, M., 1969, Behavioral and somatic effects of bradykinin injected into the cerebral ventricles of unanesthetized rabbits, *Br. J. Pharmacol.* **37:**723–732.

Hoffman, W. E., and Schmid, P. G., 1978, Separation of pressor and anti-diuretic effects of intraventricular bradykinin, *Neuropharmacology* **17:**992–1002.

Hori, S., 1968, The presence of bradykinin-like polypeptides, kinin-releasing and destroying activity in the brain, *Jpn. J. Pharmacol.* **18:**772–787.

Innis, R. B., Manning, D. C., Stewart, J. M., and Snyder, S. H., 1981, [^{3}H]Bradykinin receptor binding in mammalian tissue membranes, *Proc. Natl. Acad. Sci. U.S.A.* **78:**2630–2634.

Inoki, R., Hayashi, T., Kudo, T., and Matsumoto, K., 1978, Effects of aspirin and morphine on the release of a bradykinin-like substance into the subcutaneous perfusate of the rat paw, *Pain* **5:**53–68.

Inouye, A., Kataoka, K., and Tsujioka, T., 1961, On a kinin-like substance in the nervous tissue extracts treated with trypsin, *Jpn. J. Physiol.* **11:**319–334.

Iwata, H., Shikimi, T., and Oka, T., 1969, Pharmacological significance of peptidase and proteinase in the brain: Enzymatic inactivation of bradykinin in the rat brain, *Biochem. Pharmacol.* **18:**119–128.

Iwata, H., Shikimi, T., Iida, M., and Miichi, H., 1970, Effect of bradykinin on the central nervous system and role of the enzyme inactivating bradykinin in the mouse brain, *Jpn. J. Pharmacol.* **20:**80–86.

Iwata, T., Hashimoto, H., Hiwada, K., and Kokubu, T., 1984, Changes of plasma renin activity by intracerebroventricular administration of biologically active peptides in conscious rats, *Clin. Exp. Hypertension* **A6:**1055–1066.

Johnson, A. R., 1979, Effects of kinins in organ systems, in: *Bradykinin, Kallidin and Kallikrein. Handbook of Experimental Pharmacology,* Vol. 25, Suppl. (E. G. Erdos, ed.), Springer-Verlag, Berlin, pp. 357–399.

Juan, H., and Lembeck, F., 1974, Actions of peptides and other algesic agents on paravascular pain receptors of the isolated perfused rabbit ear, *Naunyn Schmiedebergs Arch. Pharmacol.* **290:**389–395.

Kariya, K., and Yamauchi, A., 1981, Effects of intraventricular injection of bradykinin on the EEG and the blood pressure in conscious rats, *Neuropharmacology* **20:**1221–1224.

Kariya, K., Iwaki, H., Ihda, M., Maruta, E., and Murasa, M., 1981, Electroencephalogram of bradykinin and its degradation system in rat brain, *Jpn. J. Pharmacol.* **31:**261–267.

Kariya, K., Yamauchi, A., and Chatani, Y., 1982a, Relationship between central actions of bradykinin and prostaglandins in the conscious rat, *Neuropharmacology* **21:**267–272.

Kariya, K., Yamauchi, A., Hattori, S., Tsuda, Y., and Okata, Y., 1982b, The disappearance rate of intraventricular bradykinin in the brain of the conscious rat, *Biochem. Biophys. Res. Commun.* **107:**1461–1466.

Kaufman, M. P., Coleridge, H. M., Coleridge, J. C. G., and Baker, D. G., 1980a, Bradykinin stimulates afferent vagal C-fibers in intrapulmonary airways of dogs, *J. Appl. Physiol.* **48:**511–517.

Kaufman, M. P., Baker, D. G., Coleridge, H. M., and Coleridge, J. C. G., 1980b, Stimulation by bradykinin of afferent vagal C-fibers with chemosensitive endings in the heart and aorta of the dog, *Circ. Res.* **46:**476–484.

Kellermeyer, R. W., and Graham, R. C., 1968, Kinins—possible physiologic and pathologic roles in man, *N. Engl. J. Med.* **279:**754–759, 802–807; 859–866.

Kondo, K., Okuno, T., Konishi, K., Saruta, T., and Kato, E., 1979, Central and peripheral effects of bradykinin and prostaglandin E_2 in blood pressure in conscious rats, *Naunyn Schmiedebergs Arch. Pharmacol.* **308:**111–115.

Lambert, G. A., and Lang, W. J., 1970, The effects of bradykinin and eledoisin injected into the cerebral ventricles of conscious rats, *Eur. J. Pharmacol.* **9:**383–386.

Lembeck, F., Popper, H., and Juan, H., 1976, Release of prostaglandins by bradykinin as an instrinsic mechanism of its algesic effect. *Naunyn Schmiedebergs Arch. Pharmacol.* **294:**69–73.

Lewis, G. P., and Reit, E., 1965, The action of angiotensin and bradykinin on the superior cervical ganglion of the cat, *J. Physiol. (Lond.)* **179:**538–553.

Lewis, R. E., and Phillips, M. I., 1984, Localization of the central pressor action of bradykinin to the cerebral third ventricle, *Am. J. Physiol.* **247:**R63–R68.

Lewis, R. E., Hoffman, W. E., and Phillips, M. I., 1983, Angiotensin II and bradykinin: Interactions between two centrally active peptides, *Am. J. Physiol.* **244:**R285–R291.

Lim, R. K. S., 1970, Pain, *Annu. Rev. Physiol* **32:**269–288.

Manning, D. C., and Snyder, S. H., 1983, [^{3}H]Bradykinin receptor localization in spinal cord and sensory ganglia—evidence for a role in primary afferent function, *Neurosci. Abstr.* **9:**590.

Manning, D. C., and Snyder, S. H., 1986, [^{3}H]Bradykinin binding site localization in guinea pig urinary system, in: *Kinins IV(A)* (L. M. Greenbaum and H. S. Margolius, eds.), Plenum Press, New York, pp. 561–568.

Manning, D. C., Snyder, S. H., Kachur, J. F., Miller, R. J., and Field, M., 1982, Bradykinin receptor-mediated chloride secretion in intestinal function, *Nature* **299:**256–259.

McGiff, J. C., Itskovitz, H. D., Terragno, N. A., and Wong, P. Y. K., 1976, Modulation and mediation of the action of the renal kallikrein–kinin system by prostaglandins, *Fed. Proc.* **35:**175–180.

Melo, J. C., and Graeff, F. G., 1975, Effect of intracerebroventricular bradykinin and related peptides on rabbit operant behavior, *J. Pharmacol. Exp. Ther.* **193:**1–10.

Moniuszko-Jakoniuk, J., Wisniewski, K., and Koscielak, M., 1976, Investigations of the mechanism of central action of kinins, *Psychopharmacology* **50:**181–186.

Nawa, H., Kitamura, N., Hirose, T., Asai, M., Inayama, S., and Nakanishi, S., 1983, Primary structures of bovine liver low molecular weight kininogen precursors and their two mRNAs, *Proc. Natl. Acad. Sci. U.S.A.* **80:**90–94.

Odya, C. E., Goodfriend, T. L., and Pena, C., 1980, Bradykinin receptor-like binding studied with iodinated analogues, *Biochem. Pharmacol.* **29:**175–185.

Okada, Y., Tuchiya, Y., Yagyu, M., Kozawa, S., and Kariya, K., 1977, Synthesis of bradykinin fragments and their effects on pentobarbital sleeping time in mouse, *Neuropharmacology* **16:**381–383.

Oliveira, E. B., Martino, A. R., and Camargo, A. C. M., 1976, Isolation of brain endopeptidases: Influence of size and sequences of substrates structurally related to bradykinin, *Biochemistry* **15:**1967–1974.

Pearson, L., Lambert, G. A., and Lang, W. J., 1969, Centrally mediated cardiovascular and EEG responses to bradykinin and eledoisin, *Eur. J. Pharmacol.* **8:**153–158.

Pela, I. R., Gardey-Levassort, C., Lechat, P., and Rocha e Silva, M., 1975, Brain kinins and fever induced by bacterial pyrogens in rabbits, *J. Pharm. Pharmacol.* **27:**793–794.

Perry, D. C., and Snyder, S. H., 1984, Identification of authentic bradykinin in mammalian brain, *J. Neurochem.* **43:**1072–1080.

Phillips, M. I., Summers, C., Lewis, R., Hoffman, L., and Casto, R., 1982, Cardiovascular effects of neuropeptides in brain and periphery, *Fed. Proc.* **41:**1661.

Phillis, J. W., and Limacher, J. J., 1974, Excitation of cerebral cortical neurons by various polypeptides, *Exp. Neurol.* **43:**414–423.

Pisano, J. J., and Austen, K. F. (eds.), 1974, *Chemistry and Biology of the Kallikrein–Kinin System in Health and Disease,* DHEW Publication No. NIH76-791, U.S. Government Printing Office, Washington.

Regoli, D., and Barabe, J., 1980, Pharmacology of bradykinin and related kinins, *Pharmacol. Rev.* **32:**1–46.

Reiser, G., and Hamprecht, B., 1982, Bradykinin induces hyperpolarizations in rat glioma cells and in neuroblastoma × glioma hybrid cells, *Brain Res.* **239:**191–199.

Reiser, G., Walter, U., and Hamprecht, B., 1984, Bradykinin regulates the level of guanosine 3′,5′-cyclic monophosphate (cyclic GMP) in neural cell lines, *Brain Res.* **290:**367–371.

Ribeiro, S. A., and Rocha e Silva, M., 1973, Antinociceptive action of bradykinin and related kinins of larger molecular weight by the intraventricular route, *Br. J. Pharmacol.* **47:**517–528.

Ribeiro, S. A., DaSilva, G. R., Camargo, A. C. M., and Corrado, A. P., 1970, Hyperglycemia induced in the rabbit by intraventricular injection of bradykinin, *Adv. Exp. Med. Biol.* **8:**635–639.

Ribeiro, S. A., Corrado, A. P., and Rocha e Silva, M., 1971, Antinociceptive action of intraventricular bradykinin, *Neuropharmacology* **10:**725–731.

Rocha e Silva, M., Beraldo, W. T., and Rosenfeld, G., 1949, Bradykinin, a hypotensive and smooth muscle stimulating factor released from plasma globulin by snake venoms and by trypsin, *Am. J. Physiol.* **156:**261–273.

Roscher, A. A., Manganiello, V. C., Jelsema, C. L., and Moss, J., 1984, Autoregulation of bradykinin receptors and bradykinin-induced prostacyclin formulation in human fibroblasts, *J. Clin. Invest.* **74:**552–558.

Shikimi, T., Kema, R., Matsumoto, M., Yamahata, Y., and Miyata, S., 1973, Studies on kinin-like substances in brain, *Biochem. Pharmacol.* **22:**567–573.

Simson, J. A. V., Dom, R., Chao, J., Chao, L., and Margolius, H. S., 1985, Immunocytochemical localization of tissue kallikrein in brain ventricular epithelium and hypothalamic cell bodies, *J. Histochem. Cytochem.* **33:**951–953.

Snider, R. M., and Richelson, E., 1984, Bradykinin receptor-mediated cyclic GMP formation in a nerve cell population (murine neuroblastoma clone N1E-115), *J. Neurochem.* **43:**1749–1754.

Stewart, J. M., and Vavreck, R. J., 1985, Competitive antagonists of bradykinin, *Peptides* **6:**161–164.

Stoner, J., Manganiello, V. C., and Vaughan, M., 1973, Effects of bradykinin and indomethacin on cyclic GMP and cyclic AMP in lung slices, *Proc. Natl. Acad. Sci. U.S.A.* **70:**3830–3833.

Takahashi, H., and Bunag, R. D., 1981, Centrally induced cardiovascular and sympathetic nerve responses to bradykinin in rats, *J. Pharmacol. Exp. Ther.* **216:**192–197.

Thomas, G. R., Thibodeaux, H., Margolius, H. S., and Privitera, P. J., 1984, Cerebrospinal fluid kinins and cardiovascular function, *Hypertension* **6**(Suppl. I):I46–I50.

Tomita, K., and Pisano, J. J., 1984, Binding of [^{3}H]bradykinin in isolated nephron segments of the rabbit, *Am. J. Physiol.* **246:**F732–F737.

Ueda, N., Muramatsu, I., and Fujiwara, M., 1984, Capsaicin and bradykinin-induced substance P-ergic responses in the iris sphincter muscle of the rabbit, *J. Pharmacol. Exp. Ther.* **230:**469–473.

Unger, T., Rockhold, R. W., Kaufman-Buhler, I., Hubner, D., Schull, B., and Ganten, D., 1980, Effects of angiotensin concerting enzyme inhibitors on the brain, in: *Angiotensin Converting Enzyme Inhibitors* (Z. P. Horovitz, ed.), Urban & Schwarzenberg, Baltimore, pp. 55–79.

Wahlestedt, C., Bynke, G., and Hakanson, R., 1985, Pupillary constriction by bradykinin and capsaicin: Mode of action, *Eur. J. Pharmacol.* **106:**577–583.

Werle, V. E., and Zack, P., 1970, Verteilung von Kininogen in Serum und Geweben bei Ratten und anderen Sangetieren, *Z. Klin. Chem. Klin. Biochem.* **8:**186–189.

Wilk, S., and Orlowski, M., 1979, Degradation of bradykinin by isolated neutral endopeptidases of brain and pituitary, *Biochem. Biophys. Res. Commun.* **90:**1–6.

Yang, H. Y. T., and Neff, N. H., 1972, Distribution and properties of angiotensin converting enzyme of rat brain, *J. Neurochem.* **19:**2443–2450.

Yano, K., Higashida, H., Inoue, R., and Nozawa, Y., 1984, Bradykinin-induced rapid breakdown of phosphatidylinositol 4,5-bisphosphate in neuroblastoma × glioma hybrid NG108-15 cells, *J. Biol. Chem.* **259:**10201–10207.

Yano, K., Higashida, H., Hattori, H., and Nozawa, Y., 1985, Bradykinin-induced transient accumulation of inositol trisphosphate in neuron-like cell line NG108-15 cell, *FEBS Lett.* **181:**403–406.

Yasujima, M., Matthews, P. G., and Johnston, C. I., 1981, Regulation of uterine smooth muscle bradykinin receptors by bradykinin and angiotensin converting enzyme inhibitor in the rat, *Clin. Exp. Pharmacol. Physiol.* **8:**515–518.

Zenser, T. V., Rapp, N. S., Spry, L. A., and Davis, B. B., 1984, Independent effects of bradykinin on adenosine 3′,5′-monophosphate and prostaglandin E_2 metabolism by rabbit renal medulla, *Endocrinology* **114:**541–544.

26

The Neuroanatomic Correlates of the Behavioral Effects of Bombesin

S. JOHNSTON, P. PARMASHWAR, and Z. MERALI

1. INTRODUCTION

Bombesin (BN), a tetradecapeptide initially isolated from anuran skin (Anastasi *et al.*, 1971), has become the standard of reference for comparison of the effects of a series of BN-like peptides recently isolated from several other species including rat, pig, chicken, dog, and human. These BN-like peptides share structural homology, particularly at the carboxyl-terminal heptapeptide region (McDonald *et al.*, 1979, 1980; Reeve *et al.*, 1983; Orloff *et al.*, 1984; Spindel *et al.*, 1984; Minamino *et al.*, 1984).

Central administration of BN induces pronounced behavioral activation (Pert *et al.*, 1980; Kulkosky *et al.*, 1982a,b; Merali *et al.*, 1983, 1985; Gmerek and Cowan, 1983, 1984; Schulz *et al.*, 1984; Cowan *et al.*, 1985). In addition, BN alters a variety of physiological and secretory functions, producing hypothermia in cold-exposed rats (Brown *et al.*, 1977a,b; Tache *et al.*, 1979), suppression of feeding (Parrot and Baldwin, 1982; Kulkosky *et al.*, 1982a,b; Stuckey and Gibbs, 1982; Gibbs, 1985), alteration of gastric secretion (Tache and Collu, 1982; Tache, 1982; Tache and Brown, 1982; Tache *et al.*, 1984), induction of hyperglycemia (Brown *et al.*, 1977c, 1979), elevation of mean arterial pressure, reduction of heart rate (Fisher and Brown, 1984), suppression of prolactin release (Collu *et al.*, 1983), and increased dopamine function in the rat brain (Widerlov *et al.*, 1984).

These potent effects coupled with the wide distribution of BN-like peptides throughout the mammalian gastrointestinal tract (Dockray *et al.*, 1979) and nervous system (Brown *et al.*, 1978; Moody and Pert, 1979; Moody *et al.*, 1980, 1981; Panula *et al.*, 1982; Soveny *et al.*, 1984) support the role of BN as an important neuroregulatory peptide.

S. JOHNSTON, P. PARMASHWAR, and Z. MERALI • School of Psychology, University of Ottawa, Ottawa, Ontario K1N 6N5, Canada.

The behavioral effects of BN injected ICV are time and dose dependent (Brown and Vale, 1980; Pert *et al.*, 1980; Kulkosky *et al.*, 1982a,b; Gmerek and Cowan, 1983; Schulz *et al.*, 1984; Merali *et al.*, 1983, 1985). Administration of BN (0.01–1.0 μg ICV) significantly and promptly increases locomotion, floor activity, and rearing for up to 2.5 hr (Merali *et al.*, 1983). A conspicuous behavioral alteration in the grooming pattern elicited by BN (ICV) is a syndrome of vastly increased scratching (Katz, 1980; Kulkosky *et al.*, 1982a; Gmerek and Cowan, 1983; Merali *et al.*, 1983; Schulz *et al.*, 1984; Cowan *et al.*, 1985), which is distinct from the grooming pattern induced by the prototypic groom-inducing peptide adrenocorticotropic hormone (ACTH). With ACTH the length of individual bouts and not the proportion of time spent per grooming element is enhanced (Gispen and Isaacson, 1981; Spruijt and Gispen, 1983). Another difference is their time course; BN-induced grooming begins immediately, whereas ACTH-induced grooming does not occur until 10–15 min following ICV administration (Merali *et al.*, 1983; Spruijt and Gispen, 1983; Gmerek and Cowan, 1983, 1984; Isaacson, 1984; Cowan *et al.*, 1985). Although both peptides require intact dopamine pathway(s) for their behavioral expression (Isaacson, 1984; Merali *et al.*, 1985), BN-induced effects (unlike those of ACTH) are not naloxone reversible (Gmerek and Cowan, 1983). Thus, marked pharmacological and behavioral differences exist between these two groom-inducing peptides.

Recently, several studies have attempted to delineate neuroanatomic loci of some of the biological effects of BN. Bombesin microinjected at the preoptic area produces hypothermia (Pittman *et al.*, 1980; Wunder *et al.*, 1980), whereas at the dorsal hypothalamus it produces a specific and rapid rise in plasma epinephrine (Brown, 1983). When microinjected into the lateral hypothalamus, BN appears to reduce food intake (Stuckey and Gibbs, 1982; Gibbs, 1985), and at the paraventricular nucleus, it produces a marked rise in gastric pH values and a decrease in secretory volume (Gunion *et al.*, 1983).

Behavioral effects of BN can be elicited by central (via the ventricular, intracisternal, or intrathecal) but not peripheral administration (Kulkosky, 1982a,b; Schulz *et al.*, 1984; Gmerek *et al.*, 1983). Furthermore, BN-induced behavioral effects appear to be independent of the pituitary–adrenal axis (Gmerek and Cowan, 1983). However, little is known about the neuroanatomic locus (or loci) or the neurochemical mechanisms subserving these behavioral effects. Recent results indicated that administration of BN, either ICV or into the ventral tegmental area (VTA), equipotently stimulated the locomotor activity. However, BN infused into the nucleus accumbens (NA) was considerably more potent in inducing locomotor stimulation (Schulz *et al.*, 1984).

Although BN-induced behavior has been blocked by haloperidol (Merali *et al.*, 1983; Schulz *et al.*, 1984) benzomorphan (Gmerek and Cowan, 1984), and diazepam (Crawley and Moody, 1983), these drugs are not known to block the BN receptors directly. Recently, a substance P analogue, spantide, has been demonstrated to be a competitive BN receptor antagonist (Jensen *et al.*, 1984; Yachnis *et al.*, 1984; Folkers *et al.*, 1984). As well as inhibiting binding of BN-like peptide to central receptors, spantide (ICV) reversed the BN-induced hypothermia and grooming (Yachnis *et al.*, 1984).

The overall objective of the present series of experiments was to attempt to identify the central locus or loci involved in elicitation of the behavioral effects of BN. Hence, experiments were undertaken to examine the behavioral effects of BN microinjected intracerebroventricularly (ICV) or at selected rat brain sites endowed with a high density of BN binding sites, including the nucleus tractus solitarius (NTS), hippocampus (CA4), nucleus accumbens (NA), and the anterior olfactory nucleus (AON) (Pert *et al.*, 1980; Wolf *et al.*, 1983; Zarbin *et al.*, 1985).

2. METHODS

Adult male Sprague–Dawley rats (275–300 g) (St. Constant, Quebec) were housed individually with free access to food (Master Laboratory Chow) and water. The environment was maintained at 24°C, 60% relative humidity, and with 12 hr of light (6 a.m. to 6 p.m.).

Separate groups of rats under sodium pentobarbital (50 mg/kg, IP) anesthesia were stereotactically implanted with 24-gauge adjustable stainless-steel guide cannulae (Kinetrods, Ottawa, Canada) (unless indicated otherwise) aimed at the various sites indicated below: lateral ventricle (22 gauge; Plastic Products, Roanoke, VA), A-P −0.8 mm, lateral −1.6 mm, depth −4.2 mm; NTS, A-P −13.3 mm, lateral +0.7 mm, depth −7.8; CA4, A-P −3.3 mm, lateral ±1.8 mm, depth—3.5 mm; NA, A-P +2.2 mm, lateral ±1.8 mm, depth −6.7 mm; AON, A-P +4.2 mm, lateral ±2.2 mm, depth −5.5 mm (Paxinos and Watson, 1982). The guide cannulae were cemented with dental acrylic to four jewelers' screws placed in the calvarium. Stainless-steel obturators were in the guide cannulae at all times except during injections. The animals were allowed a minimum of 5 postoperative recovery days prior to commencement of experiments. At the termination of the experiments, cannula placements were verified using cresyl violet dye substitution followed by standard histological procedures.

Monitoring of the locomotor, floor, and rearing activity was conducted by the procedure described previously (Merali *et al.*, 1983, 1985). Each behavioral observation chamber consisted of an inner clear polycarbonate cage (43 × 23 × 15 cm; identical to the rats' home cage) and an outer array of nine strategically placed infrared light beams (IR beams). A custom-designed Z-80 microprocessor-based controller performed most of the timing and scoring functions. The system consisted of eight chambers, and each of the IR beams was sampled once every second.

Simultaneous monitoring of the grooming activity by human raters was conducted from an adjacent room, through a one-way mirror, using a modification of the procedure described by Gispen and Isaacson (1981). A maximum of eight rats were monitored at a time. Each rat was observed for one 5-sec interval out of each 40-sec observation period for a total duration of 60 min. Thus, every 5 sec the observer recorded whether or not a given rat displayed the following grooming elements: facial licking, facial scratching, body licking, body scratching, and sniffing (Gispen and Isaacson, 1981; Gmerek and Cowan, 1983).

These behaviors were operationally defined as follows: facial licking, use of the forepaws either placed in the mouth before or after passing in a wiping motion over the face/head region; facial scratching, use of the digits of the hind legs to scratch the head and neck region; body licking, use of the mouth and forepaws in a wiping/gliding motion over the entire body excluding the head region; body scratching, scratching of any part of the body region below the neck with the digits of the hind legs; sniffing, vibration of the nostrils, which is often accompanied by either rearing or movement across the floor. Each rat received a score of 1 for any of each of the 5 behaviors it engaged in within the 5-sec interval for each 40-sec observation period (Gispen and Isaacson, 1981; Gmerek and Cowan, 1983). All doses of BN are expressed in micrograms per animal.

2.1. Experiment 1

In experiment 1, rats ($n = 8$) were implanted with guide cannulae aimed at the right lateral ventricle. Animals were randomly assigned to individual experimental chambers. After 1 hr of acclimatization, all rats were injected with one of the following doses of BN, in a randomized order, until all doses were tested: vehicle (0.9% saline), 0.01, 0.1, or 1.0 μg BN. All injections were administered at the rate of 3 μl/35 sec using a Harvard infusion pump. In addition, the injection cannulae (which protruded past the guide cannulae by 0.5 mm) were left in place for 30 sec after the injection. After treatment, the animals were distributed into their experimental chambers, and data collection began. All sessions commenced at 11:00 a.m. and were 60 min in duration. The behavioral effects were oberved and quantitated as described above.

2.2. Experiment 2

This experiment was designed to delineate the dose effect of BN, microinjected at the NTS, on the behavioral profile of rats.

Eight animals equipped with guide cannulae unilaterally implanted at the NTS were randomly assigned to individual behavioral chambers and microinjected with one of the following doses of BN: 0.0, 0.005, 0.05, 0.5, or 1.0 μg. All microinjections were administered at the rate of 0.5 μl/30 sec using a Harvard infusion pump. In addition, the injection cannulae (30 gauge), which protruded past the guide cannulae by 0.05 mm, were left in place for 30 sec after the injection. After treatment, behavior was monitored as detailed in Section 2.1.

2.3. Experiment 3

In experiment 3, the protocol of experiment 2 was repeated on a separate group of animals ($n = 8$) except that these rats were bilaterally implanted with guide cannulae aimed at the CA4. In addition, the doses per animal were vehicle, 0.01, 0.1, 1.0, or 2.0 μg BN.

2.4. Experiment 4

This experiment was designed to delineate the dose effect of BN microinjected at the NA on the behavioral profile of rats. Furthermore, the possible antagonism of BN-induced behaviors by the substance P analogue spantide was also tested.

Seven animals equipped with guide cannulae aimed at the NA were microinjected bilaterally with one of the following doses of BN until all doses were tested: 0.0, 0.001, 0.01, 0.1, 1.0, or 2.0 μg BN. The exception to this procedure occurred on the last injection day, when 2 μg spantide + 1.0 μg BN in 1 μl of saline was microinjected over 60 sec, and data were collected over 30 min.

2.5. Experiment 5

In experiment 5, the experiment 2 protocol was repeated on a separate group of rats ($n = 5$) except that the guide cannulae were implanted bilaterally at the AON. Effects of BN (1.0 μg) were compared to the control (saline) condition.

3. RESULTS

3.1. Experiment 1

Analysis of variance revealed a significant effect of dose of BN administered ICV on locomotion [$F(3,18) = 5.2$, $P < 0.01$], floor activity [$F(3,18) = 8.8$, $P < 0.001$], rearing [$F(3,18) = 3.7$, $P < 0.05$], facial scratching [$F(3,18) = 24.2$, $P < 0.001$], and facial licking [$F(3,18) = 20.5$, $P < 0.001$]. Comparisons of means using the least significant difference (LSD) procedure (Kirk, 1982) indicated that compared to the saline condition, all doses of BN (1.0, 0.1 or 0.01 μg) were significantly effective in stimulating locomotion, floor activity, and rearing over the 60-min period. However, for both facial scratching and facial licking, significant differences occurred only at doses of 0.1 μg BN or greater.

3.2. Experiment 2

Analyses of the data from the NTS revealed a significant effect of dose of BN for floor activity [$F(4,28) = 8.6$, $P < 0.001$], rearing [$F(4,28) = 6.9$, $P < 0.001$], facial scratching [$F(4,28) = 62.7$, $P < 0.001$], facial licking [$F(4,28) = 27.2$, $P < 0.001$], body licking [$F(4,28) = 4.8$, $P < 0.01$], and sniffing [$F(4,28) = 3.8$, $P < 0.05$]. Significant differences occurred between the control condition and groups administered BN at a dose of 0.05 μg BN or greater for floor activity and facial scratching. For both facial and body licking, significant differences occurred at doses of 0.5 μg BN or higher. Rearing and sniffing were significantly altered only at selective doses of BN, namely, 0.5 and 0.05 μg, respectively.

3.3. Experiment 3

Analyses of the data from the CA4 revealed a significant effect of dose of BN for locomotion [$F(4,28) = 2.7$, $P < 0.05$], floor activity [$F(4,28) = 9.4$, $P < 0.001$], facial scratching [$F(4,28) = 19.5$, $P < 0.001$], and facial licking [$F(4,28) = 4.38$, $P < 0.01$]. Comparison of means indicated significant differences between the saline condition and groups administered 1.0 μg BN and higher for locomotion, floor activity, facial scratching, and facial licking.

3.4. Experiment 4

One-way analysis of variance (repeated over dose) revealed a significant effect of dose of BN administered intra-NA on locomotion [$F(5,30) = 3.7$, $P < 0.01$], floor activity [$F(5,30) = 2.5$, $P < 0.05$], rearing [$F(5,30) = 4.0$, $P < 0.01$], facial scratching [$F(5,30) = 5.4$, $P < 0.001$], facial licking [$F(5,30) = 4.1$, $P < 0.01$], and sniffing [$F(5,30) = 7.9$, $P < 0.001$]. No significant effect of BN on either body licking or body scratching was found. Comparison of means between the saline condition and various BN doses revealed that stimulation of facial scratching was evident even at the lowest dose used (0.001 μg), as was suppression of facial licking. However, the stimulatory effects on other behaviors, namely, locomotion, floor activity, rearing, and sniffing, were significant only at doses of 1.0 μg or greater. Facial licking appeared to be suppressed by the 1.0-μg, but not by the 2.0-μg dose of BN.

Thus at the NA, BN differentially affected elements of grooming, increasing facial scratching, decreasing facial licking, and having no effect on body scratching or body licking. Furthermore, all measures of ambulatory activity (locomotion, floor activity, rearing, and sniffing) were significantly stimulated at doses of 1.0 μg or greater.

Analysis of variance (repeated over dose) also revealed that at the NA there was a significant effect of spantide (2 μg) when coadministered with BN (1.0 μg) as compared with BN (1.0 μg) alone on locomotion [$F(1,6) = 47.2$, $P < 0.001$], floor activity [$F(1,6) = 22.3$, $P < 0.001$], rearing [$F(1,6) = 15.9$, $P < 0.01$], facial scratching [$F(1,6) = 12.3$, $P < 0.01$], facial licking [$F(1,6) = 12$, $P < 0.01$], and sniffing [$F(1,6) = 28.5$, $P < 0.001$]. As illustrated in Fig. 5, the stimulatory effects of BN on locomotion, floor activity, rearing, facial scratching, and sniffing were significantly attenuated by spantide, whereas suppression of facial licking was reversed by spantide. A comparison of means revealed that values from the spantide + BN groups (for each behavioral element) were not significantly different from their respective control (saline) values, indicating that spantide (2 μg) antagonized all the behavioral effects of BN (1.0 μg).

3.5. Experiment 5

At the AON, one-way analysis of variance revealed no significant effect of BN (1 μg) on any of the behavioral parameters monitored, and hence lower doses were not tested.

4. DISCUSSION

Our results demonstrate a unique behavioral profile of ambulatory behaviors (locomotor activity, floor activity, rearing, and sniffing) and grooming elements (facial scratching and licking, body scratching and licking) induced by BN administered ICV or intra-NTS, -CA4, and -NA. None of the behaviors monitored was significantly stimulated by BN microinjection intra-AON.

The data from these experiments demonstrated that BN administered ICV or intra-CA4 or -NA, potently stimulated locomotion in a dose-dependent manner. The behavioral effects were prompt in onset and continuous over the 60-min observation period. At the NA, the effect of BN on locomotion was greater in magnitude relative to its effect on floor activity and rearing (Figs. 1–3). Experimental results from the NA, as compared to ICV and intra-NTS or CA4, demonstrated that the stimulation of locomotor behavior was greatest in magnitude following the intra-NA administration of BN (Fig. 1). The only two sites at which BN stimulated sniffing were the NA and NTS, with the sniffing frequency being higher following intra-NA administration.

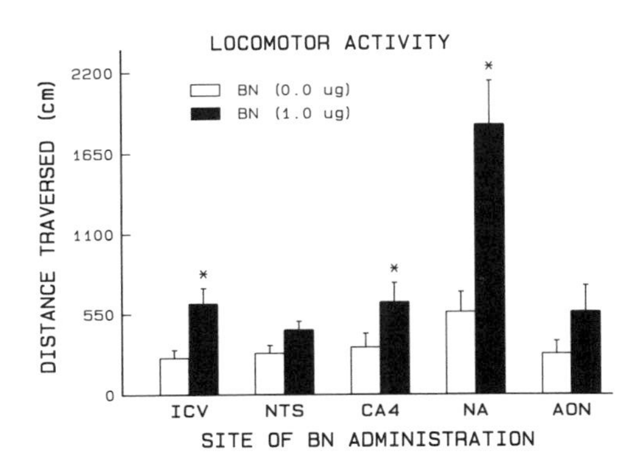

FIGURE 1. Effects of BN on locomotor activity of rats. On the ordinate, distance traversed (cm) over 60 min; on the abscissa, site of BN administration: ICV, intracerebroventricular; NTS, nucleus tractus solitarius; CA4, hippocampus; NA, nucleus accumbens; AON, anterior olfactory nucleus. The profile of vehicle-treated rats is represented by the open columns, and those of BN-treated groups are represented by the solid columns. Each value represents the mean + S.E.M. of the animals in that group (n = 5 to 8). *Significantly different from the appropriate control value at $P < 0.05$.

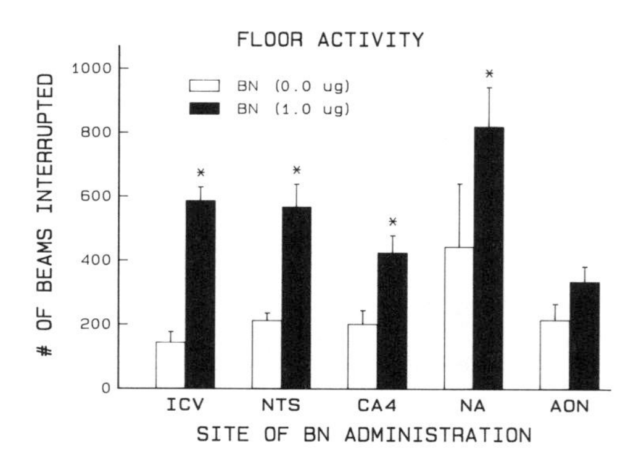

FIGURE 2. Effects of BN on floor activity of rats. On the ordinate, number of beams broken over 60 min; on the abscissa, site of BN administration: ICV, intracerebroventricular; NTS, nucleus tractus solitarius; CA4, hippocampus; NA, nucleus accumbens; AON, anterior olfactory nucleus. The profile of vehicle-treated rats is represented by the open columns, and those of BN-treated groups are represented by the solid columns. Each value represents the mean + S.E.M. of the animals in that group (n = 5 to 8). *Significantly different from appropriate control value at $P < 0.05$.

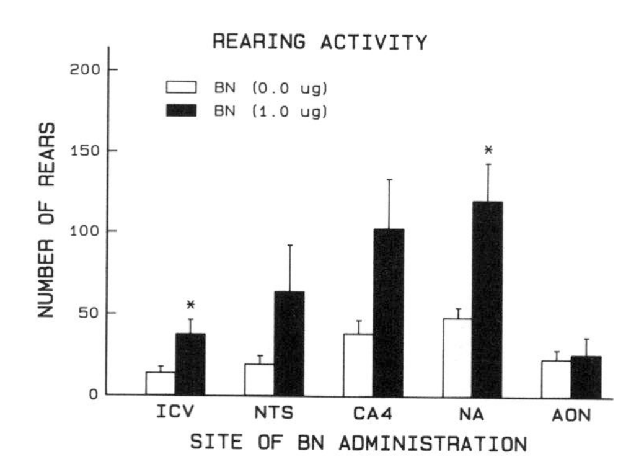

FIGURE 3. Effects of BN on the frequency of rearing activity of rats. On the ordinate, number of rears during 60 min; on the abscissa, site of BN administration: ICV, intracerebroventricular; NTS, nucleus tractus solitarius; CA4, hippocampus; NA, nucleus accumbens; AON, anterior olfactory nucleus. The profiles of vehicle-treated rats are represented by the open columns, and those of BN-treated groups are represented by the solid columns. Each value represents the mean + S.E.M. of the animals in that group (n = 5 to 8). *Significantly different from appropriate control value at $P < 0.05$.

In reference to the grooming profile at each locus, the grooming elements appeared to vary in a site-dependent manner. Our data indicated that the grooming profile observed following intra-NTS administration of BN resembled most closely that observed following ICV administration of BN in that there was a very high stimulation of facial scratching and moderately high stimulation of facial licking, whereas at the other sites, there was a moderate increase in facial scratching and a small increase or decrease in facial licking. Unique to the NTS site was the significant stimulation of body licking by BN microinjection (Fig. 4). Schulz *et al.* (1984) have similarly reported increased scratching but not sniffing following ICV BN (0.5 μg).

In contrast to the above behavioral profiles following ICV and NTS administration of BN, our data demonstrated that at the NA, BN (1.0 μg)-induced sniffing occurred with a greater frequency than scratching. Furthermore, BN (0.001 μg and 1.0 μg) intra-NA resulted in a significant decrease in facial licking as compared to the saline condition.

Gmerek and Cowan (1983) reported that the periaqueductal gray was not a critical site for BN-induced scratching, although BN did induce scratching at this site (A_{50} = 0.78 μg). Furthermore, Gmerek *et al.* (1983) have shown that intrathecally administered BN was slightly more potent than BN ICV (A_{50} = 0.004 μg and 0.013 μg, respectively) to elicit scratching. Consequently, it is feasible that a brainstem nucleus such as the NTS may be important in triggering the BN-induced scratching response induced by intrathecal BN. Another nucleus that may also be of importance in the above effect is the substantia gelatinosa of the spinal trigeminal

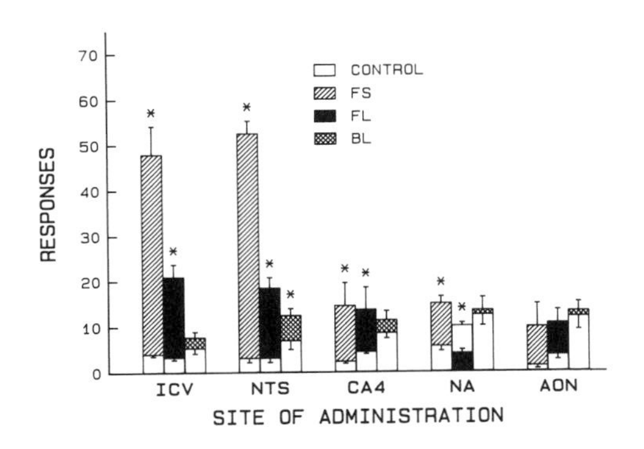

FIGURE 4. Effects of BN (1 μg) on the frequency of facial scratching (FS, hatched columns), facial licking (FL, solid columns), and body licking (BL, cross-hatched columns) when administered at various brain loci. On the ordinate, the magnitude of the quantified response over 60 min; on the abscissa, site of BN administration: ICV, intracerebroventricular; NTS, nucleus tractus solitarius; CA4, hippocampus; NA, nucleus accumbens; AON, anterior olfactory nucleus. Each value represents the mean and the S.E.M. of the animals in the respective groups (n = 5 to 8). *Significantly different from the appropriate control value (open columns) at $P < 0.05$.

nucleus (O'Donohue *et al.*, 1984), since both nuclei appear to be involved in the processing of primary sensory input from the neck and face (Carpenter, 1985), and since both nuclei contain BN-like immunoreactive peptides (Moody *et al.*, 1981; Panula *et al.*, 1982; Soveny *et al.*, 1984) as well as BN binding sites (Wolf *et al.*, 1983; Zarbin *et al.*, 1985).

Recently, it has been reported that dorsal rhizotomy in the cat resulted in a marked decrease in BN immunoreactivity in the dorsal horn, indicating that BN-like peptide(s) may be contained in primary sensory afferents. Furthermore, it was demonstrated that BN injected into the spinal cord caused a biting and scratching response indicative of sensory stimulation and thus may be a neurotransmitter of primary sensory afferents to the spinal cord (O'Donohue *et al.*, 1984). Similarily, it is possible that BN may cause grooming and scratching by stimulating second-order cranial sensory neurons, which may mimic the stimulation of the skin of the face and neck regions (O'Donohue *et al.*, 1984). Furthermore, the wide distribution of BN-like immunoreactivity (Moody *et al.*, 1981; Panula *et al.*, 1982; Soveny *et al.*, 1984) as well as BN receptors (Pert *et al.*, 1980; Wolf *et al.*, 1983; Zarbin *et al.*, 1985) in limbic and diencephalic nuclei, the cortical layers V and VI, and the dorsal horn of the spinal cord (O'Donohue *et al.*, 1984) suggest a role for BN in sensory and motor behaviors of a nonspecific (associative) nature.

Grooming in rodents has been compared to stress-induced displacement behaviors in humans, such as "nail biting" (Bolles, 1960). Since ICV administration of BN causes alteration of body temperature (Brown *et al.*, 1977a; Tache and Brown, 1982), another cause of grooming could be that rats increase grooming rates for thermoregulatory purposes. Grooming at different sites in the brain may have different behavioral significance, namely, stress reduction, temperature control, body maintenance, itching, or satiety. Certainly the different behavioral profiles at different sites are compatible with this suggestion.

At present two peptides known to induce satiety, namely, BN and cholecystokinin (CCK), also induce changes in exploratory behavior (Crawley and Schwaber, 1983; Merali *et al.*, 1983; Kulkosky *et al.*, 1982a,b; Gibbs, 1985; Cowan *et al.*, 1985). The CCK-induced satiety syndrome induced reductions in exploratory behaviors in rats and mice at doses of CCK that inhibited food consumption (Crawley and Schwaber, 1983). Recently, it has been reported that radiofrequency lesions of the nucleus tractus solitarius abolished the effects of peripherally administered CCK on exploratory behaviors (Crawley and Schwaber, 1983). Thus, the NTS may represent a critical relay loop mediating the behavioral actions of CCK (Crawley and Schwaber, 1983). This caudal and medial solitary nucleus receives mainly visceral afferent fibers from the vagus nerve along with some facial and glossopharyngeal fibers (Carpenter, 1985). The NTS gives rise to ascending projections directly to the hypothalamus, amygdala, and NA and, via a single relay, to visceral and taste cortices (Chronister *et al.* 1981; Crawley and Schwaber, 1983). Our data demonstrated that microinjection of BN at the NTS did not significantly alter the locomotor activity. In sharp contrast, however, BN at the NA caused a profound stimulation of locomotor activity. It is of interest to note that CCK, although ineffective alone, has been reported to potentiate dopamine-induced hyperlocomotion when injected intra-NA (Crawley, 1985). Furthermore, at the NTS, BN markedly stimulated grooming, whereas at the NA, grooming activity was enhanced only moderately. Since grooming typically follows satiety (Gibbs, 1985), the highly significant grooming following intra-NTS administration is one further indication that BN at the NTS may play a crucial role in satiety-related behavioral sequence.

Although BN-induced grooming is distinct from the grooming pattern induced by the prototypic groom-inducing peptide ACTH, one apparent similarity appears to be the involvement of the dopamine system(s) in the mediation of these peptide-induced behaviors. Both grooming and locomotion appear to involve the dopamine system(s), since 6-OHDA blocks locomotor simulation (Kelly and Iversen, 1976; Fink and Smith, 1980; Merali *et al.*, 1985) and BN-induced ambulatory activities (Merali *et al.*, 1985). In addition, it has been reported that the dopamine antagonist haloperidol and fluphenazine significantly attenuate the behavioral effects of BN (Merali *et al.*, 1983, 1985; Schulz *et al.*, 1984). Isaacson (1984) has reached similar conclusions on the importance of dopamine pathways in ACTH-induced grooming from lesion studies. Thus, it is of interest to note that the ventricular walls as well as the NA, where BN delivery induced potent behavioral changes, are also very rich in dopaminergic terminations (Lindvall and Bjorklund, 1983). Consequently, these results add further support to our contention that BN may be mediating some of its behavioral effects through the dopamine system(s). Recent studies that found that BN centrally administered increased tyrosine hydoxylase activity and dopamine metabolites in the rat brain (Babu and Vijayan, 1983; Widerlov *et al.*, 1984) also support such a contention.

Of particular interest is the blockade of BN-induced behavior by spantide, a competitive BN receptor antagonist, at the NA. Yachnis *et al.* (1984) have reported that the IC_{50} value of spantide was 1 μM to inhibit radiolabeled BN in the CNS. Furthermore, they report that spantide (2 μg ICV), although ineffective alone, reversed the BN-induced (1 μg; ICV) hypothermia and grooming. Our results indicated that spantide (2 μg intra-NA) blocked all the BN-induced (1 μg intra-NA) behavioral changes (Fig. 5).

In summary, of all the central loci tested, the locomotor stimulatory effect of BN was most pronounced when this peptide was microinjected at the NA. This observation is consistent with the fact that this nucleus is involved in motor responses and that it is also endowed with a very high density of BN binding sites. Although the AON is also endowed with very high density of BN binding sites, local administration of BN there failed to elicit significant behavioral changes. The behavioral pattern most conspicuous on administration at the NTS was the increase in grooming activity, particularly the increase in facial scratching, facial licking, and body licking. The effects of BN administered ICV included both components, namely, increased locomotor as well as grooming activity, and may have resulted from the diffusion of BN to various active loci such as the NA and NTS. These data indicate that some of the BN-binding sites recently identified autoradiographically (Zarbin *et al.*, 1985), may represent functionally important receptor sites and that the

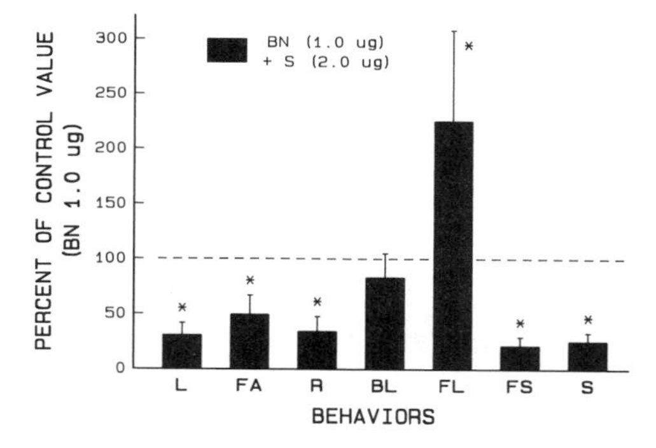

FIGURE 5. Effect of spantide on BN-induced stimulation of ambulatory activities and grooming elements at NA. On the ordinate, the magnitude of the quantified behavior over 30 min expressed as a percentage of control value (rats administered BN alone = 100%); on the abscissa; each column represents the mean + S.E.M. of rats administered spantide (2 μg) + BN (1 μg) expressed as a percentage of the control value for each behavior: L, locomotor activity; FA, floor activity; R, rearing; BL, body licking; FL, facial licking; FS, facial scratching; S, sniffing. *Significantly different from the appropriate control value at $P < 0.05$.

endogenous BN-like peptide(s) may subserve a physiological role in behaviors associated with exploration, satiety, stress, and care of body surfaces.

5. SUMMARY

Adult male Sprague–Dawley rats were microinjected with bombesin (BN) intracerebroventricularly (ICV) or at the nucleus tractus solitarius (NTS), hippocampus (CA4), nucleus accumbens (NA), and anterior olfactory nucleus (AON). Locomotion, floor activity, and rearing were monitored by a computerized system utilizing infrared beam grids, whereas the grooming elements (facial licking, facial scratching, body licking, and body scratching) and sniffing were recorded by a human rater. Bombesin differentially altered the behavioral profile in a site dependent manner.

Intraventricular administration of BN significantly stimulated locomotion, floor activity, and rearing at doses of 0.01 μg or greater and stimulated facial scratching and licking at doses of 0.1 μg or greater. However, sniffing, body scratching, and body licking were not significantly altered by ICV BN. Local microinjection of BN at the NTS significantly stimulated floor activity and facial scratching at a dose of 0.05 μg BN or greater and enhanced facial and body licking at doses of 0.5 μg or greater. Rearing was significantly stimulated only at the 0.5-μg dose, whereas sniffing was stimulated at the 0.05-μg dose. Administration of BN at CA4 significantly stimulated locomotion, floor activity, facial scratching, and facial licking at doses of 1.0 μg BN or greater. Microinjection of BN at the NA significantly stimulated facial scratching at doses as low as 0.001 μg and locomotion, floor activity, rearing, and sniffing at doses of 1.0 μg or greater. In contrast, BN (0.001 and 1.0 μg) at the same locus suppressed facial licking and failed to affect body licking and body scratching.

Spantide (2.0 μg), a BN antagonist, when coadministered with BN (1.0 μg), antagonized all the behavioral effects of BN intra-NA. Intra-AON, BN induced no significant change in behavior. In summary, of the five loci tested, the behavioral effects of BN (1 μg) were most pronounced ICV and intra-NTS and -NA. The locomotor stimulatory effects of BN were most pronounced when it was injected at the NA. Effects on grooming, however, were most pronounced when BN was administered ICV or intra-NTS. Bombesin-induced increase in body licking was unique to intra-NTS administration. The different patterns of behavioral changes evoked by BN administration at particular brain loci endowed with BN-binding sites indicate that some of these sites are pharmacologically functional receptor sites and that BN-like peptide(s) may be physiologically important in modulation of behavior.

ACKNOWLEDGMENTS. The expert technical assistance of Mr. H. van den Bergen and Mr. M. Makasare in the development of the computer-assisted behavioral monitoring apparatus and Kinetrods for the adjustable reusable cannulae are greatly appreciated. The authors wish to thank Dr. D. Coulombe for statistical assistance. This work was supported by grants from the Medical Research Council (MRC) and the Natural Sciences and Engineering Research Council of Canada (NSERC).

REFERENCES

Anastasi A., Erspamer, V., and Bucci, M., 1971, Isolation and structure of bombesin and alytesin, two analogous active peptides from the skin of the European amphibians *Bombina* and *Alytes*, *Experientia* **27:**166–167.

Babu, G. N., and Vijayan, E., 1983, Plasma gonadotropin, prolactin levels and hypothalamic tyrosine hydroxylase activity following intraventricular bombesin and secretin in ovariectomized conscious rats, *Brain Res. Bull.* **11:**25–29.

Bolles, R. C., 1960, Grooming behavior in the rat, *J. Comp. Physiol. Psychol.* **53:**306–310.

Brown, M. R., 1983, Central nervous system sites of action of bombesin and somatostatin to influence plasma epinephrine levels, *Brain Res.* **276:**253–257.

Brown, M., and Vale, W., 1980, in: *Thermoregulatory Mechanisms and Their Therapeutic Implications* (B. Cox, P. Lomax, A. S. Milton, and E. Schonbaum, eds.), S. Karger, Basel, pp. 186–194.

Brown, M., Rivier, J., and Vale, W., 1977a, Actions of bombesin, thyrotropin releasing factor, prostaglandin E and naloxone on thermoregulation in the rat, *Life Sci.* **20:**1681–1688.

Brown, M., Rivier, J., and Vale, W., 1977b, Bombesin: Potent effects on thermoregulation in rat, *Science* **196:**998–1000.

Brown, M., Rivier, J., and Vale, W., 1977c, Bombesin affects the central nervous system to produce hyperglycemia in rats, *Life Sci.* **21:**1729–1734.

Brown, M., Allen, R., Villareal, J., Rivier, J., and Vale, W., 1978, Bombesin-like activity: Radioimmunologic assessment in biological tissues, *Life Sci.* **23:**2721–2728.

Brown, M., Tache, Y., and Fisher, D., 1979, Central nervous system action of bombesin: Mechanism to induce hyperglycemia, *Endocrinology* **105:**660–665.

Carpenter, M. B., 1985, *Core Text of Neuroanatomy,* third edition, Williams & Wilkins, Baltimore, pp. 125–127.

Chronister, R. B., Sikes, R. W., Trow, T. W., and DeFrance J. F., 1981, The organization of the nucleus accumbens, in: *The Neurobiology of the Nucleus Accumbens* (R. B. Chronister and J. F. DeFrance, eds.), Haer Institute, Brunswick, ME, pp. 97–146.

Collu, R., Marchisio, A.-M., and Tache, Y., 1983, Inhibitory action of bombesin on prolactin release induced by dopamine antagonists, *Neuroendocrinol. Lett.* **5:**265–272.

Cowan, A., Khunawat, P., Zu Zhu, X, and Gmerek, D. E. IV, 1985, Effects of bombesin on behaviour, *Life Sci.* **37:**135–145.

Crawley, J. N., and Moody, T. W., 1983, Anxiolytics block excessive grooming behaviour induced by ACTH 1–24 and bombesin, *Brain Res. Bull.* **10:**399–401.

Crawley, J. N., and Schwaber, J. S., 1983, Nucleus tractus solitarius lesions block the behavioural actions of cholecystokinin, *Peptides* **4:**743–747.

Dockray, G. J., Valliant, C., and Walsh, J. H., 1979, The neuronal origin of bombesin-like immunoreactivity in the rat gastrointestinal tract, *Neuroscience* **4:**1561–1568.

Fink, J. S., and Smith, G. P., 1980, Mesolimbicocortical dopamine terminal fields are necessary for normal locomotor and investigatory exploration in rats, *Brain Res.* **199:**359–384.

Fisher, L. A., and Brown, M. R., 1984, Bombesin-induced stimulation of cardiac parasympathetic innervation, *Regul. Peptides* **8:**335–343.

Folkers, K., Hakanson, R., Horig, J., Jie-Cheng, X., and Leander, S., 1984, Biological evaluation of substance P antagonists, *Br. J. Pharmacol.* **83:**449–456.

Gibbs, J., 1985, Effect of bombesin on feeding, *Life Sci.* **37:**147–153.

Gispen, W. H., and Isaacson, R. L., 1981, ACTH-induced excesive grooming in the rat, *Pharmacol. Ther.* **12:**206–246.

Gmerek, D. E., and Cowan, A., 1983, Studies on bombesin-induced grooming in rats, *Peptides* **4:**907–913.

Gmerek, D. E., and Cowan, A., 1984, *In vivo* evidence for benzomorphan-selective receptors in rats, *J. Pharmacol. Exp. Ther.* **230:**110–115.

Gmerek, D. E., Cowan, A., and Vaught, J. L., 1983, Intrathecal bombesin in rats: Effects on behaviour and gastrointestinal transit, *Eur. J. Pharmacol.* **94:**141–143.

Gunion, M., Tache, Y., and Walsh, J. H., 1983, Gastric hyposecretion and hyperglycemia induced by bombesin injection near the paraventricular nucleus, *Soc. Neurosci. Abstr.* **3:**442.

Isaacson, R. L., 1984, Hippocampal damage: Effects on dopaminergic systems of the basal ganglia, in: *International Review of Neurobiology,* Vol. 25 (J. Smythes and R. Bradley, eds.), Academic Press, New York, pp. 339–359.

Jensen, R. T., Jones, S. W., Folkers, K., and Gardner, J. D., 1984, A synthetic peptide that is a bombesin receptor antagonist, *Nature* **309:**61–63.

Katz, R., 1980, Grooming elicited by intraventricular bombesin and eledoisin in the mouse, *Neuropharmacology* **19:**143–146.

Kelly, P. H., and Iversen, S. D., 1976, Selective 6-OHDA-induced destruction of mesolimbic dopamine neurons: Abolition of psychostimulant-induced locomotor activity in rats, *Eur. J. Pharmacol.* **40:**45–56.

Kirk, R. E., 1982, *Experimental Design: Procedures for the Behavioral Sciences,* Brooks/Cole, Monterey, CA, p. 115.

Kulkosky, P. J., Gibbs, J., and Smith, G. P., 1982a, Behavioural effects of bombesin administration in rats, *Physiol. Behav.* **28:**505–512.

Kulkosky, P. J., Gibbs, J., and Smith, G. P., 1982b, Feeding suppression and grooming repeatedly elicited by intraventricular bombesin, *Brain Res.* **242:**194–196.

Lindvall, O., and Bjorklund, A., 1983, Dopamine and norepinephrine-containing neuron systems: Their anatomy in the rat brain, in: *Chemical Neuroanatomy* (P. C. Emson, ed.), Raven Press, New York, pp. 229–255.

McDonald, T. J., Jornvall, H., Nilsson, G., Vagne, M., Ghatei M., Bloom, S. R., and Mutt, V., 1979, Characterization of a gastrin releasing peptide from porcine non-antral gastric tissue, *Biochem. Biophys. Res. Commun.* **90:**227–233.

McDonald, T. J., Jornvall, H., Ghatei, M., Bloom, S. R., and Mutt, V., 1980, Characterization of an avian (proventricular) peptide having sequence homology with porcine gastrin-releasing peptide and the amphibian peptides bombesin and alystensin, *FEBS Lett.* **122:**45–48.

Merali, Z., Johnston, S., and Zalcman, S., 1983, Bombesin-induced behavioural changes: Antagonism by neuroleptics, *Peptides* **4:**693–697.

Merali, Z., Johnston, S., and Sistek, J., 1985, Role of dopaminergic system(s) in mediation of the behavioural effects of bombesin, *Pharmacol. Biochem. Behav.* **23:**243–248.

Minamino, N., Kangawa, K., and Matsuo, H., 1984, Neuromedin B is a major bombesin-like peptide in rat brain: Regional distribution of neuromedin B and neuromedin C in rat brain, pituitary and spinal cord, *Biochem. Biophys. Res. Commun.* **124:**925–932.

Moody, T. W., and Pert, C. B., 1979, Bombesin-like peptides in rat brain: Quantitation and biochemical characterization, *Biochem. Biophys. Res. Commun.* **90:**7–14.

Moody, T. W., Thoa, N. B., O'Donohue, T. L., and Pert, C. B., 1980, Bombesin-like peptides in rat brain: Localization in synaptosomes and release from hypothalamic slices, *Life Sci.* **26:**1707–1712.

Moody, T. W., O'Donohue, T. L., and Jacobowitz, D. M., 1981, Biochemical localization and characterization of bombesin-like peptides in discrete regions of rat brain, *Peptides* **2:**75–79.

Moody, T. W., Crawley, J. N., and Jensen, R. T., 1982, Pharmacology and neurochemistry of bombesin-like peptides, *Peptides* **3:**559–563.

O'Donohue, T. L., Massari, J., Pazoles, C. J., Chronwall, B. M., Shults, C. W., Quirion, R., Chase, T. N., and Moody, T. W., 1984, A role for bombesin in sensory processing in the spinal cord, *J. Neurosci.* **412:**2956–2962.

Orloff, M. S., Reeve, J. R., Jr., Ben-Avram, C. M., Shively, J. E., and Walsh, J. H., 1984, Isolation and sequence analysis of human bombesin-like peptides, *Peptides* **5:**865–870.

Panula, P., Yang, H.-Y. T., and Costa, E., 1982, Neuronal location of the bombesin-like immunoreactivity in the central nervous system of the rat, *Regul. Peptides* **4:**275–283.

Parrott, R. F., and Baldwin, B. A., 1982, Centrally-administered bombesin produces effects unlike short term satiety in operant feeding pigs, *Physiol. Behav.* **28:**521–524.

Paxinos, G., and Watson, C., 1982, *The Rat Brain in Stereotaxic Coordinates,* Academic Press, New York.

Pert, A., Moody, T. W., Pert, C. B., DeWald, L. A., and Rivier, J., 1980, Bombesin: Receptor distribution in brain and effects on nocioception and locomotor activity, *Brain Res.* **193:**209–220.

Pittman, Q., Tache, Y., and Brown, M., 1980, Bombesin acts in preoptic areas to produce hypothermia in rats, *Life Sci.* **26:**725–730.

Reeve, J. R., Jr., Walsh, J. H., Chew, P., Clark, B., Hawke, D., and Shively, J. E., 1983, Amino acid sequences of three bombesin-like peptides from canine intestine extracts. *J. Biol. Chem.* **258:**5582–5588.

Schulz, D. W., Kalivas, P. W., Nemeroff, C. B., and Prange, A. J., Jr., 1984, Bombesin-induced locomotor hyperactivity: Evaluation of the involvement of mesolimbic dopamine system, *Brain Res.* **304:**377–382.

Soveny, C., Mercuri, J., and Hansky, J., 1984, Distribution of bombesin- and cholecystokinin-like immunoreactivity in rat and dog brain and gastrointestinal tract, *Regul. Peptides* **9:**61–68.

Spindel, E. R., Chin, W. W., Price, J., Rees, L. H., Besser, G. M., and Habener, J. F., 1984, Cloning and characterization of cDNAs encoding human gastrin-releasing peptide, *Proc. Natl. Acad. Sci. U.S.A.* **81:**5699–5703.

Spruijt, B., and Gispen, W. H., 1983, ACTH and grooming behaviour in the rat, in: *Hormones and Behaviour in Higher Vertebrates* (J. Balthazart, E. Prove, and R. Gilles, eds.), Springer-Verlag, Berlin, Heidelberg, pp. 118–136.

Stuckey, J. A., and Gibbs, J., 1982, Lateral hypothalamic injection of bombesin decreases food intake in rats, *Brain Res. Bull.* **8:**617–621.

Tache, Y., 1982, Bombesin: Central nervous system action to increase gastric mucus in rats, *Gastroenterology* **83:**75–80.

Tache, Y., and Brown, M., 1982, On the role of bombesin in homeostatis, *Trends Neurosci. Res.* **5:**431–433.

Tache, Y., and Collu, R., 1982, CNS mediated inhibition of gastric secretion by bombesin: Independence from interaction with brain catecholaminergic and serotoninergic pathways and pituitary hormones, *Regul. Peptides* **3:**51–59.

Tache, Y., Brown, M., and Collu, R., 1979, Effects of neuropeptides on adenohypophyseal hormone response to acute stress in male rats, *Endocrinology* **105:**220–224.

Tache, Y., Grijalva, C. V., Gunion, M. W., Cooper, P. H., Walsh J. H., and Novin, D., 1984, Lateral hypothalamic mediation of hypergastrinemia induced by intracisternal bombesin, *Neuroendocrinology* **39:**114–119.

Widerlov, E., Mueller, R. A., Frye, G. D., and Breese, G. R., 1984, Bombesin increases dopamine function in rat brain areas, *Peptides* **5:**523–528.

Wolf, S. S., Moody, T. W., O'Donohue, T. L., Zarbin, M. A., and Kuhar, M. J., 1983, Autoradiographic visualization of rat brain binding sites for bombesin-like peptides, *Eur. J. Pharmacol.* **87:**163–164.

Wunder, B. A., Hawkins, M. F., Avery, D. D., and Swan, H., 1980, The effects of bombesin injected into the anterior and posterior hypothalamus on body temperature and oxygen consumption, *Neuropharmacology* **19:**1095–1097.

Yachnis, A. T., Crawley, J. N., Jensen, R. T., McGrane, M. W., and Moody, T. W., 1984, The antagonism of bombesin in the CNS by substance P analogues, *Life Sci.* **35:**1963–1969.

Zarbin, M. A., Kuhar, M. J., O'Donohue, T. L., Wolf, S. S., and Moody, T. W., 1985, Autoradiographic localization of (^{125}I-Tyr^{4})bombesin-binding sites in rat brain, *J. Neurosci.* **5**(2)**:**429–437.

IV

Peptides in Normal and Malignant Cells

27

Physiological Effects of Transforming Growth Factor α

JAMES P. TAM

1. INTRODUCTION

From the conditioned medium of transformed cells, two types of mitogenic polypeptides (α and β) known as transforming growth factors were isolated and characterized. Because of their origin and their abilities to confer phenotypic transformation of cells in culture, much of the studies of transforming growth factors have been focused on their roles in malignancy and their relationship with the oncogenes (for a review, see Brown and Blakeley, 1984; Sporn and Roberts, 1985). However, because of the scarcity of these growth factors from natural sources, little is known about their physiological effects in animals. Recently, our laboratory has synthesized highly purified rat and human transforming growth factor (TGFα). These synthetic materials have been vigorously characterized and found to be indistinguishable from those obtained from the natural source. The availability of these growth factors has made it possible for the first time to study the physiological effects in whole animals. Here, we report our results using the synthetic TGFα.

1.1. Transforming Growth Factor α

Transforming growth factor α (De Larco and Todaro, 1978a; Roberts *et al.*, 1980) is released extracellularly by tumor cells of various origins into the conditioned medium in culture. These include cell lines derived from established tumors and those from viral (Todaro *et al.*, 1976; Kaplan *et al.*, 1982; Massague, 1983) or chemical transformation (Moses *et al.*, 1981; Proper *et al.*, 1982). Transforming

JAMES P. TAM • The Rockefeller University, New York, New York 10021.

growth factor α is characterized by its ability to confer reversible phenotypic transformation of untransformed, normal indicator cells in culture. However, this transforming ability is modulated by transforming growth factor type β (TGFβ). One of the most often used indicator cell lines are normal rat kidney (NRK) fibroblasts (De Larco and Todaro, 1978b). The NRK cells, when grown in the presence of both types of TGF, lose their density-dependent inhibition, assume transformed morphology, and overgrow in monolayer culture. Furthermore, NRK cells do not normally grow in soft agar, but in the presence of TGFα in culture, these cells acquire anchorage independence and respond to form colonies in soft agar. Since these properties are closely associated with transformation, they have led to the hypothesis that TGFα is produced as an autocrine growth factor by the transformed cells to sustain self-stimulation and proliferation (Sporn and Todero, 1980; Kaplan *et al.*, 1982; Sporn and Roberts, 1985). More recent studies have shown that TGFα activity is found also in nonneoplastic tissue of embryonic (Proper *et al.*, 1982; Twardzik *et al.*, 1982b) and adult fluids, including serum, urine, and colostrum (Twardzik *et al.*, 1982a). However, the results of these findings are neither conclusive nor definitive, since the biological activities of TGFα are closely similar to those of another growth factor, epidermal growth factor (EGF), which is present in biological fluids and tissues in normal states, and since none of the functional assays can distinguish between these two growth factors unequivocally, the development of a specific antibody against TGFα may help to resolve these uncertainties.

1.2. Transforming Growth Factor β

Transforming growth factor β is a homodimer of 22,000 daltons (Roberts *et al.*, 1982; Sporn and Roberts, 1985). Unlike TGFα, TGFβ is found both in the conditioned medium of virally transformed cells in culture (Roberts *et al.*, 1980; Anzano *et al.*, 1983) and in normal tissues and physiological fluid. Blood platelets have been found to be a rich source of TGFβ. Similarly, fetal calf serum also contains TGFβ. The latter finding offers an explanation of the early confusions concerning the sole ability of TGFα in soft agar to induce colony formation. It is now clear that TGFα requires the presence of either TGFβ or fetal calf serum for the synergistic action of large colony formations. Transforming growth factor β, when tested in NRK cells, has no mitogenic capacity, but recent studies have shown that TGFβ is a mitogen to other cell types. More interestingly, TGFβ also acts as an inhibitor to the growth and proliferation of certain cell lines in culture (Tucker *et al.*, 1984). The biochemical actions of TGFβ are mediated through its own cell surface receptor, which is distinctly different from the EGF receptor. So far, it has not been shown that the TGFβ receptor possesses any tyrosyl-specific protein kinase activity. Undoubtedly, TGFβ represents a distinct growth factor inhibitor and does not belong to the EGF–TGFα family. A summary of the differences between TGFα and TGFβ is shown in Table I.

TABLE I. Comparison of EGF, TGFα, and TGFβ

	EGF	TGFα	TGFβ
Amino acids	53	50	224
pI	4.5	6.8	—
Peptide chain	Single	Single	Dimeric
Precursor	1200	160	—
Soft agar growth	Yes	Yes	No
Receptor	EGF	EGF	TGFβ
Tumor cell	No	Yes	Yes

2. STRUCTURAL SIMILARITIES WITH EPIDERMAL GROWTH FACTOR

The earliest indication that TGFα is structurally and functionally similiar to EGF derives from the observation that conditioned medium containing TGFs is capable of competing with EGF for the EGF receptor (Todaro *et al.*, 1976, 1980). Furthermore, Reynolds *et al.* (1981) also found that TGFs from conditioned medium also induce EGF-receptor-associated protein kinase to autophosphorylation of its own receptor in a manner similiar to EGF. However, it is only recently that conclusive evidence from sequence determination confirms that TGFα does share considerable structural and sequence similarities with EGF. However, it is also clear that sequence homology between TGFα and EGF is not as close as anticipated and that TGFα is distinctly different in sequence.

The primary structure of TGFα from a retrovirus-transformed Fisher rat embryo fibroblast (rTGFα, Fig. 1), deduced from Edman degradation (Marquardt *et al.*, 1983, 1984; Massague, 1983) and the predicted cDNA sequence (Lee *et al.*, 1985),

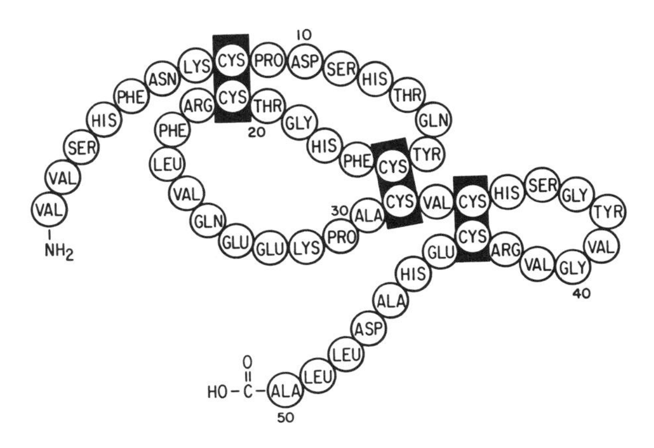

FIGURE 1. Structure of rTGFα.

contains 50 amino acid residues (Fig. 1) and differs from that of the predicted cDNA sequence of the human TGFα in four amino acid residues (Derynck *et al.*, 1984). Both the rat and human TGFα sequences share between 30 to 40% sequence homology with human (Gregory, 1975) and mouse EGF (Fig. 2). More interestingly, all four sequences can be aligned so that all six cysteine residues display homologous positions. Furthermore, amino acid residues (total four) related to conformational requirements for β-bends such as glycine and proline can also be aligned to homologous positions. Thus, it is likely that the sequence homologies between TGFs and EGFs serve to provide a common unique secondary structure for these growth factors to elicit binding to the same receptor. From the evolutionary point of view, it is also interesting to note that the sequence of TGFs is better conserved than EGFs between species. The sequence homology of TGFα between rat and human is 92% (46 out of 50 residues), but EGF between mouse and human is only 70% (37 out of 53 residues).

Another significant difference between EGFs and TGFαs lies in their biosynthetic processing. The TGFαs are derived from a precursor polypeptide of 160 amino acids, and the cleavage of the 50-amino-acid TGFα from the larger form occurs at alanine and valine residues at both amino and carboxyl termini. However, EGFs are derived from a precursor polypeptide of about 1200 amino acids, and the cleavage of the 53-amino-acid EGF occurs at paired basic residues, similar to those of other peptide hormones. Thus, structural comparison and biosynthetic processing suggest that TGFα is similar to but distinct from members of the EGF-growth factor family.

Another unexpected protein that shared similar sequence homology with TGFα is a peptide (residues 45–85) of a 140-residues polypeptide encoded by one of the early genes of vaccinia virus (Blomquist *et al.*, 1984; Brown *et al.*, 1985). Vaccinia viral peptide (VVP) shares 15 residues with both rTGFα and hTGFα and 19 residues with both mEGF and hEGF (Fig. 2). Thus, VVP is as similar to EGF as is TGF. More interestingly, several blood platelet factors, blood coagulation factor IX, factor

```
         1             10                20
rTGFα  VVSHFNKC  PDS  HTQYC  FH—GTC   RFLVQ
hTGFα  VVSHFNDC  PDS  HTQFC  FH—GTC   RFLVQ
mEGF     NSYPGC  PSS  YDGYC  LNGGVC   MHIES
hEGF     NSDSEC  PLS  HDGYC  LHDGVC   MYIEA
VVP          LC  GPE  GDGYC  LH—GDC   IHARD

     30               40            50
   EEKPAC  VC  HSGY  VGVRC  EHADL  LA
   EDKPAC  VC  HSGY  VGARC  EHADL  LA
   LDSYTC  NC  VIGY  SGDRC  QTRDL  RWWELR
   LDKYAC  NC  VVGY  IGERC  QYRDL  KWWELR
   IDGMYC  RC  SHGY  TGIRC  QHVVL  VDYQR
```

Common Sequence

CxxxxxxCxxGxCxxxxxxxxxxCxCxxGxxGxxC

FIGURE 2. Sequence homology of TGFα, EGF, and VVP (vaccina viral protein).

X, and other proteins such as low-density lipoprotein receptor and protein C also share distant sequence homology with the EGF and TGF family. The conserved sequence homology (as shown in Fig. 2), as the consensus sequence, mainly corresponds to the cysteinyl and glycyl residues, which are important for the structural and conformational role of the molecule. The structural similarities of these peptides and growth factors may represent evolutionary conservation but may also indicate functional importance. Nevertheless, future experimental confirmation will be needed to settle this issue.

3. CHEMICAL SYNTHESES OF RAT AND HUMAN TGFα

One of the impediments in the study of TGFα is the difficulty to obtain pure and sufficient quantity of material from tissue culture. From 150 liters of the conditioned medium, only 1.5 μg of purified TGFα could be isolated. Consequently, none of the physiological studies has been carried out. In addition, the scarcity of TGFα also hampers the effort to prepare specific antibodies for the study of occurrence and localization of TGFα in tissues. Our laboratory has recently undertaken the total synthesis of both rat and human TGFα with the objective of providing sufficient material to enable more extensive studies on their mode of action, particularly in whole animals. Highly purified synthetic rat and human TGFαs have been prepared by the solid-phase peptide synthesis method (Tam *et al.*, 1984; Tam, 1985, 1986; Sheikh and Tam, 1985). The homogeneity of these materials has been confirmed by reversed-phase high-pressure liquid chromatography (HPLC) (Fig. 3)

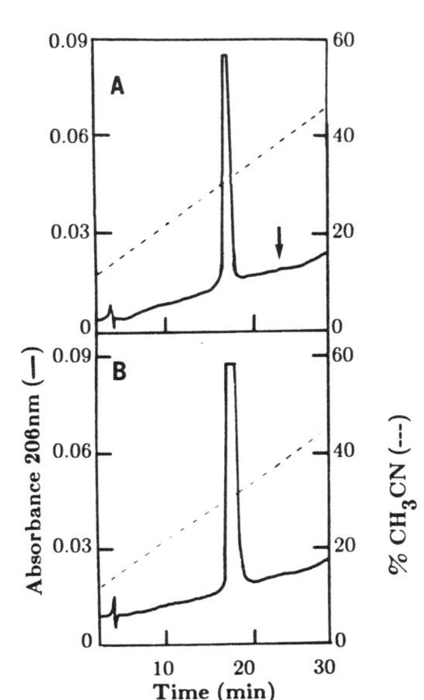

FIGURE 3. Results of HPLC analysis of synthetic rat (A) and human (B) TGFα. The arrow indicates the position where mEGF is eluted under the same conditions.

and other physical characterization. Furthermore, rat TGFα coelutes with the natural TGFα as a single symmetrical peak under similar chromatographic conditions.

Both synthetic rat and human TGFαs were compared with natural rat TGFα and mEGF for biological properties of the putative growth factor properties *in vitro* (Table II). In the mitogen assay, the stimulation of growth of serum-deprived normal rat kidney cells by the growth factor was measured by the incorporation of $[^{125}I]$iododeoxyuridine. In the soft agar assay and in the presence of TGFβ, the colony formation in soft agar was measured. Furthermore, the binding of these growth factors to the A-431 human carcinoma cells, whose cell surfaces are rich in EGF receptors, was also tested. As shown in Table II, all of the growth factors showed comparable activities and were active at the nanomolar level. A consequence of binding to the EGF receptor is the stimulation of the receptor-associated tyrosine protein kinase. Both EGF and TGFα were found to stimulate the phosphorylation of the exogenous synthetic angiotensinyl peptide substrate with similar half-maximal activities. These results support previous findings and provide convincing evidence that the synthetic TGFαs are as active as the natural rTGFα and mEGF *in vitro*.

4. EFFECTS OF TGFα ON THE SOMATIC DEVELOPMENT OF NEWBORN MOUSE

Epidermal growth factor causes marked effects on somatic development when injected into a newborn mouse. Daily subcutaneous administration of microgram quantities accelerated tooth eruption and eyelid opening (Cohen, 1962) but retarded rate of body growth and inhibited hair growth (Cohen, 1962; Moore *et al.*, 1981; Panaretto *et al.*, 1982). To determine whether TGFα possesses these physiological effects on newborn mouse, synthetic rat TGFα (0.05 to 4 μg/g body weight) was injected daily subcutaneously into newborn NCS mouse to exmaine its effect on the somatic development.

TABLE II. Comparisons of Synthetic TGFα, Natural rTGFα, and EGF by Different Growth and Transforming Assays

Assay	Half maximal activity (nM)		
	mEGF	nTGFα[a]	sTGFα[b]
Stimulation DNA synthesis	1.4	1.7	2.3
EGF-radioreceptor	1.7	3.7	4.2
Phosphotyrosine kinase	0.1	0.4	0.3
Soft agar growth	0.6	0.5	0.3

[a] Natural rat TGFα.
[b] Synthetic rat and human TGF.

4.1. Precocious Opening of Eyelid and Incisor Eruption

Early eruption of lower incisor teeth and opening of eyelids were observed in a dose-dependent manner with the TGFα-treated mouse compared with littermate controls (Fig. 4).

Doses lower than 0.25 μg TGFα/g body weight produced no significant effects. At 2.7 μg TGFα/g body weight, incisor tooth eruption and eyelid opening were observed on day 6 to day 7 instead of day 9 to 10 for tooth eruption and day 12 to 14 for eyelid opening seen in control animals. When EGF was compared to TGFα on both of these effects in a dosage of 2.7 μg/g body weight, no significant difference was observed between TGFα and EGF. The eruption of upper incisors usually occurred a day after the lower incisor eruption. Thus, on day 8, eruption of both incisors of the EGF- or TGFα-treated mouse was clearly visible, whereas none was observed with the controls (Fig. 2).

4.2. Retardation of Hair Growth and Body Weight Gain

The growth rate of the first haircoat in male neonatal mouse has been found to be significantly retarded with daily injections of 4 μg mEGF/g body weight for

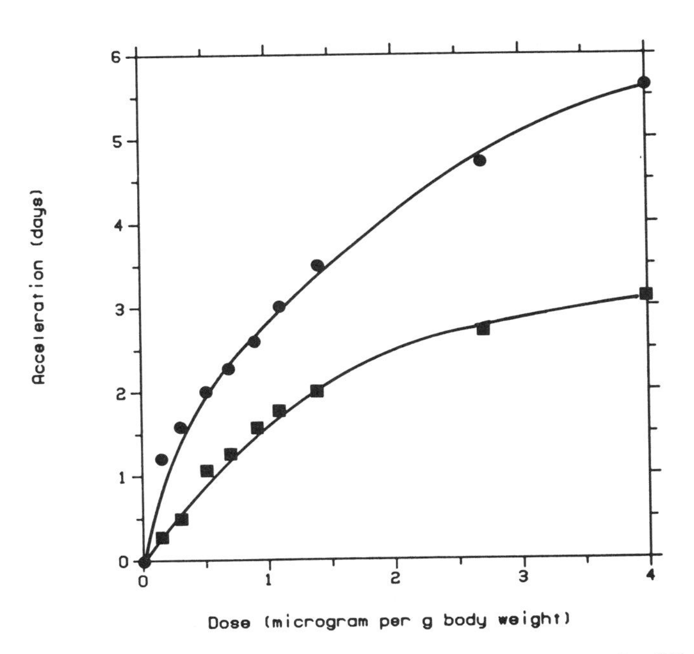

FIGURE 4. Dose–response curve of acceleration of precocious eyelid opening (●) and incisor eruption (■) in newborn mouse.

14 consecutive days from birth (Moore *et al.*, 1981). More interestingly, infusing sheep intravenously with 0.12 mg mEGF/kg body weight results not only in depression of wool growth but also in complete casting of the fleeces, leaving the sheep nude on the wool-growing regions of the body (Panaretto *et al.*, 1984). Thus, at high doses, EGF has an inhibitory effect on DNA synthesis in the dermal skin sections that contain proliferation cells of hair or wool follicles. Furthermore, EGF also retarded rate of body growth (Cohen, 1962; Moore *et al.*, 1981). Although the biochemical events leading to inhibiting effects have not been clearly established, the overall physiological effects nevertheless provide a source for comparison between EGF and TGFα.

Similar to EGF (Carpenter and Cohen, 1979), treatment with TGFα also produced inhibition of hair growth correlating to dose administered (Table III). Examination of the overall rates of hair growth visually and by scanning electron microscope on glutaraldehyde-fixed skin at high dosages revealed that EGF or TGFα produced a significantly finer and shorter coat of montrichs compared to control animals. At 2.7 μg TGFα/g body weight per day, approximately a 30% reduction of hair length and diameter was observed compared to control animals.

Transforming growth factor α was found to retard the overall growth rate of newborn mouse (Table II). The growth rates of the first 10 days of birth were examined, and in control animals, doubling and tripling of body weights were usually seen on days 5 and 9, respectively. However, the growth of the TGFα-treated animals was stunted at doses higher than 0.3 μg/g body weight per day. Again, the inhibition of rate of growth correlated well with the dose administered. At 4 μg/g body weight per day, the growth rates of either TGFα- or EGF-treated animals were 25% slower than the control animals.

Epidermal growth factor has been found to be of critical importance to the well-being of immature mouse during pregancy and the nursing period. This has led to the postulate that EGF may play an important role in physiological development and tissue differentiation. Sialoadenectomy of female mice decreases milk production and increases offspring mortality during the lactation period (Okamoto

TABLE III. Effect of Synthetic TGFα on Body Weight and Hair Growth on Newborn Mouse

Treatment	Dose (μg/g body weight per day)	Number of animals	Mean (%)	
			Body weight	Hair growth
Control	No injection or saline	22	100	100
EGF	22	2	75	63
TGF	0.5	20	112	107
	0.5–1	8	94	87
	1–3	6	85	76
	4	3	76	65

and Oka, 1984). It is possible that EGF or substances produced by EGF are lacking in sialoadenectomized mothers to pass along to the newborn mouse in milk during lactation. The similarities in physiological responses between TGFα and EGF are consistent with their role as growth factors and further suggest that TGFα is likely to be found in biological fluids during pregnancy and the nursing period.

5. STIMULATION OF ORNITHINE DECARBOXYLASE ACTIVITY

A remarkable physiological effect of EGF is its strong induction of ornithine decarboxylase activity in testes of neonatal mouse (Stastny and Cohen, 1970) and in the digestive tract (Feldman *et al.*, 1978). Since ornithine decarboxylase is a key enzyme in the biosynthetic pathway of the polyamines putrescine, spermidine, and spermine, whose production is closely related to mitogenic activities, the *in vivo* induction of TGFα will provide good evidence of the physiological similarity between these two mitogens. The level of the ornithine decarboxylase activity in a number of tissues was studied by subcutaneous injection of homogeneous synthetic TGFα or natural EGF into the 8-day-old mouse (Fig. 5). As a control, human growth hormone was also administered into the neonatal mouse under similar conditions. The responses elicited by TGFα or EGF from different tissues (three- to 24-fold) were found to be quite similar. A 24-fold increase in activity was seen in the testes by TGFα compared with a ten-fold increase by EGF. Significant but

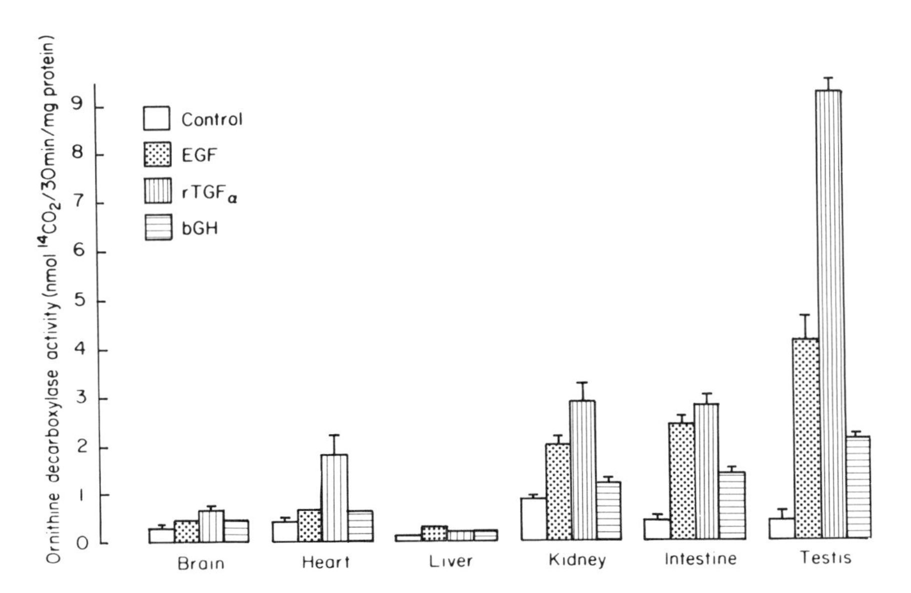

FIGURE 5. The TGFα stimulation of ornithine decarboxylase activity in tissues of an 8-day-old mouse.

lesser increases (three- to fivefold) were also seen in the intestine and kidney. Little effect was observed in the spleen and brain.

When the time course of the induction by TGFα in these tissues was also examined, the decarboxylase activity in the testes reached a maximum in 4 hr and rapidly declined. However, the maximum level of induction in kidney was reached in 2 hr. Such responses were very similar to those effects elicited by EGF (Stastny and Cohen, 1970), which also caused a maximal response of ornithine decarboxylase activity in 4 hr.

To confirm that the induction of ornithine decarboxylase activity is accompanied by protein synthesis, [^{3}H]leucine was administered to the TGFα- or EGF-treated neonatal mouse. The effect of TGFα or EGF on protein synthesis in different organs was examined by the incorporation of [^{3}H]leucine into trichloroacetic-precipitable proteins (Fig. 6). The results confirmed that TGFα induced new protein synthesis two- to tenfold compared with control. The greatest increase was seen in intestine (sixfold), and some increases were seen in spleen, kidney, and liver, but little effect was evident on brain, heart, or lung.

The marked but transient increase of ornithine decarboxylase activity is usually found in rapidly growing tissues or after the administration of growth-promoting hormones, peptides, or drugs. The elevated enzyme activity is accompanied by a net increase in polyamine synthesis and other cellular activities characteristic of growth and differentiation. In mammalian tissue, the highest concentration of polyamine is found in the testes, and one of the most potent stimulators of ornithine

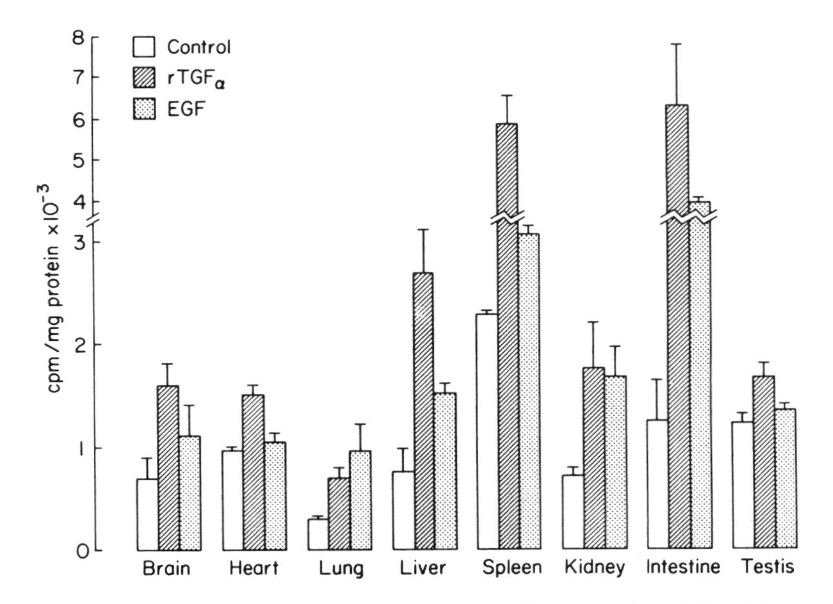

FIGURE 6. The TGFα stimulation of protein synthesis in tissues of an 8-day-old mouse.

decarboxylase in testes is EGF, which induces the enzymatic activity 20-fold in immature mouse. In this study, rat TGFα and EGF in similar dose level were found to stimulate the testicular ornithine decarboxylase activity of the immature mouse 24- and tenfold, respectively. Furthermore, the time course of the induction of enzyme activity by TGFα was similar to that by EGF. In comparison and under the same condition, human growth hormone produced only a threefold increase in activity. Thus, our findings put TGFα into the unique category of EGF, and no other peptide hormone known to date produced such a strong response of testicular ornithine decarboxylase stimulation as TGFα or EGF.

6. INHIBITION OF HISTAMINE-STIMULATED GASTRIC ACID SECRETION

Epidermal growth factor has been shown to inhibit gastric secretion stimulated by histamine, pentagastrin, or carbachol and insulin-induced hypoglycemia when given experimentally to animals. Recently, it has also been shown that an *in vitro* preparation of mammalian gastric mucosa from adult guinea pig is also a feasible model for the study of the inhibitory effects of EGF. However, the mechanism of this antisecretory activity is not known, but such antisecretory effect will provide another comparison for EGF and TGFα. When gastric mucosae were maintained in Ussing chambers in Ringer's solution, Murphy *et al.* (1985) and Rhodes *et al.* (1986) showed that TGFα was effective in the maximal reduction of the rate of secretagogue-induced acid secretion to the serosal gastric surface at a concentration of 120 ng of TGFα/ml. Similar maximal reduction of induced gastric secretion by mEGF in the matched tissue of the control experiment was also observed. The rate of inhibition was dose responsive and required 30 ng of TGFα/ml to be effective. Furthermore, both TGFα and EGF were found to produce the same electrophysiological parameters following the growth factor treatments. These included increased resistance and decreased short-circuit current. However, both EGF and TGFα were ineffective as acid secretion inhibitors from the luminal side of the mucosa at the low doses that inhibited secretion from the serosal side. These data therefore suggest that TGFα behaves similarly to EGF in the secretagogue-induced acid secretion.

7. CONCLUSION

Although the biochemical events in the present studies of TGFα on newborn mouse remain unresolved, these results clearly support the results obtained from *in vitro* studies and provide the first direct *in vivo* evidence that TGFα is a member of the EGF family. In the present study, TGFα markedly accelerates precocious eyelid opening and incisor eruption and retards hair and weight growth in the newborn mouse. TGFα also stimulates ornithine decarboxylase activity and protein synthesis but significantly inhibits gastric acid secretion. These results also provide a further stimulus for the search for the difference in the biosynthesis and physiological occurrence between TGFα and EGF. Unlike EGF, which is required to maintain the normal state of the animals, TGFα is found in states of rapid growth such as those found in pregnancy and malignancy and perhaps serves as the additional concurrent growth factor to meet such needs.

ACKNOWLEDGMENTS. This investigation was supported by PHS 36544 awarded by the National Cancer Institute, DHHS. I thank M. Sheikh, A. Nakhla, D. Rosberger, and L. Tsai for their contributions and technical assistance during the course of this work.

REFERENCES

Anzano, M. A., Roberts, A. B., Smith, J. M., Sporn, M. B., and De Larco, J. E., 1983, Sarcoma growth factor from conditioned medium of virally transformed cells is composed of both type α and type β transforming growth factors, *Proc. Natl. Acad. Sci. U.S.A.* **80:**6264–6268.

Blomquist, M. C., Hunt, L. T., and Barker, W. C., 1984, Vaccinia virus 19-kilodalton protein: Relationship to several mammalian proteins, including two growth factors, *Proc. Natl. Acad. Sci. U.S.A.* **81:**7363–7367.

Brown, J. P., Twardzik, D. R., Marquardt, H., and Todaro, G. J., 1985, Vaccinia virus encodes a polypeptide homologous to epidermal growth factor and transforming growth factor, *Nature* **313:**491–492.

Brown, K. D., and Blakeley, D. M., 1984, Transforming growth factors: Sources, properties and possible roles in normal and malignant cell growth control, *Biochem. Soc. Trans.* **12:**168–173.

Carpenter, G., and Cohen, S., 1979, Epidermal growth factor, *Annu. Rev. Biochem.* **48:**193–216.

Cohen, S., 1962, Isolation of a mouse submaxillary gland protein accelerating incisor eruption and eyelid opening in the new-born animal, *J. Biol. Chem.* **237:**1555–1562.

De Larco, J. E., and Todaro, G. J., 1978a, Growth factors from murine sarcoma virus-transformed cells, *Proc. Natl. Acad. Sci. U.S.A.* **75:**4001–4005.

De Larco, J. E., and Todaro, G. J., 1978b, Epithelioid and fibroblastic rat kidney cell clones: Epidermal growth factor (EGF) receptors and the effect of mouse sarcoma virus transformation, *J. Cell. Physiol.* **94:**335–342.

Derynck, R., Roberts, A. B., Winkler, M. E., Chen, E. Y., and Goeddel, D. V., 1984, Human transforming growth factor-α: Precursor structure and expression in *E. coli, Cell* **38:**287–297.

Feldman, E. J., Aures, D., and Grossman, M. I., 1978, Epidermal growth factor stimulates ornithine decarboxylase activity in the digestive tract of mouse (40357), *Proc. Soc. Exp. Biol. Med.* **159:**400–402.

Gregory, H. 1975, Isolation and structure of urogastrone and its relationship to epidermal growth factor, *Nature* **257:**325–327.

Kaplan, P. L, Anderson, M., and Ozanne, B., 1982, Transforming growth factor(s) production enables cells to grow in the absence of serum: An autocrine system, *Proc. Natl. Acad. Sci. U.S.A.* **79:**485–489.

Lee, D. C., Rose, T. M., Webb, N. R., and Todaro, G. J., 1985, Cloning and sequence analysis of a cDNA for rat transforming growth factor-α, *Nature* **313:**489–491.

Marquardt, H., Hunkapiller, M. W., Hood, L. E., Twardzik, D. R., De Larco, J. E., Stephenson, J. R., and Todaro, G. J., 1983, Transforming growth factors produced by retrovirus-transformed rodent fibroblasts and human melanoma cells: Amino acid sequence homology with epidermal growth factor, *Proc. Natl. Acad. Sci. U.S.A.* **80:**4684–4688.

Marquardt, H., Hunkapiller, M. W., Hood, L. E., and Todaro, G. J., 1984, Rat transforming growth factor type 1; structure and relation to epidermal growth factor, *Science* **223:**1079–1081.

Massague, J., 1983, Epidermal growth factor-like transforming growth factor, *J. Biol. Chem.* **258:**13606–13612.

Moore, G. P. M., Panaretto, B. A., and Robertson, D., 1981, Effects of epidermal growth factor on hair growth in the mouse, *J. Endocrinol.* **88:**293–299.

Moses, H. L., Branum, E. L., Proper, J. A., and Robinson, R. A., 1981, Transforming growth factor production by chemically transformed cells, *Cancer Res.* **41:**2842–2848.

Murphy, R. A., Rhodes, J. A., Tam, J., Finke, U., Saunders, M., and Silen, W., 1985, Transforming growth factor (αTGF) inhibits histamine-induced acid secretion in guinea pig gastric mucosa, *Fed. Proc.* **44:**183.

Nakhla, A. M., and Tam, J. P., 1985, Transforming growth factors stimulation of ornithine decarboxylase in neonatal mouse tissues, *Biochem. Biophys. Res. Comm.* **132:**1180–1186.

Okamoto, S., and Oka, T., 1984, Evidence for physiological function of epidermal growth factor: Pregestational sialoadenectomy of mice decreases milk production and increases offspring mortality during lactation period, *Proc. Natl. Acad. Sci. U.S.A.* **81:**6059–6063.

Panaretto, B. A., Moore, G. P. M., and Robertson, D. M., 1982, Plasma concentrations and urinary excretion of mouse epidermal growth factor associated with the inhibition of good consumption and of wool growth in Merino wethers, *J. Endocrinol.* **94:**101–102.

Panaretto, B. A., Leish, Z., Moore, G. P. M., and Robertson, D. M., 1984, Inhibition of DNA synthesis in dermal tissue of Merino sheep treated with depilatory doses of mouse epidermal growth factor, *J. Endocrinol.* **100:**25–31.

Proper, J. A., Bjornson, C. L., and Moses, H. L., 1982, Mouse embryos contain polypeptide growth factor(s) capable of inducing a reversible neoplastic phenotype in nontransformed cells in culture, *J. Cell. Physiol.* **110:**169–174.

Reynolds, F. H., Jr., Todaro, G. J., Fryling, C., and Stephenson, J. R., 1981, Human transforming growth factors induce tyrosine phosphorylation of EGF receptors, *Nature* **292:**259.

Rhodes, J. A., Tam, J. P., Finke, U., Saunders, M., Bernanke, J., Silen, W., and Murphy, R. A., 1986, Transforming growth factor alpha (TGFα) inhibits secretion of gastric acid, *Proc. Natl. Acad. Sci. U.S.A.* **83:**3844–3846.

Roberts, A. B., Lamb, L. C., Newton, D. L., Sporn, M. B., De Larco, J. E., and Todaro, G. J., 1980, Transforming growth factors: Isolation of polypeptides from virally and chemically transformed cells by acid/ethanol extraction, *Proc. Natl. Acad. Sci. U.S.A.* **77:**3494–3498.

Roberts, A. B., Anzano, M. A., Lamb, L. C., Smith, J. M., Frolik, C. A., Marquardt, H., Todaro, G. J., and Sporn, M. B., 1982, Isolation from murine sarcoma cells of novel transforming growth factors potentiated by EGF, *Nature* **295:**417.

Sheikh, M. A., and Tam, J. P., 1985, Total synthesis of biologically-active human transforming growth factor type alpha, in: *Proceedings of the Ninth American Peptide Symposium* (K. Kopple, C. M. Deber, and V. Hruby, eds.), Pierce Chemical, Rockford, IL, pp. 305–308.

Sporn, M. B., and Roberts, A. B., 1985, Autocrine growth factors and cancer, *Nature* **313:**745–747.

Sporn, M. B., and Todaro, G. J., 1980, Autocrine secretion and malignant transformation of cells, *N. Engl. J. Med.* **303:**878–880.

Stastny, M., and Cohen, S., 1970, Epidermal growth factor IV. The induction of ornithine decarboxylase, *Biochim. Biophys. Acta* **204:**578–589.

Tam, J. P., 1985, Physiological effects of transforming growth factor in the newborn mouse, *Science* **229:**673–675.

Tam, J. P., 1986, Synthesis of alpha transforming growth factor, *Methods Enzymol.* (in press).

Tam, J. P., Marquardt, H., Rosberger, D. F., Wong, T. W., and Todaro, G. J., 1984, Synthesis of biologically active rat transforming growth factor I, *Nature* **309:**376–378.

Todaro, G. J., De Larco, J. E., and Cohen, S., 1976, Transformation by murine and feline sarcoma viruses specifically blocks binding of epidermal growth factor to cells, *Nature* **264:**26–30.

Todaro, G. J., Fryling, C., and De Larco, J. E., 1980, Transformation growth factors produced by certain human tumor cells: Polypeptides that interact with epidermal growth factor receptors, *Proc. Natl. Acad. Sci. U.S.A.* **77:**5258–5262.

Tucker, R. F., Shipley, G. D., Moses, H. L., and Holley, R. W., 1984, Growth inhibitor from BSC − 1 cells closely related to platelet type β transforming growth factor, *Science* **226**:705–707.
Twardzik, D. R., Ranchalis, J. E., and Todaro, G. J., 1982a, Mouse embryonic transforming growth factors related to those isolated from tumor cells, *Cancer Res.* **42**:590–593.
Twardzik, D. R., Sherwin, S. A., Ranchalis, J. E., and Todaro, G. J., 1982b, Transforming growth factors in the urine of normal, pregnant, and tumor-bearing humans, *J. Natl. Cancer Inst.* **69**:793–798.

28

Peptides and Their Receptors in Oat Cell Carcinoma

TERRY W. MOODY, LOUIS Y. KORMAN, and DESMOND N. CARNEY

1. INTRODUCTION

Oat cell carcinoma or small-cell lung cancer (SCLC) comprises about 25% of the 110,000 new cases of malignant lung cancer each year (Bunn *et al.*, 1977). Many of the patients are responsive to chemo- and/or radiation therapy, but many of the diagnosed cases are far advanced, with the cancer having undergone metastases from the lung to the liver, lymph nodes, bone, and/or brain.

Morphologically, oat cells are similar to the endocrine Kultschitzky cells and have dense-core neurosecretory granules (Gazdar *et al.*, 1979). They are enriched in their neuroendocrine properties and contain the neuronal enzymes dopamine decarboxylase and neuron-specific enolase (Marangos *et al.*, 1982). In addition, oat cells produce transmitters such as serotonin (Pettengill *et al.*, 1982) and ectopic hormones such as ACTH, calcitonin, and vasopressin (Yesner, 1978). Recently, high concentrations of neurotensin (Wood *et al.*, 1981; Moody *et al.*, 1985a) and bombesinlike peptides (Moody *et al.*, 1981; Wood *et al.*, 1981; Erisman *et al.*, 1982) were detected in SCLC cells. Here the presence of neuropeptides and their receptors in oat cell carcinoma was investigated.

TERRY W. MOODY • Department of Biochemistry, The George Washington University School of Medicine and Health Sciences, Washington, D.C. 20037. LOUIS Y. KORMAN • Section on Gastroenterology, Veterans Administration Medical Center, Washington, D.C. 20422. DESMOND N. CARNEY • Mater Hospital, Dublin 7, Ireland.

TABLE I. Specificity of Antiserum for BN-like Peptides

Peptide	Cross reactivity[a] (%)
BN	100
GRP	110
Neuromedin C	82
GRP$_{21-27}$	64
Alytesin	52
GRP$_{22-27}$	<0.1
Litorin	<0.1
Neuromedin B	<0.1
Ranatensin	<0.1

[a] The mean value of three determinations is indicated.

2. OAT CELL PEPTIDES

2.1. Peptide Radioimmunoassays

Various human tumor cell lines were cultured in serum-free or serum-supplemented medium as described previously (Carney *et al.*, 1981). The tumor cells were harvested in log-growth phase 2 days after a medium change, and peptides were extracted in boiling 2 N acetic acid. After lyophilization, the cell extract was resuspended and assayed by radioimmunoassay.

The radioimmunoassay for bombesinlike peptides was performed as described previously (Moody *et al.*, 1983). Table I shows that the rabbit antiserum used cross reacts strongly with bombesin (BN), alytesin, gastrin-releasing peptide (GRP), GRP$_{14-27}$, Ac-GRP$_{20-27}$, GRP$_{21-27}$, and neuromedin C, all of which have identical C-terminal heptapeptides. In contrast, the antiserum does not cross react with ranatensin, litorin, and neuromedin B, which have a similar C terminal except for a phenylalanine instead of a leucine at the penultimate position. The structures of these analogues are shown in Table II.

TABLE II. Structure of BN-like Peptides[a]

GRP	Ala-Pro-Val-Ser-Val-Gly-Gly-Gly-Thr-Val-Leu-Ala-Lys-Met-Tyr-Pro-Arg-*Gly-Asn*-His-*Trp-Ala-Val-Gly-His-Leu-Met-NH*$_2$
Neuromedin C	*Gly-Asn*-His-*Trp-Ala-Val-Gly-His-Leu-Met-NH*$_2$
BN	*Pyr*-Gln-*Arg-Leu-Gly-Asn-Gln-Trp-Ala-Val-Gly-His-Leu-Met-NH*$_2$
Alytesin	*Pyr*-Gly-*Arg-Leu-Gly*-Thr-*Gln-Trp-Ala-Val-Gly-His-Leu-Met-NH*$_2$
Ranatensin	Pyr-Val-Pro-*Gln-Trp-Ala-Val-Gly-His*-Phe-*Met-NH*$_2$
Litorin	Pyr-*Gln-Trp-Ala-Val-Gly-His*-Phe-*Met-NH*$_2$
Neuromedin B	Gly-Asn-Leu-*Trp-Ala*-Thr-*Gly-His*-Phe-*Met-NH*$_2$

[a] Sequence homologies relative to BN are in italics.

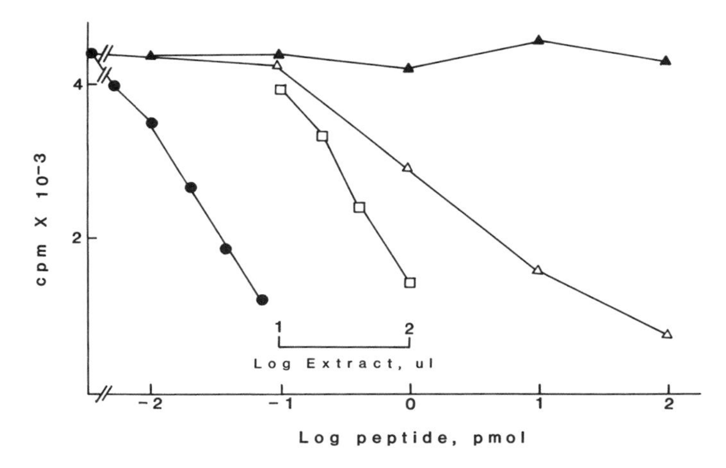

FIGURE 1. Neurotensin radioimmunoassay. The dose–response curves for neurotensin (●), neurotensin$^{8-13}$ (△), neurotensin$^{1-8}$ (▲), and an oat cell extract (□) are shown. Each value represents the mean of three determinations.

The radioimmunoassay for neurotensin was conducted as described previously (Moody *et al.*, 1985a). Figure 1 shows that this rabbit antiserum recognizes the C but not the N terminal of neurotensin. The limit of sensitivity of the antiserum was 3 fmol of neurotensin. The neurotensin standard and an extract derived from SCLC cell line NCI-N592 showed parallel dose–response curves.

2.2. Peptide Content and Characterization

The density of neurotensin and bombesinlike peptides was determined in numerous human tumor cell lines. Table III shows that the density was greatest in classic SCLC cell lines. High densities of bombesinlike peptides were detected in all classic SCLC cell lines examined (0.1–18.3 pmol/mg protein), whereas neurotensin was present in high density in 50% of classic SCLC cell lines tested (0.06–5.1 pmol/ml protein). Bombesinlike peptides and neurotensin were not abundant in six variant SCLC cell lines examined and were not detected in seven non-

TABLE III. Density of Neurotensin and BN-like Peptides[a]

Cell line	Neurotensin	BN-like peptides
Classic SCLC	0.06–5.1 (11/22)	0.1–18.3 (22/22)
Variant SCLC	<0.01 (0/6)	0.1–0.3 (3/6)
Non-SCLC lung cancer	<0.01 (0/7)	<0.01 (0/7)
Other cancer	<0.01 (0/6)	<0.01 (0/6)

[a] The range of peptide concentrations is shown. The number in parentheses shows the number of cell lines that had significant peptide levels relative to the number of cell lines examined.

SCLC cell lines including adenocarcinoma, large-cell carcinoma, and mesothelioma. Recently, high levels of BN-like peptides were detected in 45% of the bronchial carcinoid biopsy specimens examined (Yamiguchi *et al.*, 1983). Also, neurotensin or bombesinlike peptides were not found in six other human tumor cell lines including breast carcinoma, renal cell carcinoma, and melanoma. In summary, high levels of BN-like peptides and neurotensin are present in classic but not variant SCLC.

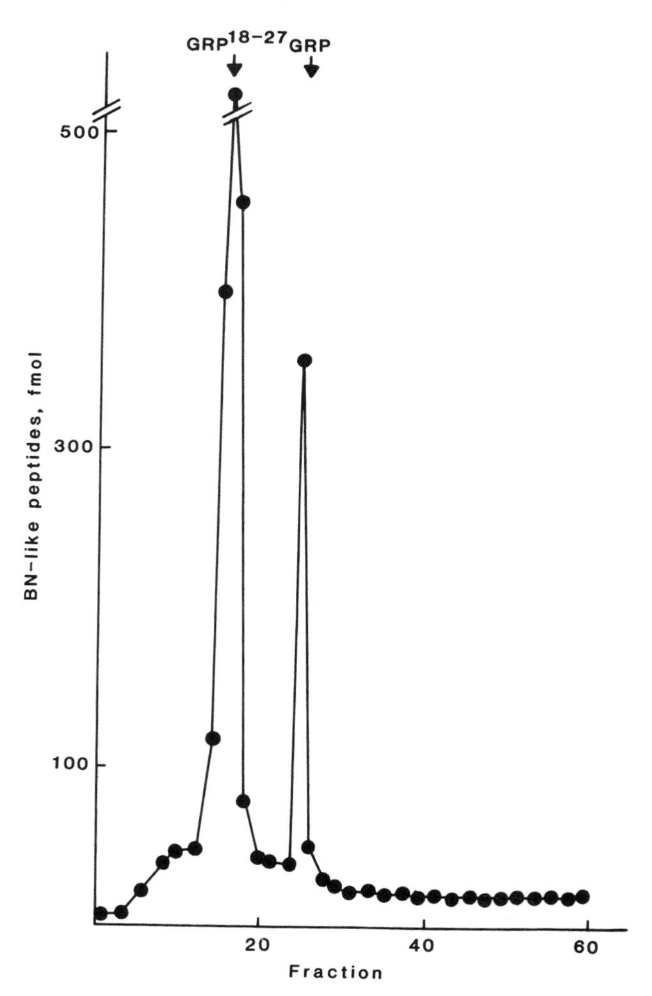

FIGURE 2. High-performance liquid chromatography profile. An extract derived from cell line NCI-N592 was fractionated using an acetonitrile gradient and 0.25 N triethylammonium phosphate, pH 3.0, and C_{18} Bondapak™ column as described previously (Moody *et al.*, 1983). The fractions were lyophilized and assayed for immunoreactive BN. The elution positions of synthetic GRP and neuromedin C (GRP_{18-27}) are shown.

The neuropeptides present in classic SCLC cell lines were characterized biochemically using cell line NCI-N592, which has a high concentration of neurotensin (5.1 pmol/mg protein) and moderate concentrations of bombesinlike peptides (1.1 pmol/mg protein). Immunoreactive neurotensin had a molecular weight and HPLC profile similar to that of synthetic standard. Thus, authentic neurotensin may be present in classic SCLC cells. The gel filtration and HPLC profile for the bombesinlike peptides were more complex. With gel filtration techniques, a minor peak and major peak of immunoreactivity were found that coeluted with GRP and neuromedin C, respectively. This was verified using HPLC techniques (Fig. 2). Thus, the bombesinlike peptides present in classic SCLC cells may be derived from GRP and its metabolites such as neuromedin C. Because each of these peptides has a C-terminal heptapeptide identical to BN, it is strongly recognized by our antiserum (Table I).

2.3. Peptide Secretion

The ability of neurotensin and bombesinlike peptides to be secreted from SCLC cells was investigated. High concentrations of extracellular K^+ (75 mM) or theophylline (10 mM) approximately doubled the secretion of neurotensin and bombesinlike peptides from classic SCLC cells. In contrast, dibutryl cAMP (1 mM) or VIP (1 μM) increased the secretion of neurotensin and bombesinlike peptides approximately 50%. These data indicate that neurotensin and bombesinlike peptides are present in and secreted from SCLC cells. Although neurosecretory granules are present in these cells, the events accompanying secretagogue-stimulated peptide secretion are undefined.

The peptide levels in patients were investigated. Bombesinlike peptides were present in low levels in normal patients or patients with non-SCLC (50 pM). Similarly, the levels in patients with limited disease were low, but the levels in patients with extensive disease were elevated up to 40-fold (Pert and Schumacher, 1982; Wood *et al.*, 1982). Thus, in patients with extensive disease, the plasma peptide levels exceeded the normal clearance and/or metabolism rate.

3. PEPTIDE RECEPTORS

3.1. Binding of Radiolabeled Peptides

Receptors for peptides were characterized. For bombesinlike peptides, [^{125}I-Tyr^4]BN bound with high affinity (K_d = 0.5 nM) to a single class of sites (B_{max} = 2000 sites/cell) in cell line NCI-H446 or NCI-H345. Binding was specific, reversible, and saturable. Pharmacology studies indicated that the C terminal of BN or GRP was important for high-affinity binding activity (Table IV). In particular, [Tyr^4]BN, BN, GRP, neuromedin C, and Ac-GRP$_{20-27}$ were potent inhibitors of specific binding activity. In contrast, [D-Trp^8]BN, [D-Val^{10}]BN, and [D-Leu^{13}]BN were not

TABLE IV. Pharmacology of Binding to SCLC Cell Line NCI H-345[a]

Peptide	IC_{50} (nM)	Relative potency (%)
[Tyr^4]BN	0.7	100
Ac-GRP_{20-27}	1.0	70
BN	1.5	47
Neuromedin C	2.0	35
GRP	5.0	14
GRP_{21-27}	300	0.23
[D-Leu^{13}]BN	1,300	0.05
[D-Val^{10}]BN	2,000	0.04
GRP_{22-27}	10,000	0.01
[D-Trp^8]BN	15,000	0.01
[des(Leu^{13},Met^{14})]BN	>20,000	<0.01

[a] Cells (2×10^6) were incubated with [^{125}I-Tyr^4]BN (1 nM) and varying doses of competitor. The concentration of competitor required to inhibit 50% of the specific [^{125}I-Tyr^4]BN bound (IC_{50}) was calculated.

potent inhibitors of binding activity. These data indicate that substitution of D- for the natural L-amino acids at the 8, 10, or 13 positions greatly reduces biological potency by over three orders of magnitude. Similarly, deletion of amino acids at the 13 and 14 positions of BN, [des-Leu^{13}, Met^{14}]BN, reduces biological potency by over four orders of magnitude. Also, deletion of the amino acid at the 20 position of GRP (GRP_{21-27}) reduces biological potency by over two orders of magnitude. Further deletion of an additional amino acid at the 21 position (GRP_{22-27}) reduces biological potency by an additional two orders of magnitude. These data indicate that the C terminal of octapeptide BN or GRP is essential for high-affinity binding activity. Because specific binding activity was greatly reduced if SCLC cells were pretreated with trypsin, radiolabeled [Tyr^4]BN may bind to plasma membrane protein. Additional data indicate that this high-affinity binding component may have a subunit molecular weight of 78,000 (Moody *et al.*, 1985b).

Recent data indicate that [^{125}I]VIP binds with high affinity to SCLC cell line NCI-N592. Also, [^{125}I]neurotensin binds with high affinity (K_d = 4 nM) to a single class of sites (2300/cell) in cell line NCI-H209. Thus, in addition to receptors for bombesinlike peptides, SCLC cells may have receptors for neurotensin and VIP.

3.2. Events after Receptor Activation

VIP receptors, when occupied, stimulate adenylate cyclase, resulting in increased intracellular levels of cAMP (fourfold) after 5 min. This increase in cAMP is accompanied by increased secretion of BN-like peptides from SCLC cells.

Occupancy of receptors for BN-like peptides causes redistribution of cellular Ca^{2+}, increased cGMP levels, and amylase release in pancreatic acinar cells (Jensen *et al.*, 1978). Recent data indicate that when SCLC receptors are activated by BN or GRP, the rate of Ca^{2+} influx increases twofold (D.G. Yoder and T.W. Moody, unpublished data). Bombesin has no effect on the cAMP levels. Also, BN stimulates the growth of SCLC cells in a concentration-dependent manner (Carney *et al.*, 1983). Thus, BN-like peptides may function as growth factors in SCLC cells. The role of neurotensin in SCLC remains unknown. Neurotensin has no effect on cellular cAMP levels.

4. DISCUSSION

The neuroendocrine nature of classic SCLC is firmly established. In addition to possessing neurotransmitter metabolites and enzymes, classic SCLC produces neuropeptides and possesses neuropeptide receptors. The peptide levels are markedly reduced in variant SCLC, which has amplification and increased expression of the c-*myc* oncogene (Little *et al.*, 1983). Neurotensin and BN-like peptides are not present in adenocarcinoma, mesothelioma, or large-cell carcinoma cell lines; however, they are present in pulmonary carcinoids (Tamai *et al.*, 1983; Roth *et al.*, 1983). Recently, the gene that codes for BN-like peptides was cloned from a lung carcinoid tumor (Spindel *et al.*, 1984). These data indicate that BN-like peptides are synthesized in the form of a high-molecular-weight precursor protein composed of 148 amino acid residues. This precursor protein is metabolized rapidly to a biologically active 27- (GRP) or 10- (neuromedin C) amino-acid form by trypsinlike proteases. Similarly, the BN-like peptides in SCLC cells are likely to be made in the form of a high-molecular-weight precursor and then metabolized into smaller biologically active peptides such as GRP or neuromedin C.

After synthesis, processing, and packaging, the peptides may then be secreted from SCLC cells. Bombesinlike peptides have been detected in the medium exposed to SCLC cells (Sorenson *et al.*, 1983; Moody *et al.*, 1983). The secretion rate can be increased *in vitro* using high concentrations of extracellular K^+ (Moody *et al.*, 1983), theophylline, or VIP; these latter two agents elevate the intracellular cAMP levels. Also, the secretion rate of BN-like peptides may be increased using muscarinic cholinergic agonists (Sorenson *et al.*, 1983). Thus, various external stimuli may regulate the secretion of neuropeptides from SCLC cells. This secretion is carefully controlled, however, in that only small amounts of peptide (5–10% of the total present in the cells) may be secreted.

When secreted, the BN peptides may interact with receptors present on the cell surface. We have characterized these receptors using radiolabeled [Tyr^4]BN, a potent BN analogue (Rivier and Brown, 1978). Binding was specific, saturable, and reversible. Pharmacological studies indicated that BN and the structurally related GRP bound with high affinity and that the C terminal of BN or GRP was important for high-affinity binding activity. Thus, the BN-like peptides produced and secreted from SCLC cells may bind to and activate receptors for BN-like peptides.

When activated, receptors for BN-like peptides stimulate growth of SCLC cells. Similarly, BN or GRP stimulates growth of 3T3 cells (Rozengurt and Sinnet-Smith, 1983) and normal human bronchial epithelial cells (Willey *et al.*, 1984). Numerous other events may occur between receptor activation and growth, such

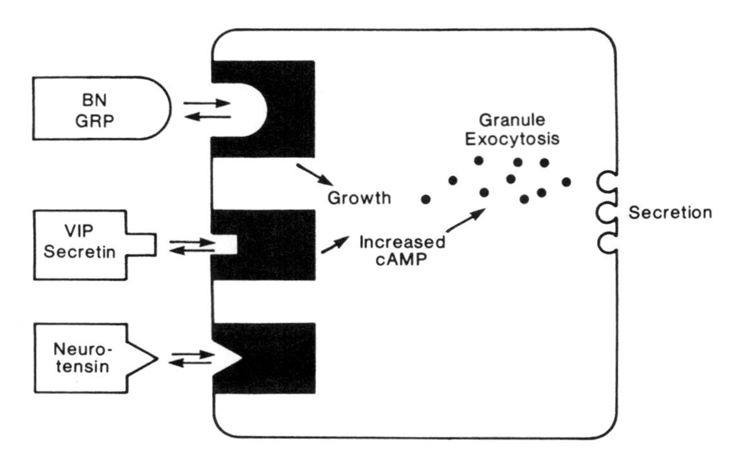

FIGURE 3. Schematic illustration of SCLC cells.

as increased Ca^{2+} flux and phosphatidylinositol turnover (Berridge *et al.*, 1984). Because SCLC tumor growth is inhibited *in vitro* and *in vivo* by an anti-BN monoclonal antibody (Cuttitta *et al.*, 1985), BN-like peptides may function as important autocrine growth factors in SCLC (Fig. 3).

It is possible that other growth factors are produced by SCLC cells. In this regard, neurotensin is produced by and secreted from SCLC cells, and neurotensin receptors are present on SCLC cells. It remains to be determined, however, if neurotensin is an additional growth factor for oat cell carcinoma. Also, VIP receptors are present in SCLC cells (Fig. 3). When activated, VIP receptors induce secretion of neurotensin and BN-like peptides.

5. SUMMARY

High levels of neurotensin and BN-like peptides are produced by classic SCLC cells. These peptides may be secreted in response to various stimuli such as VIP, which elevates the intracellular cAMP levels. Also, receptors for neurotensin and BN-like peptides are present on the surface of SCLC cells. When activated, these receptors may stimulate growth of SCLC. Thus, peptides may function as important regulatory factors in oat cell carcinoma.

ACKNOWLEDGMENTS. The authors thank Drs. F. Cuttitta, A. Gazdar, and J. Minna for helpful discussions. This research was supported by NCI grant CA 33767.

REFERENCES

Berridge, M. J., 1984, Inositol triphosphate and diacylglycerol as second messengers, *Biochem. J.* **220:**345–360.

Bunn, P. A., Cohen, M. H., Ihde, D. C., Fossieck, B. E., Matthews, M. J., and Minna, J. D., 1977, Advances in small cell bronchogenic carcinoma, *Cancer Treat. Rep.* **61:**333–342.

Carney, D., Bunn, P., Gazdar, A., Pagan, J., and Minna, J., 1981, Selective growth in serum-free hormone-supplemented medium of tumor cells obtained by biopsy from patients with small cell carcinoma of the lung, *Proc. Natl. Acad. Sci. U.S.A.* **78:**3185–3189.

Carney, C. N., Oie, H., Moody, T., Gazdar, A., Cuttitta, F., and Minna, J. D., 1983, Bombesin: An autocrine growth factor for human small cell lung cancer, *Clin. Res.* **31:**404A.

Cuttitta, F., Carney, D. N., Mulshine, J., Moody, T. W., Fedorko, J., Fischler, A., and Minna, J. D., 1985, Bombesin-like peptides can function as autocrine growth factors in human small cell lung cancer, *Nature* **316:**823–826.

Erisman, M. R., Linnoila, O., Hernandez, R., DiAugustine, R., and Lazarus, L., 1982, Human lung small-cell carcinoma contains bombesin, *Proc. Natl. Acad. Sci. U.S.A.* **79:**2379–2383.

Gazdar, A., Carney, D., Russel, E., Sims, H., Baylin, S., Bunn, P., Guccion, J., and Minna, J., 1979, Establishment of continuous, clonable, cultures of small cell carcinoma of the lung which have amine precursor uptake and decarboxylation cell properties, *Cancer Res.* **40:**3502–3507.

Jensen, R., Moody, T., Pert, C., Rivier, J., and Gardner, J., 1978, Interaction of bombesin and litorin with specific membrane receptors on pancreatic acinar cells, *Proc. Natl. Acad. Sci. U.S.A.* **75:**6139–6143.

Little, C. D., Nan, M. N., Carney,, D. N., Gazdar, A. F., and Minna, J. D., 1983, Amplification and expression of the c-*myc* oncogene in human lung cancer cell lines, *Nature* **306:**194–196.

Marangos, P., Gazdar, A., and Carney, D., 1982, Neuron specific enolase in human small cell carcinoma cultures, *Cancer Lett.* **15:**67–71.

Moody, T., Pert, C., Gazdar, A., Carney, D., and Minna, J., 1981, High levels of intracellular bombesin characterize human small cell lung carcinoma, *Science* **214:**1246–1248.

Moody, T., Russel, E., O'Donohue, T., Linden, C., and Gazdar, A., 1983, Bombesin-like peptides in small cell lung cancer: Biochemical characterization and secretion from a cell line, *Life Sci.* **32:**487–493.

Moody, T. W., Carney, D. N., Korman, L. Y., Gazdar, A. F., and Minna, J. D., 1985a, Neurotensin is produced by and secreted from classic cell lung cancer cells, *Life Sci.* **36:**1727–1732.

Moody, T. W., Carney, D. N., Cuttitta, F., Quattrocchi, K., and Minna, J. D., 1985b, High affinity receptors for bombesin/GRP-like peptides on human small cell lung cancer, *Life Sci.* **37:**105–113.

Pert, C. B., and Schumacher, U. K., 1982, Plasma bombesin concentrations in patients with extensive small cell carcinoma of the lung, *Lancet* **1:**509.

Pettengill, O. S., Bacopoulos, N. G., and Sorenson, G. D., 1982, Biogenic amine metabolites in human in tumor cells: Histochemical and mass-spectrographic demonstration, *Life Sci.* **30:**1355–1360.

Rivier, J., and Brown, M., 1978, Bombesin, bombesin analogues and related peptides: Effects on thermoregulation, *Biochemistry* **17:**1766–1771.

Roth, K. A., Evans, C. J., Weber, E., Barchas, J. D., Bostwick, D. G., and Bensch, K. G., 1983, Gastrin-releasing peptide-related peptides in a human malignant lung carcinoid tumor, *Cancer Res.* **43:**5411–5415.

Rozengurt, E., and Sinnet-Smith, J., 1983, Bombesin stimulation of DNA synthesis and cell division in cultures in Swiss 3T3 cells, *Proc. Natl. Acad. Sci. U.S.A.* **80:**2936–2940.

Sorenson, G. D., Pettengill, O. S., Cate, C. C., Ghatei, M. A., Molyneux, K. E., Gosselin, E. J., and Bloom, S. R., 1983, Bombesin and calcitonin secretion by pulmonary carcinoma is modulated by cholinergic receptors, *Life Sci.* **33:**1939–1944.

Spindel, E. R., Chin, W. W., Price, J., Rees, L. H., Besser, G. M., and Habener, J. F., 1984, Cloning and characterization of cDNAs encoding human gastrin-releasing peptide, *Proc. Natl. Acad. Sci. U.S.A.* **81:**5699–5703.

Tamai, S., Kameya, T., Yamaguchi, K., Yanai, N., Abe, K., Yanaihara, N., Yamazaki, N., Yamazaki, H., and Kageyama, K., 1983, Peripheral lung carcinoid tumor producing predominantly gastrin-releasing peptide (GRP), *Cancer* **52:**273–281.

Willey, J. G., Lechner, J. F., and Harris, C. C., 1984, Bombesin and the C-terminal tetradecapeptide of gastrin-releasing peptide are growth factors for normal human bronchial epithelial cells, *Exp. Cell Res.* **153:**245–248.

Wood, S., Wood, J., Ghatei, M., Lee, U., Shaughnessy, D., and Bloom, S., 1981, Bombesin, 5somatostatin and neurotensin-like immunoreactivity in bronchial carcinoma, *J. Clin. Endocrinol.* **53:**1310–1312.

Wood, S. M., Wood, J. R., Ghatei, M. A., Sorenson, G. D., and Bloom, S. R., 1982, Is bombesin a tumour marker for small cell carcinoma? *Lancet* **1:**690–691.

Yamiguchi, K., Abe, K., Kameya, T., Adachi, I., Taguchi, S., Otsuba, K., and Yanaihara, N., 1983, Production and molecular size heterogeneity of immunoreactive gastrin-releasing peptide in fetal and adult lungs and primary lung tumors, *Cancer Res.* **43:**3932–3937.

Yesner, R., 1978, Hormone production in small cell lung cancer, *Pathol. Ann.* **13:**207–240.

29

Bombesin/Gastrin-Releasing Peptide Stimulation of Small-Cell Lung Cancer Clonal Growth

DESMOND N. CARNEY, TERRY W. MOODY, and FRANK CUTTITTA

1. INTRODUCTION

Small-cell lung cancer (SCLC) accounts for 20–25% of all new cases of primary lung cancer. Unlike the other major histological subtypes of lung cancer (collectively referred to as non-SCLC), SCLC is highly sensitive to both chemotherapy and radiation therapy. With combination chemotherapy a clinical response will be observed in up to 95% of all patients, and complete disappearance of all known tumor is seen in 40–50% of patients (Morstyn *et al.*, 1984). However, in spite of the early dramatic responses observed in most patients, the vast majority will relapse with SCLC, and fewer than 10% of all patients will achieve long-term disease-free survival. The median survival for all patients is 10–11 months. Thus, there is a need to develop alternative approaches to the treatment of this tumor type.

Many factors may influence the response of a patient to cytotoxic therapy, including age, sex, extent of disease, and performance status (Ihde *et al.*, 1981), and it is also possible that the inherent biological properties of individual tumor cells may be of prognostic importance. In recent years major advances have been made in our ability to establish continuous cell lines of SCLC direct from patient biopsy specimens (Carney *et al.*, 1985). The availability of panels of cell lines of both SCLC and non-SCLC has greatly facilitated studies of the biological properties, including growth regulation, of these tumor cells. Among many characteristics

---

DESMOND N. CARNEY • Mater Hospital, Dublin 7, Ireland. TERRY W. MOODY • Department of Biochemistry, The George Washington University School of Medicine and Health Sciences, Washington, D. C. 20037. FRANK CUTTITTA • NCI–Navy Medical Oncology Branch, National Naval Medical Center, Bethesda, Maryland 20014.

identified in SCLC was the demonstration of significant levels of bombesinlike immunoreactivity (BLI) in both fresh specimens and established cell lines of SCLC (Moody *et al.*, 1981; Yamaguchi *et al.*, 1983). Further studies demonstrated that BLI was actively secreted by SCLC cell lines and that SCLC cell lines express a single class of high-affinity, saturable binding receptors for BLI that are similar if not identical to those previously demonstrated on nonmalignant tissue (Moody *et al.*, 1983b, 1985). In contrast to SCLC, BLI or BLI receptors have not been identified in cell lines of other types of lung cancer or in a variety of other non-APUD tumor cell lines. Because of the production of BLI by SCLC and the identification of BLI receptors in SCLC, it is possible that BLI has some important role in the growth of this tumor type. For many years it has been speculated that the "uninhibited" growth of malignant cells *in vivo* is related to the ability of these cells to produce self-stimulatory (autocrine) growth factors. The recent demonstration that some oncogenes regulate cell growth by the modulation of either growth factors or growth factor receptors stresses the importance of identifying the specific growth factors of individual tumor types. In this chapter the possible role of bombesin and its homologue, gastrin-releasing peptide (GRP), as putative growth factors for SCLC is discussed.

2. ESTABLISHMENT AND CHARACTERIZATION OF SMALL-CELL LUNG CANCER CELL LINES: GROWTH IN DEFINED MEDIUM

Initial attempts to establish continuous SCLC cell lines using serum-supplemented medium (SSM) were associated with success rates of only 10–20% (Gazdar *et al.*, 1980). These data suggested that either those cells lines that were established represented only the more "malignant" cell lines or that essential nutrients required for SCLC growth were lacking in the SSM. It is also possible that serum contained substances inhibitory to the growth of SCLC. Since culture systems are of potential value for both clinical drug sensitivity testing of fresh biopsy specimens and in the screening of new agents of potential clinical value, attempts were made to define a serum-free nutrient-supplemented medium that would support the continual growth of tumor cells from the majority of patients with SCLC. With previously established cell lines of SCLC, the influence of many growth factors on the proliferation of SCLC was evaluated.

Although numerous factors were evaluated, a defined medium, HITES, containing hydrocortisone, insulin, transferrin, estradiol, and selenium, was shown to support the continual growth of other cell lines of SCLC.

We next tested the ability of HITES to support the growth of fresh tumor cells from patient biopsy specimens including both SCLC and non-SCLC tumors. In a preliminary study of 14 fresh tumor specimens, it was demonstrated that growth of SCLC cells in defined HITES medium was superior to growth in serum-containing medium (Carney *et al.*, 1981a). In a prospective study over a 2-year period, the ability to establish continuous cell lines of SCLC was assessed in both HITES and SSM (Carney *et al.*, 1985). During this period, 43 tumor-containing specimens were received. Continuous SCLC cell lines were established in 31 (72%). Of these 31 cell lines, 22 were successfully established as continuous lines in HITES medium, and nine additional cell lines could only be established in serum-supplemented medium. Thus, although major improvements have been achieved in the growth of SCLC lines, as some cell lines could not be established without serum supplementation, it is clear that further growth factors in addition to HITES will be required for 11 fresh SCLC specimens to be cultured in defined medium.

Data from these studies suggest that the ease of establishment of cell lines in defined medium may be related to the ability of the tumor cells to secrete autocrine growth factors. Cell lines could be more readily established if they were maintained at a heavy cell density and, when passaged, were fed with 50% conditioned medium from the flasks in which they were cultured. Furthermore, studies of conditioned media from several cell lines maintained in HITES medium clearly demonstrated their ability to stimulate the clonal growth of SCLC cell lines *in vitro* (Carney *et al.*, 1984). The isolation and characterization of these autostimulatory factors in SCLC may greatly help our understanding of the regulation of tumor growth in patients.

3. BIOLOGICAL CLASSES OF SMALL-CELL LUNG CANCER

The biological properties of SCLC cell lines have recently been published (Carney *et al.*, 1985; Gazdar *et al.*, 1985). Results from these studies clearly indicate that considerable heterogeneity exists in this tumor type. Based on a variety of properties, SCLC cell lines can be subdivided into two distinct classes that may be of prognostic significance, namely, classic and variant cell lines. Classic cell lines have the typical cytological and histological characteristics of SCLC, have a relatively slow growth rate in culture, are radiosensitive, and express elevated levels of L-dopa decarboxylase (DDC), bombesinlike immunoreactivity (BLI), neuron-specific enolase (NSE), and the BB isozyme of creatine kinase (CK-BB). In contrast, variant cell lines have a much more aggressive growth behavior *in vivo* and *in vitro*, are radioresistant, and have undetectable levels of DDC and BLI but continue to express elevated levels of NSE and CK-BB. In addition, many of these variant lines have amplification and expression of the C-*myc* oncogene (Little *et al.*, 1983). Clinical correlates of these properties have been noted in that patients with the variant phenotype in their diagnostic biopsy specimens have a poorer prognosis than patients with pure SCLC (Radice *et al.*, 1982). Since classic and variant cell lines have significantly different *in vitro* growth properties, the identification of the specific growth factors for each subclass may ultimately lead to the development of new methods in the treatment of this disease.

TABLE I. Influence of Bombesin (50 nM) on the Clonal Growth of Human Tumor Cell Lines

	Cell type	
	Small cell	Nonsmall cell
Number of cell lines	10	7
Number stimulated	9	0
Fold stimulation	7–150	—

4. BOMBESIN: AN AUTOCRINE GROWTH FACTOR FOR SMALL-CELL LUNG CANCER

The demonstration that the majority of SCLC cell lines express high levels of BLI and also express a single class of high-affinity receptors for BLI suggests that BLI may have an important physiological role in SCLC. In addition, other studies have shown that bombesin and related peptides can stimulate the proliferation of 3T3 mouse fibroblasts, normal bronchial epithelium, and pancreatic hyperplasia (Rozenqurt and Sinnet-Smith, 1983; McDonald *et al.*, 1983). These data prompted studies of the effects of bombesin on SCLC growth *in vitro*.

The influence of bombesin and related hormones on the growth of small-cell lung cancer cell lines was tested using a soft agarose clonogenic assay as previously described (Carney *et al.*, 1981b, 1983). The effects of hormones on growth were determined in both SSM and in serum-free defined HITES medium. When bombesin was added to SCLC cells, a dramatic stimulation of clonal growth was observed in HITES medium (Table I). When exogenous added bombesin was tested over a wide concentration range, optimal colony stimulatory effects were observed at an exogenously added BN concentration of 50 nM. Among nine of ten individual SCLC cell lines tested, colony formation in HITES medium was stimulated seven- to 150-fold above that observed in control cultures. In contrast, and over a concentration range of 1–1000 nm, the addition of BN in the presence of serum had little or no effect on colony growth of SCLC cells. The number of colonies in

TABLE II. Comparison of the Clonal Growth of SCLC Line NCI N592 in Serum-Supplemented Medium, Serum-free HITES Medium, and HITES Medium plus Bombesin

Culture medium	Number of colonies
Serum supplemented	814 (100%)
HITES	32 (4%)
HITES + bombesin	664 (81%)

HITES medium supplemented with 50 nM BN for the most part approximated that scored in serum-supplemented medium (Table II). In contrast to SCLC cultures, no stimulation by BN of colony formation was observed in nine non-SCLC cell lines including a variety of other lung cancers, malignant melanoma, and hypernephnoma cell lines tested in both serum-supplemented medium and HITES medium (Table I).

Although high-affinity saturable receptors for BN have been identified on some (9/34) cell lines, no obvious correlation was observed in the SCLC lines among the observed *in vitro* responses to BN, the production of BN by the cell lines, and the presence of cell surface receptors for BN. This is in conflict with data demonstrated in other cell lines (Westendorf and Schonbrunn, 1983). These SCLC data suggest that the mitogenic effects of BN may be mediated through mechanisms other than binding to receptors, e.g., through stimulation of the release of other endogenous hormones by the tumor cells. In other model systems, BN has been shown to stimulate gastric, pancreatic, and pituitary hormone secretion including gastrin, prolactin, growth hormone, and insulin. In a recent report of a patient with pituitary ACTH-dependent Cushing's syndrome, and a medullary carcinoma of the thyroid, the excessive pituitary ACTH secretion was thought to be secondary to ectopic production of a bombesinlike peptide (Howlett *et al.*, 1985).

An alternative hypothesis for the lack of correlation between the mitogenic effect of BN and the presence or absence of receptors may be that the BN receptor, like those for other polypeptide hormones, may exhibit ligand-induced "down-regulation" of cell surface BN binding capacity through the constant release of BN into the culture medium. In studies of isolated pancreatic acini, preincubation of cells with BN induced densitization of enzyme secretion by reducing the number of active receptors available to interact with BN and stimulate secretion (Pandol *et al.*, 1982).

Gastrin-releasing peptide (GRP), a 27-amino-acid peptide with an affinity for BN receptors similar to that of BN alone, has also been tested for stimulation of SCLC. By using both a clonogenic assay (Carney *et al.*, 1983), and a [^{3}H]thymidine incorporation assay (Weber *et al.*, 1985), exogenously added GRP has been shown to be a potent mitogen for SCLC. The mitogenicity of GRP has been isolated to the carboxy-terminal fragment (GRP^{14-27}), which is that part of GRP homologous to bombesin. The amino-terminal fragment, GRP^{1-16}, which is not homologous to BN, was nonmitogenic. Furthermore analogues of BN with poor affinity for BN receptors (e.g., des-Leu^{13}-Met^{14}-BN) have little or no effect on the clonal growth of SCLC cells.

The results of these studies of added exogenous BN/GRP on the growth of SCLC cells suggest that this hormone may have an important role *in vivo* in the growth of this tumor type. In addition, the amphibian peptide bombesin has been demonstrated to act as a mitogen for both Swiss 3T3 fibroblasts and normal human bronchial mucosal cells (Rosenqurt and Sinnet Smyth, 1983; Willey *et al.*,1984) and thus may play a role in normal growth regulatory processes in the pulmonary tree.

5. MODIFICATION OF BOMBESIN-STIMULATED SMALL-CELL GROWTH *IN VITRO*

Although the above studies suggest that bombesin and related peptides may function as autocrine growth factors for SCLC, further support for this hypothesis comes from recent data demonstrating that a monoclonal antibody to bombesin (2A11) can inhibit the clonal growth of SCLC *in vitro* and the growth of SCLC xenografts *in vivo*. With standard culture techniques a monoclonal antibody to bombesin was generated. This antibody binds to the C-terminal region of bombesin. This portion of bombesin and GRP competes for the same class of high-affinity receptors on pituitary cells and SCLC membranes and is required for receptor binding and physiological activity (Cuttitta *et al.,* 1985). Although this antibody detects BLI in extracts of SCLC, it does not bind to live, intact SCLC cells. The antibody blocks bombesin receptor interaction in a dose-related fashion.

In a clonogenic assay, 2A11 was tested for its effect on the growth of SCLC cell lines. The clonogenic growth of lines was markedly inhibited by 3 μg/ml or greater concentration of this antibody. The addition of excess exogenous bombesin prevents the 2A11 growth inhibition, demonstrating the specificity of the 2A11 effect. When 2A11 was tested for its effects on the growth *in vivo* of SCLC tumor heterotransplants in nude mice, suppression of SCLC growth was observed following 2A11 treatment intraperitoneally, whereas no effect was observed on non-SCLC lines (Cuttitta *et al.,* 1985). These findings strongly support the hypothesis that bombesin or related peptides may function as autocrine growth factors for SCLC.

6. CONCLUSION

Although advances have been made in the diagnosis, staging, and treatment of patients with SCLC, the majority of patients still die from their disease. Data generated from studies of continuous cell lines of this tumor have clearly indicated the considerable heterogeneity that exists in this tumor type and that may be of prognostic importance to individual patients. The need for new approaches in the therapy of this disease has been emphasized by the recognition of a plateau in treatment results over the past 5 to 10 years. As discussed above, the growth of SCLC appears to be regulated by autocrine growth factor(s). The data suggest that bombesin/GRP may be one such factor. As *in vitro* and *in vivo* data suggest that disruption of the binding of BLI/GRP to its receptors may suppress the growth of SCLC, clinical trials of the use of specific antibodies or analogues of BLI (Jensen *et al.,* 1984) in patients with SCLC will indicate the role of BLI/GRP for SCLC growth *in vivo*.

REFERENCES

Carney, D. N., Bunn P. A., Gazdar, A. F., Pagan, J. A., and Minna, J. D., 1981a, Selective growth in serum-free hormone supplemented medium of tumor cells obtained by biopsy from patients with small cell carcinoma of the lung, *Proc. Natl. Acad. Sci. U.S.A.* **78**:3185–3189.

Carney, D. N., Gazdar, A. F., Bunn, P. A., and Guccion, J. G., 1981b, Demonstration of the stem cell nature of clonogenic cells in lung cancer specimens, *Stem Cells* **1**:149–164.

Carney, D. N., Oie, H., Moody, T. W., Gazdar, A. F., Cuttitta, F., and Minna, J. D., 1983, Bombesin stimulation of small cell lung cancer growth, *Clin. Res.* **31**:404.

Carney, D. N., Brower, M., Bertness, V., and Oie, H. K., 1984, The selective growth of human small cell lung cancer cell lines and clinical specimens in serum-free medium, in: *Methods in Molecular and Cell Biology* (G. Sato and D. Barnes, eds.), Alan R. Liss, New York, pp. 247–262.

Carney, D. N., Gazdar, A. G., Bepler, G., Guccion, J., Moody, T. W., Marangos, P. J., Zweig, M. H., and Minna, J. D., 1985, Establishment and identification of small cell lung cancer cell lines having classic and variant features, *Cancer Res.* **45**:2913–2923.

Cuttitta, F., Carney, D. N., Mulshine, J., Moody, T. W., Fedorko, J., Fischler, A., and Minna, J., 1985, Bombesin-like peptides can function as autocrine growth factors in human small cell lung cancer, *Nature* **316**:823–826.

Gazdar, A. F., Carney, D. N., Russel, E. K., Simms, H. L., Baylin, S. B., Bunn, P. A., Guccion, J. G., and Minna, J. D., 1980, Small cell carcinoma of the lung: Establishment of continuous clonable cell lines having APUD properties, *Cancer Res.* **40**:3502–3507.

Gazdar, A. F., Carney, D. N., Nau, M., and Minna, J. D., 1985, Characterization of variant subclasses of cell lines derived from small cell lung cancer having distinctive biochemical, morphological and growth properties, *Cancer Res.* **45**:2924–2930.

Howlett, T. A., Price, J., Hale, A. C., Donlack, I., Rees, L. H., Wass, J. A. H., and Besser, G. M., 1985, Pituitary ACTH dependent Cushing's syndrome due to ectopic production of a bombesin-like peptide by a medullary carcinoma of the thyroid, *Clin. Endocrinol.* **22**:91–101.

Ihde, D. C., Makuch, R. W., Carney, D. N., Bunn, P. A., Cohen, M. H., Matthews, M. J., and Minna, J. D., 1981, Prognostic implications of stage of disease and sites of metastases in patients with small cell carcinoma of the lung treated with intensive chemotherapy, *Am. Rev. Respir. Dis.* **123**:500–507.

Jensen, R. T., Jones, S. W., Folkers, K., and Gardner, J. D., 1984, A synthetic peptide that is a bombesin receptor antagonist, *Nature* **309**:61–63.

Little, C. D., Nau, M. M., Carney, D. N., Gazdar, A. F., and Minna, J. D., 1983, Amplification and expression of the C-*myc* oncogene in human lung cancer cell lines, *Nature* **306**:194–196.

McDonald, T. G., Ghatei, M. A., Bloom, S. R., Adrian, T. E., Mochizuk, T., Yanaihara, C., and Yanihara, N., 1983, Dose response comparisons of canine plasma gastroenteropancreatic hormone response to bombesin and the porcine gastrin releasing peptide, *Regul. Peptides* **5**:125–137.

Moody, T. W., Pert, C. B., Gazdar, A. F., Carney, D. N., and Minna, J. D., 1981, High levels of intracellular bombesin characterize human small cell lung cancer, *Science* **214**:1246–1248.

Moody, T. W., Russell, E. K., O'Donohoe, T. L., Linden, C. D., and Gazdar, A. F., 1983a, Bombesin-like peptides in small cell lung cancer; biochemical characterization and secretion from a cell line, *Life Sci.* **32**:487–493.

Moody, T. W., Bertness, V., and Carney, D. N., 1983b, Bombesin-like peptides and receptors in human tumours, *Peptides* **4**:683–686.

Moody, T. W., Carney, D. N., Cuttitta, F., Quattrocchi, K., and Minna, J. D., 1985, High affinity receptors for bombesin/GRP-like peptides on human small cell lung cancer, *Life Sci.* **37**: 105–113.

Morstyn, G., Ihde, D. C., Lichter, A. S., Bunn, P. A., Carney, D. N., Glatstein, E., and Minna, J. D., 1984, Small cell lung cancer 1973–1983: Early progress and recent obstacles, *Int. J. Radiat. Oncol.* **10**:515–539.

Pandol, S. J., Jensen, R. T., and Gardner, J. D., 1982, Mechanism of [Tyr^4]bombesin-induced desensitization in dispersed acini from guinea pig pancreas, *J. Biol. Chem.* **20:**12024–12029.

Radice, P. A., Matthews, M. J., Ihde, D. C., Gazdar, A. F., Carney, D. N., Bunn, P. A., Cohen, M. H., Fossieck, B. E., Makuch, R. W., and Minna, J. D., 1982, The clinical behaviour of mixed small cell/large cell bronchogenic carcinoma compared to pure small cell subtypes, *Cancer* **50:**2894–2902.

Rozengurt, E., and Sinnett-Smith, J., 1983, Bombesin stimulation of DNA synthesis and cell division in cultures of Swiss 3T3 cells, *Proc. Natl. Acad. Sci. U.S.A.* **80:**2936–2940.

Weber, S., Zuckerman, J. E., Bostwick, D. G., Bensch, K. G., Sikic, B. I., and Raffin, T. A., 1985, Gastrin releasing peptide is a selective mitogen for small cell lung carcinoma *in vitro, J. Clin. Invest.* **75:**306–309.

Westendorf, J. M., and Schonbrunn, A., 1983, Characterization of bombesin receptors in a rat pituitary cell line, *J. Biol. Chem.* **258:**7527–7535.

Willey, J., Lechner, J. F., and Harris, C. C., 1984, Bombesin and the C-terminal tetradecapeptide of gastrin-releasing peptide are growth factors for normal human bronchial epithelial cells, *Exp. Cell Res.* **153:**245–248.

Yamaguchi, K., Abe, K., Kameya, T., Adachi, I., Taguchi, S., Otsubo, K., and Yancuhara, N., 1983, Production and molecular size heterogeneity of immunoreactive gastrin releasing peptide in fetal and adult lungs and primary lung tumours, *Cancer Res.* **43:**3932–3939.

30

A CRF-Producing and -Secreting Tumor of the Lung

G. KAHALY, T. STRACK, U. KRAUSE, J. SCHREZENMEIR, W. ORMANNS, J. DÄMMRICH, C. MASER-GLUTH, P. VECSEI, and J. BEYER

1. INTRODUCTION

Corticotropin-releasing factor (CRF) has recently been isolated from ovine hypothalamus, characterized chemically, and synthesized. Upton and Amatruda (1971) first suggested that CRF-like activity occurred in neoplastic tissues, and Hashimoto *et al.* (1980) subsequently detected CRF-like activity in an extract of an ACTH-producing nephroblastoma. Recently, Carey *et al.* (1984) and Asa *et al.* (1984) used immunocytochemical techniques to demonstrate CRF in postmortem tumor material obtained from patients with a carcinoma of the prostate and an intrasellar gangliocytoma.

We report here on a patient with signs of Cushing's syndrome who had ectopic production and secretion of CRF from a metastatic, nondifferentiated adenocarcinoma of the lung. The clinical, biochemical, and histopathological findings of this case are presented.

2. METHODS

2.1. Case Report

The 43-year-old patient was admitted to a peripheral hospital in February, 1984 with suspicion of a lung tumor. A centrally located, large, nondifferentiated

G. KAHALY, T. STRACK, U. KRAUSE, J. SCHREZENMEIR, and J. BEYER • Department of Endocrinology, University of Mainz, Mainz, Federal Republic of Germany. W. ORMANNS and J. DÄMMRICH • Department of Pathology, University of Würzburg, Würzburg, Federal Republic of Germany. C. MASER-GLUTH and P. VECSEI • Department of Pharmacology, University of Heidelberg, Heidelberg, Federal Republic of Germany.

adenocarcinoma of the lung and intraoperatively detected lymph node metastases were totally resected. No further lymphogenic or hematogenous spreading of the tumor was detected by imaging procedures. The patient was discharged with no further therapy. A symptom-free interval lasted until June, 1984, when the first signs of Cushing's syndrome occurred. Simultaneously, symptoms of a tumor relapse appeared. In August, 1984 a cutaneous metastasis located on the left occipital skull was removed, and histology confirmed the diagnosis of a metastasis of the previously resected pulmonary carcinoma. The patient was admitted to our department of endocrinolgy and metabolism with the tentative diagnosis of ectopic Cushing's syndrome.

Physical examination on admission revealed a cachectic man with pharyngeal mycosis, slender extremities, atrophic muscles, a plethoric face, dyspnea on effort, ankle edema, diastolic hypertension, and emotional lability. Truncal obesity and cutaneous striae were not seen. High-resolution computed axial tomographic scanning of the thoracic and abdominal organs revealed two small peripheral lung metastases on the right side, a large metastasis on the lower portion of the left kidney, and massive hyperplasia of both adrenal glands. The pituitary fossa was not enlarged. Thyroid and parathyroid sonography showed no suspicious nodules.

The results of laboratory studies performed on admission were as follows: mild neutrophilic leukocytosis (15,000/nl), lymph and eosinopenia, hypokalemia (2.1 mEq/liter), hypochloremia (89 mEq/liter), metabolic alkalosis (pH 7.5), hyperglycemia (BG 370 mg/dl), elevated triglycerides (430 mg/dl), and hypogammaglobulinemia (4.6 g/liter). After endocrinological evaluation, we initiated chemotherapy with Cysplatin and Vepesid. The patient recovered well and was discharged after two chemotherapeutic courses. In December, 1984 the patient died suddenly, most likely as a result of pulmonary embolism. An autopsy was refused.

2.2. Analysis of Plasma Hormones

Corticotropin-releasing factor was extracted from plasma using Sep-Pak® C-18 cartridges (Waters Associates, Königstein, F.R.G.) as described by Maser-Gluth *et al.* (1984). A radioimmunoassay for CRF was performed with antiserum raised to a human CRF–ovine serum albumin complex in rabbits and iodinated Tyr-CRF (NEN, Dreieich, F.R.G.). Human CRF (Bachem, Bubendorf, Switzerland) was used as a standard. The minimum detectable dose of CRF was 7 pg/tube using a final incubation volume of 200 μl.

2.3. Pathology Analysis

The resected tissues of the original tumor and cutaneous metastasis as well as a needle biopsy of the metastasis in the left kidney were evaluated by immunohistochemical methods. Tissue sections of the lung tumor were fixed in 4% formaldehyde, embedded in paraffin, and stained with hematoxylin, PAS (with and without diastasic treatment), Alcian blue, van Gieson's stain, and Grimelius. The histological classification of the tumor was made in accordance with the 1981 WHO guidelines. The avidin–biotin–peroxidase complex method was used for light microscopic immunohistochemistry as follows. Deparaffinated 4-μm tissue sections were incubated in normal goat serum and treated overnight with primary antibodies. Antisera against the following peptides were used: human $ACTH_{1-39}$, human CRF, bombesin, calcitonin, CCK, glucagon, insulin, Leu-enkephalin, neurotensin, serotonin, somatostatin, substance P, VIP (Immunonuclear Inc., Stillwater MN), rat neuron-specific enolase, bovine neuron-specific enolase (Polysciences, Warrington, PA), S-100 (Protein Dakopats, Denmark), GIP (Novo, Bagsvaerd, Denmark), and PP (Milat, Sweden). The tissue sections were incubated for 30 min at room temperature with the biotinylated second antibody. Endogenous peroxidase activity was blocked with methanol/H_2O_2. Samples were then incubated with ABC complex (Vector, Burlingame, CA), and DAB was added to detect immunoreactivity.

3. RESULTS

3.1. Hormone Studies

The following data for plasma hormone levels are presented as observed values followed by normal range in parentheses. Tissue polypeptide antigen, 123 IU/liter (0–100); calcitonin, 1800 pg/ml (0–40); β-endorphin, 200 pmol/liter (3–10); and glucagon, 297 pg/ml (50–250, C-terminal antibody) were elevated. The plasma levels of carcinoembryonic antigen, human β-choriogonadotrophin, bombesin, α-fetoprotein, mid-parathyroid-hormone, gastrin, aldosterone, epinephrine, and norepinephrine were normal. Serum and urine osmolalities were likewise normal.

Diurnal variation of ACTH and cortisol levels were not observed (Table I).

TABLE I. ACTH and Steroid Levels Determined[a]

Dexamethasone suppression test (2 + 8 mg)	Day 1	Day 2	Day 3	
ACTH (pg/ml)	363	303	301	
Cortisol (μg/dl)	63	77	101	
17-OH (mg/24 hr)	45	68	60	
17-KS (mg/24 hr)	54	60	50	
Cortisol levels as a function of time	8 a.m.	12 a.m.	4 p.m.	
ACTH (pg/ml)	242	260	244	
Cortisol (μg/dl)	100	98	89	
Testosterone (ng/dl)	129	127	143	
CRF challenge (100 μg i.v.)	0 Min	15 Min	30 Min	60 Min
ACTH (pg/ml)	289	233	229	N.D.
Cortisol (μg/dl)	118	N.D.	95	89

[a] N.D. indicates not determined. There was no diurnal variation of ACTH and cortisol secretion. High doses of dexamethasone could not suppress ACTH and cortisol secretion. ACTH and cortisol plasma levels decreased after CRF administration.

Oral administration of dexamethasone (2 mg on day 1, 8 mg on day 2) did not suppress the ACTH and cortisol secretion or reduce the excretion of urinary steroids (Table I). Intravenous administration of 100 μg of synthetic CRF_{1-44} resulted in a paradoxical decrease of plasma ACTH and cortisol concentrations (Table I). The insulin hypoglycemia test reduced blood glucose from 91 mg/dl to 24 mg/dl without provoking an increase in ACTH and cortisol. There was no deficiency in other hypophyseal hormones as established by the insulin tolerance, TRH, LHRH, and GRF tests.

Selective blood sampling revealed elevated hormone levels at different locations. We found a CRF peak in the right innominate vein (192 pg/ml), where it is

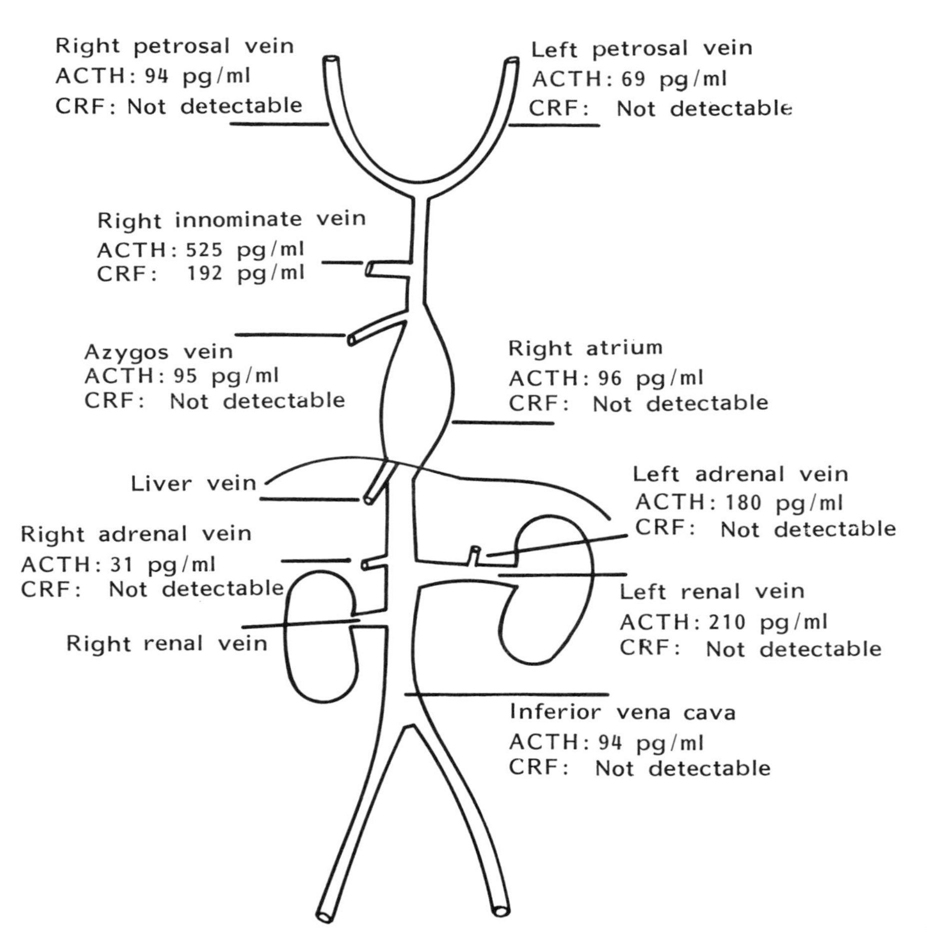

FIGURE 1. Results of the selective blood sampling. Normal range: ACTH, 0–50 pg/ml; CRF, not detectable.

normally not present in measurable levels (Fig. 1). All other peripheral serum concentrations of CRF were undetectably low. The highest ACTH concentration was in the right innominate vein (525 pg/ml, normal range 0–50), supporting the diagnosis from computed tomography of metastasis in the right lung. Furthermore, elevated ACTH levels were observed in the left renal and adrenal veins. Pituitary ACTH secretion was not suppressed, as shown by the elevated concentrations in both petrosal veins. As expected, we found maximal plasma levels of cortisol in both adrenal veins (700 and 800 μg/dl). All other peripheral serum cortisol concentrations were nearly the same (approximately 100 μg/dl, normal range 5–25).

3.2. Pathology Studies

The invasively growing tumor was classified histologically as a low-differentiated adenocarcinoma of the lung. Tumor cells were arranged in solid formations with occasional glandular lumina. The PAS stain revealed intracytoplasmic mucous substances in the tumor cells. Intracellular vacuoles were visualized in a few areas with the Alcian blue stain. The Grimelius stain showed an argyrophil reaction in numerous tumor cell complexes. Many cells of the original tumor and its metastases were immunochemically positive for various peptides, as were the solid parts of the tumor (Table II). Neurotensin (Fig. 2), calcitonin (Fig. 3), and ACTH were detected frequently. On a smaller scale, immunoreactive ACTH (Fig. 4), bombesin, and porcine CRF (Fig. 5) were found in the cytoplasm of a few cells. Positive immunoreactivity to neuron-specific enolase was observed in numerous tumor cells. S-100 protein was detected only in nerve slices and not in the surrounding tumor cells. No serotonin immunoreactivity was observed.

TABLE II. Immunostaining Results on Tumor Tissues

Peptide	Staining intensity[a]
Neurotensin	+ + +
ACTH	+ + +
Calcitonin	+ + +
NSE	+ +
CRF	+
Bombesin	+
Serotonin	—
Somatostatin	—
Leu-Enkephalin	—
GIP	—
VIP	—

[a] + + +, almost every cell positive; + +, many cells positive; +, few cells positive; —, no staining.

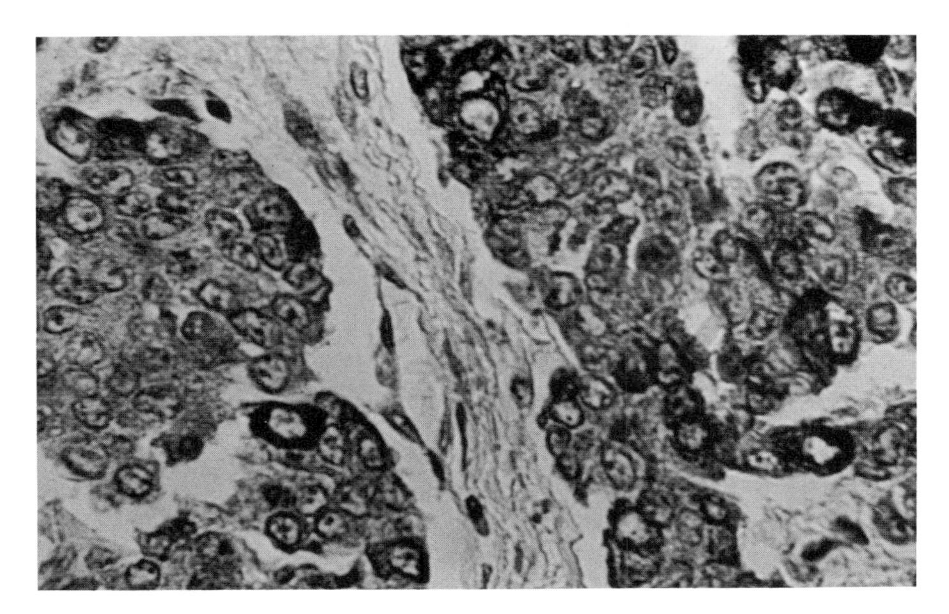

FIGURE 2. Hormone-active tumor of the lung. The immunoperoxidase technique reveals intense positive staining for neurotensin in the tumor cells. (Original magnification ×160.)

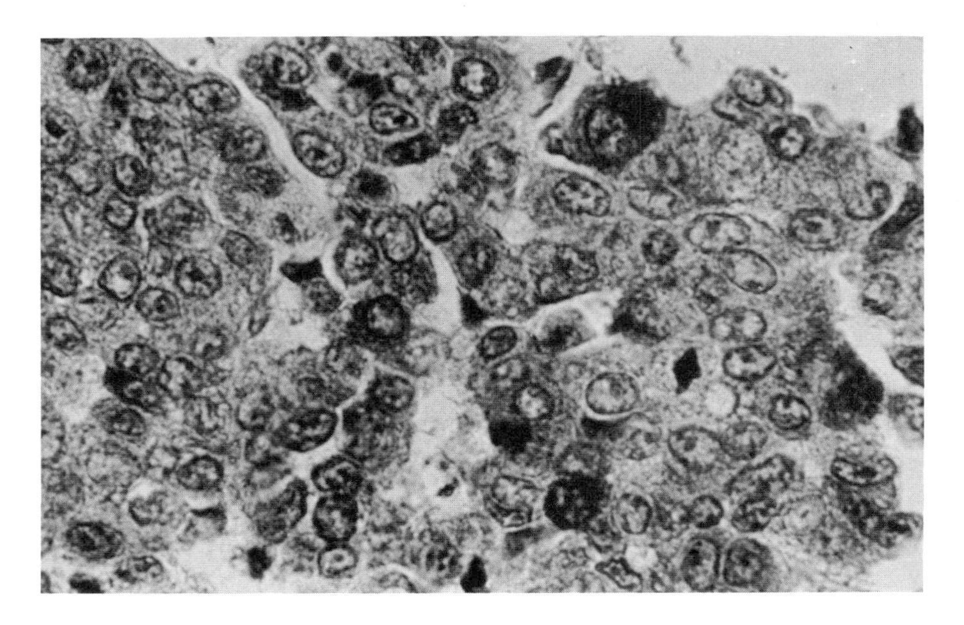

FIGURE 3. Hormone-active tumor of the lung. Numerous cells show positive immunostaining for calcitonin. (Original magnification ×200.)

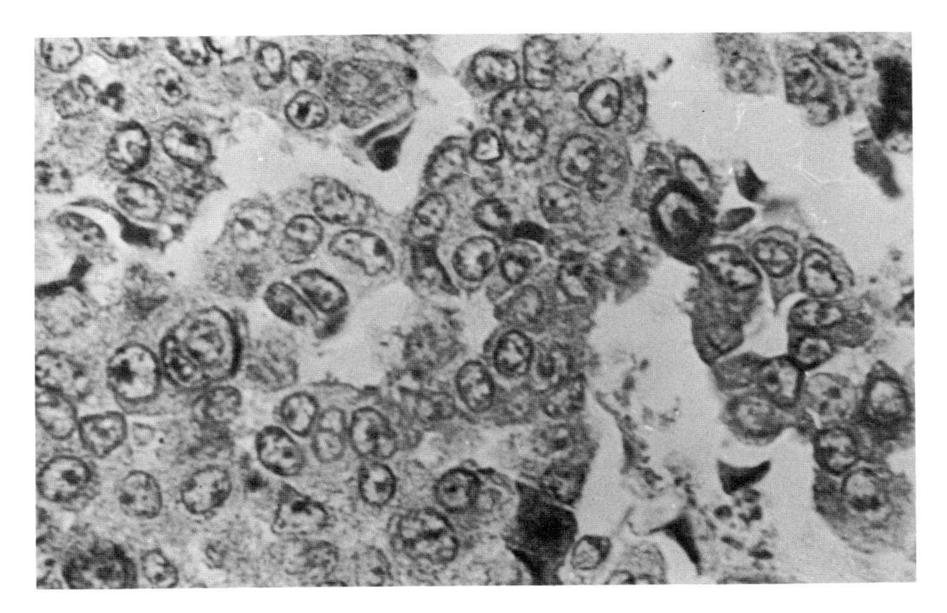

FIGURE 4. Hormone-active tumor of the lung. Numerous cells show positive immunostaining for ACTH using the immunoperoxidase method for adrenocorticotropin. (Original magnification $\times 200$.)

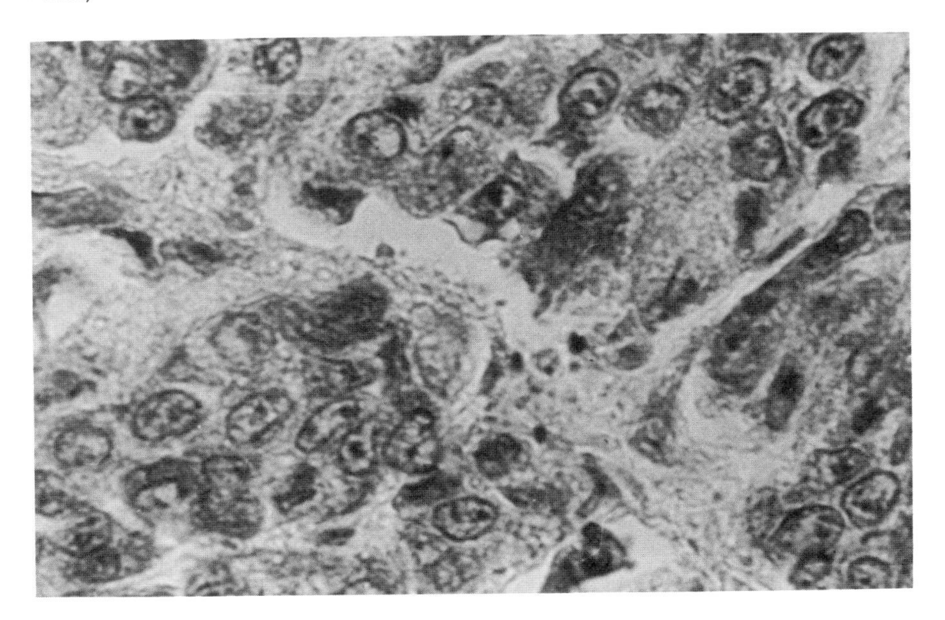

FIGURE 5. Hormone-active tumor of the lung. A few cells show positive immunostaining for CRF using the immunoperoxidase technique. (Original magnification $\times 320$.)

4. DISCUSSION

A CRF-like activity in tumors was described over a decade ago (Upton and Amatruda, 1971). Carey *et al.* (1984) and Asa *et al.* (1984) described CRF in postmortem tumor material derived from patients with ectopic Cushing's syndrome. In the case presented here, both production and secretion of CRF and ACTH were detected *in situ* in a patient with a paraneoplastic syndrome.

It is well documented that pluripotent tumors produce a variety of hormones. The cell population of the lung tumor and metastases described here was morphologically heterogeneous. The carcinoma and its metastases produced and secreted ACTH, CRF, calcitonin, bombesin, and neurotensin. The presence of these hormones and neuron-specific enolase suggests that this was a neuroendocrine tumor.

The rapid disease progression and high, noncircadian values of cortisol and ACTH observed during high doses of dexamethasone suggest a paraneoplastic genesis of the Cushing's syndrome symptoms. The paradoxical decrease of ACTH and cortisol after CRF administration are also indicative of Cushing's syndrome, since variable responses were observed in these cases. In central Cushing's disease, CRF administration would have provoked an increase in ACTH and cortisol (Chrousos *et al.*, 1984). Primary hypercortisolism caused by adrenal adenoma or carcinoma is usually accompanied by more pronounced elevation of urinary steroids and suppressed ACTH.

Because both CRF and ACTH were detected in the tumor masses, it is important to determine which hormone caused the Cushing's syndrome. Selective blood sampling showed CRF in the right innominate vein, which drains the right lung. The hormone was undetectable in venous samples from other sites, possibly because of lowered secretion by cells of the renal metastasis. ACTH concentrations were elevated at all sampling locations of the venous system except for the right renal vein. The slightly higher ACTH concentration in the petrosal veins could be explained by a constant ectopic CRF stimulation of the pituitary gland, although the difference in levels was not very pronounced and may be caused by ectopic ACTH production by the tumor itself. It is well known that tumor secretion varies considerably and may result in different hormone concentrations during the time course of blood sampling. The immunocytochemical results, which show a major production of ACTH and only minor production of CRF, further support the possibility that ACTH plays a major pathogenic role in the development of Cushing's syndrome symptoms. However, the possibility that ACTH-producing cells developed under paracrine stimulation of neighboring CRF-producing cells cannot be excluded. In the case presented here, CRF may have had an additional or potentiating effect on the symptoms, generating an interesting variation of ectopic Cushing's syndrome.

5. SUMMARY

In this study, we report on a patient with ectopic production and secretion of CRF from a metastatic, nondifferentiated adenocarcinoma of the lung. Selective blood sampling revealed elevated hormone levels at different locations. The CRF levels were elevated in the right innominate vein but not detectable at other peripheral sites. Maximal ACTH concentrations were found in the right innominate vein and left renal vein. Plasma concentrations of ACTH, calcitonin, cortisol, β-endorphin, and glucagon were elevated. Intravenous administration of 100 μg of CRF resulted in a decrease in plasma ACTH and cortisol concentrations. Immunohistochemical analysis of tissue derived from the original tumor and its metastases revealed ACTH, CRF, calcitonin, neurotensin, and neuron-specific enolase.

REFERENCES

Asa, S., Kovacs, K., Tindall, G. T., Barrow, D. L., Horvath, E., and Vecsei, P., 1984, Cushing's disease associated with an intrasellar gangliocytoma producing corticotropin-releasing factor, *Ann. Intern. Med.* **101**:789–793.

Carey, R. M., Varma, S. K., Drake, C., Thorner, M. O., Kovacs, K., Rivier, J., and Vale, W., 1984, Ectopic secretion of corticotropin-releasing factor as a cause of Cushing's syndrome: A clinical, morphologic and biochemical study, *N. Engl. J. Med.* **311**:13–20.

Chrousos, G. P., Shulte, H. M., Oldfield, E. H., Gold, P. W., Cutler, G. B., and Loriaux, D. L., 1984, The corticotropin releasing factor stimulation test, *N. Engl. J. Med.* **310**:622–626.

Hashimoto, K., Takahara, J., Ogawa, N., Yunoki, S., Ofugi, T., Arata, A., Kanda, S., and Terada, K., 1980, Adrenocorticotropin, β-lipotropin, β-endorphin and corticotropin releasing factor-like activity in an adrenocorticotropin producing nephroblastoma, *J. Clin. Endocrinol. Metab.* **50**:461–465.

Maser-Gluth, C., Toygar, A., and Vecsei, P., 1984, Time course of plasma corticosterone, 18-hydroxy-corticosterone and aldosterone concentration following CRF administration in the rat: A fact of corticotrophin inhibition, *Life Sci.* **35**:879–884.

Nicholosn, W., DeCherney, G., Jackson, R., DeBold, R., Uderman, H., Alexander, A., Rivier, J., Vale, W., and Orth, D., 1983, Plasma distribution, disappearance half-time, metabolic clearance rate and degradation of synthetic ovine corticotropin releasing factor in man, *J. Clin. Endocrinol. Metab.* **57**:1263–1269.

Upton, G. V., and Amatruda, T., 1971, Evidence for the presence of tumor peptides with corticotropin releasing factor-like activity in the ectopic ACTH syndrome, *N. Engl. J. Med.* **285**:419–423.

31

Vasoactive Intestinal Peptide Increases Tyrosine Hydroxylase Activity in Cultures of Normal and Neoplastic Chromaffin Cells

ARTHUR S. TISCHLER, ROBERT L. PERLMAN, JAMES E. JUMBLATT, DONNA COSTOPOULOS, and JOEL HORWITZ

1. INTRODUCTION

The adrenal medulla is innervated by cholinergic preganglionic sympathetic nerve fibers from the splanchnic nerve. These fibers synapse on chromaffin cells and stimulate catecholamine release. They also participate in both short-term and long-term regulation of catecholamine production through their effects on tyrosine hydroxylase, the major regulatory enzyme in catecholamine synthesis. These effects consist of rapid increases in tyrosine hydroxylase activity produced by alterations of the enzyme protein and more delayed increases produced by synthesis of new enzyme protein. They are known respectively as "activation" (Masserano and Weiner, 1979) and "transsynaptic induction" (Guidotti and Costa, 1977) of tyrosine hydroxylase, and both can be initiated by synaptic release of acetylcholine.

In addition to cholinergic innervation, recent studies have demonstrated nerve fibers in the adrenal medulla that contain regulatory peptides (Carmichael, 1983). In the adrenal medulla of both rats (Hökfelt *et al.*, 1981) and humans (Linnoila *et al.*, 1980), some nerve fibers have been shown to contain vasoactive-intestinal-peptide-like immunoreactivity (VIP-IR). In contrast to these nerve fibers, little or

ARTHUR S. TISCHLER and DONNA COSTOPOULOS • Department of Pathology, Tufts University School of Medicine, Boston, Massachusetts 02111. JAMES E. JUMBLATT • Department of Biochemistry, Tufts University School of Medicine, Boston, Massachusetts 02111. ROBERT L. PERLMAN and JOEL HORWITZ • Department of Physiology and Biophysics, University of Illinois College of Medicine, Chicago, Illinois 60680.

no VIP-IR has been demonstrated *in vivo* in normal chromaffin cells in any mammals, although chromaffin cells of frogs do contain VIP-IR (Leboulenger *et al.*, 1983). However, normal human (Tischler *et al.*, 1985) and bovine (Eiden *et al.*, 1982, 1983) chromaffin cells in cell culture produce abundant VIP. Vasoactive intestinal peptide is also present in neoplastic human chromaffin cells from pheochromocytomas both *in vivo* and *in vitro*. Evidence from our laboratories suggests that one of the roles of VIP in the adrenal medulla may be to maintain catecholamine stores by activation of tyrosine hydroxylase in chromaffin cells. This finding suggests that VIP may also function as an autocrine regulator of catecholamine production in pheochromocytomas.

2. THE DISTRIBUTION OF VIP-LIKE IMMUNOREACTIVITY IN THE ADRENAL MEDULLA

Current knowledge of the precise distribution of VIP-IR-containing nerve fibers is based primarily on two immunohistochemical studies, one dealing with rat (Hökfelt *et al.*, 1981) and the other with human (Linnoila *et al.*, 1980) tissue. In rat adrenals, nerve fibers that contain VIP-IR persist after splanchnic denervation and appear to originate from an endogenous neuron system in the adrenal medulla. They are distributed predominantly in the subcapsular region of the adrenal cortex, often around blood vessels, and to a lesser extent in a patchy fashion in the adrenal medulla itself. The distribution of VIR-IR in human adrenals appears to differ from that in rats. Numerous axonal varicosities that stain for VIP-IR are observed in close proximity to most of the chromaffin cells in the adrenal medulla. Staining is not observed in adrenal medullary neurons or in the adrenal cortex.

In comparing the above studies, it must be borne in mind that some of the differences in distribution of VIP-IR may have resulted from differences in methods of fixation and staining. The study of rat adrenals employed an immunofluorescence method with cryostat sections of tissue fixed by formaldehyde perfusion, whereas the study of human adrenals employed an immunoperoxidase method with freeze-dried, paraffin-embedded sections of tissue fixed with formaldehyde vapor. It is important to note that chromaffin cells themselves did not stain for VIP-IR in either study. However, an earlier report (Bryant *et al.*, 1976) illustrated a small group of cells that stained for VIP-IR in the human adrenal medulla. The study was conducted with freeze-dried, paraffin-embedded sections of tissue fixed with diethylpyrocarbonate vapor. Immunoreactive nerve fibers were not commented on. However, it is likely that the immunoreactive cells were adrenal medullary neurons rather than chromaffin cells.

Immunohistochemical studies of the localization of VIP-IR in the adrenal glands of other mammals have not been reported. Little or no VIP-IR is detectable by radioimmunoassay in extracts of bovine adrenal medulla (Eiden *et al.*, 1982), but a small amount of VIP-IR in nerve endings might be overlooked by this method.

Immunohistochemical studies of bovine adrenal glands and confirmatory studies of rat and human adrenal glands are needed.

In contrast to the mammalian adrenal medulla, almost all of the chromaffin cells in sections of perfusion-fixed adrenal medulla of the frog *Rana ribunda* stain for VIP-IR, and staining is not observed in nerve fibers (Leboulanger *et al.*, 1983).

Vasoactive-intestinal-peptide-like immunoreactivity is demonstrable in a high proportion of human pheochromocytomas, by both radioimmunoassay (Said, 1976; Eiden *et al.*, 1982; Gozes *et al.*, 1983; Tischler *et al.*, 1984) and immunohistochemical (Hassoun *et al.*, 1984) techniques. In a series of 14 such tumors, VIP-IR was demonstrated immunohistochemically in six (43%). In these tumors, VIP-IR was demonstrated in typical chromaffin cells without morphological evidence of neuronal differentiation (Hassoun *et al.*, 1984). In another study of tumors with mixed chromaffin-cell and neuronal features, however, the presence of VIP-IR did correlate with neuronal morphology (Mendelsohn *et al.*, 1979). Moreover, in cultures of human pheochromocytoma cells, factors that favor neuronal differentiation also tend to favor VIP production (Tischler *et al.*, 1984). It thus appears that VIP production and morphological evidence of neuronal differentiation of pheochromocytomas cells are correlated under some circumstances but that the two phenomena may be dissociated (Tischler *et al.*, 1984).

The commitment of neural crest cells in the adrenal medulla to become either neurons or chromaffin cells appears to occur late in development, since the phenotype of chromaffin cells remains plastic after birth and appears, at least to some extent, to be determined by environmental influences. Cultured normal chromaffin cells of neonatal (Unsicker *et al.*, 1978) and adult (Doupe and Patterson, 1983) rats, adult bovines (Livett *et al.*, 1978), and adult humans (Tischler *et al.*, 1980) are known to acquire morphological characteristics of neurons in culture. Further, normal human chromaffin cells resemble their neoplastic counterparts in pheochromocytomas in that conditions that favor the transition from chromaffin cell to neuronal phenotype also favor production of VIP (Tischler *et al.*, 1985). The production of VIP by pheochromocytoma cells without neuronal morphology may represent a partial switch to a neuronal phenotype.

The VIP production by pheochromocytoma cells at present appears to be confined to human tumors. Although initial evidence suggested that PC12 rat pheochromocytoma cells might produce VIP similarly to their human counterparts, this has not been confirmed (Y. C. Lee, S. R. Bloom, and A. S. Tischler, unpublished data).

3. POSSIBLE ROLES OF VIP IN THE ADRENAL MEDULLA

Because of their distribution within the gland, VIP-IR-containing nerve fibers in the rat adrenal have been postulated to play a role in controlling the flow of blood from the peripheral toward the central parts of the gland and/or in regulating secretion of corticosteroids (Hökfelt *et al.*, 1981). The VIP-IR-containing nerve fibers in the human adrenal, on the other hand, have been postulated to modulate the functions either of chromaffin cells or of splanchnic nerve endings themselves (Linnoila *et al.*, 1980).

Evidence for an effect on blood flow is derived from studies of salivary glands, where VIP which is colocalized with acetylcholine in nerve terminals has been shown to play a role in regulating both secretion and blood flow (Lundberg *et al.*, 1980). Evidence for an effect on the adrenal cortex exists in the form of data demonstrating that VIP increases adenylate cyclase activity and corticosteroid synthesis in mouse adrenal cortical carcinoma cells in cell cultures (Morera *et al.*, 1979; Kowal, 1982); VIP also increases corticosteroid secretion from perfused frog adrenal glands (Leboulanger *et al.*, 1983). It is of interest that in the frog adrenal there is a more intimate intermingling of chromaffin and steroid-producing cells than in mammals (Leboulanger *et al.*, 1983), so that the VIP that is stored in chromaffin cells themselves in this species could reach the steroid-producing cells without the need to be transported via neuronal processes. Our own findings (Tischler *et al.*, 1985) constitute the first evidence for a direct effect of VIP on chromaffin cells.

4. THE EFFECTS OF VIP ON TYROSINE HYDROXYLASE ACTIVITY AND cAMP IN NORMAL AND NEOPLASTIC CHROMAFFIN CELLS *IN VITRO*

Ip *et al.* (1982) recently reported that VIP increases tyrosine hydroxylase activity in the superior cervical ganglion (SCG) of the rat *in vitro*. The SCG is a sympathetic ganglion developmentally homologous to the adrenal medulla, and, like the adrenal medulla, it is innervated by both cholinergic and VIP-containing nerve fibers (Hökfelt *et al.*, 1977). These parallelisms prompted us to study the effects of VIP on tyrosine hydroxylase activity in cultures of normal and neoplastic chromaffin cells.

Normal rat chromaffin cells were obtained from 14-week-old male Long–Evans rats. The animals were killed by exposure to dry ice vapor, and the adrenal glands were removed aseptically. Medullary tissue was dissected away from the cortex under a dissecting microscope and dissociated in collagenase followed by trypsin plus collagenase, as previously reported (Tischler *et al.*, 1982b). Pooled chromaffin cells from four animals were plated in collagen-coated (Bornstein, 1958) 16-mm microwells. Cultures were maintained for 4 days and then assayed for tyrosine hydroxylase activity. The approximate number of chromaffin cells per well at the time of assay was estimated from counts of fluorescent cells on 2–3% of the surface area of representative wells treated with glyoxylic acid to demonstrate catecholamine storage (DeLaTorre and Surgeon, 1976). Neoplastic rat chromaffin cells were from the PC12 pheochromocytoma cell line (Greene and Tischler 1976, 1982). PC12 cells are known to respond to treatment with nerve growth factor (NGF) both by

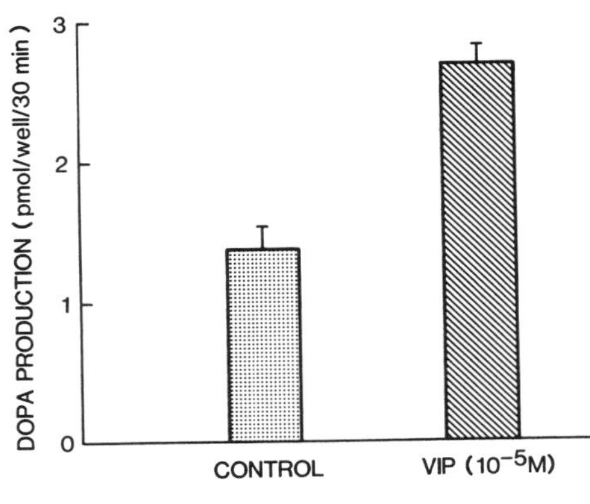

FIGURE 1. Effects of VIP (10^{-5} M) on dopa production by normal rat chromaffin cells. Cultures were maintained for 3 days in RPMI 1640 medium with 10% horse serum, 5% fetal bovine serum, penicillin, and streptomycin. Intact cultures were then incubated at 37°C for 30 min in Krebs–Ringer HEPES buffer, pH 7.3, containing 100 μM tyrosine and 100 μM brocresine, and accumulated dopa was measured by liquid chromatography with electrochemical detection (Erny *et al.*, 1981). Synthetic VIP (Sigma) was added from a 10^{-3} M stock solution at the start of the incubation. Values represent mean ± S.E.M. of four culture wells, each containing approximately 2000 chromaffin cells at the time of assay.

forming neurites (Greene and Tischler, 1976, 1982) and by increasing their number of receptors for certain ligands (Inoue and Hatanaka, 1982; Jumblatt and Tischler, 1982). In some experiments, PC12 cells were therefore treated with 2.5 S mouse salivary gland NGF for 14 days prior to assay. All cultures were maintained in a water-saturated atmosphere of 95% air and 5% CO_2 at 37°C with medium changes three times per week.

Tyrosine hydroxylase activity was determined by measuring the accumulation of 3,4-dihydroxyphenylalanine (dopa) in intact cells incubated for 30 min with 100 μM L-tyrosine in the presence of brocresine, an inhibitor of L-aromatic amino acid decarboxylase (Erny *et al.*, 1981). VIP caused two- to threefold increases in dopa production by normal rat chromaffin cells (Fig. 1) and by PC12 cells (Fig. 2). The effect of VIP on PC12 cells was detectable at 0.1 μM and increased markedly between 1 and 10 μM. This dose dependence is similar to that in the SCG (Ip *et*

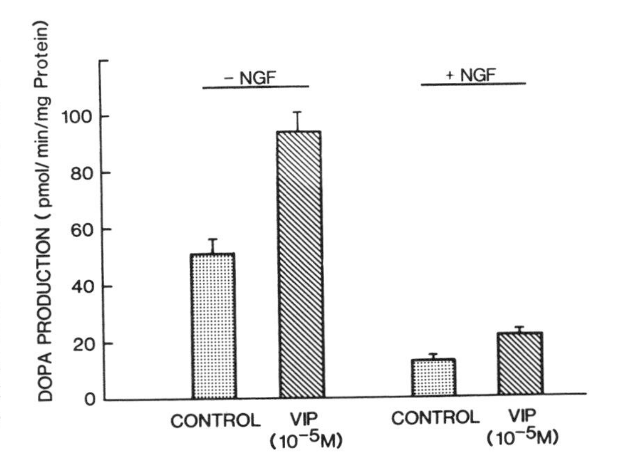

FIGURE 2. Effects of VIP (10^{-5} M) on dopa production by PC12 cells treated or untreated with NGF for 14 days *in vitro.* Values represent mean ± S.E.M. of four culture wells, each containing approximately 100 μg of cell protein. As previously reported, NGF appears to cause a large decrease in tyrosine hydroxylase activity in PC12 cells when expressed as enzyme activity per milligram protein. This apparent decrease results from the fact that NGF greatly increases the amount of protein per cell but has little effect on the amount of TH per cell (Greene and Tischler, 1976).

al., 1982). VIP caused similar increases in TH activity in PC12 cells treated or untreated with NGF (Fig. 2).

To determine whether VIP stimulates increases in cAMP in PC12 cells, the rate of conversion of [^{3}H]adenosine to [^{3}H]cAMP was measured in cultures incubated with [^{3}H]adenosine (Salomon *et al.*, 1974) in the presence or absence of VIP. The assays were performed in the presence of isobutylmethylxanthine (IBMX), a phosphodiesterase inhibitor. VIP caused a dose-dependent increase in [^{3}H]cAMP similar to that for tyrosine hydroxylase activation (Fig. 3). Since phosphodiesterase activity in all cultures was inhibited by IBMX, the increase in [^{3}H]cAMP appears to have been caused by increased synthesis rather than decreased degradation. Comparable dose–response curves have been reported for effects of VIP on adenylate cyclase activity in a number of different types of tissue *in vitro* (Robberecht *et al.*, 1982). VIP has also been shown to increase the content of cAMP in PC12 cells (Tischler *et al.*, 1985).

Other agents that increase intracellular cAMP also increase tyrosine hydroxylase activity in PC12 cells (Tischler *et al.*, 1983; Erny and Wagner, 1984). Relatively low concentrations of cholera toxin (CT), an activator of adenylate cyclase, produce increases in tyrosine hydroxylase activity similar in magnitude to those

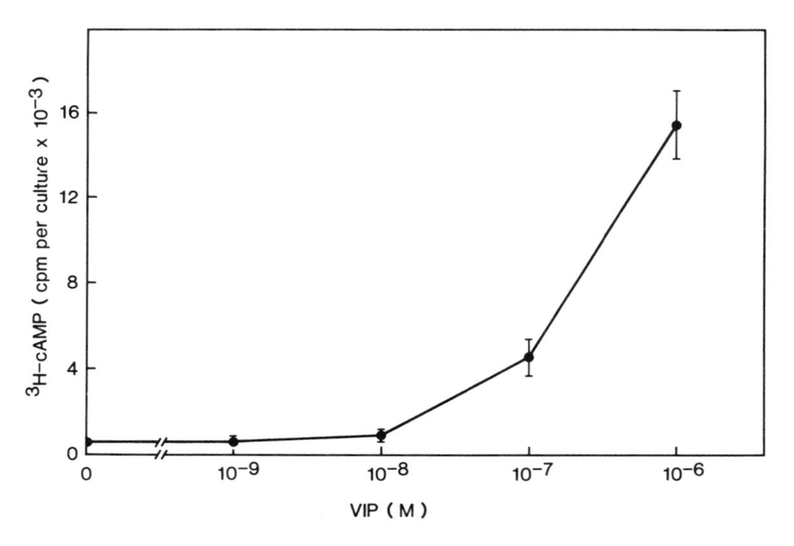

FIGURE 3. Stimulation of cAMP accumulation by VIP. PC12 cell cultures treated with NGF (100 ng/ml) for 7 days and labeled for 60 min in RPMI medium containing 1 μCi/ml [^{3}H]adenosine, washed, and treated for an additional 10 min with VIP at the indicated concentrations in medium containing 0.5 mM isobutylmethylxanthine (IBMX). All steps were performed at 37°C. Radiolabeled adenine nucleotides were extracted with 5% trichloracetic acid, and [^{3}H]cAMP content was determined by Dowex and alumina column chromatography (Salomon, 1974). Data are expressed as cpm [^{3}H]cAMP/culture dish. Values represent mean ± S.E.M. of four cultures.

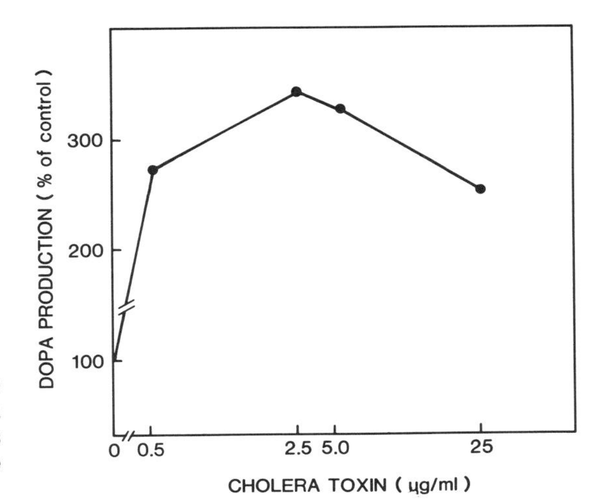

FIGURE 4. Dose–response curve for effects of cholera toxin on dopa production by PC12 cells. Values represent mean ± S.E.M. of three culture dishes.

produced by VIP, but higher concentrations of CT produce greater increases than VIP (Fig. 4). These findings suggest that tyrosine hydroxylase activity might be regulated by additive effects of multiple agents that act via cAMP. One such agent, adenosine, is probably derived from adenosine triphosphate released from PC12 cell secretory granules (Erny and Wagner, 1984), and may be responsible for plating density-dependent increases in tyrosine hydroxylase activity (Lucas *et al.*, 1979). We have observed that the basal activity of tyrosine hydroxylase in PC12 cell cultures is lowered in the presence of adenosine deaminase (Type 6, Sigma, 0.8 units/ml). This finding suggests that tyrosine hydroxylase in PC12 cell cultures is partially activated by released endogenous adenosine, and that adenosine deaminase prevents this activation by removal of released endogenous adenosine. As a result of the lowered basal activity of tyrosine hydroxylase, we have observed that VIP in the presence of adenosine deaminase produces relative increases in tyrosine hydroxylase activity that are approximately 30% greater than those in the absence of the adenosine deaminase (unpublished data). The increases in tyrosine hydroxylase activity produced both by VIP and by CT (Tischler *et al.*, 1983) occur within 30 min, suggesting that they are the result of enzyme activation rather than induction. The CT-mediated increases persist for at least 48 hr if CT is left in the culture medium (Fig. 4).

The activation of tyrosine hydroxylase by VIP in the superior cervical ganglion (Ip *et al.*, 1982) appears to involve an increase in ganglionic cAMP (Volle and Patterson, 1982), the activation of a cAMP-dependent protein kinase, and the phosphorylation of tyrosine hydroxylase by this kinase (Cahill and Perlman, 1984). A similar sequence of events may account for the activation of tyrosine hydroxylase in chromaffin cells by VIP.

5. EFFECTS OF OTHER NEUROPEPTIDES ON TYROSINE HYDROXYLASE ACTIVITY

Vasoactive intestinal peptide is a member of a family of peptides that also includes secretin, glucagon, PHI, and GRF (Miller, 1984). One of these, PHI, is synthesized with VIP from the same precursor (Miller, 1984). Because other members of this family are known to mimic the effects of VIP to different degrees in a variety of tissues (Miller, 1984), we have tested the abilities of all of the above peptides to increase tyrosine hydroxylase activity in PC12 cells. Synthetic porcine PHI-27, glucagon, GRF^{1-40} (all from Peninsula Laboratories), and secretin (Sigma Chemical Co.) were tested at concentrations from 10^{-9} to 10^{-5} M. Of these, only PHI and secretin produced detectable effects. The dose–response curve for PHI was virtually indistinguishable from that for VIP (Fig. 5). Approximately equal increases in dopa production occurred in the presence of 10^{-5} M VIP, PHI, or secretin. Ip *et al.* (1984) reported that PHI and secretin also increases tyrosine hydroylase activity in the SCG. PC12 cells might therefore be a useful model for studying mechanisms by which neuropeptides related to VIP affect sympathetic ganglia as well as the adrenal medulla. In contrast to our findings and to those of Ip *et al.* (1982), functional receptors for glucagon have been reported in one human pheochromocytoma (Levey *et al.*, 1975). It remains to be determined whether the

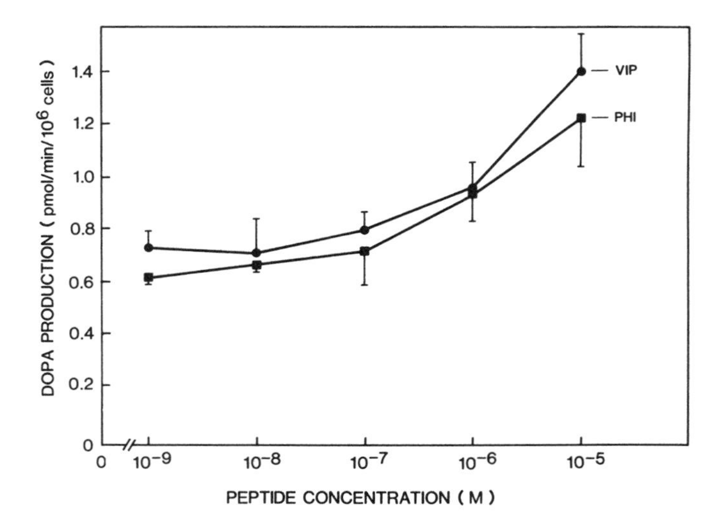

FIGURE 5. Dose–response curves for effects of VIP and PHI on dopa production in suspensions of PC12 cells without NGF. Cells were scraped from the culture dishes, rinsed, suspended in KRH buffer, counted, and assayed in microcentrifuge tubes similarly to the attached cells in Fig. 2. Approximately 100 μg of cell protein are present in 10^6 PC12 cells. Values represent mean $\pm$ S.E.M. of three replicate tubes.

apparent differences in glucagon responsiveness reflect species differences, variability between individual tumors, or other factors.

PC12 cells are known to produce at least two endogenous neuropeptides, neurotensin (Tischler *et al.*, 1982a) and neuropeptide Y (Allen *et al.*, 1984). We have therefore tested both of these agents (10^{-5} M, Sigma Chemical Co.) as possible autocrine regulators of tyrosine hydroxylase. Neither peptide produced any changes in PC12 cell tyrosine hydroxylase activity at the concentration tested.

6. PHYSIOLOGICAL AND DEVELOPMENTAL IMPLICATIONS

Our findings suggest that one role of VIP or closely related peptides in the adrenal medulla is to increase acutely tyrosine hydroxylase activity, most likely by an activation mechanism that involves phosphorylation of tyrosine hydroxylase by a cAMP-dependent protein kinase. The possibility of an additional, longer-term role in transsynaptic induction of tyrosine hydroxylase has not been ruled out. This possibility is consistent with the suggestion that cAMP might mediate transsynaptic induction as well as activation of tyrosine hydroxylase (Guidotti and Costa, 1977). A direct effect of VIP on the adrenal medulla is not inconsistent with additional effects on the cortex or on blood flow. In fact, effects at all three loci would provide a means for an integrated response of cortex and medulla to stress or other stimuli. In the rat, for example, VIP that is released from nerve endings in the cortex could both stimulate corticosteroid secretion and be carried into the medulla with the resultant steroid-enriched blood. The steroids themselves might then participate in long-term maintenance of tyrosine hydroxylase activity (Tischler *et al.*, 1982). In addition, VIP might act synergistically with acetylcholine to increase tyrosine hydroxylase activity, as reported by Ip *et al.* (1985) in the SCG. Splanchnic nerve stimulation has been reported to cause a small increase in adrenal blood flow (Wright, 1963), and this increase might well be mediated by VIP.

It has been noted by Linnoila *et al.* (1980) that many of the effects of VIP administration *in vivo,* such as increased glycogenolysis, bronchial relaxation, and vasodilation, resemble effects of epinephrine and might in fact be mediated by epinephrine. We have found that VIP does not itself stimulate catecholamine release from PC12 cells but that it facilitates depolarization-mediated release (J. E. Jumblatt, unpublished data). Further, both the dose dependence and time courses for this effect parallel those for cAMP accumulation. Similar enhancement of stimulated release without effects on basal release occurs when cAMP levels are increased by a direct activator of adenylate cyclase (Rabe *et al.*, 1982; Baizer and Weiner, 1985).

In addition to the adrenal medulla and sympathetic ganglia, VIP-containing nerve fibers are present in the carotid body, a developmentally related catecholamine-storing extraadrenal paraganglion associated with the parasympathetic rather than sympathetic nervous system (Lundberg *et al.*, 1979; Wharton *et al.*, 1980). Like the adrenal medulla, the carotid body receives predominantly cholinergic innervation. Our findings and those of Ip *et al.* (1984, 1985) suggest that a general role of VIPergic innervation in catecholamine-producing neural and endocrine tissues might be to act cooperatively with acetylcholine (Ip *et al.*, 1985), to activate tyrosine hydroxylase, and to replenish catecholamine stores that are depleted after cholinergic stimulation. One advantage of such an effect of VIP might be longer persistence of VIP than of acetylcholine in the synaptic region, since acetylcholine is rapidly degraded by acetylcholinesterase. Further, autocrine regulation of tyrosine hydroxylase activity by VIP in human pheochromocytomas might be a clinically important effect contributing to the overproduction of catecholamines by these tumors.

ACKNOWLEDGMENTS. Supported by NIH grants CA 27808 and HL 29025. The authors thank Ms. S. Tsymbalov for technical assistance and Mrs. Carol Ostrum for secretarial assistance.

REFERENCES

Allen, J. M., Tischler, A. S., Lee, Y. C., and Bloom, S. R., 1984, Neuropeptide Y (NPY) in PC12 pheochromocytoma cultures: Responses to dexamethasone and nerve growth factor, *Neurosci. Lett.* **46:**291–296.

Baizer, L., and Weiner, N., 1985, Regulation of dopamine release from PC12 pheochromocytoma cells during stimulation with elevated potassium or carbachol, *J. Neurochem.* **44:**495–501.

Bornstein, M. B., 1958, Reconstituted rat-tail collagen used as a substrate for tissue cultures on coverslips in Maximow slides and roller tubes, *Lab. Invest.* **7:**134–137.

Bryant, M. G., Bloom, S. R., Polak, J. M., Albuquerque, R. H., Madlin, I., and Pearse, A. G. E., 1976, Possible role for vasoactive intestinal peptide as gastrointestinal hormone and neurotransmitter substance, *Lancet* **1:**991–993.

Cahill, A. L., and Perlman, R. L., 1984, Phosphorylation of tyrosine hydroxylase in the superior cervical ganglion, *Biochim. Biophys. Acta* **805:**217–226.

Carmichael, S. W., 1983, *The Adrenal Medulla,* Vol. 3, Eden Press, Quebec.

DeLaTorre, J. C., and Surgeon, J. W., 1976, A methodological approach to rapid and sensitive monoamine histofluorescence using a modified glyoxylic acid technique: The SPG method, *Histochemistry* **49:**81–93.

Doupe, A. S., and Patterson, P., 1983, A neurite-promoting activity present in heart cell conditioned medium enhances the NGF-induced conversion of chromaffin cells to neurons, *Soc. Neurosci. Abstr.* **9:**263.

Eiden, L. E., Giraud, P., Hotchkiss, A., and Brownstein, M. J., 1982, Enkephalins and VIP in human pheochromocytomas and bovine adrenal chromaffin cells, in: *Regulatory Peptides: From Molecular Biology to Function* (E. Costa and M. Trabucci, eds.), Raven Press, New York, pp. 387–395.

Eiden, L. E., Eskay, R. L., Scott, J., Pollard, H., and Hotchkiss, A. J., 1983, Primary cultures of bovine chromaffin cells synthesize and secrete vasoactive intestinal peptide (VIP), *Life Sci.* **33:**687–693.

Erny, R., and Wagner, J. A., 1984, Adenosine-dependent activation of tyrosine hydroxylase is defective in adenosine kinase-deficient PC12 cells, *Proc. Natl. Acad. Sci. U.S.A.* **81:**4974–4978.

Erny, R. E., Berezo, M. W., and Perlman, R. L., 1981, Activation of tyrosine 3-monooxygenase in pheochromocytoma cells by adenosine, *J. Biol. Chem.* **256:**1335–1339.

Gozes, I., O'Connor, D. T., and Bloom, F. E., 1983, A possible high molecular weight precursor to vasoactive intestinal polypeptide sequestered into pheochromocytoma chromaffin granules, *Regul. Peptides* **6:**111–119.

Greene, L. A., and Tischler, A. S., 1976, Establishment of a noradrenergic clonal line of rat adrenal pheochromocytoma cells which respond to nerve growth factor, *Proc. Natl. Acad. Sci. U.S.A.* **73**:2424–2428.

Greene, L. A., and Tischler, A. S., 1982, PC12 pheochromocytoma cultures in neurobiological research, in: *Advances in Cellular Neurobiology,* Vol. 3 (S. Fedoroff and L. Hertz, eds.), Academic Press, New York, pp. 373–415.

Guidotti, A., and Costa, E., 1977, Trans-synaptic regulation of tyrosine 3-monooxygenase biosynthesis in rat adrenal medulla, *Biochem. Pharmacol.* **26**:817–823.

Hassoun, J., Manges, G., Girand, P., Henry, J. F., Charpin, C., Payan, H., and Toga, M., 1984, Immunohistochemical study of pheochromocytomas. An investigation of methionine-enkephalin, vasoactive intestinal peptide, somatostatin, corticotropin, β-endorphin, and calcitonin in 16 tumors, *Am. J. Pathol.* **114**:56–63.

Hökfelt, T., Elvin, L.-G., Schultzberg, M., Fuxe, K., Said, S. I., Mutt, V., and Goldstein, M., 1977, Immunohistochemical evidence of vasoactive intestinal peptide-containing neurons and nerve-fibers in sympathetic ganglia, *Neuroscience* **2**:885–896.

Hökfelt, T., Lundberg, J. M., Schultzberg, M., and Fahrenkrug, J., 1981, Immunohistochemical evidence for a local VIPergic neuron system in the adrenal gland of the rat, *Acta Physiol. Scand.* **113**:575–576.

Inoue, N., and Hatanaka, H., 1982, Nerve growth factor induces specific enkephalin binding sites in a nerve cell line, *J. Biol. Chem.* **237**:9238–9241.

Ip, N. Y., Ho, C. K., and Zigmond, R. E., 1982, Secretin and vasoactive intestinal peptide acutely increase tyrosine 3-monoxygenase activity in the rat superior cervical ganglion, *Proc. Natl. Acad. Sci. U.S.A.* **79**:7566–7569.

Ip, N. Y., Baldwin, C., and Zigmond, R. E., 1984, Acute stimulation of ganglion tyrosine hydroxylase activity by secretin, VIP and PHI, *Peptides* **5**:309–312.

Ip, N. Y., Baldwin, C., and Zigmond, R. E., 1985, Regulation of the concentration of adenosine 3′,5′-cyclic monophosphate and the activity of tyrosine hydroxylase in the rat superior cervical ganglion by three neuropeptides of the secretin family, *J. Neurosci.* **5**:1947–1954.

Jumblatt, J. E., and Tischler, A. S., 1982, Regulation of muscarinic ligand binding sites by nerve growth factor in PC12 pheochromocytoma cells, *Nature* **297**:152–159.

Kowal, J., 1982, VIP effects on adrenal cortical cell functions, in: *Vasoactive Intestinal Peptide* (S. I. Said, ed.), Raven Press, New York, pp. 277–284.

Leboulenger, F., Leroux P., Delarue, C., Tonon, M. C., Charney, Y., Dubois, P. M., Coy, D. H., and Vandry, H., 1983, Co-localization of vasoactive intestinal peptide (VIP) and enkephalins in chromaffin cells of the adrenal gland of amphibia. Stimulation of corticosteroid production by VIP, *Life Sci.* **32**:375–383.

Levey, G. S., Weiss, S. R., and Ruiz, E., 1975, Characterization of the glucagon receptor in a pheochromocytoma, *J. Clin. Endocrinol. Metab.* **40**:720–723.

Linnoila, R. I., DiAugustine, R. P., Hervonen, A., and Miller, R. J., 1980, Distribution of (LEU^5) and (MET^5) enkephalin, VIP and substance P-like immunoreactivity in human adrenal glands, *Neuroscience* **5**:2247–2259.

Livett, B. G., Dean, D. M., and Bray, G. M., 1978, Growth characteristics of isolated adrenal medullary cells in culture, *Abstr. Soc. Neurosci.* **4**:592.

Lucas, C. A., Edgar, D., and Thoenen, H., 1979, Regulation of tyrosine hydroxylase and choline acetyltransferase activities by cell density in the PC12 pheochromocytoma clonal cell line, *Exp. Cell. Res.* **121**:79–86.

Lundberg, J. M., Hökfelt, T., Fahrenkrug, J. F., Nilsson, G., and Terenius, L., 1979, Peptides in the cat carotid body (glomus caroticum): VIP-, enkephalin-, and substance P-like immunoreactivity, *Acta Physiol. Scand.* **107**:179–281.

Lundberg, J. M., Anggard, A., Fahrenkrug, J., Hökfelt, T., and Mutt, V., 1980, Vasoactive intestinal polypeptide in cholinergic neurons of exocrine glands. Functional significance of coexisting transmitters for vasodilation and secretion, *Proc. Natl. Acad. Sci. U.S.A.* **72**:1651–1655.

Masserano, J. M., and Weiner, N., 1979, The rapid activation of adrenal tyrosine hydroxylase by decapitation and its relationship to a cyclic AMP-dependent phosphorylating mechanism, *Molec. Pharmacol.* **16:**513–528.

Mendelsohn, G., Eggleston, J. C., Olson, J. L., Said, S. I., and Baylin, S. B., 1979, Vasoactive intestinal peptide and its relationship to ganglion cell differentiation in neuroblastic tumors, *Lab. Invest.* **41:**144–149.

Miller, R. J., 1984, PHI and GRF: Two new members of the glucagon secretin family, *Med. Biol.* **61:**159–162.

Morera, A. M., Laburthe, M., and Saez, J. M., 1979, Interaction of vasoactive intestinal peptide (VIP) with a mouse adrenal cell line (Y-1): Specific binding and biological effects, *Biochem. Biophys. Res. Commun.* **90:**78–85.

Rabe, C. S., Schneider, J. E., and McGee, R., Jr., 1982, Enhancement of depolarization-dependent neurosecretion from PC12 cells by forskolin-induced elevation of cyclic AMP, *J. Cyclic Nucleotide Res.* **8:**371–384.

Robberecht, P., Chatelain, P., Waelbroeck, M., and Christophe, J., 1982, Heterogeneity of VIP-recognizing binding sites in rat tissues, in: *Vasoactive Intestinal Peptide* (S. I. Said, ed.), Raven Press, New York, pp. 323–332.

Said, S. I., 1976, Evidence for secretion of vasoactive intestinal peptide by tumors of pancreas, adrenal medulla, thyroid and lung: Support for the unifying APUD concept, *J. Clin. Endocrinol.* **5:**201S.

Salomon, Y., Lordos, C., and Rodbell, M., 1974, A highly sensitive adenylate cyclase assay, *Anal. Biochem.* **58:**541–554.

Tischler, A. S., DeLellis, R. A., Biales, B., Nunnemacher, G., Carabba, V., and Wolfe, H. J., 1980, Nerve growth factor-induced neurite outgrowth from normal human chromaffin cells, *Lab. Invest.* **43:**399–409.

Tischler, A. S., Lee, A. K., Slayton, V. W., and Bloom, S. R., 1982a, Content and release of neurotensin in PC12 pheochromocytoma cell cultures. Modulation by dexamethasone and nerve growth factor, *Regul. Peptides* **3:**415.

Tischler, A. S., Perlman, R. L., Nunnemacher, G., Morse, G. M., DeLellis, R. A., Wolfe, H. J., and Sheard, B. E., 1982b, Long-term effects of dexamethasone and nerve growth factor on adrenal medullary cells cultures from young adult rats, *Cell Tissue Res.* **225:**525–542.

Tischler, A. S., Perlman, R. L., Morse, G. M., and Sheard, B. E., 1983, Glucocorticoids increase catecholamine synthesis and storage in PC12 pheochromocytoma cell cultures, *J. Neurochem.* **40:**364–370.

Tischler, A. S., Lee, Y. C., Perlman, R. L., Costopoulos, D. Slayton, V. W., and Bloom, S. R., 1984, Production of "ectopic" vasoactive intestinal peptide-like and neurotensin-like immunoreactivity in human pheochromocytoma cell cultures, *J. Neurosci.* **4:**1398–1404.

Tischler, A. S., Perlman, R. L., Costopoulos, D., and Horwitz, J., 1985, Vasoactive intestinal peptide activates tyrosine hydroxylase in normal and neoplastic chromaffin cell cultures, *Neurosci. Lett.* **61:**141–146.

Unsicker, K., Krisch, B., Otten, U., and Thoenen, H., 1978, Nerve growth factor-induced fiber outgrowth from isolated rat adrenal chromaffin cells: Impairment by glucocorticoids, *Proc. Natl. Acad. Sci. U.S.A.* **75:**3498.

Volle, R. L., and Patterson, B. A., 1982, Regulation of cyclic AMP accumulation in a rat sympathetic ganglion: Effects of vasoactive intestinal peptide, *J. Neurochem.* **39:**1195–1197.

Wharton, J., Polak, J. M., Pearse, A. G. E., McGregor, G. P., Bryant, M. G., Bloom, S. R., Emson, P. C., Bisgard, G. E., and Will, J. A., 1980, Enkephalin VIP- and substance P-like immunoreactivity in the carotid body, *Nature* **284:**269–271.

Wright, R. D., 1963, Blood flow through the adrenal gland, *Endocrinology* **72:**418–428.

V

Peripheral Peptides and Receptors

Localization of Neural and Endocrine Peptides

J. M. POLAK and S. R. BLOOM

1. INTRODUCTION

The realization that a massive number of active peptides are present in the innervation and endocrine tissue of most organ systems of the body has opened up a new facet of endocrinology, that of the study of the so-called "diffuse neuroendocrine system" (Polak and Bloom, 1983). The concept of a "diffuse endocrine system" is not new. Feyrter (1938, 1953) described a series of cells that reacted poorly with conventional histological staining techniques. These cells were dispersed throughout the body, intermingled with strongly stained cells. They also became dark after treatment of the tissue with silver salts (argyrophilia). The terms "clear cells" or "diffuse endocrine system" were the proposed. The advent of more appropriate technology led Pearse to identify these cells under the umbrella of the APUD (amine precursor uptake and decarboxylation) system (see review by Pearse, 1984) and Fujita (1977) to describe them as the "paraneuronal" system. Both Pearse and Fujita highlighted the close association between these specialized endocrine cells and neural elements.

The discovery of a large number of active or "regulatory" peptides found throughout the body and localized in neural and endocrine tissue of the diffuse neuroendocrine system gave a further functional meaning to this system. It is now well recognized that this regulatory peptide-containing system (Polak and Bloom, 1983) is present in Feyrter's diffuse neuroendocrine system, Pearse's APUD system, and Fujita's paraneuronal system.

J. M. POLAK and S. R. BLOOM • Departments of Histochemistry and Medicine, Royal Postgraduate Medical School, Hammersmith Hospital, London W12 OHS, England.

2. TECHNOLOGY

2.1. Visualization of Neural and Endocrine Peptides (Diffuse Neuroendocrine System)

A number of antibodies to specific cytoplasmic components have now been shown to be capable of staining endocrine cells, the innervation, or both components of the "diffuse neuroendocrine system" (see Fig. 1).

2.1.1. Neuron-Specific Enolase: A Marker for Both Endocrine Cells and the Innervation

Neuron-specific enolase is an isozyme of the glycolytic enzyme enolase, which was originally extracted from the brain and found by immunocytochemistry to be localized to neurons. Subsequently, antibodies to neuron-specific enolase were noted to immunostain all components of the "diffuse neuroendocrine system" (see Polak and Marangos, 1984). Neuron-specific enolase is a cytoplasmic enzyme unrelated to the number of secretory granules. The staining pattern of neuron-specific enolase is therefore unique, since the antibodies mark both endocrine cells and the innervation. Because of its putative involvement in metabolism and its nongranular cytoplasmic localization, it may be indicative of the functional state of a nerve or an endocrine cell and may also be useful for the identification of poorly granulated tissue.

It is very important to be selective in choosing antibodies to neuron-specific enolase. Neuron-specific enolase is a large protein of 78 kD, and therefore antibodies are likely to react to many separate epitopes of this molecule. The protein, which has not yet been synthetically produced, is extracted naturally from animal or human brain. Therefore, it is likely that the antigen will be contaminated with material

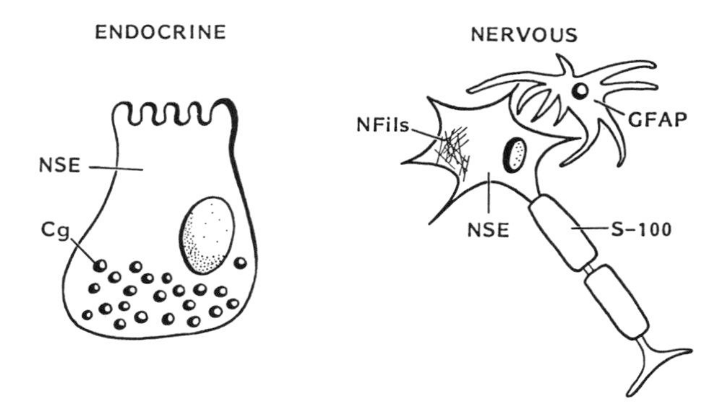

FIGURE 1. Schematic representation of different types of neuroendocrine markers. NSE, neuron-specific enolase; Cg, chromogranin; NFils, neurofilaments; S-100, protein S-100; GFAP, glial fibrillary acidic protein.

other than pure NSE, and thus spurious staining to nonneuronal elements could be expected (Dhillon *et al.*, 1982). Monoclonal antibodies to neuron-specific enolase are now becoming available (Hayano *et al.*, 1983), and it is predicted that a good mixture of high-quality monoclonal antibodies that can recognize several epitopes of the molecule will be of excellent value for morphologists.

2.1.2. Chromogranin

The chromogranins are a group of proteins first shown to be present in adrenal medullary cells by Blaschko *et al.* (1967). A major molecular form was later characterized by Schneider and co-workers (Schneider *et al.*, 1967) as a 68-kD protein. Chromogranin coexists with catecholamines and peptides in the storage granules of adrenal medullary cells and in sympathetic nerves (O'Connor *et al.*, 1983). It has been reported that chromogranin is released with catecholamines, and its concentration in cells has been used as a means to monitor sympathoadrenal secretion. Although the functions of chromogranin have yet to be fully established, it is thought to stabilize the intragranular complexes. Polyclonal antisera (staining sympathetic nerves, adrenal medullary cells, and other peptide-producing endocrine cells) and monoclonal antibodies (staining primarily peptide-producing endocrine cells) have been raised to this protein. Lloyd's group (Wilson and Lloyd, 1984) has carried out an extensive investigation to assess the potential of chromogranin monoclonal antibodies as a marker for peptide-producing endocrine cells. Studies carried out so far indicate that chromogranin antibodies stain all endocrine cell types diffusely distributed, including those of the gastrointestinal tract (Facer *et al.*, 1986) (Fig. 2) and pancreas [in the latter, primarily the glucagon-containing cells (Varndell *et al.*, 1985b), although other cell types are also weakly immunostained] and those of the thyroid and adrenal medulla. Ultrastructurally, chromogranin is present in the matrix of dense-cored secretory granules, and thus chromogranin immunostaining is related to the presence of secretory granules within the specialized cells of the diffuse neuroendocrine system.

2.1.3. Markers for Nerves and Supporting Elements

The variety of nerves containing not only classical neurotransmitters such as acetylcholine, norepinephrine, 5-hydroxytryptamine, and γ-aminobutyric acid but also putative peptide neurotransmitters can now be fully and extensively visualized by the use of polyclonal antisera and monoclonal antibodies to neurofilaments (Trojanowski, 1986; Dahl and Bignami, 1986). Neurofilaments belong to the group of cytoskeletal proteins known as intermediate filaments and have been shown to be present in different molecular weights of approximately 68,000, 150,000, and 200,000. Antibodies to neurofilaments have been extensively used in immunocytochemistry to visualize the entire innervation, as they stain both neurons and nerve fibers (Bishop *et al.*, 1985; Hacker *et al.*, 1985). Antibodies to proteins localized in both glial and Schwann cells, such as glial fibrillary acidic protein (GFAP) and

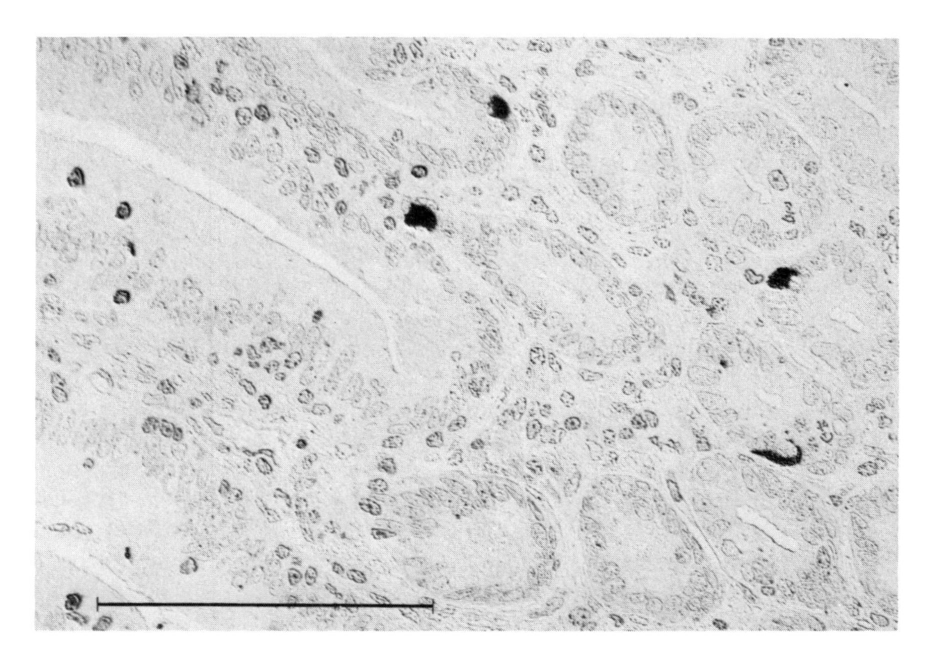

FIGURE 2. Chromogranin-immunoreactive endocrine cells in human jejunum. Freeze-dried, formaldehyde-vapor-fixed semithin section of resin-embedded tissue. Immunoperoxidase (PAP) method counterstained with hematoxylin. Scale bar, 100 μm.

S-100, are also available. Antisera to GFAP are known to stain astrocytes, in particular in the central nervous system, whereas S-100 is a good marker for both Schwann and glial cells of peripheral tissues (Jessen and Mirsky, 1980; Ferri *et al.*, 1982). Thus, antibodies to neurofilaments, to GFAP, and to S-100 are recommended to determine the pattern of innervation.

Neurofilaments are part of the cytoskeleton and are large polypeptide triplets. Monoclonal antibodies are likely to recognize restricted epitopes, and therefore, although these are useful for the demonstration of determined structures, e.g., free sensory endings, they may not pick up the entire innervation if the specific epitope is altered by fixation, etc.

Polyclonal antisera of excellent quality are available, but a mixture of high-quality monoclonal antibodies recognizing the various components of the polypeptide triplet of neurofilaments may be the ideal route for staining.

2.2. Peptide Localization

Antibodies to regulatory peptides are readily available (see Van Noorden and Polak, 1985). These antibodies work on conventionally fixed and embedded tissue.

For the visualization of peptide-containing innervation, the most appropriate techniques use cryostat sections (sections of conventional thickness), thick cryostat/Vibratome free-floating sections, or whole-mount preparations when the material is, for instance, a thin wall of a blood vessel or separated layers of the intestine (Van Noorden, 1986). A number of "enhancement" procedures have recently been proposed, some of which, e.g., immunogold silver staining, permit the visualization of neuropeptides in paraffin sections (Fig. 3) (Springall *et al.*, 1984). Antibodies are now raised to various regions of the prepropeptide molecule, thus giving a wider range of localization possibilities in the study of regulatory peptides.

Increasing evidence indicates that peptides may coexist with amines or other substances within a single cell or neuron. Several immunocytochemical methods have lately been developed to determine this coexistence accurately, including the use of the mirror-image sectioning method ("flip-flop" technique) (Ali Rachedi *et al.*, 1984), the elution methods of Tramu (Tramu *et al.*, 1978) and Nakane (Nakane, 1968), the use of paraffin- or resin-embedded 2-μm sections (Bishop *et al.*, 1982b), and the use of antibodies raised in different species tagged with different color dyes (e.g., fluorescein, rhodamine).

Electron microscopy has been instrumental in demonstrating the existence of dense-cored secretory granules with specific morphological features, e.g., size,

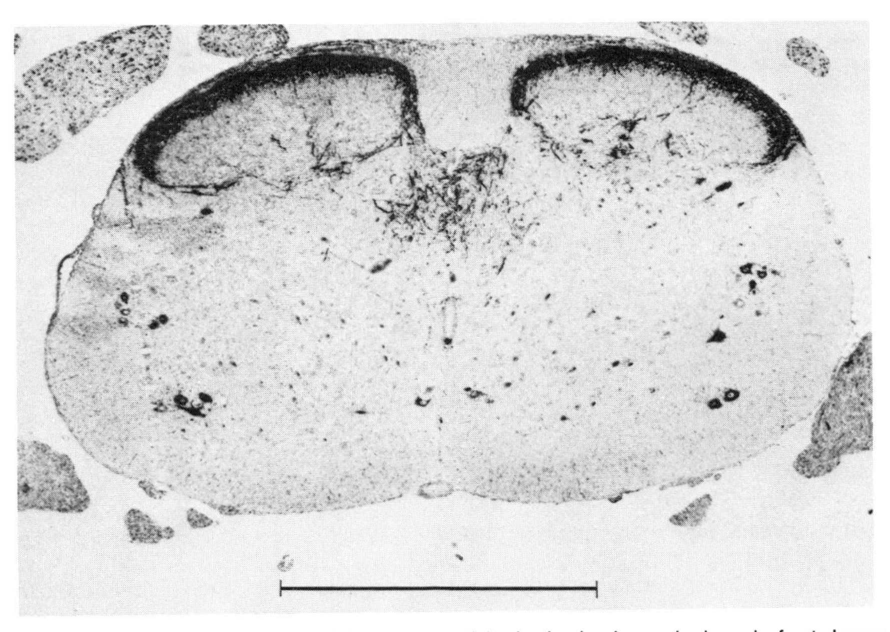

FIGURE 3. Calcitonin gene-related immunoreactivity in the lumbar spinal cord of rat. Immunoreactive fibers are concentrated in the dorsal horn, and some immunoreactive motoneurons are present in the ventral spinal cord. Bouin's fixed 5-μm wax section. Immunogold silver staining method. Scale bar, 1 mm.

shape, appearance of the limiting membrane, and electron density. The presence of a particular peptide occurring in a given morphologically characterizable secretory granule can thus be predicted with reasonable accuracy. Immunocytochemistry has been applied at the electron microscopic level with increasing effectiveness in recent years (Polak and Varndell, 1984). The techniques can be broadly divided into preembedding and postembedding methods. The well-known peroxidase–antiperoxidase (PAP) preembedding method is useful when synaptic contact or membrane visualization is required, as the material can subsequently be osmicated. For postembedding procedures a variety of gold-labeling methods are preferred. Gold particles of various sizes can be attached by noncovalent bonds to immunoglobulins. Different antibodies can be labeled with different sizes of gold particules, thus enabling multiple immunostaining procedures to be carried out (Fig. 4). These techniques have advantages over the PAP preembedding methods, since the gold particles do not obscure the fine structure of the electron-dense neurosecretory granules. Until now, the immunogold methods have only been accurately and reproducibly used on osmicated tissue with a high peptide content, e.g., the pituitary or pancreatic islets. However, it can be predicted that improvements in

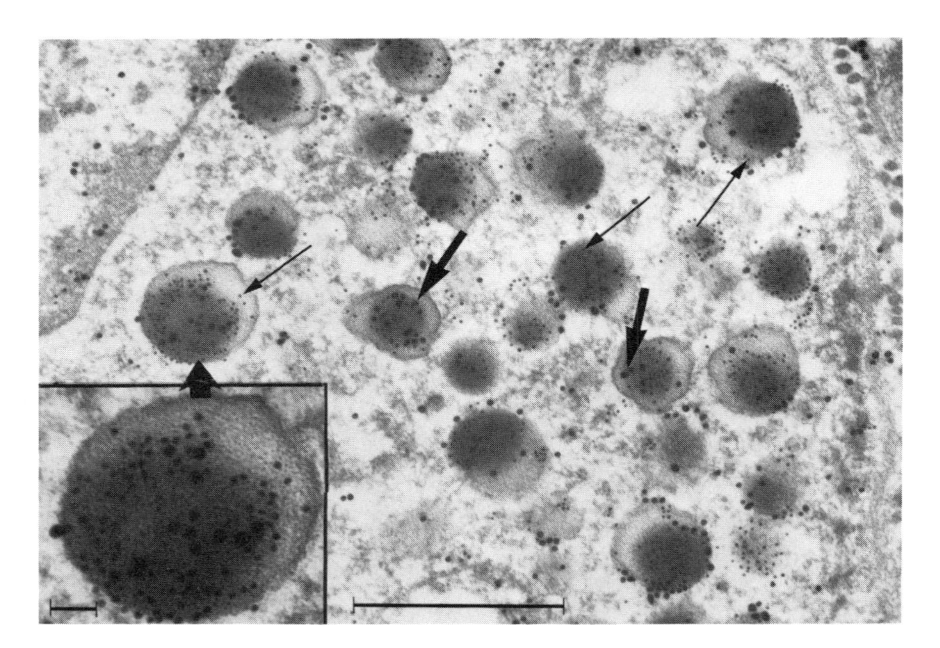

FIGURE 4. Glucagon (large arrows) and glucagonlike peptide (small arrows) immunoreactivity in human pancreatic A cell secretory granules. Both immunoreactivities are localized to the electron-dense core of the secretory granules. Glutaraldehyde fixation, Araldite-embedded ultrathin section. Double immunogold staining procedure using 20-nm (large arrows) and 10-nm (small arrows) gold particles. Uranyl acetate and lead citrate counterstains. Inset scale bar, 0.1 μm. Main scale bar, 1 μm.

the quality of antibodies and on a variety of "enhancement" immunocytochemical procedures will permit the use of gold-labeling methods on all kinds of osmicated tissue.

2.3. Determination of the Cellular Origin and Nature of Neural Peptides

Immunocytochemistry provides excellent information about the morphological network of nerve fibers (see Heimer and Robards, 1981; Bjorklund and Hökfelt, 1983; Emson, 1983). Sometimes peptide immunocytochemistry applied to untreated tissue is unable to demonstrate neuronal cell bodies because of the limitationns of the sensitivity of the technique and the low peptide storage in cell bodies. Peptide storage can, however, be enhanced by pharmacological procedures, e.g., colchicine, or surgical procedures, e.g., nerve ligation, leading to a subsequent build-up of a neurotransmitter in the cell body and depletion in the peripheral postligated segment. Determination of the nature of peptides in nerve fibers (e.g., sensory, norepinephrinergic) can also be achieved by the use of pharmacological agents including

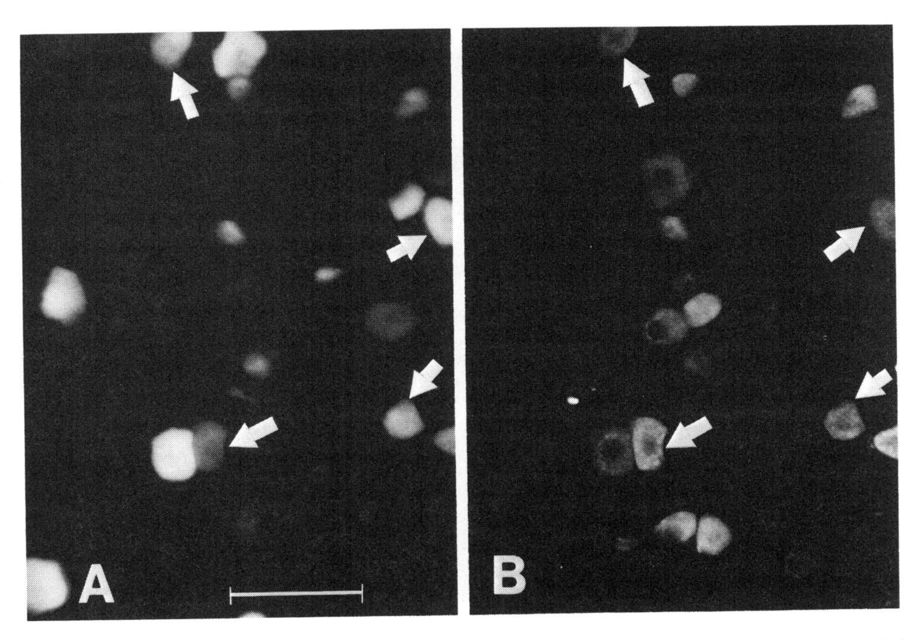

FIGURE 5. (A) Primary sensory neurons in the L4 dorsal root ganglion of the rat, labeled retrogradely with the fluorescent tracer true blue following injection of the dye into the sciatic nerve. (B) The same section processed for immunocytochemistry using antisera directed to calcitionin gene-related peptide. Some of these neurons coexist with true-blue-labeled cells (arrowed). Paraformaldehyde fixation, cryostat section (20 μm). Indirect immunofluorescence method. Scale bar, 100 μm.

capsaicin, an extract of red pepper capable of depleting sensory neurons of their neurotransmitter content, and 6-hydroxydopamine and reserpine, which are both capable of depleting (though in different ways) neurotransmitter content in norepinephrinergic neurons. The technique of retrograde tracing combined with immunocytochemistry is providing increasing information on the novel neurochemical anatomy (Fig. 5).

2.4. Receptor Visualization

Improved techniques for the autoradiographic visualization of receptors have recently entered the literature (see Chapter 18). However, another method that is also being used for visualization of receptors involves immunocytochemical means (e.g., the use of antibodies to receptors). A novel immunocytochemical technique has recently been proposed (Lackie *et al.*, 1985). This makes use of divalent forms of peptides (e.g., bombesin), one active site of which reacts with a receptor on the surface of bombesin-producing cells (e.g., small-cell carcinomas), while the other active site of the divalent peptide remains readily available for reaction with labeled antibodies, e.g., gold-labeled monoclonal antibombesin (see Fig. 6). The receptor site can then be visualized in this instance at the electron microscopic level by observing gold deposits on the surface of bombesin-producing tumor cells. Small-cell carcinoma cells are known, from biochemical analysis, to have bombesin receptors on their surface (see Chapter 29).

2.5. Visualization of Messenger RNA Directing Peptide Synthesis (Hybridization Histochemistry)

Advances in molecular biology have now made possible the construction of specific DNA probes complementary (cDNA) to determined messenger RNA sequences encoding specific peptides (see Roberts and Wilcox, 1986). These cDNA probes, which may be double or single stranded, can be labeled radioactively or with chromogens (Varndell *et al.*, 1984), and therefore the hybridized reaction product can subsequently be visualized. The technique, originally termed "*in situ* hybridization histochemistry," has been used by many workers throughout the world, and early apprehensions as to the quality of the signal have proved groundless in view of its successful use by many different centers.

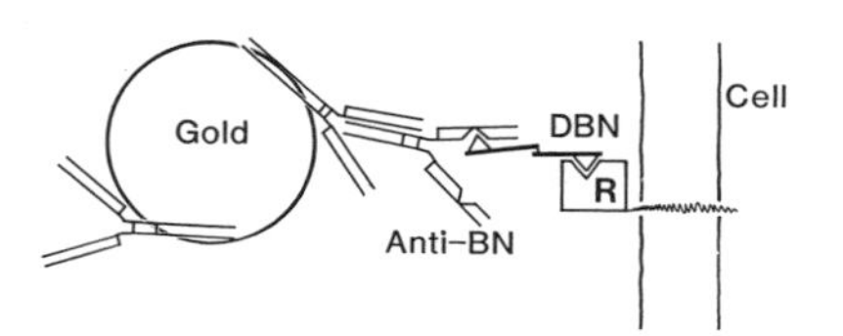

FIGURE 6. Diagrammatic representation of visualization of divalent ligand using 2A11 monoclonal antibody (anti-BN) and a colloidal gold label.

3. MORPHOLOGY OF ENDOCRINE CELLS AND NERVES

Peptide-containing endocrine cells are found scattered throughout the body, and cells located in hollow organs frequently contact the lumen by means of characteristic microvilli. The latter can be better visualized at the electron microscopic level. Endocrine cells are furnished with intracytoplasmic dense-cored secretory granules, the morphology of which permits separation into various cell types. Electron immunocytochemistry has now expanded the classification of these cell types into separate "functional" classes. The latest international agreement was reached in Cambridge in 1980 (Solcia *et al.*, 1981). Secretory granules tend to be located in the basal pole of the cells, close to the basement membrane and to the organ blood supply. Endocrine cells frequently display a cytoplasmic elongation along the basement membrane.

These anatomic features, in particular microvilli, secretory granules, and cytoplasmic extensions, invite speculation on a functional level: receptor functions are attributed to microvilli, peptide products are stored in secretory granules, and the cytoplasmic extensions may indicate a possible local or paracrine role for some of the peptide-containing endocrine cells.

Numerous neurotransmitters have now been identified in the nervous system, and these include not only the classical neurotransmitters, acetylcholine and norepinephrine, but also amino acid neurotransmitters and, in particular, peptide neurotransmitters. All of these products can now be identified by the use of specific antibodies in cell bodies, nerve fibers, and terminals. Under light microscopy, the specifically stained nerve is characterized in general by its classical beaded appearance.

4. INDIVIDUAL ORGANS

Under this subheading an attempt is made to highlight the principal distribution characteristics of peptides in some selected peripheral organs in which regulatory peptides occur in abundance.

4.1. The Gastrointestinal Tract and Pancreas

By analogy with the central nervous system, the gastrointestinal tract contains a large proportion of regulatory peptides (Polak and Bloom, 1982). These are present in typical mucosal endocrine cells (Fig. 7) and in the enteric innervation (Fig. 8). Peptide-containing endocrine cells are intermingled with nonendocrine cells, and certain areas of the gut, for instance, the gastric antrum, are considerably richer in endocrine cells compared with other areas. Certain endocrine cells, e.g., somatostatin-containing cells of the lower bowel, display the classical cytoplasmic elongation, which suggests that these cells may play a paracrine role (Larsson *et al.*,

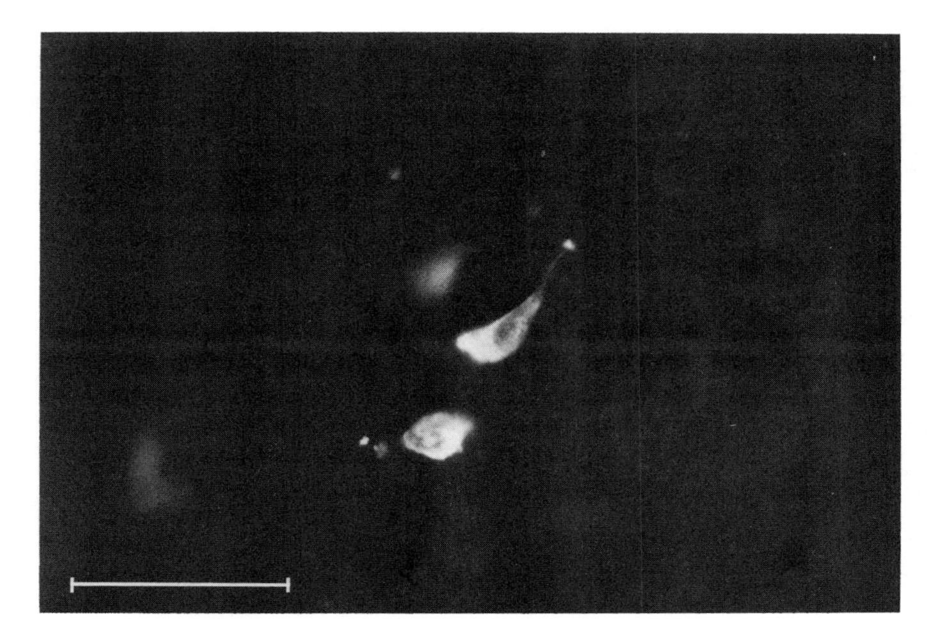

FIGURE 7. Enteroglucagon-immunoreactive cell in the mucosal epithelium of human colon. Note the luminal elongation with an aggregation of immunoreactivity at the luminal surface. The base of a similar cell can be seen in an adjacent gland. *p*-Benzoquinone solution-fixed. 10-μm cryostat section. Indirect immunofluorescence method. Scale bar, 100 μm.

1981). A good proportion of the endocrine cells in the gastrointestinal tract are connected with the lumen by means of microvilli, but others (e.g., the enterochromaffinlike cells of the fundic mucosa) are mainly located towards the basement membrane, and no luminal microvillar connection has been identified. The peptide-containing endocrine cells of the gastrointestinal mucosa are also characterized by the morphology of the electron-dense secretory granules, which, in combination with electron immunocytochemistry, permits an accurate classification of gut endocrine cells.

The gut has a rich innervation composed of nerves containing the classical neurotransmitters and a large peptidergic component (Bishop *et al.*, 1982a). The majority of peptide nerves originate from local cell bodies present in the two main ganglionated plexuses, in the submucosa and lamina propria, and are in close contact with epithelial endocrine and nonendocrine cells. Some of the peptide-containing nerves, e.g., those with neuropeptide Y (NPY), calcitonin gene-related peptide (CGRP), and vasoactive intestinal polypeptide (VIP), are closely associated with the gut vasculature. Others (e.g., galanin) are well represented in the muscle layers. The minor components of the enteric nervous system, which originate from extrinsic neurons, include a small population of NPY- and of substance P/CGRP-containing nerves. Two of the most abundant peptides present in the bowel innervation are

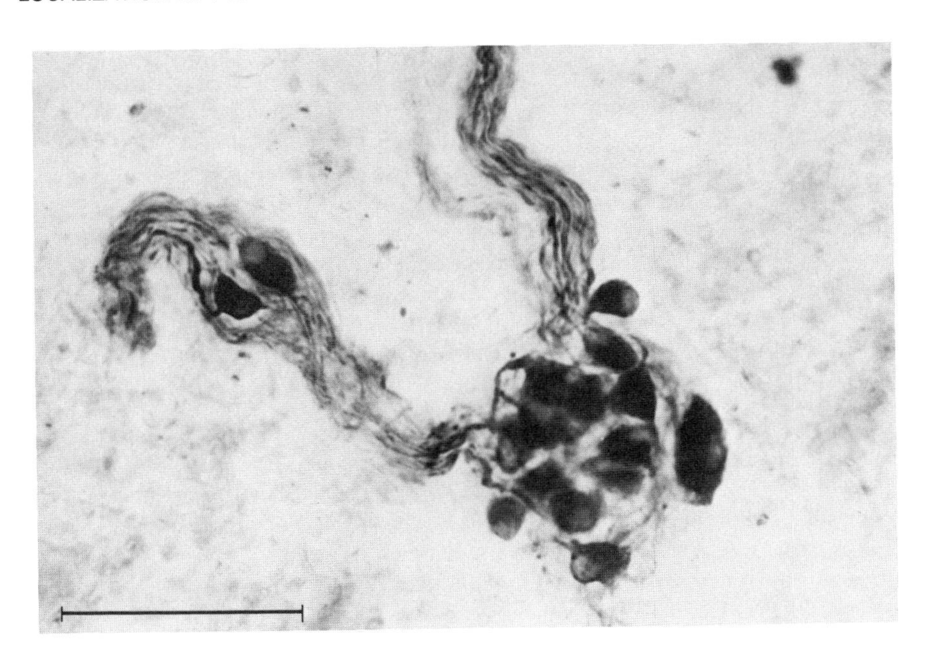

FIGURE 8. VIP-immunoreactive ganglion cells and nerve fibers in the submucosal plexus of human large bowel. *p*-Benzoquinone solution-fixed, whole-mount preparation, immunoperoxidase (PAP) technique. Scale bar, 100 μm.

VIP and peptide histidine (isoleucine/methionine) (PHI/PHM), and all of them show a distinct distribution along the length of the gastrointestinal tract.

In the pancreas, peptides are also present in endocrine cells, in particular within the pancreatic islets (Klöppel and Lenzen, 1984) and in the pancreatic innervation. Four peptides are known to be present in pancreatic islet endocrine cells, namely, insulin, glucagon, somatostatin, and pancreatic polypeptide (PP). Insulin is contained in the B cells of the pancreas, often characterized at the electron microscopic level by the existence of a crystalline bar, a broad halo, and a distinct membrane. Glucagon and the glucagonlike immunoreactants are all localized in the A cells of the pancreas and are characterized at the electron microscopic level by the presence of large, highly electron-dense granules with a distinct core and halo. Components of the preproglucagon molecule are compartmentalized in the A cells of the pancreas (Varndell *et al.*, 1985). Somatostatin cells are more abundant in the fetal and neonatal pancreas than in the adult pancreas. In the former, they form almost 25% of the endocrine cell population, whereas in the adult pancreas, no more than 10–12% is composed of somatostatin (D) cells. These D cells, like A cells, are predominantly localized in the periphery of the islets. A cells show regional differences, being abundant in the tail of the pancreas and rather more scattered in the head. The PP cells, by contrast, are more prolific and are closely related to pancreatic islets in the head of the pancreas, whereas they are scattered and intermingled within the exocrine pancreas in the tail. An ultrastructurally identified cell type exists in the pancreas containing small (D_1) granules, whose product remains to be determined.

The pancreatic innervation contains classical and peptidergic neurotransmitters, in particular VIP (Bishop *et al.*, 1980b) and PHI(PHM) (around blood vessel and pancreatic islets), CGRP (exocrine pancreas and around intrainsular capillaries), and in some mammalian species substance P and the enkephalins. A good proportion of these peptidergic nerves originate from local ganglion cells.

4.2. The Respiratory Tract

Like the gut, the respiratory tract contains peptides in mucosal endocrine cells (Feyrter clear cells or Kulchitzky cells) and in nerves (see Polak and Bloom, 1984). Substance P, VIP/PHI(PHM), galanin, and NPY are present in the lung innervation. Bombesin/gastrin-releasing peptide (GRP) is mostly present in mucosal endocrine cells, in particular in the human fetal and neonatal respiratory tract. In rat, CGRP is present both in mucosal endocrine cells and in the lung innervation. In other species CGRP is exclusively localized to nerves. Substance P and CGRP nerves originate from primary sensory neurons of the nodose ganglion, NPY from the sympathetic chain, and VIP/PHI(PHM) and galanin from local cell bodies, particularly prominent in the upper trachea. Substance P and CGRP nerves are found beneath the bronchial epithelium and, particularly in man, in close contact with bronchial smooth muscle and around blood vessels. Galanin and VIP are found principally in the upper respiratory tract in close contact with seromucous glands and blood vessels. Galanin is also found around bronchial smooth muscle. Bombesin- and CGRP-containing mucosal endocrine cells are present singly or in groups in the bronchial epithelium. Neuropeptide Y is particularly associated with the lung vasculature.

4.3. Cardiovascular System

Increasing evidence indicates that the whole cardiovascular system of man and other mammals contains a large number of powerfully active peptides (Fig. 9), in particular CGRP (Mulderry *et al.*, 1985), substance P (Wharton *et al.*, 1981), and NPY (Gu *et al.*, 1983a, 1984b). Somatostatin and VIP are present in lesser concentrations. Neuropeptide Y is the principal neuropeptide of the human heart and is also found in peripheral blood vessels. Neuropeptide Y immunoreactivity partly coexists with norepinephrine in the same neurons, although an independent NPY-containing cardiac plexus has recently been described. Neuropeptide Y nerves are found around coronary and other blood vessels, including those of the circle of Willis. The afferent and efferent arterioles of the juxtaglomerular apparatus of the

FIGURE 9. Neuropeptide-Y-immunoreactive nerve fibers in the adventitia–media border of normal guinea pig abdominal aorta. *p*-Benzoquinone-fixed whole-mount preparation, immunoperoxidase (PAP) technique. Scale bar, 100 μm.

kidney as well as the interlobular and renal arteries are heavily innervated by NPY-containing fibers, which are present in the catecholaminergic system.

Primary afferent fibers containing CGRP and substance P are found throughout the cardiovascular system. In the heart, these sensory nerve fibers are found in the conduction system and around the coronary arteries. All classes of blood vessels, arteries, veins, and capillaries, are richly innervated by primary afferents that contain sensory neuropeptides, in particular CGRP.

The cardiac atrium also contains large quantities of a family of peptides grouped under the heading of atrial natriuretic peptide (ANP) (Palluk *et al.*, 1985). The existence of a factor that is released from a distended atrium in order to regulate water and sodium load from the kidney has long been postulated. This hypothesis fitted in well with the granules in atrial cardiocytes. Application of molecular biology has now permitted analysis of the nucleotide sequence and deduction of the amino acid sequence of the ANP family of peptides. It is possible to extract considerable amounts of ANP from the right atrium, and immunocytochemistry has enabled localization of ANP to electron-dense secretory granules within specialized cardioendocrine cells.

5. PATHOLOGY

5.1. Nontumor Pathology

Under this heading a number of disease situations are discussed in which an abnormal pattern of regulatory peptide distribution has been detected.

5.1.1. Endocrine Cell Abnormalities

5.1.1a. Hyperplasia. Hyperplasia of endocrine cells has been found in a number of situations.

Gastrin (G) Cells. This entity was first discovered using immunocytochemistry (Polak *et al.,* 1972). Subsequently, functional studies were carried out, and the entity has now been recognized to be G-cell hyperplasia/hyperfunction (Lewin *et al.,* 1984). G-cell hyperplasia can be found in severe cases of duodenal ulcer with symptoms closely resembling those of gastrinoma (although no tumor is found) or in atrophic gastritis with or without pernicious anemia (Polak *et al.,* 1973a).

Celiac Disease. The presence of numerous immunoreactive endocrine cells in the affected area has frequently been reported. The finding of so-called secretin- and cholecystokinin (CCK)-cell hyperplasia is particularly relevant in celiac disease (Polak *et al.,* 1973b; Bloom *et al.,* 1978). It has been recognized for some time that anatomic abnormalities of the duodenal mucosa lead to an impairment of hormone release by intraluminal stimuli. Consequently, pancreatic bicarbonate and enzyme secretion is defective. In contrast, if secretin and CCK are administered intravenously, the good pancreatic response is obtained with dynamic studies that indicate an impairment of hormone release after normal intraduodenal acid stimulus. Other endocrine cells have also been shown to be hyperplastic in this condition (Sjolund *et al.,* 1982).

Enterochromaffinlike Cells. Since 1969 a characteristic endocrine cell type has been reported to be present in the fundic mucosa (Hakanson and Owman, 1969). This cell, with a nonluminal elongated shape, was named an enterochromaffinlike (EC-like/ECL) cell. The peptide product of these cells, which contain characteristic secretory granules, is as yet unknown, but EC-like cells in rodents have been shown to produce histamine. Histamine can now be demonstrated using specific antibodies or indirectly with antibodies to the converting enzyme histidine decarboxylase. Enterochromaffinlike cells are specifically stained by Sevier–Munger silver impregnation staining (Sevier and Munger, 1965). These cells have been shown to be sensitive to elevated gastrin levels (Hakanson *et al.,* 1984) resulting either from hyperfunction of G cells or from the existence of a gastrinoma. In addition, an overswing of gastrin release from G cells for lack of gastric acid inhibition (e.g., in pernicious anemia) can lead to EC-like hyperplasia with subsequent micronodule formation and carcinoids (Larsson *et al.,* 1978; Solcia, 1984). These findings are becoming increasingly common as drugs capable of producing maximum gastric acid suppression (e.g., new generation of H_2 receptor blocker, proton pump inhibitors) are now available for the treatment of duodenal ulcer and other hypersecretory conditions. These drugs, taken long term, produce a complete blockage of gastric acid secretion and subsequent oversecretion of G cells, leading to chronically elevated circulating gastrin levels.

Glucagon-Immunoreactive Cells. Hyperplasia and hyperactivity of EG cells have been found in situations of gut repair, i.e., intestinal resection (Buchan *et al.*, 1985).

Insulin-Containing (B) Cells of the Pancreas. A condition known in the early days as "nesidioblastosis" or intractable neonatal hypoglycemia has been found to be associated with the so-called B-cell hyperplasia (Heitz *et al.*, 1977).

5.1.1b. Hypoplasia. Two cell types in particular have been reported to undergo hypoplastic changes. These are somatostatin cells, in both the pancreas and the stomach and bombesin-containing cells of the respiratory tract. Somatostatin cells in certain cases of intractable hypoglycemia have been found to be markedly hypoplastic (Bishop *et al.*, 1981). Somatostatin is known to suppress the release of other hormones, a lack of somatostatin leading to an overactivity and a hyperplasia of B cells with subsequent hyperinsulinemia and hypoglycemia. Somatostatin cells have been shown to be decreased in certain patients with excess gastric acid secretion and duodenal ulceration (Arnold *et al.*, 1982).

It has been shown that bombesin cells present in the fetal and neonatal lung are present in lower concentration in hypoplastic lungs associated with respiratory distress syndrome (Ghatei *et al.*, 1983). These findings fit in well with the putative trophic role of bombesin.

5.1.2. Neural Abnormalities

5.1.2a. Hyperplasia

Inflammatory Bowel Disease/Crohn's Disease. Neuronal hyperplastic changes have frequently been reported to occur in the bowel of patients with Crohn's disease (Bishop *et al.*, 1980a; O'Morain *et al.*, 1984). These changes have now been shown to be related to abnormalities of the VIPergic component of the enteric nervous system. The VIP-containing nerves appear markedly hyperplastic, distorted, and brightly immunostained. Changes are particularly evident in the mucosa and lamina propria. This has indicated that VIP immunostaining could be a useful tool for diagnosing inflammatory bowel disease in endoscopic biopsies.

Diabetic Eye and Bladder. Diabetic Chinese hamsters show a marked VIPergic nerve hyperplasia around blood vessels of the choroid compared with age-matched normoglycemic controls (Diani *et al.*, 1985).

5.1.2b. Hypoplasia

Unstable Bladder. This condition of unknown etiology was, until recently, thought to be psychosomatically caused. However, VIP immunostaining has shown marked VIPergic nerve hypoplasia, in particular in the muscle layer. This depletion is also quantifiable by specific radioimmunoassay (Gu *et al.*, 1983a,b). Since VIP is known to be a potent smooth muscle relaxant in the bladder, lack of VIP leads to an overreactive, "unstable" bladder.

Impotence. In impotent men the VIPergic innervation around the erectile tissue of the corpus cavernosum is markedly depleted or absent (Gu *et al.*, 1984a). As VIP is known to be a potent vasodilator, its absence may lead to an inability to dilate arteries of the erectile tissue with consequent impotence.

Hypertension. The potent vasoconstrictor peptide NPY is found in the innervation of the juxtaglomerular apparatus of the kidney, coexisting with catecholamines (Ballesta *et al.*, 1984). Experimental malignant hypertension, produced in rats by ligation of the abdominal aorta, induces a marked depletion of NPY-containing nerves not only of the juxtaglomerular apparatus but also of the interlobular arteries.

Chronic, Intractable Constipation. Peptidergic nerves, in particular those containing VIP and substance P, have been shown to be present in abnormally low levels in diseases of the bowel associated with intractable chronic constipation and absence or degeneration of intrinsic neuronal cell bodies, e.g., Chagas' disease, Hirschsprung's disease, and grass sickness disease in the horse. In contrast, normal levels of peptidergic nerves appear to be present in a generalized autonomic neuropathy like that of the Shy–Drager syndrome, which shows no involvement of the gut neuronal cell bodies (Bishop *et al.*, 1982a).

5.2. Tumor Pathology

Although endocrine tumors (APUDomas) can be regarded as rare, our understanding of them has advanced greatly since the advent of immunocytochemistry and electron microscopy (Polak and Bloom, 1985). Among these tumors, the most frequently found are those of pancreatic origin (Sabate *et al.*, 1985). At least five characteristic clinical syndromes are now well recognized as caused by tumors supplying markedly increased circulating levels of the corresponding hormone. They are the insulinoma syndrome, the glucagonoma syndrome, the VIPoma [Verner–Morrison or watery diarrhea hypokalemia and achlorhydria (WDHA)] syndrome, the gastrinoma or Zollinger–Ellison syndrome, and the syndrome associated with a tumor producing growth hormone-releasing factor (GRF). Although many of these tumors produce more than one hormone (and are routinely called islet cell tumors), one regulatory peptide is predominantly produced, as shown by immunocytochemistry. These features are further validated by electron microscopy. Although several endocrine cell types can be distinguished in a mixed tumor, one of them, usually corresponding to the hormone responsible for the clinical features, is in the majority. In addition, immunocytochemistry for neuron-specific enolase (NSE) can be of excellent diagnostic value.

Neuroendocrine tumors of the lung are quite commonly found, in particular the small-cell carcinoma of the lung (Sheppard *et al.*, 1985). Small-cell carcinomas have been shown to produce and release several regulatory peptides, in particular the trophic peptide bombesin or gastrin-releasing peptide (GRP). They are poorly granulated tumors, and thus the general neuroendocrine marker neuron-specific enolase is of excellent diagnostic value. Chromogranin is also a good marker for neuroendocrine differentiation but is especially useful in highly granulated tumors. The features of the medullary carcinoma can be found in Williams (1985).

Pheochromocytomas, like normal adrenal medullary tissue, frequently produce regulatory peptides, including NPY, the enkephalins, and ACTH. The presence of neurotensinlike material has lately been shown in a pheochromocytoma cell line maintained in culture. The endocrine nature of the primary neuroendocrine tumor of the skin (the Merkel cell tumor), like the small-cell carcinoma of the lung, is best demonstrated by the use of antibodies to neuron-specific enolase (Gu *et al.*, 1983b). These tumors are also frequently found to be poorly granulated; argyrophilic silver impregnation, in consequence, is of little or unpredictable value. Neuroendocrine tumors of other regions, including the esophagus, larynx and uterine cervix, have frequently been described (Ibrahim *et al.*, 1985). Peptides have been demonstrated by immunocytochemistry to be produced in extrapulmonary small-cell carcinomas, but again neuron-specific enolase has been shown to be the best marker for diagnosis of these tumors.

6. CONCLUSION

The existence of a novel endocrinology, that of the diffuse neuroendocrine system, is becoming increasingly recognized. This system, containing active peptides, is present throughout the body of all animals. In evolutionary terms, the system spans more than 600 million years and is functionally very potent. Until recently, techniques for its visualization consisted of either capricious histological staining or partial localization by specific peptide immunocytochemistry. However, there now exist general neuroendocrine markers that permit the visualization of the entire regulatory peptide system, thus bypassing the need for mental reconstructions. The quite distinct distribution of many of these peptides fits in well with their various roles. Many antibodies are now available and suitable for routinely fixed tissue. The increasing awareness of regulatory peptide abnormalities in disease allows for further understanding of their physiological role and gives new insight into the pathophysiology of a given disease. Luckily, this new endocrinology has now arrived. The future is envisaged with eagerness and interest.

REFERENCES

Ali-Rachedi, A., Varndell, I. M., Adrian, T. E., Gapp, D. A., Van Noorden, S., Bloom, S. R., and Polak, J. M., 1984, Peptide YY (PYY) immunoreactivity is co-stored with glucagon-related immunoreactants in endocrine cells of the gut and pancreas, *Histochemistry* **80:**487–491.

Arnold, R., Hülst, M. V., Neuhof, Ch., Schwarting, H., Becker, H. D., and Creutzfeldt, W., 1982, Antral gastrin-producing G-cells and somatostatin-producing D-cells in different states of gastric acid secretion, *Gut* **23:**285–291.

Ballesta, J., Polak, J. M., Allen, J. M., and Bloom, S. R., 1984, The nerves of the juxtaglomerular apparatus of man and other mammals contain the potent peptide NPY, *Histochemistry* **80:**483–485.

Bishop, A. E., Polak, J. M., Bryant, M. G., Bloom, S. R., and Hamilton, S., 1980a, Abnormalities of VIP-containing nerves in Crohn's disease, *Gastroenterology* **79:**853–860.

Bishop, A. E., Polak, J. M., Green, I. C., Bryant, M. G., Bloom, S. R., 1980b, The location of VIP in the pancreas of man and rat, *Diabetologia* **18:**73–78.

Bishop, A. E., Polak, J. M., Garin Chesa, P., Timson, C. M., Bryant, M. G., and Bloom, S. R., 1981, Decrease of pancreatic somatostatin in neonatal nesidioblastosis, *Diabetes* **30:**122–126.

Bishop, A. E., Ferri, G. L., Probert, L., Bloom, S. R., and Polak, J. M., 1982a, Peptidergic nerves, in: *Structure of the Gut. Basic Science in Gastroenterology* (J. M. Polak, S. R. Bloom, N. A. Wright, and M. J. Daly, eds.), Glaxo Group Research Limited, London, pp. 221–231.

Bishop, A. E., Polak, J. M., Facer, P., Ferri, G.-L., Marangos, P. J., and Pearse, A. G. E., 1982b, Neuron-specific enolase: A common marker for the endocrine cells and innervation of the gut and pancreas, *Gastroenterology* **83:**902–915.

Bishop, A. E., Carlei, F., Lee, C., Trojanowski, J., Marangos, P. J., Dahl, D., and Polak, J. M., 1985, Combined immunostaining of neurofilaments, neuron specific enolase, GFAP and S-100, *Histochemistry* **82:**93–97.

Björklund, A., and Hökfelt, T. (eds.), 1983, *Methods in Chemical Neuroanatomy,* Vol. 1, Elsevier, Amsterdam.

Blaschko, H., Comline, R. S., Schneider, F. H., Silver, M., and Smith, A. D., 1967, Secretion of a chromaffin granule protein, chromogranin, from the adrenal gland after splanchnic stimulation, *Nature* **215:**58–59.

Bloom, S. R., Polak, J. M., and Besterman, H. S., 1978, Gut hormone profile in coeliac disease: A characteristic pattern of pathology, in: *Perspectives in Coeliac Disease* (B. McNichol, C. E. McCarthy, and P. F. Fottrell, eds.), MTP Press, Lancaster, pp. 399–410.

Buchan, A. M. J., Griffiths, C. J., Morris, J. F., and Polak, J. M., 1985, Enteroglucagon cell hyperfunction in rat small intestine after gut resection, *Gastroenterology* **88:**8–12.

Dahl, D., and Bignami, A., 1986, Intermediate filaments and differentiation in the central nervous system, in: *Immunocytochemistry—Modern Methods and Applications,* 2nd ed. (S. Van Noorden and J. M. Polak, eds.), John Wright and Sons, Bristol, pp. 401–411.

Dhillon, A. P., Rhode, J., and Leatham, A., 1982, Neurone specific enolase: An aid to the diagnosis of melanoma and neuroblastoma, *Histopathology* **6:**81–92.

Diani, A. R., Peterson, T., Sawada, G. A., Wyse, B. M., Blanks, M. C., Gerritsen, G. C., Terenghi, G., Varndell, I. M., Polak, J. M., Blank, M. A., and Bloom, S. R., 1985, Elevated levels of vasoactive intestinal peptide in the eye and urinary bladder of diabetic and prediabetic Chinese hamsters, *Diabetologia* **28:**302–307.

Emson, P. C. (ed.), *Chemical Neuroanatomy,* Raven Press, New York.

Facer, P., Bishop, A. E., Lloyd, R. V., Wilson, B. S., Hennessy, R. J., and Polak, J. M., 1986, Chromogranin A: A newly recognised marker for endocrine cells of the human gastrointestinal tract, *Gastroenterology* **89:**1366–1373.

Ferri, G. L., Probert, L., Cocchia, D., Michetti, F., Marangos, P. J., and Polak, J. M., 1982, Evidence for the presence of S-100 protein in the glial component of the human enteric nervous system, *Nature* **297:**409–410.

Feyrter, F., 1938, *Uber Diffuse Endokrine Epitheliale Organe,* J. A. Barth, Leipzig.

Feyrter, F., 1953, *Uber die Peripheren Endokrinen (Parakrinen) Drüsen des Menschen,* Maudrich, Vienna.

Fujita, T., 1977, Concept of paraneurons, *Arch. Histol. Jpn.* **40:**1–12.

Ghatei, M. A., Sheppard, M. N., Henzen-Logman, S., Blank, M. A., Polak, J. M., and Bloom, S. R., 1983, Bombesin and VIP in the developing lung: Marked changes in acute respiratory distress syndrome, *J. Clin. Endocrinol. Metab.* **57:**1226–1232.

Gu, J., Polak, J. M., Adrian, T. E., Allen, J. M., Tatemoto, K., and Bloom, S. R., 1983a, Neuropeptide tyrosine (NPY)—A major cardiac neuropeptide, *Lancet* **1:**1008–1010.

Gu, J., Polak, J. M., Van Noorden, S., Pearse, A. G. E., Marangos, P. J., and Azzopardi, J. G., 1983b, Immunostaining of neuron-specific enolase as a diagnostic tool for Merkel cell tumours, *Cancer* **52:**1039–1043.

Gu, J., Restorick, J. M., Blank, M. A., Huang, W. M., Polak, J. M., Bloom, S. R., and Mundy, A. R., 1983c, Vasoactive intestinal polypeptide in the normal and unstable bladder, *Br. J. Urol.* **55:**645–647.

Gu, J., Lazarides, M., Pryor, J. P., Blank, M. A., Polak, J. M., Morgan, R., Marangos, P. J., and Bloom, S. R., 1984a, Decrease of vasoactive intestinal polypeptide (VIP) in the penises from impotent men, *Lancet* **2:**315–318.

Gu, J., Polak, J. M., Allen, J. M., Huang, W. M., Sheppard, M. N., Tatemoto, K., and Bloom, S. R., 1984b, High concentrations of a novel peptide, neuropeptide Y, in the innervation of mouse and rat heart, *J. Histochem. Cytochem.* **32:**467–472.

Hacker, G. W., Polak, J. M., Springall, D. R., Ballesta, J., Cadieux, A., Gu, J., and Trojanowski, Q., 1985, Antibodies to neurofilament protein and other brain proteins reveal the innervation of peripheral organs, *Histochemistry* **82:**581–593.

Hakanson, R., and Owman, C., 1969, Argyrophilic reaction of histamine-containing epithelial cells in murine gastric mucosa, *Experientia* **25:**625–626.

Hakanson, R., Ekelund, M., and Sundler, F., 1984, Activation and proliferation of gastric endocrine cells, in: *Evaluation and Tumour Pathology of the Neuroendocrine System* (S. Falkmer, R. Hakanson, and F. Sundler, eds.), Elsevier, Amsterdam, pp. 371–383.

Hayano, T., Kato, K., Shimizu, A., Ariyoshi, Y., Yamada, Y., and Yamamoto, R., 1983, Production and characterisation of a monoclonal antibody to human nervous system-specific gamma gamma enolase, *J. Biochem.* **93:**1457–1460.

Heimer, L., and RoBards, M. J. (eds.), 1981, *Neuroanatomical Tract-Tracing Methods,* Plenum Press, New York.

Heitz, P., Kloppel, G., Hacki, W. H., Polak, J. M., and Pearse, A. G. E., 1977, Nesidioblastosis: The pathological basis of persistent hyperinsulinaemic hypoglycaemia in infants. Morphological and quanitative analysis of seven cases based on specific immunostaining and electron microscopy, *Diabetes* **26:**632–642.

Ibrahim, N. B. N., Briggs, J. C., and Corbishley, C. M., 1984, Extrapulmonary oat cell carcinoma, *Cancer* **54:**1645–1661.

Jessen, K. R., and Mirsky, R., 1980, Glial cells in the enteric nervous system contain glial fibrillary acidic protein, *Nature* **286:**736–737.

Klöppel, G., and Lenzen, S., 1984, Anatomy and physiology of the endocrine pancreas, in: *Pancreatic Pathology* (G. Klöppel and P. U. Heitz, eds.), Churchill Livingstone, Edinburgh, p. 133.

Lackie, P. M., Cuttitta, F., Minna, J. D., Bloom, S. R., and Polak, J. M., 1985, Localisation of receptors using a dimeric ligand and electron immunocytochemistry, *Histochemistry* **83:**57–59.

Larsson, L. I., Rehfeld, J. F., Stockbrugger, R., Blohme, G., Schöön, I. M., Lundquist, G., Kindblom, L. G., Save-Soberberg, J., Grimelius, L., Olbe, L., 1978, Mixed endocrine gastric tumours associated with hypergastrinaemia of antral origin, *Am. J. Pathol.* **93:**53–68.

Larsson, L. I., Rehfeld, J. F., Stockbrugger, R., Blohme, G., Schöön, I. M., Lundquist, G., Kindblom, L. G., Save-Soberberg, J., and Larsson, K.-I., 1981, Somatostatin cells, in: *Gut Hormones,* 2nd ed. (S. R. Bloom and J. M. Polak, eds.), Churchill Livingstone, Edinburgh, pp. 350–359.

Lewin, K. J., Yang, K., Ulich, T., Elashoff, J. D., and Walsh, J., 1984, Primary gastrin cell hyperplasia, *Am. J. Surg. Pathol.* **8:**821–832.

Mulderry, P. K., Ghatei, M. A., Rodrigo, J., Allen, J. M., Rosenfeld, M. G., Polak, J. M., and Bloom, S. R., 1985, Calcitonin gene-related peptide in cardiovascular tissues of the rat, *Neuroscience* **14:**947–954.

Nakane, P. K., 1968, Simultaneous localization of multiple tissue antigens using the peroxidase-labeled antibody method: A study in pituitary glands of the rat, *J. Histochem. Cytochem.* **16:** 557–560.

O'Connor, D. T., Frigon, F. P., and Sokoloff, R. L., 1983, Human chromogranin A: Purification and characterization from catecholamine storage vesicles of pheochromocytoma, *Hypertension* **6:**2–12.

O'Morain, C., Bishop, A. E., McGregor, G. P., Levi, A. J., Bloom, S. R., Polak, J. M., and Peters, T. J., 1984, Vasoactive intestinal peptide concentrations and immunocytochemical studies in rectal biopsies from patients with inflammatory bowel disease, *Gut* **25:**57–61.

Palluk, P., Gaida, W., and Hoefke, W., 1985, Atrial natriuretic factor, *Life Sci.* **36:**1415–1427.

Pearse, A. G. E., 1984, The diffuse neuroendocrine system: Historical review, in: *Interdisciplinary Neuroendocrinology,* Vol. 12 (M. Ratzenhofer, H. Höfler, and G. F. Walter, eds.), S. Karger, Basel, pp. 1–7.

Polak, J. M., and Bloom, S. R., 1982, Localization of regulatory peptides in the gut, *Br. Med. Bull.* **38:**303–308.

Polak, J. M., and Bloom, S. R., 1983, Regulatory peptides: Key factors the control of bodily functions, *Br. Med. J.* **286:**1461–1466.

Polak, J. M., and Bloom, S. R., 1984, Regulatory peptides of the respiratory tract: A newly discovered control system, in: *Frontiers in Neuroendocrinology* Vol. 8 (L. Martini and W. E. Ganong, eds.), Raven Press, New York, pp. 199–205.

Polak, J. M., and Bloom, S. R., 1985a, Neuroendocrine neoplasms and regulatory peptides. An introduction, in: *Endocrine Tumours* (J. M. Polak and S. R. Bloom, eds.), Churchill Livingstone, Edinburgh, pp. 1–20.

Polak, J. M., and Bloom, S. R., 1985b, Pathology of peptide-producing neuroendocrine tumours, *Br. J. Hosp. Med.* **1985:**78–88.

Polak, J. M., and Marangos, P. J., 1984, Neuron specific enolase, a marker for neuroendocrine cells, in: *Evolution and Tumour Pathology the Neuroendocrine System,* Vol. 4 (S. Falkmer, R. Hakanson, and F. Sundler, eds.), Elsevier, Amsterdam, pp. 433–452.

Polak, J. M., and Varndell, I. M. (eds.), 1984, *Immunolabelling for Electron Microscopy,* Elsevier, Amsterdam.

Polak, J. M., Stagg, B., and Pearse, A. G. E., 1972, The two types of Zollinger–Ellison syndrome. Immunofluorescent, cytochemical and ultrastructural studies of the antral and pancreatic gastric cells in different clinical states, *Gut* **13:**510–512.

Polak, J. M., Hoffbrand, V., Reed, P. I., and Pearse, A. G. E., 1973a, Qualitative and quantitative studies of antral and fundic G cells in pernicious anaemia, *Scand. J. Gastroenterol.* **8:**361–367.

Polak, J. M., Pearse, A. G. E., Van Noorden, S., Bloom, S. R., and Rossiter, M. A., 1973b, Secretin cells in coeliac disease, *Gut* **14:**870–874.

Roberts, J. Z., and Wilcox, J. N., 1986, Hybridization histochemistry: Analysis of specific mRNAs in individual cells, in: *Immunocytochemistry. Modern Methods and Applications,* 2nd ed. (J. M. Polak and S. Van Nooren, eds.), John Wright & Sons Ltd., Bristol, pp. 198–203.

Sabate, M. I., Carlei, F., Bloom, S. R., and Polak, J. M., 1985, Endocrine tumours of the gut and pancreas, in: *Endocrine Tumours* (J. M. Polak and S. R. Bloom, eds.), Churchill Livingstone, Edinburgh, pp. 193–208.

Schneider, F. H., Smith, A. D., and Winkler, H., 1967, Secretion from the adrenal medulla: Biochemical evidence for exocytosis, *Br. J. Pharm. Chemother.* **31:**94–104.

Sevier, A. C., and Munger, B. L., 1965, A silver method for paraffin sections of neural tissue, *J. Neuropathol. Exp. Neurol.* **24:**130–135.

Sheppard, M. N., Corrin, B., Bloom, S. R., and Polak, J. M., 1985, Lung endocrine tumours, in: *Endocrine Tumours* (J. M. Polak and S. R. Bloom eds.), Churchill Livingstone, Edinburgh, pp. 209–228.

Sjolund, K., Alumets, J., Berg, N. O., Hakanson, R., and Sundler, F., 1982, Enteropathy of coeliac disease in adults: Increased number of enterochromaffin cells in the duodenal mucosa, *Gut* **23:**42–48.

Solcia, E., 1984, Cytology of tumours in the gastroenteropancreatic and diffuse neuroendocrine system, in: *Evaluation and Tumour Pathology of the Neuroendocrinc System* (S. Falkmer, R. Hakanson, and F. Sundler, eds.), Elsevier, Amsterdam, pp. 453–457.

Solcia, E., Polak, J. M., Larsson, L.-I., Buchan, A. M. J., and Capella, C., 1981, Update on Lausanne classification of endocrine cells, in: *Gut Hormones,* 2nd ed. (S. R. Bloom and J. M. Polak, eds.), Churchill Livingstone, Edinburgh, pp. 96–101.

Springall, D. R., Hacker, G. W., Grimelius, L., and Polak, J. M., 1984, The potential of the immunogold-silver staining method for paraffin sections, *Histochemistry* **81:**603–608.

Tramu, G., Pillez, A., and Leonardelli, J., 1978, An efficient method of antibody elution for the successive or simultaneous localisation of two antigens by immunocytochemistry, *J. Histochem. Cytochem.* **78**(26):322–324.

Trojanowski, J. Q., 1986, Neurofilaments and glial filaments in neuropathology, in: *Immunocytochemistry, Modern Methods and Applications,* 2nd ed. (S. Van Noorden and J. M. Polak, eds.), John Wright & Sons, Bristol, pp. 413–424.

Van Noorden, S., 1986, Tissue preparation and immunocytochemical staining techniques, in: *Immunocytochemistry. Modern Methods and Applications,* 2nd ed. (S. Van Noorden and J. M. Polak, eds.), John Wright & Sons, Bristol, pp. 26–27.

Van Noorden, S., and Polak, J. M., 1985, Immunocytochemistry of regulatory peptides, in: *Techniques in Immunocytochemistry,* Vol. III (G. R. Bullock and P. Petrusz, eds.), Academic Press, London, pp. 116–154.

Varndell, I. M., Polak, J. M., Sikri, K. L., Minth, C. D., Bloom, S. R., and Dixon, J. E., 1984, Visualisation of messenger RNA directing peptide synthesis by *in situ* hybridisation using a novel single-stranded cDNA probe: Potential for the investigation of gene expression and endocrine cell activity, *Histochemistry* **81:**597–601.

Varndell, I. M., Bishop, A. E., Sikri, K. L., Uttenthal, L. O., Bloom, S. R., and Polak, J. M., 1985a, Localization of glucagon-like peptide (GLP) immunoreactants in human gut and pancreas using light and electron microscopial immunocytochemistry, *J. Histochem. Cytochem.* **33:**1080–1086.

Wharton, J., Polak, J. M., McGregor, G. P., Bishop, A. E., and Bloom, S. R., 1981, The distribution of substance P-like immunoreactive nerves in the guinea-pig heart, *Neuroscience* **6:**2193–2204.

Williams, E. D., 1985, Medullary carcinoma of the thyroid. in: *Endocrine Tumours* (J. M. Polak and S. R. Bloom, eds.), Churchill Livingstone, Edinburgh, pp. 229–240.

Wilson, B. S., and Lloyd, R. V., 1984, Detection of chromogranin in neuroendocrine cells with a monoclonal antibody, *Am. J. Pathol.* **115:**458–468.

33

Structure–Function Studies of Agonists and Antagonists for Gastrointestinal Peptide Receptors on Pancreatic Acinar Cells

ROBERT T. JENSEN and JERRY D. GARDNER

1. INTRODUCTION

With the development of techniques to prepare dispersed pancreatic acini that respond to secretagogues, it became possible to investigate the cellular basis of action of the various peptides that stimulate pancreatic enzyme secretion. Various peptides differed in efficacy, potency, and configuration of their dose–response curves for stimulating enzyme secretion (Fig. 1). Pancreatic acini have receptors for different secretagogues, and with the development of techniques to prepare biologically active radiolabeled ligands for each group of structurally similar peptides, the existence of multiple classes of receptors was demonstrated directly (Fig. 2). Figure 2 summarizes the seven different classes of receptors that mediate the action of various secretagogues on pancreatic acini. Also summarized in Fig. 2 is the biochemical basis of action of each of these different receptors.

There are two functionally distinct pathways by which secretagogues can stimulate pancreatic enzyme secretion (for review see Gardner and Jensen, 1981). One pathway involves activation of adenylate cyclase and increased cellular cAMP and mediates the action of secretagogues such as VIP, secretin, calcitonin gene-related peptide (CGRP), and cholera toxin. Another pathway involves phosphoinositide breakdown and mobilization of cellular calcium and mediates the action of cholinergic agents; cholecystokinin (CCK), gastrin, and structurally related peptides;

ROBERT T. JENSEN and JERRY D. GARDNER • Digestive Diseases Branch, National Institute of Arthritis, Diabetes, and Digestive and Kidney Diseases, National Institutes of Health, Bethesda, Maryland 20205.

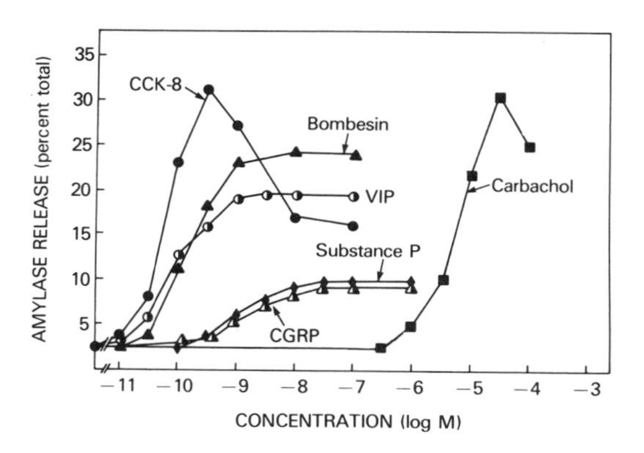

FIGURE 1. Ability of various peptides to stimulate amylase release from dispersed pancreatic acini from the guinea pig. Incubations were for 30 min at 37°C. Amylase release is expressed as the percentage of the total cellular amylase at the start of the incubation released into the extracellular medium during the incubation.

bombesin, gastrin-releasing peptide, and structurally related peptides; and physalaemin, substance P, and structurally related peptides. The secretagogues that cause release of cellular calcium also increase cellular cGMP and cause depolarization and decreased resistance of the plasma membrane of the acinar cell (Gardner and Jensen, 1981; Petersen, 1981). It is clear that neither changes in cGMP nor membrane electrical changes mediate the action of secretagogues on enzyme secretion but are, instead, secondary to the mobilization of cellular calcium (Gardner and Jensen, 1981; Petersen, 1981).

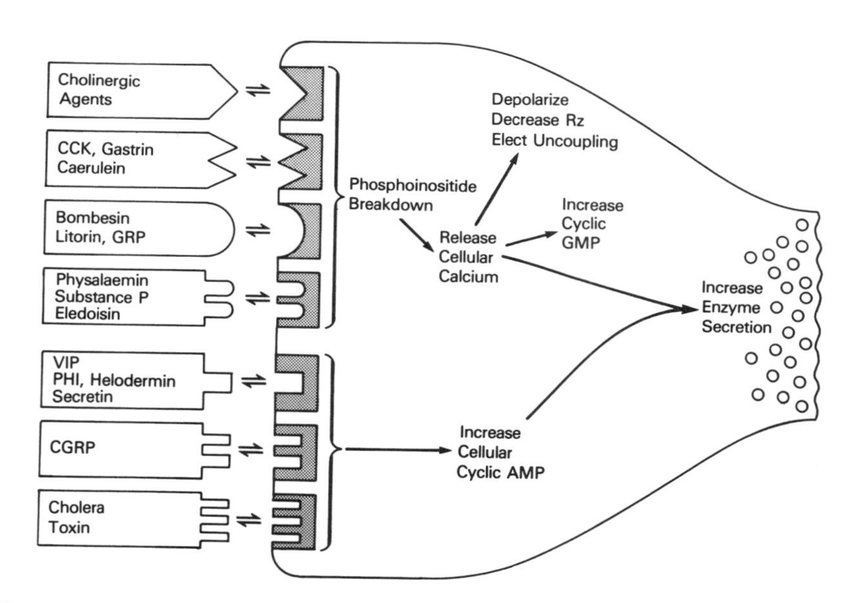

FIGURE 2. Receptors for secretagogues on pancreatic acini and their biochemical basis of action.

Because of the existence of multiple receptors on pancreatic acini and the marked responsiveness of the acini to various secretagogues, pancreatic acini have proved to be a particularly good system for detailed structure–function studies of various agonists and antagonists.

This chapter, because of limited space, concentrates primarily on the results of studies done in our laboratories, and the discussion of receptor agonists is primarily confined to results from studies of peptide fragments and excludes results from studies that analyzed the effects of various peptide analogues.

2. RECEPTORS FOR SECRETAGOGUES THAT MOBILIZE CELLULAR CALCIUM

2.1. Receptors for Cholecystokinin and Structurally Related Peptides

Specific receptors on pancreatic acini interact with the naturally occurring peptides CCK, CCK-39, CCK-8 (the C-terminal octapeptide of cholecystokinin), gastrin, and caerulein (a decapeptide isolated from the skin of the Australian frog *Hyla caerulea* that shares seven of eight C-terminal amino acids with CCK) (Anastasi, 1968; Jensen *et al.*, 1980). There is a good correlation between the concentration at which CCK and structurally related peptides inhibit binding of [^{125}I]CCK and the concentrations at which these same peptides produce changes in acinar cell function (Jensen *et al.*, 1980; Jensen and Gardner, 1981). For all CCK-related peptides, their dose–response curves for inhibition of binding of [^{125}]CCK are broad in that for a given peptide, its maximal effective concentration is at least 1000 times greater than its threshold concentration. The dose response for the ability of CCK-8 and related peptides to stimulate amylase secretion is also broad and biphasic (see Fig. 1), and the range of concentrations over which CCK-8 binds to receptors is the same as that over which the secretagogue produces changes in amylase secretion. Occupation of up to 40% of the receptors by CCK-8 causes progressive stimulation of amylase secretion, and occupation of the remaining 60% causes a progressive reduction in stimulated amylase secretion. Occupation of up to 75% of the receptors by CCK-8 causes a progressive increase in calcium outflux, and occupation of the remaining 25% causes no further change in calcium outflux. Thus, pancreatic acinar cells possess spare receptors for CCK in terms of the secretagogue-induced changes in calcium outflux but not in terms of the changes in amylase secretion (Jensen and Gardner, 1981; Jensen *et al.*, 1982).

Two hypotheses·have been proposed to account for the broad dose–inhibition curves for inhibition of binding of [^{125}I]CCK. One hypothesis, developed by Williams and his collaborators (Sankaran *et al.*, 1982), proposes that acinar cells possess two distinct classes of receptors—one class with a low capacity and a high affinity for CCK and another class with a high capacity and a low affinity for CCK. The major problem with this hypothesis is that it will not account for all the biological actions of CCK on pancreatic acinar cells (for example, see Collins *et al.*, 1981a,b). Another hypothesis, developed by Gardner and co-workers (Collins *et al.*, 1981a,b), proposes that pancreatic acinar cells possess receptors each of which have two classes of binding sites for CCK and that occupation of one binding site influences the affinity of the other binding site for CCK. The major problem with this hypothesis is that it does not account for the broad dose–inhibition curve in a quantitative manner. Thus, at the present time there is no hypothesis that adequately describes the binding of [^{125}I]CCK as well as the biological actions of CCK on pancreatic acinar cells.

The relative potencies for CCK and its naturally occurring analogues for occupying the CCK receptor and stimulating changes in cellular function (amylase release, cGMP generation, calcium outflux) are caerulein > CCK-8 > CCK > gastrin (Table I) (Jensen *et al.*, 1980; Jensen and Gardner, 1981).

Structure–function studies of various fragments and analogues of CCK indicate that C-terminal fragments as small as the dipeptide amide are capable of interacting with the CCK receptor on pancreatic acinar cells (Jensen *et al.*, 1982, 1983b). In general, the following can be stated concerning the relationship between receptor occupation and the accompanying change in cellular function. The desulfated C-terminal heptapeptide amide of CCK ([des-SO_3H]CCK-7) is the smallest C-terminal fragment of CCK with the full spectrum of biological activity of CCK. Adding a sulfate ester to the tyrosine residue of [des-SO_3H]CCK-7 increases the apparent affinity of the peptide by approximately 1000-fold. Adding an N-terminal aspartyl residue to CCK-7, thereby producing CCK-8, causes a three- to tenfold increase in the apparent affinity of the peptide for the CCK receptors. Extending the N-terminus of CCK-8 either does not change or reduces by as much as tenfold the apparent affinity of the peptide for CCK receptors. Shorter peptides, CCK-6, CCK-5, CCK-4, interact with CCK receptors but do not reproduce the full spectrum of the dose–response curve for stimulation of enzyme secretion. In particular, CCK-6, CCK-5, and CCK-4 cause the same maximal stimulation of enzyme secretion as CCK-7 or longer peptides but at supramaximal concentrations do not cause the same decrement in the dose–response curve. Moreover, CCK-6, CCK-5, and CCK-4 inhibit the downstroke of the dose–response curve for enzyme secretion caused by supramaximal concentrations of CCK-7 or longer peptides. CCK-3 and CCK-2 interact with CCK receptors on pancreatic acinar cells but do not possess agonist activity; therefore, these peptides function as CCK receptor antagonists.

Structure–function studies demonstrate that not only the position but the distance of the sulfate ester from the peptide backbone is also important in determining the potency of CCK-8 and caerulein. Evidence that the distance of the sulfate ester from the peptide backbone is important was demonstrated by a study in which the tyrosine O-sulfate in the 27th residue of CCK was replaced by ξ-hydroxynorleucine O-sulfate. This change resulted in only a tenfold decrease in activity, whereas replacement by serine sulfate and removal of the sulfate ester from the tyrosine resulted in a 1000-fold and a 600-fold decrease in activity, respectively (Bodanszky *et al.*, 1978). Studies of caerulein analogues demonstrate that the relationship of

TABLE I. Secretagogue Receptors on Acinar Cells from Guinea Pig Pancreas: Agents That Mobilize Cellular Calcium[a]

Class of receptors	Ligand used for binding studies	Naturally occurring agonist (EC_{50})	Competitive antagonist
CCK	[^{125}I]CCK	Caerulein (0.2 nM)	Derivatives of cyclic nucleotides
	[^{125}I]CCK-8	CCK (2 nM)	Derivatives of amino acids
		Gastrin (2 μM)	Partial sequences of the C-terminal region of CCK
Bombesin	[^{125}I-Tyr^{4}]Bombesin	Bombesin (4 nM)	
		Ranatensin (12 nM)	[D-Arg^{1}, D-Pro^{2}, D-Trp^{7,9}, Leu^{11}]Substance P
		Alytesin (12 nM)	
		Litorin (40 nM)	
Physalaemin	[^{125}I]Physalaemin	Physalaemin (2 nM)	[D-Pro^{2}, D-Trp^{7,9}]Substance P
	[^{125}I]Substance P	Substance P (5 nM)	[D-Pro^{2}, D-Phe^{7}, D-Trp^{9}]Substance P
		Eledoisin (100 nM)	[D-Arg^{1}, D-Pro^{2}, D-Trp^{7,9}, Leu^{11}]Substance P
		Kassinin (250 nM)	
Muscarinic cholinergic	[^{3}H]QNB [^{3}H]NMS	Muscarinic cholinergic agonists	Muscarinic cholinergic antagonists

[a] EC_{50} refers to the concentration required to occupy 50% of the receptors. CCK, cholecystokinin; QNB, quinuclidinyl benzilate; NMS, N-methylscopolamine.

the distance of the C terminus to the sulfate ester is important, because moving the tyrosyl-O-sulfate eight or six amino acids from the C terminus yielded peptides that were no more active than the desulfated analogue (de Castiglione, 1977).

Since 1979 three classes of CCK receptor antagonists have been described: (1) derivatives of cyclic nucleotides, (2) derivatives of amino acids, and (3) partial sequences of the C-terminal region of CCK (for review see Gardner and Jensen, 1984). The abilities of these various agents to antagonize the actions of CCK have in every instance been first described in dispersed acini prepared from guinea pig pancreas. These antagonists share several common features. Each antagonist inhibits CCK-stimulated amylase secretion and CCK-stimulated outflux of calcium, and the dose–response curves for those inhibitory actions correlate closely with the dose–response curve for the ability of the antagonist to inhibit binding of [^{125}I]CCK to its cell surface receptors on pancreatic acinar cells. The antagonist-induced inhibition of the action of CCK is fully and rapidly reversible, is competitive in nature, and is specific for those secretagogues that interact with the CCK receptor. In particular, the various CCK receptor antagonists do not alter the actions of muscarinic cholinergic agents, bombesin, physalaemin, vasoactive intestinal peptide, secretin, cholera toxin, 8Br-cAMP, or A23187.

In pancreatic acini that have first been incubated with relatively high concentrations of CCK and then washed, there is significant residual stimulation of enzyme secretion (Collins *et al.*, 1981a,b), and this CCK-induced residual stimulation of enzyme secretion can be reversed immediately by adding a CCK receptor antagonist (Collins *et al.*, 1981b; Gardner and Jensen, 1984). Even though their action is specific, reversible, and competitive in nature, each of the classes of CCK receptor antagonists is of relatively low affinity (Gardner and Jensen, 1984). Structure–function studies demonstrate that for the cyclic nucleotides, various derivatives of cIMP, cGMP, and cAMP all function as CCK receptor antagonists, and of this class, dibutyryl cGMP is the most potent (K_i 0.1 mM) (Barlas *et al.*, 1982; Gardner and Jensen, 1984).

The second class of CCK receptor antagonists comprises amino acid derivatives. This class of CCK receptor antagonists including proglumide, a glutaramic acid derivative (K_i 1.0 mM), benzotript, a derivative of tryptophan (N-*p*-chlorobenzoyltryptophan) (K_i 0.3 mM) (Hahne *et al.*, 1981), as well as other amino acid derivatives (Jensen *et al.*, 1983B; Maton *et al.*, 1985). Structure–function studies have demonstrated that both the nature of the amino acid side chain and the hydrophobicity of the N-aminoacyl moiety are important determinants of potency (Maton *et al.*, 1985; Jensen *et al.*, 1985). Of the amino acid derivatives tested, phenoxyacetylproglumide [(N-phenoxyacetyl)-L-glutamic acid-1-di-*n*-propylamide] is the most potent (K_i 4.5 μM) and has been shown to function as a specific CCK receptor antagonist of the action of CCK in rat, mouse, and guinea pig pancreas (Jensen *et al.*, 1985).

The third class of CCK receptor antagonists consists of various fragments of the C-terminus of CCK. Structure–function studies have demonstrated that both C-terminal fragments, BOC-Met-Asp-Phe-NH_2 (K_i 0.3 mM), BOC-Asp-Phe-NH_2 (K_i

1 mM), and BOC-Phe-NH_2 (K_i 3 mM), and various N-terminal fragments of CCK-7 or CCK-8 function as CCK receptor antagonists (Jensen *et al.*, 1983c; Gardner and Jensen, 1984). Of the N-terminal fragments, CBZ-CCK_{27-32}-NH_2 [CBZ-Tyr-(SO_3H)-Met-Gly-Trp-Met-Asp-NH_2] is the most potent (K_i 3 μM) at inhibiting the action of CCK in guinea pig pancreas (Gardner and Jensen, 1984). In a recent study, however, CBZ-CCK_{27-32}-NH_2 had partial agonist activity in pancreatic acini from mouse or rat (Howard *et al.*, 1984), suggesting that COOH-terminal analogues of CCK may not function as antagonists in all species (Howard *et al.*, 1984).

2.2. Receptors for Bombesin and Structurally Related Peptides

Specific receptors on pancreatic acini interact with the naturally occurring peptides bombesin, gastrin-releasing peptide (GRP), neuromedin B, neuromedin C, litorin, alytesin, and ranatensin (Jensen *et al.*, 1978; Jensen and Gardner, 1984). There is a good correlation between the concentration at which bombesin and related peptides inhibit the binding of [^{125}I-Tyr^4]-bombesin and their abilities to stimulate amylase secretion. The range of concentrations over which bombesin binds to its receptors is the same as that over which bombesin produces changes in amylase secretion. Occupation of up to approximately 50% of the receptors by bombesin causes progressive stimulation of amylase secretion, and occupation of the remaining 50% causes a small reduction (approximately 10%) in stimulated amylase secretion. Occupation of up to 75% of the receptors by bombesin causes a maximal increase in calcium outflux, and occupation of the remaining 25% causes no further change in calcium outflux. Thus, as occurs with CCK receptors, pancreatic acinar cells possess spare receptors for bombesin in terms of the secretagogue-induced changes in calcium outflux but not in terms of the changes in amylase secretion (Jensen and Gardner, 1981).

The relative potencies for bombesin and its naturally occurring analogues for occupying the bombesin receptor and stimulating changes in cellular function are bombesin > neuromedin C > alytesin, ranatensin, GRP > litorin > neuromedin B (Table I) (Jensen *et al.*, 1978; Jensen and Gardner, 1984).

As is the case with CCK, the intrinsic biological activity of bombesin and structurally related peptides is a property of the C-terminal portion of the molecule, and the C-terminal nonapeptide of bombesin has the same potency and efficacy as does the native tetradecapeptide (Erspamer and Melchiorri, 1973). Shorter C-terminal fragments of bombesin still possess full intrinisic biological activity, but their potencies are less than that of native bombesin. The region of the bombesin molecule that possesses intrinsic biological activity also determines the affinity of the peptide for its receptors. That is, the variation in potency among various fragments and analogues of bombesin reflects variation in their receptor affinities (Jensen *et al.*, 1979).

Of the various fragments and analogues of bombesin that have been tested, none have been found to occupy the receptor and not cause a full biological response (Jensen *et al.*, 1978, 1979). Recently, a substituted analogue of substance P, [D-

Arg[1], D-Pro[2], D-Trp[7,9], Leu[11]]substance P, which functions as a substance P receptor antagonist, has been found also to function as a bombesin receptor antagonist (Jensen *et al.*, 1984a). This substituted analogue of substance P inhibited binding of [^{125}I-Tyr[4]]bombesin and bombesin-stimulated enzyme secretion over the same concentrations. The inhibition was competitive in nature and was specific for agents that interacted with the bombesin receptor. [D-Arg[1], D-Pro[2], D-Trp[7,9], Leu[11]]Substance P was fivefold more potent as a substance P receptor antagonist (K_i 0.6 μM) than as a bombesin receptor antagonist (K_i 3 μM) but had 150-fold less affinity than substance P for the substance P receptor and 750 times less affinity than bombesin for the bombesin receptor (Jensen *et al.*, 1984a).

2.3. Receptors for Substance P and Structurally Related Peptides

Specific receptors on pancreatic acini have been shown to interact with physalaemin, substance P, eledoisin, and kassinin. Either [^{125}I]physalaemin or [^{125}I]substance P can be used to identify the class of binding sites that interacts with physalaemin and structurally related peptides but not with other pancreatic secretagogues (Table I) (Jensen *et al.*, 1979, 1984b; Sjodin *et al.*, 1980). There is a good correlation between the concentrations at which physalaemin and structurally related peptides inhibit binding of [^{125}I]physalaemin and concentrations at which these same peptides produce changes in acinar cell function.

The dose–response curve for physalaemin-stimulated amylase secretion is the same as that for the physalaemin-induced increase in calcium outflux (Jensen and Gardner, 1981). Occupation of approximately 50% of the receptors by physalaemin causes maximal stimulation of amylase release and calcium outflux, and occupation of the remaining receptors causes no further change in these functions (Jensen and Gardner, 1981). Thus, as occurs with secretagogues whose actions are mediated by CCK receptors and with those whose actions are mediated by bombesin receptors, pancreatic acinar cells possess spare receptors for physalaemin in terms of the secretagogue-induced changes in calcium outflux. In contrast to results with CCK or with bombesin, there is also spareness at the receptor level in terms of the ability of physalaemin to stimulate amylase secretion.

The relative potencies for physalaemin, substance P, and their naturally occurring analogues for occupying the substance P receptor and causing changes in cellular function are physalaemin > substance P > eledoisin > kassinin (Table I) (Jensen and Gardner, 1984).

As occurs with CCK-related peptides and with bombesin-related peptides, the intrinsic biological activity of physalaemin, substance P, and structurally related peptides is a property of the C-terminal portion of the molecule. The smallest C-terminal fragment of physalaemin that has been examined and found to retain biological activity is the pentapeptide (Espamer and Melchiorri, 1973). Eledoisin is approximately 50% more effective than physalaemin or substance P in stimulating pancreatic enzyme secretion (Jensen and Gardner, 1979). This finding suggests that with substance P and related peptides, the portion of the molecule that possesses intrinsic biological activity is not congruent with the region of the molecule that influences the affinity of the peptide for its receptor. The finding is further supported by the recent description of D-amino acid-substituted derivatives and C-terminal fragments of substance P that function as competitive inhibitors of the action of substance P and related peptides (Table I) (see below).

Recently three analogues of substance P have been synthesized and found to antagonize competitively the actions of substance P and structurally related peptides on pancreatic acinar cells, [D-Pro^2, D-Trp^{7,9}]substance P (K_i 6 μM), [D-Pro^2, D-Phe^7, D-Trp^9]substance P (K_i 1.3 μM), and [D-Arg^1, D-Pro^2, D-Trp^{7,9}, Leu^{11}]substance P (K_i 0.6 μM) (Jensen *et al.*, 1984b; Table I). These antagonists share several common features. Each antagonist inhibits the stimulation of amylase secretion and calcium outflux caused by physalaemin, substance P, and structurally related peptides, and the dose–response curves for these inhibitory actions correlate closely with the dose–response curve for the ability of the antagonist to inhibit binding of [^{125}I]physalaemin to its cell surface receptors on pancreatic acinar cells. The antagonist-induced inhibition of the action of physalaemin is fully and rapidly reversible, is competitive in nature, and is specific for those secretagogues that interact with the substance P receptors. In particular, the various substance P receptor antagonists do not alter the actions of CCK, bombesin, muscarinic cholinergic agonists, vasoactive intestinal peptide, secretin, cholera toxin, 8Br-cAMP, or A23187 (Jensen, 1984b).

3. RECEPTORS FOR SECRETAGOGUES THAT INCREASE CELLULAR CYCLIC AMP

3.1. Receptors for Vasoactive Intestinal Peptide and Secretin

Measurement of binding of [^{125}I]VIP to acinar cells from guinea pig pancreas indicates that these cells have two functionally distinct classes of receptors, each of which interact with VIP and secretin (for review see Gardner and Jensen, 1981; Jensen and Gardner, 1981). One class has a high affinity for VIP and a low affinity for secretin; the other class has a high affinity for secretin and a low affinity for VIP. Measurement of binding of [^{125}I]secretin to acinar cells from guinea pig pancreas detects only the sites with a high affinity for secretin, because the sites having a high affinity for VIP are small in number and have a low affinity for secretin and therefore do not bind significant amounts of [^{125}I]secretin (Jensen *et al.*, 1983a). There is a close correlation between occupation of the VIP-preferring receptors and the secretin-preferring receptors by VIP, secretin, or a structurally related peptide and the accompanying increase in cellular cAMP (Gardner and Jensen, 1981; Raufman *et al.*, 1982).

Occupation of the VIP-preferring receptors by VIP, secretin, PHI, or gila monster venom causes a significant increase in cellular cAMP and an accompanying increase in enzyme secretion (Jensen and Gardner, 1981; Jensen *et al.*, 1981, 1983a;

Raufman *et al.*, 1982; Table II). The dose–response curve for VIP binding to receptors is the same as the dose–response curve for the action of VIP on cellular cAMP. Both the dose–response curve for VIP binding and that for the VIP-induced increase in cellular cAMP are to the right of the dose–response curve for VIP-stimulated amylase secretion. These findings indicate that there are no spare VIP-preferring receptors in terms of the increase in cellular cAMP but that there is "spare mediator" in the sense that relatively small increases in cellular cAMP are sufficient to produce maximal stimulation of amylase secretion (Jensen and Gardner, 1981).

Occupation of the secretin-preferring receptors by secretin, VIP, PHI, or gila monster venom (Jensen and Gardner, 1981; Jensen *et al.*, 1981, 1983a; Raufman *et al.*, 1982; Table II) causes a pronounced increase in cellular cAMP without an accompanying increase in enzyme secretion. These findings indicate that in acinar cells from guinea pig pancreas, cAMP is compartmentalized, and only the cAMP that is produced in response to occupation of the VIP-preferring receptors is in the appropriate compartment to cause stimulation of enzyme secretion (Gardner *et al.*, 1982). Measurements of binding of [^{3}H]ouabain suggest that the increase in cellular cAMP that occurs in response to occupation of secretin-preferring receptors may be responsible for stimulation of fluid and electrolyte secretion from pancreatic acinar cells (Hootman *et al.*, 1983). The dose–response curve for binding of secretin is to the right of that for the secretin-induced increase in cellular cAMP. In contrast to the results with the VIP-preferring receptor, these results indicate that there are spare secretin-preferring receptors with respect to the increase in cellular cAMP and that occupation of only a fraction of the secretin-preferring receptors is sufficient to produce a maximal increase in cellular cAMP.

TABLE II. Secretagogue Receptors on Acinar Cells from Guinea Pig Pancreas: Agents that Increase AMP[a]

Class of receptors	Ligand used for binding studies[b]	Naturally occurring agonist (EC_{50})	Competitive antagonist
VIP-preferring	[^{125}I]VIP	VIP (0.9 nM)	$Secretin_{5-27}$
		PHI (100 nM)	$Secretin_{9-27}$
		Secretin (7 μM)	$Secretin_{14-27}$
		Gila monster venom (3 μg/ml)	
Secretin-preferring	[^{125}I]Secretin	Secretin (7 nM)	$Secretin_{5-27}$
	[^{125}I]VIP	VIP (6 μM)	$Secretin_{9-27}$
		PHI (6 μM)	$Secretin_{14-27}$
		Gila monster venom (30 μg/ml)	
Cholera toxin	[^{125}I]Cholera toxin	Cholera toxin (3 nM)	Choleragenoid
CGRP	[^{125}I]CGRP	Rat CGRP (100 nM)	None
		Human CGRP (300 nM)	None

[a] EC_{50} refers to the concentration required to occupy 50% of the receptors.
[b] VIP, vasoactive intestinal peptide; PHI, peptide histidine isoleucine; CGRP, calcitonin gene-related peptide.

The foregoing discussion pertains to results obtained using acinar cells from guinea pig pancreas. A much different pattern occurs with acinar cells from rat pancreas, in which four classes of receptors are required to account for the actions of VIP and secretin (Bissonnette *et al.*, 1984). One class has a high affinity for VIP and does not interact with secretin, and occupation of this class of receptors causes increased cellular cAMP and stimulation of amylase secretion. A second class has a low affinity for VIP and for secretin, and occupation of these receptors does not cause changes in cAMP or amylase secretion. A third class of receptors has a high affinity for secretin and does not interact with VIP, and occupation of these receptors causes increased cellular cAMP and stimulation of amylase secretion. A fourth class of receptors has a low affinity for secretin and does not interact with VIP, and occupation of these receptors causes stimulation of amylase secretion by a non-cAMP-mediated mechanism. Not only do rat acinar cells differ from guinea pig acinar cells in terms of the number and type of receptors that interact with VIP and secretin, but in rat acinar cells all of the cellular cAMP appears to be coupled to stimulation of enzyme secretion. In addition, rat acinar cells possess a mechanism for secretin-induced stimulation of enzyme secretion that is not mediated by cAMP. In every other system studied to date, occupation of secretin receptors by secretin caused activation of adenylate cyclase and increased cAMP.

As indicated in Table II, VIP, secretin, PHI, and gila monster venom each interact with VIP-preferring receptors and with secretin-preferring receptors on acinar cells from guinea pig pancreas; VIP, PHI, and gila monster venom have higher affinities for the VIP-preferring receptors than for the secretin-preferring receptors, whereas secretin has a higher affinity for secretin-preferring receptors than for VIP-preferring receptors.

In contrast to those peptides that produce mobilization of calcium from pancreatic acinar cells and have their intrinsic biological activities in the C-terminal portion of the molecule, in VIP and secretin the intrinsic biological activity resides in the N-terminal portion of the molecule (Christophe *et al.*, 1976; Robberecht *et al.*, 1976). For example, secretin$_{1-14}$ has an efficacy that is equal to that of native secretin, whereas secretin$_{5-27}$ and secretin$_{14-27}$ have efficacies that are less than 2% of that of native secretin (Christophe *et al.*, 1976; Gardner *et al.*, 1979; Robberecht *et al.*, 1976). Another feature that distinguishes VIP and secretin from those peptides that cause mobilization of cellular calcium is that, to date, no fragment of VIP or secretin has been found to be as potent as the native molecule. For example, although secretin$_{1-14}$ has the same efficacy as native secretin, the apparent affinity of secretin$_{1-14}$ for secretin-preferring receptors on pancreatic acinar cells is approximately 1000 times less than the affinity of native secretin for these same receptors (Jensen and Gardner, 1981).

C-terminal fragments of secretin such as secretin$_{5-27}$, secretin$_{9-27}$, and secretin$_{14-27}$ function as specific, competitive antagonists of the interaction of secretagogues with VIP-preferring receptors and with secretin-preferring receptors on acinar cells from guinea pig pancreas (Table II). The antagonist-induced inhibition is fully and rapidly reversible and is specific for those secretagogues that interact with VIP-preferring and secretin-preferring receptors. In particular, C-terminal fragments of secretin do not alter the actions of CCK, bombesin, muscarinic cholinergic agonists, physalaemin, cholera toxin, 8Br-cAMP, or A23187. Studies of secretin$_{5-27}$ have also provided insight into the role of the N-terminal tetrapeptide of secretin in influencing the apparent affinity of secretin for VIP-preferring receptors and secretin-preferring receptors on acinar cells from guinea pig pancreas. Deleting the N-terminal tetrapeptide from secretin does not alter the apparent affinity of the peptide for VIP-preferring receptors but causes a 1000-fold decrease in the apparent affinity of the peptide for secretin-preferring receptors (for review see Jensen and Gardner, 1981).

Originally the interpretation that acinar cells from guinea pig pancreas possess two functionally distinct classes of receptors for VIP and secretin was based primarily on the dose–response curves for the abilities of VIP and secretin to inhibit binding of [^{125}I]VIP and the corresponding changes in cellular cAMP and amylase secretion. Subsequent studies using secretin$_{5-27}$ and two analogues of secretin$_{5-27}$ showed that the structural requirements for occupation of VIP-preferring receptors differ from those for occupation of secretin-preferring receptors (Gardner *et al.*, 1979). For example, in secretin$_{5-27}$, replacing aspartic acid in position 15 by lysine increased the affinity of the peptide for VIP-preferring receptors but decreases the affinity of the peptide for secretin-preferring receptors.

The apparent affinities of secretin$_{5-27}$ and its analogues for receptors on acinar cells from guinea pig pancreas also provide insight into the role of the charge of the amino acid in position 15 of secretin and of VIP in influencing the interaction of these two peptides with their receptors. The amino acid in position 15 is of potential interest because secretin has an acid aspartyl residue in this position, whereas VIP has a basic lysyl residue in this position. In secretin$_{5-27}$, replacing the aspartic acid residue in position 15 by its carboxamide derivative ([Asn^{15}]secretin$_{5-27}$) or by lysine ([Lys^{15}]secretin$_{5-27}$) causes an eightfold increase in the apparent affinity of the peptide for the VIP-preferring receptors (Gardner *et al.,* 1979).

Thus, the absence of an acidic residue in position 15 is sufficient to increase the apparent affinity of the peptide for the VIP-preferring receptors, and the presence of a basic lysyl residue does not cause a further increase in the affinity of the peptide. Because VIP has lysine in position 15, one might anticipate that in secretin$_{5-27}$ replacing aspartic acid in position 15 by lysine would make the peptide more VIP-like. That is, in fact, what occurs; however, the increased affinity of the substituted peptide appears to result from the removal of the negatively charged aspartic acid rather than from the addition of the positively charged lysine (Gardner *et al.,* 1979). In secretin$_{5-27}$, replacing the aspartic acid residue in position 15 by asparagine does not alter the affinity of the peptide for the secretin-preferring receptors, whereas replacement by lysine causes a sixfold reduction in the affinity of the peptide for these receptors (Gardner *et al.,* 1979). These findings indicate that the absence of a basic lysyl residue in position 15 of secretin$_{5-27}$ is sufficient to increase the affinity of the peptide for the secretin-preferring receptors, and the presence of an acidic aspartyl residue does not cause a further increase in the apparent affinity of the peptide.

Thus, although VIP-preferring receptors differ from secretin-preferring receptors in terms of their abilities to distinguish among $secretin_{5-27}$ and its analogues (Gardner *et al.*, 1979), each class of receptors displays the somewhat surprising characteristic that the affinities for $secretin_{5-27}$ and its analogues are influenced primarily by the absence of a particular charge in position 15 but not by the presence of the opposite charge (Gardner *et al.*, 1979).

3.2. Receptors for Cholera Toxin

[^{125}I]Cholera toxin can be used to identify binding sites on pancreatic acinar cells that interact with cholera toxin and its "aggregated B subunit," choleragenoid (Gardner and Rottman, 1979). There is a close correlation between the ability of cholera toxin to bind to receptors and its ability to produce changes in acinar cell function. As is the case for the relationship between occupation of VIP-preferring receptors on guinea pig pancreatic acinar cells and cellular cAMP, the dose–response curve for cholera toxin binding is the same as the dose–response curve for cholera-toxin-induced increase in cellular cAMP. Both the dose–response curve for binding of cholera toxin and that for the cholera-toxin-induced increase in cAMP are to the right of the dose-response curve for cholera-toxin-stimulated amylase secretion. These findings indicate that there are no spare receptors for cholera toxin in terms of the increase in cellular cAMP but that there is "spare mediator" in the sense that relatively small increases in cellular cAMP are sufficient to produce maximal stimulation of amylase secretion. Thus, in acinar cells from guinea pig pancreas the behavior of receptors for cholera toxin is similar to that of VIP-preferring receptors and different from that of secretin-preferring receptors.

Choleragenoid or the "aggregated B subunit" of cholera toxin (Gill, 1977) is a competitive antagonist of the interaction of cholera toxin with its receptors on pancreatic acinar cells (Gardner and Rottman, 1979; Jensen and Gardner, 1981) (Table II).

3.3. Receptors for Calcitonin Gene-Related Peptide

At present no detailed studies of the action of CGRP have been published, but one recent preliminary report indicates that pancreatic acini possess specific receptors for CGRP (Zhou *et al.*, 1985). Rat CGRP caused a fourfold increase in cAMP and amylase release with a half-maximal effect at 2 nM. CGRP did not affect binding of [^{125}I]VIP, demonstrating that CGRP did not interact with VIP-preferring receptors. With the phosphodiesterase inhibitor Ro-20-1724, maximally effective concentrations of secretin or natural glucagon caused a 145-fold and 71-fold increase in cAMP, respectively. In contrast, high concentrations of CGRP (1 μM) caused a fivefold increase in cAMP but did not affect the increase caused by secretin or natural glucagon, indicating that CGRP does not interact with the secretin-preferring receptors or with the receptors with which natural glucagon interacts to stimulate amylase release (Zhou *et al.*, 1985). These results have recently been confirmed by direct binding studies, which show that [^{125}I]CGRP binds to high-affinity sites on pancreatic acini (K_i 20 nM) and that the binding is not affected by secretin, VIP, or natural glucagon (Z.-C. Zhou, M. Noguchi, J. D. Gardner, and R. T. Jensen, unpublished data). There are no structure–function studies of agonists of the CGRP receptor on pancreatic acini and no reports of a CGRP receptor antagonist.

4. SUMMARY

Guinea pig pancreatic acini possess specific receptors for a number of gastrointestinal peptides that are able to stimulate secretion of digestive enzymes. Specific receptors have been identified that interact with cholecystokinin (CCK), caerulein, and gastrin; bombesin, neuromedin B and C, and related peptides; substance P, physalaemin, and related peptides; vasoactive intestinal peptide (VIP), PHI, and related peptides; secretin and related peptides; calcitonin gene-related peptide (CGRP) and related peptides; and cholera toxin.

For each receptor, specific radiolabeled ligands have been developed, which have allowed detailed structure–function studies to be done, studies of the relationship of receptor occupation to intracellular events and studies evaluating possible receptor antagonists. The results from some of these studies are briefly reviewed. Three groups of specific CCK receptor antagonists have been identified, which include cyclic nucleotide analogues such as dibutyryl cGMP, amino acid derivatives such as proglumide and various glutamic acid derivatives, which are more potent, and C-terminal fragments of CCK such as CCK_{27-32}-NH_2. Various C-terminal fragments of VIP and secretin act as receptor antagonists of the action of VIP and secretin. Various amino-acid-substituted substance P and SP_{4-11} analogues act as receptor antagonists of the action of substance P. No amino-acid-substituted analogues of bombesin or partial fragments of bombesin have been identified that function as partial agonists or receptor antagonists; however, [D-Arg^1-Pro^2, D-Trp^{7,9}, Leu^{11}]substance P, a substance P receptor antagonist, also functions as a bombesin receptor antagonist.

REFERENCES

Anastasi, A., Erspamer, V., and Endean, R., 1968, Isolation and amino acid sequence of caerulein, the active decapeptide of the skin of *Hyla caerulea*, *Arch. Biochem. Biophys.* **125:**57–68.

Barlas, N., Jensen, R. T., Beinfeld, M. C., and Gardner, J. D., 1982, Cyclic nucleotide antagonists of cholecystokinin: Structural requirements for interaction with the cholecystokinin receptor, *Am. J. Physiol.* **242:**G161–167.

Bissonnette, B. M., Collen, M. J., Adachi, H., Jensen, R. T., and Gardner, J. D., 1984, Receptors for vasoactive intestinal peptide and secretin on rat pancreatic acini, *Am. J. Physiol.* **246:**G710–G717.

Bodanszky, M., Martinez, J., Priestley, G. P., Gardner, J. D., and Mutt, V., 1978, Cholecystokinin (pancreozymin). 4′. Synthesis and properties of a biologically active analogue of the C-terminal heptapeptide with ξ-hydroxynorleucine sulfate replacing tyrosine sulfate, *J. Med. Chem.* **21:**1030–1035.

Christophe, J. P., Conlon, T. P., and Gardner, J. D., 1976, Interaction of porcine vasoactive intestinal peptide with dispersed pancreatic acinar cells from the guinea pig. Binding of radioiodinated peptide, *J. Biol. Chem.* **251**:4629–4634.

Collins, S. M., Abdelmoumene, S., Jensen, R. T., and Gardner, J. D., 1981a, Cholecystokinin-induced persistent stimulation of enzyme secretion from pancreatic acini, *Am. J. Physiol.* **240**:G459–G465.

Collins, S. M., Abdelmoumene, S., Jensen, R. T., and Gardner, J. D., 1981b, Reversal of cholecystokinin-induced persistent stimulation of pancreatic enzyme secretion by dibutyryl cyclic GMP, *Am. J. Physiol.* **240**:G466–G471.

de Castiglione, R., 1977, Structure-function relationships in ceruletide-like peptides, in: *First International Symposium on Hormonal Receptors in Digestive Tract* (S. Bonfils, P. Fromagot, and G. Rosselin, eds.), Elsevier/North Holland Biomedical Press, Amsterdam, pp. 33–42.

Erspamer, V., and Melchiorri, P., 1973, Active polypeptides of the amphibian skin and their synthetic analogs, *Pure Appl. Chem.* **35**:463–494.

Gardner, J. D., and Jensen, R. T., 1981, Regulation of pancreatic enzyme secretion *in vitro*, in: *Physiology of the Gastrointestinal Tract* (L. R. Johnson, ed.), Raven Press, New York, pp. 831–871.

Gardner, J. D., and Jensen, R. T., 1984, Cholecystokinin receptor antagonists, *Am. J. Physiol.* **246**:G471–G476.

Gardner, J. D., and Rottman, A. J., 1979, Action of cholera toxin on dispersed acini from guinea pig pancreas, *Biochim. Biophys. Acta* **585**:250–265.

Gardner, J. D., Rottman, A. J., Natarajan, S., and Bodanszky, M., 1979, Interaction of secretin$_{5-27}$ and its analogues with hormone receptors on pancreatic acini, *Biochim. Biophys. Acta* **583**:491–503.

Gardner, J. D., Korman, L., Walker, M., and Sutliff, V. E., 1982, Effects of inhibitors of cyclic nucleotide phosphodiesterase on the actions of vasoactive intestinal peptide and secretin on pancreatic acini, *Am. J. Physiol.* **242**:G547–G551.

Gill, D. M., 1977, Mechanism of action of cholera toxin, in: *Advances in Cyclic Nucleotide Research*, Vol. 8 (P. Greengard and G. A. Robison, eds.), Raven Press, New York, pp. 85–118.

Hahne, W. F., Jensen, R. T., Lemp, G. F., and Gardner, J. D., 1981, Proglumide and benzotript: Members of a different class of cholecystokinin receptor antagonists, *Proc. Natl. Acad. Sci. U.S.A.* **78**:6304–6308.

Hootman, S. R., Ernst, S. A., and Williams, J. A., 1983, Secretagogue regulation of Na^+–K^+ pump activity in pancreatic acinar cells, *Am. J. Physiol.* **245**:G339–G346.

Howard, J. M., Knight, M., Jensen, R. T., and Gardner, J. D., 1984, Discovery of an analogue of cholecystokinin with partial agonist activity, *Am. J. Physiol.* **247**:G261–G269.

Jensen, R. T., and Gardner, J. D., 1979, Interaction of physalaemin, substance P and eledoisin with specific membrane receptors on pancreatic acinar cells, *Proc. Natl. Acad. Sci. U.S.A.* **76**:5679–5683.

Jensen, R. T., and Gardner, J. D., 1981, Identification and characterization of receptors for secretagogues on pancreatic acinar cells, *Fed. Proc.* **40**:2486–2496.

Jensen, R. T., and Gardner, J. D., 1984, Receptors for secretagogues on pancreatic acinar cells, *J. Pediatr. Gastroenterol. Nutr.* **3**(Suppl. 1):524–535.

Jensen, R. T., Moody, T., Pert, C., Rivier, J. E., and Gardner, J. D., 1978, Interaction of bombesin and litorin with specific membrane receptors on pancreatic acinar cells, *Proc. Natl. Sci. U.S.A.* **75**:6139–6143.

Jensen, R. T., Rivier, J. E., and Gardner, J. D., 1979, Structural requirements for interaction of various peptides with the bombesin receptor on pancreatic acinar cells, *Gastroenterology* **76**:1160.

Jensen, R. T., Lemp, G. F., and Gardner, J. D., 1980, Interaction of cholecystokinin with specific membrane receptors on pancreatic acinar cells, *Proc. Natl. Acad. Sci. U.S.A.* **77**:2079–2083.

Jensen, R. T., Tatemoto, K., Mutt, V., Lemp, G. F., and Gardner, J. D., 1981, Actions of a newly isolated intestinal peptide, PHI, on dispersed acini from guinea pig pancreas, *Am. J. Physiol.* **241**:G498–G502.

Jensen, R. T., Lemp, G. F., and Gardner, J. D., 1982, Interactions of COOH-terminal fragments of cholecystokinin with receptors on dispersed acini from guinea pig pancreas, *J. Biol. Chem.* **257**:5554–5559.

Jensen, R. T., Charlton, C. G., Adachi, H., Jones, S. W., O'Donohue, T. L., and Gardner, J. D., 1983a, Use of ^{125}I-secretin to identify and characterize high-affinity secretin receptors on pancreatic acini, *Am. J. Physiol.* **245:**G186–G195.

Jensen, R. T., Jones, S. W., and Gardner, J. D., 1983b, COOH-terminal fragments of cholecystokinin: A new class of cholecystokinin receptor antagonists, *Biochim. Biophys. Acta* **757:**250–258.

Jensen, R. T., Jones, S. W., and Gardner, J. D., 1983c, Structure–function studies of N-acyl derivatives of tryptophan that function as specific cholecystokinin receptor antagonists, *Biochim. Biophys. Acta* **761:**269–277.

Jensen, R. T., Jones, S. W., Folkers, K., and Gardner, J. D., 1984a, A synthetic peptide that is a bombesin receptor antagonist, *Nature* **309:**61–63.

Jensen, R. T., Jones, S. W., Lu, Y.-A., Xu, J.-C., Folkers, K., and Gardner, J. D., 1984b, Interaction of substance P antagonists with substance P receptors on dispersed pancreatic acini, *Biochim. Biophys. Acta* **804:**181–191.

Jensen, R. T., Murphy, R. B., Trampota, M., Schneider, L. H., Jones, S. W., Howard, J. M., and Gardner, J. D., 1985, Proglumide analogues: Potent cholecystokinin receptor antagonists, *Am. J. Physiol.* **249:**G214–G220.

Maton, P. N., Sutliff, V. E., Jensen, R. T., and Gardner, J. D., 1985, Carbobenzoxy amino acids: Structural requirements for cholecystokinin receptor antagonist activity, *Am. J. Physiol.* **248:**G479–G484.

Petersen, O. H., 1981, Electrophysiology of exocrine gland cells, in: *Physiology of the Gastrointestinal Tract* (L. R. Johnson, ed.), Raven Press, New York, pp. 749–772.

Raufman, J.-P., Jensen, R. T., Sutliff, V. E., Pisano, J. J., and Gardner, J. D., 1982, Actions of gila monster venom on dispersed acini from guinea pig pancreas, *Am. J. Physiol.* **242:**G470–G474.

Robberecht, P., Conlon, T. P., and Gardner, J. D., 1976, Interaction of porcine vasoactive intestinal peptide with dispersed acinar cells from the guinea pig: Structural requirements for effects of VIP and secretin on cellular cyclic AMP, *J. Biol. Chem.* **251:**4635–4639.

Sankaran, H., Goldfine, I. D., Bailey, H., Licko, V., and Williams, J. A., 1982, Relationship of cholecystokinin receptor binding to regulation of biological functions in pancreatic acini, *Am. J. Physiol.* **242:**G250–G257.

Sjodin, L., Brodin, E., Nilsson, G., and Conlan, T. P., 1980, Interaction of substance P with pancreatic acinar cells from guinea pig: Binding of radiolabeled peptide sites, *Acta Physiol. Scand.* **109:**97–105.

Zhou, Z.-C., Villanueva, M. L., Gardner, J. D., and Jensen, R. T., 1985, Pancreatic acini possess a distinct receptor for calcitonin gene-related peptide (CGRP), *Gastroenterology* **88:**1643.

34

Antagonists for Substance P and Other Neurokinins

D. REGOLI, J. MIZRAHI, P. d'ORLEANS-JUSTE, S. DION, G. DRAPEAU, and E. ESCHER

1. INTRODUCTION: NEUROKININ ANTAGONISTS

Because of the numerous roles of neurokinins in physiology and physiopathology, antagonists for these peptides represent a new type of promising pharmacological tool (Rosell and Folkers, 1982). During the last 5 years, a fairly large number of compounds have been developed by various investigators (see references in Table I), and some prototypes of the various categories (undeca-, octa-, hepta-, or hexapeptides) are presented in Table I. From these results, it is evident that (1) all antagonists contain at least two D-amino acids (generally D-Trp in the positions 7 and 9), which are needed to confer to the peptide its antagonistic properties, (2) the N-terminal residues (either the first or the second) are replaced with D isomers (D-Arg^1, D-Pro^2 in the undecapeptide, D-Pro^4 in the octapeptides) or with L isomers (Arg or Pro in the hepta and hexapeptides) in order to protect the antagonist from degradation or improve its solubility, and (3) the C-terminal Met has generally been replaced with an aliphatic (Leu, Nle) or aromatic (Phe) residue to increase the affinity of undeca-, octa-, and heptapeptide antagonists.

All compounds of Table I have been tested against substance P in the most commonly used pharmacological assay, the guinea pig ileum. The recent identification of new neurokinins in mammals (neurokinin A by Nawa *et al.*, 1983; neurokinin B by Kangawa *et al.*, 1983), raises important questions about the distribution of the three neurokinins, their possible specific receptors in target organs, and their physiological roles. Some of the antagonists (undeca- or octapeptides) have also been tested against neurokinin A and have been found to be slightly more active

D. REGOLI, J. MIZRAHI, P. d'ORLEANS-JUSTE, S. DION, G. DRAPEAU, and E. ESCHER • Department of Pharmacology, Medical School, University of Sherbrooke, Sherbrooke, Quebec J1H 5N4, Canada.

TABLE I. Tachykinin Antagonists

Compound[a]	Guinea pig ileum (pA_2) Against NKP	Against NKA	Reference
Undecapeptide			
[arg^1,pro^2,trp^{7,9},Leu^{11}]SP	6.13	6.30	Hunter and Maggio (1984)
[arg^1,trp^{7,9},Leu^{11}]SP	7.10	—	Folkers *et al.* (1984)
[pro^2,trp^{7,9,10}]SP	6.20	6.44	Regoli *et al.* (1984a)
[trp^{7,9,10}]SP	6.40	6.65	D. Regoli *et al.* (unpublished)
Octapeptide			
[pro^4,trp^{7,9,10}]SP_{4-11}	6.60	6.65	Regoli *et al.* (1984a)
[pro^4,trp^{7,9},Nle^{11}]SP_{4-11}	7.10	7.25	Regoli *et al.* (1984a)
[tyr^4,trp^{7,9},Nle^{11}]SP_{4-11}	6.72	6.82	D. Regoli *et al.* (unpublished)
[pro^4,trp^{7,9,10},Phe^{11}]SP_{4-11}	7.02	7.25	Regoli *et al.* (1985)
[pro^4,Lys^6,trp^{7,9,10},Phe^{11}]SP_{4-11}	6.11	7.41	Regoli *et al.* (1985)
Heptapeptide			
[Arg^5,trp^{7,9},Nle^{11}]SP_{5-11}	6.71	—	Hörig and Schultheiss (1984)
[Arg^5,trp^{7,9}]SP_{5-11}	5.91	—	Hörig and Schultheiss (1984)
[Arg^5,trp^{7,9},Phe^{11}]SP_{5-11}	5.70	—	Hörig and Schultheiss (1984)
[Arg^5,trp^{7,9},Leu^{11}]SP_{5-11}	5.55	—	Hörig and Schultheiss (1984)
Hexapeptide			
[Pro^6,phe^7,trp^9]SP_{6-11}	5.60	—	Baizman *et al.* (1983)
[Pro^6,phe^{7,9}]SP_{6-11}	5.40	—	Baizman *et al.* (1983)
[Pro^6,trp^{7,9}]SP_{6-11}	5.70	—	Baizman *et al.* (1983)

[a] Lower case indicates D-amino acids, e.g., pro is D-Pro.

against this agent than against substance P (Table I). However, the use of a single pharmacological preparation, in particular, the guinea pig ileum, appears to be insufficient and inadequate for evaluating the usefulness and the potential of new neurokinin antagonists. We have therefore performed a series of basic pharmacological experiments in order to identify several pharmacological preparations that will permit a fair evaluation of new compounds as a first step towards the development of potent, selective, and competitive antagonists for each of the three neurokinins identified in mammals.

2. CRITICAL ANALYSIS OF THE PHARMACOLOGICAL PREPARATIONS

All experiments reported in this chapter have been performed in isolated smooth muscles taken from various animal species (rat, guinea pig, hamster, dog) and from different organs (intestine, trachea, urinary bladder, arterial vessels) suspended *in vitro* under the experimental conditions generally used for pharmacological assays. Peptides structurally and functionally related to neurokinins (the naturally occurring peptides; some homologues, fragments, or analogues) as well as other peptides utilized for comparative purposes have been characterized as agonists, partial agonists, and antagonists by measuring their activities in isolated organs. Correlation between changes of chemical structure and biological activities has been established for several series of compounds in an attempt to identify antagonists specific for each of the three naturally occurring neurokinins. It must be remembered, however, that isolated organs are complex bioassay systems, since they contain various tissues including endothelia, epithelia, exocrine glands, autonomic nerves and ganglia, and smooth muscle fibers. Naturally occurring agents generally exert their biological effects by acting on more than one structure, for instance, on autonomic nerve terminals and smooth muscles, which contain pre- and postsynaptic receptor sites that may not necessarily be of the same type. Moreover, various agents may act directly on receptors located on the smooth muscle membrane or indirectly by promoting the release of endogenous agents. Whereas direct actions are expected to remain stable, exhaustion of endogenous mediators or other factors may interfere with the stability of the indirect ones.

Similarly to other agents, neurokinins exert both pre- and postsynaptic effects and direct or indirect actions by promoting the release of endogenous mediators. How can we measure and evaluate quantitatively these various components of a peptide's pharmacological effect? This can be tested by using antagonists in the frame of specificity studies such as that schematically summarized in Table II. Two preparations are analyzed in Table II, and the effects of agonists (II A) and antagonists (II B) have been shown to be different in untreated guinea pig ileums (G.P.I.) than in those treated with atropine to prevent the action of acetylcholine, whose

TABLE IIA. Affinities of Agonists (pD_2)

Agonist	G.P.I.	G.P.I.[a]	DF	R.M.V.	R.M.V.[a]	Difference
Physalaemin	9.04	8.80	−0.24	7.55	7.59	+0.04
Substance P	8.78	8.48	−0.30	7.55	7.55	0
SP_{4-11}	9.30	8.86	−0.44	7.75	7.72	−0.03
SP_{5-11}	9.11	9.51	−0.60	7.30	7.32	+0.02
Eledoisin	9.21	8.60	−0.61	7.88	7.90	+0.02
Kassinin	9.35	8.78	−0.57	8.08	8.02	−0.06

[a] Guinea pig ileum and rabbit mesenteric vein treated with atropine, 5.1 × 10^{-6} M.

TABLE IIB. Affinities of Antagonists (pA_2)

Antagonist	G.P.I.	G.P.I.[a]	Difference
$[pro^4,trp^{7,9}]SP_{4-11}$	5.65	5.85	+0.20
$[pro^4,trp^{7,9},Leu^{11}]SP_{4-11}$	6.40	6.67	+0.27
$[arg^1,pro^2,trp^{7,9},Leu^{11}]SP$	6.15	6.40	+0.25
$[pro^4,trp^{7,9,10}]SP_{4-11}$	6.30	6.60	+0.30
$[pro^4,trp^{7,9,10},Leu^{11}]SP_{4-11}$	6.25	6.75	+0.50

[a] Guinea pig ileum treated with atropine, 5 × 10^{-6} M. Lower case indicates D-amino acids.

release (from the intramural autonomic nerves) is stimulated by substance P and related peptides (Fosbraey *et al.*, 1984; Regoli *et al.*, 1984b). Similar experiments have been performed on the rabbit mesenteric vein for comparison.

As shown in Table II, the affinities of the various peptides are higher in the G.P.I. not treated with atropine, but the differences are smaller (−0.25, −0.30) for physalaemin and substance P (SP) than for eledoisin and kassinin (−0.61, −0.57). Surprisingly, the effect of the fragment SP_{5-11} is also markedly reduced (−0.60). No changes were found for the same compounds in rabbit mesenteric veins from treatment with atropine. Why is there such a difference? Antagonists have also been tested (Table IIB), and some compounds, for instance, $[pro^4, trp^{7,9,10}, Leu^{11}]SP_{4-11}$, show higher affinity ($pA_2$ 6.75 compared to 6.25) in the presence of atropine.

Thus, atropine eliminated the presynaptic component and affects eledoisin and kassinin more than physalaemin and substance P. Moreover, the affinities of some antagonists (selective for SP-P receptors?) are increased more than those of others (nonselective?). Thus, differences between preparations and the presence of multiple receptor types can modify the apparent affinity constants (pD_2 or pA_2) of both agonists and antagonists.

From the above, it emerges that the guinea pig ileum (and not the rabbit mesenteric vein) has at least two receptor types for neurokinins, the presynaptic, particularly sensitive to eledoisin and kassinin (SP-E type?), and the postsynaptic, particularly sensitive to physalaemin and substance P (SP-P type?).

We therefore performed a series of basic pharmacological studies on various isolated organs to determine the type of action (direct, indirect) of several neurokinins, including neurokinin A and neurokinin B. The results have been described in detail in a recent paper (Mizrahi *et al.*, 1985a), and a few are summarized in Table III.

The results obtained with the neurokinins separate the preparations into three groups, showing the following order of potency:

I SP $\geqslant$ Phys > Eled > NKA > Kass > NKB
II NKA > Kass > Eled > NKB > Phys > SP
III Kass > NKA > Eled > Phys > SP

3. THE NEUROKININS AND THEIR RECEPTORS

3.1. Critical Analysis

In addition to showing three different patterns, the results summarized in Table III indicate that the G.P.I. shows an order of potency of agonists that does not fit into any of the three other patterns. The G.P.I. is also the only preparation in which the six substance P homologues utilized in the present study show similar affinities. As shown in Table IIA, pD_2 values of physalaemin, substance P, eledoisin, and

TABLE III. Order of Potency of Neurokinins[a] (Relative Affinities Compared to SP = 1.0)

G.P.I. (guinea pig ileum) SP-P	Phys. 2.1	Kass. 2.0	Eled. 1.3	SP 1.0	NKA 0.6	NKB 0.15
D.C.A. (dog carotid artery) SP-P	SP 1.0	Phys. 1.0	Eled. 0.3	NKA 0.25	Kass. 0.15	NKB 0.09
R.D. (rat duodenum) SP-E	NKA 53	Kass. 32	Eled. 17	NKB 9	Phys. 1.1	SP 1.0
G.P.T. (guinea pig trachea) SP-E	NKA 45	Kass. 13	NKB 5	Eled. 4	Phys. 1.3	SP 1.0
H.U.B. (hamster urinary bladder) SP-E	Kass. 214		NKA 69	Eled. 27	Phys. 2.1	SP 1.0
D.U.B. (dog urinary bladder) SP-E	Kass. 28		NKA 13	Eled. 6	Phys. 1.4	SP 1.0

[a] Phys., physalaemin; Kass., kassinin; Eled., eledoisin; SP, substance P; NKA, neurokinin A; NKB, neurokinin B.

kassinin in preparations not treated with atropine correspond to 9.04, 8.78, 9.21, 9.35, and those of neurokinin A and B are 8.68 and 8.65, respectively.

All these values are very similar; maximum differences are 0.5 log units or less. In this respect, the G.P.I. (much less the D.C.A.) corresponds to the description of the SP-P receptor (Lee *et al.*, 1982) as the one that is equally sensitive to the various tachykinins. But on a closer analysis, it appears that the various tachykinins are equally active because some of them activate more efficiently the pre- and the others the postsynaptic receptor, and in this way the pD_2 value of, for instance, neurokinin A is comparable to that of substance P. If that is the case, then the term SP-P indicates a mixture of SP-E and SP-X receptors rather than a single receptor entity.

The designation SP-E needs also to be revised. In fact, eledoisin occupies the central position in Table III, and neurokinin A and kassinin are the most active stimulants of the two SP-E receptor subtypes.

3.2. Proposal of New Designations

Based on the above data and considerations, new designations are proposed here for neurokinins and their receptors. First, we propose to designate substance P and its naturally occurring homologues under the general term (peptide family name) of neurokinins, since the term tachykinins (in opposition to bradykinin) is based on a peripheral and rather irrelevant pharmacological effect of these peptides. In accord with the suggestion of a group of experts at the last Substance P International Symposium in Maidstone, Kent, we propose the following designation for the three mammalian neurokins and their receptors:

Neurokinins		Receptors
Neurokinin P (NKP) for substance P	NK-P	instead of SP-P
Neurokinin A (NKA) for substance K	NK-A	} instead of SP-E
Neurokinin B (NKB) for neuromedin K	NK-B*	

In the above scheme, three different receptor types for the neurokinins are proposed on the basis of (1) differences in the order of potency of agonists determined with biological assays (see Table II), (2) differences in the affinities measured by Buck *et al.* (1984) with binding assays, (3) differences in the relative activities of SP fragments (D. Regoli, J. Mizrahi, P. d'Orléans-Juste, S. Dion, G. Drapeau, and E. Escher, unpublished data), and (4) differences in the affinities of some antagonists (see Table IV). This new classification takes also into account much published and unpublished data showing different distributions (central and peripheral) and different physiological actions for the three mammalian neurokinins.

4. NEUROKININ ANTAGONISTS

4.1. Selectivity

The suggestion of three different receptor types for neurokinins that has been emerging from data obtained with agonists has been validated by findings with antagonists. A large number of new compounds (most of them octapeptides and some undecapeptides) have been tested to identify the three compounds presented in Table IV. These compounds show some selectivity for one or the other receptor site and have been selected for their pharmacological characteristics. Numerous other antagonists do not show such selectivity. Also noteworthy is the fact that in assembling Table IV, we selected three preparations out of eight with the purpose of emphasizing the differences among the three antagonists. The first compound of Table IV appears to be selective for the NK-P receptor type since its affinity for this receptor is 1.1 to 1.3 log units higher than that for the NK-A. On the other hand, the compound is inactive both as agonist and antagonist in preparations containing NK-B receptors.

The second compound acts as an agonist in preparations containing receptors of the NK-P type: it shows a fairly good affinity (pA_2 6.25) against NK-A in a

*Note: Kassinin was used instead of neurokinin B for characterizing the NK-B receptor site because the preparations of neurokinin B available to us were found to be unstable and did not provide reproducible biological effects.

TABLE IV. Selectivity of Neurokinin Antagonists (pA_2)[a]

Antagonist		D.C.A. (NK-P)	R.D. (NK-A)	H.U.B. (NK-B)
[pro^4,trp^{7,9,10},Phe^{11}]SP_{4-11}	NKP	6.7	5.6	In.
	NKA	7.0	5.7	In.
[pro^4,Lys^6,trp^{7,9,10},Phe^{11}]SP_{4-11}	NKP	Ag.	5.2	5.7
	NKA	Ag.	6.2	5.1
[tyr^4,trp^{7,9},Nle^{11}]SP_{4-11}	NKP	Ag.	5.1	5.9
	Kassinin	Ag.	6.1	6.5

[a] Ag., agonist; In., inactive; NKP, neurokinin P; NKA, neurokinin A; D.C.A., dog carotid artery; R.D., rat duodenum; H.U.B., hamster urinary bladder. Lower case indicates D-amino acids.

preparation (the R.D.) of the NK-A group, and its affinity is 1.15 log unit higher than that observed in preparations containing the NK-B receptor type.

The third compound of Table IV, [tyr^4,trp^{7,9},Nle^{11}]SP_{4-11}, is an agonist on the D.C.A. and an antagonist in the other two preparations. However, its affinity is higher in the H.U.B. (NK-B receptor type) than in the R.D. (NK-A receptor), the difference being between 0.4 and 0.8 log units.

Thus, data obtained with antagonists support the suggestion that tachykinins may exert their peripheral actions by activating two and possibly three different receptor types. Moreover, the data reported above indicate that a new classification of tachykinin receptors is required to cope with the data recently obtained in various laboratories with the two tachykinins recently identified in mammals.

4.2. Specificity

In order to determine the specificity of neurokinin antagonists, several compounds were tested in the six preparations reported in Table III and in others (see Mizrahi *et al.*, 1985a, for details) against one or more tachykinins as well as against one of the following agents: angiotensin II, bradykinin (BK), desArg^9-BK, bombesin, vasopressin, histamine, acetylcholine, 5-hydroxytryptamine, and norepinephrine. The myotropic effects (stimulatory or inhibitory) of all these agents, except bombesin, were not affected by neurokinin antagonists. The antagonistic effect of two compounds against bombesin was studied in two preparations, the rat and guinea pig urinary bladders (R.U.B. and G.P.U.B.) and the results are shown in Table V.

Other data and the technical details of the study summarized in Table V can be found in a recent paper (Mizrahi *et al.*, 1985b).

The first compound of Table V is more active against bombesin than against substance P, whereas the second shows lower but similar affinity against the two peptides. Our results on smooth muscle preparations confirm the recent observations of Jensen *et al.* (1984a,b) on a secretory system, the pancreatic acinar cell.

TABLE V. Neurokinin Antagonists Active against Bombesin (pA_2)

Antagonist	Agonist	Preparation	
		R.U.B.	G.P.U.B.
[pro[4],Lys[6],trp[7,9,10],Phe[11]]SP_{4-11}	Bombesin	6.36 (1.4)[a]	6.18 (1.3)
	Neurokinin P	5.33 (1.1)	5.41 (1.0)
	Bradykinin	Inactive	Inactive
[pro[4],trp[7,9,10],Phe[11]]SP_{4-11}	Bombesin	5.10	5.15
	Neurokinin P	5.10	5.65
	Bradykinin	Inactive	Inactive

[a] In parentheses: the difference $pA_2 - pA_{10}$. Lower case indicates D-amino acids.

4.3. Competitivity

In addition to determining pA_2 and pA_{10} values and calculating the difference, as in Table V, a series of dose–response curves for NKP and NKA were measured in various preparations. The data were analyzed according to Schild (Arunlakshana and Schild, 1959; Schild, 1973). Results obtained with three antagonists, each tested in two preparations, are illustrated in Fig. 1. As shown by the value of the slope (on the right of Fig. 1), the first and second compounds are competitive, whereas the third one shows slope values too low for an antagonism of a competitive type.

Some neurokinin antagonists, particularly the undecapeptides, show stimulatory effects in some isolated organs such as the guinea pig trachea. This effect has been attributed to the ability of antagonists containing the N-terminal positively charged portion of substance P to release histamine (Mizrahi *et al.*, 1984). Such an interpretation is supported by the data summarized in Table VI.

The role of histamine in the contractile effects of various undecapeptides is demonstrated by the reduction of the effect in the presence of diphenhydramine, a fairly specific antagonist of the H_1 receptors. Moreover, undecapeptides are more potent than substance P as histamine releasers from rat mast cells. The data reported in Table VI have been obtained by Devillier *et al.* (1985) and confirm previous findings by Foreman *et al.* (1982, 1983). Octapeptide antagonists are 20 to 500 times less active than the undecapeptides (Regoli *et al.*, 1984a).

5. CONCLUSIONS

1. Isolated organs suspended *in vitro* are among the most useful pharmacological assays for characterizing new synthetic peptides related to neurokinins, particularly antagonists.
2. The discovery of new neurokinins (neurokinin A and B) in mammals has prompted a reassessment of the pharmacological tests. A critical analysis of the neurokinin myotropic effects on the guinea pig ileum (G.P.I., the

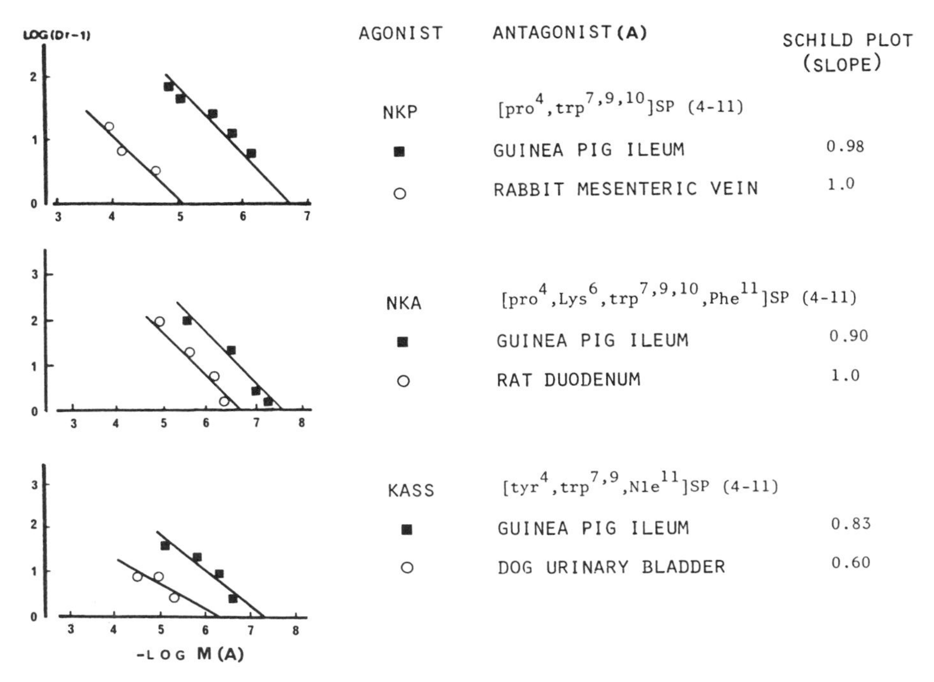

FIGURE 1. Schild plots obtained with three octapeptide antagonists against the myotropic effects of substance P (NKP), neurokinin A (NKA), and kassinin (KASS) on the guinea pig ileum, the rabbit mesenteric vein, the rat duodenum, and the dog urinary bladder. Each point is the mean of at least five experiments. Abscissa, negative -iog of molar (M) concentrations of antagonist; ordinate, log of (D_r − 1).

TABLE VI. Substance P Antagonists and Histamine Release

Agent	α^E	Guinea pig trachea: Control contraction (mm)	Guinea pig trachea: Anti-H_1[a] contraction (mm)	Rat mast cells: ED_{50} (M)	Rat mast cells: Relative activity
[pro^2,trp^{7,9}]SP	1.0	20.1 ± 3.3	8.0 ± 1.0*		
[pro^2,trp^{7,9,10}]SP	0.8	13.0 ± 2.0	6.1 ± 0.8*	2.4×10^{-6}	2.0
[trp^{7,9,10}]SP	1.0	29.0 ± 3.1	10.5 ± 1.2*	6.2×10^{-8}	54.1
[pro^2,trp^{7,9},Leu^{11}]SP	1.0	35.0 ± 3.2	13.4 ± 1.5*		
[arg^1,pro^2,trp^{7,9},Leu^{11}]SP	0.9	25.5 ± 3.5	7.6 ± 0.9*		
[pro^4,trp^{7,9},Leu^{11}]SP_{4-11}	0.1	—		3.8×10^{-5}	0.1
Substance P	1.0	—		4.3×10^{-6}	1.0

[a] Anti-H_1: diphenhydramine 3.3×10^{-6} M; $*P < 0.001$. Lower case indicates D-amino acids.

most commonly used assay organ) indicates that these effects are produced by the activation of (at least) two functionally different sites.

3. Five other tissues (in addition to the G.P.I.) have been characterized, and three major types (patterns) of biological responses have been identified. Thus, the dog carotid artery shows high sensitivity to NKP and Phys; the rat duodenum and guinea pig trachea are particularly sensitive to NKA and Kass; Kass appears to be very active on the hamster and dog urinary bladders. We suggest that some of these preparations should be used in parallel with the G.P.I. in the characterization of any new synthetic or endogenous neurokinin and their antagonists.
4. Antagonists showing some selectivity for one or the other mammalian neurokinin receptors in the three types of tissues have been identified. These are [pro^4,$trp^{7,9,10}$,Phe^{11}[SP_{4-11} (**1**) in the dog carotid artery, [pro^4,Lys^6, $trp^{7,9,10}$,Phe^{11}[SP_{4-11} (**2**) in the rat duodenum, and [tyr^4,$trp^{7,9}$,Nle^{11}] SP_{4-11} (**3**) in the hamster urinary bladder.
5. Compounds **1** and **2** appear to be competitive (Schild plot slope near 1.0); compound **3** is noncompetitive. Compounds **1** and **2** are also active against bombesin, and the affinity of compound **2** for the bombesin receptors is higher than that of compound **1.** When tested against other peptides (angiotensin, bradykinin) or nonpeptide agents (acetylcholine, histamine, serotonin), the neurokinin antagonists are inactive.
6. When compared for their effects on histamine release, undecapeptide antagonists are definitely more active than the octapeptides.
7. Based on data obtained with agonists and antagonists, a new receptor classification, including three different receptor types for neurokinins, is proposed. We also suggest that the term tachykinin should be replaced by neurokinin.

ACKNOWLEDGMENTS. We thank Mrs. Cécile Théberge for typing the manuscript and the tables. All experimental work reported in this paper has been performed with the financial help of the Medical Research Council of Canada (MRCC) and of the Canadian Heart Foundation (CHF). D.R. is a Career Investigator of the MRCC, E.E. is a Scholar of the CHF, and J.M. is a fellow of the Fonds de la Recherche en Santé du Québec.

REFERENCES

Arunlakshana, O. and Schild, H. O., 1959, Some quantitative uses of drug antagonists, *Br. J. Pharmacol.* **14:**48–58.

Baizman, E. R., Gordon, T. D., Hansen, P. E., Kiefer, D., Lopresti, D. M., McKay, F. C., Morgan, B. A., and Perrone, W. H., 1983, Structure and antagonist activity in a series of hexapeptide substance P analogs, in: *Peptides: Structure and Function* (J. V. Hruby, and D. H. Rich, eds.) Pierce Chemical, Rockford, IL, pp. 437–440.

Buck, S. H., Burcher, E., Shults, C. W., Lovenberg, W., and O'Donohue, T. L., 1984, Novel pharmacology of substance K-binding sites: A third type of tachykinin receptor, *Science* **226:**987–989.

Devillier, P., Renoux, M., Giroud, J.-P., and Regoli, D., 1985, Peptides and histamine release from rat peritoneal mast cells, *Eur. J. Pharmacol.* **117:**89–96.

Folkers, K., Häkanson, R., Hörig, J., Xu, J. C. and Leander, S., 1984, Biological evolution of substance P antagonists, *Br. J. Pharmacol.* **83:**449–456.

Foreman, J.-C., and Jordan, C. C., 1983, Structure–activity relationships for some substance P-related peptides that cause wheal and flare reactions in human skin, *J. Physiol. (Lond.)* **335:**449–465.

Foreman, J.-C., Jordan, C.-C., and Piotrowski, W., 1982, Interaction of neurotensin with the substance P receptor mediating histamine release from rat mast cells and the flare in human skin, *Br. J. Pharmacol.* **77:**531–539.

Fosbraey, P., Featherstone, R.-L., and Morton, I. K. M., 1984, Comparison of potency of substance P and related peptides on [^{3}H]-acetylcholine release and contractile actions in the guinea pig ileum, *Naunyn Schmiedebergs Arch. Pharmacol.* **326:**111–115.

Hörig, J. A., and Schultheiss, H., 1984, Structure–activity relationship of C-terminal hexa- and heptapeptide substance P antagonists as studied in the guinea pig ileum, *Eur. J. Pharmacol.* **105:**65–72.

Hunter, J. C., and Maggio, J. E., 1984, A pharmacological study with substance K: Evidence for multiple types of tachykinin receptors, *Eur. J. Pharmacol.* **105:**149–153.

Jensen, R. T., Jones, S. W., Folkers, K., and Gardner, J. D., 1984a, A synthetic peptide that is a bombesin receptor antagonist, *Nature* **309:**61–63.

Jensen, R. T., Jones, S. W., Lu, Y. A., Xu, J. C., Folkers, K., and Gardner, J. D., 1984b, Interaction of substance P antagonists with substance P receptors on dispersed pancreatic acini, *Biochim. Biophys. Acta* **804:**181–191.

Kangawa, K., Minamino, N., Fukuda, A., and Matsuo, H., 1983, Neuromedin K: A novel mammalian tachykinin identified in porcine spinal cord, *Biochem. Biophys. Res. Commun.* **114:**533–540.

Lee, C. M., Iversen, L. L., Hanley, M. R., and Sandberg, B. E. B., 1982, The possible existence of multiple receptors for substance P, *Nauyn Schmiedebergs Arch. Pharmacol.* **318:**281–287.

Mizrahi, J., Escher, E., D'Orléans-Juste, P., and Regoli, D., 1984, Undeca- and octa-peptide antagonists for substance P; a study on the guinea pig trachea, *Eur. J. Pharmacol.* **99:**193–202.

Mizrahi, J., Dion, S., D'Orléans-Juste, P., Escher, E., and Regoli, D., 1985a, Tachykinin receptors in smooth muscles: A study with agonists (substance P, neurokinin A) and antagonists, *Eur. J. Pharmacol.* **118:**25–36.

Mizrahi, J., Dion, S., D'Orléans-Juste, P., and Regoli, D., 1985b, Activities and antagonism of bombesin on urinary smooth muscles, *Eur. J. Pharmacol.* **111:**339–345.

Nawa, H., Dotenchi, M., Igano, K., Inouye, K., and Nakanishi, S., 1983, Substance K: A novel mammalian tachykinin that differs from substance P in its pharmacological profile, *Life Sci.* **34:**1153–1160.

Regoli, D., Escher, E., and Mizrahi, J., 1984a, Structure–activity studies and the development of antagonists, *Pharmacology* **28:**301–320.

Regoli, D., Mizrahi, J., D'Orléans-Juste, P., and Escher, E., 1984b, Receptors for Substance P. II. Classification by agonists fragments and analogues, *Eur. J. Pharmacol.* **97:**171–177.

Regoli, D., Mizrahi, J., D'Orléans-Juste, P., Dion, S., Drapeau, G., and Escher, E., 1985, Substance P antagonists showing selectivity for different receptor types, *Eur. J. Pharmacol.* **109:**121–125.

Rosell, S., and Folkers, K., 1982, Substance P antagonists: A new type of pharmacological tool, *Trends Pharmacol. Sci.* **3:**211–212.

Schild, O. H., 1973, Receptor classification with special reference to beta-adrenergic receptors, in: *Drug Receptors* (H. P. Rang, ed.), University Park Press, Baltimore, pp. 29–36.

35

The Enkephalin-Containing Peptides

SIDNEY UDENFRIEND

There are three known genes that code for enkephalin-containing (EC) peptides (Udenfriend and Kilpatrick, 1983). The one coding for proenkephalin is expressed in many tissues. Having characterized many of the proenkephalin products and developed methods for their assay (Udenfriend *et al.*, 1983), and having cloned bovine (Gubler *et al.*, 1982) and rat (Howells *et al.*, 1984) proenkephalin cDNA, we had the tools to investigate the biology of this EC peptide precursor.

One of the interesting observations that we have investigated in detail is the effects of denervation on adrenal proenkephalin EC peptides. Schultzberg *et al.* (1978) had used immunohistochemical methods to show that the adrenal medullas of many animals, including guinea pigs, cows, and rabbits, were stained when exposed to fluorescent-labeled [Met]enkephalin antibody. On the other hand, when adrenal medullas of rats were used, there was little uptake of fluorescent antibody by the tissues. However, when rat adrenals were denervated several days in advance, [Met]enkephalin antibody was taken up. This indicated that in some manner denervation increased the [Met]enkephalin content of the adrenal medulla. Since free [Met]enkephalin can only be derived from proenkephalin, we realized that this could represent an interesting example of regulation of the proenkephalin gene or its products. We repeated the experiments of Schultzberg *et al.* (1978) and monitored the various products of proenkephalin processing as a function of time after denervation. As shown in Fig. 1, after a several-hour lag there was a rapid and large increase in total EC peptides, maximal values being reached 72 to 96 hr after denervation.

It is interesting to examine the composition of EC peptides at each time point. The normal rat adrenal medulla, prior to denervation, contains much smaller amounts

SIDNEY UDENFRIEND • Roche Institute of Molecular Biology, Roche Research Center, Nutley, New Jersey 07110.

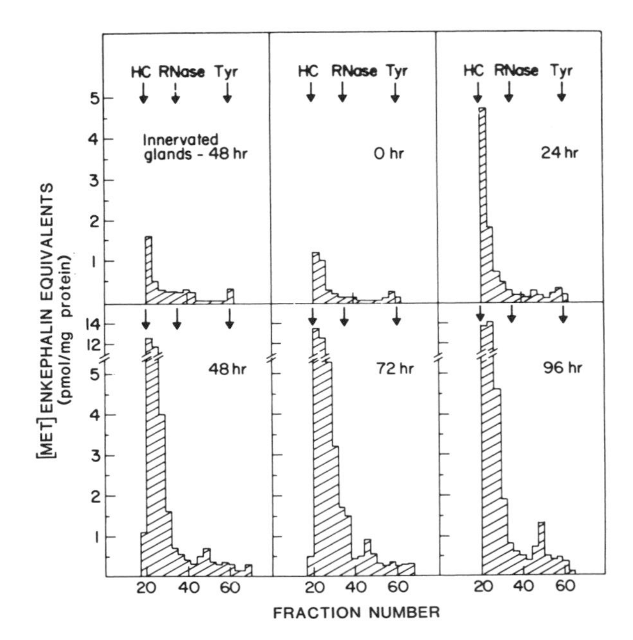

FIGURE 1. Sephadex G-75 chromatography of [Met]EC peptides in adrenal extracts at various times after denervation. Aliquots (20–200 μl) from each fraction were lyophilized, treated with trypsin and carboxypeptidase B, and assayed for [Met]enkephalin. The column markers are hemocyanin (HC), RNase, and tyrosine (Tyr) (after Fleminger *et al.,* 1984).

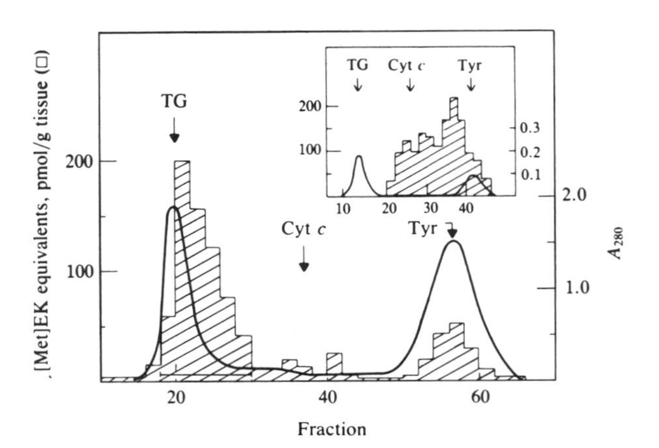

FIGURE 2. Sephadex G-75 chromatography of an extract of 150 denervated rat adrenal glands. Fractions (3.7 ml) were collected, and 50-μl aliquots were taken for [Met]enkephalin assay. Fractions 20 to 30 were pooled for further purification and identification of proenkephalin. The column markers were (TG) thyroglobulin (M_r 669,000), (Cyt *c*) cytochrome *c* (M_r 13,000), and (Tyr) tyrosine (M_r 181), respectively. (*Inset*) Sephadex G-75 chromatography of an extract of bovine chromaffin granules shown for comparison (after Fleminger *et al.,* 1984).

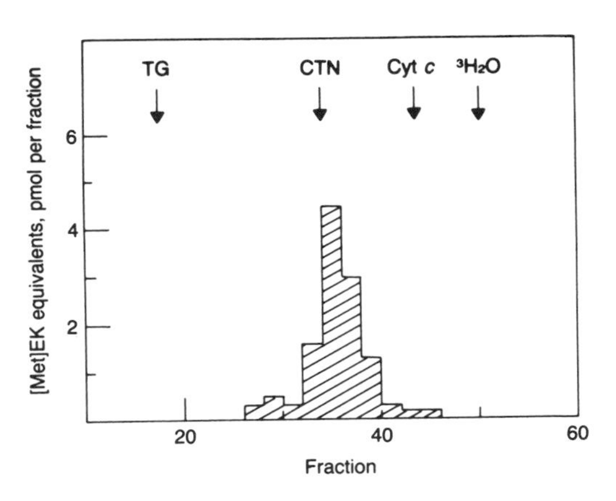

FIGURE 3. Estimation of molecular weight of "proenkephalin." An aliquot of the high-molecular-weight material from the Sephadex G-75 column (Fig. 2) was applied to a Fractogel TSK HW-50(F) column. One-milliliter fractions were collected and assayed for [Met]-enkephalin after treatment with trypsin and carboxypeptidase B. TG, CTN, and Cyt represent the elution volumes of thyroglobulin (M_r 669,000), chymotrypsinogen (M_r 25,000), and cytochrome *c* (M_r 13,000), respectively (after Fleminger *et al.*, 1984).

of EC peptides than medullas of other animals (Fleminger *et al.*, 1984). Total EC peptides in guinea pig medulla are over 300 pmol/g, compared to about 25 to 50 pmol/g in the rat. Furthermore, medullas from guinea pigs, cows, and rabbits contain a spectrum of EC peptides ranging from the largest characterized products of proenkephalin (12.6 and 18.6 kDa) to those of intermediate size to the free pentapeptides (Fig. 2). In those animals, the intermediate-size EC peptides predominate. By contrast, rat medullas contain mainly free pentapeptides and the largest EC peptides, with negligible amounts of intermediate-size EC peptides. Following denervation of rat adrenals, it is mainly the largest EC peptides that increase in amount. Only after 48 to 72 hr do some small amounts of intermediate-size EC peptides appear. Increases in free pentapeptides occur even later and are small and variable.

Because there was so little processing of the newly formed large EC peptides, it seemed possible that intact proenkephalin might actually be the product that accumulated. When the large EC peptide fraction from denervated medullas was subjected to more careful analysis by size exclusion chromatography, most of it eluted in the 29-kDa region, consistent with proenkephalin (Fig. 3). The material

TABLE I. Enkephalin-Containing Peptides Generated from Proenkephalin by Lys-C Endoproteinase and Carboxypeptidase B

	Theoretical[a]	Found[b]
[Met]Enkephalin	1	0.71
Arg^0 [Met]Enkephalin	2	3.1
Arg^0 [Leu]Enkephalin	1	0.92
Arg^0 [Met]Enkephalin-Arg^6-Gly^7-Leu^8	1	0.71
Arg^0 [Met]Enkephalin-Arg^6-Phe^7	1	0.92
$Proenkephalin_{187-206}$	1	0.95

[a] Based on sequence of rat proenkephalin.
[b] Each number represents the average of two independent experiments. One equivalent of proenkephalin is defined as the amount of [Met]enkephalin released by the enzymes divided by 6 (after Fleminger *et al.*, 1984).

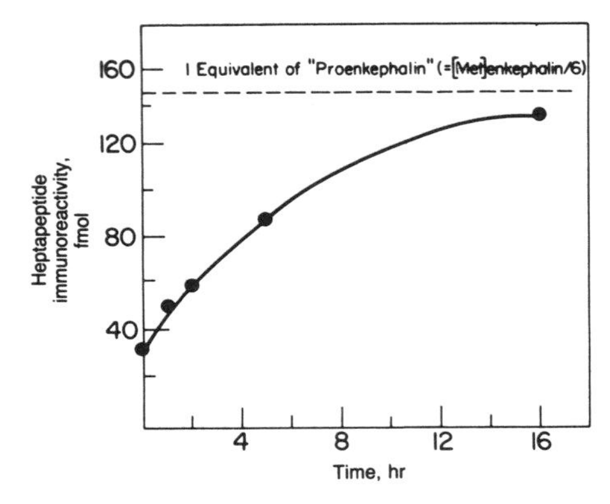

FIGURE 4. Stoichiometric release of [Met]-enkephalin-Arg^{6}-Phe^{7} from "proenkephalin" by Lys-C endoproteinase. An aliquot of the pooled fractions that had been purified on RP-8 HPLC was incubated with Lys-C endoproteinase at 37° C. Aliquots were removed at various times and assayed for the carboxyl-terminal heptapeptide by immunoassay. The dashed line represents the calculated amount of proenkephalin in the reaction mixture based on the [Met]enkephalin assay, assuming six copies of [Met]enkephalin per proenkephalin molecule (after Fleminger *et al.*, 1984).

was definitely identified as proenkephalin through the products released by Lys-C endoproteinase action (Table I) and by demonstrating the stoichiometric appearance of the characteristic heptapeptide sequence Tyr-Gly-Gly-Pro-Met-Arg-Phe at its carboxyl terminus (Fig. 4). Thus, denervation of the rat adrenal induces a ten- to 20-fold increase in intact proenkephalin with only a small amount of processing

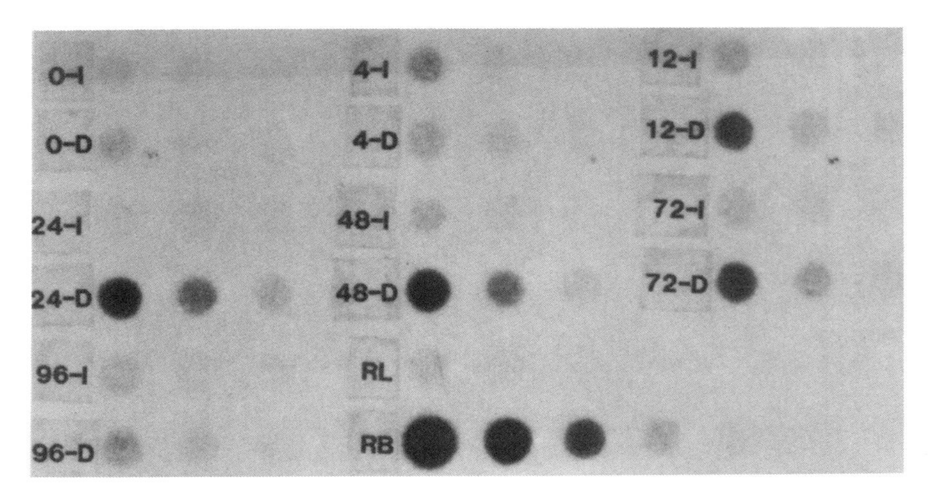

FIGURE 5. Time course of the effect of unilateral denervation on rat adrenal preproenkephalin mRNA. Poly(A)$^{+}$ RNA (12 μg) from innervated (I) and denervated (D) rat adrenal glands, rat liver (RL), and rat brain (RB) was denatured by heating to 60°C for 15 min in the presence of 7.4% formaldehyde/0.9 M NaCl/0.09 M sodium citrate (denaturing buffer). Serial dilutions of each sample were made up in denaturing buffer and applied to nitrocellulose (8.0, 2.7, and 0.9 μg of adrenal samples and 8.0–0.1 μg of liver and brain samples, from left to right). Wells were rinsed with denaturing buffer, and the filter was baked, prehybridized, and hybridized with nick-translated human proenkephalin cDNA (10^{7} cpm) at 42°C. Numbers refer to the hours after denervation when adrenal glands were removed. The data are representative of four independent experiments (after Kilpatrick *et al.*, 1984).

occurring over the 96 hr of the experiment. This apparent limitation in processing enzymes might explain the normal pattern of EC peptides in the rat medulla.

The increased amount of proenkephalin following denervation could represent induction at either the translational or transcriptional level. To distinguish between these two possible sites of regulation, proenkephalin cDNA was used to monitor changes in the corresponding mRNA brought on by denervation (Kilpatrick *et al.*, 1984). As shown in Fig. 5, proenkephalin mRNA also increased manyfold. As would be expected, increases in the mRNA preceded those in proenkephalin. It is worth noting that total mRNA actually decreased as proenkephalin mRNA was increasing (Table II). It is likely, therefore, that the observed increase represents newly synthesized proenkephalin mRNA and that denervation acts at the transcriptional level.

A possible explanation of these findings is that the cholinergic input to the adrenal medulla regulates proenkephalin synthesis by a negative control at the transcriptional level. If that were true, then administration of ganglionic blocking agent should also induce adrenal proenkephalin synthesis. Several such drugs were administered. However, in our hands even the most potent of them, when administered in amounts that produced overt pharmacological effects, did not increase adrenal enkephalin much if at all. Dr. C. Inturissi (unpublished observations) has

TABLE II. Effect of Denervation on Adrenal EC Peptide and Nucleic Acid Levels[a]

Time (hr)	Total RNA (μg)	Poly(A)$^+$ RNA (μg)	Proenkephalin mRNA (pg equivalents)	EC peptides (pmol/mg protein)
0 Innervated	56	4.9	0.0	5
Denervated	53	4.9	0.1	5
4 Innervated	56	5.5	0.3	5
Denervated	51	4.8	0.2	6
12 Innervated	59	6.3	0.3	6
Denervated	28	2.6	1.4	10
24 Innervated	62	5.8	0.0	7
Denervated	21	1.9	1.7	15
48 Innervated	68	7.4	0.2	6
Denervated	29	2.6	2.8	35
72 Denervated	66	7.6	0.2	7
Denervated	38	2.6	1.7	49
96 Innervated	66	7.2	0.3	6
Denervated	46	3.4	0.6	55

[a] The RNA values are given per adrenal gland. Picogram equivalents of nick-translated human proenkephalin cDNA hybridized per adrenal gland. One microgram of rat brain poly(A)$^+$RNA hybridized with 1.8 pg of cDNA (after Kilpatrick *et al.*, 1984).

observed up to a twofold increase in rat adrenal enkephalins on administration of ecolid. We have occasionally seen such increases, but the variability was large enough to question their statistical significance. In any event, ganglionic blocking agents do not produce the ten- to 20-fold increase in rat EC peptides that is seen after adrenal denervation.

There are other findings that question the universality of the phenomenon. Unlike rat adrenals, guinea pig adrenals, when denervated, do not show this increased synthesis of proenkephalin peptides. It should be noted that guinea pig adrenals normally contain much larger amounts of EC peptides than do rat adrenals. In addition, the composition of proenkephalin-derived adrenal peptides is quantitatively different in guinea pigs than in rats. There is very little, if any, intact proenkephalin in the guinea pig gland. It contains mainly large and intermediate-sized EC peptides with very small proportions of free enkephalins (Fig. 5). By contrast, rat adrenals contain about 60% of what appears to be intact proenkephalin, little, if any, intermediate-sized EC peptides, and about 30% free enkephalins and other small EC peptides. Thus far, denervation has not been attempted in any other species. It should be noted that adrenals from beef, rabbits, and humans resemble those of the guinea pig in their EC peptide content.

Other rat tissues such as the pineal gland and superior cervical ganglion were denervated, but no effect on their total EC peptide content was observed. Thus, although the induction of adrenal proenkephalin by denervation is an interesting and dramatic effect, it has not yet been duplicated in any other tissue.

Of the three genes that code for enkephalin-containing peptides, the one for proenkephalin appears to be the most ubiquitous. Products of proenkephalin have been reported in adrenal medulla, brain, sympathetic nerves, pineal gland, anterior pituitary, posterior pituitary, pancreas, intestine, and heart. The availability of the rat cDNA probe made it possible for us to examine tissues for proenkephalin mRNA with some surprising findings. We found that rat heart contains large amounts of proenkephalin mRNA (Howells *et al.*, 1985). This mRNA was identified by its hybridization with rat proenkephalin cDNA as well as with a synthetic deoxynucleotide that is complementary to another part of the molecule. Both probes are in the coding region of the mRNA. On dissection, almost all of the mRNA was found in the ventricles. The mRNA content of rat ventricles was found to be about twice as high as that in the brain. The latter had heretofore been considered the richest source of proenkephalin mRNA. Localization in the ventricles was surprising since, unlike the atria, they contain no secreting organelles. Cellular localization of proenkephalin mRNA in rat heart ventricles is being investigated by *in situ* hybridization.

The very high content of proenkephalin mRNA in rat heart ventricles is not paralled by the EC peptide content, which is less than 1 pmol/g. By contrast, rat brain, with half as much proenkephalin mRNA as the ventricles, contains over 50 pmol/g of proenkephalin-derived EC peptides. Hamster, bovine, and human heart contain only traces of proenkephalin mRNA in venticles and atria. Their EC peptide content is also low (Howells *et al.*, 1985).

Proenkephalin mRNA was also found to be high in both male and female gonads of the rat (Kilpatrick *et al.*, 1985). The amounts of specific mRNA in testes and ovaries were about half of that found in rat brain. On further study, it was found that hamster testes and ovaries also contain appreciable amounts of proenkephalin mRNA. As in the case of the rat heart ventricle, products of translation were not found in amounts commensurate with the amount of specific mRNA in the gonads of the three species examined.

An interesting finding was that the preproenkephalin mRNA in rat testes is about 500 nucleotides larger than the preproenkephalin mRNA derived from other sources, which contains 1450 nucleotides. This larger mRNA was identified by its hybridization with a large restriction fragment from rat preproenkephalin cDNA cloned from rat brain, which is within the coding region of the mRNA. The structure and significance of this large form of proenkephalin mRNA in rat testes are under investigation.

It is worth noting that the rat has long been used as the standard laboratory animal for investigating opiate drugs, and much of our knowledge of opiate pharmacology is based on *in vivo* studies in this animal. We see, however, that in our studies alone there are several significant differences in the biosynthesis and regulation of proenkephalin-derived peptides between rats and the other species investigated. Those noted by us are:

1. Rat adrenal proenkephalin EC peptides differ qualitatively and quantitatively from those in other species, suggesting differences in the processing of proenkephalin.
2. The stimulatory effect of adrenal denervation on adrenal proenkephalin transcription and translation has not been observed in other species.
3. The very large amounts of proenkephalin mRNA present in rat heart ventricles are not observed in ventricles of other species.
4. The larger species of proenkephalin mRNA is found only in rat testes. The proenkephalin mRNA in other species is of the same size that is found in brain of all the species examined.

With so many differences, one wonders whether the rat should continue to be used as the model laboratory animal for studying opioid peptide pharmacology.

ACKNOWLEDGMENTS. This report is based on studies carried out in my laboratory by a number of colleagues. Dr. R. Howells together with Dr. A. Noe are continuing investigations on regulation of proenkephalin biosynthesis. Dr. K. Kilpatrick is continuing investigations on proenkephalin in testes and ovaries in his new laboratories at the Worcester Foundation.

REFERENCES

Fleminger, G., Howells, R. D., Kilpatrick, D. L., and Udenfriend, S., 1984, Intact proenkephalin is the major enkephalin-containing peptide-produced in rat adrenal glands after denervation, *Proc. Natl. Acad. Sci. U.S.A.* **81:**7985–7988.

Gubler, U., Seeburg, P., Hoffmann, B. J., Gage, L. P., and Udenfriend, S., 1982, Molecular cloning establishes proenkephalin as precursor of enkephalin-containing peptides, *Nature* **295**:206–208.
Howells, R. D., Kilpatrick, D. L., Bhatt, R., Monahan, J. J., Poonian, M., and Udenfriend, S., 1984, Molecular cloning and sequence determination of rat preproenkephalin cDNA: Sensitive probe for studying transcriptional changes in rat tissues, *Proc. Natl. Acad. Sci. U.S.A.* **81**:7651–7655.
Howells, R. D., Kilpatrick, D. L., Noe, A., Fleminger, G., Bailey, L. C., and Udenfriend, S., 1985, Products of proenkephalin transcription and translation in mammalian heart, *Fed. Proc.* **44**:420.
Kilpatrick, D. L., Howells, R. D., Fleminger, G., and Udenfriend, S., 1984, Denervation of rat adrenal glands markedly increases preproenkephalin mRNA, *Proc. Natl. Acad. Sci. U.S.A.* **81**:7221–7223.
Kilpatrick, D. L., Howells, R. D., Noe, A., Bailey, L. C., and Udenfriend, S., 1985, Expression of preproenkephalin mRNA and its peptide products in mammalian testis and ovary, *Proc. Natl. Acad. Sci. U.S.A.* **81**:7221–7223.
Schultzberg, M., Hokfelt, T., Lundberg, J., Terenius, L., Elfirm, L.-G., and Elde, R., 1978, Enkephalin-like immunoreactivity in nerve terminals in sympathetic ganglia and adrenal medulla and in adrenal medullary gland cells, *Acta Physiol. Scand* **103**:475–477.
Udenfriend, S., and Kilpatrick, D. L., 1983, Biochemistry of the enkephalins and enkephalin-containing peptides, *Arch. Biochem. Biophys.* **221**:309–323.

36

Endocrine and Neuroendocrine Actions of Cardiac Peptides

WILLIS K. SAMSON and ROBERT L. ESKAY

1. CARDIAC PEPTIDES: ISOLATION AND CHARACTERIZATION

The presence of secretory granules in atrial myocytes (Jamieson and Palade, 1964) suggested that the heart, in addition to its muscular function, might be a secretory organ. More recently, the function of these secretorylike granules became apparent when deBold (1979) demonstrated that water deprivation resulted in an increase in granularity of the atria, suggesting a role for these specific granules, or their contents, in the regulation of water and electrolyte balance. Later, his group (deBold *et al.*, 1981) demonstrated that saline extracts of atrial tissues, when injected intravenously (i.v.) into anesthetized rats, caused a rapid and potent increase in urinary sodium and chloride excretion and urine flow. They further speculated that atrial extracts contained a substance that inhibited renal tubular sodium chloride reabsorption and that the atria, a tissue known to monitor intravascular volume (Gauer *et al.*, 1961), could be the source of hormonal agents that controlled fluid volume via renal mechanisms. There soon followed a virtual flood of reports detailing the renal and vascular effects of partially purified atrial extracts, culminated by the reports by several groups (Flynn *et al.*, 1983; Atlas *et al.*, 1984; Geller *et al.*, 1984; Misono *et al.*, 1984; and Seidah *et al.*, 1984) of the isolation, purification, sequencing, and synthesis of a family of peptides from atrial extracts that possess the natriuretic/diuretic and/or spasmolytic effects of the crude extracts.

Since these various groups were working ostensively independently of one another, it is not peculiar that quite a variety of names are commonly used: atrial

WILLIS K. SAMSON • Department of Physiology, University of Texas Health Science Center at Dallas, Dallas, Texas 75235. ROBERT L. ESKAY • Section of Neurochemistry, Laboratory of Clinical Studies, Division of Intramural, Clinical and Biologic Research, National Institute on Alcohol Abuse and Alcoholism, National Institutes of Health, Bethesda, Maryland 20892.

natriuretic factor (ANF) or cardionatrin (Flynn *et al.*, 1983), auriculin (Atlas *et al.*, 1984), atriopeptin (Geller *et al.*, 1984), and atrial natriuretic peptide (Misono *et al.*, 1984). Essentially, they are members of the same family of peptides, differing only in the amount C- and N-terminal extension of a common 17-membered ring formed by an internal cystine disulfide (Fig. 1). Multiple precursor forms of atrial natriuretic factors also have been identified using cDNA (Yamanaka *et al.*, 1984; Maki *et al.*, 1984; Oikawa *et al.*, 1984) and amino-acid sequence (Lazure *et al.*, 1984) analysis. However, the central 17-membered internal disulfide ring is essential for biological activity (Misono *et al.*, 1984), and the degree of C- and N-terminal extension determines relative efficacy in a variety of bioassays (Geller *et al.*, 1984). For example, the 21-amino-acid peptide atriopeptin I, which differs from atriopeptin II and atriopeptin III only in the extent of C-terminal extension, lacks the *in vitro* vasorelaxant activity of the 23- and 24-membered peptides (Geller *et al.*, 1984), which each share equal biopotency; yet, all three members of this family of peptides share equipotent natriuretic–diuretic activities.

2. REGULATION OF RELEASE

An awareness of the stimuli that regulate the release of ANF is just beginning to emerge as a result of the development of radioimmunoassays (RIAs) with which to quantify ANF levels. The overall role of these peptides is thought to be an orchestrated means by which the body rids itself of a volume load; therefore, increased extracellular fluid volume acts as a stimulus for ANF release as a consequence of the atrial stretching induced by enhanced venous return. Tanaka *et al.*, (1984) have reported a significant increase in circulating immunoreactive (ir) ANF in rats provided 1.0% NaCl as drinking water. In addition, Lang *et al.* (1985) and Eskay *et al.* (1985) have demonstrated that volume loading with either 0.9% NaCl

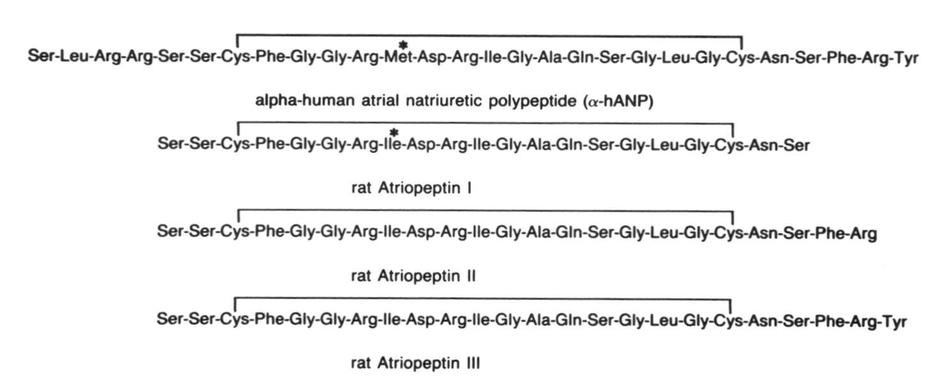

FIGURE 1. Amino acid sequences of the family of cardiac peptides known as atrial natriuretic factors (ANFs). The human form differs from that of the rat at only one amino acid locus (*).

or 5.0% glucose resulted in rapid increases in irANF in rat plasma. Hyperosmotic challenge (1 ml containing 2.5 mEq NaCl), as well, resulted in a four- to fivefold increase in rat plasma ANF content within minutes (Eskay *et al.*, 1985), which may have resulted simply from the rapid transfer of cellular fluid to the extracellular fluid space and subsequent activation of atrial stretch receptors. Thus, atrial-stretch-induced diuresis (Ledsome and Linden, 1968; Kappagoda *et al.*, 1979; Ackermann and Rudolph, 1984) is now thought to be mediated at least in part via ANF release.

Complicating an understanding of the potential multiple stimuli that may alter ANF release are a number of methodological considerations, which to date have been overlooked for the most part. As a consequence, reported basal ANF values in plasma vary 15-fold (Gutkowska *et al.*, 1984; Lang *et al.*, 1984; Tanaka *et al.*, 1984). In our studies, it appears that circulating basal irANF levels obtained in the conscious rat range from 50 to 100 pg/ml plasma (Eskay *et al.*, 1985). Anesthetics such as pentobarbital, halothane, and ether, which to date have been used by various investigators, result in artifactual elevation, by three- to eightfold, of circulating irANF levels (D'Souza and Eskay, 1986; Horky *et al.*, 1985). Clearly, caution should be exercised in the interpretation of results obtained in anesthetized animals because of the unexplored possibility of the differential effects of anesthetics on myocardial versus neuronal tissue with regard not only to the magnitude but also the direction of experimentally induced changes in plasma ANF.

Furthermore, volume expansion alone probably does not account for the full spectrum of stimuli that release ANF. Indeed, *in vitro* studies have demonstrated the direct action of acetylcholine, epinephrine, and vasopressin (AVP) to alter the release of ANF-like bioassayable material (Sonnenberg *et al.*, 1984; Sonnenberg and Veress, 1984). Others have been unable to determine any change in ANF secretion *in vitro* using the same secretagogues and incubation protocol; however, infusions of these agents *in vivo* do result in enhanced ANF release (D'Souza and Eskay, 1985). Acute infusion of microgram per kilogram body weight doses of carbachol or epinephrine or nanogram per kilogram doses of AVP in conscious rats with indwelling catheters resulted in three- to fivefold increases in irANF levels within 1 to 3 min. Pretreatment with atropine completely abolished the carbachol-induced release of ANF, suggesting the involvement of muscarinic cholinergic receptors in the regulation of ANF release (D'Souza and Eskay, 1986), although the exact site or mechanism of action of these secretagogues remains unclear.

Finally, germane to the study of the stimuli that regulate circulating ANF levels is the clarification of the forms of ANF released. There is general agreement that a large precursor molecule is stored *in situ,* and recent evidence suggests that one or more precursor-derived peptides (25 to 33 amino acids in length) are secreted (Tanaka *et al.*, 1984; Lang *et al.*, 1985; Eskay *et al.*, 1985; Vuolteenaho *et al.*, 1985) from the atria and thus are able to exert the various renal, adrenal, and vascular smooth muscle endocrine effects previously ascribed to the ANFs.

3. ENDOCRINE ACTIONS

Initial studies employing atrial extracts elucidated the natriuretic and diuretic (deBold *et al.*, 1981; Sonnenberg *et al.*, 1982; Briggs *et al.*, 1982; Borenstein *et al.*, 1983; Camargo *et al.*, 1984) as well as spasmolytic (Currie *et al.*, 1983; Camargo *et al.*, 1984) and adrenal (Atarashi *et al.*, 1984) effects of ANF. The availability of synthetic peptides resulted in a reexamination of the bioactions of ANF, and, naturally, the seemingly clear-cut bioactions have now come under close scrutiny, and some controversy has been generated.

3.1. Renal

The original description of a natriuretic–diuretic effect of atrial extracts (deBold *et al.*, 1981) reported increases in urinary sodium, chloride, and, to a lesser degree, potassium excretion and urine flow in anesthetized, nondiuretic rats. The authors observed no change in glomerular filtration rate (GFR) and therefore suggested that the effects of the extract were exerted specifically on NaCl transport in the ascending limb of Henle or medullary collecting duct. Subsequent micropuncture studies (Briggs *et al.*, 1982; Sonnenberg *et al.*, 1982) also implicated the medullary collecting duct and tubules as the site of action of the atrial extract; however, in the study of Briggs *et al.* (1982), high doses of extract resulted in increased GFR. This observation of increased GFR was corroborated by Borenstein *et al.* (1983) and Camargo *et al.* (1984), raising the possibility that the natriuretic effect of the atrial extracts resulted in part from hemodynamic and not tubular actions. Indeed, Atlas *et al.* (1984) furthered their concept of hemodynamic control when they demonstrated that synthetic ANF had both renal vasoconstrictive and vasorelaxant effects. An increase in GFR subsequent to ANF infusion was proposed by these authors to result from constriction of the efferent ateriole, and the natriuretic effect accounted for by the resulting increase in GFR and filtered sodium load.

The ability of ANF to increase renal blood flow was established by microsphere studies (Koike *et al.*, 1984; Garcia *et al.*, 1985) in conscious rats and using implantable flow probes in anesthetized dogs (Katsube *et al.*, 1985; Wakitani *et al.*, 1985; Ishihara *et al.*, 1985). However, others (Lappe *et al.*, 1985) have reported failure of ANF to cause vasodilation in the conscious rat and the anesthetized cat (Wendt *et al.*, 1985). Although the differences in results between some studies might be ascribed to the use of anesthetics or species variation, Maack *et al.* (1985) obtained increases in GFR with ANF infusion in conscious and anesthetized dogs in the face of only transient increases in renal blood flow. This led them to speculate that the decrease in urine osmolality after ANF infusion in conscious dogs, during which time free water clearance remained stable (Maack *et al.*, 1984), resulted from increased inner medullary blood flow and subsequent medullary washout. Data from conscious rats seem to support this possibility (Huang *et al.*, 1985).

Even though convincing data for the hemodynamic basis of the natriuretic effect of ANF exist, a tubular effect of ANF cannot be ruled out (Maack *et al.*,

1985). Atrial-natriuretic factor receptor binding is present, albeit at low levels, in the collecting tubules (DeLean *et al.*, 1985) as well as on membranes from rabbit and rat kidney cortex (Napier *et al.*, 1984), and data from conscious rats (Gellai *et al.*, 1985) and isolated rat kidneys (Murray *et al.*, 1985) strongly suggest a dissociation of the hemodynamic and natriuretic–diuretic effects.

We have approached this problem differently (Vanatta and Samson, 1985). Although others (Pamnani *et al.*, 1984) had failed to observe an effect of atrial extract on short-circuit currents in the toad bladder, we reasoned that the diuretic effect of ANF might not be secondary to its natriuretic action but instead via inhibition of vasopressin-stimulated water absorption. Indeed, we have demonstrated that atriopeptin III, at equimolar concentrations ranging from 10^{-12} to 10^{-11} M, significantly inhibits vasotocin-induced water reabsorption in the toad bladder (Vanatta and Samson, 1985). These results suggest that earlier micropuncture studies (Briggs *et al.*, 1982; Sonnenberg *et al.*, 1982) did in fact identify a tubular action of atrial extracts and that the observed natriuretic–diuretic actions of ANF are mediated through many sites of action, not the least of which may be the distal nephron (Sonnenberg, 1985).

3.2. Vascular

The vasorelaxant effect of ANF was also first identified in atrial extracts (Currie *et al.*, 1983; Grammer *et al.*, 1983; Kleinert *et al.*, 1984). In rabbit aortic and chick rectum strips precontracted *in vitro* with norepinephrine and carbachol, respectively, dose-dependent relaxation was observed (Currie *et al.*, 1983). Furthermore, both hormonally (angiotensin) and nonhormonally (potassium) induced contractions of aortic strips were reversed by atrial extracts (Kleinert *et al.*, 1984). Synthetic ANF was later shown to relax norepinephrine-contracted aortic strips (Currie *et al.*, 1984; Misono *et al.*, 1984), carbachol-contracted chick rectum (Currie *et al.*, 1984; Kangawa and Matsuo, 1984), and histamine-contracted aortic rings (Atlas *et al.*, 1984).

Although concordance exists over the effect of ANF on *in vitro* smooth muscle relaxation, the literature on *in vivo* vasodilatory actions, particularly the effects on renal vascularity (see above), is discordant. The hypotensive effects of atrial exracts (deBold *et al.*, 1981) were first demonstrated in anesthetized rats. Synthetic ANF also induces hypotension in conscious as well as anesthetized dogs (Ishihara *et al.*, 1985), an effect theorized by the authors to result from selective vasodilation of the renal artery bed and fluid loss secondary to diuresis. More recently, however, the hypotensive effect of ANF has been explained not by a decrease in total peripheral resistance but by a reduction in cardiac output (Smits and Lappe, 1985; Kleinert *et al.*, 1985).

It is of interest, and perhaps this concept will also develop in studies on the neuroendocrine actions of ANF, that the vascular action of the cardiac peptides is expressed in the presence of hormonal agents that by themselves contract smooth muscle. Thus, the concept of hormonal antagonism is central to the analysis of

ANF's vascular effects. This too might be the case for its diuretic effect (Vanatta and Samson, 1985) and its adrenal actions (see below).

3.3. Adrenal/Renal

The ability of atrial stretching to inhibit aldosterone secretion (Anderson *et al.*, 1959) suggested a role for ANF in the control of adrenal function. Atrial extracts subsequently were shown to reduce basal and inhibit ACTH- and angiotensin-II-stimulated aldosterone release from cultured adrenal capsular cells (Atarashi *et al.*, 1984), and synthetic ANF (Goodfriend *et al.*, 1984) inhibited basal and angiotensin-, cAMP-, and potassium-stimulated aldosterone production in suspensions of bovine adrenal glomerulosa cells. Infusion (i.v.) of synthetic ANF blocks the aldosterone-releasing action of angiotensin II in conscious rats (Chartier *et al.*, 1984) and in anesthetized dogs (Maack *et al.*, 1984) and decreases plasma aldosterone levels as well as renin secretory rate and plasma renin activity. This effect of ANF on renin secretion was also observed by Burnett *et al.* (1984). Both groups suggested that the reduced renin secretion was caused by an increased sodium load delivered to the macula densa as a result of increased GFR but could not entirely rule out the direct action of ANF on the juxtaglomerular apparatus. Recently, Inagami's group (Naruse *et al.*, 1985) has demonstrated that synthetic ANF decreases plasma renin activity in anesthetized rats and, further, has identified a direct action on the juxtaglomerular cells, since ANF inhibits renin release from kidney cortex slices *in vitro*. Therefore, ANF can act directly at the level of renal juxtaglomerular cells and adrenal glomerulosa cells to inhibit the release of renin and angiotensin. This, in addition to its ability to alter blood pressure (Maack *et al.*, 1984a; Tang *et al.*, 1984) and promote natriuresis and diuresis, suggests an important role for ANF in the physiological control of fluid volume and vascular homeostasis.

4. NEUROENDOCRINE ACTIONS

The smooth muscle and renal actions of ANF suggested a role for the cardiac peptides in the control of extracellular fluid volume, electrolyte homeostasis, and vascular function quite contrary to that of vasopressin (AVP). Therefore, initial neuroendocrine investigations were directed toward possible ANF–AVP interactions.

4.1. Posterior Pituitary

The inhibitory action of ANF on aldosterone (Chartier *et al.*, 1984) and renin secretion (Naruse *et al.*, 1985) suggested that, in addition to antagonizing the action of AVP in the kidney (Vanatta and Samson, 1985), ANF might inhibit AVP release from the posterior pituitary. Since peripheral plasma levels of AVP in conscious rats are very low, however, it would be difficult to detect accurately any inhibitory

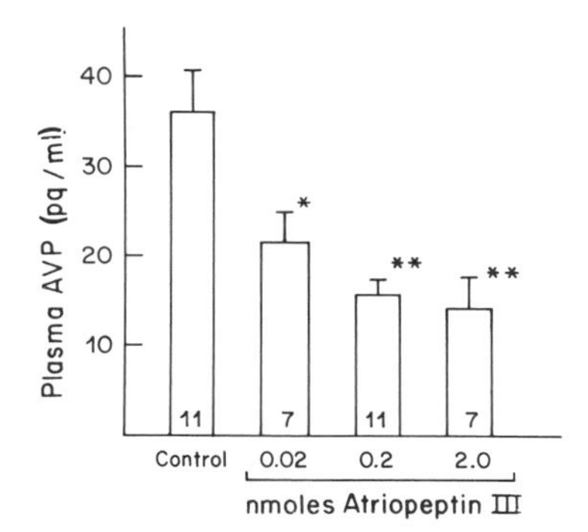

FIGURE 2. Effects of intravenous infusion of control saline (0.1 ml) or saline containing ANF on plasma AVP in water-deprived conscious rats. Group size is indicated in bars. *$P < 0.05$, **$P < 0.005$. (Reprinted by permission of S. Karger, Basel, from Samson, 1985a.)

action of ANF on circulating AVP. Therefore, rats were manipulated such that circulating levels of AVP would be elevated prior to examining the effects of ANF (Samson, 1985a). In the first series, conscious male rates subjected to a 72-hr water deprivation were infused i.v. with 0.1 ml of either saline or saline-containing atriopeptin III. Fifteen minutes later, rats were sacrificed by decapitation, trunk bloods were collected, and plasma AVP levels were determined by specific RIA. Significant dose-related inhibition by ANF of dehydration-elevated AVP levels was observed (Fig. 2).

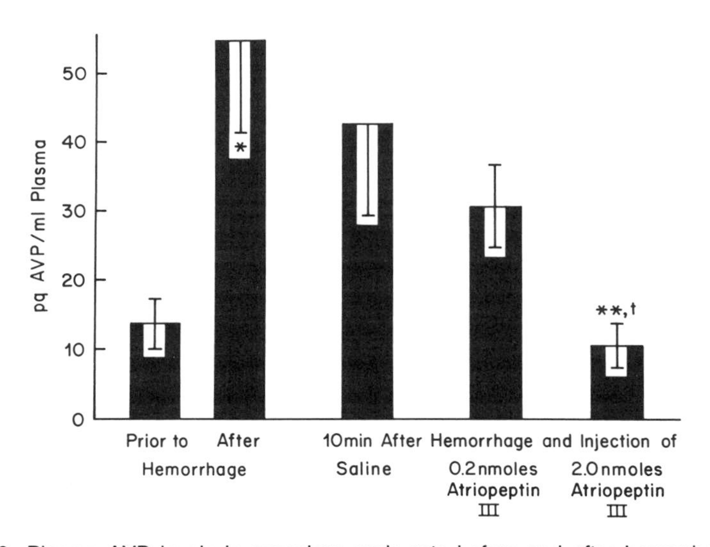

FIGURE 3. Plasma AVP levels in conscious male rats before and after hemorrhage (5 ml blood/300 g body weight) and 10 min after i.v. infusion of saline (0.1 ml) or saline containing ANF. *$P < 0.05$ versus nonhemorrhaged levels; **$P < 0.05$ versus saline-infused hemorrhaged controls; $^{+}P < 0.02$ versus posthemorrhage levels. (Reprinted by permission of S. Karger, Basel, from Samson, 1985a.)

In a second experiment, conscious male rats were hemorrhaged by removal of blood (5 ml/300 g body weight) over a 5-min period via indwelling jugular cannulae. At the end of this period, saline alone (0.1 ml) or saline containing atriopeptin III was infused via the cannulae, and rats were sacrificed 10 min later. A dose-related inhibition of hemorrhage-induced AVP release was observed, with significant inhibition evident after infusion of 2 nmol ANF (Fig. 3).

The results of this series of experiments indicated that in the conscious rat, ANF could inhibit AVP release, but the site of action of i.v. infused ANF was unclear. In addition to a direct neural lobe effect, ANF could have acted at the osmoreceptor sites in the hypothalamus (Ramsay *et al.*, 1983) to overcome the dehydration-induced stimuli for AVP. A reduction in AVP release secondary to a central action of ANF (if it indeed crossed the blood–brain barrier) on parasympathetic and sympathetic afferent input to the hypothalamus (Schrier *et al.*, 1979; Sladek, 1983) or by interaction with peptidergic systems in the brain (Sladek, 1983) could also explain these results. The direct neural lobe site of action was attractive for a variety of reasons; thus, neurointermediate lobe explants were incubated *in vitro* in the presence of ANF, and both AVP and oxytocin (OT) releases were monitored by RIA.

Initial studies involved static incubation of neural lobe explants in medium 199. These studies demonstrated the inhibitory action of high doses of ANF on AVP release when release was expressed either in terms of picograms hormone released or picograms hormone released per nanogram hormone content (Fig. 4). No significant effect on OT release was observed. Since these static studies poorly mimic the *in situ* situation, perifusion studies were undertaken, which would have

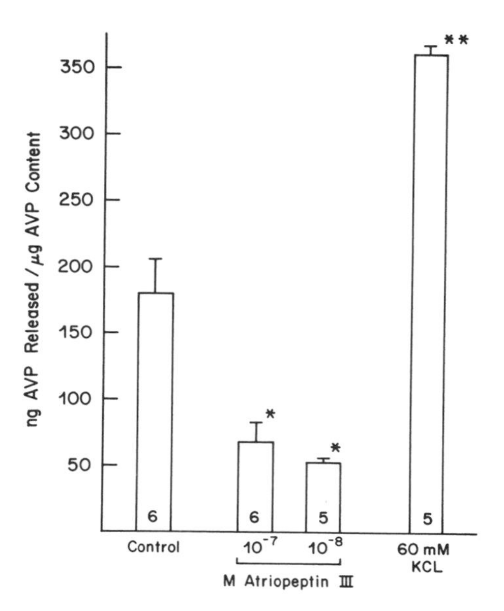

FIGURE 4. Effects of ANF or a depolarizing concentration of potassium chloride on AVP release from rat neurointermediate lobe explants incubated *in vitro*.

TABLE I. The Effect of Atrial Natriuretic Factor or a Depolarizing Concentration of KCl on Vasopressin Release from Perifused Rat Neurointermediate Lobes[a]

	Preexposure	Exposure
Control	1.28 ± 0.06	1.33 ± 0.07
10^{-11} M ANF	1.02 ± 0.03	0.89 ± 0.02*
60 mM KCl	1.34 ± 0.06	1.92 ± 0.11**

[a] Data are expressed as nanograms AVP per milliliter perfusate. $^*P < 0.002$; $^{**}P < 0.001$.

the advantage of simulating the *in vivo* constant perifusion situation and allowing initial kinetic observations. For these studies, four neural lobes were loaded onto perifusion columns supported by a gel matrix (Biogel, P-2) and perifused with culture medium for an initial stabilization period. Thereafter, the perifusate was collected automatically at 1-min intervals prior to, during, and following exposure to medium alone or medium containing 10^{-11} M atriopeptin III or depolarizing concentrations of KCl (Table I). The inhibitory action of ANF displayed a brief latency (3–4 min) to onset and a slightly longer latency to extinction (not shown). However, the action of ANF was specific to AVP, since OT levels in perifusate medium were unaffected by ANF (Table II).

These preliminary studies suggest a direct action of the cardiac peptides to inhibit AVP release selectively from the neural lobe. Some controversy does, however, exist. Although Januszewicz *et al.* (1985) failed to observe any significant effect of ANF on whole neural lobe explants in static incubation, they reported a significant stimulatory action of what they called physiological doses (3×10^{-11} and 3×10^{-10} M) on minced neural lobes. Oxytocin release was not determined. Their AVP release results have not been duplicated in further studies in this laboratory; however, the differing experimental protocols might explain the conflicting results. Yet, a stimulatory action of ANF on AVP release would seem physiologically contradictory, and even Januszewicz *et al.* (1985) allow in their discussion that, despite their *in vitro* data, "the overall effect of atrial distension, as possibly mediated both by ANF release and by nervous reflexes, may well be a decrease in AVP release." Certainly, the demonstration of ANF binding sites in neural lobe membranes (Januszewicz *et al.*, 1985) and the ability of ANF to inhibit basal and stimulated adenylate cyclase activity in posterior pituitary homogenates (Anand-Srivasta *et al.*, 1985) would imply an inhibitory action of ANF on AVP release,

TABLE II. Oxytocin Release from Perifused Rat Neurointermediate Lobes

	Preexposure	Exposure
Control	143.1 ± 13.3	122.8 ± 13.5
10^{-11} M ANF	152.2 ± 12.1	181.7 ± 11.5
60 mM KCl	138.4 ± 6.1	242.7 ± 12.3[a]

[a] $P < 0.001$.

since adenylate cyclase activation has been linked to release of AVP (Ruoff *et al.,* 1976). Indeed, recently Genest's group reported unpublished observations (Anand-Srivasta *et al.,* 1985) claiming, unlike their earlier findings (Januszewicz *et al.,* 1985), to have seen an inhibitory action of ANF on AVP release.

4.2. Hypothalamic

Since ANF-like immunoreactivity (henceforth called simply ANF) has been detected in the hypothalamus (see Section 5.1), it is possible that cardiac peptides act centrally, perhaps on the cell bodies in the paraventricular (PVN) or supraoptic (SOP) nuclei, to alter AVP release. To test this hypothesis, synthetic atriopeptin III was slowly infused (2 nmol in 2 μl saline) via an indwelling cannula into the third cerebroventricle of conscious, unrestrained male rats, and trunk blood was collected 15 min later. In two experiments, central infusion of ANF resulted in significant reductions in basal AVP release (Fig. 5). This implies that ANF can act either directly on the AVP-producing cells of the hypothalamus or on other cellular elements that provide afferent input to these neurons and suggests a neuromodulatory role for ANF. The anatomic substrate for such a role for ANF has been demonstrated by immunocytochemistry (Jacobowitz *et al.*, 1985; Saper *et al.*, 1985; Skofitsch *et al.,* 1986) and the visualization of ANF binding sites in the CNS (Quirion *et al.,* 1984).

In addition to hypothalamic actions to alter hormone release, behavioral effects of the cardiac peptides can be predicted. Of obvious interest is the possible effect of ANF on thirst mechanisms. We are examining this possibility and have observed a significant inhibitory action of centrally administered ANF on water intake in 18-hr water-deprived conscious rats, on angiotensin-II-induced intake in normally hydrated conscious rats (Antunes-Rodrigues *et al.,* 1985), and on salt preference in salt-depleted, normally hydrated rats (Antunes-Rodrigues *et al.,* 1986).

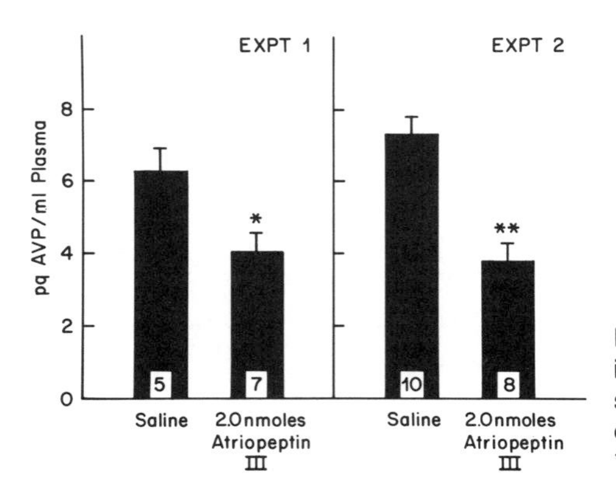

FIGURE 5. Effect of intracerebroventricular infusion of saline (2 μl) or 2.0 nmol ANF in saline on plasma AVP levels in normally hydrated, conscious, unrestrained male rats. $^{*}P < 0.05$, $^{**}P < 0.01$.

4.3. Anterior Pituitary

The presence of ANF (Samson, 1985b) in anterior pituitary extracts and ANF binding in the gland (Quirion *et al.*, 1985) strongly suggest a role for ANF in the control of anterior pituitary function. Indeed, its presence in median eminence extracts (Samson, 1985b; Zamir *et al.*, 1986) and in nerve terminals in the external layer of the median eminence (Jacobowitz *et al.*, 1985; Saper *et al.*, 1985; Skofitsch *et al.*, 1985) suggests its presence in the hypophyseal portal plasma and its possible role as a regulator of anterior pituitary function. The ability of ANF not only to antagonize the action (Vanatta and Samson, 1985) of AVP but also to inhibit its release (Samson, 1985a) suggests that in addition to antagonizing the effect of ACTH in the adrenal (Goodfriend *et al.*, 1984), ANF might interfere with ACTH release from the anterior pituitary gland. Indeed, ANF has been shown to inhibit basal and corticotropin-releasing-factor-stimulated adenylate cyclase activity in the anterior pituitary (Anand-Srivasta *et al.*, 1985). In a subsequent study (J. Dave, W. K. Samson, and R. L. Eskay, personal observations), we were not able to observe ANF inhibitory effects on basal adenylate cyclase activity in anterior pituitary membranes, but ANF clearly blocked CRF-stimulated adenylate cyclase activity. Changes in cAMP levels in the anterior pituitary have been linked causally to ACTH release (Bilezikjian and Vale, 1983).

Although initial studies using cultured, dispersed anterior pituitary cells from male rats suggested such an inhibitory effect of ANF (Fig. 6) at a dose of 1 nM, subsequent studies have failed to demonstrate the ability of 10 mM ANF to inhibit basal ACTH release or CRF-stimulated β-endorphin or ACTH release in similar cell cultures. Therefore, although atrial stretching not only stimulates ANF release but also inhibits circulating ACTH levels (Plotsky *et al.*, 1985), it does not necessarily follow that ANF inhibits ACTH release directly. Instead, ANF released on atrial pulsation might be responsible for the resultant decrease in portal plasma AVP levels observed (Plotsky *et al.*, 1985) and thereby decrease ACTH release, since AVP has been implicated in the control of ACTH secretion (Vale *et al.*, 1983).

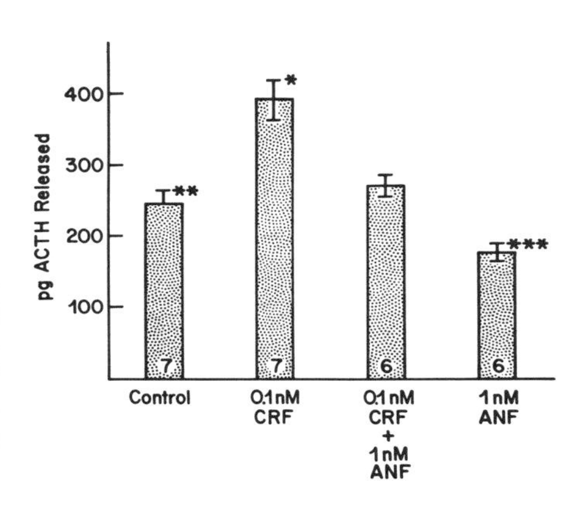

FIGURE 6. Effect of ANF on basal and corticotropin-releasing factor (CRF)-stimulated ACTH release from cultured dispersed anterior pituitary cells *in vitro*. $^*P < 0.001$, CRF alone versus CRF + ANF; $^{**}P < 0.001$ control versus CRF; $^{***}P < 0.05$ control versus ANF alone.

TABLE III. Lack of Significant Effect of Atriopeptin III on Basal Release of Anterior Pituitary Hormones *in Vitro*[a]

	Control	Atriopeptin III 0.1 nM	Atriopeptin III 1.0 nM	Atriopeptin III 10 nM
Prolactin (PRL)	134.2 ± 4.3	129.7 ± 5.6	144.8 ± 4.5	132.3 ± 5.6
Luteinizing hormone (LH)	3.5 ± 0.2	3.3 ± 0.2	3.2 ± 0.2	3.3 ± 0.1
Growth hormone (GH)	1382 ± 69	1505 ± 77	1498 ± 37	1384 ± 63
Adrenocorticotropin (ACTH × 10^{-3})	304.3 ± 12.6	304.9 ± 8.0	287.6 ± 13.5	299.4 ± 8.3

[a] Means ±S.E.M. (ng), $n = 7$.

The possibility that ANF might alter the release of other anterior pituitary hormones is presently under investigation. Preliminary results failed to demonstrate (Table III) any significant effect of ANF on basal release of luteinizing hormone, prolactin, growth hormone, or ACTH. However, it is possible, in spite of the apparent lack of effect on hormone release, that ANF may alter the synthesis of certain anterior-pituitary-derived hormones without altering hormone release *per se*.

5. CENTRAL NERVOUS SYSTEM LOCALIZATION

The presence of natriuretic/diuretic substances within the brain was suggested during studies examining the cholinergic mediation of fluid intake and urinary sodium excretion in conscious rats (Dorn *et al.*, 1970). Later studies by the same group (Morris *et al.*, 1976) implicated the hypothalamus as the source of humoral agents that exert potent natriuretic effects subsequent to intraventricular hypertonic saline infusion. The isolation and characterization of natriuretic substances in cardiac extracts suggested the nature of the proposed hypothalamic natriuretic substance, since many hormones originally isolated in extracts of non-CNS tissue (e.g., gastrointestinal hormones) subsequently were localized in the CNS, particularly in hypothalamic structures, and were demonstrated either to exert hypothalamic actions (Samson *et al.*, 1982) or to be released from the hypothalamus into the circulation (Shimatsu *et al.*, 1981). Therefore, once antisera to the cardiac peptides became available, several groups endeavored to demonstrate the presence of ANF in the CNS.

5.1. Regional Distribution

Tanaka *et al.* (1984) first reported the presence of ANF in rat hypothalamic extracts and further identified both low- and high-molecular-weight forms of the immunocrossreactivity detected. Levels of ANF in hypothalamic extracts were considerably less than those present in atrial extracts; however, nanogram quantities were present per gram tissue (Tanaka *et al.*, 1984). Soon thereafter, the first immunohistochemical mapping of CNS neurons containing ANF was published (Jacobowitz *et al.*, 1985). These authors identified ANF neurons in distinct hypothalamic regions including the organum vasculosum lamina terminalis (OVLT), bed nucleus of the stria terminalis, and several hypothalamic nuclei. Brainstem centers containing ANF immunoreactivity included the mesencephalic central gray substance, dorsal tegmental nucleus, and dorsal parabrachial nucleus. Nerve fibers containing ANF were observed in the lateral septum, diagonal band, ventral bed nucleus, interpeduncular nucleus, dorsal raphe, and median eminence. This localization led the authors to speculate that ANF-like peptides are not only produced in the brain but also that their presence in regions known to be important in the regulation of fluid and electrolyte homeostasis indicated a possible role for them in the CNS control of fluid balance (Jacobowitz *et al.*, 1985). Shortly thereafter, Saper *et al.* (1985) reported a similar distribution of ANF in hypothalamic and pontine structures and likewise speculated on a neuromodulatory role for ANF in the control of blood volume and composition. Certainly, the presence of ANF binding sites in the subfornical organ, area postrema, median preoptic nucleus, median eminence, and nucleus tractus solitarius (Quirion *et al.*, 1984) justified such a proposal.

More recently, detailed mapping by immunohistochemistry (Skofitsch *et al.*, 1985) and radioimmunoassay (Zamir *et al.*, 1986) of the regional distribution of central ANF has been accomplished. These extensive studies revealed a concentration of ANF-containing neuronal elements in the telencephalon (nucleus interstitialis stria terminalis and amygdala), the region of the anteroventral third ventricle (AV3V), basal hypothalamus (including the external layer of the median eminence), posterior pituitary, interpeduncular nucleus, caudal mesencephalic reticular nuclei, nucleus tractus solitarius, and the dorsal and ventral horns of the spinal cord. Radioimmunoassay of extracts of microdissected brain regions revealed a quantitative distribution that agreed well with the immunohistochemical mapping. By far, the highest concentrations (560–580 fmol/mg protein) of ANF were found in the paraventricular nucleus of the hypothalamus and the median preoptic nucleus. High levels (360–470 fmol/mg protein) were also detected in the interpenduncular nucleus, preoptic and hypothalamic periventricular nuclei, median eminence, and the OVLT (Zamir *et al.*, 1986). Further characterization of irANF in the brain on the basis of reversed-phase high-performance liquid chromatography revealed that an extract of the preoptic hypothalamic area of the rat contained two peaks. The primary peak, which constituted two-thirds of the total immunoreactivity, coeluted with atriopeptin III, whereas the remaining material eluted much later, suggesting a substantially larger peptide (Zamir *et al.*, 1986).

5.2. Experimentally Induced Alterations in ANF Content

On solely anatomic grounds, a role for ANF in the hypothalamic control of salt and water intake and extracellular fluid volume homeostasis can be hypothesized. Therefore, it is not unreasonable to propose that alterations of fluid balance might alter cardiac and/or CNS content of ANF. Indeed the initial report of hypothalamic ANF (Tanaka *et al.*, 1984) also revealed that whereas rat atrial and plasma ANF content rose after high sodium intake, hypothalamic levels fell. This suggested a central action of ANF quite different from the peripheral response.

Since dehydration-induced AVP release can be reversed by ANF infusion (Samson, 1985a), and centrally administered ANF can alter basal AVP release, the effect of water deprivation on levels of ANF and AVP in discrete structures associated with the hypothalamo–neurohypophyseal system was examined (Samson, 1986). In addition to confirming the presence of ANF immunoreactivity within basal forebrain structures, this study revealed that although 72-hr water deprivation resulted in increased heart content of ANF, levels of the peptide in several hypothalamic structures actually fell significantly or remained unchanged (Fig. 7). Significant reductions in ANF content of some (OVLT, posterior pituitary), but not all (anterior pituitary, median eminence) of the tissues examined that do not lie behind the blood–brain barrier argue against these changes merely being a reflection of altered circulating ANF levels. Furthermore, the reduction of ANF content in the supraoptic and suprachiasmatic, but not the paraventricular, nuclei suggests differential processing of central ANF-containing neuronal circuitry. These dehydration-induced alterations in ANF content of discrete brain regions known to be involved in behavioral as well as neuroendocrine mechanisms regulating vasopressin secretion and extracellular fluid balance suggest a neuromodulatory role for the cardiac peptides in hypothalamic function.

	ANF	AVP
Neural Lobe	↓	↓
Supraoptic Nucleus	↓	n.c.
Paraventricular Nucleus	n.c.	n.c.
Median Eminence	n.c.	↑
Suprachiasmatic Nucleus	↓	n.c.
OVLT	↓	n.c.
Anterior Pituitary	n.c.	↓
Plasma	(↓)	↑
Heart	↑	?

FIGURE 7. Summary of the effects of dehydration on tissue AVP and ANF content in the rat; n.c., no change. (From Samson, 1985a.)

6. CONCLUSIONS AND PROSPECTIVES

It is clear that the cardiac peptides possess potent endocrine and neuroendocrine actions. The physiological significance of these effects is now and will continue to be a fertile, rapidly expanding field of both fundamental and clinical research. The major circulating forms of ANF are being identified, and the biospecificity of the various family members established. Their presence in tissue other than the heart certainly indicates the potential for a wider spectrum of action than originally hypothesized. In addition to established sites of actions, the effects of these peptides on central nervous system function are being discovered; and, as with the developing literature dealing with the renal, vascular, and adrenal sites of action of ANF, surely controversy will develop over the exact mechanisms and sites of action of the cardiac peptides within the CNS.

We have not only added to the controversy over the renal site of action of ANF (Vanatta and Samson, 1985) with our demonstration of a possible mode of action of ANF to block AVP-induced antidiuresis but also have entered into an as yet unresolved dispute with the Genest group (Januszewicz *et al.*, 1985) over the ability of ANF to alter AVP release (Samson, 1985b). However, since the studies on the neuroendocrine actions of ANF are still in their nascent stages, these discrepancies in results can be expected. Already our findings on the neuromodulatory action of centrally administrated ANF to alter water intake (Antunes-Rodrigues *et al.*, 1985) have been corroborated by others (Nakao *et al.*, 1985). Furthermore, our studies detailing the adenylate-cyclase-inhibited effect of ANF in anterior- and posterior-pituitary-derived membranes have been duplicated by others (Anand-Srivasta *et al.*, 1985).

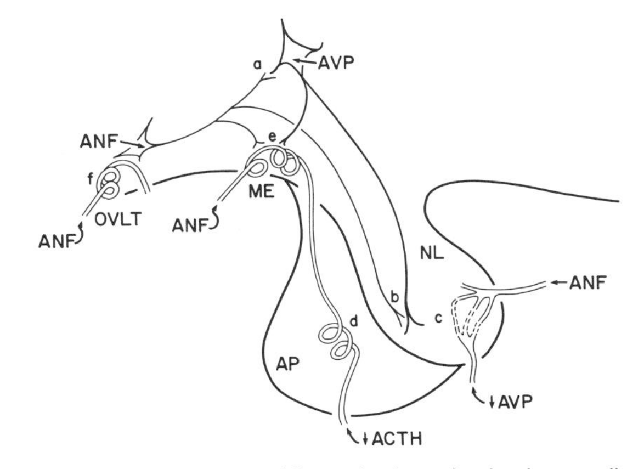

FIGURE 8. Possible sites of action of ANF within the hypothalamo–neurohypophyseal and pituitary units. ANF of peripheral origin might alter AVP release by an action at the OVLT or subfornical organ (not shown) or by an action on AVP-containing nerve terminals in the neural lobe (c). Alternatively, ANF of CNS origin might act as a neuromodulatory agent directly on AVP-containing cell bodies (a) and nerve terminals (e) or via interneurons within the hypothalamus or axo-axonally (b) in the neural lobe (NL). An anterior pituitary (AP) site of action of ANF to alter hormone release (d) can be hypothesized, as well as an action of ANF of central (a) or peripheral (e,f) origin to alter neuronally derived behavioral events such as fluid intake.

Thus, both neuromodulatory and neuroendocrine effects of the cardiac peptides now can be hypothesized. Our studies indicate a CNS site of action of ANF to mediate the control of fluid and electrolyte balance in a number of important ways. First, ANF can alter AVP release both *in vivo* (Samson, 1985a) and *in vitro* (Samson, 1985b) and thereby not only alter the renal handling of water directly (Vanatta and Samson, 1985) but also modulate the normal hormonal mechanisms regulating water conservation. Second, ANF can alter the neurogenic mechanisms by which the intake of salt and water is controlled (Antunes-Rodrigues *et al.*, 1985, 1986). Lastly, an action of ANF to alter in some way anterior pituitary hormone secretion is suggested not only on anatomic grounds (Jacobowitz *et al.*, 1985; Saper *et al.*, 1985; Skofitsch *et al.*, 1985; Zamir *et al.*, 1986; Samson 1986) but also by the fact that atrial pacing reduces plasma ACTH levels (Plotsky *et al.*, 1985). Certainly, the ability of ANF to inhibit adenylate cyclase activity in anterior pituitary homogenates suggests such a site of action.

In summary, we feel that important neuroendocrine actions of the cardiac peptides soon will be acknowledged. The sites of action of ANF in the hypothalamo–hypophyseal unit can be many (Fig. 8). Elucidation of the cellular sites of action of these peptides is the current goal of several laboratories.

REFERENCES

Ackermann, U., and Rudolph, J. R., 1984, Control of right atrial pressure at constant cardiac output suppresses volume natriuresis in anesthetized rats, *Can. J. Physiol. Pharmacol.* **62:**798–801.

Anand-Srivasta, M. B., Cantin, M., and Genest, J., 1985, Inhibition of pituitary adenylate cyclase by atrial natriuretic factor, *Life Sci.* **36:**1873–1879.

Anderson, C. H., McCally, M., and Farrell, G. L., 1959, The effects of atrial stretch on aldosterone secretion, *Endocrinology* **64:**202–207.

Antunes-Rodrigues, J., McCann, S. M., Rogers, L. C., and Samson, W. K., 1985, Third cerebroventricular infusion of atrial natriuretic factor inhibits water intake in conscious rats, *Proc. Natl. Acad. Sci. U.S.A.* **82:**8720–8723.

Antunes-Rodrigues, J., McCann, S. M., and Samson, W. K., 1986, Atrial natriuretic factor infusion inhibits salt intake in salt-depleted, normally hydrated rats *Endocrinology* **118:**1726–1728.

Atarashi, K., Mulrow, P. J., Franco-Saenz, R., Snajdar, R., and Rapp, J., 1984, Inhibition of aldosterone production by an atrial extract, *Science* **224:**992–994.

Atlas, S. A., Kleinert, H. D., Camargo, M. J., Januszewicz, A., Sealy, J. E., Laragh, J. H., Schilling, J. W., Lewicki, J. A., Johnson, L. K., and Maack, T., 1984, Purification, sequencing, and synthesis of natriuretic and vasoactive rat atrial peptide, *Nature* **309:**717–719.

Bilezikjian, L. M., and Vale, W. W., 1983, Glucocorticoids inhibit corticotropin-releasing factor-induced production of adenosine 3′,5′-monophosphate in cultured anterior pituitary cells, *Endocrinology* **113:**657–662.

Borenstein, H. B., Cupples, W. A. Sonnenberg, H., and Veress, A. T., 1983, The effect of natriuretic atrial extract on renal hemodynamics and urinary excretion in anesthetized rats, *J. Physiol. (Lond.)* **334:**133–140.

Briggs, J. P., Steipe, B., Schubert, G., and Schnermann, J., 1982, Micropuncture studies of the renal effects of atrial natriuretic substance, *Pfleugers Arch.* **395:**271–276.

Burnett, J. C., Granger, J. P., and Opgenorth, T. J., 1984, Effects of synthetic atrial natriuretic factor on renal function and renin release, *Am. J. Physiol.* **247:**F863–F866.

Camargo, M. J. F., Kleinert, H. D., Atlas, S. A., Sealey, J. E., Laragh, J. H., and Maack, T., 1984, Ca-dependent hemodynamic and natriuretic effects of atrial extract in isolated rat kidney, *Am. J. Physiol.* **246:**F447–F456.

Chartier, L., Schiffrin, E., Thibault, G., and Garcia, R., 1984, Atrial natriuretic factor inhibits the stimulation of aldosterone secretion by angiotensin II, ACTH and potassium *in vitro* and angiotensin II-induced steroidogenesis *in vivo, Endocrinology* **115:**2026–2028.

Currie, M. G., Geller, D. M., Cole, B. R., Boylan, J. G., Yusheng, W., Holmberg, S. W., and Needleman, P., 1983, Bioactive cardiac substances: Potent vasorelaxant activity in mammalian atria, *Science* **221:**71–73.

Currie, M., Geller, D. M., Cole, B. R., Siegel, N. R., Fok, K. F., Adams, S. P., Eubanks, S. R., Galluppi, G. R., and Needleman, P., 1984, Purification and sequence analysis of bioactive atrial peptides (atriopeptins), *Science* **223:**67–69.

deBold, A. J., 1979, Heart atria granularity effects of changes in water electrolyte balance, *Proc. Soc. Exp. Biol. Med.* **161:**508–511.

deBold, A. J., Borenstein, H. B., Veress, A. T., and Sonnenberg, H., 1981, A rapid and potent natriuretic response to intravenous injection of atrial myocardial extract in rats, *Life Sci.* **28:**89–94.

DeLean, A., Thibault, G., Seidah, N. G., Lazure, C., Gutkowska, J., Vinay, P., Chretien, M., and Genest, J., 1985, Pharmacological characteristics of adrenal and renal receptors for atrial natriuretic factor, *Fed. Proc.* **44:**1571.

Dorn, J., Antunes-Rodrigues, J., and McCann, S. M., 1970, Natriuresis in the rat following intraventricular carbachol, *Am. J. Physiol.* **219:**1292–1298.

D'Souza, N., and Eskay, R. L., 1986, Adrenergic, cholinergic, and vasopressinergic-induced release of atrial natriuretic peptides *in vivo, Biochem. Biophys. Res. Commun.* (in press).

Eskay, R. L., Zukowska-Grojec, Z., Haass, M., and Zamir, N., 1985, Characterization of circulating atrial natriuretic peptides in conscious rats: regulation of release by multiple factors, *Science* **232:**636–639.

Flynn, T. G., deBold, M. L., and DeBold, A. J., 1983, The amino acid sequence of an atrial peptide with potent diuretic and natriuretic properties, *Biochem. Biophys. Res. Commun.* **117:**859–865.

Garcia, R., Thibault, G., Gutkowska, J., Cantin, M., and Genest, J., 1985, Changes of regional blood flow induced by atrial natriuretic factor (ANF) in conscious rats, *Life Sci.* **36:**1687–1692.

Gauer, O. H., Henry, J. P., and Sieker, H. O., 1961, Cardiac receptors and fluid volume control, *Prog. Cardiovasc. Dis.* **4:**1–26.

Gellai, M., Kinter, L. B., and Beeruwkes, R., 1985, Cardiovascular and renal effects of atrial natriuretic factor (ANF) in the spontaneously hypertensive (SHR) rat, *Kidney Int.* **27:**295.

Geller, D. M., Currie, M. G., Wakitani, K., Cole, B. R., Adams, S. P. Fok, K. F., Siegel, N. R., Eubanks, S. R., Galluppi, G. R., and Needleman, P., 1984, Atriopeptins: A family of potent biologically active peptides derived from mammalian atria, *Biochem. Biophys. Res. Commun.* **120:**333–338.

Goodfriend, T. L., Elliott, M. E., and Atlas, S. A., 1984, Actions of synthetic atrial natriuretic factor on bovine adrenal glomerulosa, *Life Sic.* **35:**1675–1682.

Grammer, R. T., Fukumi, H., Inagami, T., and Misono, K. S., 1983, Rat atrial natriuretic factor: Purification and vasorelaxant activity, *Biochem. Biophys. Res. Commun.* **116:**696–703.

Gutkowska, J., Horky, K., Thibault, G., Januszewicz, P., Cantin, M., and Genest, J., 1984, Atrial natriuretic factor is a circulating hormone, *Biochem. Biophys. Res. Commun.* **125:**315–323.

Horky, K., Gutkowska, J., Garcia, R., Thibault, G., Genest, J., and Cantin, M., 1985, Effect of different anesthetics on immunoreactive atrial natriuretic factor concentrations in rat plasma, *Biochem. Biophys. Res. Commun.* **129:**651–657.

Huang, C. L., Lewicki, J., Johnson, L. K., and Cogan, M. G., 1985, Renal mechanism of action of rat atrial natriuretic factor, *J. Clin. Invest.* **75:**770–773.

Ishihara, T., Aisaka, K., Hattori, K., Hamasaki, S., Morita, M., Noguchi, T., Kangawa, K., and Matsuo, H., 1985, Vasodilatory and diuretic actions of α-human atrial natriuretic polypeptide (α-hANP), *Life Sci.* **36:**1205–1215.

Jacobowitz, D. M., Skofitsch, G., Keiser, H. R., Eskay, R. L., and Zamir, N., 1985, Evidence for the existence of atrial natriuretic factor-containing neurons in the rat brain, *Neuroendocrinology* **40:**92–94.

Jamieson, J. D., and Palade, G. E., 1964, Specific granules in atrial muscle cells, *J. Cell Biol.* **23:**151–172.

Januszewicz, P., Gutkowska, J., DeLean, A., Thibault, G., Garcia, R., Genest, J., and Cantin, M., 1985, Synthetic atrial natriuretic factor induces release (possibly receptor mediated) of vasopressin from rat posterior pituitary, *Proc. Soc. Exp. Biol. Med.* **178:**321–325.

Kangawa, K., and Matsuo, H., 1984, Purification and complete amino acid sequence of α-human natriuretic polypeptide (α-hANP), *Biochem. Biophys. Res. Commun.* **118:**131–139.

Kappagoda, C. T., Knapp, M. F., Linden, R. J., Pearson, M. J., and Whittaker, E. M., 1979, Diuresis from left atrial receptors: Effects of plasma on the secretion of the malpighian tubules of *Rhodnius prolixus, J. Physiol. (Lond.)* **291:**381–391.

Katsube, N., Wakitani, K., Fok, K. F., Tjoeng, F. S., Zupec, M. E., Eubanks, S. R., Adams, S. P., and Needleman, P., 1985, Differential structure–activity relationships of atrial peptides as natriuretics and renal vasodilators in the dog, *Biochem. Biophys. Res. Commun.* **128:**325–330.

Kleinert, H. D., Maack, T., Atlas, S. A., Januszewicz, A., Sealey, J. E., and Laragh, J. H., 1984, Atrial natriuretic factor inhibits angiotensin-, norepinephrine-, and potassium-induced vascular contractility, *Hypertension* **6** (Suppl. I):1-143–1-147.

Kleinert, H. D., Volpe, M., Camargo, M. J. F., Atlas, S. A., Laragh, J. H., and Maack, T., 1985, Cardiovascular effects of synthetic atrial natriuretic factor (ANF) in normal dogs, *Fed. Proc.* **44:**1729.

Koike, H., Sada, T., Miyamoto, M., Oizuma, K., Sugiyama, M., and Inagami, T., 1984, Atrial natriuretic factor selectively increases renal blood flow in conscious spontaneously hypertensive rats, *Eur. J. Pharmacol.* **104:**391–392.

Lang, R. E., Tholken, H., Ganten, D., Luft, F. C., Ruskoaho, H., and Unger, T., 1985, Atrial natriuretic factor—a circulating hormone stimulated by volume loading, *Nature* **314:**264–266.

Lappe, R. W., Smits, J. F. M., Todt, J. A., Debets, J. J. M., and Wendt, R. L., 1985, Failure of atriopeptin II to cause arterial vasodilation in the conscious rat, *Circ. Res.* **56:**606.

Lazure, C., Seidah, N. C., Chretien, M., Thibault, G., Garcia, R., Cantin, M., and Genest, J., 1984, Atrial pronatriodilatin: A precursor for natriuretic factor and cardiodilatin, *FEB Lett.* **172:**80–86.

Ledsome, J. R., and Linden, R. J., 1968, The role of the left atrial receptors in the diuretic response to left atrial distension, *J. Physiol. (Lond.)* **198:**487–503.

Maack, T., Marion, D. N., Camargo, M. J. F., Kleinert, H. D., Laragh, J. H., Vaughn, E. D., and Atlas, S. A., 1984, Effects of auriculin (atrial natriuretic factor) on blood pressure, renal function, and the renin–aldosterone system in dogs, *Am. J. Med.* **77:**1069–1075.

Maack, T., Camargo, M. J. F., Kleinert, H. D., Laragh, J. H., and Atlas, S. A., 1985, Atrial natriuretic factor: Structure and functional properties, *Kidney Int.* **27:**607–615.

Maki, M., Takayanagi, R., Misono, K. S., Pandey, K. N., Tibbetts, C., and Inagami, T., 1984, Structure of rat atrial natriuretic factor precursor deduced from cDNA sequence, *Nature* **309:**722–724.

Misono, K. S., Fukumi, H., Grammer, R. T., and Inagami, T., 1984, Rat atrial natriuretic factor: Complete amino acid sequence and disulfide linkage essential for biological activity, *Biochem. Biophys. Res. Commun.* **119:**524–529.

Morris, M., McCann, S. M., and Orias, R., 1976, Evidence for hormonal participation in the natriuretic and kaliuretic responses to intraventricular hypertonic saline and norepinephrine, *Proc. Soc. Exp. Biol. Med.* **152:**95–98.

Murray, R. D., Carretero, O. A., Scicli, A. G., and Inagami, T., 1985, Dissociation of hemodynamic from natriuretic effects of atrial natriuretic factor (ANF), *Kidney Int.* **27:**317.

Nakao, K., Moui, N., Sugawara, A., Sakamoto, M., Suida, M., Nakamura, M., Katsuura, G., Kawata, M., Sano, Y., and Imura, H., 1985, Atrial natriuretic polypeptide in rat brain, *Endocr. Soc. Abstr.* **629:**158.

Napier, M. A., Vandlen, R. L., Albers-Schonberg, G., Nutt, R. F., Brady, S., Lyle, T., Winquist, R., Faison, E. P., Heinel L. A., and Blaine, E. H., 1984, Specific membrane receptors for atrial natriuretic factor in renal and vascular tissues, *Proc. Natl. Acad. Sci. U.S.A.* **81:**5946–5950.

Naruse, M., Obana, K., Naruse, K., Demura, H., and Inagami, T., 1985, Synthetic atrial natriuretic factor inhibits *in vivo* and *in vitro* renin secretion in rats, *Endocr. Soc. Abstr.* **679:**170.

Oikawa, S., Imai, M., Ueno, A., Tanaka, S., Noguchi, T., Nakayato, H., Kangawa, K., Fukuda, A., and Matsuo, H., 1984, Cloning and sequence analysis of cDNA encoding a precursor for human atrial natriuretic polypeptide, *Nature* **309:**724–726.

Pamnani, M. B., Clough, D. L., Chan, J. S., Link, W. T., and Haddy, F. J., 1984, Effects of rat atrial extract on sodium transport and blood pressure in the rat, *Proc. Soc. Exp. Biol. Med.* **176:**123–131.

Plotsky, P. M. Bruhn, T. O., and Vale, W., 1985, Evidence for multifactor regulation of the adrenocorticotropin secretory response to hemodynamic stimuli, *Endocrinology* **116:**633–639.

Quirion, R., Dalpe, M., DeLean, A., Gutkowska, J., Cantin, M., and Genest, J., 1984, Atrial natriuretic factor (ANF) binding sites in brain and related structures, *Peptides* **5:**1167–1172.

Ramsey, D. J., Thrasher, T. N., and Keil, L. C., 1983, The organum vasculosum laminae terminalis: A critical area for osmoreception, in: *The Neurohypophysis* (B. A. Cross and G. Leng, eds.), Elsevier, Amsterdam, pp. 91–98.

Ruoff, H. J., Mathison, R., and Lederis, K., 1976, Cyclic 3′5′-adenosine monophosphate in the hypothalamo–neurohypophysial system of normal, NaCl-treated and lactating rats, *Neuroendocrinology* **22:**18–29.

Samson, W. K., 1985a, Atrial natriuretic factor inhibits dehydration and hemorrhage-induced vasopressin release, *Neuroendocrinology* **40:**277–279.

Samson, W. K., 1985b, Atrial natriuretic factor inhibits vasopressin release, *Fed. Proc.* **44:**1200.

Samson, W. K., 1986, Dehydration-induced alterations in rat brain vasopressin and atrial natriuretic factor immunoreactivity, *Endocrinology* **117:**1279–1281.

Samson, W. K., Vijayan, E., Lumpkin, M. D., and McCann, S. M., 1982, Gastrointestinal hormones: Central nervous system localization and sites of neuroendocrine action, *Endocrinol. Exp.* **16:**117–121.

Saper, C. B., Standaert, D. G., Currie, M. G., Schwartz, D., Geller, D. M., and Needleman, P., 1985, Atriopeptin-iommunoreactive neurons in the rat brain: Presence in cardiovascular regulatory areas, *Science* **227:**1047–1049.

Schrier, R. W., Berl, T., and Anderson, R. J., 1979, Osmotic and nonosmotic control of vasopressin release, *Am. J. Physiol.* **236:**321–322.

Seidah, N. G., Lazure, C., Chretien, M., Thibault, G., Garcia, R., Cantin, M., Genest, J., Nutt, R. F., Brady, S. F., Lyle, T. A., Palveda, W. J., Colton, C. D., Ciccarone, T. M., and Veber, D. F., 1984, Amino acid sequence of homologous rat atrial peptides: Natriuretic activity of native and synthetic forms, *Proc. Natl. Acad. Sci. U.S.A.* **81:**2640–2644.

Shimatsu, A., Kato, Y., Matsushita, N., Katakami, H., Yanaihara, N., and Imura, H., 1981, Immunoreactive vasoactive intestinal polypeptide in rat hypophysial portal blood, *Endocrinology* **108:**935–940.

Skofitsch, G., Jacobowitz, D. M., Eskay, R. L., and Zamir, N., 1986, Distribution of atrial natriuretic factor-like immunoreactive neurons in the rat brain, *Neuroscience* **16:**917–918.

Sladek, C. D., 1983, Regulation of vasopressin release by neurotransmitters, neuropeptides and osmotic stimuli, in: *The Neurohypophysis* (B. A. Cross and G. Leng, eds.), Elsevier, Amsterdam, pp. 71–90.

Smits, J. F. M., and Lappe, R. W., 1985, Systemic hemodynamic effects of synthetic atriopeptin II (AP II) in conscious SHR, *Fed. Proc.* **44:**1555.

Sonnenberg, H., 1985, Renal tubular effects of atrial natriuretic factor (ANF), *Kidney Int.* **27:**105.

Sonnenberg, H., and Veress, A. T., 1984, Cellular mechanism of release of atrial natriuretic factor, *Biochem. Biophys. Res. Commun.* **124:**443–449.

Sonnenberg, H., Cupples, W. A., de Bold, A. J., and Veress, A. T., 1982, Intrarenal localization of the natriuretic effect of cardiac atrial extract, *Can. J. Physiol. Pharmacol.* **60:**1149–1152.

Sonnenberg, H., Krebs, R., and Veress, A. T., 1984, Release of atrial natriuetic factor from incubated rat heart atria, *IRCS Med. Sci.* **12:**783–784.

Tanaka, I., Misono, K. S., and Inagami, T., 1984, Atrial natriuretic factor in rat hypothalamus, atria, and plasma: Determination by specific radioimmunoassay, *Biochem. Biophys. Res. Commun.* **124:**663–668.

Tang, J., Webber, R. J., Chang, P., Chang, J. K., Kiang, J., and Wei, E. T., 1984, Depressor and natriuretic activities of several atrial peptides, *Regul. Peptides* **9:**53–59.

Vale, W., Vaughan, J., Smith, N., Yamamoto, G., Rivier, J., and Rivier, C., 1983, Effects of synthetic ovine corticotropin-releasing factor, glucocorticoids, catecholamines, neurohypophysial peptides, and other substances on cultured corticotropic cells, *Endocrinology* **113:**1121–1131.

Vanatta, J. C., and Samson, W. K., 1985, Atrial natriuretic factor produces an inhibition of vasotocin-induced water absorption in the toad urinary bladder, *Fed. Proc.* **44:**1731.

Vuolteenaho, O., Arjamaa, O., and Ling, N., 1985, Atrial natriuretic polypeptides (AVP): Rat atria store high molecular weight precursor but secrete processed peptides of 25–35 amino acids, *Biochem. Biophys. Res. Commun.* **129:**82–88.

Wakitani, K., Oshima, T., Loewy, A. D., Holmberg, S. W., Cole, B. R., Adams, S. P., Fok, K. F., Currie, M. G., and Needleman, P., 1985, Comparative vascular pharmacology of the atriopeptins, *Circ. Res.* **56:**621–627.

Wendt, R. L., Shropshire, A. T., and Lappe, R. W., 1985, Cardiovascular effects of synthetic atriopeptin II (AP II) in the anesthetized cat, *Fed. Proc.* **44:**1728.

Yamanaka, M., Greenberg, B., Johnson, L., Seilhamer, J., Brewer, M., Friedemann, T., Miller, J., Atlas, S., Laragh, J., Lewicki, J., and Fiddles, J., 1984, Cloning and sequence analysis of the cDNA for the rat atrial natriuretic factor precursor, *Nature* **309:**719–722.

Zamir, N., Skofitsch, G., Eskay, R. L., and Jacobowitz, D. M., 1986, Distribution of atrial natriuretic peptide in the rat brain, *Brain Res.* **365:**105–111.

37

[Norleucine^{3,6}]-Substituted Cholecystokinin Octapeptide Analogues

Design, Synthesis, and Comparative Structure–Activity Relationships in Guinea Pig Brain, Lung, and Pancreas Tissues

TOMI K. SAWYER, DOUGLAS J. STAPLES, ROBERT A. LAHTI, PEGGY J. K. D. SCHREUR, ANITA E. WILKERSON, HENRY H. HOLZGREFE, and STUART BUNTING

1. INTRODUCTION

Cholecystokinin octapeptide (CCK-8, Fig. 1) is a neurogastric peptide hormone and neurotransmitter that possesses multiple biological activities (Table I). Cholecystokinin octapeptide is one of several molecular variants (e.g., CCK-39, CCK-33, CCK-12, and CCK-8) existing within a family of CCK peptides that have been identified in both the central and peripheral nervous systems as well as in the gastrointestinal tract. The specific details related to the discovery, distribution, biosynthesis, metabolism, biological activities *in vitro* and *in vivo,* and mechanisms of action of CCK peptides have been excellently reviewed recently (Mutt, 1980; Kelley and Dodd, 1981: Williams, 1982; Morley, 1982; Beinfeld, 1983; Dockray, 1983).

In this chapter we describe the comparative biological properties *in vitro* of

TOMI K. SAWYER and DOUGLAS J. STAPLES • Biopolymer Chemistry/Biotechnology, The Upjohn Company, Kalamazoo, Michigan 49001. ROBERT A. LAHTI, PEGGY J. K. D. SCHREUR, and ANITA E. WILKERSON • CNS Diseases Research, The Upjohn Company, Kalamazoo, Michigan 49001. HENRY H. HOLZGREFE and STUART BUNTING • Lipids Research, The Upjohn Company, Kalamazoo, Michigan 49001.

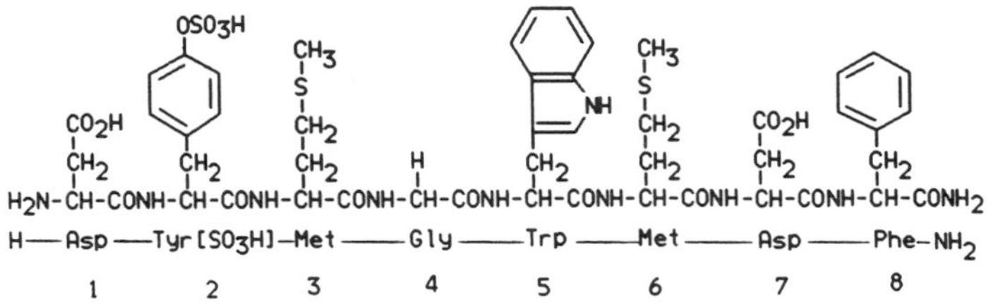

H—Asp—Tyr[SO_3H]—Met—Gly—Trp—Met—Asp—Phe—NH_2

1 2 3 4 5 6 7 8

FIGURE 1. Chemical structure of CCK-8.

CCK-8 and several of its [$Nle^{3,6}$]-substituted analogues in the guinea pig brain, pancreas, and lung tissue. We discovered CCK-8 to be a highly potent stimulant of tracheal contraction and provide preliminary results on the characterization of a CCK peptidergic receptor system in the guinea pig lung as based on structure–activity studies. Our design strategy was focused on developing a facile synthesis of CCK-8 analogues that would be chemically more stable than the native peptide toward oxidative conditions. The [$Nle^{3,6}$]-substituted synthetic congeners of CCK-8 described herein may provide superior molecular probes for further pharmacological analysis of CCK-8 *in vitro* and *in vivo*.

2. STRUCTURE–ACTIVITY RELATIONSHIPS OF CCK-8 ANALOGUES IN RETROSPECT

Cholecystokinin octapeptide has been examined in numerous structure–activity investigations (Morley, 1968, 1977; Plusec *et al.*, 1970; Gardner *et al.*, 1975, 1984; Castiglione, 1977; Bodansky *et al.*, 1977, 1978; Kaminski *et al.*, 1977; Meyer *et*

TABLE I. A Synopsis of Several Different Bioactivities Reported for CCK-8

Peripherally mediated effects
Gallbladder contraction
Pancreatic enzyme secretion and Ca^{2+} efflux
Trophic effect on pancreas
Gastric acid secretion (weak/partial agonism)
Satiety
Centrally mediated effects
Sedation, catalepsy, and ptosis
Neurolepticlike, antipsychoticlike activities
Reduction of exploratory behavior
Elevation of threshold for chemically induced convulsions
Biphasic effect related to antinociceptive activity (analgesia at high dose and antagonism of opiate-induced analgesia at low dose)
Stimulation of somatotropin and adrenocorticotropin release
Inhibition of thyrotropin release
Thermoregulation (hypothermia in the rat and mouse; hyperthermia in the guinea pig)
Satiety

al., 1980; Adachi *et al.*, 1981; Holland *et al.*, 1982; Martinez *et al.*, 1982; Jensen *et al.*, 1983; Magous *et al.*, 1983; Gaudreau *et al.*, 1983a; Spanarkel *et al.*, 1983; Penke *et al.*, 1984) to determine which of its amino acid constituents are important for CCK-like agonism, intrinsic bioactivity, and/or potency *in vitro* as related to pancreatic amylase secretion or Ca^{2+} efflux, gallbladder contraction, ileum contraction, and gastric acid secretion. Generally, these studies have shown that the C-terminal tetrapeptide sequence Trp-Met-Asp-Phe-NH_2 is the "active site" for CCK-like biological activities. For example, H-Trp-Met-Asp-Phe-NH_2 is about 1/1000 as potent as CCK-8 in stimulating pancreatic amylase secretion or Ca^{2+} outflux and is about 1/10,000 as potent as CCK-8 in stimulating gallbladder contraction. The relationship of sequential N-terminal extension and Tyr^2 sulfation of the H-Trp-Met-Asp-Phe-NH_2 tetrapeptide to yield CCK-8 and guinea pig gallbladder-contracting activity has been clearly shown by Gaudreau and co-workers (1983a) and is summarized in Table II. Their studies reiterate the well-known structural requirement for $Tyr[SO_3H]^2$ within the heptapeptide sequence of H-Tyr[SO_3H]-Met-Gly-Trp-Met-Asp-Phe-NH_2 for providing high CCK-like potency.

Spectroscopic investigations (Schiller *et al.*, 1978; Penke *et al.*, 1983; Durieux *et al.*, 1983) have also been performed on CCK-8 derivatives. However, structure–conformation–activity studies are yet devoid of data derived for cyclic congeners of any of the previously reported linear CCK-8 analogues. Nevertheless, several heptapeptide analogues of Ac-Tyr[SO_3H]-Met-Gly-Trp-Met-Asp-Phe-NH_2 (Ac-CCK-7) having singular stereostructural modifications such as D-Ala^4, D-Trp^5, D-Met^6, or Hyp^7 (hydroxyproline) and effecting potent gallbladder-contracting activities *in vitro* as reported by Penke and co-workers (1984) have supported, in part, the possible existence of reverse-turn secondary structures (e.g., β-turn or γ-turn) within CCK-8, which may be important to the expression of CCK-like bioactivities.

More recently, CCK-8 has been evaluated for its central and/or behavioral effects in several structure–activity investigations (Fekete *et al.*, 1981; Halmy *et al.*, 1982; Cohen *et al.*, 1982; Rogawski, 1982; Kadar *et al.*, 1983; Faris *et al.*,

TABLE II. A Summary of CCK-8 Structure–Activity Studies on Guinea Pig Gallbladder Contraction *in Vitro* as Reported by Gaudreau and Co-workers (1983a)

Peptide		Relative potency
1. H-Asp-Tyr[SO_3H]-Met-Gly-Trp-Met-Asp-Phe-NH_2	(CCK-8)	1.0
2. H-Asp-Tyr-Met-Gly-Trp-Met-Asp-Phe-NH_2		0.0004
3. H-Tyr[SO_3H]-Met-Gly-Trp-Met-Asp-Phe-NH_2	(CCK-7)	0.4
4. H-Tyr-Met-Gly-Trp-Met-Asp-Phe-NH_2		0.0002
5. H-Met-Gly-Trp-Met-Asp-Phe-NH_2	(CCK-6)	<0.0001
6. H-Gly-Trp-Met-Asp-Phe-NH_2	(CCK-5)	<0.0001
7. H-Trp-Met-Asp-Phe-NH_2	(CCK-4)	<0.0001
8. H-Met-Asp-Phe-NH_2	(CCK-3)	Inactive
9. H-Asp-Phe-NH_2	(CCK-2)	Inactive

1983; van Ree *et al.*, 1983; Penke *et al.*, 1984; Crawley *et al.*, 1984; Zetler, 1982, 1983, 1984) in regard to satiety, conditioned feeding behavior, exploratory activity, anticonvulsant activity, antipsychotic activity, analgesia, thermoregulation, and others. In the human being, CCK-8, CCK-33, and the decapeptide ceruletide (<Glu-Gln-Asp-Tyr[SO_3H]-Thr-Gly-Trp-Met-Asp-Phe-NH_2) have been investigated (Stacher *et al.*, 1982a,b; Pi-Sunyer *et al.*, 1982; Nair *et al.*, 1982; Moroji *et al.*, 1982) in relation to satiety, analgesia, and antipsychotic-like activities. In addition, CCK-based pharmaceuticals have evolved in the forms of both diagnostic and therapeutic agents (de Castiglione, 1983).

Comparative structure–activity and biochemical characterization studies of CCK–receptor interactions (i.e., binding and signal transduction) at different target tissues as well as correlation of these results to the aforementioned biological effects *in vitro* and *in vivo* have rapidly become an area of intensive investigation (Sankaran *et al.*, 1980; Saito *et al.*, 1980, 1981; Hays *et al.*, 1980; Innis and Snyder, 1980; Praissman *et al.*, 1982, 1983; Gaudreau *et al.*, 1983b; Sakamoto *et al.*, 1984; Fourmy *et al.*, 1984a,b; Beglinger *et al.*, 1984; Praissman and Walden, 1984; Finkelstein *et al.*, 1984; Wennogle *et al.*, 1985; Szecowka *et al.*, 1985; Chang *et al.*, 1983, 1985).

To accommodate the requirements of CCK receptor analysis performed in these studies, several radiolabeled CCK ligands have been developed such as [^{125}I-Bolten–Hunter]CCK-33, [^{125}I-ρ-hydroxybenzimidoyl]-CCK-8, [^{125}I-Bolten–Hunter]CCK-8, [^{125}I-Bolten–Hunter, Thr^4, Nle^7]CCK-9, and [^{3}H]CCK-5. Several highlights of the above comparative studies include (1) identification of CCK binding site(s) in brain and pancreas tissues that are saturable, reversible, specific, and high affinity (i.e., K_d values in the nanomolar range); (2) evidence of a less stringent requirement for Tyr[SO_3H] versus Tyr in CCK-8 binding to brain tissue (i.e., [Tyr^2]CCK-8 typically ranges in potency from 1 to 0.1 relative to CCK-8) as compared to CCK-8 binding to pancreatic tissue (i.e., [Tyr^2]-CCK-8 typically ranges in potency from 10^{-2} to 10^{-3} relative to CCK-8); (3) implications for the existance of heterogeneous (i.e., at least two subtypes) populations of CCK receptors in both brain and pancreas tissues as based on detailed examinations of saturation binding isotherms and dissociation kinetics as well as differential binding selectivities of various CCK peptides to such tissues; and (4) direct physiochemical analysis of the structural organization (i.e., size and subunit) of brain and pancreatic CCK receptors.

3. DESIGN AND SYNTHESIS OF [Nle^{3,6}]-SUBSTITUTED CCK-8 ANALOGUES

Based on the preceding structure–activity relationships for CCK-8 *in vitro* and *in vivo,* our design strategy was to develop a facile synthesis of CCK-8 analogues employing solid-phase methodology followed by a single-step O-sulfation procedure to yield the desired Tyr[SO_3H]-containing derivatives. To accomplish these objec-

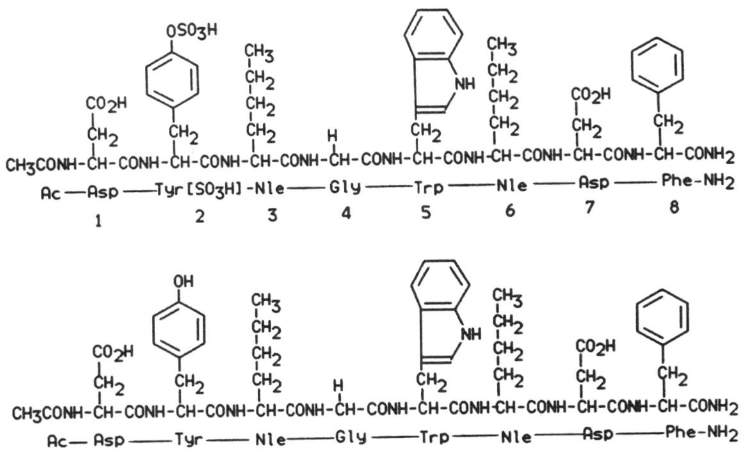

CH_3CONH-CH-CONH-CH-CONH-CH-CONH-CH-CONH-CH-CONH-CH-CONH-CH-CONH-CH-$CONH_2$

Ac—Asp—Tyr[SO_3H]-Nle—Gly—Trp—Nle—Asp—Phe-NH_2

1 2 3 4 5 6 7 8

CH_3CONH-CH-CONH-CH-CONH-CH-CONH-CH-CONH-CH-CONH-CH-CONH-CH-CONH-CH-$CONH_2$

Ac—Asp—Tyr—Nle—Gly—Trp—Nle—Asp—Phe-NH_2

FIGURE 2. Chemical structures of Ac-[Nle^{3,6}]-CCK-8 and Ac-[Tyr^2, Nle^{3,6}]-CCK-8.

tives, the two Met residues at positions 3 and 6 were each substituted by Nle, and the N-terminal Asp residue was acetylated. The resultant title compounds (Fig. 2) were then prepared as illustrated in Fig. 3. These [Nle^{3,6}]-substituted CCK-8 analogues are expected to be resistant to either chemical or biological oxidative conditions, which could otherwise effectively transform Met to the undesirable Met (O) within the native peptide, CCK-8, and structurally related derivatives. In addition, the Ac-[Tyr^2,Nle^{3,6}]CCK-8 may provide a useful template for direct radiolabeling followed by a similar single-step O-sulfation procedure to yield an Ac-[Nle^{3,6}]CCK-8 radioligand.

4. COMPARATIVE BIOLOGICAL PROPERTIES *IN VITRO* OF [Nle^{3,6}]-SUBSTITUTED CCK-8 ANALOGUES

The Ac-[Nle^{3,6}]CCK-8 and Ac-[Tyr^2, Nle^{3,6}]CCK-8 were tested on guinea pig brain binding *in vitro* (Table III). The Ac-[Nle^{3,6}]CCK-8 was about equipotent to CCK-8 and effected tenfold higher binding compared to its Tyr^2 correlate. These results are consistent with earlier studies discussed above.

In the guinea pig pancreas assay, the [Nle^{3,6}]-substituted CCK-8 analogues were tested for their dose–response relationships as amylase enzyme secretagogues (Table IV). The Ac-[Nle^{3,6}]CCK-8 was slightly more potent ($EC_{50} \approx 10^{-10}$ M) than CCK-8. As predicted, the Tyr^2 correlates were about 1000-fold less potent than the Tyr[SO_3H]-containing peptides.

We have recently discovered (Bunting *et al.*, 1985) a CCK-8-specific sensitive peptidergic receptor system in guinea pig lung tissue. These CCK receptors have been characterized by examination of the comparative trachea-contracting

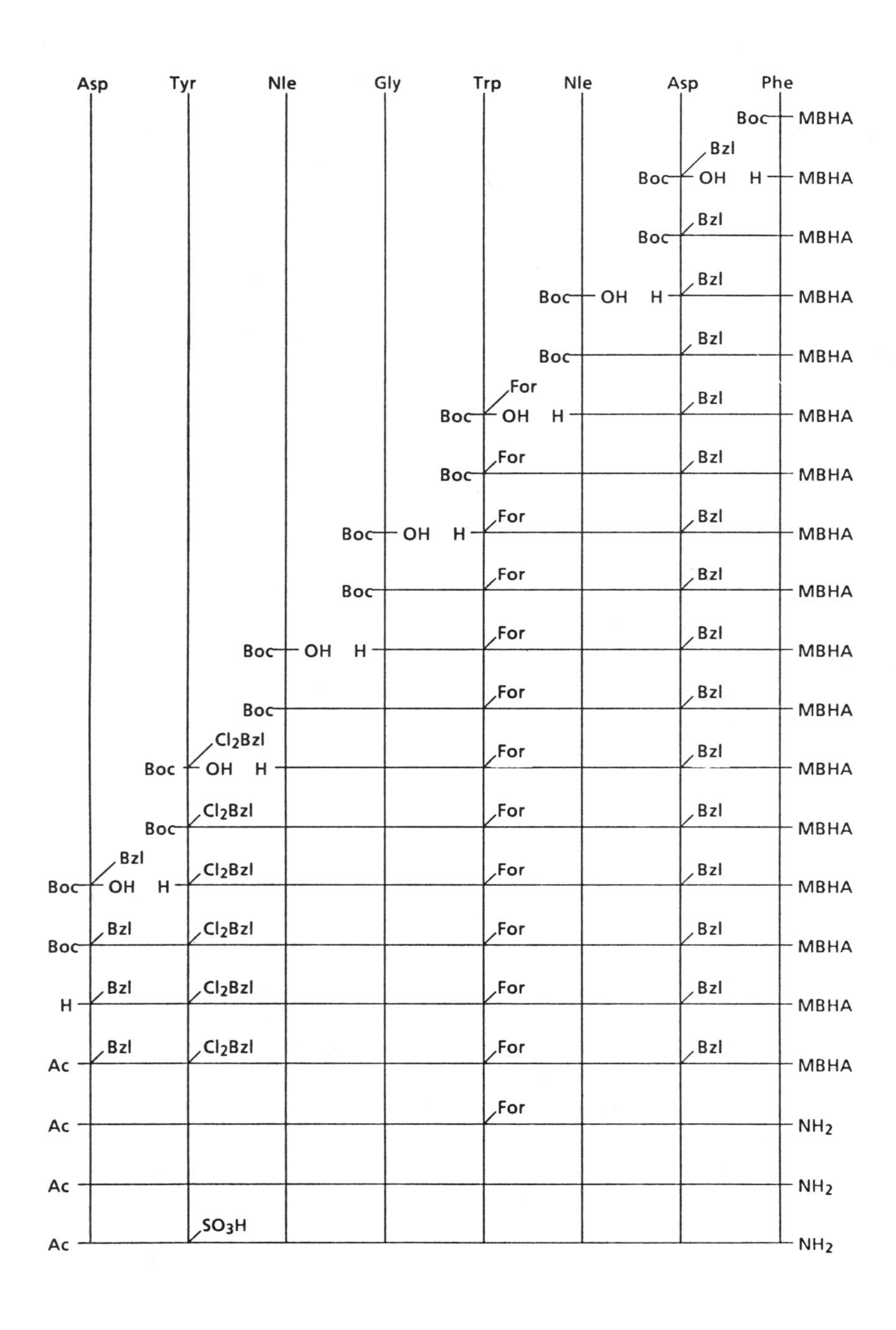

Asp
Tyr
Nle
Gly
Trp
Nle
Asp
Phe
Boc
MBHA
Bzl
OH
H
For
Cl2Bzl
Ac
NH2
SO3H

TABLE III. Structure–Activity Studies of [$Nle^{3,6}$]-Substituted CCK-8 Analogues on Guinea Pig Brain Binding *in Vitro*

Peptide	D_{50} (nM)[a]	Relative potency
1. H-Asp-Tyr[SO_3H]-Met-Gly-Trp-Met-Asp-Phe-NH_2	0.44	1.0
2. H-Asp-Tyr-Met-Gly-Trp-Met-Asp-Phe-NH_2	3.5	0.1
3. Ac-Asp-Tyr[SO_3H]-Nle-Gly-Trp-Nle-Asp-Phe-NH_2	0.59	0.7
4. Ac-Asp-Tyr-Nle-Gly-Trp-Nle-Asp-Phe-NH_2	21.9	0.02

[a] Concentration of CCK-8 analogue effecting 50% displacement of [^3H-propionyl]-CCK-8 (purchased from Amersham) on guinea pig brain (less cerebellum) semipurified homogenate based on the methods described by Lahti *et al.* (1983). All determinations (in triplicate) were conducted at 32°C for 30 min incubation time. Nonspecific binding was determined using CCK-8. The D_{50} values were calculated using linear regression techniques. The details of this assay will be published elsewhere.

dose–response relationships of numerous CCK-8-based analogues and fragments. The Ac-[$Nle^{3,6}$]CCK-8 was a potent stimulant of tracheal contraction (Table V), and the comparative bioactivities of CCK-8, [Tyr^2]CCK-8, the [$Nle^{3,6}$]-substituted analogues as well as other CCK-8 derivatives (Bunting *et al.*, 1985) are consistent with earlier pharmacological studies on CCK receptor systems of pancreas and gallbladder tissues.

TABLE IV. Structure–Activity Studies of [$Nle^{3,6}$]-Substituted CCK-8 Analogues on Guinea Pig Pancreatic Amylase Secretion *in Vitro*

Peptide	EC_{50} (nM)[a]	Relative potency
1. H-Asp-Tyr[SO_3H]-Met-Gly-Trp-Met-Asp-Phe-NH_2	0.09	1.0
2. H-Asp-Tyr-Met-Gly-Trp-Met-Asp-Phe-NH_2	20.0	0.005
3. Ac-Asp-Tyr[SO_3H]-Nle-Gly-Trp-Nle-Asp-Phe-NH_2	0.05	2.0
4. Ac-Asp-Tyr-Nle-Gly-Trp-Nle-Asp-Phe-NH_2	>100.0	<0.001

[a] Concentration of CCK-8 analogue effecting 50% maximal amylase secretion *in vitro* using dispersed pancreatic acini prepared from albino guinea pigs based on the methods described by Peikin *et al.* (1978) and modifications thereof obtained from Gardner and co-workers. Amylase secretion was quantitated by known spectrophotometric procedures. The details of this assay will be published elsewhere.

←

FIGURE 3. Synthesis of Ac-[Tyr^2, $Nle^{3,6}$]-CCK-8 and Ac-[$Nle^{3,6}$]-CCK-8 by sequential solid-phase and solution-phase methods. The intermediate peptide resin, Ac-Asp[Bzl]-Tyr[Cl_2Bzl]-Nle-Gly-Trp[For]Nle-Asp[Bzl]-Phe-MBHA (MBHA, *p*-methylbenzhydrylamine), was cleaved and deprotected, except for Trp[For], by HF to yield Ac-[Tyr^2, $Nle^{3,6}$, Trp(For)^5]CCK-8. The preceding Trp[For]-substituted intermediate was subsequently deformylated by NaOH and O-sulfated at Tyr^2 by pyridine-SO_3 in a stepwise manner to yield the two title CCK-8 analogues. Each title compound was purified to homogeneity by preparative reverse-phase HPLC and characterized for its physicochemical properties by amino acid analysis, FAB-mass spectroscopy, IR-spectroscopy, optical rotation, and analytical reverse-phase HPLC. The details of the chemistry of these CCK-8 analogues will be published elsewhere.

TABLE V. Structure–Activity Studies of [$Nle^{3,6}$]-Substituted CCK-8 Analogues on Guinea Pig Tracheal Contraction *in Vitro*

Peptide	EC_{50} (M)[a]	Relative potency
1. H-Asp-Tyr[SO_3H]-Met-Gly-Trp-Met-Asp-Phe-NH_2	5.6×10^{-8}	1.0
2. H-Asp-Tyr-Met-Gly-Trp-Met-Asp-Phe-NH_2	1.6×10^{-5}	0.0035
3. Ac-Asp-Tyr[SO_3H]-Nle-Gly-Trp-Nle-Asp-Phe-NH_2	9.0×10^{-9}	1.28
4. Ac-Asp-Tyr-Nle-Gly-Trp-Nle-Asp-Phe-NH_2	5.0×10^{-5}	0.0024

[a] Concentration of CCK-8 analogue effecting contractile activity on ziz-zag strips of guinea pig trachea prepared according to a modification of the method of Emerson and MacKay (1979). The EC_{50} values were calculated from combined dose–response curves, and all compounds generated similar maximal contractile force. Relative potencies were calculated by comparison of CCK-8 and its analogue in the same experiment. The details of this assay will be published elsewhere.

Recently, CCK-8 has been reported (Gillis *et al.*, 1983) to stimulate cardiorespiratory function when administered into selected regions of the central nervous system. However, no direct effects of CCK-8 or other CCK peptides on lung tissue have previously been described. Our data suggest that a new regulatory function may exist for CCK-8 within the respiratory system; however, the physiological relevance of a CCK peptide in this system has yet to be ascertained. Interestingly, extractable CCK-like immunoreactive material in lung tissue has been described (Polak and Bloom, 1982). The application of improved CCK peptide-specific radioimmunoassay methods (Chang and Chey, 1983; Rehfeld and Morley, 1983; Bacarese-Hamilton *et al.*, 1985; Adrian *et al.*, 1985) may provide further insight towards characterization and exploration of such CCK–target tissue interactions.

5. FUTURE RESEARCH DIRECTIONS

In this report, we have described the design, synthesis, and comparative structure–activity relationships of [$Nle^{3,6}$]-substituted CCK-8 analogues on guinea pig brain, pancreas, and lung tissue systems *in vitro*. The *in vivo* anorexigenic activities of these CCK-8 analogues in the rat have also recently been investigated (Schreur *et al.*, 1985), and the Ac-[$Nle^{3,6}$]CCK-8 was found to be significantly more potent than the native peptide in effecting satiety. Thus, such [$Nle^{3,6}$]-substituted CCK-8 analogues may provide excellent molecular probes for future pharmacological studies related to the multitude of peripheral and/or central bioactivities known for CCK peptides.

In the human brain, CCK-8 has recently been identified and extensively characterized by physiochemical analyses (Miller *et al.*, 1984). In the rat brain, the modulatory effects of CCK-8 on the binding of dopamine, spiperone, benzodiazepines, and β-adrenergic agents have recently been investigated (Murphy and Schuster, 1982; Agnathi *et al.*, 1983; Bradwejn and Montigny, 1984; Kochman *et al.*, 1984). Thus, the molecular mechanism(s) of CCK peptide action underlying the many known biological effects of this neurogastric peptide hormone and neurotransmitter remains an area of intensive multidisciplinary research. Based on our initial studies on Ac-[Nle^{3,6}]CCK-8 described above, we are further examining CCK-8 by structure–conformation–activity studies and evaluating CCK-8 analogues for therapeutic application.

REFERENCES

Adachi, H., Rajh, H. M., Tesser, G. I., dePont, J. J. H. M. M., Jensen, R. T., and Gardner, J. D., 1981, Interaction of tryptophan-modified analogues of cholecystokinin-octapeptide with cholecystokinin receptors on pancreatic acini, *Biochim. Biophys. Acta* **678:**358–363.

Adrian, T. E., Bacarese-Hamilton, A. J., and Bloom, S. R., 1985, Measurement of cholecystokinin octapeptide using a new specific radioimmunoassay, *Peptides* **6:**11–16.

Agnathi, L. F., Fuxe, K., Benfenati, F., Celani, M. F., Battistini, N., Mutt, V., Cavicchioli, L., Galli, G., and Hokfelt, T., 1983, Differential modulation by CCK-8 and CCK-4 of [^{3}H]-spiperone binding sites linked to dopamine and 5-hydroxytryptamine receptors in the brain of the rat, *Neurosci. Lett.* **35:**179–183.

Bacarese-Hamilton, A. J., Adrian, T. E., Chohan, P., Antony, T., and Bloom, S. R., 1985, Oxidation/reduction of methionine residues in CCK: A study by radioimmunoassay and isocratic reverse phase high pressure liquid chromatography, *Peptides* **6:**17–22.

Beglinger, C., Solomon, T. E., Gyr, K., Moroder, L., and Wunsch, E., 1984, Exocrine pancreatic secretion in response to a new CCK-analog, CCK-33 and caerulein in dogs, *Regul. Peptides* **8:**291–296.

Beinfeld, M. C., 1983, Cholecystokinin in the central nervous system. A minireview, *Neuropeptides* **3:**411–427.

Bodansky, M., Natarajan, S., Hahne, W., and Gardner, J. D., 1977, Cholecystokinin (pancreozymin). 3. Synthesis and properties of an analogue of the C-terminal heptapeptide with serine sulfate replacing tyrosine sulfate, *J. Med. Chem.* **20:**1047–1050.

Bodansky, M., Martinez, J., Priestly, G. P., Gardner, J. D., and Mutt, V., 1978, Cholecystokinin (pancreozymin). 4. Synthesis and properties of a biologically active analogue of the C-terminal heptapeptide with ε-hydroxynorleucine sulfate replacing tyrosine sulfate, *J. Med. Chem.* **21:**1030–1035.

Bradwejn, J., and de Montigny, C., 1984, Benzodiazepines antagonize cholecystokinin-induced activation of rat hippocampal neurones, *Nature* **312:**363–364.

Bunting, S., Holzgrefe, H. H., deVaux, A. E., Staples, D. J., and Sawyer, T. K., 1985, Evidence for a cholecystokinin-octapeptide receptor on guinea pig trachea, in: *Peptides: Structure and Function, Proceedings of the Ninth American Peptide Symposium* (C. M. Deber, V. J. Hruby, and K. D. Kopple, eds.), Pierce Chemical, Rockford, IL, pp. 579–582.

Castiglione, R. de, 1977, Structure–activity relationships in ceruletide-like peptides, in: *First International Symposium of Hormonal Receptors in Digestive Tract Physiology* (S. Bonfils, P. Fromageot, and G. Rosselin, eds.), Elsevier/North-Holland Biomedical Press, Amsterdam, pp. 33–42.

Castiglione, R. de, 1983, Exploitation and exploration of ceruletide and eledoisin, two peptides of nonmammalian origin, *Biopolymers* **22:**507–515.

Chang, T.-M., and Chey, W. Y., 1983, Radioimmunoassay of cholecystokinin, *Dig. Dis. Sci.*, **28:**456–468.

Chang, R. S. L., Lotti, V. J., Martin, G. E., and Chen, T. B., 1983, Increase in brain ^{125}I-cholecystokinin (CCK) receptor binding following chronic haloperidol treatment, intracisternal 6-hydroxydopamine or ventral tegmental lesions, *Life Sci.* **32:**871–878.

Chang, R. S. L., Lotti, V. J., and Chen, T. B., 1985, Cholecystokinin receptor mediated hydrolysis of inositol phospholipids in guinea pig gastric glands, *Life Sci.* **36:**965–971.

Cohen, S. L., Knight, M., Tamminga, C. A., and Chase, T. N., 1982, Cholecystokinin-octapeptide effects on conditioned-avoidance behavior, stereotypy and catalepsy, *Eur. J. Pharmacol.* **83:**213–222.

Crawley, J. N., St. Pierre, S., and Gaudreau, P., 1984, Analysis of the behavioral activity of C- and N-terminal fragments of cholecystokinin octapeptide, *J. Pharmacol. Exp. Ther.* **230**:438–444.

Dockray, G. J., 1983, Cholecystokinin, in: *Brain Peptides* (D. T. Krieger, M. J. Brownstein, and J. B. Martin, eds.), John Wiley & Sons, New York, pp. 851–869.

Durieux, C., Belleney, J., Lallemand, J.-Y., Roques, B. P., and Fournie-Zaluski, M.-C., 1983, ^{1}H-NMR conformational study of sulfated and non-sulfated cholecystokinin fragment CCK_{27-33}: Influence of the sulfate group on the peptide folding, *Biochem. Biophys. Res. Commun.* **114**:705–712.

Emerson, J., and MacKay, D., 1979, The zig-zag tracheal strip, *J. Pharm. Pharmacol.* **31**:798.

Faris, P. L., Komisaruk, B. R., Watkins, L. R., and Mayer, D. J., 1983, Evidence for the neuropeptide cholecystokinin as an antagonist of opiate analgesia, *Science* **219**:310–312.

Fekete, M., Balázs, M., Penke, B., and Telegdy, G., 1981, Influence of sulfated and unsulfated cholecystokinin octapeptide on conditioned feeding behavior in rats, *Peptides* **2**:385–388.

Finkelstein, J. A., Steggles, A. W., Martinez, P. A., and Praissman, M., 1984, Cholecystokinin receptor binding levels in genetically obese rat brain, *Peptides* **5**:11–14.

Fourmy, D., Pradayrol, L., Vaysse, N., Susini, C., and Ribet, A., 1984a, ^{125}I-[Thr^{34}, Nle^{37}]-CCK_{31-39}: A nonoxidizable tracer for the characterization of CCK receptor on pancreatic acini and radioimmunoassay of C-terminal CCK peptides, *J. Immunoassay* **5**:99–120.

Fourmy, D., Zahidi, A., Pradayrol, L., Vayssette, J., and Ribet, A., 1984b, Relationship of CCK/gastrin receptor binding to amylase release in dog pancreatic acini, *Regul. Peptides* **10**:57–68.

Gardner, J. D., Conlon, T. P., Klaeveman, H. L., Adams, T. D., and Ondetti, M. A., 1975, Action of cholecystokinin and cholinergic agents on calcium transport in isolated pancreatic acinar cells, *J. Clin. Invest.* **56**:366–375.

Gardner, J. D., Knight, M., Sutliff, V. E., Tamminga, C. A., and Jensen, R. T., 1984, Derivatives of CCK-(26–32) as cholecystokinin receptor antagonists in guinea pig panreatic acini, *Am. J. Physiol.* **246**:G292–G295.

Gaudreau, P., Morell, J. L., St. Pierre, S., Quirion, R., and Pert, C., 1983a, Cholecystokinin octapeptide fragments: Synthesis and structure–activity relationship, in: *Peptides: Structure and Function* (V. J. Hruby and D. H. Rich, eds.), Pierce Chemical, Rockford, IL, pp. 441–444.

Gaudreau, P., Quirion, R., St. Pierre, S., and Pert, C. B., 1983b, Characterization and visualization of cholecystokinin receptors in rat brain using [^{3}H]pentagastrin, *Peptides* **4**:755–762.

Gillis, R. A., Quest, J. A., Pagani, F. D., Souza, J. D., Taveira da Silva, A. M., Jensen, R. T., Garvey, T. Q. III, and Hamosh, P., 1983, Activation of central nervous system cholecystokinin receptors stimulates respiration in the cat, *J. Pharamcol. Exp. Ther.* **224**:408–414.

Halmy, L., Nyakas, C., and Walter, J., 1982, The C-terminal tetrapeptide of cholecystokinin decreases hunger in rats, *Experientia* **38**:873–874.

Hays, S. E., Beinfeld, M. C., Jensen, R. T., Goodwin, F. K., and Paul, S. M., 1980, Demonstration of a putative receptor site for cholecystokinin in rat brain, *Neuropeptides* **1**:53–62.

Holland, J., Hirst, B. B., and Shaw, B., 1982, Structure–activity studies with cholecystokinin on gastric secretion in the cat, *Peptides* **3**:891–895.

Innis, R. B., and Snyder, S. H., 1980, Cholecystokinin receptor binding in brain and pancreas: Regulation of pancreatic binding by cyclic and acyclic guanine nucleotides, *Eur. J. Pharmacol.* **65**:123–124.

Jensen, R. T., Jones, S. W., and Gardner, J. D., 1983, COOH-Terminal fragments of cholecystokinin: A new class of cholecystokinin receptor antagonists, *Biochim. Biophys. Acta.* **757**:250–258.

Kadar, T., Pesti, A., Penke, B., Toth, G., Zarandi, M., and Telegdy, G., 1983, Structure–activity and dose–effect relationships of the antagonism of picrotoxin-induced seizures by cholecystokinin, fragments and analogues of cholecystokinin in mice, *Neuropharmacology* **22**:1223–1229.

Kaminski, D. L., Ruwart, M. J., and Jellinek, M., 1977, Structure–function relationships of peptide fragments of gastrin and cholecystokinin, *Am. J. Physiol.* **233**:E286–E292.

Kelley, J. S., and Dodd, J., 1981, Cholecystokinin and gastrin as transmitters in the mammalian central nervous system, in: *Neural Peptides and Neuronal Communication* (J. B. Martin, S. Reichlin, and K. L. Bick, eds.), Raven Press, New York, pp. 133–144.

Kochman, R. L., Grey, T. R., and Hirsch, J. D., 1984, Cholecystokinin *in vivo* reduces binding to rat hypothalamic β-adrenergic sites, *Peptides* **5**:499–502.

Lahti, R. A., Sethy, V. H., Barsuhn, C., and Hester, J. B., 1983, Pharmacological profile of the antidepressant adinazolam, a triazolobenzodiazepine, *Neuropharmacology* **22:**1277–1282.
Magous, R., Martinez, J., Lignon, M. F., and Bali, J. P., 1983, The role of the Asp-32 residue of cholecystokinin in gastric acid secretion and gastrin receptor recognition, *Regul. Peptides* **5:**327–332.
Martinez, J., Winternitz, F., Bodansky, M., Gardner, J. D., Walker, M. D., and Mutt, V., 1982, Synthesis and some pharmacological properties of Z-Tyr(SO_3H)-Met-Gly-Trp-Met-Asp(Phe-NH_2)-OH, a 32 β-aspartyl analogue of cholecystokinin (pancreozymin) 27–33, *J. Med. Chem.* **25:**589–593.
Meyer, F. D., Gyr, K., Kayasseh, L., Jeker, L., Wall, M., Trzeciak, A., and Gillesen, D., 1980, Biological activity of the C-terminal octapeptide of cholecystokinin, of three of its analogues and of caerulein in the dog, *Experientia* **36:**434–436.
Miller, J. L., Jardine, I., Weissman, E., Go, V. L. W., and Speicher, D., 1984, Characterization of cholecystokinin from the human brain, *J. Neurochem.* **43:**835–840.
Morley, J. S., 1968, Structure–function relationships in gastrin-like peptides, *Proc. R. Soc. Lond. [Biol.]* **170:**97–111.
Morley, J. S., 1977, Information about peptide hormone receptors from structure–activity studies, in: *First International Symposium of Hormonal Receptors in Digestive Tract Physiology* (S. Bonfils, P. Fromageot, and G. Rosselin, eds.), Elsevier/North-Holland Biomedical Press, Amsterdam, pp. 3–11.
Morley, J. E., 1982, The ascent of cholecystokinin (CCK) from gut to brain, *Life Sci.* **30:**479–493.
Moroji, T., Watanabe, N., Aoki, N., and Itoh, S., 1982, Antipsychotic effects of ceruletide (cerulein) on chronic schizophrenia, *Arch. Gen. Psychiatry* **39:**485.
Murphy, R. B., and Schuster, D. I., 1982, Modulation of [^{3}H]-dopamine binding by cholecystokinin octapeptide (CCK-8), *Peptides* **3:**539–543.
Mutt, V., 1980, Cholecystokinin: Isolation, structure, and functions, in: *Gastrointestinal Hormones* (G. B. J. Glass, ed.), Raven Press, New York, pp. 169–221.
Nair, N. P. V., Bloom, D. M., and Nestoros, J. N., 1982, Cholecystokinin appears to have antipsychotic properties, *Prog. Neuropsychopharmacol. Biol. Psychiatry* **6:**509–512.
Peikin, S. R., Rottman, A. J., Batzri, S., and Gardner, J. D., 1978, Kinetics of amylase release by dispersed acini prepared from guinea pig pancreas, *Am. J. Physiol.* **235:**E743–E749.
Penke, B., Zarandi, M., Toth, G. K., Kovacs, K., Fekete, M., Telegdy, G., and Pham, P., 1983, The active centres of gastrin and cholecystokinin: Syntheses, conformational problems, correlations between chemical structures and biological activity, in: *Peptides 1982* (K. Blaha and P. Malon, eds.), Walter DeGruyter, Berlin, pp. 569–575.
Penke, B., Hajnal, F., Lonovics, J., Holzinger, G., Kadar, T., Telegdy, G., and Rivier, J., 1984, Synthesis of potent heptapeptide analogues of cholecystokinin, *J. Med. Chem.* **27:**845–849.
Pi-Sunyer, X., Kissileff, H. R., Thornton, J., and Smith, G. P., 1982, C-Terminal octapeptide of cholecystokinin decreases food intake in obese men, *Physiol. Behav.* **29:**627–630.
Plusec, J., Sheehan, J. T., Sabo, E. F., Williams, N., Kocy, O., and Ondetti, M. A., 1970, Synthesis of analogs of the C-terminal octapeptide of cholecystokinin–pancreozymin. Structure–activity relationship, *J. Med. Chem.* **13:**349–352.
Polak, J. M., and Bloom, S. R., 1982, Regulatory peptides and neuron-specific enolase in the respiratory tract of man and other mammals, *Exp. Lung Res.* **3:**313–328.
Praissman, M., and Walden, M., 1984, The binding characteristics of ^{125}I-gastrin and ^{125}I-CCK-8 to guinea pig fundic gastric glands differ: Is there more than one binding site for peptides of the CCK-gastrin family? *Biochem. Biophys. Res. Commun.* **123:**641–647.
Praissman, M., Izzo, R. S., and Berkowitz, J. M., 1982, Modification of the C-terminal octapeptide of cholecystokinin with a high-specific-activity iodinated imidoester: Preparation, characterization, and binding to isolated pancreatic acinar cells, *Anal. Biochem.* **121:**190–198.
Praissman, M., Martinez, P. A., Saladino, C. F., Berkowitz, J. M., Steggles, A. W., and Finkelstein, J. A., 1983, Characterization of cholecystokinin binding sites in rat cerebral cortex using a ^{125}I-CCK-8 probe resistant to degradation, *J. Neurochem.* **40:**1406–1413.
Rehfeld, J. F., and Morley, J. S., 1983, Residue-specific radioimmunoanalysis: A novel analytical tool. Application to the C-terminal of CCK/gastrin peptides, *J. Biochem. Biophys. Methods* **7:**161–170.

Rogawski, M. A., 1982, Cholecystokinin octapeptide: Effects on the excitability of cultured spinal neurons, *Peptides* **3**:545–555.

Saito, A., Sankaran, H., Goldfine, I. D., and Williams, J. A., 1980, Cholecystokinin receptors in the brain: Characterization and distribution, *Science* **208**:1155–1156.

Saito, A., Williams, J. A., and Goldfine, I. D., 1981, Alterations in brain cholecystokinin receptors after fasting, *Nature* **289**:599–600.

Sakamoto, C., Williams, J. A., and Goldfine, I. D., 1984, Brain CCK receptors are structurally distinct from pancreas CCK receptors, *Biochem. Biophys. Res. Commun.* **124**:497–502.

Sankaran, H., Goldfine, I. D., Deveney, C. W., Wong, K.-Y., and Williams, J. A., 1980, Binding of cholecystokinin to high affinity receptors on isolated rat pancreatic acini, *J. Biol. Chem.* **255**:1849–1853.

Schiller, P. W., Natarajan, S., and Bodansky, M., 1978, Determination of the intramolecular tryosine–tryptophan distance in a 7-peptide related to the C-terminal sequence of cholecystokinin, *Int. J. Peptide Protein Res.* **12**:139–142.

Schreur, P. J. K. D., Sawyer, T. K., Ruwart, M. J., Collins, R. J., Staples, D. J., Rush, B. D., Nichols, N. F., and Russell, R. R., 1985, Satiety effects of CCK-8 in rats: Studies on peptide delivery, structure–activity, and subchronic treatment, in: *Fifth International Washington Spring Symposium: Neural and Endocrine Peptides and Receptors* (abstract).

Spanarkel, M., Martinez, J., Briet, C., Jensen, R. T., and Gardner, J. D., 1983, Cholecystokinin-27–33-amide: A member of a new class of cholecystokinin receptor antagonists, *J. Biol. Chem.* **258**:6746–6749.

Stacher, G., Steinringer, H., Schmierer, G., Schneider, C., and Winklehner, S., 1982a, Cholecystokinin octapeptide decreases intake of solid food in man, *Peptides* **3**: 133–136.

Stacher, G., Steinringer, H., Schmierer, G., Winklehner, S., and Schneider, C., 1982b, Ceruletide increases threshold and tolerance to experimentally induced pain in healthy man, *Peptides* **3**:955–962.

Szecowka, J., Goldfine, I. D., and Williams, J. A., 1985, Solubilization and characterization of CCK receptors from mouse pancreas, *Regul. Peptides* **10**:71–83.

Van Ree, J. M., Gaffori, O., and de Wied, D., 1983, In rats, the behavioral profile of CCK-8 related peptides resembles that of antipsychotic agents, *Eur. J. Pharmacol.* **93**:63–78.

Wennogle, L. P., Steel, D. J., and Petrack, B., 1985, Characterization of central cholecystokinin receptors using a radioiodinated octapeptide probe, *Life Sci.* **36**:1485–1492.

Williams, J. A., 1982, Cholecystokinin: A hormone and a neurotransmitter, *Biomed. Res.* **3**:107–121.

Zetler, G., 1982, Cholecystokinin octapeptide, caerulein and caerulein analogues: Effects on thermoregulation in the mouse, *Neuropharmacology* **21**:795–801.

Zetler, G., 1983, Cholecystokinin octapeptide (CCK-8), ceruletide and analogues of ceruletide: Effects on tremors induced by oxotremorine, harmine and ibogaine. A comparison with prolyl-leucylglycine amide (MIF), anti-parkinsonian drugs and clonazepam, *Neuropharmacology* **22**:757–766.

Zetler, G., 1984, Ceruletide, ceruletide analogues and cholecystokinin octapeptide (CCK-8): Effects on isolated intestinal preparations and gallbladders of guinea pigs and mice, *Peptides* **5**:729–736.

Characterization of Tachykinin Receptors by Ligand Binding Studies and by Utilization of Conformationally Restricted Tachykinin Analogues

MARGARET A. CASCIERI, GARY G. CHICCHI, ROGER M. FREIDINGER, CHRISTIANE DYLION COLTON, DEBRA S. PERLOW, BRIAN WILLIAMS, NEIL R. CURTIS, DANIEL F. VEBER, and TEHMING LIANG

1. INTRODUCTION

Tachykinins are a family of peptides that share the carboxyl-terminal sequence -Phe-X-Gly-Leu-Met-NH_2 and that cause contraction of various smooth muscle systems. Substance P (Chang and Leeman, 1970) and the newly discovered neurokinin A (Maggio *et al.*, 1983; Minamino *et al.*, 1984a; Kimura *et al.*, 1983) and neurokinin B* (Kimura *et al.*, 1983; Kangawa *et al.*, 1983) are mammalian tachykinins. Several nonmammalian tachykinins have been isolated, which include eledoisin, physalaemin, kassinin, and phyllomedusin (for a review, see Erspamer and Melchiorri, 1983). The amino acid sequences of these peptides are shown in Fig. 1.

Various studies have shown that substance P is widely located in synaptic vesicles in both the central and peripheral nervous systems, that it can be released by agents causing membrane depolarization, and that it causes excitation of certain

* Neurokinin A is also substance K, neuromedin L, and neurokinin α; neurokinin B is also neuromedin K and neurokinin β.

MARGARET A. CASCIERI, GARY G. CHICCHI, and TEHMING LIANG • Merck Sharp and Dohme Research Laboratories, Rahway, New Jersey 07065. ROGER M. FREIDINGER, CHRISTIANE DYLION COLTON, DEBRA S. PERLOW, and DANIEL F. VEBER • Merck Sharp and Dohme Research Laboratories, West Point, Pennsylvania 19486. BRIAN WILLIAMS and NEIL R. CURTIS • Merck Sharp and Dohme Research Laboratories, Neuroscience Research Centre, Terlings Park, Essex, United Kingdom.

Mammalian	
Substance P	Arg-Pro-Lys-Pro-Gln-Gln-Phe-Phe-Gly-Leu-Met-NH_2
Neurokinin A	His-Lys-Thr-Asp-Ser-Phe-Val-Gly-Leu-Met-NH_2
Neurokinin B	Asp-Met-His-Asp-Phe-Phe-Val-Gly-Leu-Met-NH_2
Nonmammalian	
Physalaemin	pGlu-Ala-Asp-Pro-Asn-Lys-Phe-Tyr-Gly-Leu-Met-NH_2
Kassinin	Asp-Val-Pro-Lys-Ser-Asp-Gln-Phe-Val-Gly-Leu-Met-NH_2
Eledoisin	pGlu-Pro-Ser-Lys-Asp-Ala-Phe-Ile-Gly-Leu-Met-NH_2
Phyllomedusin	pGlu-Asn-Pro-Asn-Arg-Phe-Ile-Gly-Leu-Met-NH_2

FIGURE 1. Structure of tachykinins.

neuron populations (Nicoll *et al.*, 1980). Recent studies have shown that both neurokinin A and neurokinin B are discretely distributed in the rat brain and spinal cord (Minamino *et al.*, 1984b). Thus, it is likely that these three peptides function as neurotransmitters or modulators in the mammalian central and peripheral nervous systems. In this chapter, we summarize our studies describing the pharmacology, function, and distribution of tachykinin receptors. We also describe a series of conformationally restricted hexapeptide analogues that are selective for one class of tachykinin receptors.

2. THE SUBSTANCE P RECEPTOR OF PERIPHERAL AND CENTRAL TISSUES

Substance P is a potent stimulator of salivation in rats and increases the rate of amylase secretion in a preparation of isolated parotid acinar cells (Liang and Cascieri, 1979). Methods developed by other investigators to demonstrate substance P receptor binding in isolated cells or membranes using [^{3}H]substance P (Nakata *et al.*, 1978; Hanley *et al.*, 1980b) or [^{125}I]physalaemin (Jensen and Gardner, 1979; Putney *et al.*, 1980) were inadequate because of low specific activity or high nonspecific binding. Since chloramine T oxidation of substance P could cause oxidation of the carboxyl-terminal methionine residue, resulting in a peptide with reduced biological activity (Floor and Leeman, 1980), we attempted to prepare a labeled ligand under less harsh conditions. We conjugated substance P with [^{125}I]Bolton–Hunter reagent and showed that this ligand ([^{125}I]BH-SP) binds to parotid cells with high specific activity (Liang and Cascieri, 1980). Nonradioactive BH-SP is three times more potent than substance P in stimulating salivation. Thus, this ligand is both fully biologically active and useful as a receptor probe. The binding of [^{125}I]BH-SP to parotid cells is reversible and to a single class of high-affinity (K_d = 4 nM) binding sites (Liang and Cascieri, 1981). The binding affinity of 13 fragments and analogues of substance P correlates well (correlation coefficient = 0.92) with their relative potency in stimulating salivation (Fig. 2). As amino acids are removed from the amino terminal of substance P, there is a progressive loss in both biological activity and binding affinity. Fragments smaller than

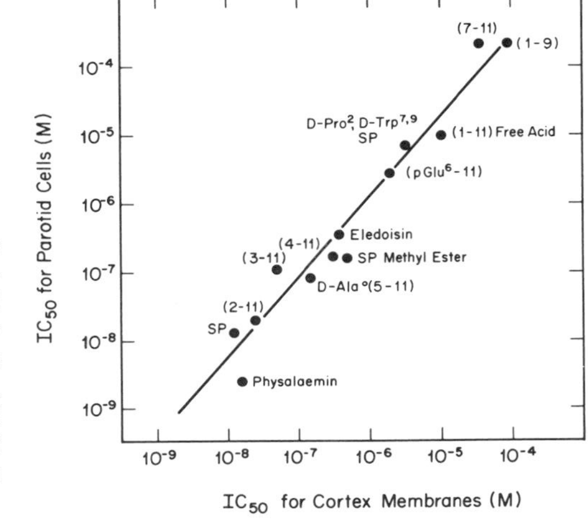

FIGURE 2. Correlation between the relative binding affinity to the [^{125}I]BH-SP binding site on rat parotid cells and the relative saliva-stimulating activity for substance P and related peptides. The line is the theoretical correlation of 1.0. The correlation coefficient calculated from these data is 0.92 with a slope of 1.01. (Data reprinted with permission from Liang and Cascieri, 1981.)

the carboxyl-terminal hexapeptide are poor inhibitors of the binding of [^{125}I]BH-SP and weak stimulators of salivation. Substance P_{1-9} and substance P free acid are less than 0.1% as potent as substance P both as stimulators of rat salivation and as inhibitors of [^{125}I]BH-SP binding. Various peptides structurally unrelated to substance P do not inhibit the binding. Thus, [^{125}I]BH-SP binds to a physiologically relevant receptor on parotid acinar cells.

[^{125}I]BH-Substance P also binds with high affinity (K_d = 12 ± 4 nM) to a single binding site in gradient-purified rat brain cortex membranes (Cascieri and Liang, 1983, 1984). This binding is both reversible and saturable (0.5 ± 0.1 pmol/mg protein). The proteolytic degradation of [^{125}I]BH-SP was effectively inhibited

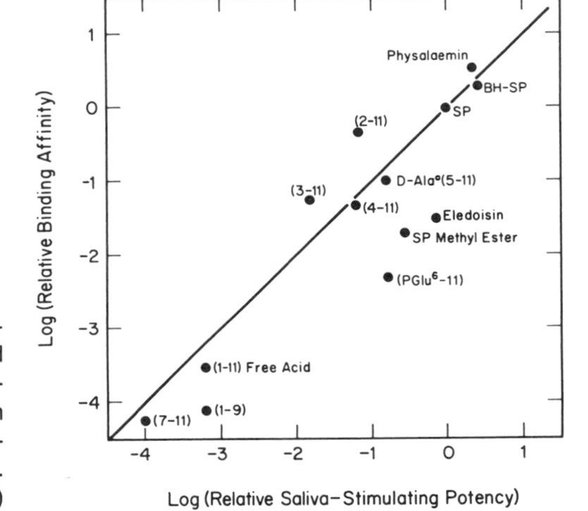

FIGURE 3. Correlation between the relative binding affinities of substance P and related peptides for the [^{125}I]BH-SP binding sites on rat parotid cells and rat brain cortex membranes. The correlation coefficient is 0.978, and the slope is 1.19. (Reprinted from Cascieri and Liang, 1983.)

by the presence of chymostatin in the incubation. In addition, [^{125}I]BH-SP extracted from cortex membranes after labeling and extensive washing coeluted with authentic [^{125}I]BH-SP on reverse-phase HPLC chromatography. The relative binding affinities of substance P fragments and analogues for the cortex membrane binding site are nearly identical with the relative binding affinities of these peptides for the parotid cell substance P receptor (Fig. 3). Thus, the substance P receptors of both the peripheral and central nervous systems are similar with regard to peptide affinity.

Our data describing the pharmacological properties of this receptor have been confirmed using both [^{125}I]BH-SP (Beaujouan *et al.*, 1982; Viger *et al.*, 1983) in mouse mesencephalic cells and rat brain synaptosomes and [^{3}H]substance P (Lee *et al.*, 1983) in salivary gland membranes. In collaboration with the laboratories of Candace Pert and Miles Herkenham, we have studied the distribution of [^{125}I]BH-SP binding sites in rat brain slices by autoradiography after *in vitro* incubation with ligand (Rothman *et al.*, 1984b). The [^{125}I]BH-SP binding to brain slices is inhibited by substance P fragments and analogues with the same specificity that we observed previously in cortex membranes. The [^{125}I]BH-SP binding sites are widely but discretely localized throughout the sensory, limbic, and cortical areas of rat brain. The distribution of these sites is similar to that observed by others using [^{3}H]substance P (Quirion *et al.*, 1983) and [^{125}I]BH-SP (Shults *et al.*, 1984).

These studies all show that the distribution of substance P receptors does not correlate well with the distribution of immunoreactive substance-P-containing nerve tracts. Whereas in many areas of the brain both substance P receptors and immunoreactive substance P coexist, in many areas only receptors (e.g., cerebellum) or peptide (e.g., substantia nigra) can be found. These data and the recent demonstration that neurokinins A and B are discretely located in the brain and the spinal cord (Minamino *et al.*, 1984b) suggest that there may be more than one type of tachykinin binding site.

3. [^{125}I]BH-ELEDOISIN BINDING TO CORTEX MEMBRANES

As first recognized in the laboratory of Leslie Iversen, although the spectrum of activity of the tachykinins is similar, their rank order of potency in various pharmacological preparations differs (Lee *et al.*, 1982). For example, substance P is more potent than eledoisin in stimulating salivation and lowering blood pressure, but it is less potent in stimulating smooth muscle contraction in the urogenital system (Erspamer *et al.*, 1980). This led them to propose that at least two types of tachykinin receptors exist: a P type in which substance P is equipotent to or more active than eledoisin and an E type in which eledoisin is 10 to 100 times more potent (Lee *et al.*, 1982). In addition, the potencies of substance P fragments to stimulate salivation (Liang and Cascieri, 1981) and lower blood pressure (Couture and Regoli, 1982) decrease with decreasing peptide length. However, in many smooth muscle preparations (Couture and Regoli, 1982; Teichberg *et al.*, 1981)

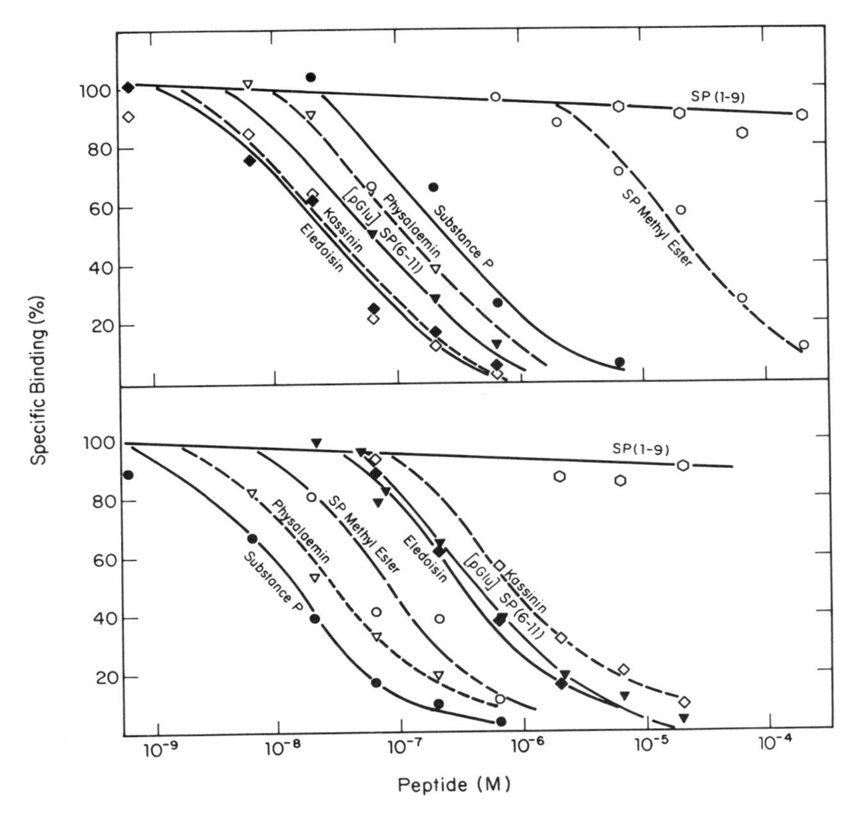

FIGURE 4. Displacement of [^{125}I]BH-eledoisin (top) and [^{125}I]BH-SP (bottom) binding to rat brain cortex membranes by tachykinins. (Reprinted with permission from Cascieri *et al.*, 1985b.)

and in frog spinal cord neurons (Otsuka and Konishi, 1975), substance P_{4-11} and substance P pyroglutamyl$_{6-11}$ are as potent or more potent than substance P.

Eledoisin conjugated with [^{125}I]Bolton–Hunter reagent ([^{125}I]BH-eledoisin) specifically binds to rat brain cortex membranes (Cascieri and Liang, 1984). Eledoisin and kassinin are 6.8 times more potent than substance P as inhibitors of this binding, whereas they are only 3% and 1.5%, respectively, as potent as substance P as inhibitors of [^{125}I]BH-SP binding. Scatchard analysis of the binding of [^{125}I]BH-eledoisin is curvillinear, suggesting that the ligand may bind to two distinct sites with K_d = 0.9 ± 0.7 nM and K_d = 20 ± 10 nM (Cascieri *et al.*, 1985). The inhibition of [^{125}I]BH-SP and [^{125}I]BH-eledoisin binding by various tachykinins is shown in Fig. 4. In addition to the differences noted above, substance P pyroglutamyl$_{6-11}$ is only 3% as potent as substance P as an inhibitor of [^{125}I]BH-SP binding, whereas it is 2.5 times as potent as substance P as an inhibitor of [^{125}I]BH-eledoisin binding. Substance P methyl ester in which the carboxyl terminal

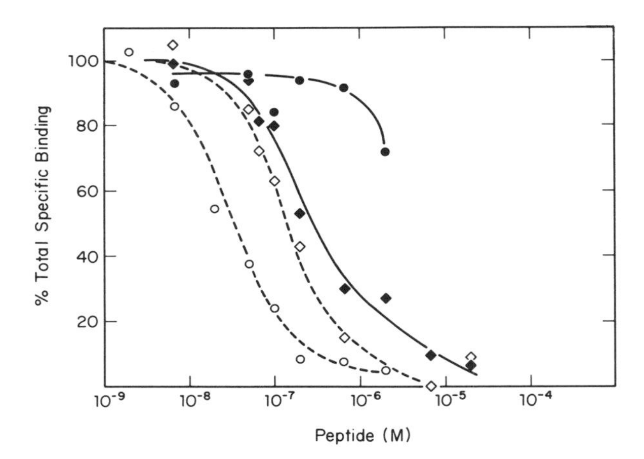

FIGURE 5. Displacement of [^{125}I]BH-eledoisin (open symbols) and [^{125}I]BH-SP (closed symbols) binding to rat brain cortex membranes by neurokinins A (◇, ◆) and B (○, ●). (Reprinted with permission from Cascieri *et al.*, 1985b.)

amide is replaced with a methyl group (Cascieri *et al.*, 1981) has a 1200-fold higher affinity for the [^{125}I]BH-SP binding site than for the [^{125}I]BH-eledoisin binding site. Substance P methyl ester is a potent agonist in muscle contraction assays of the P type but is less than 1% as active as substance P in E-type tissues (Watson *et al.*, 1983). Thus, [^{125}I]BH-eledoisin binds to a high-affinity site that is pharmacologically similar to the E receptor subtype proposed by Iversen and associates.

Neurokinin B is at least 58 times more potent as an inhibitor of [^{125}I]BH-eledoisin binding than as an inhibitor of [^{125}I]BH-SP binding (Fig. 5). In contrast, neuromedin A is nearly equipotent as an inhibitor of the binding of both ligands. This indicates that neuromedin B may be the endogenous mammalian ligand for the [^{125}I]BH-eledoisin binding site. Table I lists the IC_{50} of these peptides and others as inhibitors of the binding of both ligands. It is interesting that carboxyl-terminal fragments of substance P larger than the pentapeptide are more potent relative to substance P as inhibitors of [^{125}I]BH-eledoisin binding than as inhibitors of [^{125}I]-BH-SP binding. In various smooth muscle preparations, the relative potencies of carboxyl-terminal fragments vary greatly (Couture and Regoli, 1982; Teichberg *et al.*, 1981). These tissues may contain mixtures of the two types of binding sites, and the relative order of potency of peptides in a given tissue may depend on the ratio of the two sites present.

The difference in the carboxyl-terminal sequences of substance P and eledoisin is the substitution of an isoleucine for phenylalanine at position 8 (Fig. 1). In addition, kassinin, which is a better inhibitor of [^{125}I]BH-eledoisin binding than of [^{125}I]BH-SP binding, also contains a nonaromatic residue (valine) at this position. Lee *et al.* (1982) also observed that tachykinins with nonaromatic residues at this position were more active in E-type pharmacological preparations than substance

TABLE I. Inhibition of [^{125}I]BH-Eledoisin and [^{125}I]BH-SP Binding to Brain Cortex Membranes by SP and Related Peptides

Peptide	[^{125}I]BH-Eledoisin binding IC$_{50}$ (M)[a,b]		Relative affinity	[^{125}I]BH-SP binding IC$_{50}$ (M)[a,b]		Relative affinity
SP	$2.1 \pm 1.6 \times 10^{-7}$	(4)	1	$1.0 \pm 0.5 \times 10^{-8}$	(7)	1
Eledoisin	$3.1 \pm 2 \times 10^{-8}$	(4)	6.8	$3.6 \pm 0.5 \times 10^{-7}$	(3)	0.03
Physalaemin	$1.1 \pm 0.07 \times 10^{-7}$	(2)	1.9	$1.4 \pm 0.7 \times 10^{-8}$	(3)	0.71
Kassinin	$3.1 \pm 1.7 \times 10^{-8}$	(2)	6.8	$6.7 \pm 2 \times 10^{-7}$	(3)	0.015
Neurokinin A	$1.3 \pm 0.4 \times 10^{-7}$	(2)	1.6	$2.6 \pm 0.4 \times 10^{-7}$	(2)	0.038
Neurokinin B	$3.4 \pm 0.9 \times 10^{-8}$	(2)	6.2	$>2 \times 10^{-6}$	(2)	<0.005
BH-Eledoisin	$4.2 \pm 0.3 \times 10^{-9}$	(2)	50.0	$>5 \times 10^{-7}$	(1)	<0.02
Eledoisin-related peptide	3.4×10^{-6}	(1)	0.06	2.8×10^{-7}	(1)	0.04
Phyllomedusin	8.8×10^{-8}	(1)	2.3	1.8×10^{-8}	(1)	0.5
SP methyl ester	$2.4 \pm 0.5 \times 10^{-5}$	(3)	0.008	$2.0 \pm 0.9 \times 10^{-8}$	(3)	0.5
SP free acid	$>2 \times 10^{-4}$	(2)	<0.001	$1.1 \pm 0.5 \times 10^{-5}$	(2)	0.001
SP_{2-11}	$2.7 \pm 1.8 \times 10^{-7}$	(2)	0.77	$2.7 \pm 1.3 \times 10^{-8}$	(2)	0.37
SP_{3-11}	$4.3 \pm 3.3 \times 10^{-7}$	(2)	0.49	$5.3 \pm 0.3 \times 10^{-8}$	(2)	0.19
SP_{4-11}	$3.0 \pm 0.3 \times 10^{-7}$	(2)	0.67	$4.9 \pm 0.4 \times 10^{-8}$	(2)	0.2
SP_{5-11}	$7.5 \pm 7 \times 10^{-7}$	(2)	0.28	$2.5 \pm 1.2 \times 10^{-7}$	(2)	0.04
SP_{6-11}	$1.6 \pm 0.6 \times 10^{-7}$	(2)	1.3	$6.0 \pm 1.2 \times 10^{-7}$	(2)[c]	0.02
[pGlu]SP_{6-11}	$8.5 \pm 3.5 \times 10^{-8}$	(2)	2.5	$3.7 \pm 1 \times 10^{-7}$	(3)	0.03
SP_{7-11}	$9.7 \pm 6 \times 10^{-6}$	(2)	0.02	$4.1 \pm 0.9 \times 10^{-5}$	(2)	0.0003
SP_{8-11}	1.3×10^{-4}	(1)	0.002	$>2 \times 10^{-5}$	(2)	<0.0006
SP_{9-11}	$>2 \times 10^{-4}$	(2)	<0.001	$>2 \times 10^{-5}$	(2)	<0.0006
SP_{1-9}	$>2 \times 10^{-4}$	(2)	<0.001	$>6 \times 10^{-5}$	(2)	<0.0002
$SP_{1-9}NH_2$	$>1 \times 10^{-4}$	(1)	<0.002	$>1 \times 10^{-4}$	(1)	<0.0001
[D-Ala^0]SP_{5-11}	5.0×10^{-7}	(1)	0.42	$1.5 \pm 0.3 \times 10^{-7}$	(2)	0.07
[D-Pro^2,D-Trp^{7,9}]SP	1.3×10^{-5}	(1)	0.02	$3.5 \pm 0.03 \times 10^{-6}$	(2)	0.003
[D-Pro^4,D-Trp^{7,9}]SP_{4-11}	7.9×10^{-5}	(1)	0.003	4.8×10^{-5}	(1)	0.0003

[a] Mean ± standard deviation; number of experiments shown in parentheses.
[b] Data reprinted with permission from Cascieri *et al.*, 1985b.
[c] SP_{6-11} is a mixture of cyclized and noncyclized peptides.

P. A comparison of the activities of the carboxyl-terminal hexapeptides containing phenylalanine, isoleucine, and valine is shown in Table II. The substitution of a nonaromatic residue does not increase the affinity of the peptide for the [^{125}I]BH-eledoisin receptor. However, all of the hexapeptides are poor inhibitors of [^{125}I]BH-SP binding. In addition, eledoisin-related peptide and phyllomedusin are better inhibitors of [^{125}I]BH-SP binding than of [^{125}I]BH-eledoisin binding (Table II). These two peptides are only 5–20% as active as eledoisin in E-type pharmacological preparations (Watson *et al.*, 1983). These data indicate that the nonaromatic residue in position 8 of eledoisin is not responsible for increased affinity to the [^{125}I]BH-eledoisin receptor or for increased selectivity for this receptor subtype. Also, the data in Tables I and II clearly show that the amino-terminal residues of substance P are important in maintaining high potency at the [^{125}I]BH-SP receptor subtype.

TABLE II. Effect of Substitution of Branched-Chain Amino Acids for Phenylalanine in Tachykinin Peptides

	IC_{50} (M)	
	[^{125}I]BH-SP	[^{125}I]BH-Eledoisin
pGlu-Phe-Phe-Gly-Leu-Met-NH_2	3.4×10^{-7}	1.8×10^{-8}
pGlu-Phe-Val-Gly-Leu-Met-NH_2	2.2×10^{-6}	2.0×10^{-8}
pGlu-Phe-Ile-Gly-Leu-Met-NH_2	5.0×10^{-7}	1.2×10^{-8}
Lys-Phe-Ile-Gly-Leu-Met-NH_2 (eledoisin related peptide)	2.8×10^{-7}	3.4×10^{-6}
Phyllomedusin	1.8×10^{-8}	8.8×10^{-8}

[^{125}I]BH-Eledoisin binding to rat parotid cells cannot be demonstrated even at protein concentrations seven times higher than required to demonstrate specific binding of [^{125}I]BH-SP (Cascieri *et al.*, 1985b). These data and the fact that the pharmacological specificity of the [^{125}I]BH-SP receptor correlates well with the specificity of tachykinin-induced salivation indicate that the parotid contains a nearly homogeneous population of [^{125}I]BH-SP receptors. The distribution of [^{125}I]BH-eledoisin binding sites in the rat brain (Rothman *et al.*, 1984a) and spinal cord (Ninkovic *et al.*, 1985) is distinct from that of [^{125}I]BH-SP. The pattern of binding in the cortical layers is very different for the two ligands (Rothman *et al.*, 1984a). In contrast to [^{125}I]BH-SP, [^{125}I]BH-eledoisin densely labels the interpeduncular nucleus and sparsely labels the substantia nigra. In the rat spinal cord, [^{125}I]BH-SP binding is present in layers I and II of the dorsal horn and on motor neurons in the ventral horn. In contrast, [^{125}I]BH-eledoisin labels only layer I and the outer part of layer II and does not label any motor neurons (Ninkovic *et al.*, 1985). Thus, by both pharmacological and distribution criteria, [^{125}I]BH-SP and [^{125}I]BH-eledoisin bind to distinct tachykinin receptor subtypes.

4. COUPLING OF TACHYKININ RECEPTORS TO GUANINE NUCLEOTIDE BINDING PROTEINS

GTP, GDP, and Gpp(NH)p inhibit the binding of [^{125}I]BH-SP and [^{125}I]BH-eledoisin to cortex membranes (Cascieri and Liang, 1984; Cascieri *et al.*, 1985a), whereas other nucleotides have no effect. These data suggest that these tachykinin receptors may be coupled to a guanine nucleotide binding component. Although there are some reports that substance P stimulates adenylate cyclase in human brain homogenates (Duffy *et al.*, 1975) and rat and guinea pig spinal cord (Northam and Jones, 1984), we and others (Quik *et al.*, 1978) see no stimulation of adenylate cylcase in cortex membranes or in rat brain slices, respectively. In rat small intestine, substance P and eledoisin stimulate turnover of phosphoinositides, but they have no effect on levels of cAMP (Watson, 1984). Substance P also stimulates phosphoinositide turnover in rat salivary glands (Hanley *et al.*, 1980a) and in rat brain slices (Mantyh *et al.*, 1984). The relationship between these processes and the inhibition of tachykinin binding by guanine nucleotides has not been elucidated.

5. DEVELOPMENT OF A CONFORMATIONALLY RESTRICTED AGONIST SELECTIVE FOR [^{125}I]BH-ELEDOISIN RECEPTORS

Since the presence of a glycine residue may favor a turn in the peptide backbone, we tested several conformationally restricted analogues of substance P hexapeptide to see if such a turn is important in the recognition of the peptide by its receptor (Cascieri *et al.*, 1986). The properties of a series of analogues in which the nitrogen of the leucine residue is linked by a two-carbon bridge to the α-carbon of glycine to form a five-membered lactam ring are shown in Table III. Compound IA, in which the stereochemistry at the α-carbon is *R*, is 500 times more potent as an inhibitor of [^{125}I]BH-eledoisin binding than as an inhibitor of [^{125}I]BH-SP binding. In contrast, its stereoisomer IB is a poor inhibitor of the binding of both ligands. The substitution of isoleucine (IC) or valine (ID) for phenylalanine in the carboxyl-terminal sequence of these analogues does not significantly affect the selectivity of these peptides for the two binding sites. IA, IC, and ID are potent agonists in the guinea pig ileum contraction assay but are at least 500 times less potent than substance P as stimulants of rat salivation (Table III). Thus, these compounds are potent [^{125}I]BH-eledoisin receptor-selective agonists.

Preliminary NMR studies suggest that the preferred conformation of IA in solution (DMSO) has a turn that does not fit any of the classic definitions for the

TABLE III. Receptor Affinities and Biological Potencies of Lactam-Containing Tachykinin Hexapeptides

<Glu-Phe-X-N–α(lactam)–N–δ–Met NH$_2$

	X	α	δ	IC$_{50}$ (nM) [^{125}I]BH-SP	IC$_{50}$ (nM) [^{125}I]BH-Eledoisin	ED$_{50}$ Guinea pig ileum (nM)	ED$_{50}$ Salivation (nmol/rat)
IA	Phe	*R*	*S*	16,000	27	1.2	>100
IB	Phe	*S*	*S*	>60,000	32,000	≥1000	>65
IC	Ile	*R*	*S*	140,000	150	16.4	>100
ID	Val	*R*	*S*	120,000	72	10.0	>100
Substance P				10	210	0.44	0.2
Eledoisin				360	30	1.6	0.4

various types of peptide turns. The peptide backbone of IB would be expected to have a type II′ β turn. This is inferred from studies of various constrained analogues of LHRH including the *S*-lactam (Freidinger *et al.*, 1980). Therefore, we conclude that the [^{125}I]BH-SP and [^{125}I]BH-eledoisin receptor bound conformations do not have X-Gly-Leu-Met-NH_2 type II′ β turns.

6. DISTINCT TACHYKININ RECEPTORS IN THE GUINEA PIG ILEUM

We have used two selective agonists, substance P methyl ester for the [^{125}I]BH-SP receptor and IA for the [^{125}I]BH-eledoisin receptor, to study the function of the two receptors in the guinea pig ileum. The substance P-induced contraction of guinea pig ileum is not inhibited by atropine. However, Holzer and Lembeck (1980) showed that after induction of a maximal longitudinal contraction by high levels of substance P, the contraction is not followed by complete relaxation. This remaining contraction could be enhanced by the cholinesterase inhibitor physostigmine and was decreased by the anticholinergic tropicamide. Subsequently, Yau and Youther (1982) were able to demonstrate directly the release of acetylcholine from the myenteric plexus by substance P. Recently, Fosbraey *et al.* (1984) showed that eledoisin is 26 times more active than substance P in eliciting acetylcholine release.

The contraction of the guinea pig ileum induced by 1 nM IA is completely attenuated by preincubation of the tissue with atropine (2 μM, Fig. 6). In contrast, the activity of substance P methyl ester is unchanged by atropine treatment. Atropine causes a rightward shift in the dose–response curve for IA. The ED_{50} for IA in the

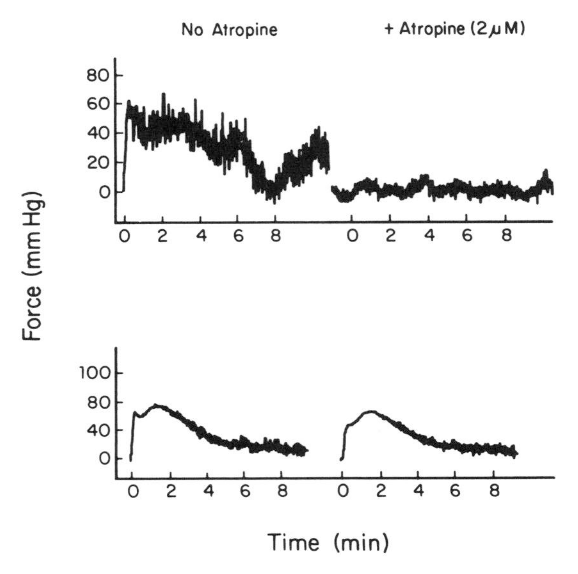

FIGURE 6. Differential effect of atropine on the stimulation of guinea pig ileum contraction by IA (top panel) and substance P methyl ester (bottom panel). Atropine was added 10 min prior to addition of peptide. (Reprinted from Cascieri *et al.*, 1986.)

absence or presence of atropine is 1.2 nM or 6.7 nM, respectively (significantly different at $P < 0.05$). These data indicate that the IA-induced contraction of guinea pig ileum is largely dependent on the release of acetylcholine. Thus, the tachykinin receptors in the myenteric plexus may be largely [^{125}I]BH-eledosin receptors, whereas those in the smooth muscle may be largely [^{125}I]BH-SP receptors.

7. CONCLUSIONS

Our data indicate that there are at least two distinct tachykinin receptor subtypes in the central and peripheral nervous systems. This statement is supported by the following: (1) the rank orders of potencies of peptides to inhibit [^{125}I]BH-SP and [^{125}I]BH-eledoisin binding to brain membranes are different, (2) autoradiography of brain sections shows different patterns of distribution for [^{125}I]BH-SP and [^{125}I]BH-eledoisin binding sites, and (3) tachykinin agonists that are selective for either [^{125}I]BH-SP or [^{125}I]BH-eledoisin binding sites have been identified. Since neurokinin B is a mammalian tachykinin selective for the [^{125}I]BH-eledoisin binding site, this peptide may be the endogenous ligand for the [^{125}I]BH-eledoisin binding site. The pharmacological specificities of these two binding sites indicate that the ratio of sites in a given tissue may explain the differences observed in the relative potencies of tachykinin fragments and analogues in various tissues. Thus, the parotid gland contains a nearly homogeneous population of [^{125}I]BH-SP binding sites, and the rank order of potencies of peptides to inhibit its binding and to stimulate salivation correlates well. In tissues in which eledoisin is 50 to 100 times more potent than substance P (i.e., E-type tissues; Lee *et al.*, 1982), the [^{125}I]BH-eledoisin binding site may predominate. However, tissues in which substance P fragments are equipotent to substance P may contain nearly equal amounts of the two sites.

The specificity for the [^{125}I]BH-eledoisin binding site is determined by the conformation of the carboxyl-terminal sequence. The selective, potent, and conformationally restricted agonists IA, IC, and ID contain a newly identified type of turn conformation, which appears to be required for recognition at the [^{125}I]BH-eledoisin binding site. Since binding affinity for the [^{125}I]BH-SP receptor is lost with the successive removal of amino acids from the amino terminal, and since the hexapeptide sequences are relatively poor inhibitors of [^{125}I]BH-SP binding, as yet undefined determinants in the amino-terminal sequence must contribute to the conformation required for recognition at the [^{125}I]BH-SP receptor.

REFERENCES

Beaujouan, J. C., Torrens, Y., Herbet, A., Daguet, M.-C., Glowinski, J., and Prochiantz, A., 1982, Specific binding of an immunoreactive and biologically active ^{125}I-labeled substance P derivative to mouse mesencephalic cells in primary culture, *Mol. Pharmacol.* **22:**48–55.

Cascieri, M. A., and Liang, T., 1983, Characterization of the substance P receptor in rat brain cortex membranes and the inhibition of radioligand binding by guanine nucleotides, *J. Biol. Chem.* **258:**5158–5164.

Cascieri, M. A., and Liang, T., 1984, Binding of ^{125}I Bolton Hunter conjugated eledoisin to rat brain cortex membranes—evidence for two classes of tachykinin receptors in the mammalian central nervous system, *Life Sci.* **35:**179–184.

Cascieri, M. A., Goldenberg, M. M., and Liang, T., 1981, Biological activity of substance P methyl ester, *Mol. Pharmacol.* **20:**457–459.

Cascieri, M. A., Chicchi, G. G., and Liang, T., 1985, Demonstration of two distinct tachykinin receptors in rat brain cortex, *J. Biol. Chem.* **260:**1501–1507.

Cascieri, M. A., Chicchi, G. G., Freidinger, R. M., Colton, C. D., Perlow, D. S., Williams, B., Curtis, M. R., McKnight, A. T., Maguire, J. J., Veber, D. F., and Liang, T., 1986, Conformationally constrained tachykinin analogs which are selective ligands for the eledoisin binding site, *Mol. Pharmacol.* **29:**34–38.

Chang, M. M., and Leeman, S. E., 1970, Isolation of a sialogogic peptide from bovine hypothalamic tissue and its characterization as substance P, *J. Biol. Chem.* **245:**4784–4790.

Couture, R., and Regoli, D., 1982, Mini review: Smooth muscle pharmacology of substance P, *Pharmacology* **24:**1–25.

Duffy, M. J., Wong, J., and Powell, D., 1975, Stimulation of adenylate cyclase activity in different areas of human brain by substance P, *Neuropharmacology* **14:**615–618.

Erspamer, V., and Melchiorri, P., 1983, Actions of amphibian skin peptides on the central nervous system and the anterior pituitary, In: *Neuroendocrine Perspectives,* Vol. 2 (L. E. Muller and R. M. MacLeod, eds.), Elsevier, Amsterdam, pp. 37–106.

Erspamer, G. F., Erspamer, V., and Piccinelli, D., 1980, Parallel bioassay of physalaemin and kassinin, a tachykinin dodecapeptide from the skin of the African frog *Kassina senegalensis, Naunyn Schmiedebergs Arch. Pharmacol.* **311:**61–65.

Floor, E., and Leeman, S. E., 1980, Substance P sulfoxide: Separation from substance P by high-pressure liquid chromatography, biological and immunological activities, and chemical reduction, *Anal. Biochem.* **101:**498–503.

Fosbraey, P., Featherstone, R. L., and Morton, I. K. M., 1984, Comparison of potency of substance P and related peptides on ^{3}H -acetylcholine release, and contractile actions, in the guinea pig ileum, *Naunyn Schmiedebergs Arch. Pharmacol.* **326:**111–115.

Freidinger, R. M., Veber, D. F., Perlow, D. S., Brooks, J. R., and Saperstein, R., 1980, Bioactive conformation of luteinizing hormone releasing hormone: Evidence from a conformationally constrained analog, *Science* **210:**656–658.

Hanley, M. R., Lee, C. M., Jones, L. M., and Michell, R. H., 1980a, Similar effects of substance P and related peptides on salivation and on phosphatidylinositol turnover in rat salivary glands, *Mol. Pharmacol.* **18:**78–83.

Hanley, M. R., Sandberg, B. E. B., Lee, C. M., Iversen, L. L., Brundish, D. E., and Wade, R., 1980b, Specific binding of ^{3}H-substance P to rat brain membranes, *Nature* **286:**810–812.

Holzer, P., and Lembeck, F., 1980, Neurally mediated contraction of ileal longitudinal muscle by substance P, *Neurosci. Lett.* **17:**101–105.

Jensen, R. T., and Gardner, J. D., 1979, Interaction of physalaemin, substance P, and eledoisin with specific membrane receptors on pancreatic acinar cells, *Proc. Natl. Acad. Sci. U.S.A.* **76:**5679–5683.

Kangawa, K., Minamino, N., Fukuda, A., and Matsuo, H., 1983, Neuromedin K: A novel mammalian tachykinin identified in porcine spinal cord, *Biochem. Biophys. Res. Commun.*, **114:**533–540.

Kimura, S., Okado, M., Sugita, Y., Kanazawa, I., and Munekata, E., 1983, Novel neuropeptides, neurokinin α and β, isolated from porcine spinal cord, *Proc. Jpn. Acad.* **59:**101–104.

Lee, C.-M., Iversen, L. L., Hanley, M. R., and Sandberg, B. E. B., 1982, The possible existence of multiple receptors for substance P, *Naunyn Schmiedebergs Arch. Pharmacol.* **318:**281–287.

Lee, C.-M., Javitch, J. A., and Snyder, S. H., 1983, ^{3}H-Substance P binding to salivary gland membranes, *Mol. Pharmacol.* **23:**563–569.

Liang, T., and Cascieri, M. A., 1979, Substance P stimulation of amylase release by isolated parotid cells and inhibition of substance P induction of salivation by vasoactive peptides, *Mol. Cell. Endocrinol.* **15:**151–162.

Liang, T., and Cascieri, M. A., 1980, Specific binding of an immunoreactive and biologically active ^{125}I-labeled N(1)acylated substance P derivative to parotid cells, *Biochem. Biophys. Res. Commun.* **96:**1793–1799.

Liang, T., and Cascieri, M. A., 1981, Substance P receptor on parotid cell membranes, *J. Neurosci.* **1:**1133–1141.

Maggio, J. E., Sandberg, B. E. B., Bradley, C. V., Iversen, L. L., Santikarn, S., Williams, B. H., Hunter, J. C., and Hanley, M. R., 1983, Substance K: A novel tachykinin in mammalian spinal cord, in: *Substance P: Dublin, 1983* (P. Skrabanek and D. Powell, eds.), Boole Press, Dublin, pp. 20–21.

Mantyh, P. W., Pinnock, R. D., Downes, C. P., Goedert, M., and Hunt, S. P., 1984, Correlation between inositol phospholipid hydrolysis and substance P receptors in rat CNS, *Nature* **308:**795–797.

Minamino, N., Kangawa, K., Fukuda, A., and Matsuo, H., 1984a, Neuromedin L: A novel mammalian tachykinin identified in porcine spinal cord, *Neuropeptides* **4:**157–166.

Minamino, N., Masuda, H., Kangawa, K., and Matsuo, H., 1984b, Regional distribution of neuromedin K and neuromedin L in rat brain and spinal cord, *Biochem. Biophys. Res. Comm.* **124:**731–738.

Nakata, Y., Kusaka, Y., Segawa, T., Yajima, H., and Kitagawa, K., 1978, Substance P: Regional distribution and specific binding to synaptic membranes in rabbit central nervous system, *Life Sci.* **22:**259–268.

Nicoll, R. A., Schenker, C., and Leeman, S. E., 1980, Substance P as a transmitter candidate, *Annu. Rev. Neurosci.* **3:**227–268.

Ninkovic, M., Beaujouan, J. C., Torrens, Y., Saffroy, M., Hall, M. D., and Glowinski, J., 1985, Differential localization of tachykinin receptors in rat spinal cord, *Eur. J. Pharmacol.* **106:**463–464.

Northam, W. J., and Jones, D. J., 1984, Comparison of capsaicin and substance P induced cyclic AMP accumulation in spinal cord tissue slices, *Life Sci.* **35:**293–302.

Otsuka, M., and Konishi, S., 1975, Substance P an excitatory transmitter of primary sensory neurons, *Cold Spring Harbor Symp. Quant. Biol.* **40:**135–143.

Putney, J. W., Van de Walle, C. M., and Wheeler, C. S., 1980, Binding of ^{125}I-physalaemin to rat parotid acinar cells, *J. Physiol. (Lond.)* **301:**205–212.

Quik, M., Iversen, L. L., and Bloom, S. R., 1978, Effect of vasoactive intestinal peptide and other peptides on cAMP accumulation in rat brain, *Biochem. Pharmacol.* **27:**2209–2213.

Quirion, R., Shults, C. W., Moody, T. W., Pert, C. B., Chase, T. N., and O'Donohue, T. L., 1983, Autoradiographic distribution of substance P receptors in rat central nervous system, *Nature* **303:**714–716.

Rothman, R. B., Danks, J. A., Herkenham, M., Cascieri, M. A., Chicchi, G. G., Liang, T., and Pert, C. B., 1984a, Autoradiographic localization of a novel peptide binding site in rat brain using the substance P analog, eledoisin, *Neuropeptides* **4:**343–349.

Rothman, R. B., Herkenham, M., Pert, C. B., Liang, T., and Cascieri, M. A., 1984b, Visualization of rat brain receptors for the neuropeptide, substance P, *Brain Res.* **309:**47–54.

Shults, C. W., Quirion, R., Chronwall, B., Chase, T. N., and O'Donohue, T. L., 1984, A comparison of the anatomical distribution of substance P and substance P receptors in the rat central nervous system, *Peptides* **5:**1097–1128.

Teichberg, V. I., Cohen, S., and Blumberg, S., 1981, Distinct classes of substance P receptors revealed by a comparison of the activities of substance P and some of its segments, *Regul. Peptides* **1:**327–333.

Viger, A., Beaujouan, J. C., Torrens, Y., and Glowinski, J., 1983, Specific binding of a ^{125}I-substance P derivative to rat brain synaptosomes, *J. Neurochem.* **40:**1030–1039.

Watson, S. P., 1984, The action of substance P on contraction, inositol phospholipids and adenylate cyclase in rat small intestine, *Biochem. Pharmacol.* **33:**3733–3737.

Watson, S. P., Sandberg, B. E. B., Hanley, M. R., and Iversen, L. L., 1983, Tissue selectivity of substance P alkyl esters: Suggesting multiple receptors, *Eur. J. Pharmacol.* **87:**77–84.

Yau, W. H., and Youther, M. L., 1982, Direct evidence for a release of acetylcholine from the myenteric plexus of guinea pig small intestine by substance P, *Eur. J. Pharmacol.* **81:**665–668.

Conformationally Restricted Cyclic Penicillamine Analogues with High Selectivity for δ- and μ-Opioid Receptors

K. GULYA, J. T. PELTON, D. R. GEHLERT, J. K. WAMSLEY, S. P. DUCKLES, V. J. HRUBY, and H. I. YAMAMURA

1. INTRODUCTION

The concept that there are subtypes of the opiate receptor was originally suggested by Martin *et al.* (1976) almost a decade ago and is widely accepted based on both *in vitro* (Martin *et al.*, 1976; Tyers, 1980) and *in vivo* (Lord *et al.*, 1977; Schulz *et al.*, 1980) studies. The demonstration of separate opioid target sites such as the brain (Herz *et al.*, 1970; Jacquet and Lajtha, 1974; Pert and Yaksh, 1974; Wei *et al.*, 1975) and spinal cord (Yaksh and Rudy, 1976, 1977) has led to the suggestion that a specific effect may be mediated by different opioid receptors at different central nervous system sites (Ling and Pasternak, 1983; Porreca and Burks, 1983). Although numerous pharmacological and biochemical studies of classical (nonpeptide) opiates and opioid peptide analogues have revealed the existence of several subclasses of receptors (e.g., μ, κ, and δ, Martin *et al.*, 1976; Gilbert and Martin, 1976; Lord *et al.*, 1977; Chang and Cuatrecasas, 1979; Wolozin and Pasternak, 1981), it is well documented that the vast majority of opioid ligands available interact extensively with the different types of receptors, making it difficult to define receptor roles.

The endogenous opioid pentapeptides, [Met[5]]enkephalin and [Leu[5]]enkephalin, interact with several subtypes of opiate receptors that mediate different biological

K. GULYA, S. P. DUCKLES, and H. I. YAMAMURA • Department of Pharmacology, University of Arizona, Tucson, Arizona 85724. J. T. PELTON and V. J. HRUBY • Department of Chemistry, University of Arizona, Tucson, Arizona 85724. D. R. GEHLERT and J. K. WAMSLEY • Department of Psychiatry, University of Utah, Salt Lake City, Utah 84142. *Permanent address of K.G.:* Central Research Laboratory, Medical University, Szeged 6720, Hungary.

responses (Lord *et al.*, 1977; Wolozin and Pasternak, 1981; Chang and Cuatrecasas, 1977). In a series of δ-receptor agonists, for example, the widely used [D-Ala^2, D-Leu^5]enkephalin (DADLE) has shown cross reactivity with the μ binding site in the guinea pig brain (Gillan *et al.*, 1980) or rat brain (Gillan and Kosterlitz, 1982). Obviously, an opioid receptor ligand with a high degree of selectivity for one subtype of opioid receptor is needed and is essential to the elucidation of the function of opioid receptor subtypes that have been proposed.

2. CONFORMATIONALLY RESTRICTED PEPTIDES

Small linear oligopeptides like enkephalins are flexible molecules, and their biologically active conformations may only be determined from an interaction with their receptors (Schiller *et al.*, 1981). Appropriate structural modifications can provide insight into steric and electronic requirements for ligand–receptor interactions. However, conformational insights generally require the design of conformationally constrained analogues. Conformational restriction can be achieved either by methylation of α-carbon atoms or amide nitrogens of the peptide backbone (Marshall *et al.*, 1974) or by synthesis of analogues containing cyclic elements.

In the case of enkephalins, the latter approach recently led to the synthesis of cyclic analogues of enkephalins (Schiller *et al.*, 1981; Schiller and DiMao, 1982, Mosberg *et al.*, 1982, 1983a–c). These analogues have a greatly reduced number of possible solution conformations and, if sufficiently rigid, must maintain a similar conformation when bound to the receptor.

Conformational restrictions of enkephalin analogue by cyclic elements have generally been of two types. Two half-cystine residues are substituted in positions 2 and 5, and ring closure is achieved by oxidative disulfide bond formation (Schiller *et al.*, 1981). Increased rigidity can be conferred on a disulfide-containing peptide if half-cystine is replaced by half-penicillamine (β,β-dimethyl-half-cystine; Pen) because of the effect of *gem*-dimethyl substitution in medium-sized rings (Eliel *et al.*, 1965; Meraldi *et al.*, 1977).

2.1. Conformationally Restricted Enkephalin Analogues for δ-Opioid Receptors

A pioneering approach to the design of a ligand selective for the δ-opioid receptors was implemented by Mosberg *et al.* (1982), who demonstrated that the cyclic conformationally restrained enkephalin analogues [D-Pen^2, L-Cys^5]enkephalinamide and [D-Pen^2, D-Cys^5]enkephalinamide displayed substantial δ-opioid receptor selectivity. Although the corresponding COOH-terminal analogues [D-Pen^2, L-Cys^5]enkephalin and [D-Pen^2, D-Cys^5]enkephalin were found to possess even more pronounced δ-opioid receptor selectivity (Mosberg *et al.*, 1983a), the same group later found that further rigidity imposed by two *gem*-dialkyl substituents incorporated into the medium rings of the cyclic penicillamine-disubstituted enkephalin analogues H-Tyr-D-Pen-Gly-Phe-L-Pen-OH ([D-Pen[2], L-Pen[5]]enkephalin) and H-Tyr-D-Pen-Gly-Phe-D-Pen-OH ([D-Pen[2], D-Pen[5]]enkephalin resulted in additional improvement in δ-receptor selectivity (Mosberg *et al.*, 1983b,c).

Fournie-Zaluski *et al.* (1981) has recently proposed that favorable interaction with μ receptor requires compact formation with the ligand, whereas favorable interaction with δ receptors requires a more extended ligand conformation. The very high δ-receptor selectivity of the *bis*-Pen analogues casts serious doubt on this suggestion (Mosberg *et al.*, 1983c).

2.2. Conformationally Restricted Somatostatin Analogues for μ-Opioid Receptors

The tetradecapeptide somatostatin, originally identified as a somatotropin release-inhibiting factor, has since been demonstrated to be a potent modulator of numerous endocrine functions. It has been shown to be distributed throughout the central and peripheral nervous system and to exert a wide range of biological actions (Brownstein *et al.*, 1975; Kastin *et al.*, 1978; Patel and Reichlin, 1978) over a broad concentration range. High-affinity binding with a receptor system is generally thought to have physiological significance, but when the affinity is weak (micromolar), the physiological significance is not clear. The weak affinity that somatostatin displays for the opioid receptor is one such example. It has been reported, for instance, that micromolar concentrations of somatostatin can inhibit the binding of [^{3}H]naloxone and [^{3}H]DADLE to rat brain homogenates as well as give rise to an *in vivo* analgesic response in mice (Terenius, 1976; Rezek *et al.*, 1978). Recently, it has been reported that the somatostatin analogue [D-Phe[5], Cys[6], D-Trp[8], Cys[11]]somatostatin$_{5-12}$-ol, which we refer to as [D-Phe-Cys-Phe-D-Trp-Lys-Thr-Cys-Thr(ol), with an IC_{50} value of 38 nM at the μ-opiate receptor, was capable of antagonizing the excitatory effects of the stable enkephalin analogue [D-Ala[2], MePhe[4], Gly-ol[5]]enkephalin (DAGO) in electrophysiological experiments (Maurer *et al.*, 1982). Our interest in somatostatin stems from the reported antagonist activity of this analogue to opiates as well as the apparent lack of structural similarity of somatostatin and its analogues to either the rigid opiates or the enkephalin compound.

Recently we have prepared and tested a number of highly constrained cyclic analogues of somatostatin containing penicillamines in order to probe the structural and conformational features important for somatostatin's activity at opioid receptors (Pelton *et al.*, 1985a,b). These studies have led to the development of the Pen-containing analogue [D-Phe[5], Cys[6], Tyr[7], D-Trp[8], Pen[11]]somatostatin$_{5-12}$-amide, which we refer to as D-Phe-Cys-Tyr-D-Trp-Lys-Thr-Pen-Thr-NH_2 or [Cys[2], Tyr[3], Pen[7]]SA. Other Pen analogues are [Cys[2], Pen[7]]SA, [Cys[2], Tyr[3], Orn[5], Pen[7]]SA, and [Cys[2], Tyr[3], Arg[5], Pen[7]]SA, which display high μ-opiate receptor affinity and selectivity in the rat brain radioreceptor assay.

3. δ-OPIOID RECEPTOR-SELECTIVE CYCLIC ENKEPHALIN ANALOGUES

3.1. *In Vivo* Effects

[D-Pen[2], L-Cys[5]]- and [D-Pen[2], D-Cys[5]]enkephalinamide (DPLCEA and DPDCEA) were both effective (3.3. μg, i.c.v.) in increasing the hot-plate (52–55°C) latencies in mice, which is an indication of the analgesic property of these peptides (Mosberg *et al.*, 1982). This analgesic effect was abolished by pretreatment of the animals with naloxone. The analgesic effect in rats treated with either peptide alone lasted 20–30 min. Higher doses of the peptide (10 μg) prolonged the duration of analgesia.

[D-Pen[2], L-Cys[5]]Enkephalin (DPLCE) produced a dose-related analgesic effect after i.c.v. administration with peak effects occurring within 5 min and lasting up to 20 min. Maximum analgesia occurred with 10 μg of DPLCE administered i.c.v. Intrathecally administered peptide (1.0 μg) produced analgesia with maximal effects seen within 5–10 min (Porreca *et al.*, 1983). Intrathecal administration (1, 3, or 10 μg) of DPLCE effectively inhibited gastrointestinal transit at all doses tested in mice (Porreca and Burks, 1983). However, in rats transit was delayed only at the highest dose (100 μg) administered.

[D-Pen[2],L-Pen[5]]- and [D-Pen[2], D-Pen[5]]enkephalin (DPLPE and DPDPE) did not affect small intestinal transit at doses up to 50 μg (DPLPE) and 125 μg (DPDPE) (Galligan *et al.*, 1984). The thermal analgesic effects of these peptides in rats could be antagonized by pretreatment with naloxone (2.0 mg/kg i.p.) when evaluated at peak effect, usually 10 min after peptide administration (Galligan *et al.*, 1984). Naloxone antagonism indicates that the analgesic effects of these conformationally constrained enkephalin analogues are produced by binding to opioid receptors regardless of their subtype or anatomic location. The finding that highly selective δ-receptor agonists induce analgesia, although they are less potent than the μ-selective or nonselective compounds (Galligan *et al.*, 1984), indicating a role of brain δ receptors in mediating analgesia is consistent with previous suggestions to this effect (Fredrickson *et al.*, 1981; Ward and Takemori 1983). According to the experiments using highly selective δ agonists (DPLPE, DPDPE), hot-plate analgesia occurring through receptors in the spinal cord is mediated mainly by δ receptors, whereas inhibition of transit is mediated through both δ and μ receptors (Porreca *et al.*, 1984).

3.2. *In Vitro* Effects

3.2.1. Guinea Pig Ileum and Mouse Vas Deferens Assays

[D-Pen[2], L-Cys[5]]- or [D-Pen[2], D-Cys[5]]enkephalinamide analogues (DPLCEA and DPDCEA) were effective in inhibiting electrically evoked muscle contractions in both guinea pig ileum (GPI) and mouse vas deferens (MVD) (Mosberg *et al.*, 1982). These responses were sensitive to naloxone antagonism (10 and 200 nM).

Because these enkephalinamide analogues were more potent in the MVD assay, where the functional receptors are believed to be of the δ subtype (Lord *et al.*, 1977; Kosterlitz *et al.*, 1980), than in the GPI assay, these results indicated a marked preference of these analogues for δ receptors (Table I).

Similarly, DPLPE and DPDPE were able to inhibit the electrically induced contractions of GPI and MVD preparations (Mosberg *et al.*, 1983b), and these effects were naloxone reversible. Both DPLPE and DPDPE were more potent in the MVD system (2.5 and 2.2. nM, respectively) than in the GPI system (2720 and 6930 nM, respectively), in which the μ receptor is thought to mediate the effect (Lord *et al.*, 1977; Kosterlitz *et al*, 1980). The GPI(IC_{50})/MVD(IC_{50}) was 1088 for DPLPE and 3164 for DPDPE, showing that DPDPE is a highly selective ligand for δ receptors. For comparison, we examined the selectivity of DADLE, a postulated prototypical δ-receptor ligand. We found that this ligand was among the least selective analogues tested, with a GPI(IC_{50})/MVD(IC_{50}) value of 90 (Mosberg *et al.*, 1983b).

These results indicate that the cyclic enkephalin analogues are highly selective for the δ-opioid receptors. Recently, Corbett *et al.* (1984) also reported that these enkephalin analogues (DPLPE and DPDPE) were potent in the MVD assay but were not active in the rat vas deferens assay (mainly but not exclusively μ receptors) or in the rabbit vas deferens system (where κ receptors predominate), thus again illustrating the selectivity of these compounds.

3.2.2. Radioreceptor Binding Assay

Attempts to synthesize a δ-opioid receptor-selective analogue were successful (Table II). Both DPDCEA and DPLCEA were able to inhibit [^{3}H]DADLE binding

TABLE I. Inhibitory Effects (IC_{50}, nM) of Several Enkephalin Analogues on Electrically Evoked Contractions of the Myenteric Plexus–Longitudinal Muscle of Guinea Pig (GPI) and Vasa Deferentia of Mouse (MVD) and Rat (RVD)

Analogues	GPI	MVD	RVD	GPI/MVD	RVD/MVD	References[a]
[Leu]Enkephalin	35.6	1.73		20.6		4
DADLE	24.3	0.27		90.0		1
	8.9	0.73		12.2		4
DALLE	21.1	1.37		15.4		4
DPDCEA	117.0	16.80		7.0		2
DPLCEA	118.0	3.60		32.7		2
DPLCE	213.0	0.32		665.6		2
DPDCE	1250.0	6.27		215.3		2
DPLPE	2720.0	2.50		1088.0		3
	2350.0	2.77	10,000	848.4	3610	4
DPDPE	6930.0	2.19		3164.0		3
	3000.0	4.14	10,000	724.6	2415	4

[a] 1, Mosberg *et al.* (1982); 2, Mosberg *et al.* (1983a); 3, Mosberg *et al.* (1983b); 4, Corbett *et al.* (1984).

TABLE II. Inhibitory Effects (IC_{50}, nM) of Several Enkephalin Analogues on the Binding of Different Labeled Putative δ- and μ-opioid Receptor Ligands in Rat Brain Membrane Preparations

Analogue	[^{3}H]Naloxone	[^{3}H]DADLE	[^{3}H]DPDPE	References[a]
[Leu]Enkephalin	35.00	1.15	1.3	1,4
[Met]Enkephalin	27.00	1.75		1
DADLE	16.00	3.90	1.3	3,4
DPLCEA	73.40	3.35		1
DPDCEA	162.40	7.20		1
DPLCE	177.90	11.90	0.9	2,4
DPDCE	156.80	26.00		2
DPLPE	3710.00	10.00	2.8	3,4
DPDPE	2840.00	16.00	2.3	3,4

[a] 1, Mosberg *et al.* (1982); 2, Mosberg *et al.* (1983a); 3, Mosberg *et al.* (1983c); 4, Akiyama *et al.* (1985).

in rat homogenate (Mosberg *et al.*, 1982) with IC_{50} values of 7.2 and 3.35 nM, respectively, but both analogues were less potent in inhibiting [^{3}H]naloxone binding (162.4 and 73.4 nM, respectively). Their $IC_{50}(\mu)/IC_{50}(\delta)$ values were about 22, somewhat better than those of [Leu^5]enkephalin (Porreca *et al.*, 1982). The corresponding COOH-terminal analogues DPLCE and DPDCE showed somewhat less δ-opioid receptor selectivity with ratios $IC_{50}(\mu)/IC_{50}(\delta)$ of 15 and 6 nM, respectively (Mosberg *et al.*, 1983b) in the radioreceptor assay. Further rigidity imposed by two *gem*-dialkyl substituents produced DPLPE and DPDPE, which resulted in additional improvement in δ-receptor selectivity (Mosberg *et al.*, 1983c). Inhibition experiments in rat brain homogenates showed the DPLPE inhibited [^{3}H]DADLE binding with an IC_{50} value of 10 nM, whereas DPDPE had an IC_{50} value of 16 nM (Mosberg *et al.*, 1983c). The $IC_{50}(\mu)/IC_{50}(\delta)$ value was 371 and 175 for DPLPE and DPDPE, respectively. The δ-opioid selectivities of these *bis*-Pen-containing analogues were substantially higher than those reported for the dimeric enkephalin analogue (H-Tyr-D-Ala-Gly-Phe-NH-$)_2$-(-CH_2-$)_{12}$, for which receptor selectivity was assessed by comparing binding to δ receptors of neuroblastoma–glioma cell membranes with binding to μ sites in rat brain preparations (Shimohigashi *et al.*, 1982).

Corbett *et al.* (1984) also concluded that DPDPE and DPLPE are the most selective δ-opioid agonists. They examined the binding spectrum in guinea pig brain homogenate of six enkephalin analogues with preference for the δ-opioid site. The K_i values of δ-receptor binding of the [D-Pen^2, L-Pen^5]- and [D-Pen^2, D-Pen^5]enkephalin were 2.8 nM and 2.7 nM, respectively. The K_i values at the μ binding site showed a much lower affinity (K_i value of 659 and 713 nM). Thus, [Pen^2, Pen^5]enkephalin analogues are the ligands of choice for binding at the δ receptor site (Mosberg *et al.*, 1983c; Corbett *et al.*, 1984).

Recently [D-Pen^2, D-Pen^5]enkephalin has become available in a tritiated form, and two reports have recently been published characterizing the binding properties in guinea pig and rat brain (Cotton *et al.*, 1985) and in rat brain and the neuro-

TABLE III. Dissociation Constants and Binding Capacities of [^{3}H]DPDPE in Homogenates of Rat and Guinea Pig Brains and in Membranes of NG 108-15

	B_{max} (fmol/mg protein)	K_D (nM)	Reference[a]
Rat brain	2.47[b]	5.42	2
Cerebral cortex	117	3.3	1
Hippocampus	65	4.4	1
Pons–medulla	39	3.3	1
Cerebellum	32	5.2	1
Guinea pig brain	4.19[b]	1.61	2
NG 108-15	235.3	1.24	1

[a] 1, Akiyama *et al.* (1985); 2, Cotton *et al.* (1985).
[b] Picomoles per gram wet tissue.

blastoma–glioma cell line NG 108-15 (Akiyama *et al.* 1985, see Table III). Cotton *et al.* (1985) demonstrated that the maximum binding capacity (B_{max}) of [^{3}H]DPDPE was 4.19 pmol/g wet tissue and the K_D was 1.61 nM in the guinea pig brain. In the rat brain the corresponding values were 2.47 pmol/g wet tissue and 5.42 nM. In both species, the B_{max} and the affinity were not altered when the μ-receptor binding was suppressed with a potent μ agonist [D-Ala^2, MePhe^4, Gly-ol^5]enkephalin. The K_i values for a number of opioids at the δ site were also determined in homogenates of guinea pig brain using [^{3}H]DPDPE or [^{3}H]DADLE as ligands with suppression of μ-receptor sites. The K_i values obtained with the two tritiated ligands were of similar nature. Binding at the δ site was selectively inhibited by unlabeled DPDPE (K_i = 2.56 nM) and DADLE (K_i = 1.35 nM). A similar pattern of selectivity was noticed in homogenates of rat brain.

Akiyama *et al.* (1985) have recently characterized the specific binding properties of [^{3}H]DPDPE in rat brain and in the neuroblastoma–glioma cell line (NG 108-15). The kinetic data showed that association of [^{3}H]DPDPE to specific opioid receptor sites in these two preparations proceeded at a slow rate. At least 8–9 hr (NG 108-15 membranes) and 3 hr (rat brain membranes) were necessary to allow the specific binding to reach steady state at 25°C, probably because of the low receptor and ligand concentrations used in the assay. Saturation isotherms of [^{3}H]DPDPE binding gave an apparent K_D value of 1.24 nM in neuroblastoma–glioma cell membranes, whereas the K_D values in different rat brain regions varied between 3.3. and 5.2 nM. These values were in good agreement with the K_D value obtained from kinetic studies. The B_{max} value in NG 108-15 membranes was 235.3 fmol/mg protein. In the cerebral cortex of the rat brain, the maximal binding capacity was 117.0 fmol/mg protein, whereas the hippocampus and medulla–pons region had B_{max} values of 64.6 and 39.0 fmol/mg protein, respectively. The lowest B_{max} was seen in the cerebellum (32.2 fmole/mg protein). A number of δ-specific enkephalin analogues inhibited [^{3}H]DPDPE binding with high affinity in both receptor systems (0.5–5.0 nM). Putative μ-receptor-selective ligands such as morphine, [D-Ala^2, MePhe^4, Gly-ol^5]enkephalin, (MePhe^3, D-Pro^4)morphiceptin, and naloxone were less effective inhibitors of [^{3}H]DPDPE binding in both systems tested in this study (Akiyama *et al.*, 1985).

3.2.3. Light Microscopic Autoradiographic Localization of [^{3}H]DPDPE Binding

Gulya *et al.* (1985, 1986) have used light microscopic autoradiography to visualize the neuroanatomic distribution of [^{3}H]DPDPE binding sites in tissue-mounted sections of rat brain, providing the first characterization and distribution of this highly selective δ-opioid agonist. The appropriate rinse time required for the highest specific [^{3}H]DPDPE binding (83%) to tissue-mounted sections of rat brain appeared to be 30 min at 0–4°C. At least 1 hr was necessary to allow the specific binding of [^{3}H]DPDPE (5 nM) to reach equilibrium at 25°C. Saturation isotherms of [^{3}H]DPDPE binding to tissue-mounted sections of forebrain gave a maximal number of binding sites of 79.9 fmol/mg protein and an apparent dissociation constant of 6.25 nM. This K_D value was slightly greater than that found in membrane studies of different brain regions of rat (Akiyama *et al.*, 1985). The B_{max} value derived from tissue-mounted experiments was in good agreement with values from homogenate binding studies (Akiyama *et al.*, 1985). Both DPDPE and [Met^5]enkephalin inhibited [^{3}H]DPDPE binding with high affinity (IC_{50} values of 5.8 and 14.1 nM, respectively), whereas putative μ-opioid selective ligands (morphine, [D-Ala^2, MePhe^4, Gly-ol^5]enkephalin and [MePhe^3, D-Pro^4]morphiceptin) were less potent inhibitors of [^{3}H]DPDPE binding.

The rat brain areas containing the highest densities of δ-opioid receptors were the claustrum, basolateral amygdaloid nucleus, caudate–putamen, nucleus accumbens, the external plexiform layer of the olfactory bulb, and the olfactory tubercle (Table IV). Moderate receptor density was characteristic of the hippocampal formation, where grains were seen over the molecular layer of the dentate gyrus and stratum oriens (CA1), and the different layers of cerebral cortex. Generally, a low density of δ-opioid receptors was found over the thalamus and the septal nuclei. Low specific [^{3}H]DPDPE binding was also seen in the cerebellum, medulla oblongata, and in the dorsal horn of the spinal cord. There was very low specific binding over the white matter areas (Gulya *et al.*, 1985, 1986).

4. μ-OPIOID RECEPTOR-SELECTIVE CYCLIC SOMATOSTATIN ANALOGUES

Recently, Pelton *et al.* (1985a,b) have prepared a number of a highly constrained cyclic somatostatin analogues. As can be seen from Table V, somatostatin$_{1-14}$ and the [Cys^2, Cys^7] analogue showed little or no preference for either the μ- or δ-opioid receptors as determined by their ability to inhibit the binding of the radiolabeled opioid ligands. The introduction of penicillamine at the 7 position [Cys^2, Pen^7]S, resulted in an approximately fourfold shift toward the μ-opioid receptor

TABLE IV. Specific [^{3}H]DPDPE Binding in Slide-Mounted Rat Brain Sections

Brain area	Specific binding[a] (fmol/mg tissue, mean ± S.E.M.)
Amygdala	
Basolateral nucleus	18.2 ± 0.7
Caudate–putamen	19.7 ± 1.1
Claustrum	23.9 ± 1.2
Corpus callosum	1.62 ± 0.32
Cortex	
Cingulate	9.93 ± 0.87
Retrospenial	4.65 ± 0.26
Temporal (anterior)	
Lamina I–III	11.3 ± 0.2
Lamina IV	9.15 ± 0.26
Lamina V	8.00 ± 0.53
Lamina VI	11.1 ± 0.55
Globus pallidus	2.30 ± 0.27
Hippocampal formation	
Dentate gyrus, molecular layer	6.05 ± 0.69
Stratum lacunosum moleculare	1.68 ± 0.19
Stratum oriens (CA1)	6.05 ± 0.57
Subiculum, ventral	7.98 ± 0.89
Interpeduncular nucleus	13.9 ± 0.8
Medial geniculate	2.65 ± 0.28
Nucleus accumbens	17.8 ± 0.2
Olfactory bulb	
External plexiform layer	25.1 ± 0.4
Internal granular layer	10.2 ± 0.8
Olfactory tubercle	21.4 ± 0.9
Periaquaductal gray	3.23 ± 0.19

[a] The [^{3}H]DPDPE autoradiographic images were quantitatively analyzed using computer-assisted microdensitometry (Gulya *et al.*, 1986).

binding affinity (Pelton *et al.*, 1985b). When Pen was substituted for Cys at the second position, a decrease of 66-fold potency in μ-opioid receptor affinity was observed.

The substitution of Tyr for Phe at position 3 resulted in an interesting increase of μ-opioid binding affinity. The Tyr-containing analogue [Pen^2, Tyr^3, Cys^7]S was 130-fold more potent at the μ receptor than the corresponding phenylalanyl peptide. The increase in receptor binding activity associated with the tyrosyl analogue may be related to the well-known requirement for a phenolic hydroxyl moiety in the rigid opiate and enkephalin systems. On the other hand, introduction of tyrosine at the third position resulted in a decreased affinity for somatostatin receptors as detected by [^{125}I]CGP 23,996 binding (Pelton *et al.*, 1985b) (data not shown). It was concluded that an introduction of Tyr to the more potent [Cys^2, Pen^7]S analogue would result in a somatostatin analogue with binding activity similar to that of

TABLE V. Inhibitory Effects (IC_{50}, nM) of Various Cyclic Somatostatin Analogues on the Binding of Different Labeled Putative δ- and μ-Opiod Receptor Ligands in Rat Brain Membrane Preparations

Analogue	[^{3}H]Naloxone[a]	[^{3}H]DADLE[a]	[^{3}H]DPDPE	IC_{50} ([^{3}H]DADLE) / IC_{50} ([^{3}H]naloxone)	IC_{50} ([^{3}H]DPDPE) / IC_{50} ([^{3}H]naloxone)
[Cys^2, Cys^7]S	2,600 ± 260	3,100 ± 720	20,000 ± 8,000	1.2	7.7
[Cys^2, Pen^7]S	930 ± 72	3,800 ± 610	31,000 ± 2,600	4.1	33.3
[Pen^2, Cys^7]S	61,000 ± 17,500	38,100 ± 3,000	16,600 ± 8,000	0.62	0.27
[Pen^2, Tyr^3, Cys^7]S	470 ± 10	2,600 ± 410	20,660 ± 3,100	5.5	44.0
[Cys^2, Tyr^3, Pen^7]S	290 ± 58	3,800 ± 610	8,330 ± 1,250	13.1	28.7
[Cys^2, Tyr^3, Pen^7]SA	3.5 ± 0.2	950 ± 210	7,100 ± 600	271.4	2029
[Cys^2, Pen^7]SA	9.9 ± 1.6	1,100 ± 120	24,250 ± 3,100	111.1	2450
[Cys^2, Tyr^3, Orn^5, Pen^7]SA	5.8 ± 1.4	—	12,350 ± 660	—	2129
[Cys^2, Tyr^3, Arg^5, Pen^7]SA	2.8 ± 0.8	—	3,850 ± 300	—	1375
Somatostatin$_{1-14}$	27,400 ± 4,200	16,400 ± 8,500	—	0.6	—

[a] Data are referred from Pelton *et al.* (1985a,b).

morphine. The nature of this newly synthesized somatostatin analogue agreed with the expectations. [Cys^2, Tyr^3, Pen^7]Somatostatin and its amidated form [Cys^2, Tyr^3, Pen^7]SA had IC_{50} values of 290 and 3.5 nM, respectively, compared to the IC_{50} value of 930 nM in the case of the [Cys^2, Pen^7]S analogue. These compounds were not potent in inhibiting [^{3}H]DADLE binding. The IC_{50}(DADLE)/IC_{50}/(naloxone) value was 271 in the case of [Cys^2, Tyr^3, Pen^7]SA and 111 in the case of [Cys^2, Pen^7]SA, indicating a highly improved selectivity of these analogues toward μ-opioid receptors, being 7830 and 2770 times more potent, respectively, than somatostatin.

In radioreceptor studies using 1 and 10 nM [^{3}H]naloxone, a parallel shift to the right in the inhibition curve of the analogue [Cys^2, Tyr^3, Pen^7]SA was obtained, suggesting that the compound was acting at the μ-receptor site in a competititve manner (data not shown).

The most potent and selective somatostatin analogues for μ-opioid receptor sites were also examined on [^{3}H]DPDPE binding in rat brain homogenate, as this radioligand is more selective than [^{3}H]DADLE at the δ-opioid receptor. The [^{3}H]DPDPE binding was inhibited by [Cys^2, Tyr^3, Pen^7]SA, [Cys^2, Pen^7]SA, and [Cys^2, Pen^7]S with IC_{50} values of 7,100, 24,250, and 31,000 nM, respectively (Gulya *et al.*, 1986). [Cys^2, Tyr^3, Pen^7]- and [Cys^2, Pen^7]SA had IC_{50}([^{3}H]DPDPE)/IC_{50}([^{3}H]naloxone) ratios of 2029 and 2450, respectively. The introduction of Orn in the fifth position into the potent [Cys^2, Tyr^3, Pen^7]SA analogue resulted in an additional improvement in inhibition of μ receptor binding (IC_{50} = 5.8 nM), whereas the introduction of Arg into the fifth position of the [Cys^2, Tyr^3, Pen^7]SA molecule created the most potent, although not the most selective, cyclic analogue of somatostatin in inhibiting [^{3}H]naloxone binding ([Cys^2, Tyr^3, Arg^5, Pen^7]SA; IC_{50} = 2.8 nM, IC_{50}[^{3}H]DPDPE/IC_{50}[^{3}H]naloxone = 1375). The results of our studies show that the somatostatin octapeptide analogues can bind to μ-opioid receptors with high affinity and selectivity.

5. CONCLUSIONS

In summary, we have developed a series of very potent and highly selective analogues of enkephalin and somatostatin that exhibit exceptionally high δ- and μ-opioid receptor selectivity, respectively.

A series of conformationally restricted *bis*-penicillamine-containing cyclic analogues of enkephalin were synthesized and evaluated in *in vivo* and *in vitro* bioassays as well as in radioreceptor assays. The highest selectivities observed were for the cyclic *bis*-penicillamine analogues DPDPE and DPLPE, which showed (IC_{50}[^{3}H]naloxone)/IC_{50}([^{3}H]DPDPE) ratios of 1235 and 1325, respectively. Tritiated DPDPE was recently custom synthesized and evaluated for its selectivity toward the δ-opioid receptors in different membrane systems. Saturable high-affinity binding occurred with this radioligand in every system examined, with dissociation constants of about 1–4 nM and B_{max} values of about 150–250 fmol/mg protein.

Tissue-mounted experiments gave similar results to membrane studies. Autoradiographic studies showed a distinct regional distribution of [^{3}H]DPDPE binding sites in the rat brain. A number of δ-selective enkephalin analogues inhibited [^{3}H]DPDPE binding with high affinity (low nanomolar range), whereas μ-opioid receptor-selective ligands were significantly less effective (micromolar concentrations), thus illustrating [^{3}H]DPDPE as the ligand of choice for further δ-opioid receptor research.

A series of conformationally restricted bis-Pen-containing cyclic somatostatin analogues were also synthesized and evaluated for their opioid receptor-binding inhibitory activity in radioreceptor assays. Somatostatin octapeptide analogues such as [Cys^2, Pen^7]SA, [Cys^2, Tyr^3, Pen^7]SA, [Cys^2, Tyr^3,Orn^5, Pen^7]SA, and [Cys^2, Tyr^3, Arg^5, Pen^7]SA were synthesized and were very potent at the μ-opioid receptor (IC_{50} values of 9.9, 3.5., 5.8, and 2.8 nM, respectively). These compounds exhibited a 1375- to 2450-fold selectivity (IC_{50}[^{3}H]DPDPE/IC_{50}[^{3}H]naloxone) toward the μ- versus δ-opioid receptors with minimal inhibition at the somatostatin receptor.

ACKNOWLEDGMENTS. We would like to thank our colleagues K. Akiyama, K. Gee, H. Mosberg, and T. Burks for making this review possible. Supported by NIH grants NS-19972, MH-27257, HL-30956, HL-36042, and MH-30626. H.I.Y. is a recipient of a RSDA from the NIMH (MH-00095). S. P. Duckles is a recipient of an Established Investigator Award from the A.H.A.

REFERENCES

Akiyama, K., Gee, K., Mosberg, H. I., Hruby, V. J., and Yamamura, H. I., 1985, Characterization of [^{3}H][2-D-penicillamine, 5-D-penicillamine]enkephalin binding to delta opiate receptors in the rat brain and neuroblastoma–glioma hybrid cell line (NG 108-15), *Proc. Natl. Acad. Sci. U.S.A.* **82:**2543–2547.

Brownstein, M., Arimura, A., Sato, H., Schally, A. V., and Kizer, J. S., 1975, The effect of hypothalamic deafferentiation on somatostatin-like activity in the rat brain, *Endocrinology* **96:**1456–1461.

Chang, K.-J., and Cuatrecasas, P., 1979, Multiple opiate receptors: Enkephalins and morphine bind to receptors of different specificity, *J. Biol. Chem* **254:**2610–2618.

Corbett, A. D., Gillan, M. G. C., Kosterlitz, H. W., McKnight, A. T., Paterson, S. J., and Robson, L. E., 1984, Selectivities of opioid peptide analogues as agonists and antagonists at the delta receptor, *Br. J. Pharmacol.* **83:**271–279.

Cotton, R., Kosterlitz, H. W., Paterson, S. J., Rance, M. J., and Traynor, J. R., 1985, The use of [^{3}H]-[D-Pen^2,D-Pen^5]enkephalin as a highly selective ligand for the delta binding site, *Br. J. Pharmacol.* **84:**927–932.

Eliel, E. L., Allinger, N. L., Angyl, S. J., and Morrison, G. A., 1965, *Conformational Analysis*, Interscience Publishers, New York.

Fournie-Zaluski, M.-C., Gacel, G., Maigret, B., Premilat, S. and Roques, B. P., 1981, Structural requirements for specific recognition of mu or delta opiate receptors, *Mol. Pharmacol.* **20:**484–491.

Frederickson, R. C. A., Smithwick, E. L., Shuman, R., and Bemis, K. B., 1981, Metkephamid, a systemically active analog of methionine enkephalin with potent delta receptor selectivity, *Science* **211:**603–605.

Galligan, J. J., Mosberg, H. I., Hurst, R., Hruby, V. J., and Burks, T. F., 1984, Cerebral delta opioid receptors mediate analgesia but not the intestinal motility effects of intracerebroventricularly administered opioids, *J. Pharmacol. Exp. Ther.* **229:**641–648.

Gilbert, P. E., and Martin, W. R., 1976, The effects of morphine and nalorphine-like drugs in the nondependent, morphine-dependent and cyclazocine-dependent chronic spinal dog, *J. Pharmacol. Exp. Ther.* **198:**66–82.

Gillan, M. G. C., and Kosterlitz, H. W., 1982, Spectrum of the mu, delta and kappa binding sites in homogenates of rat brain, *Br. J. Pharmacol.* **77:**461–469.

Gillan, M. G. C., Kosterlitz, H. W., and Paterson, S. J., 1980, Comparison of the binding characteristics of tritiated opiates and opioid peptides, *Br. J. Pharmacol.* **70:**481–490.

Gulya, K., Gehlert, D. R., Wamsley, J. K., Mosberg, H. I., Hruby, V. J., Duckles, S. P., and Yamamura, H. I., 1985, Autoradiographic localization of delta opioid receptors in the rat brain using a highly selective bis-penicillamine cyclic enkephalin analog, *Eur. J. Pharmacol.* **111:**285–286.

Gulya, K., Gehlert, D., Wamsley, J. K., Mosberg, H. I., Hruby, V. J., Duckles, S. P., and Yamamura, H. I., 1986, Light microscopic autoradiographic localization of delta opioid receptors in the rat brain using a highly selective bis-penicillamine cyclic enkephaline analog, *J. Pharmacol. Exp. Ther.* (in press).

Gulya, K., Pelton, J. T., Hruby, V.J., and Yamamura, H.I., 1986, Cyclic somatostatin octapeptide analogues wih high affinity and selectivity toward mu opioid receptors, *Life Sci.***38:**2221–2229.

Herz, A., Albus, J., Metys, J., Schubert, P., and Teschemacher, H., 1970, On the central sites for the antinociceptive action of morphine and fentanyl, *Neuropharmacology* **9:**539–551.

Jacquet, Y., and Lajtha, A., 1974, Paradoxical effects after microinjection of morphine in the periaqueductal gray matter in the rat, *Science* **185:**1055–1057.

Kastin, A. J., Coy, D. H., Jacquet, Y., Schally, A. V., and Plotnikoff, N., 1978, CNS effects of somatostatin, *Metabolism* **27:**1347–1352.

Kosterlitz, H. W., Lord, J. A. H., Paterson, S. J., and Waterfield, A. A., 1980, Effects of changes in the structure of enkephalins and of narcotic analgesic drugs on their interactions with mu and delta receptors, *Br. J. Pharmacol.* **68:**333–342.

Ling, G. S. F., and Pasternak, G. W., 1983, Spinal and supraspinal opioid analgesia in the mouse: Role of subpopulations of opioid binding sites, *Brain Res.* **271:**152–156.

Lord, J. A., Waterfield, A. A., Hughes, J., and Kosterlitz, H. W., 1977, Endogenous opioid peptides: Multiple agonists and receptors, *Nature* **267:**495–499.

Marshall, G. R., Bossard, H. E., Vine, W. H., Glickson, J. D., and Needleman, P., 1974, Angiotensin II: Conformation and interaction with the receptor, in: *Recent Advances in Renal Physiology and Pharmacology* (L. G. Wesson and G. M. Fanelli, Jr., eds.), University Park Press, Baltimore, pp. 215–256.

Martin, W. R., Eades, C. G., Thompson, J. A., Huppler, R. E., and Gilbert, P. E., 1976, The effects of morphine and nalorphine-like drugs in the non-dependent and morphine-dependent chronic spinal dog, *J. Pharmacol. Exp. Ther.* **197:**517–523.

Maurer, R., Gaehwiler, B. H., Buescher, H. H., Hill, R. C., and Roemer, D., 1982, Opiate antagonistic properties of an octapeptide somatostatin analog, *Proc. Natl. Acad. Sci. U.S.A.* **79:**4815–4817.

Meraldi, J.-P., Hruby, V. J., and Brewster, A. I. R., 1977, Relative conformational rigidity in oxytocin and (1-penicillamine)-oxytocin: A proposal for the relationship of conformational flexibility to peptide hormone agonism and antagonism, *Proc. Natl. Acad. Sci. U.S.A.* **74:**1373–1377.

Mosberg, H. I., Hurst, R., Hruby, V. J., Galligan, J. J., Burks, T. F., Gee, K., and Yamamura, H. I., 1982, (D-Pen^2, L-Cys^5)enkephalinamide and (D-Pen^2, D-Cys^5)enkephalinamide conformationally constrained cyclic enkephalinamide analogs with delta receptor specificity, *Biochem. Biophys. Res. Commun.* **106**(2)**:**506–512.

Mosberg, H. I., Hurst, R., Hruby, V. J., Galligan, J. J., Burks, T. F., Gee, K., and Yamamura, H. I., 1983a, Conformationally constrained cyclic enkephalin analogs with pronounced delta opioid receptor agonist selectivity, *Life Sci.* **32:**2565–2569.

Mosberg, H. I., Hurst, R., Hruby, V. J., Gee, K., Akiyama, K., Yamamura, H. I., Galligan, J. J., and Burks, T. F. 1983b, Cyclic penicillamine containing enkephalin analogs display profound delta receptor selectivities, *Life Sci.* (Suppl.1)**33:**447–450.

Mosberg, H. I., Hurst, R., Hruby, V. J., Gee, K., Yamamura, H. I., Galligan, J. J., and Burks, T. F. 1983c, Bis-penicillamine enkephalins possess highly improved specificity toward delta opioid receptors, *Proc. Natl. Acad. Sci. U.S.A.* **80:**5871–5874.

Patel, Y. C., and Reichlin, S., 1978, Somatostatin in hypothalamus, extrahypothalamic brain and peripheral tissues of the rat, *Endocrinology* **102:**523–530.

Pelton, J. T., Gulya, K., Hruby, V. J., Duckles, S. P., and Yamamura, H. I., 1985a, Conformationally restricted analogs of somatostatin with high mu opiate receptor specificity, *Proc. Natl. Acad. Sci. U.S.A.* **82:**236–239.

Pelton, J. T., Gulya, K., Hruby, V. J., Duckles, S., and Yamamura, H. I., 1985b, Somatostatin analogs with affinity for opiate receptors in rat brain binding assay, *Peptides* (Suppl.)**6:**159–163.

Pert, A., and Yaksh, T., 1974, Sites of morphine-induced analgesia in the primate brain: Relation to pain pathways, *Brain Res.* **80:**135–140.

Porreca, F., and Burks, T., 1983, The spinal cord as a site of opioid effects on gastrointestinal transit in the mouse, *J. Pharmacol. Exp. Ther.* **227:**22–27.

Porreca, F., Mosberg, H. I., Hurst, R., Hruby, V. J., and Burks, T. F., 1983, A comparison of the analgesic and gastrointestinal transit effects of (D-Pen^2, L-Pen^5)enkephalin after intracerebroventricular and intrathecal administration to mice, *Life Sci.* (Suppl. 1)**33:**457–460.

Porreca, F., Mosberg, H. I., Hurst, R., Hruby, V. J., and Burks, F. T., 1984, Roles of mu, delta and kappa opioid receptors in spinal and supraspinal mediation of gastrointestinal transit effects and hot-plate analgesia in the mouse, *J. Pharmacol. Exp. Ther.* **230:**341–348.

Rezek, M. V., Havlicek, L., Leybin, F. S., LaBella, F. S., and Friesen, H., 1978, Opiate-like naloxone-reversible actions of somatostatin given intracerebrally, *Can. J. Physiol. Pharmacol.* **56:**227–231.

Schiller, P. W., and DiMao, J., 1982, Opiate receptor subclasses differ in their conformational requirements, *Nature* **297:**74–76.

Schiller, P. W., Eggiman, B., DiMao, J., Lemieux, C., and Nguyen, T. M.-D., 1981, Cyclic enkephalin analogs containing a cystine bridge, *Biochem. Biophys. Res. Commun.* **101:**337–343.

Schulz, R., Wuster, M., Kreuss, H., and Herz, A., 1980, Selective development of tolerance without dependence in multiple opiate receptors of mouse vas deferens, *Nature* **285:**242–243.

Shimohigashi, Y., Costa, T., Chen, H.-C., and Rodbard, D., 1982, Dimeric tetrapeptide enkephalins display extraordinary selectivity for the delta opiate receptor, *Nature* **297:**333–335.

Terenius, L., 1976, Somatostatin and ACTH are peptides with partial antagonist-like selectivity for opiate receptors, *Eur. J. Pharmacol.* **38:**211–213.

Tyers, M. B., 1980, A classification of opiate receptors that mediate antinociception in animals, *Br. J. Pharmacol.* **69:**503–512.

Ward, S. J., and Takemori, A. E., 1983, Relative involvement of receptor subtypes in opioid-induced inhibition of gastrointestinal transit in mice, *J. Pharmacol. Exp. Ther.* **224:**359–363.

Wei, E., Sigel, S., Loh, H., and Way, E., 1975, Central sites of naloxone-precipitated shaking in the anesthetized morphine-dependent rat, *J. Pharmacol. Exp. Ther.* **195:**480–486.

Wolozin, B. L., and Pasternak, G. W., 1981, Classification of multiple morphine and enkephalin binding sites in the central nervous system, *Proc. Natl. Acad. Sci. U.S.A.* **78:**6181–6185.

Yaksh, T. L., and Rudy, T. A., 1976, Analgesia mediated by a direct spinal action of narcotics, *Science* **192:**1357–1358.

Yaksh, T. L., and Rudy, T. A., 1977, Studies on the direct spinal action of narcotics in the production of analgesia in the rat, *J. Pharmacol. Exp. Ther.* **202:**411–428.

40

Rat Islet Endocrine Cells Contain Met- and Leu-Enkephalins in High- and Low-Molecular-Weight Forms

KIM TIMMERS, NANCY R. VOYLES, CLIFFORD KING, MICHAEL WELLS, RICHARD FAIRTILE, and LILLIAN RECANT

1. BACKGROUND

The rate of insulin secretion is known to be influenced by a large number of substances, including glucose and other substrates, ions, other hormones, and neurotransmitters. For example, Lerner and Porte (1971) showed that an epinephrine infusion profoundly inhibited the acute, first-phase insulin secretory response to glucose in normal humans. More recently, following the reports by Margules *et al.* (1978) and others (Rossier *et al.*, 1979) of increased β-endorphin in pituitary of genetically obese hyperinsulinemic mice, evidence has begun to accumulate that suggests a role for endogenous opioid peptides in the modulation of insulin secretion and possibly in the pathophysiology of the diabetic state. Types of evidence include pharmacological studies of opiate effects on islet function, a smaller number of reports of the presence in pancreas of endogenous opioids, and a few reports suggesting that islets may secrete endogenous opioid peptides.

1.1. Pharmacological Effects of Exogenous Opiates on Islet Function

Pharmacological studies may be divided into infusion studies in humans and dogs, pancreas perfusions in dogs, and studies with isolated rat islets or islet cell

KIM TIMMERS, NANCY R. VOYLES, CLIFFORD KING, MICHAEL WELLS, RICHARD FAIRTILE, and LILLIAN RECANT • Veterans Administration Medical Center, Washington, D.C. 20422.

cultures. A general problem with most if not all of these studies is the use of agents that are now known to interact with more than one opiate receptor subtype (Chang and Cuatrecasas, 1981; Wolozin and Pasternak, 1981; James *et al.*, 1982), which may result in multiple (and perhaps opposing) effects. We note that a preliminary report of opiate receptors in membranes of pancreas and islets has appeared (Barkey *et al.*, 1981).

1.1.1. Infusion Studies

Ipp *et al.* (1980) showed that morphine infused at a high dose was capable of increasing the plasma insulin and glucagon concentrations in nondiabetic dogs, whereas similar infusions in alloxan diabetic dogs only resulted in increased plasma glucagon concentrations. Two years later, Ipp *et al.* (1982) showed similar effects in the dog using the enkephalin analogue, D-Met^2-Pro^5-enkephalinamide (DMPE), with the addition of a demonstration that the effects were blocked by naloxone. In contrast, Robinson *et al.* (1982) reported a delayed decrease in plasma glucagon in diabetic dogs occurring after 45 min during a morphine infusion, whereas naloxone infusion alone resulted in a similarly delayed increase in glucagon. Werther *et al.* (1985) also found a delayed rise in glucagon during infusions of normal dogs with either naloxone or DMPE. It is difficult to interpret these conflicting results.

In studies with normal human subjects, β-endorphin infusion resulted in an increase in plasma insulin and glucagon (Reid and Yen, 1981; Feldman *et al.*, 1983); these effects were not reversed by naloxone, however (Feldman *et al.*, 1983). β-Endorphin also produced an increase in plasma glucagon in insulin-dependent (Feldman *et al.*, 1983) as well as insulin-independent (type II) diabetics (Reid *et al.*, 1984). In the type II diabetics, a rise in plasma insulin also occurred (Reid *et al.*, 1984). These reports of human studies with β-endorphin are consistent with each other. Other studies showed no change in plasma insulin or glucagon in normal humans infused with naloxone (Morley *et al.*, 1980) or with the enkephalin analogue D-Ala^2-Me-Phe^4-Met^5-(O)-ol enkephalin (DAMME) (Stubbs *et al.*, 1978), whereas administration of DMPE during oral glucose tolerance testing resulted in decreased insulin concentrations in plasma (Jeanrenaud *et al.*, 1981). In type II diabetics, Giugliano *et al.* (1982) reported a naloxone-induced increase (toward normal) in the acute, first-phase insulin release in response to intravenous glucose.

The latter two studies (Jeanrenaud *et al.*, 1981; Giugliano *et al.*, 1982) taken together are suggestive of an oversecretion of an endogenous opioid within islets of type II diabetics with consequent reduction in the rate of insulin secretion. However, infusion studies in intact individuals (rather than studies *in vitro*) cannot provide definitive evidence for events presumed to take place in the islet, since neurogenic effects on islet secretion are well known (Miller, 1981), and opioid peptides could conceivably be acting on some component of the nervous system with consequent neurogenic effects on the islet.

1.1.2. Pancreas Perfusion

The use of isolated perfused pancreas preparations eliminates possible interactions with the central nervous system while leaving intact interactions between islets and ganglia within the pancreas. Ipp *et al.* (1978) first showed that both morphine and β-endorphin were capable of stimulating secretion of insulin and glucagon and inhibiting the secretion of somatostatin in the isolated perfused dog pancreas in response to glucose plus arginine. These effects were partially or completely blocked by simultaneous perfusion with naloxone, indicating apparent specificity for opiate receptors. Hermansen (1983) found similar results using morphine, Met^5- (Met) enkephalin, and Leu^5- (Leu) enkephalin, again showing naloxone reversibility of the effects. Met-Enkephalin was more potent than Leu-enkephalin in stimulating insulin secretion. Thus, studies with the canine isolated perfused pancreas are similar to those obtained in several of the infusion studies in intact normal dogs.

1.1.3. Studies with Isolated Islets

Considerable evidence exists for effects of opiates on isolated rat islets and cultured islet cells. Kanter *et al.* (1980) first reported that Met- and Leu-enkephalin as well as D-Ala^2-Leu^5-enkephalin decreased the release of both glucagon and insulin from cultured neonatal rat pancreas cultures. In the presence of arginine, morphine stimulated release of insulin and glucagon, and enkephalin again inhibited. These disparate effects were blocked by naloxone (Kanter *et al.*, 1980). Green *et al.* (1980), in contrast, reported that Met-enkephalin, D-Ala^2-Met^5-enkephalin, and DAMME each stimulated insulin release from adult rat islets at low concentrations (10^{-10} to 10^{-8} M) while failing to stimulate at higher concentrations (10^{-5} M), all in the presence of partially stimulating concentrations of glucose (8 mM). Pierluissi *et al.* (1981) found similar effects using DAMME and further showed that the stimulation of insulin secretion was better viewed as a potentiation of the effect of submaximal glucose concentrations, since no effect of DAMME was found when tested in the presence of very low or very high glucose. The stimulatory effects could be blocked by naloxone.

More detailed comparative studies by Green *et al.* (1983a,b) have attempted to determine whether cAMP is involved in the opiate effects on insulin secretion. DAMME produced an increase in islet cAMP concentrations, which peaked a few minutes after the peak of insulin release. In addition, Met- and Leu-enkephalins stimulated insulin release at concentrations that failed to stimulate cAMP (Green *et al.*, 1983a). The rise in cAMP produced by DAMME, by Leu-enkephalin, or by α-endorphin was reversed by naloxone, and the increase in insulin release produced by DAMME and α-endorphin but not that produced by Leu-enkephalin was also reversed (Green *et al.*, 1983a). These results indicate that opioid peptides influence insulin secretion at least in part through mechanisms that do not involve increases in cellular cAMP.

Studies with dynorphin A_{1-13} (Green *et al.*, 1983b) showed, as with other opiates tested, a biphasic response with maximal stimulation of insulin release from islets at 5.8×10^{-9} M and no stimulation at 10^{-7} M. Dynorphin caused an increase in uptake of ^{45}Ca ions by islets, and the stimulatory effect on insulin release could be blocked by verapamil, indicating an important role of calcium fluxes in the stimulation of secretion by dynorphin, as for other stimuli. There was no effect of dynorphin on islet secretion of somatostatin, in contrast to secretagogues such as glucose and glyceraldehyde in the same studies (Green *et al.*, 1983b). Dynorphin also increased islet cAMP levels, but a continuous increase with increasing concentrations of opiate was found, in contrast to the biphasic response of insulin secretion to the same doses of dynorphin. These data further support the dissociation of cAMP accumulation from the stimulation of insulin secretion.

1.1.4. Discussion

Pharmacological effects of enkephalins and other opioid peptides on insulin secretion from isolated islets provide good evidence that opiate receptors of one or more types exist within the islet. In fact, the biphasic response curves suggest the presence of more than one receptor type. However, until the present time, the lack of availability of agonists and antagonists of sufficient selectivity for particular opioid receptor subtypes has made it extremely difficult if not impossible to determine which types of receptor are present and/or bioactive in complex systems such as the islet. The very recent development of several highly selective ligands may make studies feasible in the near future.

One must bear in mind, however, that the isolated islet contains at least four different endocrine cell types (one of which is the insulin-containing β cell), which may differ in their surface opiate receptor types. Further, the non-β cells secrete other peptide hormones (e.g., glucagon, somatostatin that themselves can influence the rate of insulin secretion. Thus, the intact, isolated islet may prove to be too complex to allow one to study adequately interactions with opioid peptides.

How do the opiate peptides reach the islets in order to influence insulin secretory rates? One possibility is the peripheral circulation. The concentrations of Met-enkephalin that produce increases in the rate of insulin secretion from rat islets *in vitro* (10^{-10} to 10^{-8} M) are not strikingly higher than those reported (Awoke *et al.*, 1984) for fasting peripheral plasma levels in humans (2×10^{-11} M). Furthermore, since the rate of insulin secretion in the overnight fasting state is low (not stimulated), one might expect enkephalin levels also to be low. Circulating enkephalin levels during periods of stimulated or increased insulin secretion have not yet been measured. Paracrine effects of enkephalins secreted within the islet are perhaps more likely than effects of enkephalins measured in the peripheral circulation, since the plasma half-life of these pentapeptides is extremely short (Hambrook *et al.*, 1976). Therefore, physiological effects on insulin secretion might depend on the local concentration of released enkephalins in the interstitial fluid or the capillaries of the islet rather than on the concentration produced after mixing with peripheral blood, which could represent a substantial dilution.

1.2. Content of Endogenous Opioids in Pancreas: Previous Studies

1.2.1. β-Endorphin

Grube *et al.* (1978) first described immunohistochemical studies that showed that β-endorphinlike immunoreactivity was localized to glucagon-containing cells within the rat islet. Subsequent biochemical studies (Smyth and Zakarian, 1982) showed that β-endorphinlike immunoreactive peptides in rat pancreas are "principally" of molecular weight similar to β-endorphin itself, although some material of higher molecular weight was also found. Human pancreas extracts have been variously reported (using varying methodologies) to contain immunoreactive β-endorphin of low molecular weight (Bruni *et al.*, 1979) and of higher molecular weight, i.e., β-lipotropin (Feurle *et al.*, 1980). This material appeared to be localized in somatostatin-containing cells (Bruni *et al.*, 1979), in contrast to the rat pancreas. [Feurle *et al.* (1980) reported that their material was localized in unidentified endocrine cells of the islet.] In porcine and fetal bovine pancreas, on the other hand, β-endorphinlike immunoreactive peptides consist almost entirely of peptides of higher molecular weight, i.e., 7000 or 20,000, respectively (Houck *et al.*, 1981; Tung and Cockburn, 1984). Matsumura *et al.* (1984) have also reported β-endorphinlike immunoreactivity in extracts of rat pancreas. In guinea pig pancreas, Stern *et al.* (1982) failed to find evidence for β-endorphinlike materials. These varying results with different mammalian species highlight the necessity of studying tissue content, localization, secretion, and bioeffects of particular opioid peptides within one species, if valid interpretations of physiology within that species are to be made.

1.2.2. Met- and Leu-Enkephalins

Feurle *et al.* (1982) have reported the presence of low-molecular-weight Met-enkephalinlike immunoreactivity in acid extracts of human pancreas obtained at autopsy. This immunoreactive material was destroyed by treatment with cyanogen bromide, a result that is consistent with the presence of Met-enkephalin *per se*. Stern *et al.* (1982) have combined HPLC, gel filtration chromatography, and radioimmunoassay to show that guinea pig pancreas contains Met- and Leu-enkephalins in both low- and high-molecular-weight (precursor) forms. [Met^5-Arg^6-Phe^7]Enkephalinlike immunoreactivity has been reported in extracts of pancreas of both rat and guinea pig (Tang *et al.*, 1982), although those authors dismissed it as "negligible" in amount as compared with other tissues.

Studies of localization of enkephalins within the pancreas (Larsson, 1983;

Shimosegawa *et al.*, 1983) appear to have been done solely in the cat and the dog, where enkephalins were confined to nerves and ganglia. However, given the species differences cited above for localization of β-endorphin, these reports should be interpreted narrowly and do not preclude localization of enkephalins in nonneural tissue in other species.

1.3. Are Opioid Peptides Secreted by Islets?

Evidence for secretion of opioid peptides by islets is fragmentary and largely indirect at the present time. It consists, to the best of our knowledge, of data contained in four published papers.

First, Sachse and Laube (1980) showed that Leu-enkephalinlike immunoreactivity in plasma of rats was decreased after oral or intravenous arginine administration, while plasma levels of both insulin and glucagon increased as expected. Second, the same investigators (Sachse *et al.*, 1981) also reported a decrease in release of Leu-enkephalinlike immunoreactivity into the perfusate during perfusions of isolated pancreas with arginine, while again the rates of secretion of insulin and glucagon were increased. The results contained in these two short papers, in which a Leu-enkephalin-directed antibody of uncertain specificity (K. Timmers, unpublished data) was used, have not so far been confirmed by others.

Third, *in vitro* incubations of genetically obese (*ob/ob*) mouse islets with naloxone and glucose resulted in a reduced rate of secretion of insulin compared to incubations with glucose alone (Recant *et al.*, 1980). There was no effect of naloxone on secretion by normal islets. This result implies that opiates are released from *ob/ob* islet cells and interact with islet cell opiate receptors, which can be blocked by naloxone and which are capable of influencing the rate of insulin secretion. This evidence is extremely indirect and is open to other interpretations, especially in view of possible non-opiate-receptor-mediated effects of naloxone (Sawynok *et al.*, 1979) as well as certain opioid peptides (Walker *et al.*, 1982).

Fourth, Matsumura *et al.* (1984) have reported that plasma β-endorphinlike immunoreactivity is fivefold higher in portal vein than in jugular vein, implying a source of secretion in the splanchnic bed, which includes the pancreas. Furthermore, this immunoreactivity was increased 50% in both portal and peripheral blood taken from animals that had ventromedial hypothalamic lesions, which lesions result in obesity, hyperinsulinemia, islet hypertrophy, and increased pancreatic content of β-endorphin (Matsumura *et al.*, 1984).

One interpretation of all these data is that they are consistent with, but do not clearly show, secretion of opioid peptides by islet endocrine cells. More direct *in vitro* measurements of islet secretion of opioid peptides are needed. The present studies of pancreatic and islet content of Met- and Leu-enkephalins in the rat were undertaken as necessary preliminaries to studies of secretion and bioeffects in normal and diabetic rat models.

2. METHODS

2.1. Tissue and Islet Extractions

Male Wistar–Firth (WF/N) or Sprague–Dawley rats weighing 180–300 g were killed by decapitation. The tissues to be extracted were rapidly excised, rinsed in saline, and frozen on dry ice. Within 3 hr, each tissue was placed in hot 2.0 N acetic acid, heated in boiling water for 10 min, homogenized, centrifuged at 10,000 × *g* for 10 min, and the supernatant lyophilized, as described (Recant *et al.*, 1983).

Batches of 100–800 islets, isolated by collagenase digestion (Lacy and Kostianovsky, 1967), were collected in protein-free Hanks bicarbonate buffer and immediately extracted in hot 2.0 N acetic acid. The extract was chilled and centrifuged to sediment the precipitated protein, which was measured by Lowry assay (Lowry *et al.*, 1951), while the supernatant was removed and lyophilized.

2.2. Column Chromatography and High-Performance Liquid Chromatography

Lyophilized tissue extracts were redissolved in 0.1 N acetic acid, applied to columns (1.2 × 60 cm) of Biogel P-60 (Bio-Rad Laboratories), and eluted with 0.1 N acetic acid at 4°C. Fractions were lyophilized before assay.

Selected samples (pooled P-60 column fractions) were analyzed further by reversed-phase high-performance liquid chromatography (HPLC). Aqueous samples were applied to an octadecylsilane column (0.46 × 25 cm, 5-μm particle size, Rainin Instruments), and peptides were eluted with linear gradients of acetonitrile in water (15–60% in 60 min, with 1 g/liter trifluoroacetic acid in both solvents) at a flow rate of 1.0 ml/min. One-minute fractions were collected and lyophilized. Positions of elution of enkephalin standards were determined in parallel.

Enkephalin-containing precursor peptides were detected as immunoreactivity present after trypsin and carboxypeptidase B treatment of the lyophilized fractions, essentially as described by Dandekar and Sabol (1982).

2.3. Immunochemical Techniques

Radioimmunoassays (RIA) were carried out using antibodies and tracers obtained from the following sources: Met-enkephalin, Immunonuclear Corp.; Leu-enkephalin, Peninsula Laboratories; β-endorphin, Immunonuclear Corp. The Met-enkephalin and the Peninsula Leu-enkephalin RIAs had low cross reactivities (2–4%) with Leu- and with Met-enkephalin, respectively. A second Leu-enkephalin RIA, obtained from Immunonuclear Corp., was used only for one of the experiments shown here (Fig. 8) and exhibits extensive cross reactivity (30%) with Met-enkephalin.

Methods used for immunohistochemical staining will be described in detail elsewhere (Timmers *et al.,* 1986) and may be summarized as follows. Pancreas was fixed *in situ* by vascular perfusion with 4% paraformaldehyde, postfixed, and embedded in paraffin. Blocks were sectioned at 6 μm, and sections on slides were deparaffinized and rehydrated. Sections were incubated in primary antisera for 24 hr at 4°C and processed for immunocytochemistry using the peroxidase–antiperoxidase method (Sternberger, 1974). Detection of specifically bound immunoperoxidase was accomplished using a metal-enhanced diaminobenzidine technique (Adams, 1981). The antibodies used were not the same as those used for radioimmunoassay and were obtained from ImmunoNuclear Corp. (Met-enkephalin and Leu-enkephalin), Dr. Richard Miller (Leu-enkephalin), Dr. Stuart MacLean (Leu-enkephalin), Dr. Andrew Baird (BAM-22P), and Dr. Betty Eipper (β-endorphin).

3. RESULTS

3.1. Detection of Enkephalins and Enkephalin-Containing Peptides in Extracts of Rat Pancreas

Acetic acid extracts of whole rat pancreas were studied first using low-pH gel-filtration chromatography. As shown in Fig. 1, peaks of immunoreactive Met- and Leu-enkephalins (scale on right) were detectable in the low-molecular-weight region of the chromatogram. The identity of this immunoreactive material was verified using reversed-phase HPLC. As is illustrated for Leu-enkephalin in Fig. 2, most of the material eluted at the position of Leu-enkephalin (66% of total, as shown) or Met-enkephalin (80%, not shown).

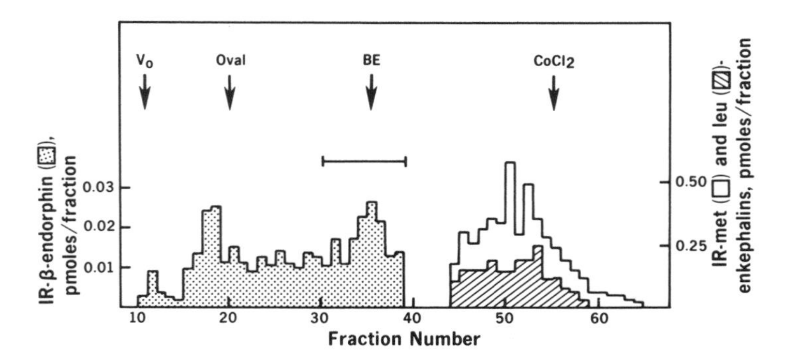

FIGURE 1. Immunoreactive β-endorphin (scale on left) and Met- and Leu-enkephalins (scale on right) in aliquots of fractions from a single representative Biogel P-60 column. Data are expressed as picomoles per column fraction, obtained from chromatography of pooled extracts of two rat pancreata. Arrows indicate elution position of ovalbumin (Oval), β-endorphin (BE), and cobalt chloride ($CoCl_2$). (Reproduced with permission from Timmers *et al.,* 1986.)

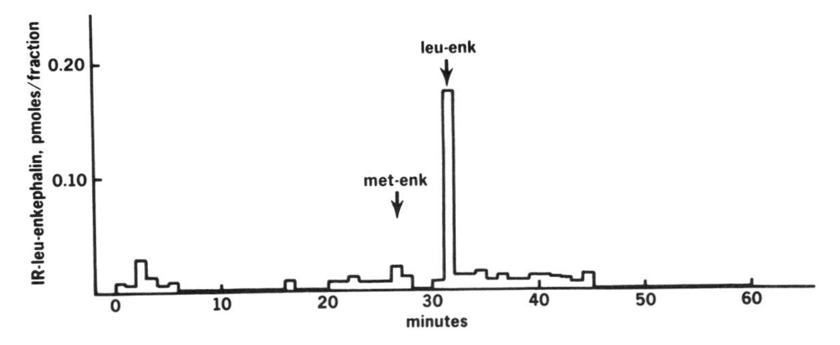

FIGURE 2. Immunoreactive Leu-enkephalin after reversed-phase HPLC of pooled low-molecular-weight fractions (45–68) from Fig. 1.

In addition, immunoreactive β-endorphin was detectable in gel filtration fractions coeluting with an iodinated β-endorphin standard (bracket in Fig. 1) as well as at higher molecular weights, suggesting the presence of β-lipotropin and/or other larger peptides derived from proopiomelanocortin. This result is in general agreement with results of Smyth and Zakarian (1982). However, the total amount of immunoreactive β-endorphin, summed across the column fractions, was only 6% that of free Met-enkephalin in the later fractions of the same columns (2.11 ± 0.18 versus 36 ± 4 fmol/mg protein, respectively; notice the difference in the left and right scales of Fig. 1).

Early gel filtration fractions (11–45) contained little or no detectable enkephalinlike immunoreactivity when assayed directly after lyophilization (not shown). However, treatment of each fraction with trypsin and carboxypeptidase B released considerable Met- and Leu-enkephalin immunoreactivity (Fig. 3), indicating the presence of high-molecular-weight enkephalin-containing peptides. Met-enkephalinlike immunoreactivity appeared as a major peak eluting just after the excluded volume with some additional material eluting later. Leu-enkephalin-containing materials were eluted in a somewhat different pattern (Fig. 3, bottom).

Enzyme-digested material from several gel filtration column fractions was subjected to HPLC analysis in order to verify the identify of these immunoreactive peptides. As is shown in Fig. 4, Met-enkephalinlike immunoreactivity was eluted in two major peaks with retention times corresponding to those of authentic Met-enkephalin ("met") and the air-oxidation product, Met-enkephalin sulfoxide ("oxidized met"). A minor peak (at 23–24 min in the figure) comprising 18% of the total immunoreactivity has not been identified. Leu-enkephalinlike immunoreactivity was eluted in a major peak (50–60% of total) at the appropriate retention time. However, a considerable proportion of the total immunoreactivity did not behave like the authentic peptide, indicating that about half of the Leu-enkephalinlike material detected after enzyme digestion of gel filtration fractions of pancreas may represent interfering peptides or perhaps N-terminally extended Leu-enkephalin-

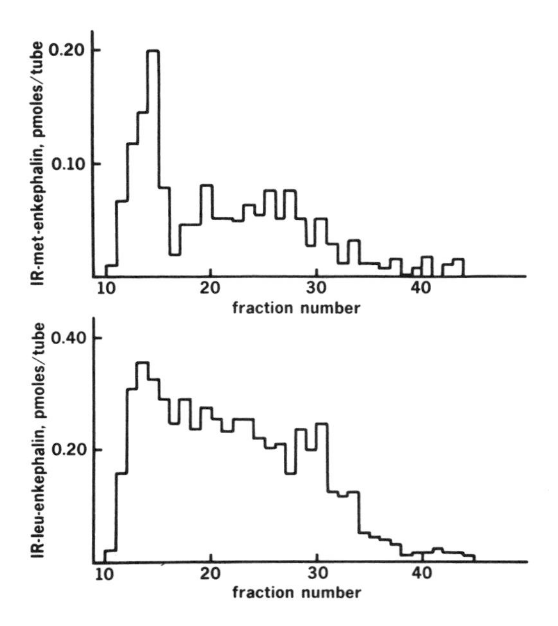

FIGURE 3. Immunoreactive Met-enkephalin- (top) and Leu-enkephalin- (bottom) containing peptides in early gel filtration (Biogel P-60) column fractions of extracts of rat pancreas. Aliquots (one-eighth) of each column fraction were treated with trypsin and carboxypeptidase B before RIA (one of three similar experiments).

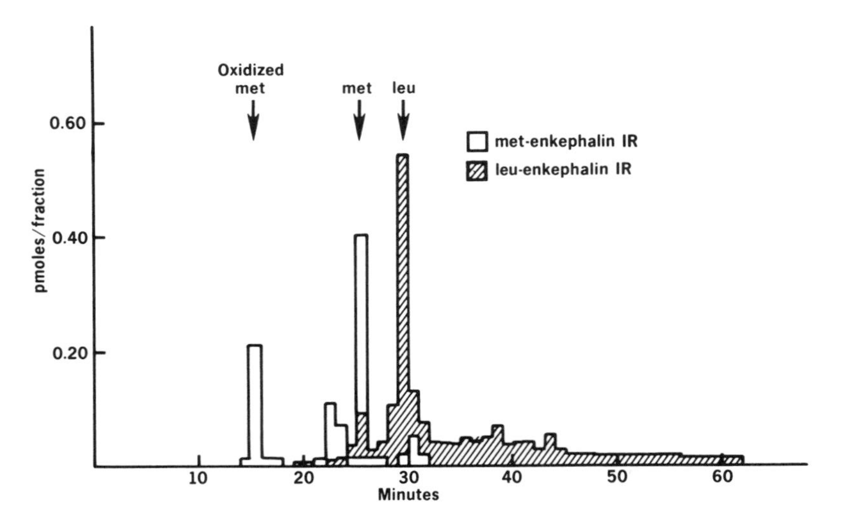

FIGURE 4. Immunoreactive Met- and Leu-enkephalins after reversed-phase HPLC of digested aliquots of fractions 15 and 16 from gel filtration shown in Fig. 3. Elution positions of Met-enkephalin sulfoxide ("oxidized met") as well as synthetic Met- and Leu-enkephalins ("met" and "leu") are shown.

containing peptides, since the antibody used is known not to cross react appreciably with C-terminally elongated Leu-enkephalin-containing peptides, e.g., dynorphins.

3.2. Localization of Enkephalins in Islets of Langerhans

To determine whether enkephalins were localized in endocrine or exocrine tissue within the pancreas, rat pancreas was fixed, embedded in paraffin, and stained by a modified peroxidase–antiperoxidase technique using several different antibodies raised against Leu-enkephalin, Met-enkephalin, and bovine adrenal medulla docosapeptide (BAM 22P), a 22-amino-acid fragment of proenkephalin A (Mizuno *et al.*, 1980). As illustrated in Fig. 5, the Leu-enkephalin-directed immunohistochemical antibodies (all of which cross react strongly with Met-enkephalin) stain the majority of the endocrine cells of the islet, i.e., the same cells that are stained using an antiinsulin antibody in an adjacent section of the same islet (Fig. 5, on the right). Similar results were obtained (Timmers *et al.*, 1986) with the antibody raised against BAM 22P, which does not recognize the Met- or Leu-enkephalin sequences (Baird *et al.*, 1984). A well-characterized antibody raised against β-endorphin ("Melinda"), on the other hand, stained only a small population of cells

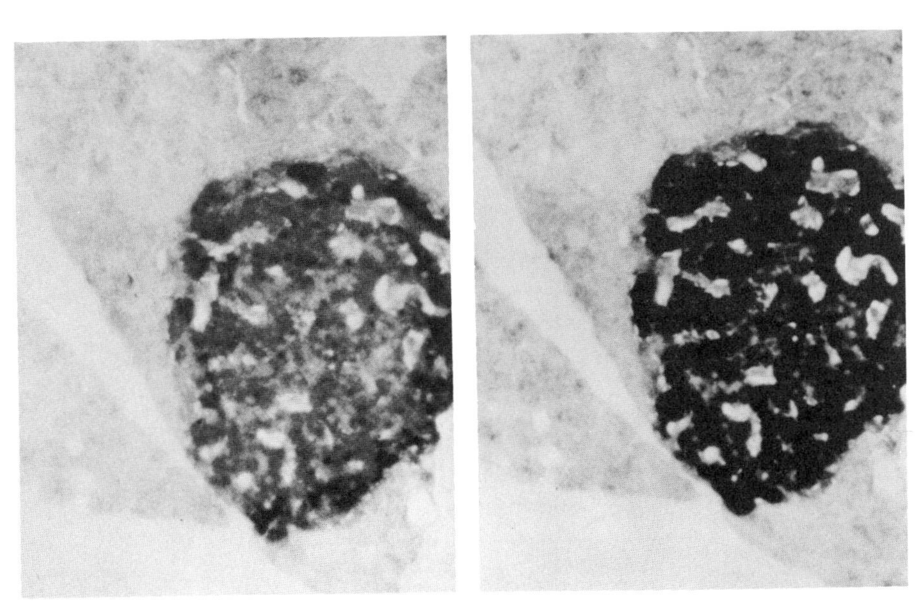

FIGURE 5. Immunohistochemical staining of adjacent sections of fixed rat pancreas for Leu-enkephalin (left) and insulin (right).

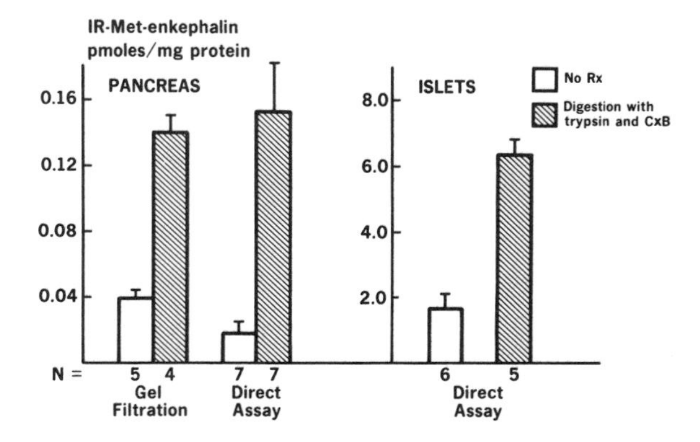

FIGURE 6. Met-enkephalinlike immunoreactivity in extracts of rat pancreas and isolated rat islets. Note that the scale for islets is 50-fold that for whole pancreas.

located on the periphery of the islet (Timmers *et al.*, 1986), as has been previously reported (Grube *et al.*, 1978).

The quantitative contribution of islets to the enkephalin content of the pancreas was determined by extracting isolated islets and comparing islet content to that of the pancreas. Free Met-enkephalin content per milligram protein was about 90-fold enriched in islets (Fig. 6; note the 50-fold difference in the vertical scales), whereas total content of Met-enkephalin-containing peptides (detected after enzyme digestion) was 40-fold enriched in islets. Leu-enkephalin content was 15- to 50-fold enriched in islets (Fig. 7; the difference in scales is tenfold). An enrichment of 50-

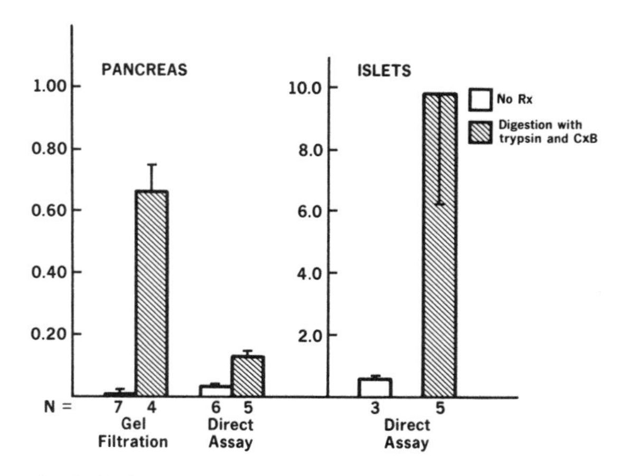

FIGURE 7. Leu-enkephalinlike immunoreactivity in extracts of rat pancreas and isolated rat islets. Note that the scale for islets is tenfold that for whole pancreas.

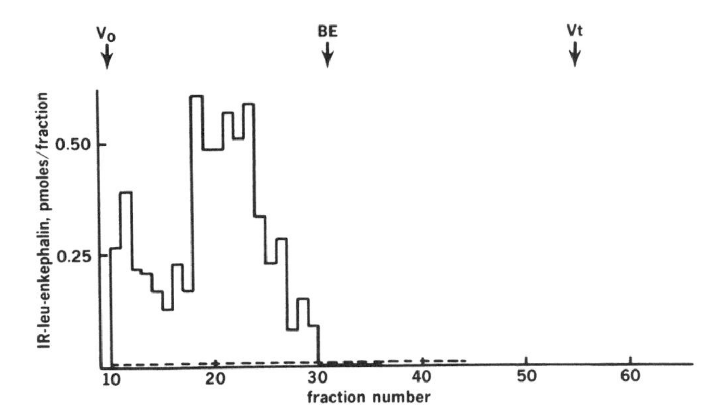

FIGURE 8. Leu-enkephalin-containing peptides in early gel filtration (P-60) column fractions of pooled extracts of 2400 isolated islets. Aliquots of each fraction were assayed before (dotted line) and after (solid line) treatment with trypsin and carboxypeptidase B.

to 100-fold would suggest that all or nearly all of the pancreatic content is contained in islets, since the islets comprise 1–2% of the total volume of the pancreas.

Data presented in Fig. 8 indicate that the increase in Leu-enkephalinlike immunoreactivity in enzyme-digested extracts of isolated islets is caused by production of enkephalinlike sequences from material of high molecular weight, since it is eluted from the gel filtration column in the early fractions.

3.3. Content of Enkephalins in Other Endocrine Tissues of the Rat

The enkephalin content of rat pancreas has sometimes been reported as negligible (Tang *et al.*, 1982), and indeed it is low compared to content in adrenals, pituitary, or hypothalamus when calculated per milligram of protein (Table I, top half). However, the pancreas as a whole is a very large organ compared to the others just mentioned, and 98–99% of its volume is comprised of nonendocrine tissue. The data presented in this report demonstrate that the endocrine cells of the pancreas, i.e., the islets of Langerhans, probably account for most (if not all) the Met-enkephalin-containing peptides in the pancreas. Islet content per milligram protein of Met-enkephalin (1.72 and 5.90 pmol/mg protein for free and precursor forms, respectively, from Fig. 6) is similar to that in the rat posterior and anterior pituitary and only six- to tenfold less than that of the rat hypothalamus (Table I, top half). Furthermore, when total Met- and Leu-enkephalinlike material is expressed per organ or per rat (Table I, bottom half), the total content in pancreas is quite similar to that in pituitary or adrenals and only twofold lower than in hypothalamus of the rat. Thus, in the rat, the endocrine cells of the islet are another potentially rich source of enkephalins.

TABLE I. Comparison of Pancreatic Content of Enkephalin-like Immunoreactivities with Content in Adrenals, Pituitary, and Hypothalamus of the Rat

	Met-Enkephalin		Leu-Enkephalin	
	Not digested[a]	Digested	Not digested	Digested
Expressed as picomoles per milligram protein				
Adrenals[b]	0.022	0.98	0.010	1.02
Pancreas[c]	0.019 ± 0.007	0.153 ± 0.032	0.033 ± 0.006	0.130 ± 0.020
Pancreas[d]	0.020 ± 0.004	No data	0.017 ± 0.002	No data
Posterior pituitary	4.76 ± 0.45	8.43 ± 2.11	7.85 ± 3.89	49.3 ± 26.0
Anterior pituitary	1.50 ± 0.13	4.42 ± 0.54	0.94 ± 0.23	7.73 ± 0.75
Hypothalamus	9.67 ± 0.49	49.8 ± 2.9	4.88 ± 0.80	52.1 ± 3.9
Expressed as total picomoles per rat				
Adrenals[b]	0.22	9.8	0.10	10.2
Pancreas[c]	1.62 ± 0.31	14.8 ± 1.2	3.38 ± 0.37	22.4 ± 5.8
Pancreas[d]	3.27 ± 0.57	No data	2.80 ± 0.25	No data
Posterior pituitary	2.61 ± 0.73	4.26 ± 0.91	3.48 ± 0.95	20.5 ± 4.8
Anterior pituitary	2.64 ± 0.25	7.38 ± 0.59	1.54 ± 0.29	13.0 ± 1.5
Hypothalamus	6.67 ± 1.28	34.2 ± 6.4	3.29 ± 0.72	36.3 ± 8.0

[a] Samples were assayed with and without digestion with trypsin and carboxypeptidase B.
[b] Data shown are for adrenals from four rats pooled (one value).
[c] Data taken from Figs. 6 and 7.
[d] Data for these pancreata and for pituitary and hypothalamus represent the mean values for three spontaneously hypertensive rats (SHR).

4. DISCUSSION

Our results indicate that all or most of the enkephalinlike immunoreactive material detectable in acid extracts of rat pancreas is localized within the endocrine cells of the islets of Langerhans. In addition to low-molecular-weight (pentapeptide) Met- and Leu-enkephalins, abundant high-molecular-weight material, detectable only after release of enkephalin sequences by treatment with trypsin and carboxypeptidase B, is also present in rat islets. The identity of the immunoreactive enkephalins was checked using the criteria of suitable molecular size in gel filtration chromatography and appropriate retention time in HPLC. The relative proportion of Met-enkephalin that is present in free, low-molecular-weight form to that in high-molecular-weight enkephalin-containing peptides in rat pancreas (12%/88%) is similar to that in bovine hypothalamus (32%/68%) (Rossier *et al.*, 1983) as well as in rat hypothalamus (19%/81%) and anterior pituitary (34%/66%) (Table I). Since such high-molecular-weight material is unlikely to be taken up intact from the circulation, these results suggest that enkephalin-containing peptides are biosynthesized within the islet cell. The presence of BAM 22P immunoreactive staining in rat islets is also consistent with biosynthesis of the proenkephalin A gene product within these cells. Nevertheless, direct evidence for biosynthesis, i.e., the presence of the appropriate messenger RNA in isolated islets, is currently being sought.

A general indication of the possible biological importance of these materials in the islets of Langerhans may be gained from comparing islet content with that in other neuroendocrine organs in which opioid peptides are thought to be important modulators of the secretion of other peptide hormones (see Pfeiffer and Herz, 1984). Preliminary data comparing both enkephalin concentration (per milligram protein) in islets as well as total content in pancreas with that in adrenals, hypothalamus, and anterior and posterior pituitary (Section 3.3) indicate that the pancreas is decidedly not an organ whose enkephalin content is negligible. Rather, within the pancreas, the endocrine cells of the islets of Langerhans are rich in enkephalins and, in the aggregate, contain nearly the same total amount of Met-enkephalin-containing peptides as does hypothalamus, adrenals, or pituitary. It is therefore quite possible, though not yet demonstrated, that islets contribute, along with other endocrine organs, to circulating peripheral levels of enkephalins and/or other proenkephalin-A-derived peptides.

We note that several intermediate-sized peptides derived from proenkephalin A such as BAM 12P, BAM 22P, and peptides E and F, which were originally isolated from bovine adrenal medulla, have also been detected in rat brain (e.g., Bloch *et al.*, 1983) and are bioactive (with varying affinities) in antinociceptive tests as well as in the guinea pig ileum (e.g., Mizuno *et al.*, 1978; Kilpatrick *et al.*, 1981). High-molecular-weight forms of Met-enkephalin have also been detected in plasma (Baird *et al.*, 1984). Verification that rat islets contain abundant enkephalin-containing larger peptides makes it necessary to characterize these materials further and to attempt to determine whether any are secreted as well as whether any are bioactive within the islet.

The presence of Met- and Leu-enkephalins in insulin-containing cells (Figs. 5 and 6) suggests the possibility of cosecretion with insulin; this remains to be examined. If endogenous opioid peptides were secreted and had bioeffects in the islet, this would make it more difficult to interpret pharmacological studies (Green *et al.*, 1980, 1983a,b; Pierluissi *et al.*, 1981) using exogenous opiate agonists, since the latter agents may be having effects that add to or even counteract the effects of endogenous ligands binding to the same or different opiate receptors. Nevertheless, the apparent stimulation of insulin secretion by certain concentrations of exogenous opiates of intermediate (i.e., physiological yet partially stimulatory) glucose concentrations suggests that opioid peptides could act as physiological potentiators of the signal provided by a rising plasma glucose level *in vivo*. This possibility, as well as possible defects in signal potentiation in type II (non-insulin-dependent) diabetes mellitus in humans, deserves further study.

ACKNOWLEDGMENT. This work was supported by the Veterans Administration and by a grant from the District of Columbia Area Affiliate of the American Diabetes Association.

REFERENCES

Adams, J. C., 1981, Heavy metal intensification of DAB-based HRP reaction product, *J. Histochem. Cytochem.* **29:**775.

Awoke, S., Voyles, N., Bhathena, S. J., Tanenberg, R., and Recant, L., 1984, Alterations of plasma opioid activity in human diabetics, *Life Sci.* **34:**1999–2006.

Baird, A., Klepper, R., and Ling, N., 1984, *In vivo* and *in vitro* evidence that the C-terminus of preproenkephalin-A circulates as an 8500-dalton molecule, *Proc. Soc. Exp. Biol. Med.* **175:**304–308.

Barkey, R. J., Wurzburger, R. J., and Spector, S., 1981, Selective opiate binding sites in the rat pancreas, *Pharmacologist* **23:**380.

Bloch, B., Baird, A., Ling, N., Benoit, R., and Guillemin, R., 1983, Immunohistochemical evidence that brain enkephalins arise from a precursor similar to adrenal preproenkephalin, *Brain Res.* **263:**251–257.

Bruni, J. F., Watkins, W. B., and Yen, S. S. C., 1979, β-Endorphin in the human pancreas, *J. Clin. Endocrinol. Metab.* **49:**649–651.

Chang, K., and Cuatrecasas, P., 1981, Heterogeneity and properties of opiate receptors, *Fed. Proc.* **40:**2729–2734.

Dandekar, S., and Sabol, S. L., 1982, Cell-free translation and partial characterization of mRNA coding for enkephalin-precursor protein, *Proc. Natl. Acad. Sci. U.S.A.* **79:**1017–1021.

Feldman, M., Kiser, R. S., Unger, R. H., and Li, C. H., 1983, Beta-endorphin and the endocrine pancreas, *N. Engl. J. Med.* **308:**349–353.

Feurle, G. E., Weber, U., and Helmsteadter, V., 1980, β-Lipotropin-like material in human pancreas and pyloric antral mucosa, *Life Sci.* **27:**467–473.

Feurle, G. E., Helmstaedter, W., and Weber, U., 1982, Met- and leu-enkephalin immuno- and bioreactivity in human stomach and pancreas, *Life Sci.* **31:**2961–2969.

Giugliano, D., Ceriello, A., Di Pinto, P., Saccomanno, F., Gentile, S., and Cappiapuoti, F., 1982, Impaired insulin secretion in human diabetes mellitus: The effect of naloxone-induced opiate receptor blockade, *Diabetes* **31:**367–370.

Green, I. C., Perrin, D., Pedley, K. C., Leslie, R. D. G., and Pyke, D. A., 1980, Effect of enkephalins and morphine on insulin secretion from isolated rat islets, *Diabetologia* **19:**158–161.

Green, I. C., Ray, K., and Perrin, D., 1983a, Opioid peptide effects on insulin release and c-AMP in islets of Langerhans, *Horm. Metab. Res.* **15:**124–128.

Green, I. C., Perrin, D., Penman, E., Yaseen, A., Ray, K., and Howell, S. L., 1983b, Effect of dynorphin on insulin and somatostatin secretion, calcium uptake, and c-AMP levels in isolated rat islets of Langerhans, *Diabetes* **32:**685–690.

Grube, D., Voigt, K. H., and Weber, E., 1978, Pancreatic glucagon cells contain endorphin-like immunoreactivity, *Histochemistry* **59:**75–79.

Hambrook, J. M., Morgan, B. A., Rance, M. J., and Smith, C. F., 1976, Mode of deactivation of the enkephalins by rat and human plasma and rat brain homogenates, *Nature* **262:**782–783.

Hermansen, K., 1983, Enkephalins and the secretion of pancreatic somatostatin and insulin in the dog: Studies *in vitro, Endocrinology* **113:**1149–1154.

Houck, J. C., Chang, C. M., and Kimball, C. D., 1981, Pancreatic beta-endorphin-like polypeptides, *Pharmacology* **23:**14–23.

Ipp, E., Dobbs, R., and Unger, R. H., 1978, Morphine and β-endorphin influence the secretion of the endocrine pancreas, *Nature* **276:**190–191.

Ipp, E., Schusdziarra, V., Harris, V., and Unger, R. H., 1980, Morphine-induced hyperglycemia: Role of insulin and glucagon, *Endocrinology* **107:**461–463.

Ipp, E., Dhorajiwala, J., Pugh, W., Moosa, A. R., and Rubenstein, A. H., 1982, Effects of an enkephalin analog on pancreatic endocrine function and glucose homeostasis in normal and diabetic dogs, *Endocrinology* **111:**2110–2116.

James, J. F., Chavkin, C., and Goldstein, A., 1982, Preparation of brain membranes containing a single type of opioid receptor highly selective for dynorphin, *Proc. Natl. Acad. Sci. U.S.A.* **79:**7570–7574.

Jeanrenaud, X., Maeder, E., Del Pozo, E., and Felber, J. P., 1981, Enkephalin-induced glucose–insulin dissociation in man, *Diabetologia* **21**:287.

Kanter, R. A., Esinck, J. W., and Fujimoto, W. Y., 1980, Disparate effects of enkephalin and morphine upon insulin and glucagon secretion by islet cell cultures, *Diabetes* **29**:84–86.

Kilpatrick, D. L., Taniguchi, T., Jones, B. N., Stern, A. S., Shively, J. E., Hullihan, J., Kimura, S., Stein, S., and Udenfriend, S., 1981, A highly potent 3200-dalton adrenal opioid peptide that contains both a [Met]- and [Leu]enkephalin sequence, *Proc. Natl. Acad. Sci. U.S.A.* **78**:3265–3268.

Lacy, P. E., and Kostianovsky, M., 1967, Method for the isolation of intact islets of Langerhans from the rat pancreas, *Diabetes* **16**:35–39.

Larsson, L., 1979, Innervation of the pancreas by substance P, enkephalin, vasoactive intestinal polypeptide and gastrin/CCK immunoreactive nerves, *J. Histochem. Cytochem.* **27**:1283–1284.

Lerner, R. L. and Porte, D., 1971, Epinephrine: Selective inhibiton of the acute insulin response to glucose, *J. Clin. Invest.* **50**:2453–2457.

Lowry, O. H., Rosebrough, N. J., Farr, A. L., and Randall, R. J., 1951, Protein measurement with the Folin phenol reagent, *J. Biol. Chem.* **193**:265–275.

Margules, D. L., Moisset, B., Lewis, M. J., Shibuya, H., and Pert, C. B., 1978, β-Endorphin is associated with overeating in genetically obese mice (*ob/ob*) and rats (*fa/fa*), *Science* **202**:988–991.

Matsumura, M., Yamanoi, A., Sato, K., Tsuda, M., Chikamori, K., Mori, H., and Saito, S., 1984, Alterations in the levels of β-endorphin-like immunoreactivity in plasma and tissues of obese rats with hypothalamic lesions, *Horm. Metab. Res.* **16**:105–106.

Miller, R. E., 1981, Pancreatic neuroendocrinology: Peripheral neural mechanisms in the regulation of the islets of Langerhans, *Endocr. Rev.* **2**:471–494.

Mizuno, K., Minamino, N., Kangawa, K., and Matsuo, H., 1980, A new family of endogenous "big" met-enkephalins from bovine adrenal medulla: Purification and structure of docosa- (BAM-22P) and eicosapeptide (BAM-20P) with very potent opiate activity, *Biochem. Biophys. Res. Commun.* **97**:1283–1290.

Morley, J. E., Baranetsky, N. G., Wingert, T. D., Carlson, H. E., Hershman, J. M., Melmed, S., Levin, S. R., Jamison, K. R., Weitzman, R., Chang, R. J., and Varner, A. A., 1980, Endocrine effects of naloxone induced opiate receptor blockade, *J. Clin. Endocrinol. Metab.* **50**:251–257.

Pfeiffer, A., and Herz, A., 1984, Endocrine actions of opioids, *Horm. Metab. Res.* **16**:386–397.

Pierluissi, R., Pierluissi, J., and Ashcroft, S. J. H., 1981, Effects of an enkephalin analogue (DAMME) on insulin release from cultured rat islets of Langerhans. *Diabetologia* **20**:642–646.

Recant, L., Voyles, N., Luciano, M., and Pert, C. B., 1980, Naltrexone reduces weight gain, alters "β-endorphin," and reduces insulin output from pancreatic islets of genetically obese mice, *Peptides* **1**:309–313.

Recant, L., Voyles, N., Wade, A., Awoke, S., and Bhathena, S., 1983, Studies on the role of opiate peptides in two forms of genetic obesity: *ob/ob* mouse and *fa/fa* rat, *Horm. Metab. Res.* **15**:589–593.

Reid, R. L., and Yen, S. S. C., 1981, β-Endorphin stimulates the secretion of insulin and glucagon in humans, *J. Clin. Endocrinol. Metab.* **52**:592–594.

Reid, R. L., Sandler, J. A., and Yen, S. S., 1984, Beta-endorphin stimulates the secretion of insulin and glucagon in diabetes mellitus, *Metabolism* **33**:197–199.

Robinson, R. P., Williams, P. E., Gooch, B. R., and Abumrad, N. N., 1982, The effect of morphine and naloxone on plasma glucagon and glucose production in the diabetic dog. *Diabetes* (Suppl. 2)**31**:161A.

Rossier, J., Rogers, J., Shibasaki, T., Guillemin, R., and Bloom, F. E., 1979, Opioid peptides and α-melanocyte-stimulating hormone in genetically obese (*ob/ob*) mice during development, *Proc. Natl. Acad. Sci. U.S.A.* **76**:2077–2080.

Rossier, J., Liston, D., Patey, G., Chaminade, M., Foutz, A. S., Cupo, A., Giraud, P., Roisin, M. P., Henry, J. P., Verbanck, P., and Vanderhaeghen, J. J., 1983, The enkephalinergic neuron: Implications of a polyenkephalin precursor, *Cold Spring Harbor Symp. Quant. Biol.* **48**:393–404.

Sachse, G., and Laube, H., 1980, Leucine-enkephalin plasma levels after an oral or intravenous arginine stimulation in rats, *Horm. Metab. Res.* **12**:416–417.

Sachse, G., Svedberg, J., and Laube, H., 1981, Effect of glucose and arginine on L-enkephalin secretion of the isolated perfused rat pancreas, *Horm. Metab. Res.* **13**:360.

Sawynok, J., Pinsky, C., and LaBella, F. S., 1979, Minireview on the specificity of naloxone as an opiate antagonist, *Life Sci.* **25**:1621–1632.

Shimosegawa, T., Kobayashi, S., Fujita, T., Mochizuki, T., Yanaihara, C., and Yanaihara, N., 1983, Nerve elements containing Met-enkephalin-Arg-Gly-Leu-like immunoreactivity in canine pancreas: A histochemical study, *Neurosci. Lett.* **42**:161–165.

Smyth, D. G., and Zakarian, S., 1982, α,N-Acetyl derivatives of β-endorphin in rat pituitary: Chromatographic evidence for processed forms of β-endorphin in pancreas and brain, *Life Sci.* **31**:1887–1890.

Stern, A. S., Wurzburger, R. J., Barkey, B., and Spector, S., 1982, Opioid polypeptides in guinea pig pancreas, *Proc. Natl. Acad. Sci. U.S.A.* **79**:6703–6706.

Sternberger, L. A., 1974, *Immunocytochemistry,* Prentice-Hall, Englewood Cliffs, NJ.

Stubbs, W. A., Delitala, G., Jones, A., Jeffcoate, W. J., Edwards, C. R., Ratter, S. J., Besser, G. M., Bloom, S. R., and Alberti, K. G., 1978, Hormonal and metabolic responses to an enkephalin analog in normal man, *Lancet* **2**:1225–1227.

Tang, J., Yang, H.-Y. T., and Costa, E., 1982, Distribution of Met-enkephalin-Arg^6-Phe^7 in various tissues of rats and guinea pigs, *Neuropharmacology* **21**:595–600.

Timmers, K. I., Voyles, N., King, C., Wells, M., Fairtile, R., and Recant, L., 1986, Opioid peptides in rat islets of Langerhans: Immunoreactive Met- and Leu-enkephalins and BAM 22P, *Diabetes* **35**:52–57.

Tung, A. K., and Cockburn, E., 1984, β-Endorphin-like immunoreactivity in extracts of the fetal bovine pancreas, *Diabetes* **33**:235–238.

Walker, J. M., Moises, H. C., Coy, D. H., Baldright, F., and Akil, H., 1982, Nonopiate effects of dynorphin and des-Tyr-dynorphin, *Science* **218**:1136–1138.

Werther, G. A., Joffe, S., Artal, R., and Sperling, M. A., 1985, Opiate modulation of glucose turnover in dogs, *Metabolism* **34**:136–140.

Wolozin, B. L., and Pasternak, G. W., 1981, Classification of multiple morphine and enkephalin binding sites in the central nervous system, *Proc. Natl. Acad. Sci. U.S.A.* **78**:6181–6185.

41

Modulation of Carboxypeptidase Processing Enzyme Activity

VIVIAN Y. H. HOOK, LEE E. EIDEN, and REBECCA M. PRUSS

1. INTRODUCTION

Regulation of peptide hormone biosynthesis can occur at several cellular levels: transcription of the peptide hormone gene, RNA processing, translation of the mature mRNA, and posttranslational processing of the precursor form of the peptide. Among these steps, the role of the posttranslational processing enzymes in the control of peptide hormone formation has not yet been determined. These enzymes represent an important potential point of regulation, since it is at this step that the biologically inactive peptide hormone precursor is converted to its final smaller biologically active form.

Peptide hormone sequences within their precursors are often characteristically flanked at their NH_2 and COOH termini by pairs of basic amino acid residues (lysine and arginine), which may serve as recognition sites for the processing enzymes. Several different proteases are thought to be involved in prohormone processing. A trypsinlike endopeptidase that cleaves the precursor at the pairs of basic amino acid residues could liberate peptide hormone(s) extended at the carboxyl terminus with lysine or arginine. The COOH-terminal basic residue can be clipped off by a carboxypeptidase-B-like processing enzyme, resulting in the formation of the small active peptide. If the endopeptidase cleaves between the pairs of basic residues, an aminopeptidase may be required to remove the NH_2-terminal basic residue extension on the peptide.

VIVIAN Y. H. HOOK, LEE E. EIDEN, and REBECCA M. PRUSS • Laboratory of Cell Biology, National Institute of Mental Health, Bethesda, Maryland 20205. *Present address of V.Y.H.H.:* Department of Biochemistry, Uniformed Services University of the Health Sciences, Bethesda, Maryland 20814.

Much progress has recently been made in the identification and characterization of the carboxypeptidase-B-like enzyme involved in prohormone processing. Carboxypeptidase processing enzyme activities have been identified in purified secretory granules from bovine adrenal medulla (Hook *et al.*, 1982; Fricker and Snyder, 1982; Fricker *et al.*, 1982; Supattapone *et al.*, 1984; Hook, 1984; Hook and Eiden, 1984), from anterior, intermediate, and neural lobes of rat pituitary (Hook and Loh, 1984), and from rat pancreatic tumor cells (Docherty and Hutton, 1983). These activities are thought to be involved in the processing of proenkephalin, proopiomelanocortin, provasopressin, and proinsulin, respectively, and they display similar properties. Such activity has also been identified in bovine pituitary and brain and has previously been referred to as "enkephalin convertase" (Fricker *et al.*, 1982; Fricker and Snyder, 1983; Supattapone *et al.*, 1984). Characterization of the purified carboxypeptidase ("enkephalin convertase") has revealed that the bovine pituitary, brain, and adrenal medulla forms of the enzyme are identical with respect to kinetics, substrate specificity, inhibitor profile, and molecular weight (Fricker and Snyder, 1983; Supattapone *et al.*, 1984). This suggests that a common carboxypeptidase enzyme may be involved in the processing of many peptide precursors.

The processing carboxypeptidase's acidic pH optimum, thiol dependence, molecular weight, and potency of Zn-metalloprotease inhibitors differentiate it (Hook *et al.*, 1982; Fricker *et al.*, 1982; Supattapone *et al.*, 1984; Hook, 1984; Hook and Eiden, 1984; Hook and Loh, 1984; Docherty and Hutton, 1983; Fricker and Snyder, 1982, 1983) from pancreatic carboxypeptidase B (CPB) and plasma carboxypeptidase N (CPN) (Folk *et al.*, 1960; Erdos *et al.*, 1964). However, the processing carboxypeptidase resembles CPB and CPN in its specificity for basic amino acid residues and stimulation by Co^{2+}. Specific rabbit polyclonal antibodies against the purified bovine pituitary processing carboxypeptidase does not cross react with carboxypeptidases B, N, Y, P, or A, suggesting that this enzyme may be a protein that is structurally distinct from other known carboxypeptidases (Hook *et al.*, 1985,b). Thus, a unique carboxypeptidase appears to be involved in processing a variety of peptide hormone precursors.

2. REGULATION OF CARBOXYPEPTIDASE PROCESSING ENZYME

2.1. Reserpine and Forskolin Treatment of Cultured Chromaffin Cells

Once a peptide hormone precursor-processing enzyme has been properly identified and characterized, it will be important to investigate how its activities may be regulated. Are the processing enzyme activities altered as cellular peptide levels are elevated or lowered, and, if so, what is the mechanism of this regulation?

We have begun to study the problem of measuring carboxypeptidase processing enzyme activity and immunoreactivity during periods of increased [Met]enkephalin

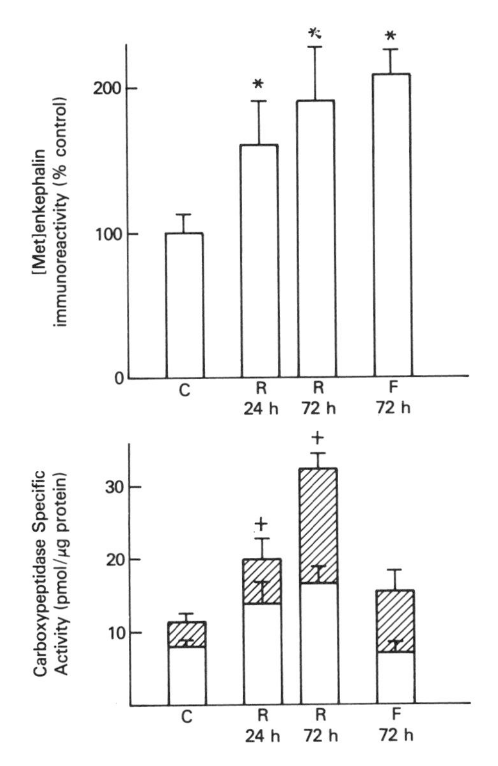

FIGURE 1. Effect of reserpine (R) and forskolin (F) treatment on [Met]enkephalin levels and carboxypeptidase activity in cultured bovine chromaffin cells. Carboxypeptidase activity was measured in the absence (□) or presence (▨) of Co^{2+}. Values were calculated as $\bar{x} \pm$ S.E.M. * $P < 0.01$, + $P < 0.05$, statistically significant compared to control using students two-tailed *t*-test ($n = 3$).

formation in primary cultures of bovine adrenal medulla chromaffin cells (Hook *et al.*, 1985.) Two agents, reserpine and forskolin, were used to induce [Met]enkephalin levels in the chromaffin cells. Reserpine inhibits catecholamine reuptake into the chromaffin granules, resulting in a depletion of granule catecholamine stores (Viveros *et al.*, 1969), and forskolin is a diterpene that elevates cellular cAMP levels through activation of adenylate cyclase (Seamon and Daly, 1981). Both agents increased enkephalin levels by approximately twofold in the chromaffin cells (Fig. 1). In the reserpine-treated cells, carboxypeptidase processing enzyme specfic activity as well as total activity (data not shown) was increased by two- to threefold over control. Forskolin treatment, on the other hand, had no significant effect on the carboxypeptidase activity. These data suggest that the carboxypeptidase enzyme may be selectively regulated during periods of elevated enkephalin biosynthesis.

In contrast to reserpine, which appears to elevate cellular [Met]enkephalin levels at least in part through stimulation of carboxypeptidase processing enzyme activity, forskolin must induce [Met]enkephalin levels through different mechanism(s). Other studies (Eiden *et al.*, 1984) have found that forskolin elicits a three- to fivefold increase in chromaffin cell preproenkephalin messenger RNA ($mRNA^{enk}$), but following exposure to reserpine, $mRNA^{enk}$ levels were reduced. Forskolin treatment also increased levels of high-molecular-weight enkephalin-containing peptides, but reserpine decreased their levels. These findings suggest a scheme in which forskolin increases preproenkephalin mRNA to result in the synthesis of greater levels of enkephalin precursors, whereas reserpine enhances the processing of existing high-molecular-weight precursors. Thus, alterations in levels of $mRNA^{enk}$ and/or changes in the processing enzyme activities may represent two different points of control in the regulation of peptide hormone biosynthesis.

2.2. Mechanism of the Reserpine-Induced Activation of Carboxypeptidase Processing Enzyme

The mechanism for the reserpine-induced increase in carboxypeptidase activity was investigated. Reserpine had no direct effect on carboxypeptidase activity (*in vitro*), and the increase in enzyme activity did not result from changes in endogenous [Met]enkephalin-Arg substrate concentration, since the increased enzyme activity was retained after removal of low-molecular-weight peptides and proteins (M_r less than 10,000) (Hook *et al.*, 1985a). Reserpine dramatically lowered the K_m for Co^{2+}-stimulated carboxypeptidase to one-fourth the value for Co^{2+}-stimulated carboxypeptidase from control cells (Table I). However, reserpine did not alter levels of carboxypeptidase immunoreactivity in the chromaffin cells, indicating that the increase in carboxypeptidase activity did not result from an elevation in the number of enzyme molecules. These data suggest that during reserpine treatment, inactive enzyme molecules may be converted to active molecules, or that less active enzyme molecules have been converted to more active molecules that display a higher affinity for [Met]enkephalin-Arg^6 substrate.

TABLE I. K_m Values for Carboxypeptidase Processing Enzyme in Chromaffin Granules from Reserpine-Treated Cells[a]

	K_m (mM [Met]enkephalin-Arg^6)	
	No Co^{2+}	+ Co^{2+}
Control	0.576 ± 0.112	0.447 ± 0.054
Reserpine	0.727 ± 0.047	0.136 ± 0.007*

[a] The K_m values for [Met]enkephalin-Arg^6 as substrate for the carboxypeptidase processing enzymes, assayed in the absence or presence of 3 mM $CoCl_2$, are expressed as the mean ± S.E.M. *Statistically significant $P < 0.005$ compared to control using student's two-tailed *t*-test.

3. SUBPOPULATIONS OF CARBOXYPEPTIDASE PROCESSING ENZYME

3.1. Activation State of Soluble and Membrane-Bound Enzyme

A hypothesis concerning the conversion of less active to more active carboxypeptidase enzyme molecules following reserpine treatment implies that populations of enzymes at different states of activation are already present in the cell. The ratio of enzyme activity/immunoreactivity can be used as an index of the level of enzyme activity per unit number of enzyme molecules. Such a ratio can distinguish between highly active and inactive populations of carboxypeptidase enzyme molecules. Comparison of this ratio for the carboxypeptidase in the soluble and membrane components of chromaffin granules (Table II) revealed that the soluble form of the enzyme was five to six times more active than the membrane-bound form (Hook, 1985).

Treatment of granule membranes with NaCl results in the removal of 20% of membrane-bound carboxypeptidase activity (Hook, 1984). The enzyme removed from the membrane by NaCl was three times more active than the enzyme remaining in the membrane (Table II). Transfer of the enzyme from the membrane to the soluble fraction could provide a mechanism for the production of greater amounts of active enzyme. It is possible that the reserpine-induced increase in chromaffin

TABLE II. Comparison of Carboxypeptidase Processing Enzyme Activity and Immunoreactivity in Different Components of Chromaffin Granules[a]

	Activity/immunoreactivity (pmol [Met]enkephalin formed per 30 min per ng carboxypeptidae)	
	no Co^{2+}	+ Co^{2+}
Chromaffin granule fraction		
Whole granules (lysed)	1.30 ± 0.03	2.65 ± 0.32
Soluble component	2.86 ± 0.26	5.41 ± 0.38
Membrane component	0.50 ± 0.04	1.06 ± 0.10
Soluble, NaCl	1.18 ± 0.25	3.43 ± 0.92
Membrane, NaCl	0.43 ± 0.11	0.96 ± 0.28
Release media of chromaffin cells in culture		
Control	—	2.99 ± 0.64
Nicotine (10 μM)	—	10.10 ± 1.02

[a] Carboxypeptidase activity and immunoreactivity were measured in each of the above fractions. Chromaffin granule fractions were purified from fresh bovine adrenal medulla tissue. Nicotine-induced release of Co^{2+}-stimulated carboxypeptidase activity (3 mM Co^{2+}) and immunoreactivity from cultured cells were computed as the difference between values from control and nicotine media, because enzyme in the nicotine media represents the sum from control and nicotine alone. Carboxypeptidase activity in the release media was very low and, therefore, was detected only in the presence of Co^{2+}. The activity/immunoreactivity ratio for the enzyme in granules purified from cultured chromaffin cells was the same as that for enzyme in granules purified from fresh adrenal medulla tissue (data not shown). Values are expressed as the mean ± S.E.M. (n = 3).

cell carboxypeptidase may involve a redistribution of active enzyme molecules between the soluble and membrane components of the granules. Future studies should investigate in more detail the relationship between soluble and membrane-bound forms of the carboxypeptidase processing enzyme.

3.2. Activation State of Released Enzyme

After the enkephalin precursor is processed to its smaller peptide products, these biologically active small enkephalin peptides in mature granules can be released to the extracellular environment upon specific receptor stimulation of the cell. Because the enkephalin peptides and the carboxypeptidase processing enzyme are both present in the soluble component of secretory granules, an obvious question to study is: are the small peptides and their processing enzymes coreleased on secretagogue stimulation of the cell? And if so, what are the functional properties of the released processing enzyme? Chromaffin cells possess nicotinic receptors, which stimulate the release of enkephalin peptides and catecholamines (Livett *et al.*, 1981). Nicotine was found to stimulate the corelease of catalytically active carboxypeptidase enzyme and [Met]enkephalin (Hook and Eiden, 1985). This release was blocked by hexamethonium, indicating that the release was specifically mediated by nicotinic receptor stimulation. The carboxypeptidase processing enzyme is also coreleased with other peptide hormones. Studies in AtT-20 mouse pituitary tumor cells have shown that carboxypeptidase processing enzyme activity and ACTH are coreleased by corticotropin-releasing factor and β-adrenergic receptor stimulation (Mains and Eipper, 1984).

When the enzyme activity/immunoreactivity ratio was determined for the released carboxypeptidase, it became apparent that nicotine stimulated the release of a highly active population of carboxypeptidase enzyme molecules (Table II). The released enzyme was approximately two times more active than enzyme in the soluble component of purified chromaffin granules. This observation can be explained on the basis of the heterogeneity of secretory granules with respect to immature and mature granules.

Proenkephalin (Eiden *et al.*, 1984; Fleminger, *et al.*, 1983) and other prohormones (Loh and Gainer, 1983) are processed to their smaller products in a time-dependent manner. Thus, immature secretory vesicles contain the large peptide hormone precursor, whereas the mature vesicles contain primarily the small biologically active forms of the peptide. Purified chromaffin granules, however, contain granules at early and late stages of maturation, since they contain both large enkephalin precursors and small enkephalin peptides. This heterogeneity with respect to immature and mature granules can be distinguished functionally during secretion. In perfused cat adrenal, nicotine stimulates the preferential release of mature enkephalin pentapeptide, whereas potassium depolarization induces the release of both enkephalin pentapeptide and larger enkephalin-containing species (Chaminade *et al.*, 1984). Differential release of fully and partially processed polypeptide from two pools of secretory vesicles is thought to reflect a recruitment of mature and /or immature secretory granules into the exocytotic process.

Therefore, the nicotine-induced release of mature secretory granule contents and release of a highly active population of carboxypeptidase enzyme indicates that the carboxypeptidase is more active in mature than in immature granules. These data suggest a preferential activation of the enzyme in the later stages of granule maturation preceding or during movement into the nicotine-releasable pool. Perhaps this activation could occur via movement of the enzyme from a less active membrane-bound form to a more active soluble form. It will be interesting in future studies to investigate how activation of this processing enzyme may be regulated to result in altered rates of prohormone processing.

4. CONCLUDING REMARKS

The carboxypeptidase processing enzyme is one of several enzymes thought to be required for the complete conversion of proenkephalin and other prohormones to their smaller biologically active peptide products. It will be important to determine how the trypsinlike endopeptidase and perhaps aminopeptidase processing enzymes may be regulated during the biosynthesis of enkephalin and other peptide hormones. These studies will also provide some insight as to how the processing enzymes may be coordinately regulated.

ADDENDUM

The carboxypeptidase peptide hormone processing enzyme has been designated "carboxypeptidase H" ("H" for hormone; E.C. number 3.4.17.10, Enzyme Nomenclature Committee of IUB) to distinguish it from other carboxypeptidases.

REFERENCES

Chaminade, M., Foutz, A. S., and Rossier, J., 1984, Co-release of enkephalins and precursors with catecholamines from the perfused cat adrenal gland *in situ*, *J. Physiol. (Lond.)* **353:**317–319.

Docherty, K., and Hutton, J. C., 1983, Carboxypeptidase activity in the insulin secretory granule, *FEBS Lett.* **162**(1)**:**137–141.

Eiden, L. E., Giraud, P., Affolter, H.-U., Herbert, E., and Hotchkiss, A. J., 1984, Alternate modes of enkephalin biosynthesis regulation by reserpine and cyclic AMP in cultured chromaffin cells, *Proc. Natl. Acad. Sci. U.S.A.* **81:**3949–3953.

Erdos, E. G., Sloane, E. M., and Wohler, I. M., 1964, Carboxypeptidase in blood and other fluids, *Biochem. Pharmacol.* **13:**893–905.

Fleminger, G., Ezra, E., Kilpatrick, D. L., and Udenfriend, S., 1983, Processing of enkephalin-containing peptides in isolated bovine adrenal chromaffin granules, *Proc. Natl. Acad. Sci. U.S.A.***80:**6418–6421.

Folk, J. E., Piez, K. A., Carroll, W. R., and Gladner, J. A., 1960, Carboxypeptidase B, purification and characterization of the porcine enzyme, *J. Biol. Chem.* **235**:2272–2277.

Fricker, S., and Snyder, S. H., 1982, Enkephalin convertase: Purification and characterization of a specific enkephalin-synthesizing carboxypeptidase localized to adrenal chromaffin granules, *Proc. Natl. Acad. Sci. U.S.A.* **79**:3886–3890.

Fricker, L. D., and Snyder, S. H., 1983, Purification and characterization of enkephalin convertase, an enkephalin-synthesizing carboxypeptidase, *J. Biol. Chem.* **258**(18):10950–10955.

Fricker, S., Supattapone, S., and Snyder, S. H., 1982, Enkephalin convertase: A specific enkephalin synthesizing carboxypeptidase in adrenal chromaffin granules, brain, and pituitary gland, *Life Sci.* **31**:1841–1844.

Hook, V. Y. H., 1984, Carboxypeptidase B-like activity for the processing of enkephalin precursors in the membrane component of bovine adrenomedullary chromaffin granules, *Neuropeptides* **4**:117–126.

Hook, V. Y. H., 1985, Differential distribution of carboxypeptidase processing enzyme activity and immunoreactivity in membrane and soluble components of chromaffin granules, *J. Neurochem.* **45**:987–989.

Hook, V. Y. H., and Eiden, L. E., 1984, Two peptidases that convert ^{125}I-Lys-Arg-(Met)enkephalin and ^{125}I-(Met)enkephalin-Arg^{6}, respectively, to ^{125}I-(Met)enkephalin in bovine adrenal medullary chromaffin granules, *FEBS Lett.* **172**:212–218.

Hook, V. Y. H., and Eiden, L. E., 1985, (Met)Enkephalin and carboxypeptidase processing enzyme are co-released from chromaffin cells by cholinergic stimulation, *Biochem. Biophys. Res. Commun.* **128**(2):563–570.

Hook, V. Y. H., and Loh, P. Y., 1984, Carboxypeptidase B-like converting enzyme activity in secretory granules of rat pituitary, *Proc. Natl. Acad. Sci. U.S.A.* **81**:2776–2780.

Hook, V. Y. H., Eiden, L. E., and Brownstein, M. J., 1982, A carboxypeptidase processing enzyme for enkephalin precursors, *Nature* **295**:341–342.

Hook, V. Y. H., Eiden, L. E., and Pruss, R. M., 1985a, Selective regulation of carboxypeptidase peptide hormone processing enzyme during enkephalin biosynthesis in cultured bovine adrenomedullary chromaffin cells, *J. Biol. Chem.* **260**:5991–5997.

Hook, V. Y. H., Mezey, E., Fricker, L. D., Pruss, R. M., Siegel, R., and Brownstein, M. J., 1985b, Immunochemical characterization of carboxypeptidase B-like peptide hormone processing enzyme, *Proc. Natl. Acad. Sci. U.S.A.* **82**:4745–4749.

Livett, B. G., Dean, D. M., Whelan, L. G., Udenfriend, S., and Rossier, J., 1981, Co-release of enkephalin and catecholamines from cultured chromaffin cells, *Nature* **289**:317–319.

Loh, Y. P., and Gainer, H., 1983, Biosynthesis and processing of neuropeptides, in: *Brain Peptides* (D. T. Krieger, M. J. Brownstein, and J. B. Martin, eds.), John Wiley & Sons, New York, pp. 79–116.

Mains, R. E., and Eipper, B. A., 1984, Secretion and regulation of two biosynthetic enzyme activities, peptidyl-glycine α-amidating monooxygenase and a carboxypeptidase, by mouse pituitary corticotropic tumor cells, *Endocrinology* **115**(6):1683–1690.

Seamon, K. B., and Daly, J. W., 1981, Forskolin: A unique diterpene activator of cyclic AMP-generating systems, *J. Cyclic Nucleotide Res.***7**:201–224.

Supattapone, S., Fricker, L. D., and Snyder, S. H., 1984, Purification and characterization of a membrane-bound enkephalin-forming carboxypeptidase, "enkephalin convertase," *J. Neurochem.* **42**:1017–1023.

Viveros, O. H., Arqueros, L., Connett, R. J., and Kirshner, N., 1969, The fate of the storage vesicles following insulin and reserpine administration, *Mol. Pharmacol.* **5**:69–82.

Neuropeptides and Adrenal Steroidogenesis

GEORGE V. VAHOUNY, CAROL HERRON, ROBERT TOMBES, and TERRY W. MOODY

1. INTRODUCTION

There has been a great deal of attention directed toward the source, regulation, and function of the peptides derived from the 31-kD proadrenocorticotropin–endorphin precursor in the pituitary (Barker, 1977; Krieger, 1980; Numa and Nakanishi, 1980). The parent molecule is metabolized to an amino-terminal segment, referred to as the 16K fragment, the 39-amino-acid hormone ACTH, and a 91-amino-acid peptide, β-lipotropin (β-LPH). In turn, ACTH and β-LPH may be metabolized to the smaller peptides, including α-melanotropin (α-MSH) from ACTH and β-MSH and β-endorphin from β-lipotropin.

Various studies have suggested that one or more of these peptides may play a role either in independent or synergistic stimulation of adrenal steroidogenesis by the fascicula or glomerulosa zones. These studies have included the 16K fragment (Pederson and Brownie, 1980), γ_3-MSH (Pederson *et al.*, 1980), β-endorphin (Shanker and Sharma, 1979), α-MSH (Seelig and Sayers, 1971; Lowry and McMartin, 1972), α-MSH (Lowry and McMartin, 1972; Vinson *et al.*, 1981), and β-LPH (Vinson *et al.*, 1981; Matsuoka *et al.*, 1980, 1981). The evidence for the independent or synergistic effects on the fascicula or glomerulosa zones of the adrenal cortex has, however, been inconsistent and has generally required concentrations of peptides several orders of magnitude greater than that of ACTH, which produces a maximal biological response.

The purpose of the present study was to compare the independent and syn-

GEORGE V. VAHOUNY, CAROL HERRON, ROBERT TOMBES, and TERRY W. MOODY • Department of Biochemistry, The George Washington University School of Medicine and Health Sciences, Washington, D.C. 20037.

ergistic effects of a number of pituitary peptides on glucocorticoid release by a suspension of rat adrenal fasciculata cells. This includes ACTH, β-LPH, β-endorphin, α-MSH, β-MSH, γ-MSH, and another neuropeptide, bombesin, which is thought to alter adrenal function (Brown *et al.*, 1979). The ability of these peptides to stimulate cAMP formation by an adrenal membrane preparation was also tested.

2. MATERIALS AND METHODS

Adrenocortical (fasciculata reticularis) cells were prepared and characterized as described previously (Vahouny *et al.*, 1978). These were resuspended in Krebs bicarbonate buffer, pH 7.4, containing 11 mM glucose, 0.08 mM albumin, and 7.65 mM Ca^{2+}. Aliquots of the cell suspension (1.9 ml containing 2–3 × 10^5 cells/ml) were incubated for 1 or 2 hr at 37°C following addition of 0.1 ml buffer alone or buffer containing the peptides indicated. Corticosterone was assayed fluorimetrically on methylene chloride extracts of the incubation media (Silber *et al.*, 1958).

Rat adrenal membranes were prepared by a modification of the procedures of Londos and Rodbell (1975) and Dazord *et al.* (1974). Incubations contained the following components in 500 μl: 0.2 mM [α-^{32}P]ATP (5 × 10^6 cpm), 5 mM Mg^{2+}, 1 mM cAMP, 10 μM GTP, 1 mM dithiothreitol, 0.1 mg bovine serum albumin, 2.5 units creatine phosphokinase, 5 mM phosphocreatine, 1 mM 3-isobutyl-1-methylxanthine, 10 μg membrane protein, 125 mM tris-HCl buffer, pH 7.6, and peptides in the concentrations indicated. Reaction was carried out for 15 min at 30°C and was terminated by addition of 100 μl of 2% sodium dodecylsulfate containing 45 mM ATP and 1.3 mM cAMP. After addition of [^{3}H]cAMP as recovery standard, samples were boiled for 3 min, cooled, and diluted with 1 ml distilled water. Labeled cAMP was separated from other labeled nucleotides by the double-column technique of Salomon *et al.* (1974) and was analyzed by liquid scintillation spectrometry. The pituitary peptides were kindly provided by Dr. T. O'Donohue, National Institute of Mental Health.

3. RESULTS

$ACTH_{1-39}$ and $ACTH_{1-24}$ gave identical results at equivalent concentrations with either the isolated cell preparation or the adrenal membrane preparation. The various peptide test preparations were included at levels of 1–1000 ng per incubation and represented concentrations either equivalent to that of the ACTH that elicited maximal responses (1 nM) or 50-fold that of ACTH. As shown in Fig. 1, there were no significant steroidogeneic responses to α-, β-, or γ-MSH or to bombesin, β-endorphin, or the lower concentrations of β-LPH. As has been demonstrated earlier (Lowry and McMartin, 1972), there was a marked steroidogenic response to the higher concentration of β-LPH (5 × 10^{-8} M).

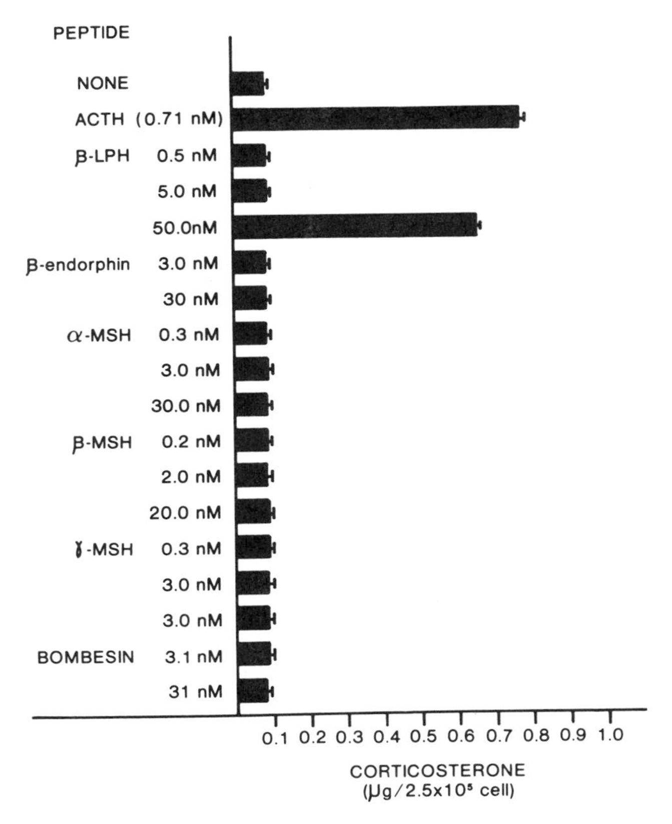

FIGURE 1. Effect of individual pituitary peptides on steroidogenesis by rat adrenal cells in suspension.

The ability of neuropeptides to synergize the effects of ACTH was tested using both optimal (0.71 nM) and suboptimal (0.28 nM) concentrations of ACTH. As shown in Table I, there was no synergism by the lower concentrations of β-endorphin, γ-MSH, or bombesin. At the highest concentration of β-LPH tested, the response to 0.28 nM ACTH was increased by only 14%; α-MSH inhibited the steroidogenic response to ACTH, whereas results with β-MSH were highly variable. In three of the seven studies, β-MSH exhibited a slight synergistic response (*ca.* 50%) in the presence of either concentration of ACTH, but the effect was not reproducible in four other studies. Thus, the composite data from seven studies demonstrate no significant response to β-MSH in the presence of ACTH.

The adrenal membrane preparation used in the present studies has been extensively characterized according to previous reports on similar preparations (Londos and Rodbell, 1975; Dazord *et al.*, 1974, 1975; Salomon *et al.*, 1974; Glynn *et al.*, 1977; Schlegal and Schwyzer, 1977). It was established that maximal re-

TABLE I. Effect of Pituitary Peptides on ACTH-Induced Steroidogenesis by Rat Adrenal Fasciculata Cells[a]

Peptide and concentration added to ACTH		Percentage maximal stimulation
None (ACTH 0.71 nM)		100 ± 1.7
β-LPH	5.0 nM	101 ± 1.4
	50.0 nM	114 ± 1.1
β-Endorphin	30.0 nM	102 ± 16
α-MSH	3.0 nM	58 ± 2.2
β-MSH	2.2 nM	108 ± 23
γ-MSH	3.2 nM	94 ± 3.2
Bombesin	31.0 nM	94 ± 1.4

[a] ACTH was included at 0.71 nM or 0.28 nM in individual studies. Incubations were carried out for 1 or 2 hr at 37°C prior to analysis for corticosterone. Figures represent normalized values relative to ACTH-stimulated corticogenesis ± S.E.M. for two to six studies.

sponse of cAMP production occurred at 2.5×10^{-6} M ACTH, as reported by others (Londos and Rodbell, 1975; Dazord *et al.*, 1975; Glynn *et al.*, 1977). As shown in Table II, cAMP production was increased by over threefold above basal levels by addition of 1 mM NaF, a direct activator of adenylate cyclase. This was essentially mimicked by 2.5×10^{-7} M $ACTH_{1-24}$ and was further increased twofold at 2.5×10^{-6} M ACTH. Because of lack of availability, only four of the neuropeptides were tested with this membrane system. When added at levels equivalent to the concentration of ACTH eliciting the half-maximal cAMP responses (e.g., $0.9–3 \times 10^{-7}$ M), none of the peptides tested significantly altered base-line levels of cAMP.

TABLE II. Effect of Selected Pituitary Peptides on cAMP Production by Rat Adrenal Membranes[a]

Peptide	Concentration	cAMP formation (pmol/min per mg)
None	—	4.0 ± 0.1
NaF	1.0 μM	13.4 ± 0.4
ACTH	0.25 μM	15.2 ± 1.3
ACTH	2.5 μM	29.2 ± 0.8
α-MSH	0.12 μM	4.4 ± 0.2
β-MSH	0.03 μM	3.7 ± 0.2
γ-MSH	1.2 μM	4.5 ± 0.3
β-Endorphin	3.0 μM	3.5 ± 0.1

[a] Contents of the incubation mixture are described in Section 2. Incubations were for 15 min at 30°C, and the production of [^{32}P]cAMP from [α-^{32}P]ATP was assayed according to Salomon *et al.* (1974).

4. DISCUSSION

The finding that the component peptides of the ACTH–β-LPH precursor are encoded by a single gene and the possibility that their biological responses may be coordinated (Numa and Nakanishi, 1981) have stimulated studies on comparative effects of these peptides on adrenocortical function. Previous evidence for independent or synergistic effects of these peptides has not been entirely consistent and, when observed, appears to require supraphysiological concentrations to elicit responses by one or more zones of the adrenal cortex.

Gasson (1979) originally reported that the intact pro-ACTH/endorphin precursor was 0.33% as effective as ACTH on steroidogenesis by rat adrenocortical cells. It was found that the 16K fragment and β-LPH were 10,000- and 250-fold less potent, respectively, than ACTH. Pederson and Brownie (1980) subsequently reported that the intact 16K fragment or its tryptic products stimulated corticosterone production by rat adrenocortical cells to a maximum of 4.2% of that elicited by ACTH at an equivalent concentration (1 nM). Both the 16K fragment and its trypic products, however, demonstrated synergism with a lower concentration of ACTH (0.01 nM) but resulted in inhibition at the higher concentration of the tropic hormone (1 nM).

Of the possible 16K products, neither the γ-MSH precursor (Vinson *et al.*, 1981; Al-Dujaili *et al.*, 1981), synthetic γ_3-MSH (Pederson *et al.*, 1980), nor γ_1-MSH (Vinson *et al.*, 1981) has been shown to exhibit an independent effect on either zone of the adrenal cortex. However, the γ-MSH precursor was reported to synergize the response of glomerulosa cells to ACTH if the cells were initially "primed" with the peptide (Al-Dujaili *et al.*, 1981). Synthetic γ_3-MSH was also reported to enhance ACTH-induced levels of plasma aldosterone and corticosterone *in vivo* (Pederson and Brownie, 1980). In the present study, γ-MSH had no effect on membrane adenylate cyclase activity, on independent stimulation of steroidogenesis by rat fascicula cells, or with respect to synergism with ACTH.

A major proopiocortin peptide, β-lipotropin, has been reported to have little independent effect on either fascicula or glomerulosa cells at concentrations below 10 nM (Vinson *et al.*, 1981). As reported earlier and confirmed in the present study, concentrations exceeding 50 nM (representing a 100- to 1000-fold excess with respect to ACTH) can elicit corticosterone production by fasciculata cells (Seeling and Sayers, 1971; Lowry and McMartin, 1972; Vinson *et al.*, 1981) and aldosterone output by the glomerulosa zone (Vinson *et al.*, 1981). However, Vinson *et al.* (1981) have presented evidence to suggest that this effect may be a result of contamination of the β-LPH preparation with ACTH at levels of 0.1% or less.

Of the β-LPH fragments, β-MSH had been reported to stimulate aldosterone production by adrenal capsular cells in a dose-dependent manner (Matsuoka *et al.*, 1981). However, the concentration of β-MSH required to elicit a response was twice that of ACTH, and the maximum stimulation was only 60% of that with ACTH. Others (Pederson and Brownie, 1980; Vinson *et al.*, 1981) have not found an effect of this peptide on cells of either zone of the adrenal cortex. In the present study, β-MSH had no effect on adrenal fasculata cells or on membrane adenylate cyclase. It was also the only peptide to exhibit an inhibitory effect on ACTH-induced steroidogenesis.

Shanker and Sharma (1979) had originally reported an effect of β-endorphin on corticosterone production by rat adrenal fascicula cells. However, the maximum stimulation was only 17% that with ACTH, and the concentration required was 1000-fold greater than that of ACTH. Subsequent studies (Vinson *et al.*, 1981; Matsuoka *et al.*, 1980), including those in the present report, suggest that at levels of 0.1 nM to 30 nM, β-endorphin does not exhibit an independent steroidogenic effect, nor does it appear to synergize the response to ACTH. In addition, it is not possible to demonstrate a direct effect of this peptide on membrane adenylate cyclase.

Finally, of the ACTH fragments, neither CLIP at 100 nM (Vinson *et al.*, 1981) nor α-MSH at 0.1–1000 nM (Pederson and Brownie, 1980; Pederson *et al.*, 1980; Shanker and Sharma, 1979; Seelig and Sayers, 1971; Lowry and McMartin, 1972; Vinson *et al.*, 1981) was shown to elicit an independent response by the adrenal fasculata cells. It also appears from the present studies that α-MSH had no effect on membrane adenylate cyclase. With respect to synergism, it had been reported (Pederson and Brownie, 1980) that α-MSH (1 nM) resulted in a 15% depression of ACTH (1 nM)-induced steroidogenesis. In the present study, it was found that 3 nM α-MSH reduced the maximal response to ACTH by almost one-half.

From the present study and those reported earlier, the peptide derivatives of proopiocortin appear to have little independent influence on steroidogenesis by the fascicula–reticularis layers of the rat adrenal. In those cases in which an independent effect has been purported (e.g., 16K fragment and β-LPH), the effects are either small compared to those of ACTH or require high concentrations (two to three orders of magnitude) compared to ACTH. In contast, the finding that β-LPH (Vinson *et al.*, 1981), γ-MSH (Al-Dujaili *et al.*, 1981), and β-MSH (Matsuoka *et al.*, 1981) can specifically enhance aldosterone production, albeit at relatively high concentrations, might have relevance, particularly during sodium restriction (e.g., Vinson *et al.*, 1981).

5. SUMMARY

Several neuropeptides, some of which are derived from proopiomelanocortin, have been tested for their ability to stimulate independently steroidogenesis by rat adrenal fasciculata cells and to activate adenylate cyclase in plasma membrane preparations. With intact adrenal cells, β-lipotropin (β-LPH) could independently stimulate corticosterone production but required 80-fold the concentration of ACTH (0.71 nM) to produce an equivalent response. In the presence of ACTH (0.28 nM), β-LPH (50 nM) exhibited a modest (16%) synergistic effect. Of the smaller proopiomelanocortin peptides, neither β-endorphin nor α-, β-, or γ-MSH nor the neuropeptide bombesin had any independent or synergistic effects on adrenal corticosterone production. The adrenal membrane adenylate cyclase was fully activated by 1 μM NaF and by 2.5 μM ACTH. Half-maximal activation occurred with 0.25 μM ACTH. With this preparation, neither α-, β-, or γ-MSH nor β-endorphin, at concentrations from 0.09 μM to 3.0 μM, had any effect on cAMP production. The present studies generally support earlier findings that proopiomelanocortin peptides other than ACTH may not exert direct effects on adrenal glucocorticoid production at biologically relevant concentrations.

ACKNOWLEDGMENTS. This work was supported by grants from USPHS: AM-32309 (G.V.V.) and MH-36498 (T.W.M.).

REFERENCES

Al-Dujaili, E. A. S., Hope, J., Estivariz, F. E., Lowry, P. J., and Edwards, C. R. W., 1981, Circulating human pituitary pro-γ-melanotropin enhances the adrenal response to ACTH, *Nature* **291:**156–159.

Barker, J. L., 1977, Physiological roles of peptides in the nervous system, in: *Peptides in Neurobiology* (H. Gainer, ed.), Plenum Press, New York, pp. 295–344.

Brown, M., Tache, Y., and Fisher, D., 1979, Central nervous system action of bombesin: Mechanism to induce hyperglycemia, *Endocrinology* **105:**660–665.

Dazord, A., Morera, A. M., Bertrand, J., and Saez, J. M., 1974, Prostaglandin receptors in human and ovine adrenal glands: Binding and stimulation of adenyl cyclase in subcellular preparations, *Endocrinology* **95:**352–359.

Dazord, A., Gallet, D., and Saez, J. M., 1975, Adenyl cyclase activity in rat, ovine and human adrenal preparations, *Horm. Metab. Res.* **7:**184–189.

Gasson, J. G., 1979, Steroidogenic activity of high molecular weight forms of corticotropin, *Biochemistry* **18:**4215–4224.

Glynn, P., Cooper, D. M. F., and Schulster, D., 1977, Modulation of the response of bovine adrenocortical adenylate cyclase to corticotropin, *Biochem. J.* **168:**277–282.

Krieger, D. T., 1980, Pituitary hormones in the brain: What is their function? *Fed. Proc.* **39:**2937–2941.

Londos, D., and Rodbell, M., 1975, Multiple inhibitory and activating effects of nucleotides and magnesium on adrenal adenylate cyclase, *J. Biol. Chem.* **250:**3459–3465.

Lowry, P. F., and McMartin, C., 1972, A study of the action of peptides with corticotropic activity on isolated adrenal cells, *J. Endocrinol.* **55:**xxxiii.

Matsuoka, H., Mulrow, P. J., and Li, C. H., 1980, β-Lipotropin: A new aldosterone-stimulating factor, *Science* **209:**307–308.

Matsuoka, H., Mulrow, P. J., Franco-Saenz, R., and Li, C. H., 1981, Stimulation of aldosterone production by β-melanotropin, *Nature* **291:**155–159.

Numa, S., and Nakanishi, S., 1981, Corticotropin–β-lipotropin precursor—a multihormone precursor and its gene, *Trends Biochem. Sci.* **6:**274–277.

Pederson, R. C., and Brownie, A. C., 1980, Adrenocortical response to corticotropin is potentiated by part of the amino-terminal region of pro-corticotropin/endorphin, *Proc. Natl. Acad. Sci. U.S.A.* **77:**2239–2243.

Pederson, R. C., Brownie, A. C., and Ling, N., 1980, Proadrenocorticotropin/endorphin-derived peptides: Coordinate action on adrenal steroidogenesis, *Science* **208:**1044–1046.

Salomon, Y., Londos, C., and Rodbell, M., 1974, A highly sensitive adenylate cyclase assay, *Anal. Biochem.* **58:**541–548.

Schlegal, W., and Schwyzer, R., 1977, Purification of bovine adrenal cortex plasma membrane vesicles containing a highly corticotropin-sensitive adenylate cyclase system and angiotensin II binding sites, *Eur. J. Biochem.* **72:**415–424.

Seelig, S., and Sayers, G., 1971, Structure activity relationship among ACTH's, *Fed. Proc.* **30:**316.

Shanker, G., and Sharma, R. K., 1979, β-Endorphin stimulates corticosterone synthesis in isolated rat adrenal cells, *Biochem. Biophys. Res. Commun.* **36:**1–5.

Silber, R. H., Busch, R., and Oslapas, R., 1958, Practical procedure for estimation of corticosterone or hydrocortisone, *Clin. Chem.* **4:**278–285.

Vahouny, G. V., Chanderbhan, R., Hinds, R., Hodges, V. A., and Treadwell, C. R., 1978, ACTH-induced hydrolysis of cholesterol esters in rat adrenal cells, *J. Lipid Res.* **19:**570–577.

Vinson, G. P., Whitehouse, B. J., Sell, A., Etienne, A. T., and Morris, H. R., 1981, Specific stimulation of steroidogenesis in rat adrenal zone glomerulosa cells by pituitary peptides, *Biochem. Biophys. Res. Commun.* **99:**65–72.

43

Substance K in Cardiovascular Control

JACOB EIMERL and GIORA FEUERSTEIN

1. INTRODUCTION

Until recently, substance P (SP) was considered to be the only tachykinin present in mammalian tissues; all other members of this family of peptides were isolated from amphibian extracts (Erspamer, 1981).

Recently, substance K (SK, neuromedin L, neurokinin α, neurokinin A; Lindefors *et al.*, 1985), a decapeptide that shares the same common C-terminal amino acid sequence (Phe-X-Gly-Leu-Met-NH_2) as all other naturally occurring tachykinins (Fig. 1), was identified in porcine spinal cord (Kimura *et al.*, 1983; Maggio *et al.*, 1983) and in rat brain and spinal cord (Minamino *et al.*, 1984). Moreover, Nawa *et al.* (1984b) have shown that one of two bovine brain SP precursors also contains the same amino acid sequence as SK. The C-terminal sequence is relatively constant within the tachykinin family and accounts for some of their common biological properties, i.e., hypotension and stimulatory action on intestinal smooth muscle and salivary secretion (Erspamer, 1971); the N terminus is considerably variable and is probably responsible for their spectrum of activity, potency, efficacy, and duration of action in different preparations and species (Erspamer, 1981; Nawa *et al.*, 1984a).

Pharmacological studies that related SP effects on smooth muscle to the amphibian tachykinins kassinin, eledoisin, and physalaemin eventually led Iversen *et al.* (1982) to classify two subtypes of SP receptors: SP-P, which presumably is activated by most tachykinins including SP and SK with similar potency, and SP-E, which is more specific for eledoisin, kassinin, and SK than SP. More recent and detailed studies, however, using SK, SP, and various SP analogues have suggested

---

JACOB EIMERL and GIORA FEUERSTEIN • Neurobiology Research Division, Department of Neurology, Uniformed Services University of the Health Sciences, Bethesda, Maryland 20814. The opinions or assertions contained herein are the private ones of the authors and are not to be construed as official or reflecting the views of the Department of Defense or the Uniformed Services University of the Health Sciences.

Substance P	Arg-Pro-Lys-Pro-Gln-Gln-Phe-Phe-Gly-Leu-Met-NH_2
Neurokinin A	
Neurokinin α	His-Lys-Thr-Asp-Ser-Phe-Val-Gly-Leu-Met-NH_2
Substance K	
Neuromedin L	
Neurokinin B	Asp-Met-His-Asp-Phe-Phe-Val-Gly-Leu-Met-NH_2
Neurokinin β	
Neuromedin K	

FIGURE 1. The amino acid sequences of the natural tachykinins identified thus far in mammalian tissues. Homology among the three tachykinins SP, SK, and neurokinin B is denoted by a solid line. Additional homology between SP and SK is denoted by a dashed line and between Sk and neurokinin B by a dotted line. Terminology has been adopted from Lindefors *et al.* (1985).

the presence of a specific SK receptor in addition to SP receptors (Beaujouan *et al.*, 1984; Buck *et al.*, 1984; Hunter and Maggio, 1984a,b; Mantyh *et al.*, 1984; Quirion and Pilapil, 1984). Subtypes of SP and SK receptors have been shown to be widely distributed not only within the CNS but also in peripheral organs (Boksa and Livett, 1985; Buck *et al.*, 1984; Gundersen *et al.*, 1985; Hunter and Maggio, 1984a,b; Ninkovic and Hunt, 1985; Regoli *et al.*, 1984a). The capability of SK to activate SP receptors with similar or even greater potency and its identification in various mammalian tissues suggested a role for SK in modulation of functions related so far to SP only. A primary function in this regard is cardiovascular regulation, a function in which SP was already implicated.

Substance P has been shown along with other tachykinins to lower blood pressure and to increase heart rate when administered systemically to several experimental animals (Nawa *et al.*, 1984a; Petty and Reid, 1982). Substance P and other tachykinins were also shown to have contractile/relaxant effects (Bayorh and Feuerstein, 1985; Regoli *et al.*, 1984a,b) that depend on the type of vascular tissue and the species. Furthermore, SP was found to modulate autonomic activity (Boksa and Livett, 1985; Couture and Regoli, 1982; Furness *et al.*, 1982; Otsuka and Konishi, 1983). In the central nervous system, SP was argued to participate in regulation of blood pressure and modulation of baroreceptor reflex (Fuxe *et al.*, 1982; Gillis *et al.*, 1980; Haeusler and Osterwalder, 1980; Helke, 1982; Petty and Reid, 1982).

The present chapter summarizes the evidence that suggests a role for SK in regulation of vascular tone, cardiac functions, and systemic hemodynamic control.

2. SYSTEMIC EFFECTS OF SUBSTANCE K

2.1. Effects on Arterial Blood Pressure

Systemic administration of SK to the conscious rat causes a rapid but transient (max. -17 ± 5 mm Hg) hypotensive response. In the pithed rat (Gillespie and Muir, 1967) in which peripheral cardiovascular responses were evaluated without

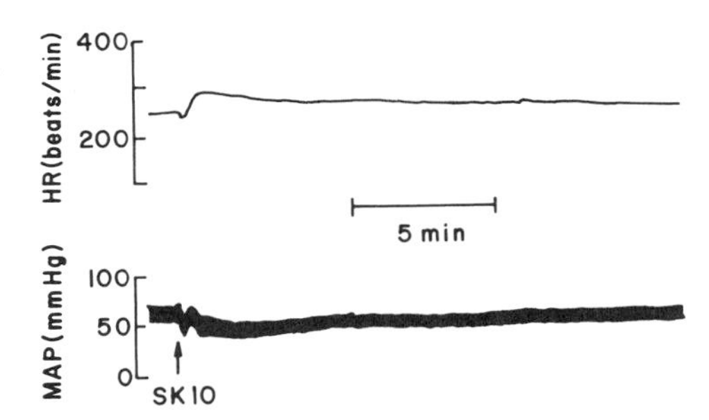

FIGURE 2. Typical recording of blood pressure and heart rate responses to administration of SK (10 nmol/kg, i.v.) in the pithed rat.

modification by central regulatory circuits or spinal reflexes, SK also produces dose-dependent decreases in mean blood pressure (Eimerl *et al.*, 1985). However, in the pithed rat, the higher dose of SK produced a typical triphasic blood pressure response (Fig. 2) composed of a depressor phase interrupted within the first minute by a short pressor phase, which then turned into a prolonged depressor phase. Both of these effects were not significantly affected by pretreatment with either phentolamine (α-adrenergic blocker) or propranolol (β-adrenergic blocker).

In the open-chest anesthetized domestic pig, the effects of SK on coronary and systemic circulation were studied by direct intracoronary injections of SK into the left anterior descending (LAD) coronary artery (Eimerl *et al.*, 1985). In this model, higher doses of SK caused a similar hypotensive response (max. -21 ± 4 mm Hg). These results show that the hypotensive property of SK *in vivo* is not confined to one species and is in accord with previous studies conducted on the anesthetized rat (Nawa *et al.*, 1984a) and guinea pig (Hua *et al.*, 1984). In this regard, SK has similar action to other related tachykinins (Erspamer, 1981). It is of interest to note that of the three tachykinins that have been identified so far in mammalian tissues, SK, SP, and neurokinin B, only SK was shown to induce hypotension in both the conscious and the pithed rat. Therefore, the depressor response to SK is probably not mediated by SP receptors; moreover, the well-documented hypotensive capacity of SP could be mediated by SK receptors in vascular tissue. Furthermore, the pressor response to very high doses of SP (Bayorh and Feuerstein, 1985) and the pressor response to moderate doses of SK in the pithed rat suggest action on a similar receptor, which mediates smooth muscle contraction.

2.2. Effect on Cardiac Function

In the conscious rat and the anesthetized pig, SK produced only small and transient increases in heart rate (HR), which accompanied the decrease in systemic blood pressure. In the pithed rat, a relatively high dose of SK (10 nmol/kg) produced significant tachycardia (41 ± 4 beats/min), which accompanied the pressor phase. The increase in heart rate produced by the higher dose of SK is completely blocked by the β blocker propranolol. This β-adrenoreceptor-mediated cardiac accelerator effect of SK corroborates well with the capacity of SK to increase circulating levels of catecholamines (Eimerl *et al.*, 1985). Though a direct effect of SK on cardiac contractility or venous return cannot be excluded, our results clearly indicate that the depressor effect of SK is independent of the heart rate response and therefore likely to be a direct vascular response. An additional explanation for the depressor response to SK might be related to its action on the microcirculation (increased permeability), a phenomenon well documented for the other tachykinins (Hua *et al.*, 1984). Reduced venous return, which leads to decrease in cardiac output, might also contribute to its overall effect.

2.3. Effects on Regional Blood Flow and Vascular Resistance

Many studies thus far have demonstrated the vasodilatory capacity of SK on various vascular preparations *in vitro*. However, the effect of SK on discrete blood vessels in different organs has not been reported as yet. In recent studies in our laboratory, we have utilized the directional pulsed Doppler technique (Haywood *et al.*, 1981) for simultaneous blood flow measurements in various organs of the conscious rat (Fig. 3).

These studies have shown differential blood flow and vascular resistance changes as follows:

1. Mesenteric artery. A dose-dependent increase in vascular resistance (max. 83 ± 20%) and decrease in blood flow (max. −49 ± 4%).
2. Hindquarter. Biphasic response: a short initial decrease in vascular resistance (max. −36 ± 6%) accompanied by increased blood flow (max. 41 ± 8%) followed later by a minor opposite response.
3. Renal artery. Minor increase in vascular resistance (max. 32 ± 13%) and decrease in blood flow (max. −21 ± 7%).

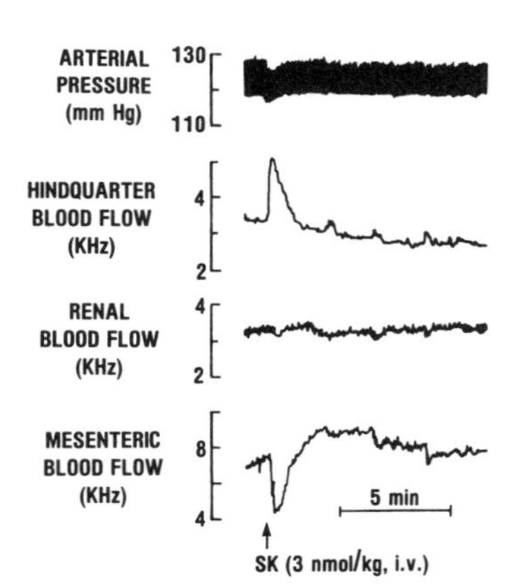

FIGURE 3. Typical simultaneous recording of blood pressure and regional blood flow responses to administration of SK (3 nmol/kg, i.v.) in the conscious rat.

In the domestic pig, SK (0.3–10 nmol) injected directly into the coronary artery produced dose-dependent decreases in coronary blood flow (max. −26 ± 3%), and the coronary dilation induced by SK was independent of systemic changes (Eimerl *et al.*, 1985).

These results taken together not only confirm the vasodilator capacity of SK *in vivo* in different species but also demonstrate its capacity to redistribute blood flow differentially to various vascular beds. Its potent constrictor effect on the mesenteric artery *in vivo* may correspond to its peripheral pressor effect as shown in the pithed rat. Catecholamines, which may produce a similar pattern of blood flow changes through their combined α- and β-adrenoreceptor activation, were excluded as potential mediators of this response in the pithed rat (Eimerl *et al.*, 1985). Other mechanisms (nonadrenergic) could have been activated by SK; however, the differential vascular responses to SK in the conscious rat might also be the result of SK receptors mediating opposite vascular responses in different vascular beds. Thus, SK receptors in the mesenteric circulation produce constriction, whereas SK receptors in skeletal muscles produce relaxation. The comparatively minor changes in renal blood flow might suggest the lack or paucity of SK receptors on renal blood vessels.

3. CENTRAL EFFECTS OF SUBSTANCE K

It was previously shown that SP injected into the cerebroventricular system of the rat produces an opposite effect on blood pressure to systemic injection: central injections of SP produce increments in blood pressure and tachycardia, which were argued to have been mediated by increased sympathetic activity (Fuxe *et al.*, 1982).

Since SP and SK might have similar actions not only in the periphery but also in the central nervous system, we decided to examine the effect of SK on cardiovascular regulation through direct injections of SK into the cerebroventricular system.

To this end, we prepared chronically instrumented rats with guide cannulas fixed on the skull to allow direct injections into the lateral cerebral ventricle. Injection of 0.9% NaCl solution produced inconsistent blood pressure and heart rate changes: −2 ± 2 mm Hg and +12 ± 5 beats/min, respectively. Substance K at 1, 10, or 100 ng/kg produced the following respective changes in blood pressure: +8 ± 2 mm Hg, +8 ± 2 mm Hg, and +18 ± 2 mm Hg. The pressor response was accompanied by moderate increases in heart rate: +30 ± 15, +22 ± 12, and +30 ± 14 beats/min for the respective doses of SK (Fig. 4). These data clearly show that SK, like SP, has opposite effects on blood pressure when injected centrally versus directly into the systemic circulation. However, it is not possible at this time to determine whether the central effects of SP and SK are mediated by different receptors or a receptor common to both tachykinins.

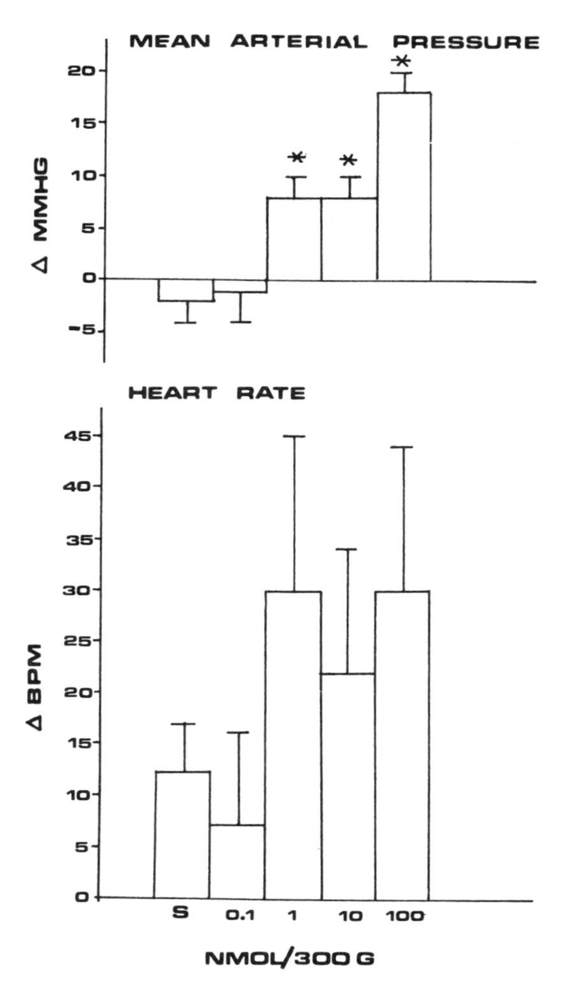

FIGURE 4. The effect of central administration (cerebroventricular) of SK on mean arterial pressure (ΔMMHG) and heart rate (ΔBPM) response in the conscious rat (n = 8–10). *$P < 0.01$ (ANOVA followed by Student–Newman–Keul's test).

4. MECHANISM OF ACTION

Unlike SP, which was shown to decrease induced catecholamine release from the adrenal (Boksa and Livett, 1985), we have found in the pithed rat that SK increases circulating levels of catecholamines through presynaptic sympathoadrenomedullary action. Also, catecholamine release induced by electrical stimulation of the spinal cord, which activates the whole spinal sympathetic outflow, is not affected by SK. These observations suggest that SK interference with catecholamine-induced vascular constriction involves the postsynaptic element, the vascular smooth muscle. Moreover, the failure of pretreatment with either propranolol or phentolamine to affect the depressor or pressor effects of SK in the pithed rat and its potent inhibitory action on various pressor stimuli (spinal cord stimulation, Arg^8-vasopressin) further support a direct mode of action on the vascular smooth muscle. Yet, peripheral nonadrenergic mechanisms or mediators cannot be excluded at this stage. In this regard, it is pertinent to mention that SP was suggested to involve several secondary mediators in its activity, i.e., ACh (Boksa and Livett, 1985; Yau *et al.*, 1985), opiates (Fuxe *et al.*, 1982; Ninkovic and Hunt, 1985), prostaglandins, leukotrienes (Regoli *et al.*, 1984a), and 5-HT (Holmgren *et al.*, 1985). Some of these mediators might also be involved in SK effects. These possibilities, however, await further investigations.

5. CONCLUSION

Of the tachykinin family of neuropeptides, SK is the most recently isolated mammalian tachykinin that was shown to affect potently cardiac activity and vascular tone. Like other members of this family of peptides, it was shown to dilate certain vascular beds or preparations (*in vivo* and *in vitro*) and to cause hypotension in several animal models. When administered centrally, it increases blood pressure and heart rate and causes a general excitement in the conscious rat. Its hypotensive property, as shown in the pithed rat, is probably related to peripheral vasodilation by direct action on the vascular smooth muscle; its constrictive and pressor capacity as shown in the conscious and pithed rat may involve activation of a different receptor or other unknown mechanisms. Likewise, SK was shown to enhance catecholamine release from the sympathoadrenomedullary elements in the periphery and thereby accelerate heart rate. However, other, nonadrenergic mechanisms might also be activated by SK. These combined data, along with the recent identification of a specific SK receptor with wide central and peripheral distribution and the possible capacity by SK to activate SP receptors, strongly suggest that SK plays a physiological role in the overall regulation of the cardiovascular system.

ACKNOWLEDGMENT. This work was supported by the Uniformed Services University of the Health Sciences Protocol No. RO9211.

REFERENCES

Bayorh, M. A., and Feuerstein, G., 1985, Bombesin and substance P modulate peripheral sympathetic and cardiovascular activity, *Peptides* (Suppl. 1)**6:**115–120.

Beaujouan, J. C., Torrens, Y., Viger, A., and Glowinski, J., 1984, A new type of tachykinin binding site in the rat brain characterized by specific binding of a labeled eledoisin derivative, *Mol. Pharmacol.* **26:**248–254.

Boksa, P., and Livett, B. G., 1985, The substance P receptor subtype modulating catecholamine release from adrenal chromaffin cells, *Brain Res.* **332**:29–38.

Buck, S. H., Burcher, E., Shults, C. W., Lovenberg, W., and O'Donohue, T. L., 1984, Novel pharmacology of substance K binding sites: A third type of tachykinin receptor, *Science* **226**:987–989.

Couture, R., and Regoli, D., 1982, Mini review: Smooth muscle pharmacology of substance P, *Pharmacology* **24**:1–25.

Eimerl, J., Bayorh, M. A., Zukowska-Grojec, Z., Faden, A. I., Ezra, D., and Feuerstein, G., 1985, Substance K: Vascular and cardiac effects in rat and pig, *Peptides* (Suppl. 2)**6**:149–153.

Erspamer, V., 1971, Biogenic amines and active polypeptides of the amphibian skin, *Annu. Rev. Pharmacol.* **11**:327–350.

Erspamer, V., 1981, The tachykinin peptide family, *Trends Neurosci.* **4**:267–269.

Furness, J., Papka, R. E., Pella, N. G., Costa, M., and Askay, R. L., 1982, Substance P-like immunoreactivity in nerves associated with vascular system of guinea pigs, *Neuroscience* **7**:477–489.

Fuxe, K., Agnati, L. F., Rossel, S., Härfstrand, A., Folkers, K., Lundberg, J. M., Anderson, K., and Hökfelt, T., 1982, Vasopressor effects of substance P and C-terminal sequences after intracisternal injection to α-chloralose-anesthetized rats: Blockade by a substance P antagonist, *Eur. J. Pharmacol.* **77**:171–176.

Gillespie, J. S., and Muir, T. C., 1967, A method of stimulating the complete sympathetic outflow from the spinal cord to blood vessels in the pithed rat, *Br. J. Pharmacol. Chemother.* **30**:78–87.

Gillis, R. A., Helke, C. J., Hamilton, B. L., Norman, W. P., and Jacobowitz, D. M., 1980, Evidence that substance P is a neurotransmitter of baro- and chemoreceptor afferents in nucleus tractus solitarius, *Brain Res.* **181**:476–481.

Gundersen, K., Øktedalen, O., and Fonnum, F., 1985, Substance P in subdivisions of the sciatic nerve, and in red and white skeletal muscles, *Brain Res.* **329**:97–103.

Haeusler, G., and Osterwalder, R., 1980, Evidence suggesting a transmitter or neuromodulatory role for substance P at the first synapse of the baroreceptor reflex, *Naunyn Schmiedebergs Arch. Pharmacol.* **314**:111–121.

Haywood, J. R., Shaffer, R. A., Fastenow, C., Fink, G. D., and Brody, M. J., 1981, Regional blood flow measurement in the conscious rat with pulsed Doppler flowmeter, *Am. J. Physiol.* **241**(10):H273–H278.

Helke, C. J., 1982, Neuroanatomical localization of substance P: Implications for central cardiovascular control, *Peptides* **3**:479–483.

Holmgren, S., Grove, D. J., and Nilsson, S., 1985, Substance P acts by releasing 5-hydroxytryptamine from enteric neurons in the stomach of the rainbow trout, *Salmo gairdneri, Neuroscience* **14**:683–693.

Hua, X., Lundberg, J. M., Theodorsson-Norheim, E., and Brodin, E., 1984, Comparison of cardiovascular and bronchoconstrictor effects of substance P, substance K and other tachykinins, *Naunyn Schmiedebergs Arch. Pharmacol.* **328**:196–201.

Hunter, J. C., and Maggio, J. E., 1984a, Pharmacological characterization of a novel tachykinin isolated from mammalian spinal cord, *Eur. J. Pharmacol.* **97**:159–160.

Hunter, J. C., and Maggio, J. E., 1984b, A pharmacological study with substance K: Evidence for multiple types of tachykinin receptors, *Eur. J. Pharmacol.* **105**:149–153.

Iversen, L. L., Hanley, M. R., Sandberg, B. E. B., Lee, C.-M., Pinnock, R. D., and Watson, S. P., 1982, Substance P receptors in the nervous system and possible receptor subtypes, in: *Substance P in the Nervous System, Ciba Foundation Symposium 91* (R. Porter and M. O'Conour, eds.), Pitman, London, pp. 186–195.

Kimura, S., Okada, M., Sugita, Y., Kanazawa, I., and Munekata, E., 1983, Novel neuropeptides, neurokinin A and B, isolated from porcine spinal cord, *Proc. Jpn. Acad. [B]* **59**:101–104.

Lindefors, N., Brodin, E., Theodorsson-Norheim, E., and Ungerstedt, U., 1985, Regional distribution and *in vivo* release of tachykinin-like immunoreactivities in rat brain: Evidence for regional differences in relative proportions of tachykinins, *Regul. Peptides* **10**:217–230.

Maggio, J. E., Sandberg, B. E. B., Bradley, C. V., Iversen, L. L., Santikarn, B. H., Williams, B. H., Hunter, J. C., and Hanley, M. R., 1983, Substance K: A novel tachykinin in mammalian spinal cord, in: *Substance P* (P. Scrabanek and D. Powell, eds.), Boole Press, Dublin, pp. 20–21.

Mantyh, P. W., Maggio, J. E., and Hunt, S. P., 1984, The autoradiographic distribution of kassinin and substance K binding sites is different from the distribution of substance P binding sites in rat brain, *Eur. J. Pharmacol.* **102:**361–364.

Minamino, N., Masuda, H., Kangawa, K., and Matsuo, H., 1984, Regional distribution of neuromedin K and neuromedin L in rat brain and spinal cord, *Biochem. Biophys. Res. Commun.* **124**(3)**:**731–738.

Nawa, H., Dotenchi, M., Igano, K., Inouye, K., and Nakanishi, S., 1984a, Substance K: A novel mammalian tachykinin that differs from substance P in its pharmacological profile, *Life Sci.* **34:**1153–1160.

Nawa, H., Hirose, T., Takashima, H., Inayama, S., and Nakanishi, S., 1984b, Nucleotide sequences of cloned cDNAs for two types of bovine brain substance P precursor, *Nature* **306:**32–36.

Ninkovic, M., and Hunt, S. P., 1985, Opiate and histamine H1 receptors are present on some substance P-containing dorsal root ganglion cells, *Neurosci. Lett.* **53:**133–137.

Otsuka, M., and Konishi, S., 1983, Substance P—the first peptide neurotransmitter? *Trends Neurosci.* **6:**317–320.

Petty, M., and Reid, J., 1982, The cardiovascular effects of centrally administered substance P in the anesthetized rabbit, *Eur. J. Pharmacol.* **82:**9–14.

Quirion, R., and Pilapil, C., 1984, Comparative potencies of substance P, substance K and neuromedin K on brain substance P receptors, *Neuropeptides* **4:**325–329.

Regoli, D., D'Orleans-Juste, P., Escher, E., and Mizrahi, J., 1984a, Receptors for substance P. I. The pharmacological preparations, *Eur. J. Pharmacol.* **97:**161–170.

Regoli, D., Escher, E., Drapeau, G., D'Orleans-Juste, P., and Mizrahi, J., 1984b, Receptors for substance P. III. Classification by competitive antagonists, *Eur. J. Pharmacol.* **97:**179–189.

Yau, W. M., Youther, M. L., and Verdun, P. R., 1985, A presynaptic site of action of substance P and vasoactive intestinal polypeptide on myenteric neurons, *Brain Res.* **330:**382–385.

VI

Clinical Applications of Neural and Endocrine Peptides

Combined Treatment with an LHRH Agonist and the Antiandrogen Flutamide in Prostate Cancer

F. LABRIE, A. DUPONT, A. BELANGER, R. ST.-ARNAUD, M. GIGUERE, Y. LACOURCIERE, J. EMOND, and G. MONFETTE

1. INTRODUCTION

The original observation that treatment of adult male rats with LHRH agonists causes a marked inhibition of testosterone secretion accompanied by a loss in secondary sex organ weight (Labrie *et al.*, 1978, 1980) opened a new era in the treatment of androgen-dependent diseases. When administered to adult men in appropriate doses by the subcutaneous route, these peptides decrease serum testosterone and dihydrotestosterone to castrated levels (Labrie *et al.*, 1980; Faure *et al.*, 1982; Jacobi and Wenderoth, 1982). Since LHRH agonists have no adverse side effects other than those related to low circulating levels of androgens, they offer an advantageous alternative to orchiectomy and treatment with estrogens in advanced carcinoma of the prostate.

However, when LHRH agonists are administered alone, the transient elevation of serum androgens seen during the first days of treatment has been reported to induce disease flare in a significant proportion of cases (Kahan *et al.*, 1984; Leuprolide Study Group, 1984). Moreover, since LHRH agonists achieve medical castration, one cannot expect any improvement in prognosis over the already well-known but limited effects of orchiectomy (Slack *et al.*, 1984).

In order to neutralize the androgens of adrenal origin, which remain at high

F. LABRIE, A. DUPONT, A. BELANGER, R. ST.-ARNAUD, M. GIGUERE, and Y. LACOURCIERE • Departments of Medicine, Molecular Endocrinology and Urology, Laval University Medical Center, Quebec GIV 4G2, Canada. J. EMOND and G. MONFETTE • Hôtel-Dieu Hospital, St-Jérôme, Quebec J7Y2T8, Canada.

levels in the prostatic tissue following castration (Geller *et al.*, 1984), as well as to prevent the possibility of disease flare during the first 2 weeks of treatment, we have developed a two-way approach for the treatment of prostate cancer that includes an LHRH agonist administered in combination with a pure antiandrogen at the start of treatment (Labrie *et al.*, 1982, 1983a,b, 1984, 1985).

Following detailed animals studies, we have applied the combined treatment with an LHRH agonist (or surgical castration) in association with a pure antiandrogen for the treatment of 88 previously untreated patients having stage D2 (bone metastases) prostate cancer. The duration of treatment extended from 5.7 to 36.7 months with an average of 17.6 months. Some patients were also randomly assigned to orchiectomy alone, but entry into this group was stopped when survival became significantly different.

The results are also compared with those obtained when the same combination therapy is applied to patients who have received previous hormonal therapy (orchiectomy, estrogens, or LHRH agonists alone).

2. PATIENTS, MATERIALS, AND METHODS

Ninety-five patients with histology-proven prostatic adenocarcinoma and bone metastases visualized by bone scintigraphy (stage D2) took part in this multicenter study after written informed consent. The criteria for inclusion and exclusion were those of the U.S. NPCP (Slack *et al.*, 1984; Labrie *et al.*, 1982, 1983a,b). Of the 88 previously untreated stage D2 patients who had combination therapy, 78 received the combined treatment with the LHRH agonist [D-Trp^6, des-Gly-NH_2^{10}]LHRH ethylamide (Peptal®) or [D-Ser(TBU)^6, des-Gly-NH_2^{10}]LHRH ethylamide in association with the pure antiandrogen 2-methyl-N-[4-nitro-3-(trifluoromethyl)pheny1]propanamide (Flutamide®, Euflex®), whereas ten had orchiectomy (instead of LHRH agonist) in association with the antiandrogen. No difference was observed between chemical or surgical castration, and the two LHRH agonists used show equipotency in both experimental animals and men. Twenty patients were originally started with the Flutamide® analogue 5,5-dimethyl-3[4-nitro-3-(trifluromethyl)phenyl]-2,4-imidazolidione (RU23908, Anandron®). However, the occurrence of visual side effects in approximately 70% of patients receiving Anandron® has led to a change from Anandron® to Flutamide® and to the exclusive use of Flutamide® in all patients.

The LHRH agonists were injected subcutaneously at the daily dose of 500 μg at 0800 hr for 1 month followed by a 250-μg daily dose while the antiandrogens were given three times daily at 0700, 1500, and 2300 hr at doses of 100 (Anandron®) or 250 (Flutamide®) mg orally. The antiandrogen was started 1 day before the first administration of the LHRH agonist or orchiectomy.

During the course of the study, seven patients were randomly assigned to orchiectomy alone. The conditions of entry were the same as described above except that patients had to accept orchiectomy and LHRH agonist treatment. It was agreed at the start that entry into the orchiectomy-alone arm would be stopped as soon as a significant difference in survival was obtained. When a lack of objective response or tumor progression was noted in the orchiectomy-alone patients, treatment with Flutamide® was started. All patients were followed to determine survival.

In addition, 56 patients (average age 66.5 years) had received previous hormonal therapy, namely, orchiectomy, DES, or the LHRH agonists Leuprolide or [D-Trp^6]LHRH alone for an average of 10.9 months. These patients received the combined antiandrogen therapy at the time of relapse. Those already orchiectomized received only Flutamide®, whereas the others received the combination of LHRH agonist plus antiandrogen.

Complete clinical, urological, biochemical, and radiological evaluation of the patients was performed before starting treatment as described (Labrie *et al.*, 1983a). The initial evaluation included history, physical examination, bone scan, transrectal and transabdominal ultrasonography of the prostate, ultrasonography of the abdomen, chest roentgenogram, skeletal survey, and sometimes computerized axial tomography (CAT) of the abdomen and pelvis as well as excretory urogram (IVP). Performance status and pain were evaluated on a scale of 0 to 4. The follow-up was as described (Labrie *et al.*, 1983a). The criteria of the U.S. National Prostatic Cancer Project were used for assessment of objective response to treatment (Slack *et al.*, 1984). Statistical significance was measured according to the Wilcoxon and log-rank techniques as well as by the Fisher's exact test (Armitage, 1971; Goldman and Elwood, 1979) when appropriate. The probability of continuing response and survival was calculated according to Kaplan and Meier (1958).

3. RESULTS

As illustrated in Fig. 1A, the first administration of the LHRH agonist caused an approximately four-fold increase in serum LH levels measured 1 to 24 hr later ($P < 0.01$). Serum testosterone, on the other hand, increased slowly during the first 24 hr to reach 50% above control at 24 hr ($P < 0.05$) (Fig. 2A). Serum radioimmunoassayable LH then decreased to 45% of control at 1 month and to approximately 30% of control between 6 and 24 months of treatment ($P < 0.01$) (Figs. 1B to 1F). Serum testosterone, on the other hand, decreased to 10% of control at 1 month and remained at this low level thereafter ($P < 0.01$) (Figs. 2B to 2F). The most important finding, however, was that no increase in either serum LH or testosterone was observed after the subcutaneous injection of 500 μg of either of the two LHRH agonists used in our study.

Biologically active serum LH exhibited the same stimulatory pattern at the beginning of treatment. However, bioactive LH levels, although parallel to radioimmunoassayable LH during the first 2 weeks of treatment, then showed a drastic inhibition to 7% of pretreatment values after 30 days of combined hormonal therapy (0.43 ± 0.04 and 0.030 ± 0.007 ng/ml on days -2 and 30, respectively) and remained low thereafter. This marked inhibition of the biological activity of serum

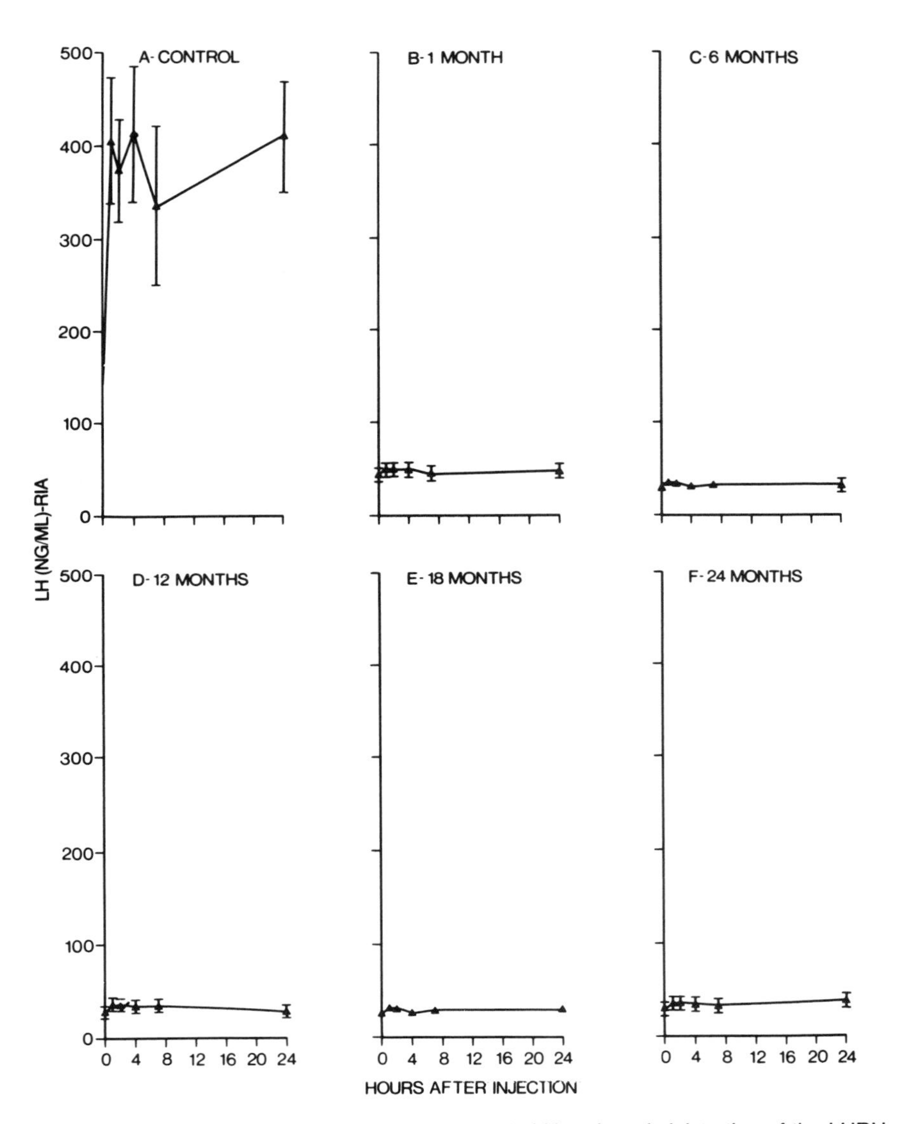

FIGURE 1. Response of serum radioimmunoassayable LH to the administration of the LHRH agonist [D-Trp^6, des-Gly-NH$_2$^{10}]LHRH ethylamide or [D-Ser(TBU)^6, des-Gly-NH$_2$^{10}]LHRH ethylamide on the first day of treatment (A) and after 1(B), 6(C), 12(D), 19(E), and 24(F) months of daily treatment (500 μg daily for 1 month followed by 250 μg daily). The basal levels of LH were 94.9 ± 12.7 ng/ml (LER-907 as standard).

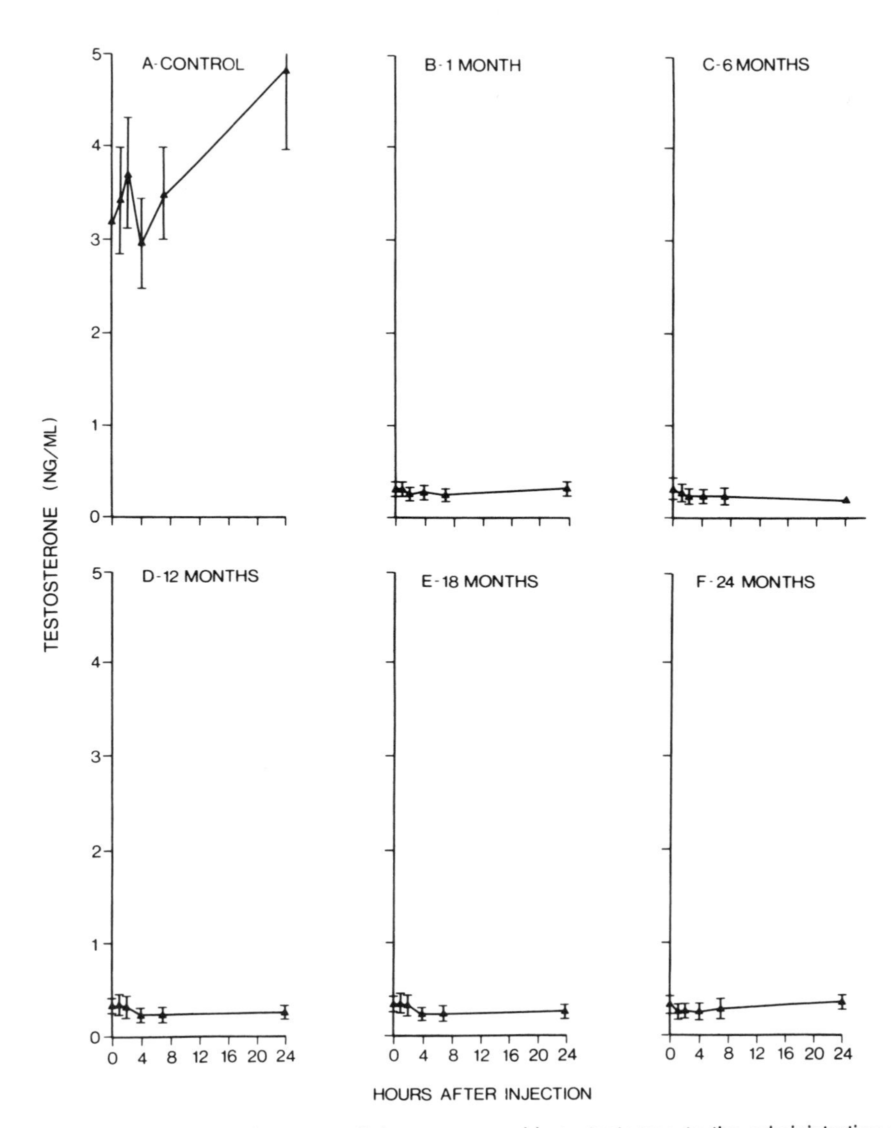

FIGURE 2. Response of serum radioimmunoassayable testosterone to the administration of the LHRH agonist [D-Trp^{6}, des-Gly-$NH_2{}^{10}$]LHRH ethylamide or [D-Ser(TBU)^{6}, des-Gly-$NH_2{}^{10}$]LHRH ethylamide on the first day of treatment(A) and after 1(B), 6(C), 12(D), 18(E), and 24(F) months of daily treatment (500 μg daily for 1 month followed by 250 μg daily). The basal levels of testosterone were 3.18 ± 0.49 ng/ml (1 nmol/liter = 0.30 ng/ml).

LH induced by the combined therapy led to a marked decrease in the bioactive/immunoactive ratio of serum LH in these patients. The bio/immuno ratio decreased rapidly from a value of 0.26 ± 0.03 before the onset of therapy (day −2) to the low value of 0.020 ± 0.003 on day 30 ($P < 0.01$). The bio/immuno ratio of 7.7% of control on day 30 remained at similarly low values thereafter up to at least 3 months (St.-Arnaud *et al.*, 1985).

It was also of great interest to see the rapid decrease in the serum concentration of the four adrenal steroids that act as precursors for the biosynthesis of testosterone and 5α-dihydrotestosterone in the prostate cancer tissue. From basal values of 915 ± 75 ng/ml, the serum concentration of dehydroepiandrosterone sulfate (DHEA-S) was already decreased to 73% of control ($P < 0.05$) between days 1 and 4 of treatment and reached 61% at 1 month ($P < 0.01$). Thereafter, the mean concentration of circulating DHEA-S remained at 60% of control or lower ($P < 0.01$).

The serum levels of dehydroepiandrosterone (DHEA) followed a pattern almost superimposable to that of DHEA-S. In fact, from basal values of 2.18 ± 0.17 ng/ml, the concentration of serum DHEA decreased progressively to reach 63% of control after 1 month of treatment ($P < 0.01$). Thereafter, the concentration of DHEA remained approximately constant at mean values ranging between 46 and 67% of control ($P < 0.01$).

An even more striking inhibitory effect was observed on the serum concentration of androst-5-ene 3β,17β-diol (Δ^5-diol) (Fig. 3). From basal values of 0.59 ± 0.07 ng/ml, the serum concentration of this adrenal steroid decreased to 44% of control at 1 month ($P < 0.01$). Thereafter, the mean serum concentration of Δ^5-diol remained at values ranging between 36 and 50% of control ($P < 0.01$) during most of the treatment period. The serum levels of androstenedione (Δ^4-dione) followed a similar pattern (data not shown). It was of great interest to see that the serum concentration of cortisol remained constant during the whole period of combined antihormonal treatment, the pretreatment value being 184 ± 6.65 ng/ml. During the whole course of treatment, the mean values of serum cortisol varied only between 170 ± 7.10 and 215 ± 11.9 ng/ml.

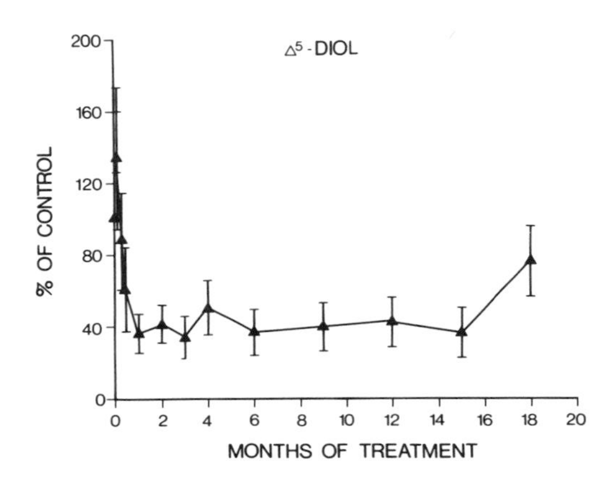

FIGURE 3. Changes in serum Δ^5-diol in previously untreated patients having clinical stage D2 prostate cancer receiving the combined therapy with a pure antiandrogen in association with orchiectomy or an LHRH agonist. The pretreatment values of serum Δ^5-diol were 0.59 ± 0.07 ng/ml (15 patients).

Starting in March 1982, 88 previously untreated patients with histology-proven prostatic carcinoma and bone metastases identified by bone scan and X-ray received the combined treatment for more than 5.7 months as first therapy. Only one patient was excluded from evaluation (interruption of treatment). Pain was present in 52 patients, the pain being moderate in 24% (21) and severe requiring antalgic positions or movements in 25% of them (22). Seven patients (8%) had pain requiring the use of a wheel chair for their transport, and two patients were totally bedridden. The pain subsided completely in all cases during the first month of treatment.

Almost all patients displayed various levels of prostatism, which was improved during the first 2 months of treatment. The rectal examination revealed an enlarged and a hard prostate in 85% of the patients. In all of them, the volume of the gland regressed, and its consistency improved to become small and soft during the first 6 months.

Of the 88 patients, 51.1% (45) had a normal activity. Twenty-three patients (26%) were symptomatic but ambulatory, whereas 11 patients (12.5%) stayed in bed for less than 50% of the time and seven (8%) stayed in bed more than 50% of the time, two of them being totally bedridden. Performance returned to normal in 64, 88, 96, and 100% of the patients after 1, 2, 3, and 4 months of treatment, respectively. The two patients who were bedridden became ambulatory after 2 months of treatment.

Figure 4 illustrates the changes in serum prostatic acid phosphatase (PAP) levels following the start of combination antihormonal treatment. The serum levels of PAP were initially elevated in 77 of the patients (87.5%), the values ranging between 1.0 and 896 ng/ml (normal < 2.0 ng/ml). In all cases, the start of treatment

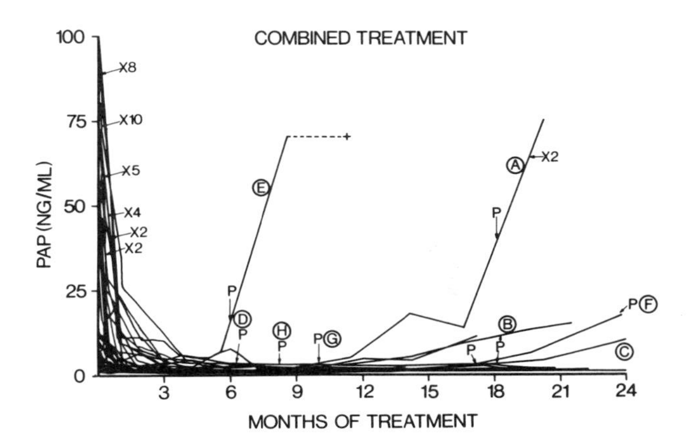

FIGURE 4. Effect of combined antihormonal treatment on serum prostatic acid phosphatase (PAP) levels in previously untreated patients having clinical stage D2 prostate cancer. In a group of 88 patients, 77 (87.5%) had serum PAP levels above 2.0 ng/ml at the start of treatment. P indicates time of objective progression of the disease by bone scan and/or ultrasonography for local progression at the prostatic level. × indicates that the value shown should be multiplied by the indicated number, and + indicates time of death. The circled letter identifies the patient.

was followed by an extremely rapid fall in serum PAP, a decrease to 41 and 24% of control being reached on days 1 to 4 and 5 to 10 ($P < 0.01$) after the start of treatment, respectively. In patients treated for 6 and 9 months, serum PAP values had returned to normal in 90 and 95% of the cases, respectively.

Bone scintigraphy performed 4 to 6 months after the start of treatment was an absolute requirement for inclusion of the patients in one of the categories of objective

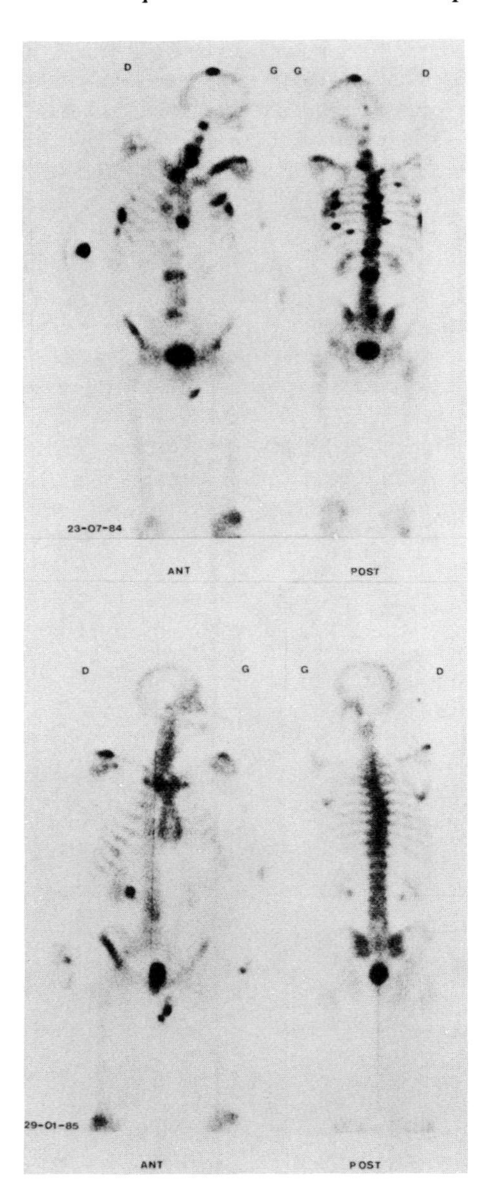

FIGURE 5. Bone scans with ^{99m}Tc-labeled methylene diphosphonate of patient treated with the pure antiandrogen Flutamide® and the LHRH agonist [D-Trp^6,des-Gly-NH_2^{10}]LHRH ethylamide. (A) Before treatment on July 23, 1984, showing disseminated bone metastases; (B) January 29, 1985, 6 months after the start of combined antihormonal therapy. Note the disappearance of all areas of increased uptake.

responses. An example of the changes in bone scintigraphy in a patient who showed a complete response at 6 months of treatment is illustrated in Fig. 5. It can be seen in Table I that with an average duration of treatment of 17.6 months (from 5.7 to 36.7), 22 of 88 patients (25%) showed a complete response with normalization of the bone scan and serum PAP as well as disappearance of any clinical symptom or sign of prostate cancer. Thirty-five patients (39.8%) showed a partial response with a decrease by more than 50% in the number of areas of increased uptake at bone scintigraphy and a return to normal of serum PAP in all cases. Twenty-seven patients (30.7%) showed an improvement or stabilization of their disease confirmed by bone scan (Table I). Despite marked subjective and objective improvement, a new lesion was seen on the bone scan at 6 months in four patients (4.6% of progression).

Figure 6 illustrates the probability of continuing response in the group of 88 patients who received the combined treatment as first therapy. Quite remarkably, the probability of having a continuing positive response at 2 years, calculated according to Kalpan and Meier (1958), is 81%.

As mentioned earlier, it was felt important to include a randomized group of patients with orchiectomy alone before such a randomized study became ethically unacceptable because of the evidence of a higher risk in the group receiving no antiandrogen. Such a study should eliminate any potential bias related to our population of patients. The finding of a rate of response to orchiectomy similar to the previous studies would then permit a comparison of the effect of combination treatment with the numerous previous and contemporary studies on the effect of orchiectomy and estrogens alone.

The changes in serum PAP levels in the group of patients who had orchiectomy alone are shown in Fig. 7. In the four patients who had serum PAP levels above 10 ng/ml at the start of treatment (43, 575, 35, and 20 ng/ml, patients 2, 3, 4, and 5, respectively), a progressive but relatively slow decrease was seen during the first 11 months. The serum PAP values did, however, remain above normal in all these four cases. Of the three patients who had slightly elevated serum PAP levels (below

TABLE I. Comparison of the Rate of Objective Response[a] to Combined Antiandrogen Treatment in Previously Untreated and Treated Patients Having Clinical Stage D2 Prostate Cancer

Previous treatment	Number of patients	Months of treatment (mean; range)	Best objective response				Relapse	Deaths	
			Complete	Partial	Stable	Progression		Prostate cancer	Other causes
Untreated	88	17.6 (5.7–36.7)	22 (25.0%)	35 (39.8%)	27 (30.7%)	4 (4.6%)	22/86 (26.2%)	4 (4.6%)	4 (4.6%)
Orch/DES/LHRH agonist alone	56	24.3 (6.1–63.2)	4 (7.1%)	8 (14.3%)	12 (21.4%)	32 (57.1%)	13/23 (54.2%)	38 (67.9%)	2 (3.6%)

[a] U.S. National Prostatic Cancer Project (NPCP) criteria.

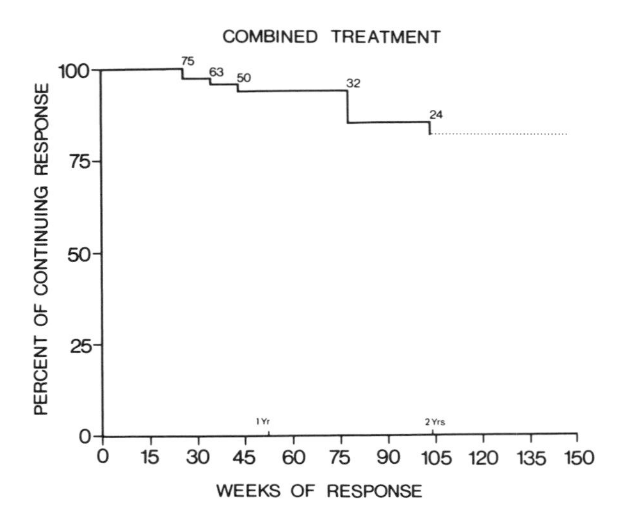

FIGURE 6. Probability of continuing positive response following combined antihormonal therapy in previously untreated patients having clinical stage D2 prostate cancer [calculated according to Kaplan and Meier (1958) for 88 patients].

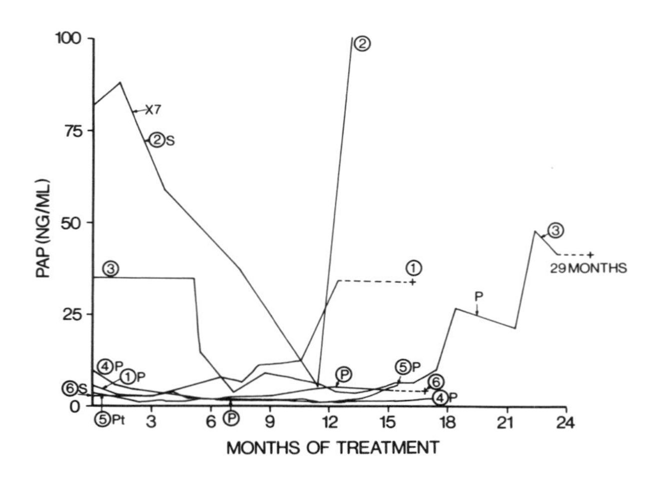

FIGURE 7. Effect of orchiectomy alone on serum PAP levels in previously untreated patients having clinical stage D2 prostate cancer. P indicates objective progression of the disease; Pt means partial response; S indicates stable disease; + indicates time of death. × indicates that the value shown should be multiplied by the following number. Circled numbers are for patient identification.

10 ng/ml) at the start of treatment, the concentration of serum PAP temporarily decreased to normal in 2 cases.

The most dramatic but expected finding in this study of the effect of orchiectomy alone is that four out of the seven patients have already died from their cancer at 11.5, 16, 17, and 29 months, respectively, and the three remaining patients show progression of the disease. Comparison of the probability of survival following orchiectomy alone and the combination therapy is illustrated (according to Kaplan–Meier method) in Fig. 8.

At each visit, the patients answered a detailed questionnaire concerning any possible symptom or sign of intolerance to the drugs. Hot flashes were described spontaneously by the patients in approximately 50% of the cases after 1 to 3 months of treatment. Usually, the severity of the hot flashes decreased with time or disappeared within 2 years. A decrease or loss of libido was observed in approximately 75% of patients. However, it should be mentioned that in 25% of the patients, libido and potency are maintained.

A side effect not related to the neutralization of androgens is that approximately 70% of patients treated with Anandron® showed a significant delay in obtaining good vision on coming from a bright area. Although the upper limit of normal of the photo stress test in patients of that age is 1 min, the delay observed in some patients treated with Anandron® was increased up to 25 min. However, on cessation of Anandron® treatment and change to Flutamide®, those symptoms rapidly disappeared in all cases.

As illustrated in Fig. 9, the probability of survival is markedly increased ($P < 0.01$) in the group of stage D2 patients who received the combined treatment with the pure antiandrogen as first treatment as compared to those who had orchiectomy or received estrogens or LHRH agonists alone prior to the administration of Flutamide®. In fact, the probability of survival at 2 years is 89.1% in the group of patients who received the combination therapy at the start of treatment as com-

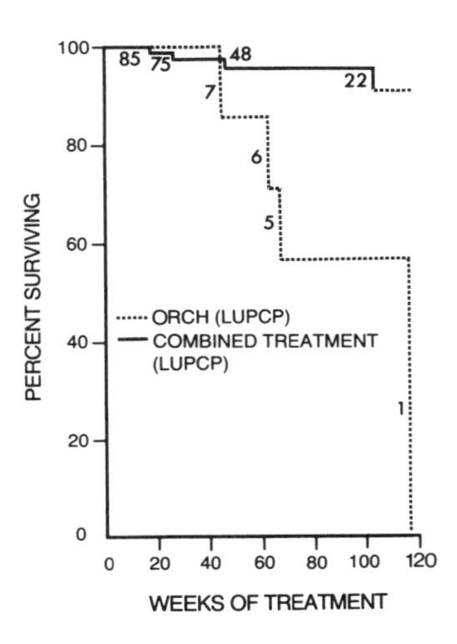

FIGURE 8. Comparison of the probability of survival (Kaplan–Meier method) following orchiectomy alone and combination therapy in previously untreated patients having clinical stage D2 prostate cancer. Entry into the orchiectomy group was stopped when the difference became significant at 18 months with three of seven patients having died from their cancer ($P < 0.05$).

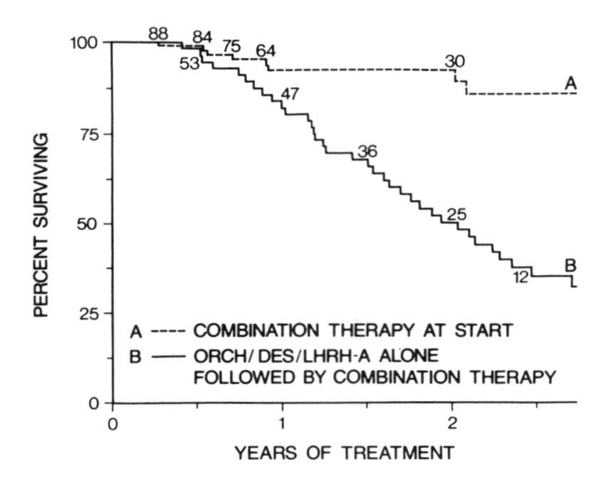

FIGURE 9. Comparison of the probability of continuing response following combination therapy and the administration of Leuprolide® alone or DES (Leuprolide Study Group, 1984).

pared to only 48.2% in those who had partial androgen blockade for various time intervals before receiving the combination therapy. Up to 2 years of treatment, the death rate is thus increased by more than fourfold by delaying Flutamide® administration (from 10.9 to 51.8%, $P < 0.01$). When only deaths from prostate cancer are considered, it can be seen in Table I that there are 38 deaths in the group of patients who received delayed combination therapy as compared to only four deaths from prostate cancer among the patients who had a blockage of both adrenal and testicular androgens at the start of treatment.

In agreement with all previous data on the effect of adrenalectomy, hypophysectomy, and antiandrogens in patients who had received previous hormonal therapy (Huggins and Scott, 1945; Sogani *et al.*, 1975; Stoliar and Albert, 1974; Robinson *et al.*, 1974), it can be seen in Fig. 10 and Table I that only 24 of 56 (42.9%) previously treated patients showed a positive objective response to the combined antiandrogen blockade. Patients previously castrated received only Flutamide®, whereas those treated with DES or an LHRH agonist alone received [D-Trp[6]]LHRH ethylamide in combination with Flutamide®. By contrast, 84 of 88 (95.4%) of

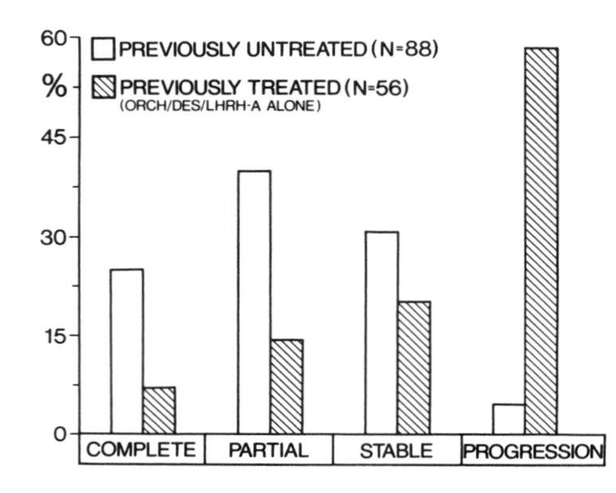

FIGURE 10. Comparison of the objective response rates (criteria of the U.S. NPCP) in previously untreated patients having clinical stage D2 prostate cancer who received the combination therapy as first treatment and those who had orchiectomy or received estrogens or LHRH agonists alone before administration of the combination therapy at the time of progression. Survival was calculated according to the Kaplan–Meier method from the time of first treatment in the two groups.

previously untreated patients showed a positive objective response at start of treatment ($P < 0.01$, Fisher's exact test) (Fig. 9, Table I).

The use of a pure antiandrogen at the start of treatment eliminates the unnecessary risks of disease flare that are known to occur in a significant proportion of patients (Kahan *et al.*, 1984; Leuprolide Study Group, 1984). It seems obvious that exposure of the tumor cells to supraphysiological levels of androgens represents an increased stimulus for the tumor to grow and to metastasize. The present data clearly show that the pure antiandrogen makes it possible to take advantage of the well-tolerated LHRH agonists as substitutes for orchiectomy and estrogen while eliminating the risk of disease flare of LHRH agonists used alone.

Since the present study rigorously applied the criteria of the U.S. NPCP (Slack *et al.*, 1984) for determining the rate of objective response, the results can be confidently compared with those obtained in the study 500 of the NPCP (Murphy *et al.*, 1983) as well as with those more recently reported by the Leuprolide Study Group (1984). In addition, the present results observed following orchiectomy alone are in agreement with all the data previously obtained by the groups that studied the effects of orchiectomy or estrogens (Murphy *et al.*, 1983; Nesbit and Baum, 1950; Jordan *et al.*, 1977; Mettlin *et al.*, 1982; Leuprolide Study Group, 1984).

Combined androgen blockade at the start of treatment in previously untreated stage D2 patients has led, so far, to a 95.4% positive objective response as compared to 81% following DES or orchiectomy in the NPCP 500 trial (Murphy *et al.*, 1983). Initial response rates of 86 and 85% have been obtained with Leuprolide® and DES alone, respectively, in the recent study of the Leuprolide Study Group (1984). In addition to the improved percentage of positive responses at the start of treatment, another most important aspect of the effect of the combination treatment is the marked increase in the duration of the positive response. In study 500 of the NPCP, only 40% of the patients were in remission after 18 months of treatment (Murphy *et al.*, 1983), thus indicating progression or relapse of the cancer in 60% of the patients as compared to 12% at the same time interval in the present study (Fig. 6).

Although the percentage of patients still in remission at 2 years is 81% with the combination therapy, it has already decreased to 0% with Leuprolide® at 22 months and to only 30% at 18 months with DES (Leuprolide Study Group, 1984). There is thus a remarkable advantage of the combination therapy not only on the percentage of initial responses but even more strikingly on the duration of the positive response. The most impressive result is that observed on survival and quality of life. In fact, only four patients have died from prostate cancer, although four have died from other causes.

A most important conclusion of this study is that previously untreated prostate cancer, even at the metastatic stage, is exquisitely sensitive to androgens. These data clearly support the direct measurements of DHT in prostatic cancer tissue (Geller *et al.*, 1984), which indicates that following DES or orchiectomy, a significant amount of androgens are left in the prostatic tissue and continue to stimulate cancer growth.

The most likely explanation for the difference observed between the present results and those of previous studies is that previous hormonal therapy was limited to the neutralization of androgens of testicular origin by surgical castration and/or estrogens, whereas the present approach achieves more complete blockade of androgens of both testicular and adrenal origin at the start of treatment. A large number of reports have shown that neutralization of adrenal androgens has beneficial effects on prostate cancer (for review, see Labrie *et al.*, 1985). However, in the past, medical or surgical adrenalectomy or hypophysectomy was never performed as a first approach in combination with blockade of testicular androgens. The neutralization of adrenal androgens was always achieved as a second step following the lack of response to castration or when relapse of the disease had occurred after a period of remission (Labrie *et al.*, 1985).

An unexpected but most important additional benefit of combined antiandrogen treatment is that it inhibits by approximately 50% the serum levels of adrenal steroids responsible for the formation of active androgens in prostatic cancer tissue, especially dehydroepiandrosterone (DHEA), DHEA sulfate (DHEA-S), androstenedione, and androst-5-ene-3β,17β-diol. This approximately 50% decrease in the serum levels of precursor steroids should lead to a similar decrease in the level of active androgens in the prostatic cancer, thus decreasing the stimulatory androgenic influence on cancer growth. This 50% decrease in local androgen concentration should facilitate the inhibitory action of the antiandrogen. It is thus quite remarkable that the combination treatment, in addition to completely blocking testicular androgen secretion as well as neutralizing the peripheral action of androgens, can also achieve a partial medical adrenalectomy limited to androgen precursors and not affecting the secretion of cortisol. In order to minimize the development of androgen-sensitive tumors that are induced by the low androgen levels remaining in the prostate following castration alone (Labrie *et al.*, 1985), it is suggested that in the future the treatment of prostate cancer should aim at a complete blockade of the secretion and/or action of androgens of both testicular and adrenal origin at the start of treatment.

In agreement with the present data, 33 to 39% of patients already castrated or treated with estrogens have shown a positive response to Flutamide® (Sogani *et al.*, 1975; Stoliar and Albert, 1974). Similarly, after adrenal androgen suppression with aminoglutethimide in patients who had become refractory to orchiectomy or estrogens, a positive response was observed in 50% of patients (Robinson *et al.*, 1974). Bilateral adrenalectomy and hypophysectomy have been associated with palliation in 20 to 60% of cases (Huggins and Scott, 1945; Labrie *et al.*, 1985).

There is convincing clinical and laboratory evidence for an important role of adrenal androgens in prostate cancer. The clinical data pertain to the observation in all studies of a 30 to 50% response to adrenalectomy, hypophysectomy, Flutamide®, or aminoglutethimide in patients who showed relapse after orchiectomy or treatment with estrogens (Huggins and Scott, 1945; Sogani *et al.*, 1975; Stoliar and Albert, 1974; Robinson *et al.*, 1974; Labrie *et al.*, 1985). Such a response can only be explained by the continuing stimulatory action of adrenal androgens on prostate cancer growth following medical or surgical castration.

As a support to the clinical data obtained following blockade of the secretion or action of adrenal androgens, all the enzymes required for the conversion of the adrenal precursors dehydroepiandrosterone (DHEA), DHEA sulfate (DHEA-S), androst-5-ene-3β,17β-diol (Δ^5-diol), and androstenedione (Δ^4-dione) have been described in prostate cancer tissue (Harper *et al.*, 1974). Moreover, castrated levels of serum testosterone at 5–10% of control (0.2–0.4 ng/ml) are still effective in androgen target tissues and can maintain androgen-dependent functions at 30 to 40% of control (Labrie *et al.*, 1985; Bartsch *et al.*, 1983). These findings are in agreement with the report that the intraprostatic concentration of the most active androgen, dihydrotestosterone (DHT), is only reduced by 50 to 70% following orchiectomy (Geller *et al.*, 1984).

There is thus ample clinical and biochemical evidence for a role of adrenal androgens in prostate cancer. The remaining question was the relative benefits of early versus late blockage of adrenal androgens. The finding of a 95.4% rate of positive objective response at start of treatment following combination therapy in previously untreated patients (Table I) indicates that prostate cancer, even at the advanced stage of metastases, remains exquisitely sensitive to androgens when continuously exposed to normal levels of androgens. However, when exposed for some time to the low level of adrenal androgens, tumors become autonomous or androgen insensitive, thus losing their property of responding to the combined antiandrogen blockade in approximately 60% of the cases.

It should be mentioned that the development of treatment resistance has been observed in some patients who had received estrogens or had orchiectomy for only 5 to 10 days, although the average pretreatment time was 10.9 months. The appearance of autonomous prostate cancer tumors in the human is analogous to the development of androgen resistance in mouse Shionogi mammary carcinoma cells, where the phenomenon occurs extremely rapidly and has been found within 2 weeks of exposure to low androgens (King *et al.*, 1977; Luthy and Veilleux, 1985). The recent finding that the antiandrogen Flutamide® can prevent the loss of androgen sensitivity (Luthy and Veilleux, 1985) in the presence of low androgen levels may partly explain the favorable results observed in the present study in patients who received the combination therapy at the start of treatment. In fact, Flutamide® not only blocks the stimulatory action of the adrenal androgens that remain present in prostatic cancer tissue following medical or surgical castration but may well block the spontaneous action of the free androgen receptor and thus prevent or delay the development of androgen-resistant tumors (Simard and Labrie, 1984).

The present data confirm that a large proportion of prostatic cancer tumors are autonomous or treatment resistant at the time of relapse following medical or surgical castration. As a consequence, the delay in administering the combination therapy has a major negative impact on the quality of life and survival, the death rate during the first 2 years being more than fourfold higher in the group of previously treated patients as compared to those who received the combination therapy as first treatment. Since the combined antiandrogen blockade provides a much higher rate of positive response at the start of treatment (95.4 versus 60–80%) and provides additional years of excellent quality of life with no side effects other than those related to the blockage of androgens (hot flashes and a decrease or loss of libido), it seems reasonable to propose that the combination therapy should be given as first treatment with no exception to all patients having advanced prostate cancer. For all those who have received previous hormonal therapy, the combination treatment remains the best alternative and can still improve their quality of life and possibly survival.

4. SUMMARY

Chronic treatment of adult men with LHRH agonists causes a rapid decrease in serum testosterone and 5α-dihydrotestosterone to castration levels. In the presence of such low levels of circulating testicular androgens, the concentration of serum LH measured by radioimmunoassay (RIA) sometimes remains normal or is only partially inhibited. The close parallelism observed between serum testosterone and bioactive LH levels suggests that the loss of biological activity of the gonadotropin is mainly, if not exclusively, responsible for the inhibition of testicular androgen secretion observed during chronic treatment with LHRH agonists.

Eighty-eight previously untreated patients with clinical stage D2 (bone metastases) prostate cancer have received the combination therapy with a pure antiandrogen and an LHRH agonist (or orchiectomy) as first treatment in a multicenter study for up to 36 months (average 17.6 months). A positive objective response assessed according to the criteria of the U.S. NPCP has been observed in 95.4% of cases. Pain disappeared in all patients within 1 month, and performance became normal in all (including two bedridden patients) within 4 months. Progression of the disease after a period of remission was observed in 22 patients. Four patients have died from prostate cancer, and four died from other causes. The probability of continuing response and survival at 2 years for the patients who received the combination treatment at start of treatment (Kaplan–Meier method) is 81 and 89.1%, respectively. By contrast, in the randomized group who had orchiectomy alone, four of seven died from prostate cancer ($P < 0.05$ as compared to combination therapy).

In addition to a marked improvement in the remission rate and survival, combination therapy maintains a good quality of life. Hot flashes and a decrease or loss of libido are the only side effects. Comparison of the efficacy of combination therapy was made in a group of 56 patients who had previous hormonal therapy (orchiectomy, estrogens, or LHRH agonists alone). Only 42.9% of patients who had previous treatment showed a positive response at start of treatment. Whereas a death rate of 10.9% was observed at 2 years in the group of patients who received the combination therapy as first treatment, a death rate of 51.8% was observed in those who had medical or surgical castration as first treatment prior to combination therapy at the time of relapse. Survival was calculated from the time of first treatment in the two comparable groups of patients. The present data show that combination therapy with an LHRH agonist or orchiectomy in association with a pure antiandrogen (Flutamide®) should be the first line treatment for all patients having advanced prostate cancer. Early combination therapy decreases by more than fourfold the death rate within the first 2 years of treatment while preserving a good quality of life.

REFERENCES

Armitage, T., 1971, *Statistical Methods in Medical Research,* Blackwell Scientific, Oxford.

Bartsch, W., Knabbe, B., and Voigt, K. D., 1983, Regulation and compartmentalization of androgens in rat prostate and muscle, *J. Steroid Biochem.* **19:**929–937.

Faure, N., Labrie, F., Lemay, A., Bélanger, A., Gourdeau, Y., Laroche, B., and Robert, G., 1982, Inhibition of serum androgen levels by chronic intranasal administration of a potent LHRH agonist in adult men, *Fertil. Steril.* **37:**416–424.

Geller, J., Albert, J. D., Nachtsheim, D. A., and Loza, D. C., 1984, Comparison of prostatic cancer tissue dehydrotestosterone levels at the time of relapse following orchiectomy or estrogen therapy, *J. Urol.* **132:**693–696.

Goldman, A. J., and Elwood, J. M., 1979, Examining survival data, *CMA J.* **121:**1065–1071.

Harper, M. E., Pike, A., Peeling, W. B., and Griffiths, K., 1974, Steroids of adrenal origin metabolized by human prostatic tissue both *in vivo* and *in vitro, J. Endocrinol.* **60:**117–125.

Huggins, C., and Scott, W. W., 1945, Bilateral adrenalectomy in prostatic cancer, *Ann. Surg.* **122:**1031–1041.

Jacobi, G. H., and Wenderoth, U. K., 1982, Gonadotropin-releasing hormone analogs for prostatic cancer: Untoward side-effects of high dose regimens acquire a therapeutical dimension, *Eur. Urol.* **8:**129–134.

Jordan, W. P., Jr., Blackard, C. E., and Byar, D. P., 1977, Reconsideration of orchiectomy in the treatment of advanced prostatic carcinoma, *South. Med. J.* **70:**1411–1413.

Kahan, A., Delrieu, F., Amor, B., Chiche, R., and Steg, A., 1984, Disease flare induced by D-Trp[6]-LHRH analogue in patients with metastatic prostatic cancer, *Lancet* **1:**971–972.

Kaplan, E. L., and Meier, P., 1958, Nonparametric estimation from incomplete observations, *Am. Statist. Assoc. J.* **53:**457–481.

King, R. J. B., Cambray, G. J., Jagus-Smith, R., Robinson, J. H., and Smith, J. A., 1977, Steroid hormone and the control of tumor growth: Studies on androgen-responsive tumor cells in culture, in: *Receptors and Mechanism of Action of Steroid Hormones* (J. R. Pasqualini, ed.), Marcel Dekker, New York, pp. 215–227.

Labrie, F., Auclair, C., Cusan, L., Kelly, P. A., Pelletier, G., and Ferland, L., 1978, Inhibitory effects of LHRH and its agonists on testicular gonadotrophin receptors and spermatogenesis in the rat, *Int. J. Androl.* [Suppl.]**2:**303–318.

Labrie, F., Bélanger, A., Cusan, L., Séguin, C., Pelletier, G., Kelly, P. A., Lefebvre, F. A., Lemay, A., and Raynaud, J. P., 1980, Antifertility effects of LHRH agonists in the male, *J. Androl.* **1:**209–228.

Labrie, F., Dupont, A., Bélanger, A., Cusan, L., Lacoursière, Y., Monfette, G., Laberge, J. G., Emond, J. P., Fazekas, A. T. A., Raynaud, J. P., and Husson, J. M., 1982, New hormonal therapy in prostatic carcinoma: Combined treatment with an LHRH agonist and an antiandrogen, *Clin. Invest. Med.* **5:**267–275.

Labrie, F., Dupont, A., Bélanger, A., Lacourcière, Y., Raynaud, J. P., Husson, J M., Gareau, J., Fazekas, A. T. A., Sandow, J., Monfette, G., Girard, J. G., Emond, J. P., and Houle, J. G., 1983a, New approach in the treatment of prostate cancer: Complete instead of only partial withdrawal of androgens, *Prostate* **4:**579–594.

Labrie, F., Dupont, A., Bélanger, A., Lefebvre, F. A., Cusan, L., Monfette, G., Laberge, J. G., Emond, J P., Raynaud, J. P., Husson, J. M., and Fazekas, A. T. A., 1983b, New hormonal treatment in cancer of the prostate: Combined administration of an LHRH agonist and an antiandrogen, *J. Steroid Biochem.* **19:**999–1008.

Labrie, F., Bélanger, A., Dupont, A., Emond, J., Lacourcière, Y., and Monfette, G., 1984, Combined treatment with an LHRH agonist and a pure antiandrogen in advanced carcinoma of the prostate, *Lancet* **1:**1090.

Labrie, F., Dupont, A., and Bélanger, A., 1985, Complete androgen blockade for the treatment of prostate cancer, in: *Important Advances in Oncology* (V. T. De Vita, Jr., S. Hellman, and S. A. Rosenberg, eds.), J. B. Lippincott, Philadelphia, pp. 193–217.

Leuprolide Study Group, 1984, Leuprolide versus diethylstilbestrol for metastatic prostate cancer, *N. Engl. J. Med.* **311:**1281–1286.

Luthy, I., and Veilleux, R., 1985, The antiandrogen Flutamide® prevents the loss of androgen responsiveness in SC-115 mouse mammary carcinoma cells, in: *Proceedings 67th Annual Meeting Endocrine Society,* p. 193.

Mettlin, C., Natarajan, N., and Murphy, G. P., 1982, Recent patterns of care of prostatic cancer patients in the United States: Results from the surveys of the American College of Surgeons Commission on Cancer, *Int. Adv. Surg. Oncol.* **5:**277–321.

Murphy, G. P., Beckley, S., Brady, M. F., Chu, M., DeKernion, J. B., Dhabuwala, C., Gaeta, J. F., Gibbons, R. P., Loening, S., McKiel, C. F., McLeod, D. G., Pontes, J. E., Prout, G. R., Scardino, P. T., Schlegel, J. U., Schmidt, J. D., Scott, W. W., Slack, N. H., and Soloway, M., 1983, Treatment of newly diagnosed metastatic prostate cancer patients with chemotherapy agents in combination with hormones versus hormones alone, *Cancer* **51:**1264–1272.

Nesbit, R. M., and Baum, W. C., 1950, Endocrine control of prostatic carcinoma: Clinical and statistical survey of 1818 cases, J.A.M.A. **143:**1317–1320.

Robinson, M. R. G., Shearer, R. J., and Fergusson, J. D., 1974, Adrenal suppression in the treatment of carcinoma of the prostate, *Br. J. Urol.* **46:**555–559.

Simard, J., and Labrie, F., 1984, Unoccupied androgen receptors are biologically active in rat pituitary gonadotrophs. *Excerpta Medica Congr. Ser.* **652:**973.

Slack, N. H., Murphy, G. D., and NPCP participants, 1984, Criteria for evaluating patient responses to treatment modalities for prostatic cancer, *Urol Clin. North Am.* **11:**337–342.

Sogani, P. C., Ray, B., and Whitmore, W. F. Jr., 1975, Advanced prostatic carcinoma: Flutamide® therapy after conventional endocrine treatment, *Urology* **6:**164–166.

St.-Arnaud, R., Lachance, R., Kelly, S. J., Bélanger, A., Dupont, A., and Labrie, F., 1986, Loss of luteinizing hormone (LH) bioactivity in patients with prostatic cancer treated with an LHRH agonist and a pure antiandrogen, *Clin. Endocrinol.* **24:**21–30.

Stoliar, B., and Albert, D. J., 1974, SCH 13521 in the treatment of advanced carcinoma of the prostate, *J. Urol.* **111:**803–807.

Relationship of Clinical to Basic Research with Peptides as Illustrated by MSH

ABBA J. KASTIN, RICHARD D. OLSON, JAMES E. ZADINA, and WILLIAM A. BANKS

1. INTRODUCTION

Many of the principles that are now influencing the field of brain peptides were established more than a decade ago with melanocyte-stimulating hormone (MSH). A series of studies with MSH demonstrated not only the multiple actions of peptides but also the complex relationship of basic to clinical investigation. The actions of MSH, therefore, can be considered a prototype for those of other peptides. The present review discusses several of these actions with examples taken from our own experience. Where possible, clinical studies are emphasized even though most of our research has involved laboratory animals.

2. CENTRAL EFFECTS OF PEPTIDES AFTER PERIPHERAL ADMINISTRATION

2.1. Electroencephalography

The first clinical trial of a brain peptide was performed in the mid-1960s with MSH. It showed changes in the electrical activity of the brain after infusion in a peripheral vein (Kastin *et al.*, 1968b). These electroencephalographic (EEG) changes were similar to those found earlier after injection of MSH in rabbits (Dyster-Aas

ABBA J. KASTIN, RICHARD D. OLSON, JAMES E. ZADINA, and WILLIAM A. BANKS • Veterans Administration Medical Center, University of New Orleans, and Tulane University School of Medicine, New Orleans, Louisiana 70146.

and Krakau, 1965). The next EEG study of MSH also involved humans. It demonstrated an increase in the averaged somatosensory cortical evoked response that was so marked it could be seen on single trials of the EEG. This increase was enhanced when the subject focused attention to a particular task (Kastin *et al.*, 1971). The subsequent studies were performed, in order, in rats (Sandman *et al.*, 1972), frogs (Denman *et al.*, 1972), and dogs (Urban *et al.*, 1974). An intriguing finding from one of the more recent EEG studies in rats showed that the longer the latency of onset of the EEG changes, the shorter the duration (Miller *et al.*, 1981).

2.2. Behavior

The EEG studies in both laboratory animals and humans were paralleled by a series of behavioral studies. In most cases, psychological tests were performed together with the EEG studies. In both rats and humans, MSH seemed to improve visual attention even in the mentally retarded (Kastin *et al.*, 1971; Sandman *et al.*, 1972, 1975, 1976, 1977; Miller *et al.*, 1974, 1976, 1980; Beckwith *et al.*, 1976). There was even more resistance to the concept of a peptide inducing behavioral changes than there was to the possibility that it could produce EEG changes. This is probably explained by the misconception that changes in the electrical activity of the brain are somehow more meaningful and scientific than are psychological studies. In the case of MSH, EEG and behavioral studies were performed in both rats and humans to demonstrate that a peripherally administered peptide could exert effects that were not confined to pigmentary effects in the periphery.

2.3. Blood–Brain Barrier

Although the demonstration of effects of MSH on the central nervous system (CNS) after peripheral administration seems to constitute evidence for passage of this peptide across the blood–brain barrier (BBB), several other possibilities could explain such effects, as has been reviewed elsewhere (Kastin *et al.*, 1979, 1981). Some of the first attempts to demonstrate direct penetration involved MSH, but there were limitations to these early studies, including lack of confirmation that the portion of the molecule identified by radioactivity or by radioimmunoassay (RIA) remained on the intact molecule (Kastin *et al.*, 1981). These studies were useful, nevertheless, in providing preliminary evidence of direct penetration of the BBB and especially in focusing attention on this previously ignored problem.

The first demonstration that peripherally injected peptides could penetrate the BBB in intact form was provided in a study with tritiated MSH. Thin-layer chromatography, as well as chromatoelectrophoresis, of homogenized brain tissue from animals injected with the labeled peptide revealed a radioactive peak at the same R_f as the MSH standard (Kastin *et al.*, 1976). Although clinical studies with MSH have contributed greatly to the general understanding of the behavioral and EEG effects of peptides like MSH, clinical studies using sophisticated scanning techniques of the penetration of peptides across the BBB have been lacking.

3. ACTIONS DEPENDENT ON SITUATIONAL REQUIREMENTS

3.1. Behavior

To some investigators, a behavorial effect of a substance like a peptide must be manifested by a change in activity. It was demonstrated many years ago in the rat, however, that MSH does not affect locomotor activity when injected by itself (Kastin *et al.*, 1973b). It can interact with other substances to alter locomotor activity (Yehuda and Kastin, 1980), but different paradigms must be examined to determine specific effects.

Most early studies with MSH involved rats in aversive situations (Greven and De Wied, 1977). These led to the conclusion that MSH affected memory. Although our first study of MSH in rats involved a nonaversive situation (Sandman *et al.*, 1969), much of the work demonstrating an action of MSH on attention rather than memory was performed in humans (Kastin *et al.*, 1971; Miller *et al.*, 1974, 1976, 1980; Sandman *et al.*, 1975, 1976, 1977). An occasional subject would spontaneously comment on his perception of improvement after MSH (Miller *et al.*, 1974, 1980), but specific testing for attention was usually required. In humans, performance on the Benton Visual Retention Test was most frequently improved by MSH (Kastin *et al.*, 1971; Sandman *et al.*, 1975, 1976; Miller *et al.*, 1980), whereas in rats the Y-maze and Thompson box involving reversal learning were most sensitive to the effects of MSH (Sandman *et al.*, 1972, 1973a, 1980a; Beckwith *et al.*, 1976). Three entirely different patterns of response for various MSH/ACTH-related compounds were found in the three phases of a learning task involving such an apparatus (Sandman *et al.*, 1980a).

Although it was a parallel series of clinical and basic studies that established that the more specific effects of MSH were on attention rather than classical memory, clinical studies established the specificity for visual in contrast to verbal attention (Kastin *et al.*, 1971; Miller *et al.*, 1976, 1980) long before it was demonstrated in the rat (Handlemann *et al.*, 1983). Evoked potentials were used in an elegant clinical study to demonstrate further the effects of MSH compounds on the process of attention (Fehm-Wolfsdorf *et al.*, 1981).

Melanocyte-stimulating hormone has been used in other studies to demonstrate different effects depending on the experimental situation. Both the level of electrical shock (Nockton *et al.*, 1972; Stratton and Kastin, 1974a) and complexity of the maze (Sandman *et al.*, 1969; Stratton and Kastin, 1974b) can contribute to different results in different situations. Moreover, just because an analogue of a compound (e.g., Org 2766) is 1000 times more potent than the parent compound (MSH) in one situation (Greven and De Wied, 1977) does not mean that it will be more potent in every situation (File, 1981).

3.2. MSH versus ACTH Release

It has been axiomatic for years that MSH release always occurs in parallel with ACTH release. More recently, similar statements have been made about β-endorphin and ACTH. Depending on the experimental situation, however, it is possible to show a dissociation of such release.

Starting with the basic differences we had found between the predominant inhibitory influence of the hypothalamus on MSH release as compared with its stimulatory influence on ACTH release (Kastin and Schally, 1972), we reasoned that other situations might be found with similar dissociations. This differential release of MSH and ACTH was found after adrenalectomy in rats (Gosbee *et al.*, 1970) and dexamethasone suppression in both rats (Dunn *et al.*, 1972) and humans (Kastin *et al.*, 1973a). The apparent titration of the degree of stress with the plasma levels of peptide that was found in one clinical study (Kastin *et al.*, 1973a) might also help to explain some of the differences in hormonal levels in two other studies, one in rats (Sandman *et al.*, 1973) and one in humans (Kastin *et al.*, 1974). Thus, both basic and clinical conditions have been found in which the release of MSH did not always parallel that of ACTH.

4. TYPE OF SUBJECT

4.1. Sex

For many years all studies with peptides like MSH in laboratory animals used only males. We found, however, a marked difference in the response of male and female rats to MSH (Beckwith *et al.*,1977b). Differences were later found in some (Veith *et al.*, 1978; Miller *et al.*, 1980) but not all (Ward *et al.*, 1979) clinical studies with MSH compounds. One of these studies involved elderly human subjects, who seemed to respond in the same manner as the younger subjects usually tested (Miller *et al.*, 1980).

4.2. Strain

Using albino and hooded rats of different strains, we also found different effects of MSH on the interaction between activity and dark preferences (Stratton *et al.*, 1973). Differences were also observed according to strain of rat for visual discrimination and reversal learning (Sandman *et al.*, 1973a).

4.3. Species

The results from almost all studies with MSH appear to be readily transferrable from rat to human and human to rat. One of the few exceptions involves an MSH/ACTH analogue that failed to influence extinction of the conditioned avoidance response in the 20 young men studied (Miller *et al.*, 1977). It is possible that the dose was too high, particularly since it is known that such peptides are more slowly degraded by human serum than by rat serum (Marks *et al.*, 1976).

5. HALF-LIFE

Although variations exist in the persistence of peptides like MSH in blood depending on the method of determination, there is a clear dissociation between their presence in blood and their actions. Onset as well as duration of effects (Miller *et al.*, 1981) are found in both rat (Kastin *et al.*, 1975) and human (Redding *et al.*, 1978) at times exceeding the presence of MSH in blood. In this situation, the human studies followed those in the rat. Despite different rates of degradation (Marks *et al.*, 1976), they show that correlations of the blood levels of peptides are not necessarily relevant for their actions. This demonstrates that rules generally accepted for other compounds need not apply to peptides.

6. PERINATAL ADMINISTRATION

Perhaps the most extreme example of the dissociation of blood levels of MSH from its length of action occurs in the perinatal situation. Rats injected during the first week of life with MSH but not tested until the adult stage show long-lasting effects of the MSH (Beckwith *et al.*, 1977a,b). Similar types of studies have now been performed with other peptides such as β-endorphin in rats. Obviously, no clinical studies of this have been performed, but analogous situations may occur naturally, such as the stressful release of excessive peptides during childbirth. It is known that MSH exists in the human being as early as 11 weeks of gestation. (Kastin *et al.*, 1968a).

7. NONLINEAR DOSE–RESPONSE PATTERNS

Another example of the unexpected results found with peptides involves the inverted-U dose–response curve and other complex relationships. Although originally described with another peptide, it has been occasionally observed in studies with MSH. These include increased footshook-induced fighting in mice (Plotnikoff and Kastin, 1976), thermoregulation in rats (Yehuda and Kastin, 1980), and workshop productivity in mentally retarded humans (Sandman *et al.*, 1980b). In mentally retarded adults paid to contour electrical resistors to fit a mold, 5 or 10 mg of an MSH/ACTH analogue significantly enhanced productivity of tasks requiring precision and concentration, whereas 20 mg depressed performance of all tasks (Sandman *et al.*, 1980b). A linear relationship between the dose of a peptide and its effect, therefore, does not always apply in laboratory animals or humans.

8. CONSTRICTED NOMENCLATURE

Both basic and clinical experiments illustrate that the name of a peptide does not define its only action. None of the effects of MSH described in this paper are known to be mediated by melanocytes. The harm done by naming a peptide for the action for which it is first described can be avoided only if investigators do not feel constrained by such nomenclature (Kastin *et al.*, 1983).

9. CONCLUSION

Melanocyte-stimulating hormone has served as a prototype for many of the principles of peptide research in both laboratory animals and humans. These principles should serve as a good illustration that new concepts can be discovered with peptides without known physiological roles. Although full explanations of each phenomenon are still lacking, MSH well illustrates the many dissociative actions that characterize the multiple actions of peptides.

ACKNOWLEDGMENTS. This work was supported in part by the Veterans Administration and the Office of Naval Research.

REFERENCES

Beckwith, B. E., Sandman, C.A., and Kastin, A. J., 1976, Influence of three short-chain peptides (α-MSH, MSH/ACTH 4–10, MIF-1) on dimensional attention, *Pharmacol. Biochem. Behav.* **5**:11–16.

Beckwith, B. E., O'Quinn K. R., Petro, M. S., Kastin, A. J., and Sandman, C. A., 1977a, The effects of neonatal injections of α-MSH on the open-field behavior of juvenile and adult rats, *Physiol. Psychol.* **5**:295–299.

Beckwith, B. E., Sandman, C. A., Hothersall, D., and Kastin, A. J., 1977b, Influence of neonatal injections of α-MSH on learning, memory and attention in rats, *Physiol. Behav.* **18**:63–71.

Denman, P. M., Miller, L. H., Sandman, C. A., Schally, A. V., and Kastin, A. J., 1972, Electrophysiological correlates of melanocyte-stimulating hormone activity in the frog, *J. Comp. Physiol. Psychol.* **80**:59–65.

Dunn, J. D., Kastin, A. J., Carrillo, A. J., and Schally, A. V., 1972, Additional evidence for dissociation of melanocyte-stimulating hormone and corticotrophin release, *J. Endocrinol.* **55**:463–464.

Dyster-Aas, H.K., and Krakau, C. E. T., 1965, General effects of α-melanocyte stimulating hormone in the rabbit, *Acta Endocrinol.* **48**:609–618.

Fehm-Wolfsdorf, G., Elbert, T., Lutzenberger, W., Rockstroh, B., Birbaumer, N., and Fehm, H. L., 1981, Effect of an ACTH 4–9 analog on human cortical evoked potentials in a two-stimulus reaction time paradigm, *Psychoneuroendocrinology* **6:**311–320.

File, S. E., 1981, Contrasting effects of Org 2766 and α-MSH on social and exploratory behavior in the rat, *Peptides* **2:**255–260.

Gosbee, J. L., Kraicer, J., Kastin, A. J., and Schally, A. V., 1970, Functional relationship between the pars intermedia and ACTH secretion in the rat, *Endocrinology* **86:**560–567.

Greven, H. M., and De Wied, D., 1977, Influence of peptides structurally related to ACTH and MSH on active avoidance behavior in rats, *Front. Horm. Res.* **4:**140–152.

Handelmann, G. E., O'Donohue, T. L., Forrester, D., and Cook, W., 1983, Alpha-melanocyte stimulating hormone facilitates learning of visual but not auditory discriminations, *Peptides* **4:**145–148.

Kastin, A. J., and Schally, A. V., 1972, MSH release in mammals, in: *Pigmentation: Its Genesis and Biologic Control* (V. Riley, ed.), Appleton-Century-Crofts, New York, pp. 215–223.

Kastin, A. J., Gennser, G., Arimura, A., Miller, M. C. III, and Schally, A. V., 1968a, Melanocyte-stimulating and corticotrophic activities in human foetal pituitary glands, *Acta Endocrinol.* **58:**6–10.

Kastin, A. J., Kullander, S., Borglin, N. E., Dahlberg, B., Dyster-Aas, K., Krakau, C. E. T., Ingvar, D. H., Miller, M. C. III, Bowers, C. Y., and Schally, A. V., 1968b, Extrapigmentary effects of melanocyte-stimulating hormone in amenorrhoeic women, *Lancet* **1:**1007–1010.

Kastin, A. J., Miller, L. H., Gonzalez-Barcena, D., Hawley, W. D., Dyster-Aas, K., Schally, A. V., Velasco De Parra, M. L., and Velasco, M., 1971, Psycho-physiologic correlates of MSH activity in man, *Physiol. Behav.* **7:**893–896.

Kastin, A. J., Beach, G. D., Hawley, W. D., Kendall, J. W., Jr., Edward, M. S., and Schally, A. V., 1973a, Dissociaton of MSH and ACTH release in man, *J. Clin. Endocrinol. Metab.* **36:**770–772.

Kastin, A. J., Miller, M. C., Ferrell, L., and Schally, A. V., 1973b, General activity in intact and hypophysectomized rats after administration of melanocyte-stimulating hormone (MSH), melatonin, and Pro-Leu-Gly-NH_2, *Physiol. Behav.* **10:**399–401.

Kastin, A. J., Hawley, W. D., Miller, M. C., Schally, A. V., and Lancaster, C., 1974, Plasma MSH and cortisol levels in 567 patients with special reference to brain trauma, *Endocrinol. Exp.* **8:**97–105.

Kastin, A. J., Nissen, C., Nikolics, K., Medzihradszky, K., Coy, D. H., Teplan, I., and Schally, A. V., 1976, Distribution of ^{3}H-α-MSH in rat brain, *Brain Res. Bull.* **1:**19–26.

Kastin, A. J., Olson, R. D., Schally, A. V., and Coy, D. H., 1979, CNS effects of peripherally administered brain peptides, *Life Sci.* **25:**401–414.

Kastin, A. J., Olson, R. D., Fritschka, E., and Coy, D. H., 1981, Neuropeptides and the blood–brain barrier, in: *Cerebral Microcirculation and Metabolism* (J. Cervos-Navarro, and E. Fritschka, eds.), Raven Press, New York, pp. 139–145.

Kastin, A. J., Banks, W. A., Zadina, J. E., and Graf, M., 1983, Brain peptides: The dangers of constricted nomenclatures, *Life Sci.* **32:**295–301.

Marks, N., Stern, F., and Kastin, A. J., 1976, Biodegradation of α-MSH and derived peptides by rat brain extracts, and by rat and human serum, *Brain Res. Bull.* **1:**591–593.

Miller, L. H., Kastin, A. J., Sandman, C. A., Fink, M., and Van Veen, W. J., 1974, Polypeptide influences on attention, memory and anxiety in man, *Pharmacol. Biochem. Behav.* **2:**663–668.

Miller, L. H., Harris, L. C., Van Riezen, H., and Kastin, A. J., 1976, Neuroheptapeptide influence on attention and memory in man, *Pharmacol. Biochem. Behav.* **5:**17–21.

Miller, L. H., Fischer, S. C., Groves, G. A., and Rudrauff, M. E., 1977, MSH/ACTH 4–10 influences on the CAR in human subjects: A negative finding, *Pharmacol. Biochem. Behav.* **7:**417–419.

Miller, L. H., Groves, G. A., Bopp, M. J., and Kastin, A. J., 1980, A neuroheptapeptide influence on cognitive functioning in the elderly, *Peptides* **1:**55–57.

Miller, L. H., Kastin, A. J., Hayes, M., Sterste, A., Garcia, J., and Coy, D. H., 1981, Inverse relationship between onset and duration of EEG effects of six peripherally administered peptides, *Pharmacol. Biochem. Behav.* **15:**845–848.

Nockton, R., Kastin, A. J., Elder, S. T., and Schally, A. V., 1972, Passive and active avoidance responses at two levels of shock after administration of melanocyte-stimulating hormone, *Horm. Behav.* **3**:339–344.

Plotnikoff, N. P., and Kastin, A. J., 1976, Neuropharmacological tests with α-melanocyte stimulating hormone, *Life Sci.* **18**:1217–1222.

Redding, T. W., Kastin, A. J., Nikolics, K., Schally, A. V., and Coy, D. H., 1978, Disappearance and excretion of labeled α-MSH in man, *Pharmacol. Biochem. Behav.* **9**:207–212.

Sandman, C. A., Kastin, A. J., 1978, Interaction of α-MSH and MIF-1 with *d*-amphetamine on open-field behavior of rats, *Pharmacol. Biochem. Behav.* **9**:759–762.

Sandman, C. A., Kastin, A. J., and Schally, A. V., 1969, Melanocyte-stimulating hormones and learned appetitive behavior, *Experientia* **25**:1001–1002.

Sandman, C. A., Denman, P. M., Miller, L. H., and Knott, J. R., 1971, Electroencephalographic measures of melanocyte-stimulating hormone activity, *J. Comp. Physiol. Psychol.* **76**:103–109.

Sandman, C. A., Miller, L. H., Kastin, A. J., and Schally, A. V., 1972, Neuroendocrine influence on attention and memory, *J. Comp. Physiol. Psychol.* **80**:54–58.

Sandman, C. A., Alexander, W. D., and Kastin, A. J., 1973a, Neuroendocrine influences on visual discrimination and reversal learning in the albino and hooded rat, *Physiol. Behav.* **11**:613–617.

Sandman, C. A., Kastin, A. J., Schally, A. C., Kendall, J. W., and Miller, L. H., 1973b, Neuroendocrine responses to physical and psychological stress, *J. Comp. Physiol. Psychol.* **81**:386–390.

Sandman, C. A., George, J., Nolan, J. D., Van Reizen, H., and Kastin, A. J., 1975, Enhancement of attention in man with ACTH/MSH 4–10, *Physiol. Behav.* **15**:427–431.

Sandman, C. A., George, J., Walker, B. B., and Nolan, J. D., 1976, Neuropeptide MSH/ACTH 4–10 enhances attention in the mentally retarded, *Pharmacol. Biochem. Behav.* **5**:23–28.

Sandman, C. A., George, J. M., McCanne, T. R., Nolan, J. D., Kaswan, J., and Kastin, A. J., 1977, MSH/ACTH 4–10 influences behavioral and physiological measures of attention, *J. Clin. Endocrinol. Metab.* **44**:884–891.

Sandman, C. A., Beckwith, B. E., and Kastin, A. J., 1980a, Are learning and attention related to the sequence of amino acids in ACTH/MSH peptides? *Peptides* **1**:277–280.

Sandman, C. A., Walker, B. B., and Lawton, C. A., 1980b, An analog of MSH/ACTH 4–9 enhances interpersonal and environmental awareness in mentally retarded adults, *Peptides* **1**:109–114.

Stratton, L. O., and Kastin, A. J., 1974a, Avoidance learning at two levels of shock in rats receiving MSH, *Horm. Behav.* **5**:149–155.

Stratton, L. O., and Kastin, A. J., 1974b, Increased acquisition of a complex appetitive task after MSH and MIF, *Pharmacol. Biochem. Behav.* **3**:901–904.

Stratton, L. O., Kastin, A. J., and Coleman, W. P. III, 1973, Activity and dark preference responses of albino and hooded rats receiving MSH, *Physiol. Behav.* **11**:907–909.

Urban, I., Lopes da Silva, F. H., Storms van Leeuwen, W., and DeWied, D., 1974, A frequency shift in the hippocampal theta activity: An electrical correlate of central action of ACTH analogues in the dog? *Brain Res.* **69**:361–365.

Veith, J. L., Sandman, C. A., George, J.M., and Stevens, V. C., 1978, Effects of MSH/ACTH 4–10 on memory, attention and endogenous hormone levels in women, *Physiol. Behav.* **20**:43–50.

Ward, M. M., Sandman, C. A., George, J. M., and Shulman, H., 1979, MSH/ACTH 4–10 in men and women: Effects upon performance of an attention and memory task, *Physiol. Behav.* **22**:669–673.

Yehuda, S. and Kastin, A. J., 1980, Interation of MIF-1 or alpha-MSH with *d*-amphetamine or chlorpromazine on thermoregulation and motor activity of rats maintained at different ambient temperatures, *Peptides* **1**:243–248.

Perspectives in the Design of Peptide Analogues in Biomedical Applications

SERGE A. ST-PIERRE

1. INTRODUCTION

In 1971, when the late Josef Rudinger wrote *The Design of Peptide Hormone Analogs* for the series *Drug Design*, edited by E. J. Ariens (Rudinger, 1971), he was the pioneer in a discipline that has become a major trend in biomedical research: the design of drugs based on biologically active peptides.

At that period, relatively few biologically active peptides were known. Some of them, such as insulin and oxytocin, had already reached the status of drugs. The study of the biological function of others such as bradykinin, angiotensin, corticotropin, gastrin, and cholecystokinin was becoming documented through the work of a limited number of investigators.

During the last 15 years, important advances in isolation and purification techniques such as affinity chromatography and HPLC, the improvement of sequencing techniques and instruments, and the development of molecular biology were mostly responsible for the discovery of more than 100 new mammalian peptides, several of which still await a physiological role to be assigned. Moreover, a host of other new peptides were found in lower living organisms such as amphibians, invertebrates, or microorganisms, and they also possess interesting biological properties.

2. BIOMEDICAL USES OF PEPTIDES

Natural peptides and peptide analogues have a large number of diagnostic and therapeutic applications in human as well as in veterinary medicine. Table I sum-

SERGE A. ST-PIERRE • INRS-Santé, Institut National de la Recherche Scientifique, Pavillon Gamelin, Montreal, Quebec H1N 3M5, Canada.

TABLE I. Major Fields of Biomedical Applications for Peptides

Pharmacotherapy	Replacement of internal deficiencies (hormones, transmitters, antibodies, etc.)
	Adjustment of functioning of organs (blood pressure, water and salt balance, behavior, digestion, reproduction, etc.)
Chemotherapy	Diseases caused by viral or bacterial infections
	Neoplasms
Immunotherapy	Promoting natural defense mechanisms (synthetic vaccines)
	Stimulation of inhibitory cells or inhibition of effector cells (immunostimulants and immunomodulants)
Diagnostic	Markers of various pathologies characterized by significant changes of peptide concentrations in biological fluids or tissues
Others	Pain, fertility control, inflammation, etc.

marizes the major fields of application for peptides, and Table II gives a list of current as well as potential peptide drugs and their biomedical utilizations.

Replacement of deficient hormones in the organism represents the most classical example of the use of natural peptides as drugs, as illustrated by the treatment of diabetes with insulin. The word natural, in our sense, is related to the sequence of a given peptide as discovered, no matter whether this peptide is commercially obtained as an extract from living organisms or as a synthetic material prepared by chemical, enzymatic, or biotechnological techniques.

As most hormone deficiences are usually caused by a malfunction of vital organs or tissues, regulatory peptides with cardiotonic, diuretic, or psychotropic properties, among others, are finding a larger number of applications in medicine. This area is the one in which the most important effort is devoted to the design and the synthesis of analogues. Peptide inhibitors of regulatory enzymes such as the converting enzyme also fall into this category of applications (Ondetti *et al.*, 1977).

Infections and cancer have become promising new areas for the use of peptides as chemotherapeutic agents. For the treatment of certain forms of hormone-sensitive tumors or infections, the therapeutic effect of peptides or peptide antagonists has been found to depend either on the control of hormone secretion from producing cells or the regulation of their action at the receptor level. The hypothalamic peptides somatostatin and LHRH and their synthetic analogues have already found important clinical applications in these areas. Recent clinical studies have demonstrated that prostate cancer can be successfully treated with synthetic analogues of LHRH (Labrie *et al.*, 1985). Another investigation has shown that tumors in the digestive system can be treated with long-acting somatostatin analogues (Maton *et al.*, 1985).

Immunostimulating or immunomodulating peptides are known to enhance lymphocytic or phagocytic functions or the production of immunostimulating hormones such as interferons or interleukins. In this case, the word "chemotherapy" should be replaced by "immunopharmacotherapy" to better describe the mode of action of this new generation of drugs. Viral infections such as herpes simplex appear to respond to treatment with immunostimulating peptides (Djawari *et al.*, 1984). Pep-

TABLE II. Current and Potential Peptide Drugs and Their Fields of Application

Peptide drug or drug candidate	Field of application
Insulin	Diabetes
Oxytocin (Pitocin and others)	Uterus contraction Lactation stimulation
Vasopressin and analogues (Ornipressin and others)	Antidiuretic Diabetes insipidus
Angiotensin II antagonists (Saralasin)	Blood pressure control
Angiotensin-converting enzyme inhibitors (Captopril, Enalapril)	Blood pressure control
ACTH and analogues (Acethropan, Synchrodyn, and others)	Stimulation of glucocorticoid production Chronization
Calcitonin (Calsynar and others)	Regulation of calcium metabolism
Parthyroid hormone (Parathorm)	Regulation of calcium metabolism
Somatostatin and analogues (Stilamin)	Inhibition of growth hormone secretion Gastroduodenal ulcers and gastritis Cancer
LHRH (Factrel), analogues (Buserelin), and antagonists	Fertility control Cancer
TRH (Antepan and others)	Stimulation of TSH synthesis and release
GRF	Promotion of growth
Atrial natriuretic factor	Blood pressure control
Opiate peptides (endorphin, dynorphin, enkephalins)	Analgesia
Thymopoietin and thymosins	Viral infections Cancer
Delta sleep-inducing peptide (DSIP)	Promotion of sleep
CCK-8 and analogues	Suppression of appetite
VIP	Bronchodilation Vasodilation Promotion of intestinal secretions Promotion of hormone secretion (pituitary and pancreas)

tide immunostimulants such as tuftsin, muramyl-dipeptide, and bestatin were shown to be beneficial for the immunotherapeutic treatment of tumors in animals. On the other hand, investigations to develop the treatment of bacterial infections using peptide antibiotics that can interfere at various steps of vital processes of microbial pathogens are being carried out mainly in Japan, where most of these peptides have been isolated from microorganisms.

Very recently, following the pioneering work of Sela in Israel, Chedid in France, and Lerner in the United States, several promising results have been obtained with new bacterial, viral, and hormonal vaccines using synthetic peptide fragments as antigens. This revolutionary approach has already given rise to clinically relevant vaccines against viral infections such as influenza (Shapira *et al.*, 1984) and hepatitis

B (Bhatnagar *et al.*, 1982) in humans and foot and mouth disease in animals (Biddle *et al.*, 1982), which are presently under clinical investigation.

As another example of the application of peptides in immunology, the measurement by immunodiagnostic methods of abnormal concentrations of discrete peptide hormone markers in biological tissues and fluids is being extensively explored as a sensitive test for the early detection of certain forms of cancers. Convincing evidence was provided in recent years that peptides such as VIP, neurotensin, CRF, GRF, calcitonin, GRP, substance P, PP, β-endorphin, and other peptides could be synthesized and secreted by various tumors of endocrine nature. This discovery has stimulated investigators to develop sensitive immunodiagnostic tests (RIA or ELISA) using specific antibodies that can detect minute concentrations of these markers of cancer cells. Since the pioneering work of Moody *et al.* (1981), the principle of peptide hormone markers has been used by several authors in their attempt to develop a test for the early detection of small-cell lung cancer, which is characterized by the production and the secretion of peptide hormones such as gastrin-releasing peptide, substance P, somatostatin, and neurotensin (Wolfsen and Odell, 1982).

Several other applications for peptides can be found in various areas of biomedical interest. The need for nonaddictive peptide analgesics that could replace classical opiates such as morphine has led to the synthesis of a large number of analogues of the so called opiate peptides. Also, the elucidation of the role of LHRH in reproductive function is at the origin of new tools for fertility control in man and animal based on LHRH antagonists. Chemical castration has been made possible by immunization with LHRH (Carelli *et al.*, 1982). Also, as the role of kinins and of the complement peptides in inflammation is better understood, antagonists of peptides and inhibitors of enzymes involved in the inflammatory process appear as potential antiinflammatory drugs.

3. SPECIAL FEATURES OF PEPTIDES AS DRUGS

Peptides differ from classical drugs in several aspects. Some of their characteristics give them important advantages over classical drugs, but others restrict their application in therapeutics. These features are summarized in Table III.

TABLE III. Special Features of Peptides as Drugs

Stereotyped structural units: the amino acids
High molecular weight
Important conformational freedom
Metabolically unstable
Among the biologically most active compounds known
May have more than one distinct biological effect
Naturally occurring peptides may differ in structure and in physiological function in different species

Because they are composed of the most diversified ensemble of building blocks among all the natural substances, an astronomical number of entirely different peptides containing two to 50 residues can be assembled using various combinations of the 20 natural amino acids. This number of combinations can be increased manyfold with the use of a host of other amino acids obtained from organic synthesis or extraction procedures as substituents in natural sequences. On the other hand, since they are usually much simpler substances, the number of possibilities for creating biologically active analogues of classical drugs is infinitely lower than it is for peptides.

Although the molecular weight zone separating peptides and proteins is not clearly defined, most biologically active peptides known contain fewer than 50 amino acids. Therefore, peptides are large molecules compared to other drugs, with molecular weights ranging from 200 to several thousands. For instance, the size of dexamethasone, a large steroid drug, is comparable to enkephalin, a small peptide. In addition to being the origin of their high cost of fabrication, the large molecular weight of most peptides limits their diffusion through biological membranes. It also complicates their transport to active sites and, consequently, restricts their therapeutic applications.

Even though the large number of bonds and side chains gives peptides an immense variety of possible biological activities, it generates a proportional number of degrees of freedom. In addition to the large number of possible values for peptide bond angles, each amino acid side chain can adopt a whole range of conformations. As a result, for a simple decapeptide, thousands of discrete theoretical conformations are possible, whereas classical drugs are a lot more rigid and have a better defined tridimensional structure.

Metabolic instability is one of the main restrictions to the use of natural peptides as drugs. Biological tissues and fluids contain a variety of proteolytic enzymes that can rapidly degrade amide bonds located either at the ends (exopeptidases) or in the middle (endopeptidases) of peptide chains, causing their inactivation. In recent years, the nature of many specific and nonspecific proteases has been determined, together with their preferred cleavage sites. These findings make possible the design of reinforced peptide bonds at these sites or, from another approach, the preparation of specific enzyme inhibitors.

Several peptides are known to act in the vicinity of their site of production. On the other hand, most biologically active peptides are produced and have receptors in various parts of the organism, where they can have a whole range of different functions. This characteristic of peptides can be at the origin of numerous side effects that cancel or surpass expected beneficial effects.

Still, peptides are among the most active biological substances known. Nanomolar doses of most peptides can given rise to a biological response, whereas classical drugs are usually active only at much higher concentrations. Therefore, from the above observations, there is generally a lesser need for peptide superagonists than for analogues with improved resistance to proteolytic enzymes, selectivity, and transport properties.

4. THE DESIGN OF PEPTIDE ANALOGUES

The synthesis of peptide analogues with some advantages over the natural hormone has been a major goal for peptide chemists ever since du Vigneaud synthesized oxytocin (du Vigneaud *et al.*, 1953). A summary of these advantages is presented in Table IV.

From their first experiences in peptide design, chemists concluded that they had first to gain knowledge of hormone action, including physiological, biochemical, and molecular mechanism of action, before obtaining significant results. Even though some interesting peptide analogues were prepared from the approach of "let's see what happens if . . . ," most successes in peptide design were obtained from more systematic strategies.

In *The Design of Peptide Hormone Analogs*, Rudinger (1971) proposed a general approach to the problem of peptide design. As a first step, he suggested that in the study of a new peptide molecule, small changes in the sequence should first be made in order to determine which portion of the chain is responsible for biological activity: fragments and analogues in which functional side chains are replaced systematically by a neutral amino acid such as Gly or Ala should be used. Once biologically important amino acids have been identified, structural changes should be made isosteric or isofunctional to separate steric from chemically functional effects. Modifications in the C and N terminals are useful to determine the importance of protected or free amino and carboxyl groups for activity.

Although the principles of drug design have remained quite the same since Rudinger, the modern approach has become more global or conformational. A greater attention is now devoted to chemical modifications that can promote the tendencies of a peptide to adopt a given tridimensional structure. Such modifications are generally suggested by conformational predictions based on the sequence of the peptide or by conformational analysis using spectroscopic studies and theoretical calculations.

Recently, a promising approach was proposed by Emil Kaiser to preserve or accentuate the original secondary structure of biologically active peptides and polypeptides by altering entire regions of the molecule rather than individual amino acids. The amphiphilic structure of a given peptide can be successfully predicted,

TABLE IV. Biological Aspects and Special Aims of Peptide Design

Simplified synthesis
Increased potency and selectivity
Suppression of immunoreactivity
Antagonists
Specific inhibitors
Modification of metabolism
Modification of distribution

and potent analogues can be designed and synthesized, as demonstrated with β-endorphin (Blanc and Kaiser, 1984) and other peptides.

The two approaches appear to complete each other, as the former is most useful for the identification and modification of the peptide active site, whereas the latter can help to reveal the conformation of peptides at the receptor and suggest multiple substitutions that will improve the interaction with the membrane of target cells.

Therefore, the strategy summarized in Table V can generally be applied to identify the biologically relevant residues of a peptide and, consequently, to design peptide analogues with improved biological properties, provided that a specific bioassay exists to evaluate the relative potency of the analogues. Several biologically relevant analogues of various peptides, among which are found superagonists and antagonists or more specific analogues, were prepared from side-chain modification of key amino acids, changes in the peptide backbone at critical positions, or various conformational restrictions enhancing a given structure.

Several hundred synthetic analogues of the hypothalamic peptides TRH, LHRH, and somatostatin, the opiate enkephalin, and others such as oxytocin, angiotensin, bradykinin, and neurotensin containing the modification described above have been prepared in recent years. The stabilization of a postulated β-turn by substituting a D-residue for naturally occurring Gly has given rise to superagonists of LHRH (Monahan *et al.*, 1973) and enkephalin (Pert *et al.*, 1976). Also, D-substitution can increase specificity or resistance to enzyme degradation, as demonstrated in the cases of somatostatin (Marks *et al.*, 1976) and enkephalin (Pert *et al.*, 1976). Several chemical transformations have been utilized to protect peptides against enzymatic degradation. The replacement of amide linkages by retro–inverso bonds, as introduced by Goodman and collaborators, appears to be a promising method to prepare long-acting analogues of peptides, as demonstrated with enkephalin (Chorev *et al.*, 1979), somatostatin (Pallai *et al.*, 1983), and others. Also, ketomethylene and N-carboxylmethyl groups can be used as enzyme-resistant isosteric replacements of peptide bonds, as demonstrated by the synthesis of orally active inhibitors of the angiotensin-converting enzyme (ACE) (Patchett *et al.*, 1980).

More recently, conformational restrictions have been successfully applied to several peptides. A reduced list of these modifications and some examples are

TABLE V. Stepwise Approach to Structure–Activity Studies of Peptides

Shortening at amino and carboxylic ends
Chain extension
Elision and intercalation of residues
Modification of side chains
Modification of terminal or side-chain reactive groups
Changes in the peptide backbone
Conformational restrictions, including cyclization

presented in Table VI. The most impressive achievements with such analogues have been obtained by the Merck group, who recently designed and synthesized potent orally active analogues of somatostatin. These compounds, whose biologically active restricted structure was deduced from painstaking spectroscopic and molecular modeling studies, contain six out of the 14 residues of native somatostatin (Veber *et al.*, 1984). Such cyclic peptides show remarkable chemical and enzymatic resistance and improved pharmacodynamic properties.

On the other hand, cyclization of naturally occurring linear peptides has been shown, in some cases, to result in valuable analogues. Cycloenkephalin (DiMaio *et al.*, 1982), cyclo-MSH (Sawyer *et al.*, 1982), and cyclobradykinin (Chipens *et al.*, 1981) are all superagonists of the respective natural peptide. Also, less rigid restrictions brought about by the introduction of residues such as penicillamine (Mosberg *et al.*, 1981) or dehydroamino acids (English and Stammer, 1978) or by the formation of lactam bridges (Freidinger *et al.*, 1980) have been used to obtain potent peptide analogues.

5. PEPTIDE DRUGS IN PRACTICE

Peptide design and synthesis have now reached a high degree of efficiency and sophistication. Synthetic analogues have been most useful to study the phys-

TABLE VI. Conformational Restrictions Introduced in Peptides by Side-Chain Modification

Conformational restriction	Peptide	Biological effect	Reference
D-Residue substitution for stabilization of turn	LH-RH	Increased activity	Monahan *et al.* (1973)
Aib or Cyl substitution for stabilization of turn	Chemotactic peptide	Activity	Iqbal *et al.* (1984)
D-Trp substitution	Neurotensin	Antagonist	Quirion *et al.* (1980)
	Neurotensin	Increased activity	Jolicoeur *et al.* (1981)
	Substance P	Antagonist	Fuxe *et al.* (1982)
	Substance P	Antagonist	Caranikas *et al.* (1982)
Penicillamine substitution for Cys	Oxytocin	Inhibitor, prolonged activity	Hruby *et al.* (1982)
Dehydro amino acid substitution	Enkephalinn	Increased potency	English and Stammer (1978)
	Bradykinin	Increased potency	Fisher *et al.* (1978)
Introduction of lactam bridge	LH-RH	Increased potency	Freidinger *et al.* (1980)
Isosteric cyclization of linear peptides	Enkephalin	Increased potency	Di Maio *et al.* (1982)
	Melanotropin	Increased potency, prolonged activity	Sawyer *et al.* (1982)
	β-Endorphin	High binding activity	Blake *et al.* (1981)
Reduction of cycle size of cyclic peptides	Somatostatin	Increased potency, oral activity	Veber *et al.* (1984)

iological role of peptides and to understand better the nature of their interaction with receptors. Although no general rule yet exists that could allow peptide chemists to prepare synthetic superagonists or antagonists out of any peptide on demand, such an impressive amount of new knowledge and experience has been gained during the recent years that we can realistically expect to witness such an accomplishment in the near future.

However, in spite of their spectacular achievements, chemists have to recognize that few among the hundreds of peptide analogues that have been described so far can be utilized as drugs or at least be compared advantageously to existing classical drugs unless the difficulties discussed above could be circumvented. Even though the problem of chemical and enzymatic stability appears to be solvable by means of appropriate chemical modifications, specificity and bioavailability of most peptide drugs are unsatisfactory.

Several different routes of administration are possible for drugs, as listed in Table VII. Except for a few small peptides such as ACE inhibitors, hydrophobic oligopeptides, or cyclic peptides such as the recently developed reduced somatostatin analogues, bioavailability of peptides following oral administration is in general too low. Their instability in the digestive system and their nonlipophilicity, which restricts their diffusion through the gastrointestinal mucosa, are two main obstacles to their enteral administration. Therefore, even though orally active drugs are highly desirable, most peptides used in the clinic are normally administered via parenteral routes (intravenously, subcutaneously, intranasally). For instance, in spite of important efforts that have been made to dispense insulin orally, subcutaneous injection still is its most current mode of administration. Another large peptide, growth hormone-releasing factor, is most conveniently delivered from subcutaneous deposits. Luteinizing hormone-releasing hormone and analogues are normally injected intravenously to chronic patients; nasal administration has also been shown possible.

Moreover, even though some peptides have been reported to exert central effects when injected systemically, it is still not clear whether they do so following a direct interaction with receptors in the brain or through unknown mediated mechanisms. The impermeability of the blood–brain barrier, which is composed of

TABLE VII. Methods for the Administration of Drugs

Conventional	Controlled delivery systems
Oral	Infusion pumps
Nasal	Liposomes
Intravenous	Microcapsules
Intramuscular	Microspheres
Subcutaneous	Antibodies
Intradermal	Macromolecules
Rectal	Biodegradable polymers
	Prodrugs

cerebrovascular endothelial cells that are connected by tight junctions, appears to be a serious obstacle to the delivery of most peptides and proteins to the CNS. In contrast, some investigators have reported that peptides are significantly permeable at the blood–brain barrier (Banks *et al.*, 1984). However, until this controversy has been settled, the delivery of peptides to the brain remains a major problem.

In the second part of Table VII is found a list of systems that have been developed during recent years for the delivery of drugs. Even though these systems are just beginning to be applied to peptides, they appear to provide a promising answer to the problems of specificity and stability that are beyond the reach of peptide design. For instance, the use of implantable insulin pumps is being investigated in diabetes patients (Rupp *et al.*, 1982). Also, new Accurel® microporous tubes have been conceived to deliver constant quantities of neuroactive drugs to the brain (Boer *et al.*, 1984). Such a device allows the constant delivery of peptides directly to the target organ without the saw-tooth pattern of subtherapeutic and toxic levels that is observed during multiple injections. On the other hand, liposomal carrying systems appear to be applicable for the delivery of anticancer peptides (Key *et al.*, 1982) and synthetic peptide vaccines. More recently, the use of subcutaneous deposits of biodogradable copolymers containing LHRH analogues has been investigated as a method for the treatment of prostate cancer patients (Sanders, 1984). Finally, as mechanisms of transport systems by microorganisms and of intracellular cleavage of peptides by proteases are becoming better known, the concept of peptide prodrugs is becoming attractive. Even though only a few examples of peptide prodrugs are found, there is every reason to believe that a fusion of chemical and biological expertise can lead to the evolution of peptide prodrugs with a high degree of target specificity and, possibly, penetration of tissue.

6. CONCLUSIONS

There is now a consensus that peptides and their derivatives are already an important part of the pharmacopoeia. Even though the emphasis in this report has been on peptide drugs obtained through design and chemical synthesis procedures, the author also wants to stress the importance of peptides in other areas of clinical interest as well as the promising applications of recombinant DNA and enzymatic techniques of peptide synthesis. Unfortunately, each of these subjects needs a whole symposium to be barely covered and, therefore, is beyond the scope of the present discussion.

Peptide chemists have reasons to be proud of their past accomplishments and the importance that their pet molecules are taking in medicine and on the markets today. Moreover, as an essential complement to new genetic and enzymatic modes of synthesis, considerable possibilities exist for them to devise simpler synthetic substitutes for natural hormones and to find stimulants or inhibitors of their secretion, potentiators of their action, or blockers of the receptors on which they act. However, they have to keep in mind that there is more to it than the scientific game of creating new analogues with attractive properties in *in vitro* systems and in laboratory animals. The ultimate aim is the human being to whom the drug is finally destined. For this reason, factors such as the mode of administration, the comfort of the patient, the possibility of side effects, the drug's shelf half-life, and the cost of production also have to be taken into consideration.

7. SUMMARY

Much of the research being conducted in the drug industry today is centered around newly discovered endogenous substances of the human body. Among these substances, peptides have become one of the most rapidly progressing fields for research and development of new drugs.

A need exists for economical, orally effective, and metabolically stable analogues of peptides, neurotransmitters, and neuromodulators. Although it is sometimes difficult to predict which area of clinical medicine will benefit the most from the discoveries being made around one given peptide following its isolation and characterization, experience has shown that most valuable therapeutic tools can be obtained from synthetic analogues or substitutes that can stimulate or inhibit their secretion, potentiate their action, or block the receptors on which they act.

Although peptides are among the biologically most active compounds known, they are generally ill-suited for drug use. Most peptides are relatively large, and their pharmacokinetics is usually complicated. They generally can act on a wide range of target tissues or organs. Moreover, they are metabolically less stable than most classes of drugs because of the ubiquity of peptidases with various specificities. On the other hand, their stereotyped structural units, the amino acids, confer on them a variety of biological properties that is found in no other class of natural substances.

Important advances have been made in recent years toward the discovery of new peptide drugs. The mapping of "active regions" and "binding regions" of a number of peptide hormones has been performed by means of structure–activity studies with synthetic analogues. Chemical manipulations of the structures and the use of improved spectroscopic tools such as high-resolution NMR and computer modeling have contributed to the generation of new peptide analogues and peptidomimetics that are orally effective. The new concept of conformationally restrained conformation and the development of stereochemically equivalent nonnatural peptide bonds is at the origin of peptide analogues with improved metabolic stability. Some of these analogues should lead to drugs with novel pharmacological profiles in therapeutically important areas including psychopharmacology, endocrinology, cardiovascular pharmacology, and gastrointestinal pharmacology.

REFERENCES

Banks, W. A., Kastin, A. J., and Coy, D. H., 1984, Evidence that [^{125}I]N-Tyr-delta sleep-inducing peptide crosses the blood–brain barrier by a noncompetitive mechanism, *Brain Res.* **301**:201–207.

Bhatnagar, P. K., Papas, E., Blum, H. E., Milich, D. R., Nitecki, D., Karels, M. J., and Vyas, G. N., 1982, Immune response to synthetic peptide analogues of hepatitis B surface antigen for the a determinant, *Proc. Natl. Acad. Sci. U.S.A.* **79**:4400–4404.

Biddle, J. L., Houghten, R. A., Alexander, H., Shinnick, T. M., Sutcliff, J. G., Lerner, R. A., Rowlands, D. J., and Brown, F., 1982, Protection against foot-and-mouth disease by immunization with a chemically synthesized peptide predicted from the viral nucleotide sequence, *Nature* **298**:30–33.

Blake, J., Ferrara, P., and Li, C. H., 1981, Beta-endorphin: Synthesis and radioligand binding activity of analogs containing cystine bridges, *Int. J. Peptide Protein Res.* **17**:239–242.

Blanc, J. P., and Kaiser, E. T., 1984, Biological and physical properties of a beta-endorphin analog containing only D-amino acids in the amphiphilic helical segment 13–31, *J. Biol. Chem.* **259**:9549–9556.

Boer, G. J., van der Woude, T. P., Kruisbrink, J., and van Heerikhuize, J., 1984, Successful ventricular application of the miniaturized controlled-delivery Accurel technique for sustained enhancement of cerebrospinal fluid peptide levels in the rat, *J. Neurosci. Methods* **11**:281–289.

Caranikas, S., Mizhari, J., D'Orléans-Juste, P., and Regoli, D., 1982, Antagonists of substance P, *Eur. J. Pharmacol.* **77**:205–206.

Carelli, C., Audibert, F., Gaillard, J., and Chedid, L., 1982, Immunological castration of male mice by a totally synthetic vaccine administered in saline, *Proc. Natl. Acad. Sci. U.S.A.* **79**:5392–5395.

Chipens, G. I., Mutulis, F. K., Batayev, B. S., Klusha, V. E., Misina, I. P., and Myshlyiakova, N. V., 1981, Cyclic analogs of bradykinin, *Int. J. Peptide Protein Res.* **18**:302–311.

Chorev, M., Shavitz, R., Goodman, M., Minick, S., and Guillemin, R., 1979, Partially modified retro–inverso enkephalinamides: Topochemical long-acting analogs *in vitro* and *in vivo, Science* **204**:1210–1212.

DiMaio, J., Nguyen, T. M. D., Lemieux, C., and Schiller, P. W., 1982, Synthesis and pharmacological characterization *in vitro* of cyclic enkephalin analogues: Effect of conformational constraints on opiate receptor selectivity, *J. Med. Chem.* **25**:1432–1438.

Djawari, D., Bolla, K., and Haneke, E., 1984, Treatment of recurrent herpes simplex with thymopoietin pentapeptide, *Deut. Med. Wochenschr.* **109**:496–498.

du Vigneaud, V., Ressler, C., Swan, J. M., Roberts, C. W., Katsoyannis, P. G., and Gordon, S., 1953, The synthesis of oxytocin, *J. Am. Chem. Soc.* **75**:4879–4885.

English, M. L., and Stammer, C. H., 1978, D-Ala^2, dehydro-Phe^4-Methionine enkephalinamide, a dehydropeptide hormone, *Biochem. Biophys. Res. Commun.* **85**:780–782.

Fisher, G. H., Malborough, D. I., Ryan, J. W., Felix, A. M., 1978, L-3,4-Dihydroproline analogues of bradykinin: Synthesis, biological activity, and solution conformation, *Arch. Biochem. Biophys.* **189**:81–85.

Freidinger, R. M., Veber, D. F., Perlow, D. S., Brooks, J. R., and Saperstein, R., 1980, Bioactive conformation of luteinizing hormone-releasing hormone: Evidence from a conformationally constrained analog, *Science* **210**:656–658.

Fuxe, K., Agnati, L. F., Rosell, S., Harfstrand, A., Folkers, K., Lundberg, J. M., Andersson, K., and Hökfelt, T., 1982, Vasopressor effects of substance P and C-terminal sequences after intracisternal injection to alpha-chloralose-anesthetized rats: Blockade by a substance P antagonist, *Eur. J. Pharmacol.* **77**:171–176.

Hrudy, V. J., and Mosberg, H. I., 1982, Structural, conformational, and dynamics considerations in the development of peptide hormone antagonists, in: *Hormone Antagonists* (M. K. Agarwal, ed.), W. DeGruder, New York, pp. 433–473.

Iqbal, M., Balaram, P., Showell, H. J., Freer, R. J., and Becker, E. L., 1984, Conformationally constrained chemotactic peptide analogs of high biological activity, *FEBS Lett.* **165**:171–174.

Jolicoeur, F. B., Barbeau, A., Rioux, F., Quirion, R., and St -Pierre, S., 1981, Differential neuro-behavioral effects of neurotensin and structural analogs, *Peptides* **2:**171–176.

Key, M. E., Talmadge, J. E., Fogler, W. E., Bucano, C., and Fidler, I. J., 1982, Isolation of tumoricidal melanoma metastases of mice treated systemically with liposomes containing a lipophilic derivative of muramyl dipeptide, *J. Natl. Cancer Inst.* **69:**1198.

Labrie, F., Dupont, A., and Belanger, A., 1985, Complete androgen blockade for the treatment of prostrate cancer, in: *Important advances in Onoclogy* (V. T. De Vita, Jr., S. Hellman, and S. A. Rosenberg, eds.), J. B., Lippincott, Philadelphia, pp.193–217.

Marks, N., Stern, F., and Benuck, M., 1976, Correlation between biological potency and biodegradation of a somatostatin analogue, *Nature* **261:**511–512.

Maton, P. N., O Dorisio, T. M., Howe, B. A., McArthur, K. E., Howard, J. M., Cherner, J. A., Malarkey, T. B., Collen, M. J., Gardner, J. D., and Jensen, R. T., 1985, Effect of a long-acting somatostatin analog (SMS 201-995) in a patient with pancreatic cholera, *N. Engl. J. Med.* **312:**17–21.

Monahan, M. W., Amoss, M. S., Anderson, H. A., and Vale, W., 1973, Synthetic analogs of the hypothalamic luteinizing hormone releasing factor with increased agonist or antagonist properties, *Biochemistry* **12:**4616–4620.

Moody, T. W., Pert, C. B., Gazdar, A. F., Carney, D. N., and Minna, J. D., 1981, High levels of intracellular bombesin characterize human small-cell carcinoma, *Science* **214:**1246–1248.

Mosberg, H. I., Hruby, V. J., and Meraldi, J. P., 1981, Conformational study of the potent peptide hormone antagonist [1-penicillamine, 2-leucine]-oxytocin in aqueous solution, *Biochemistry* **20:**2822–2828.

Ondetti, M. A., Rubin, B., and Cushman, D. W., 1977, Design of specific inhibitors of angiotensin-converting enzyme: New class of orally active antihypertensive agents, *Science* **196:**441–444.

Pallai, P., Struthers, S., and Goodman, M., 1983, Extended retro–inverso analogs of somatostatin, *Biopolymers* **22:**2523–2538.

Patchett, A. A., Harris, E., Tristram, E. W., Wyvratt, M. J., Wu, M. T., Taub, D., Peterson, E. R., Ikeler, T. J., ten Broeke, J., Payne, L. G., Ondeyka, D. L., Thorsett, E. D., Greenlee, W. J., Lohr, N. S., Hoffsommer, R. D., Joshua, H., Ruyle, W. V., Rothrock, J. W., Aster, S. D., Maycock, A. L., Robinson, F. M., Hirschmann, R., Sweet, C. S., Ulm, E. H., Gross, D. M., Vassil, T. C., and Stone, C. A., 1980, A new class of angiotensin-converting enzyme inhibitors, *Nature* **288:**280–283.

Pert, C. B., Pert, A., Chang, J. K., and Fong, B. T. W., 1976, [D-Ala^2]-Met-enkephalinamide: A potent, long-lasting synthetic pentapeptide analgesic, *Science* **194:**330–332.

Quirion, R., Rioux, F., Regoli, D., and St.-Pierre, S., 1980, Selective blockade of neurotensin induced coronary vessel constriction in perfused rat hearts by a neurotensin analog, *Eur. J. Pharmacol.* **61:**309–312.

Rudinger, J., 1971, The design of peptide analogs, in: *Drug Design,* Vol. II (J. Ariens, ed.), Academic Press, New York, pp. 319–419.

Rupp, W. M., Barbosa, J. J., Blackshear, P. J., McCarthy, H. B., Rohde, T. D., Goldberg, F. J., Rublein, T. G., Dorman, F. D., and Buchwald, H., 1982, The use of an implantable insulin pump in the treatment of type II diabetes, *N. Engl. J. Med.* **307:**265–270.

Sanders, L. M., 1984, An injectable biodegradable controlled release delivery system for LHRH analogues, in: *Proceedings of the VII International Congress of Endocrinology* (F. Labrie and L. Proulx, eds.), Excerpta Medica, Elsevier Science, Amsterdam, p. 1018.

Sawyer, T. K., Hruby, V. J., Darman, P. S., and Hadley, M. E., 1982, [half-Cys^4, half-Cys^{10}]-Alpha-melanocyte-stimulating hormone: A cyclic alpha-melanotropin exhibiting superagonist biological activity, *Proc. Natl. Acad. Sci. U.S.A.* **79:**1751–1755.

Shapira, M., Jibson, M., Muller, G., and Arnon, R., 1984, Immunity and protection against influenza virus by synthetic peptide corresponding to antigenic sites of hemagglutinin, *Proc. Natl. Acad. Sci. U.S.A.* **81:**2461–2465.

Veber, D. F., Saperstein, R., Nutt, R. F., Freidinger, R. M., Brady, S. F., Curley, P., Perlow, D. S., Palaveda, W. J., Colton, C. D., Zacchei, A. G., Tocco, D. J., Hoff, D. R., Vandlen, R. L., Gerich, J. E., Hall, L., Mandarino, K., Cordes, E. H., Anderson, P. S., and Hirschmann, R., 1984, A super active cyclic hexapeptide analog of somatostatin, *Life Sci.* **34:**1371–1378.

Wolfsen, A., and Odell, W., 1982, Advances in the use of peptide hormones as tumor markers, in: *Cancer Diagnosis: New Concepts and Techniques* (R. J. Stechel and R. A. Kagan, eds.), Grune and Stratton, Orlando, Florida, pp. 63–79.

Treatment of Precocious Puberty with LHRH Agonists

A Pilot Study of the Efficiency of Periodic Administration of a Delayed-Release Preparation of D-Trp^6-LHRH

MARC ROGER, JEAN-LOUIS CHAUSSAIN, FRANCOISE RAYNAUD, JEAN-EDMOND TOUBLANC, PIERRE CANLORBE, JEAN-CLAUDE JOB, and ANDREW V. SCHALLY

1. INTRODUCTION

A better understanding of the paradoxical inhibitory effects of LHRH agonists has led to the discovery of new therapeutic agents, thanks largely to the experimental studies of Knobil's group (Belchetz *et al.*, 1978). Certain structural analogues have proven to be powerful suppressors of the pituitary–gonadal axis, as well as devoid of toxic effects.

As reviewed by several investigators (Sandow, 1983; Bex and Corbin, 1984; Schally *et al.*, 1984), such a treatment might be of value in all situations in which gonadal suppression is needed, that is, mainly hormone-dependent cancers, endometriosis, polycystic ovaries, and precocious puberty.

MARC ROGER • Fondation de Recherche en Hormonologie, 94268 Fresnes Cedex, France. JEAN-LOUIS CHAUSSAIN, FRANCOISE RAYNAUD, JEAN-EDMOND TOUBLANC, PIERRE CANLORBE, and JEAN-CLAUDE JOB • Hôpital St. Vincent de Paul, Paris, France. ANDREW V. SCHALLY • Veterans Administration Medical Center and Tulane University School of Medicine, New Orleans, Louisiana 70146.

2. USE OF LHRH AGONISTS IN PRECOCIOUS PUBERTY: ROUTE OF ADMINISTRATION AND DOSAGE

Several different agonists have been used in the treatment of precocious puberty in boys and girls. All are structural analogues in which the amino acid in position 6 is replaced by D-tryptophan, *t*-butyl-D-serine, D-leucine, or 3-(2 naphthyl)-D-alanine. As early as 1981, D-Trp^6-LHRH was used by Laron's group in subcutaneous injections of 20 μg per day (Laron *et al.*, 1981) and then 40 μg per day (Kauli *et al.*, 1984).

Crowley *et al.* (1981) and Comite *et al.* (1981, 1984) have reported the results of treatment in both sexes using a daily subcutaneous injection of D-Trp^6-des-Gly^{10}-LHRH-ethylamide in doses of 4 μg per kilogram of body weight per day. D-Ser(tBu)^6-Gly^{10}-LHRH-ethylamide (Buserelin) has been used by several groups either in the form of an intranasal spray (Donaldson *et al.*, 1984) or by subcutaneous and intranasal routes successively (Luder *et al.*, 1984).

Although these treatments have proven successful both in idiopathic puberty and in puberty resulting from brain tumors, some uncertainty remained. Some cases of partial resistance or secondary escape have been reported (Kauli *et al.*, 1984; Donaldson *et al.*, 1984), which required an increase in the daily dosage of the drug either by injecting the same amount twice a day instead of once or by increasing the amount of drug per injection or per nasal spraying. In fact, the daily dosage commonly used in pediatrics was derived from the dosage that was proven efficient in adult patients. Only recently were pharmacological data available in children. Buserelin levels declines rapidly after subcutaneous injection (Holland *et al.*, 1985), and rather high dosages are needed in order to achieve plasma concentrations detectable thoughout a 24-hr period after a single subcutaneous injection. When the drug is given intranasally, the resulting concentrations were about 100 times lower than those achieved by subcutaneous injections. This made it necessary to administer 200 to 400 μg per spray and to use repetitive spraying in order to maintain therapeutic levels thoughout the day.

As a consequence, the necessity of administering the drug daily or several times a day makes this kind of treatment inconvenient and does not favor the psychosocial adaptation of the children under treatment.

Therefore we were interested in a different approach in which the agonist D-Trp^6-LHRH was administered intramuscularly on a periodic basis in the form of a delayed-release preparation (Decapeptyl® in microcapsules).

3. A PILOT STUDY OF INTRAMUSCULAR DECAPEPTYL: MATERIAL AND METHODS

3.1. Patients

This study is based on the first 13 children of a broad multicentric study, the complete results of which are reported elsewhere (Roger *et al.* 1986). Seven boys

TABLE I. Main Clinical Features of Boys in the Pilot Study of the Long-Acting Intramuscular Preparation of D-Trp^6-LHRH

Boys	Age at onset (years)	Age at first injection (years)	Height (cm)	Weight (kg)	Bone age (years)	Pubic hair score	Mean testis volume (ml)	Etiology	Previous treatment[a]
1	6.0	6.6	126	25.5	9.3	II	7	Idiopathic	None
2	0.5	5.2	130	33	11.0	III	3.4	Hamartoma + Klinefelter	CA + MPA for 4 years
3	9.5	10.4	138	37	12.0	II	5.7	Idiopathic	None
4	1.6	2.0	95	17	3.5	II	10	Idiopathic	None
5	7.5	10.0	154	47.8	13.0	III	12	Hamartoma	CA + MPA for 2 years
6	5.0	8.3	144	40.7	11.0	II	10	Brain calcification	CA for 2 years
7	9.5	10.5	125	28.8	10.0	II	8	Suprasellar cyst	None

[a] MPA, medroxprogesterone acetate; CA, cyproterone acetate.

TABLE II. Main Clinical Features of Girls in the Pilot Study of the Long-Acting Intramuscular Preparation of D-Trp^6-LHRH

Girls	Age at onset (years)	Age at first injection (years)	Height (cm)	Weight (kg)	Bone age (years)	Pubic hair score	Breast score	Uterus size (mm)	Ovary length (mm)	Follicles	Etiology	Previous treatment
8	5.6	8.8	129	29	10.5	III	IV	55	24	+	Idiopathic	None
9	3.0	3.3	99	18.3	4.6	II	III	35	22		Idopathic	None
10	6.0	7.6	136	32.5	10.0	II	III	45	30	+	Idiopathic	None
11	2.5	6.4	130	27	11.0	III	II	36	30	+	Idiopathic	MPA + CA for 3 years
12	0.7	0.9	76	9.6	1.7	II	II	45	18	+	Idiopathic	None
13	3.7	4.3	117	23.7	8.0	II	II	40	25		Idiopathic	None

aged from 2 to 10 years and six girls aged from 1 to 8 years (Tables I and II) have been treated for 6 to 21 months. Four of the boys had a precocious puberty associated with an organic brain lesion. Repeated brain scans did not detect brain lesions in any of the other children, and they were thus considered to have an idiopathic precocious puberty. At the beginning of treatment, all children had signs of evolutive puberty: accelerated growth velocity, advanced bone age, pubic hair, testis or breast enlargement. One girl was menstruating. All, including three boys and one girl previously treated with medroxyprogesterone acetate (MPA) and/or cyproterone acetate (CA), had a pubertal response of the pituitary to LHRH.

3.2. Product

The D-Trp^6-LHRH was synthesized by solid-phase methods (Coy *et al.*, 1976) and supplied by Debiopharm (Lausanne, Switzerland). The long-acting release formulation consisted of microcapsules prepared by a phase-separation process. The resulting product was a free-flowing powder of spherical particles consisting of Decapeptyl (2% wt./wt.) distributed in a polymeric matrix of 53 : 47 (mol%) of a biodegradable biocompatible polymer (DL-lactide-coglycolide) (98% wt./wt.). The microcapsules were loaded in disposable syringes and sterilized with a 2-Mrad dose of γ radiation. This procedure did not affect the biological activity of the peptide (Redding *et al.*, 1984). Just before intramuscular administration, the microcapsules were suspended in 1.5 ml injection vehicle containing 2% carboxymethylcellulose and 1% Tween 20 in water. Each syringe contained an equivalent of 3 mg Decapeptyl in microcapsules designed to deliver in a controlled fashion a daily dose of 100 μg D-Trp^6-LHRH for 30 days.

3.3. Protocol

The children were injected intramuscularly with an equivalent of microcapsules corresponding to a dose of 60 μg Decapeptyl per kilogram of body weight, supposedly delivered in a continuous manner at approximately 2 μg of Decapeptyl per kilogram of body weight and per day. The injection was administered on days 1 and 21 and every 28 days thereafter. As a rule, the suspension was prepared and the injection was performed by the same nurse or physician for each child. Before each injection, the children were subjected to a physical examination. The side effects were recorded when appropriate, and blood was collected for hormone assays. Blood was also collected on days 3, 7, and 14 after the first injection; LHRH stimulation tests were performed after 7, 23, and 43 weeks of therapy using 100 μg/m^2 synthetic LHRH as an intravenous bolus.

In girls, ultrasound examination of the uterus and ovaries was performed after 6 and 12 months of therapy. In children previously treated with MPA and/or CA, this treatment was continued for 7 days after the first injection of microcapsules before being interrupted. In other children, in order to prevent the effects of the initial stimulation of the pituitary–gonadal axis, cyproterone acetate (100 mg per

m^2 per day) was given for 7 days before and for 7 days after the first injection of microcapsules. No other hormonal medication was permitted during the treatment.

3.4. Methods

The increments in height were compared to French reference values computed by Sempé and Sempé (1979). Bone age was evaluated from an X-ray of the left hand and wrist according to Greulich and Pyle (1959). Pubertal stages of pubic hair and breast development were assessed according to Tanner (1962). Mean testis volume (V) was calculated from the length (L) and thickness (T) of both testes according to the formula: $V = L \times T^2 \times \pi/6$.

Plasma testosterone, estradiol, and gonadotropin levels were measured by previously described methods and compared to the reference values previously reported (Roger *et al.*, 1979, 1984; Roger, 1983).

4. EFFECTS ON PUBERTAL SYMPTOMS

4.1. Biological Effects

4.1.1. Gonadal Secretions

The effect on gonadal secretion was obvious in boys. In all except the first case, testicular inhibition was observed on day 21 after the first injection, the mean level of testerone ($\pm$S.E.M.) decreasing from 3.2 ± 0.48 to 0.17 ± 0.03 ng/ml ($P < 0.01$). It has to be noted that a stimulatory effect was observed between days 3 and 7 after the first injection, as could be expected, but suppression was already effective on day 14 (Figs. 1 and 2).

The effect on ovarian secretion was less dramatic, as estradiol levels were not very high in all cases prior to the first injection. Nevertheless, mean estradiol level ($\pm$S.E.M.) decreased within 3 weeks from 59 ± 9 to 29 ± 9 pg/ml ($P < 0.05$) and then to 14 ± 2 pg/ml at 11 weeks (Fig. 3).

4.1.2. Gonadotropins

In all children, basal levels and pituitary reserves of gonadotropins were rapidly suppressed (Figs. 4 and 5).

In girls, the mean basal level of FSH ($\pm$S.E.M.) was high before the first injection: 168 ± 27 ng LER 907/ml (3.4 ± 0.5 IU/liter). It decreased significantly after 3 weeks: 54 ± 8 ng/ml (1.1 ± 0.2 IU/liter) ($P < 0.02$). Basal immunoreactive LH level was normal before the first injection: 44 ± 6 ng LER 907/ml (2.1 ± 0.3 IU/liter) and was significantly lowered only from the 20th week: 25 ± 4 μg/ml (1.2 ± 0.2 IU/liter).

In boys, the gonadotropin pattern was not very different, mean FSH level

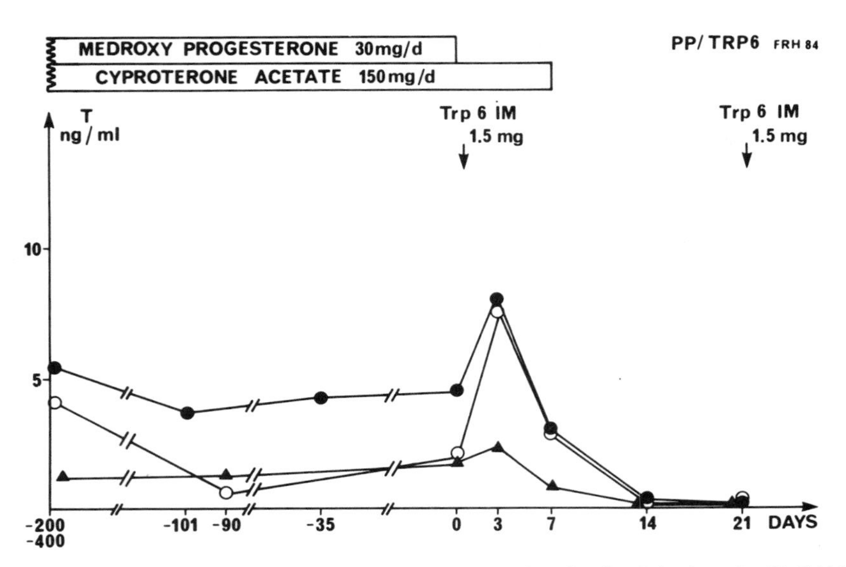

FIGURE 1. Evolution of plasma testosterone (T) levels after the first injection of D-Trp[6]-LHRH in microcapsules (Trp-6 IM) in three boys whose testosterone secretion had not been suppressed by prolonged treatment with medroxyprogesterone plus cyproterone acetate. Note the rapid suppression of testosterone after the initial flare-up.

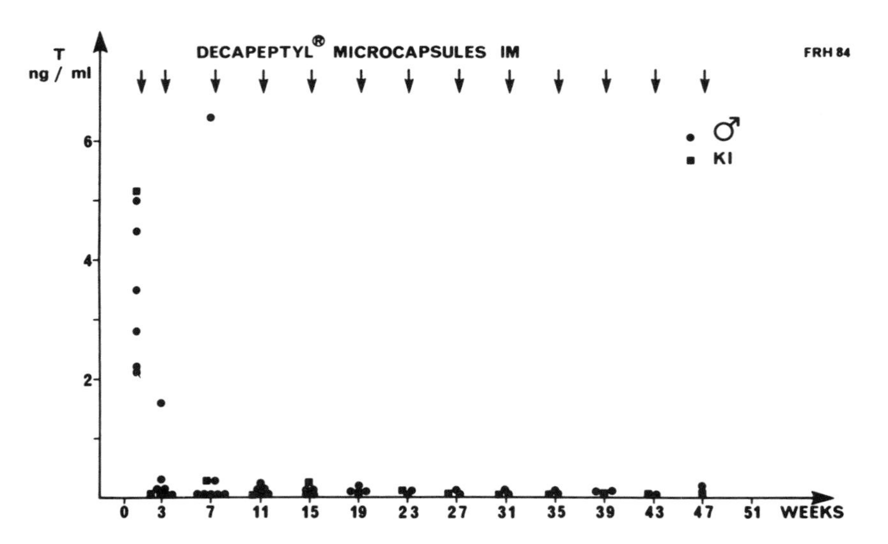

FIGURE 2. Evolution of individual plasma testosterone (T) levels in seven boys treated with the intramuscular long-acting preparation of D-Trp[6]-LHRH (Decapeptyl microcapsules IM). KI, boy with precocious puberty and Klinefelter syndrome (subject 2).

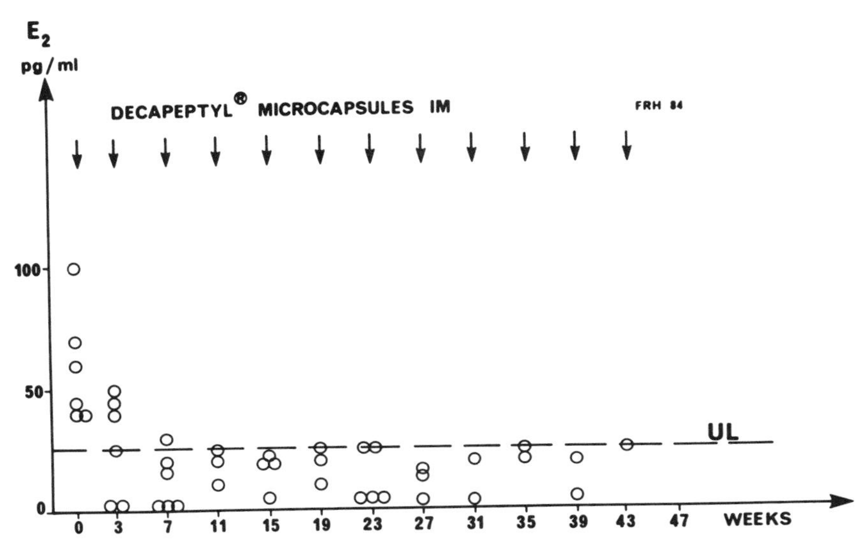

FIGURE 3. Evolution of individual plasma estradiol (E_2) levels in six girls treated with the intramuscular long-acting preparation of D-Trp^6-LHRH (Decapeptyl microcapsules IM). UL, upper limit in prepubertal girls.

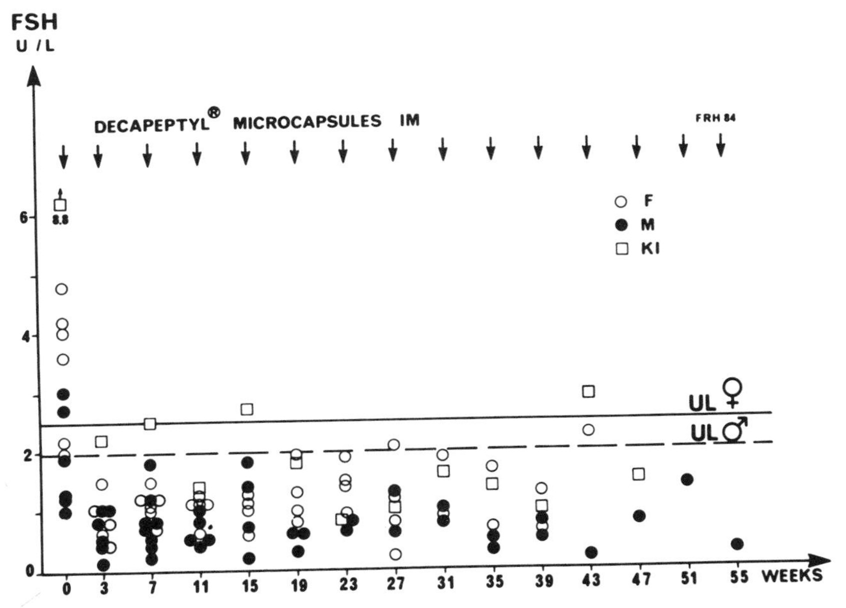

FIGURE 4. Evolution of individual plasma FSH levels in girls and boys treated with the intramuscular long-acting preparation of D-Trp^6-LHRH (Decapeptyl microcapsules IM). F, girls; M, boys; Kl, boy with Klinefelter syndrome; UL, upper limit in prepubertal boys (♂) and girls (♀).

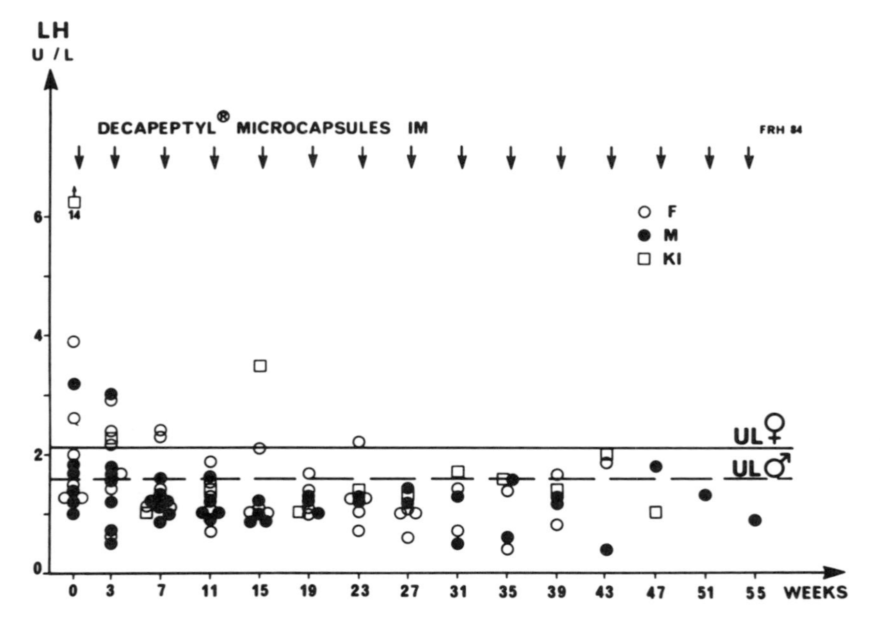

FIGURE 5. Evolution of individual plasma LH levels in girls and boys treated with the intramuscular long-acting preparation of D-Trp^6-LHRH (Decapeptyl microcapsules IM). Same symbols as in Fig. 4.

decreasing within 3 weeks from 85 ± 16 (1.7 ± 0.3 IU/liter) to 36 ± 6 ng/ml (0.7 ± 0.1 IU/liter) ($P < 0.02$). However, mean basal immunoreactive LH level was depressed more rapidly than in girls, decreasing within 7 weeks from 40 ± 4 (1.9 ± 0.2 IH/liter) to 25 ± 2 ng/ml (1.2 ± 0.1 IU/liter) ($P < 0.05$).

An effect of the intramuscular treatment on the pituitary response to the LHRH test was rapidly observed (Figs. 6 and 7). In boys and girls, FSH response was almost completely abolished within 7 weeks, whereas LH response was reduced to normal prepubertal range. Such a suppression of the pituitary response to LHRH was maintained throughout the therapeutic period in all children, even after 18 months of therapy.

These results are similar to results previously reported after therapy with the same analogue (Kauli *et al.*, 1984) or with other LHRH agonists administered daily through the subcutaneous route. It is interesting to note, however, that inhibition of gonadal suppression was observed within a few weeks in all children and that in no case did a secondary escape or partial resistance occur. Therefore, in no case was a secondary augmentation of the dosage of Decapeptyl necessary to maintain the pituitary and gonadal suppression. This was not as consistently observed in series using the subcutaneous or intranasal route. Such a rapid, constant, and permanent effect might be attributed to the mode of administration of the drug. The

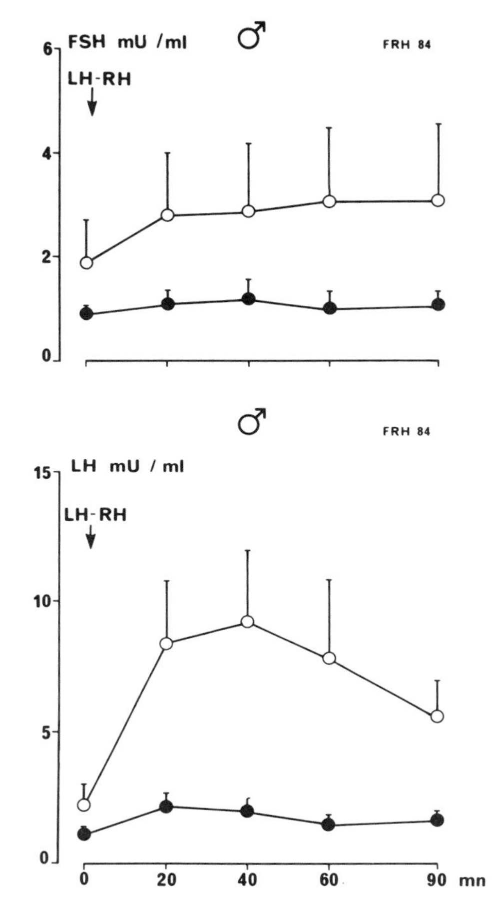

FIGURE 6. Mean and standard deviation (vertical bars) of FSH and LH levels in boys during LHRH tests (100 μg/m²) just before (open circles) and after 7 weeks (black circles) of treatment with intramuscular long-acting preparation of D-Trp^6-LHRH (Decapeptyl microcapsules).

subcutaneous injection of 100 μg Decapeptyl in adults induced a rapid rise of plasma D-Trp^6-LHRH levels, with a maximum between 30 and 90 min. However, the drug was undetectable in plasma after 24 hr (unpublished data). The intramuscular injection of Decapeptyl in microcapsules induced a more constant delivery of the drug to the bloodstream, and drug levels were easily detectable in serum for several weeks in rats (Mason-Garcia *et al.*, 1985) and in men (Roger *et al.*, 1985). In children, we were able to demonstrate as well that detectable levels were maintained for at least 21 days after a single intramuscular injection (Lahlou *et al.*, 1985).

In contrast with the complete inhibition of gonadal secretions, immunoreactive plasma levels of gonadotropins were only reduced to normal prepubertal values. Two different explanations might account for this apparent discrepancy. The fluc-

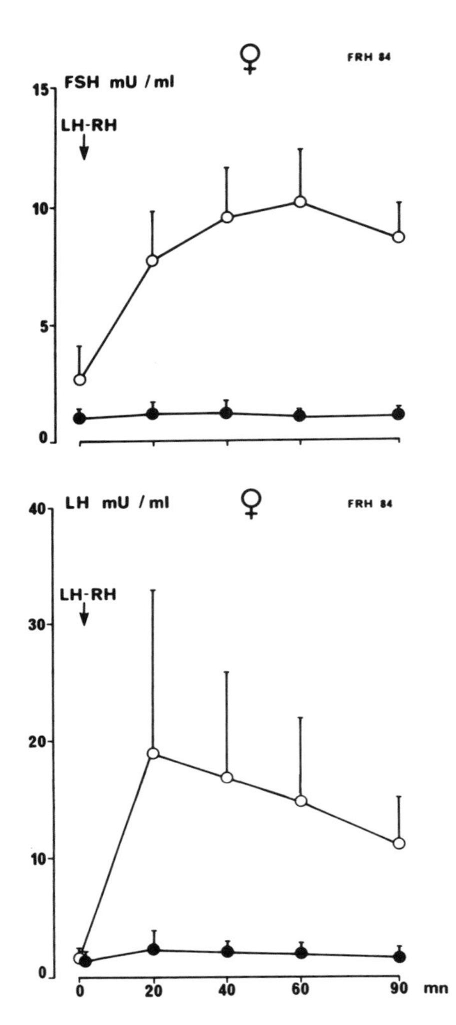

FIGURE 7. Mean and standard deviation (vertical bars) of FSH and LH levels in girls during LHRH tests (100 μg/m^2) just before (open circles) and after 7 weeks (black circles) of treatment with the intramuscular long-acting preparation of D-Trp^6-LHRH (Decapeptyl microcapsules).

tuating levels of LH are a characteristic feature of the mature pituitary and therefore the pubertal pituitary, whether or not puberty was precocious. It is well known that LHRH agonist treatments suppress the pulsatile secretion of LH (Crowley *et al.*, 1981; Mansfield *et al.*, 1983), and nonpulsatile levels of LH are able to downregulate gonadal LH receptors (Dufau *et al.*, 1984). On the other hand, there is a possibility that the LH immunoreactive activity we do measure by radioimmunoassay does not reflect the actual bioactivity, as suggested by the study of Evans *et al.* (1984).

4.2. Clinical Effects

4.2.1. Sexual Characteristics

In all publications a rapid decline of pubertal symptoms was noted in boys and in girls. This decline was noted to be more rapid and more constant than after treatment with inhibitory steroids (medroxyprogesterone or cyproterone).

In our study, a clinical improvement was obvious in all children within 6 months or less. In girls, breast enlargement decreased rapidly, and breasts lost their glandular consistency; Tanner scores were II to IV before treatment with Decapeptyl and I or II after 6 months of therapy. In the menstruating girl, bleeding disappeared after 1 month. Mean uterus size (±S.E.M.) decreased within 6 months from 43 ± 2.1 to 34 ± 1.9 mm ($P < 0.01$), and mean ovary length from 25 ± 1.9 to 16 ± 2.7 mm ($P < 0.02$).

In boys, mean testis volume (±S.E.M.) decreased within 6 months from 7.9 ± 1.3 to 6.7 ± 1.4 ml ($P < 0.05$). However, as no histological data are available in humans, it cannot be ascertained whether or not this finding is correlated with a reduction of the volume of endocrine tissue or germinal tissue. It has been reported that in rats (Lefebvre *et al.*, 1984) and in dogs (Vickery *et al.*, 1984) the numbers of germinal and Sertoli cells were reduced by LH agonist treatment, whereas Leydig cells appeared normal.

The effect of the treatment on pubic hair was more difficult to assess. In our study, as in others, pubic hair score decreased in some children and remained unchanged in others. The development of sexual hair is not a rapid process and is not dependent only on gonadal androgens. Since LHRH agonist treatments do not affect adrenal secretions (unlike treatment with medroxyprogesterone acetate or, particularly, cyproterone acetate), pubic hair, if present when agonist treatment started, might keep growing under the influence of adrenal androgens (Kauli *et al.*, 1984). Such an event should not be considered resistance to therapy.

4.2.2. Growth Velocity and Bone Maturation

Reduced adult height is one of the most severe consequences of precocious puberty, and it is well known that treatments with inhibitory steroids have not been shown to improve significantly the final height in many children.

In our study, growth velocity calculated over a 6-month or 1-year period diminished relative to the velocity recorded over the year before the first injection. This same effect as been reported in all previous publications of treatments in which a subcutaneous agonist was used. It is nonetheless difficult to attribute this evolution solely to the treatment itself, since growth velocity diminishes spontaneously after the pubertal spurt. Moreover, in children under 4 years of age, decrease in growth velocity is a spontaneous phenomenon in any event. This sign is therefore not the best indicator of therapeutic efficiency.

The evaluation of bone age is certainly a better index of the effects of treatment. In most of our 13 children, the bone age increment diminished relative to the height–age increment. The mean ratio of the height–age increment to the bone age increment was 2.0 ± 0.29 in boys and 1.3 ± 0.2 in girls. If confirmed on a larger scale, this fact could be considered evidence of an arrest in bone maturation greater than the slowing of growth velocity. This is, in fact, the primary benefit that can be expected from this treatment. It should be noted that, using the same index, treatments with inhibitory steroids have proven disappointing (Chaussain *et al.*, 1980). Mansfield *et al.* (1983) have compared with adult height prediction before and after treatment with D-Trp^6-desGly^{10}-LHRH ethylamide and found that adult height prediction increased by about 3 cm after 18 months of therapy. The accuracy of the tables for adult height prediction has been challenged when applied to pathological conditions (Zachmann *et al.*, 1978). Therefore, these different parameters should be considered only as indicative of a tendency. A definite demonstration of beneficial effects on adult height remains to be made.

5. TOLERANCE AND SIDE EFFECTS

5.1. Clinical Side Effects

Some side effects have been regularly recorded in adult patients treated with LHRH agonists for hormone-dependent cancers or endometriosis. Hot flashes, headaches, and digestive disturbances were the most commonly encountered symptoms.

Unlike adults, children tolerated high dosages of LHRH agonists without any significant immediate side effects. No adverse reactions, clinical or biological, were noted by Mansfield *et al.* (1983), Comite *et al.* (1983), or Kauli *et al.* (1984). In our study the only side effect was the occurrence in one girl of transient vaginal bleeding 9 days after the first and second intramuscular injections. Such bleedings have been observed in adult menstruating women, and they might occur without any variation in estradiol levels (Lemay *et al.*, 1984). Their pathogenesis is thus unclear.

5.2. Immune Reactions

As in other reports using subcutaneous or intranasal treatments, no allergic reactions were noted in our study. The induction of antibodies against the agonist has been carefully checked by several investigators. Mansfield *et al.* (1983) and Fraser *et al.* (1984) were unable to demonstrate a significant production of such antibodies in children treated for precocious puberty. The coupling of a small molecule, such as a small peptide, with a polymer is supposed to enhance its antigenic power. We failed, however, to demonstrate a significant binding of radioactive D-Trp^6-LHRH to patient immunoglobulins, even after 18 months of therapy. This negative result was quite consistent with the absence of secondary resistance.

5.3. Compliance

Daily treatments with subcutaneous or intranasal therapy were generally well accepted, and no families, in reports published so far, elected to discontinue the therapy. The periodic administration of intramuscular Decapeptyl every 4 weeks or once a month provided a major advantage: between two injections, both children and parents could forget the disease, as there was no therapeutic action necessary during the month. This can be considered a major improvement. By itself, the agonist suppressed the excessive emotional lability and the ill-timed sexual behavior of the children. The apparent absence of therapy with the exception of the monthly injection made easier the social integration of the children at home, at school, and during collective holidays.

5.4. Effects on Fertility

Short-term studies have shown that the inhibition of the pituitary–gonadal axis in children treated for precocious puberty with an LHRH agonist was rapidly reversible (Crowley *et al.*, 1981; Mansfield *et al.*, 1983; Comite *et al.*, 1984). Obviously, no data are available concerning the potential fertility of these children. In dogs and men, normal sperm counts were rapidly recovered after short treatments with LHRH agonists (Vickery *et al.*, 1984; Rabin *et al.*, 1984). In rats, the germinal alterations induced by a 5-month treatment are almost completely restored 5 months after discontinuation of the therapy (Lefebre *et al.*, 1984). Therefore, although we cannot state that very long-term treatments will have no effect on fertility, there is presently no evidence that fertility impairment should be expected.

6. CONCLUSION

The use of LHRH agonists has significantly modified the medical attitude regarding precocious puberty. In the recent past, the therapeutic tools have been, in some ways, so inefficient that incomplete results could be considered partial successes. At present, we know that we are able to suppress completely and permanently the functions of the gonads thanks to LHRH agonist administration.

This ability has raised new questions. Could the long-term down-regulation of pituitary and gonadal receptors induce nonreversible effects? Is the proven reversibility of the endocrine tissue inhibition a guarantee of the reversibility of the germinal cell alterations? In order to prevent eventual irreversible effects, it has been suggested that the therapy be discontinued after 1 or 2 years so that the reversibility could be assessed. Such an attitude is hardly acceptable, as the usefulness of treatment discontinuation has not been demonstrated. On the other hand, it can be stated that discontinuing the treatment can induce a secondary flare-up of the disease with disastrous consequences on psychosocial behavior and bone maturation. In any case, the desirable duration of the treatment remains an unsolved problem.

The advance in bone age was poorly controlled by treatment with inhibitory steroids, medroxyprogesterone, and/or cyproterone. The agonists of LHRH are more potent suppressors of sex steroids; thus, a stronger effect on bone maturation and a better resulting adult height could be expected. From this point of view, the perceived slowing done of bone age maturation and the improvement of adult height prediction are encouraging indices.

Peptides must be delivered directly to the bloodstream in order to retain their biological activity. Therefore, the subcutaneous and intranasal routes have long been considered the best choices. Intranasal administration several times a day or subcutaneous injection once or twice a day is a very constraining situation. There is no doubt that a long-acting delivery system would represent a significant advance. The intramuscular microcapsules of Decapeptyl dispersed within a biodegradable polymeric matrix have proven to be as efficient as daily subcutaneous treatment. It insures better child compliance than daily treatment and permits better psychosocial adaptation.

ACKNOWLEDGMENTS. The authors wish to thank Dr. R. Y. Mauvernay (Debiopharm, Lausanne, Switzerland) for his help in the pilot study of Decapeptyl microcapsules and the National Hormone and Pituitary Program (Baltimore, Maryland) for the generous gift of reagents for gonadotropin radioimmunoassays.

REFERENCES

Belchetz, P. E., Plant, T. M., Nakai, Y., Keogh, E. J., and Knobil, E., 1978, Hypophysial responses to continuous and intermittent delivery of hypothalamic gonadotropin-releasing hormone, *Science* **202:**631–633.

Bex, F. J., and Corbin, A., 1984, LHRH and analogs: Reproductive pharmacology and contraceptive and therapeutic utility, in: *Frontiers in Neuroendocrinology,* Vol. 8 (L. Martini and W. F. Ganong, eds.), Raven Press, New York, pp. 85–151.

Chaussain, J. L., Roger, M., Vassal, J., Canlorbe, P., and Job, J. C., 1980, La fonction androgénique testiculaire dans les pubertés précoces du garçon. Effets du traitement, in: *XVI International Congress of Pediatrics* (A. Ballabriga and A. Gallart, eds.), Spanish Pediatrics Association, Barcelona, p. 432.

Comite, F., Cutler, G. B., Rivier, J., Vale, W. W., Loriaux, D. L., and Crowley, W. F., Jr., 1981, Short-term treatment of idiopathic precocious puberty with a long-acting analogue of luteinizing hormone-releasing hormone. A preliminary report, *N. Engl. J. Med.* **305:**1546–1550.

Comite, F., Pescovitz, O. H., Rieth, K. G., Dwyer, A. J., Hench, K., McNemar, A., Loriaux, D. L., and Cutler, G. B., Jr., 1984, Luteinizing hormone-releasing hormone analog treatment of boys with hypothalamic harmatoma and true precocious puberty, *J. Clin. Endocrinol. Metab.* **59:**888–892.

Coy, D. H., Vilchez-Martinez, J. A., Coy, E. J., and Schally, A. V., 1976, Analogs of luteinizing hormone-releasing hormone with increased biological activity produced by D-amino acid substitutions in position 6, *J. Med. Chem.* **19:**423–425.

Crowley, W. F., Comite, F., Vale, W., Rivier, J., Loriaux, D. L., and Cutler, G. B., 1981, Therapeutic use of pituitary desensitization with a long-acting LHRH agonist: A potential new treatment for idiopathic precocious puberty, *J. Clin. Endocrinol. Metabl.* **52:**370–372.

Donaldson, M. D. C., Stanhope, R., Lee, T. J., Price, D. A., Brook, C. G. D., and Savage, D. C. L., 1984, Gonadotropin responses to GnRH in precocious puberty treated with GnRH analogue, *Clin. Endocrinol.* **21:**499–503.

Dufau, M. L., Winters, C. A., Hattori, M., Aquilano, D., Baranao, J. L. S., Nozu, K., Baukal, A., and Catt, K. J., 1984, Hormonal regulation of androgen production by the Leydig cell, *J. Steroid Biochem.* **20:**61–173.

Evans, R. M., Doelle, G. C., Lindner, J., Bradley, V., and Rabin, D., 1984, A luteinizing hormone-releasing hormone agonist decreases biological activity and modifies chromatographic behavior of luteinizing hormone in man, *J. Clin. Invest.* **73:**262–266.

Fraser, H. M., Sandow, J., and Krauss, B., 1983, Antibody production against an agonist analogue of luteinizing hormone-releasing hormone: Evaluation of immunochemical and physiological consequences, *Acta Endocrinol.* (*Kbh.*) **103:**151–157.

Greulich, N. N., and Pyle, S. I. (eds.), 1959, *Radiographic Atlas of Skeletal Development of the Hand and Wrist,* Stanford University Press, Stanford, CA.

Holland, F. J., Fishman, L., Costigan, D. C., Leeder, J. S., and Fasekas, A. T., 1985, D-Ser(TBU)^{6}EA^{9}LHRH in precocious puberty: Pharmacokinetics of SC, IV and intranasal use, *Pediatr. Res.* **19:**633.

Kauli, R., Pertzelan, A., Ben-Zeev, Z., Prager Lewin, R., Kaufman, H., Comaru-Schally, A. M., Schally, A. V., and Laron, Z., 1984, Treatment of precocious puberty with LHRH analogue in combination with cyproterone acetate—Further experience, *Clin. Endocrinol.* **20:**377–387.

Lahlou, N., Roger, M., Canlorbe, P., Chaussain, J. L., Raynaud, F., Toublanc, J. E., and Schally, A. V., 1985, Plasma levels of D-TRP-6-LH-RH (Decapeptyl) after intramuscular injection of long-acting microcapsules in children treated for precocious puberty (PP), *Pediatr. Res.* **19:**635.

Laron, Z., Kauli, R., Ben Zeev, Z., Comaru-Schally, A. M., and Schally, A. V., 1981, D-TRP^{6} analogue of luteinising hormone releasing hormone in combination with cyproterone acetate to treat precocious puberty, *Lancet* **2:**955–956.

Lefebvre, F. A., Bélanger, A., Pelletier, G., and Labrie, F., 1984, Recovery of gonadal functions in the adult male rat following cessation of five-month daily treatment with an LHRH agonist, *J. Androl.* **5:**181–192.

Lemay, A., Maheux, R., Faure, N., Jean, C., and Fazekas, A. T. A., 1984, Reversible hypogonadism induced by a luteinizing hormone-releasing hormone (LH-RH) agonist (Buserelin) as a new therapeutic approach for endometriosis, *Fertil. Steril.* **41:**863–871.

Luder, A. S., Holland, F. J., Costigan, D. C., Jenner, M. R., Wielgosz, G., and Fazekas, A. T. A., 1984, Intranasal and subcutaneous treatment of central precocious puberty in both sexes with a long-acting analog of luteinizing hormone-releasing hormone, *J. Clin. Endocrinol. Metab.* **58:**966–972.

Mansfield, M. J., Beardsworth, D. E., Loughlin, J. S., Crawford, J. D., Bode, H. H., Rivier, J., Vale, W., Kushner, D. C., Crigler, J. F., and Crowley, W. F., 1983, Long-term treatment of central precocious puberty with a long-acting analogue of luteinizing hormone-releasing hormone. Effects on somatic growth and skeletal maturation, *N. Engl. J. Med.* **309:**1286–1290.

Mason-Garcia, M., Vigh, S., Comaru-Schally, A. M., Redding, T. W., Somogyvari-Vigh, A., Horvath, J., and Schally, A. V., 1985, Radioimmunoassay for 6-D-tryptophan analog of luteinizing hormone-releasing hormone: Measurement of serum levels after administration of long-acting microcapsule formulations, *Proc. Natl. Acad. Sci. U.S.A.* **82:**1547–1551.

Rabin, D., Evans, R. M., Alexander, A. N., Doelle, G. C., Rivier, J., Vale, W., and Liddle, G. W., 1984, Heterogeneity of sperm density profiles following 20-week therapy with high-dose LHRH analog plus testosterone, *J. Androl.* **5:**176–180.

Redding, T. W., Schally, A. V., Tice, T. R., and Meyers, W. E., 1984, Long-acting delivery systems for peptides: Inhibition of rat prostate tumors by controlled release of D-Trp^6 luteinizing hormone-releasing hormone from injectable microcapsules, *Proc. Natl. Acad. Sci. U.S.A.* **81:**5845–5848.

Roger, M., 1983, La maturation des fonctions gonadotropes de la naissance à l'adolescence, *Contraception Fertil. Sexual.* **11:**1139–1151.

Roger, M., Nahoul, K., Toublanc, J. E., Castanier, M., Canlorbe, P., and Job, J. C., 1979, Les androgènes plasmatiques chez le garçon de la naissance à l'adolescence, *Ann. Pediatr.* **26:**239–245.

Roger, M., Chaussain, J. L., Evain, D., Toublanc, J. E., Canlorbe, P., and Scholler, R., 1984, Les formes habituelles et inhabituelles des précocités sexuelles féminines et leur diagnostic biologique, *Ann. Pediatr.* **31:**183–192.

Roger, M., Duchier, J., Lahlou, N., Nahoul, K., and Schally, A. V., 1985, Treatment of prostatic carcinoma with D-TRP-6-LH-RH: Plasma hormone levels after daily subcutaneous injections and periodic administration of delayed release preparations, *Prostate* **7:**271–282.

Roger, M., Chaussain, J.-L., Berlier, P., Bost, M., Canlorbe, P., Colle, M., Francois, R., Garandeau, P., Lahlou, N., Morel, Y., and Schally, A. V., 1986, Long-term treatment of male and female precocious puberty by periodic administration of a long acting preparation of D-Trp-6-LH-RH microcapsules, *J. Clin. Endocrin. Metab.* **62:**670–677.

Sandow, J., 1983, Clinical applications of LHRH and its analogues, *Clin. Endocrinol.* **18:**571–592.

Schally, A. V., Redding, T. W., and Comaru-Schally, A. M., 1984, Potential use of analogs of luteinizing hormone-releasing hormones in the treatment of hormone-sensitive neoplasms, *Cancer Treat. Rep.* **68:**281–289.

Sempé, P., and Sempé, M. (eds.), 1979, *Auxologie: Méthode et Séquences,* Théraplix, Paris.

Tanner, J. M. (ed.), 1962, *Growth at Adolescence,* Blackwell Scientific, Oxford.

Vickery, B. H., McRae, G. I., Briones, W., Worden, A., Seidenberg, R., Schanbacher, B. D., and Falvo, R., 1984, Effects of an LHRH agonist analog upon sexual function in male dogs, *J. Androl.* **5:**28–42.

Zachmann, M., Sobradillo, B., Frank, M., Frisch, H., and Prader, A., 1978, Bayley-Pinneau, Roche-Wainer-Thissen and Tanner height predictions in normal children and in patients with various pathological conditions, *J. Pediatr.* **93:**749–755.

48

Regulation of Neuroendocrine Pathways by Thymosins

NICHOLAS R. HALL, BRYAN L. SPANGELO, JOHN M. FARAH, Jr., THOMAS L. O'DONOHUE, and ALLAN L. GOLDSTEIN

1. INTRODUCTION

Historically, the thymus gland has long been considered an endocrine gland, first with a role within the reproductive axis (Anderson, 1932) and later as an organ functioning in opposition to the adrenal cortex (Martin, 1976). Within the past three decades, a crucial role during the differentiation of T lymphocytes has also been documented, a role that has greatly overshadowed the thymus gland's importance as a reproductive and/or stress-associated endocrine tissue. Evidence collected in recent years suggests that these apparently dissociated roles are not mutually exclusive and that in some instances, soluble products of thymic epithelial cells and lymphocytes are responsible for normal functioning of the cellular branch of the immune system as well as the regulation of reproductive and stress-associated neuroendocrine circuits (Goldstein *et al.*, 1981; Hall and Goldstein, 1983). These peptides with hormonelike activity are called thymosins and are able to exert their immunoregulatory influences both directly, by acting at the level of the lymphocyte, and indirectly, via pituitary–gonadal and pituitary–adrenal hormones.

The evidence in support of thymosin regulation of immunogenesis through direct pathways has been extensively reviewed elsewhere (see Goldstein *et al.*, 1972, 1981). This chapter is devoted to a discussion of the evidence that thymosin peptides are able indirectly to modulate the course of immunity by acting directly at the level of the pituitary gland or within the central nervous system. Reproductive

NICHOLAS R. HALL, BRYAN L. SPANGELO, and ALLAN L. GOLDSTEIN • The George Washington University School of Medicine and Health Sciences, Washington, D. C. 20892. JOHN M. FARAH, Jr., and THOMAS L. O'DONOHUE • Experimental Therapeutics Branch, National Institute of Neurological and Communicative Disorders and Stroke, Bethesda, Maryland 20205.

system hormones that are stimulated include luteinizing hormone (LH) and prolactin (PRL), and stress-associated hormones include adrenocorticotropin hormone (ACTH) and β-endorphin (β-END). Each of these hormones and/or target tissue products has immunoregulatory potential, as is discussed.

2. THYMOSINS AND REPRODUCTIVE PITUITARY HORMONES

At the turn of the century, thymic extracts were used extensively in the treatment of cryptorchidism and other reproductive system abnormalities (Anderson, 1932). Experimental evidence suggesting that there might be a biochemical basis for this treatment regimen began to appear in the literature in the 1970s with an early report that gonadal degeneration subsequent to thymectomy could be prevented by administration of homeostatic thymic hormone (HTH), an extract of the thymus gland (Comsa, 1979). This observation was later extended when it was found that HTH and thymosin fraction 5 (TF-5) were able to restore the decreased body, adrenal, and testis weight that was subsequent to neonatal thymectomy (Deschaux *et al.*, 1979). Plasma levels of testosterone and LH were altered following this surgical procedure, but levels of these hormones returned toward normal values following the administration of either HTH or TF-5. An involvement of the pituitary gland was suggested by the observation that the thymectomy-induced hormonal changes failed to occur when the animals were also hypophysectomized. This same conclusion had been suggested by another group of investigators following observations of abnormal LH cyclicity in athymic mice (Weinstein, 1978).

It has subsequently been found that some of these effects can be attributed to specific peptides that are components of the TF-5 preparations used in the earlier investigations. For example, in an *in vitro* superfusion system, it has been found that synthetic thymosin β_4 (Tβ_4) but not thymosin α_1 (Tα_1) stimulates luteinizing hormone-releasing hormone (LHRH) from rat medial basal hypothalami (Rebar *et al.*, 1981). Release of LH from superfused pituitary tissue occurred only when the tissue was superfused in sequence with the medial basal hypothalamus. Further evidence that Tβ_4 might be active at the level of the hypothalamus has been generated following the intracerebroventricular injection of Tβ_4 (Hall *et al.*, 1982).

This peptide was injected at a concentration of 1 μg in a volume of 2 μl of saline into the lateral cerebroventricle of chronically cannulated mice. Cannulas were fashioned from PE-50 tubing and were secured to the calvarium using dental cement. On the day of the experiment, a precisely measured needle was used to administer the peptide. Two control groups were comprised of animals that received either the saline vehicle or Tα_1 as a peptide control. The results of this study are summarized in Table I. As can be seen, Tβ_4 resulted in a significant elevation of LH, whereas the vehicle and peptide controls were without effect on this particular measure. However, Tα_1 was found to cause a significant elevation in corticosterone levels, whereas Tβ_4 had no effect on this particular measure. Based on the superfusion studies, it has been concluded that LH release that occurs subsequent to

TABLE I. Effect of Intracerebroventricular Injection of Thymosin Peptides on Serum Luteinizing Hormones and Corticosterone Levels[a]

Treatment	Corticosterone (ng/ml ± S.E.M.)	Luteinizing hormone (ng/ml ± S.E.M.)
Noninjected	118.1 ± 31.3	0.89 ± 0.45
Saline	92.2 ± 13.2	0.40 ± 0.07
$T\beta_4$, 1 μg	82.6 ± 19.6	2.82 ± 0.75[b]
$T\alpha_1$, 1 μg	225.9 ± 57.5[c]	0.34 ± 0.75

[a] Serum hormone levels were measured 3 hr following the intracerebral injection of thymosin.
[b] $P < 0.01$ vs. saline.
[c] $P < 0.02$ vs. saline.

administration of $T\beta_4$ is caused by a central-nervous-system-mediated mechanism. This is not true for another reproductive hormone, PRL, which can be released directly by TF-5 following its coincubation with a PRL-secreting pituitary tumor cell line.

The effect of PRL on the immune system has been the subject of much scrutiny in recent years. Initially, investigators proposed that PRL was an immunosuppressive agent based on the results of both *in vivo* and *in vitro* studies (Harris *et al.*, 1979; Kelly and Dineen, 1973). Subsequent to these reports, however, Berczi and colleagues (Berczi *et al.*, 1981) proposed that PRL may be an important hormone for the proper maintenance of the immune system. In these experiments, hypophysectomized female rats were shown to be immunodeficient. Chronic administration of PRL was required to restore the immune response of these animals to the normal range. Our own work (Spangelo *et al.*, 1985) has shown that PRL can potentiate the response of spleen cells to the T-cell mitogen concanavalian A (con-A) but not the B-cell mitogen lipopolysaccharide (LPS). We therefore examined the effects that PRL has on the production of antibody to the T-cell-dependent antigen sheep red blood cells (SRBC) in normal mice.

In an initial study, multiple injections of bovine PRL (bPRL) given before, during, and after sensitization to SRBC caused a significant increase in antibody titers in mice compared to control BSA-injected animals. What was not apparent from this study was the time at which PRL was most effective in the stimulation of the immune response. Normal male mice (C57Bl/6) were therefore injected with 0.2 ml of 10% SRBC (day 0) and 200 μg of BSA or bPRL on day 0, 2, or 4. All animals were sacrificed 5 days after immunization. As seen in Fig. 1, only injections of bPRL given 4 days after SRBC immunization resulted in a significant increase in antibody titers. It may be concluded from this study that PRL only stimulates an already activated immune system. This is consistent with the *in vitro* observation that bPRL has no effect on resting lymphocytes but can stimulate cells incubated with the T-cell mitogen con-A (Spangelo *et al.*, 1985).

The stimulation of the immune response by PRL suggested that the immune system itself may cause alterations in the levels of this hormone. To test this

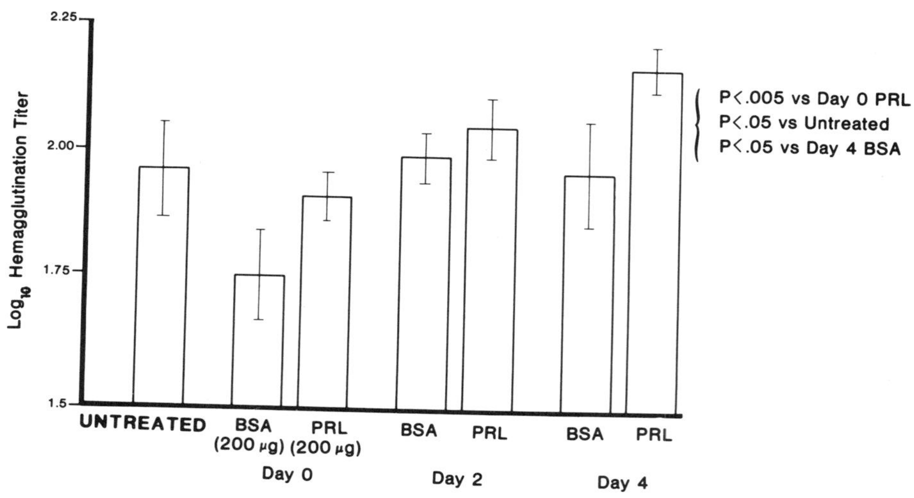

FIGURE 1. Prolactin enhances antibody production when administered during the logarithmic phase of a primary immune response.

hypothesis, the GH3 rat pituitary tumor cell line was tested for responsiveness to TNS F5. This cell line was chosen since it secretes PRL and growth hormone, which have both been shown to have immunomodulatory properties. The GH3 cells were maintained in monolayer culture in a humidified atmosphere of 95% air, 5% CO_2 and grown in Ham's F-10 medium supplemented with 15% horse serum, 5% fetal calf serum, and 50 μg/ml gentamicin.

For each experiment cells from the same parent flask were trypsinized and diluted to 50,000 cells/ml. These cells (2 ml) were subsequently plated into 35-mm multiwell plates. After 72–96 hr, medium was aspirated from each well, and

TABLE II. Release of Prolactin from GH_3 Cells Coincubated with Thymosin Fraction 5

Thymosin fraction 5 (μg/ml)	rPRL released (ng/ml)	*P* value
0.0	3.78 ± 0.71	—
0.98	4.37 ± 0.15	NS
1.95	5.03 ± 0.43	NS
3.91	5.55 ± 0.45	NS
7.81	7.75 ± 1.02	<0.025
15.63	11.7 ± 0.97	<0.005
31.25	15.5 ± 1.22	<0.001
62.50	17.9 ± 0.86	<0.001
125.00	21.9 ± 0.49	<0.001
250.00	24.3 ± 1.77	<0.001
500.00	23.2 ± 1.47	<0.001
1000.00	23.8 ± 3.05	<0.005

the cell monolayer was washed two times with serum-free Ham's F-10 buffered to pH 7.3 with 25 mM HEPES. The cells were then tested with TF-5 or media (1 ml volumes).

In an initial experiment, TF-5 from 1 pg to 10 mg was tested on the GH3 cell line. It was found that TF-5 from 10 to 1000 μg stimulated PRL release. The 10-mg dose had no apparent effect on PRL release, and doses less than 1 μg were also ineffectual. In a subsequent study, several doses from 1 to 1000 μg of TF-5 were examined. As shown in Table II, all doses of 7.81 μg TSN F5 and above were effective in stimulating release of PRL from the GH3 cells. In addition, it is evident that doses of 125 μg and above were equally effective in causing release.

3. THYMOSINS AND STRESS-ASSOCIATED PITUITARY HORMONES

Traditionally, glucocorticoids have been regarded as immunosuppressive hormones because at high concentrations they are able to diminish a number of measures of immune responsiveness, especially those that are concerned with T-lymphocyte function (see Hall and Goldstein, 1984; Monjan and Collector, 1977). Furthermore, stress-inducing events that can be correlated with elevated glucocorticoids can also be correlated with immunosuppression and increased susceptibility to opportunistic infections (Palmblad, 1985; Stein, 1985; Ader and Friedman, 1964). However, glucocorticoids are not always immunosuppressive. This class of steroids at low concentrations can exert a so-called "permissive effect" and are required at low physiological concentrations (0.01–0.11 μM) of cortisol, corticotropin, or certain related corticosteroids if lymphocytes are to be cultured successfully in serum-free media (Ambrose, 1964). Furthermore, in certain protocols involving chronic stress, immune responsiveness can be potentiated (Monjan, 1981). Consequently, this class of steroids is best described as being immunomodulatory with differential effects depending on the hormone's concentration and/or the temporal parameters associated with its release.

During the course of an immune response, the glucocorticoids have been found to be significantly elevated at the time of peak antibody formation (Besedovsky *et al.*, 1981). This rise may well be subsequent to the release of thymosin and/orlymphokines, since each class of immunopeptide has been shown to induce steroidogenesis (Besedovsky *et al.*, 1981; Hall *et al.*, 1982). This phenomenon has now been documented in a number of species including monkeys (Healy *et al.*, 1983) and rodents (McGillis *et al.*, 1985).

In the first study, premenarchial cynomolgus monkeys (*Macaca fascicularis*) were fitted with a vest and mobile tether assembly that permitted chronic cannulation of the femoral vein. It was subsequently possible to inject test substances and to withdraw blood samples from unanesthetized, freely moving animals. TF-5 at concentrations of 1 or 10 mg/kg were injected at 0700 hr, which resulted in a dose-dependent elevation of hormones associated with the pituitary–adrenal axis. Basal levels of ACTH, β-END, and cortisol were (mean ± S.E.M.) 24.06 ± 3.91 pg/ml, 37.3 ± 7.65 pg/ml, and 35.3 ± 3.16 μg/dl, respectively. Significant increases in these hormones followed the administration of the 10 mg/kg dose of TF-5, which were 46.34 ± 7.05 pg/ml, 51.3 ± 13.6 pg/ml, and 56.3 ± 8.7 μg/dl for ACTH, β-END, and cortisol. Other pituitary hormones that were measured included LH, PRL, follicle-stimulating hormone, growth hormone, and thyrotropin; however, no changes were detected following the administration of TF-5 (Healy *et al.*, 1983). The discrepancy between this observation and the previously discussed findings in which LH and PRL were found to be stimulated by thymosin peptides could represent a species difference or the fact that in the rodent studies, adult animals were used, whereas in the monkey studies, the animals were prepubertal.

If the rise in ACTH, β-END, and cortisol in response to TF-5 administration was indeed a manifestation of a regulatory circuit between the immune and nervous systems, then removal of a primary source of thymosin, the thymus gland, would be expected to result in a decrease in each of these hormones. This experiment was undertaken using thymectomized monkeys, and on its conclusion, the predicted results were observed, as summarized in Table III (Healy *et al.*, 1983). Athymic primates were found to have significantly lower plasma ACTH and β-END, and cortisol values tended to be lower than in control monkeys.

If a component peptide of thymosin plays a physiological role in modulating the release of ACTH and cortisol, then it should be possible to detect changes in either thymosin peptides or pituitary–adrenal hormones under circumstances during which either would be expected to change. A change in corticosterone during the course of an immune response when thymic peptides are thought to be active has already been discussed (Besedovsky and Sorkin, 1977). There is also evidence for a circadian periodicity of $T\alpha_1$ that can be correlated with the circadian rhythm of glucocorticoids (McGillis *et al.*, 1983a). In mice, peak levels of $T\alpha_1$ were detected at approximately 1.5 hr after the onset of light, which was temporally opposite to the peak levels of glucocorticoids. Another expression of this relationship would be that peak levels of $T\alpha_1$ preceded the peak in corticosterone. However, it is important to note that studies establishing a cause–effect relationship between these two hormones have not yet been completed.

A similar correlation has been observed following exposure to stress-inducing

TABLE III. Plasma ACTH, β-END, and Cortisol Concentrations in Athymic and Aged-Matched Control Monkeys (n = 8)

	Plasma hormone concentrations (mean ± S.E.)		
Primate group	ACTH	β-END	Cortisol
Athymic[a]	39.6 ± 5.7[b]	85.8 ± 11.3[c]	33.1 ± 4.0
Control	90.7 ± 26.6	147.1 ± 24.0	46.5 ± 7.6

[a] Primates were tested 6–10 weeks after thymectomy.
[b] $P < 0.05$.
[c] $P < 0.025$.

protocols. Male Swiss-Webster mice were subjected to restraint stress for 3 hr on two consecutive days. Following this exposure, the animals were returned to their individual cages with food and water available *ad libidum*. The animals were sacrificed approximately 24 hr after the final stress, at which time body weights, spleen weights, and thymus weights were recorded. Trunk blood was collected for the measurement of corticosterone and $T\alpha_1$ in serum. The results revealed that animals that were exposed to the stress protocol had significantly involuted thymus glands compared with the unhandled control mice (mean ± S.E.M.) (29.4 ± 2.3 mg versus 39.6 ± 1.8, $P < 0.005$). Although the spleen was reduced in size, this reduction was not significant ($P < 0.1$). There was no difference between the mean body weights of the stressed and control mice.

Consistent with the reduction in thymus weight was an increase in serum corticosterone in the mice that were stressed (181.6 ± 52.4 ng/ml) compared with the control animals (80 ± 24 ng/ml). Furthermore, serum $T\alpha_1$ was also elevated in the stressed mice (1.7000 ± 0.34 ng/ml) relative to the control animals (0.969 ± 0.128 ng/ml). The physiological relevance of this elevation of $T\alpha_1$ is not fully understood, but it might constitute a mechanism by which recovery from stress-induced immunosuppression could be potentiated. Evidence in support of this hypothesis is discussed elsewhere (Hall and Goldstein, 1983).

4. SITE OF ACTION OF THYMOSIN-INDUCED STEROIDOGENESIS

Three potential sites exist at which the thymosin peptides may exert their influence on steroidogenesis. They include a direct effect at the level of the adrenal fasciculata cell, the pituitary gland, and/or the neuronal pathways that regulate corticotropin-releasing factor (CRF) activity. Each of these has been systematically evaluated.

Direct stimulation of adrenal fasciculata cells does not occur (Vahouny *et al.*, 1983). This possibility was tested by coincubating isolated cells with various concentrations of TF-5 as well as purified component peptides. The latter included $T\alpha_1$, $T\alpha_7$, $T\beta_4$. In addition, lymphokine-containing supernatant fluids in which con-A-stimulated lymphocytes were cultured were evaluated. However, none of these preparations was effective in stimulating either corticosterone or cAMP release. Instead of direct stimulation, the possibility that thymosin might act synergistically with ACTH was considered. To test this possibility, the same concentrations of thymosin and lymphokine preparations were added to a dose of ACTH that resulted in only a suboptimal stimulation of corticosterone and cAMP. The results of this latter experiment were also negative, and so it was considered unlikely that the corticogenic effect of thymosin observed *in vivo* resulted from stimulation of the adrenal cortex.

Another potential site of action is the anterior pituitary gland, especially since ACTH and β-endorphin were elevated when TF-5 was injected into monkeys (Healy *et al.*, 1983). Initial studies revealed that at concentrations of 1 pg to 1 mg, there was no stimulation of ACTH release when pituitary tissue was superfused with TF-5 (McGillis *et al.*, 1983b; Hall *et al.*, 1985). Subsequent studies, however, suggest that direct stimulation of cultured rat pituitary cells with TF-5 does result in a dose- and time-dependent release of ACTH (McGillis *et al.*, 1985).

Similar findings have been reported following studies in which AtT-20 cells, an ACTH- and β-endorphin-secreting cell line, were coincubated with TF-5 (Farah *et al.*, 1985). An elevation in secreted β-endorphin immunoreactivity was detected within 15 min after treating the AtT-20 cells with 200 μg/ml of TF-5. The amount released was 4.4 ± 0.5 ng/ml, which was significantly higher than the control level, which was 3.1 ± 0.1 ng/ml. Lower concentrations of TF-5 had no effect on basal release of β-endorphin immunoreactivity or on the release evoked by moderate to high doses of rat CRF. However, 600 μg/ml of TF-5 in combination with 0.1 μM CRF evoked an increase in hormone release that was synergistic. This synergistic effect following treatment with doses that alone caused a near-maximal hormone secretion of β-endorphin suggests that the TF-5 may act by a mechanism that is independent of cyclic adenosine monophosphate (cAMP), the second messenger that mediates CRF stimulation of corticotrophic cells. Consistent with this interpretation was the observation that TF-5 potentiates hormone secretion elicited by forskolin, an activator of the enzyme that generates cAMP.

Although these data identify the pituitary gland as a site of action of thymosin in stimulating neuroendocrine circuits, they do not preclude the possibility of a central nervous system site of action as well. Several types of evidence suggest that this may indeed be the case. First, thymosin peptides are corticogenic when injected directly into the brains of chronically cannulated rodents. In one such study, adult male rats received injections of TF-5 into the third cerebroventricle (Hall *et al.*, 1983). The peptide preparation was injected over five consecutive days at a concentration of 10 μg per animal. The endpoint was endocrine tissue weight, with the gonads, thyroid, and adrenals being evaluated. Of these tissues, only the adrenal glands were found to undergo an increase in weight. Furthermore, only TF-5 was effective in stimulating this gland. Control preparations included a kidney fraction 5 (prepared from calf kidney tissue in a manner identical to TF-5) injected into the third ventricle, saline vehicle injected into the third ventricle, TF-5 injected i.p. to control for leakage into the peripheral circulation, and a noninjected group. A subsequent experiment was carried out using mice, and the results suggested that the increase in adrenal weight was correlated with increased production of corticosterone.

In this latter study, mice were fitted with polyethylene guide tubes that were secured in the skull overlying the lateral cerebroventricle (Hall *et al.*, 1982). The guide tubes were precisely measured so that they extended into the cerebrospinal fluid. At the time of the experiment, a microsyringe with a needle the same length as the guide tube was used to deliver the thymosin peptides in a volume of 2 μl of saline. At a concentration of 1 μg, Tα_1 administration was followed by a significant rise in corticosterone (Table I). It is important to note that when injected into monkeys, $T\alpha_1$ was not corticogenic, even though the plasma level was comparable to that observed following the injection of TF-5 (Healy *et al.*, 1983). Whether this discrepency was caused by a species difference or a failure of the $T\alpha_1$ to cross the blood–brain barrier when injected i.v. has not been established. In addition to exerting biological activity when injected into the brain, $T\alpha_1$ immunoreactivity has also been detected in discrete brain regions including areas that, when stimulated or lesioned electrically, can result in alterations of CRF activity. These include the arcuate nucleus and median eminence of the hypothalamus as well as the pituitary gland (Palaszynski *et al.*, 1983).

5. DISCUSSION

A number of hormones that are produced either by the pituitary gland or by target tissues are able to exert profound effects on the immune system. These include prolactin, β-endorphin, and adrenal corticosteroids, which have been found to modulate the immune system in both *in vitro* and *in vivo* protocols. The observation that each of these hormones can be stimulated by a product of the immune system suggests that thymosin might constitute a feedback component of a central nervous system—immune system circuit. Thus, in addition to their direct effects on lymphocytes, thymosin peptides may well exert indirect effects via the release of immunomodulatory pituitary peptides.

The physiological role that this release might play during the course of host defense can only be the subject of speculation until further studies are conducted. This is especially true with respect to prolactin, which has only recently become the subject of intense study within the domain of immunology. More is known about the effects of reproductive steroids on the immune response, but how these effects are orchestrated with the course of immunogenesis is not known. The best-characterized of the hormones stimulated by thymosin are the glucocorticoids, which can either stimulate or inhibit the immune system depending on the dose and temporal parameters involved (see Hall and Goldstein, 1984).

It has been proposed that the stimulation of glucocorticoid release during an immune response may constitute a mechanism by which the immune system limits the extent of its activation (Besedovsky and Sorkin, 1977). A basic premise of this hypothesis is that the rise in steroid is secondary to the release of thymosins and/or lymphokines. It is also important to take into account the fact that less mature T lymphocytes have a higher density of glucocorticoid receptors compared with mature T cells. Thus, as the immune response proceeds, T lymphocytes become less responsive to the potentially immunosuppressive effects of glucocorticoids. The steroids released at the time of peak antibody production would have their most deleterious effects on the immature cells, with either no or low affinity for the sensitizing immunogen. Cells actively involved in mounting a defense against the immunogen would be relatively resistant to this potential suppression because of their reduced density of receptors. Those cells most affected would constitute a

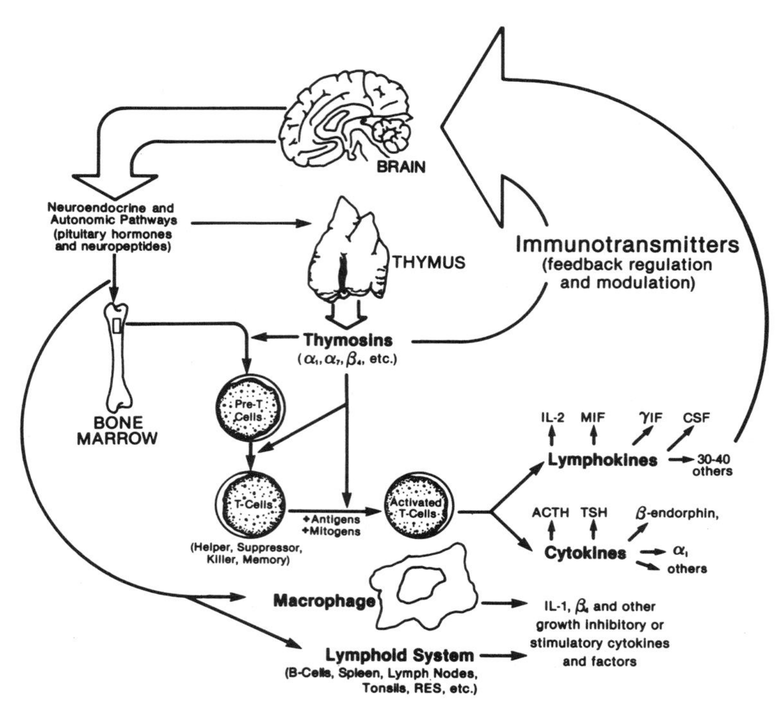

FIGURE 2. Interrelationships between thymosin peptides and neuroendocrine circuits in the regulation of immunity. See text for explanation.

population not required for the ongoing immune response but that might be capable of being stimulated by nonspecific products such as lymphokines. A modification of the original hypothesis as described by Besedovsky that incorporates the results of the experiments discussed in this review is summarized in Fig. 2.

In summary, this model could be viewed as a mechanism for eliminating background noise during an immune response. In sensory physiology, it is called lateral inhibition and involves neurons sending inhibitory collateral projections to adjacent cells. The same principle is thought to apply to the cells of the immune system.

As discussed previously, it is not clear how the other hormones might fit into this scenario. One possibility is that different component peptides of TF-5 are each responsible for the release of different pituitary peptides and that this release occurs only under conditions during which the hormone is required. The importance of pituitary hormone release subsequent to thymosin exposure and the clinical implications of this phenomena are far from clear; however, the concept of autoregulation involving the brain and pituitary gland has been documented for all of the major endocrine systems in the body. Furthermore, perturbations of the major endocrine systems can involve dysfunction at any of the levels of control. The evidence reviewed in this chapter suggests that the endocrine thymus may not be an exception.

ACKNOWLEDGMENT. These studies have been supported in part by grants and/or gifts from the NIH (CA24974), C. Hoffmann-LaRoche, Inc., and Alpha 1 Biomedicals, Inc.

REFERENCES

Ader, R., and Friedman, S. B., 1964, Social factors affecting emotionality and resistance to disease in animals: IV. Differential housing, emotionality and Walker 256 carcinosarcoma in the rat, *Psychol. Rep.* **15:**535–541.

Ambrose C. T., 1964, The requirement for hydrocortisone in antibody-forming tissue cultivated in serum-free medium, *J. Exp. Med.* **119:**1027–1049.

Ambrose, C. T., 1970, The essential role of corticosteroids in the induction of the immune response *in vitro,* in: *Hormones and the Immune Response* (G. E. W. Wolstenholme and J. Knight, eds.), J. A. A. Churchill, London, pp. 100–125.

Anderson, D. H., 1932, The relationship between the thymus and reproduction, *Physiol. Rev.* **12:**1–22.

Berczi, I., Nagy, E., Kovacs, K., and Horvath, E., 1981, Regulation of humoral immunity in rats by pituitary hormones, *Acta Endocrinol.,* **98:**506.

Besedovsky, H. O., and Sorkin, E., 1977, Network of immune–neuroendocrine interactions, *Clin. Exp. Immunol.* **27:**1–12.

Besedovsky, H. O., del Rey, A., and Sorkin, E., 1981, Lymphokine-containing supernatants from con-A stimulated cells increase corticosterone blood levels, *J. Immunol.* **126:**385–389.

Comsa, J., Leonhardt, H., and Ozminski, K., 1979, Hormonal influences on the secretion of the thymus, *Thymus* **1:**81–93.

Deschaux, P., Massengo, B., and Fontages, R., 1979, Endocrine interaction of the thymus with the hypophysis, adrenals and testes: Effects of two thymic extracts, *Thymus* **1:**95–108.

Farah, J. M., Bishop, J. F., McGillis, J. P., Hall, N. R., Goldstein, A. L., and O'Donohue, T. L., 1985, Thymosin stimulates hormone secretion from AtT20 mouse corticotropic tumor cells, *Soc. Neurosci. Abstr.* **11:**886.

Goldstein, A. L., Guha, A., Zatz, M. M., and White, A., 1972, Purification and biological activity of thymosin, a hormone of the thymus gland, *Proc. Natl. Acad. Sci. U.S.A.* **69:**1800–1803.

Goldstein, A. L., Low, T. L. K., Thurman, G. B., Zatz, M. M., Hall, N. R., Chen, C. P., Hu, S., Naylor, P. B., and McClure, J. E., 1981, Current status of thymosin and other hormones of the thymus gland, *Recent Prog. Horm. Res.* **37:**369–415.

Hall, N. R., and Goldstein, A. L., 1983, The thymus–brain connection: Interaction between thymosin and the neuroendocrine system, *Lymphokine Res.* **2:**1–6.

Hall, N. R., and Goldstein, A. L., 1984, Endocrine regulation of host immunity: The role of steroids and thymosin, in: *Immune Modulation Agents and Their Mechanisms* (R. L. Fenichel and M. A. Chirigos, eds.), Marcel Dekker, New York, pp. 533–563.

Hall, N. R., McGillis, J. P., Spangelo, B. L., Palaszynski, E., Moody, T., and Goldstein, A. L., 1982, Evidence for a neuroendocrine–thymus axis mediated by thymosin polypeptides, *Dev. Immunol.* **17:**653–660.

Hall, N. R., McGillis, J. P., and Goldstein, A. L., 1984, Activation of neuroendocrine pathways by thymosin peptides, in: *Stress, Immunity and Aging* (E. L. Cooper, ed.), Marcel Dekker, New York, pp. 209–223.

Hall, N. R., McGillis, J. P., Spangelo, B. L., Healy, D. L., Chrousos, G. P., Schulte, H. M., and Goldstein, A. L., 1985, Thymic hormone effects on the brain and neuroendocrine circuits, in: *Neural Modulation of Immunity* (R. Guillemin, M. Cohn, and T. Melnechuk, eds.), Raven Press, New York, pp. 179–193.

Harris, R. D., Kay, N. E., Seljeskog, K. J., and Douglas, S. D., 1979, Prolactin suppression of leukocyte chemotaxis *in vitro, J. Neurosurg.* **50:**462.

Healy, D. L., Hodgen, G. D., Schultz, H. M., Chrousos, G. P., Loriaux, D. L., Hall, N. R., and Goldstein, A. L., 1983, The thymus–adrenal connection: Thymosin has corticotropin-releasing activity in primates, *Science* **222:**1353–1355.

Kelly, J. D., and Dineen, J. K., 1973, The suppression of rejection of *Nippostrongylus brasiliensis* in Lewis strain rats treated with ovine prolactin, *Immunology* **24:**551.

Martin, C. R. (ed.), 1976, *Textbook of Endocrine Physiology,* Williams & Wilkins, Baltimore.

McGillis, J. P., Hall, N. R., and Goldstein, A. L., 1983a, Circadian rhythm of thymosin α_1 in normal and thymectomized mice, *J. Immunol.* **131:**146–151.

McGillis, J. P., Hall, N. R., and Goldstein, A. L., 1983b, Effect of superfused thymosin fraction 5 and thymosin α_1 on *in vitro* pituitary ACTH release, *Soc. Neurosci. Abstr.* **343:**116.

McGillis, J. P., Hall, N. R., Vahouny, G. V., and Goldstein, A. L., 1985, Thymosin fraction 5 causes increased serum corticosterone in rodents *in vivo. J. Immunol* **134:**3952–3955.

McGillis, J. P., Hall, N. R., and Goldstein, A. L., 1986, Biphasic effects of thymosin fraction 5 on ACTH release from cultured anterior pituitary cells, *Fed. Proc.* (in press).

Monjan, A. A., 1981, Stress and immunologic competence: Studies in animals, in: *Psychoneuroimmunology* (R. Ader, ed.), Academic Press, New York, pp. 185–228.

Monjan, A. A., and Collector, M. I., 1977, Stress-induced modulation of the immune response, *Science* **196:**307–308.

Palaszynski, E. W., Moody, T. W., O'Donohue, T. L., and Goldstein, A. L., 1983, Thymosin-like peptides: Localization and biochemical characterization in the rat brain and pituitary gland, *Peptides* **4:**463–467.

Palmblad, J. E. W., 1985, Stress and human immunologic competence, in: *Neural Modulation of Immunity* (R. Guillemin, M. Cohn, and T. Melnechuk, eds.), Raven Press, New York, pp. 45–53.

Rebar, R. W., Miyake, A., Low, T. L. K., and Goldstein, A. L., 1981, Thymosin stimulates secretion of luteinizing hormone-releasing factor, *Science* **214:**669–671.

Spangelo, B. L., Hall, N. R., and Goldstein, A. L., 1985, Evidence that prolactin is an immunomodulatory hormone, in: *Prolactin. Basic and Clinical Correlates* (R. M. MacLeod, M. O. Thorner, and U. Scapangnini, eds.), Liviana Press, Padova, p. 343.

Stein, M., 1985, Bereavement, depression, stress and immunity, in: *Neural Modulation of Immunity* (R. Guillemin, M. Cohn, and T. Melnechuk, eds.), Raven Press, New York, pp. 29–44.

Vahouny, G. V., Kyeyune-Nyombi, E., McGillis, J. P., Tare, N. S., Huang, K. Y., Tombes, R., Goldstein, A. L., and Hall, N. R., 1983, Thymosin peptides and lymphokines do not directly stimulate adrenal corticosteroid production *in vitro. J. Immunol.* **130:**791–794.

Weinstein, Y., 1978, Impairment of the hypothalamo–pituitary–ovarian axis of the athymic "nude" mouse, *Mech. Ageing Dev.* **8:**63–69.

Participants

ABDUL-BADI ABOU-SAMRA
Clinique Endocrinologique Hôpital de l'Antiquaille
69321 Lyon Cedex 5, France
Present address:
Endocrinology and Reproduction Research Branch
National Institute of Child Health and Human Development
National Institutes of Health
Bethesda, Maryland 20892

L. G. ALLEN
Department of Obstetrics and Gynecology
University of Florida College of Medicine
Gainesville, Florida 32610

M. C. AQUILA
Department of Physiology
The University of Texas Health Science Center at Dallas
Southwestern Medical School
Dallas, Texas 75235

WILLIAM A. BANKS
Veterans Administration Medical Center
University of New Orleans, and Tulane University School of Medicine
New Orleans, Louisiana 70146

J. BEDRAN DE CASTRO
Department of Physiology
The University of Texas Health Science Center at Dallas
Southwestern Medical School
Dallas, Texas 75235

A. BELANGER
Departments of Medicine, Molecular Endocrinology and Urology
Laval University Medical Center
Quebec G1V 4G2, Canada

J. BEYER
Department of Endocrinology
University of Mainz
Mainz, Federal Republic of Germany

J. EDWIN BLALOCK
Department of Microbiology
University of Texas Medical Branch
Galveston, Texas 77550
Present address:
Departments of Physiology and Biophysics
University of Alabama
Birmingham, Alabama 35294

S. R. BLOOM
Departments of Histochemistry and Medicine
Royal Postgraduate Medical School
Hammersmith Hospital
London W12 OHS, England

FRANCOISE BORSON
Clinique Endocrinologique Hôpital de l'Antiquaille
69321 Lyon Cedex 5, France

KENNETH L. BOST
Department of Microbiology
University of Texas Medical Branch
Galveston, Texas 77550
Present Address:
Departments of Physiology and Biophysics
University of Alabama
Birmingham, Alabama 35294

STUART BUNTING
Lipids Research
The Upjohn Company
Kalamazoo, Michigan 49001

REN-ZHI CAI
Veterans Administration Medical Center and
Tulane University School of Medicine
New Orleans, Louisiana 70146

PIERRE CANLORBE
Hôpital St. Vincent de Paul
Paris, France

DESMOND N. CARNEY
Mater Hospital
Dublin 7, Ireland

MARGARET A. CASCIERI
Merck Sharp and Dohme Research Laboratories
Rahway, New Jersey 07065

DHIRENDRA N. CHATURVEDI
Vega Biotechnologies, Inc.
Tucson, Arizona 85706

JEAN-LOUIS CHAUSSAIN
Hôpital St. Vincent de Paul
Paris, France

GARY G. CHICCHI
Merck Sharp and Dohme Research Laboratories
Rahway, New Jersey 07065

J. T. CLARK
Department of Obstetrics and Gynecology
University of Florida College of Medicine
Gainesville, Florida 32610

JOHN COLALUCA
Veterans Administration Medical Center and
Tulane University School of Medicine
New Orleans, Louisiana 70146

CHRISTIANE DYLION COLTON
Merck Sharp and Dohme Research Laboratories
West Point, Pennsylvania 19486

PATRICIA C. CONTRERAS
Experimental Therapeutics Branch
National Institute of Neurological and
Communicative Disorders and Stroke
National Institutes of Health
Bethesda, Maryland 20205

DONNA COSTOPOULOS
Department of Pathology
Tufts University School of Medicine
Boston, Massachusetts 02111

JACQUELINE N. CRAWLEY
Clinical Neuroscience Branch
National Institute of Mental Health
National Institutes of Health
Bethesda, Maryland 20205

W. R. CROWLEY
Department of Pharmacology
University of Tennessee College of Medicine
Memphis, Tennessee 38163

NEIL R. CURTIS
Merck Sharp and Dohme Research Laboratories
Neuroscience Research Centre
Terlings Park, Essex, United Kingdom

FRANK CUTTITTA
NCI-Navy Medical Oncology Branch
National Naval Medical Center
Bethesda, Maryland 20014

J. DÄMMRICH
Department of Pathology
University of Würzburg
Würzburg, Federal Republic of Germany

HENRI DECHAUD
Clinique Endocrinologique Hôpital de l'Antiqualille
69321 Lyon Cedex 5, France

BARBARA M. DeLUSTRO
Institutes of Bio-Organic Chemistry and Biological Sciences
Syntex Research
Palo Alto, California 94304

DEBORA A. DiMAGGIO
Experimental Therapeutics Branch
National Institute of Neurological and Communicative Disorders and Stroke
National Institutes of Health
Bethesda, Maryland 20205

S. DION
Department of Pharmacology
Medical School
University of Sherbrooke
Sherbrooke, Quebec J1H 5N4, Canada

P. d'ORLEANS-JUSTE
Department of Pharmacology
Medical School
University of Sherbrooke
Sherbrooke, Quebec J1H 5N4, Canada

G. DRAPEAU
Department of Pharmacology
Medical School
University of Sherbrooke
Sherbrooke, Quebec J1H 5N4, Canada

S. P. DUCKLES
Department of Pharmacology
University of Arizona
Tucson, Arizona 85724

CAROL A. DUDLEY
Department of Physiology
The University of Texas Health Science Center at Dallas
Dallas, Texas 75235

A. DUPONT
Departments of Medicine, Molecular Endocrinology and Urology
Laval University Medical Center
Quebec GIV 4G2, Canada

LEE E. EIDEN
Laboratory of Cell Biology
National Institute of Mental Health
National Institutes of Health
Bethesda, Maryland 20205

JACOB EIMERL
Neurobiology Research Division
Department of Neurology
Uniformed Services University of the Health Sciences
Bethesda, Maryland 20814

PETER J. ELLIOTT
Department of Psychiatry
Duke University Medical Center
Durham, North Carolina 27710
Present address:
Department of Psychiatry
Yale University School of Medicine
New Haven, Connecticut 06508

J. EMOND
Hôtel-Dieu Hospital
Quebec St. Jérôme, J7Y2T8, Canada

JOHN ENG
Solomon A. Berson Laboratory
Veterans Administration Medical Center
Bronx, New York 10468

E. ESCHER
Department of Pharmacology
Medical School
University of Sherbrooke
Sherbrooke, Quebec J1H 5N4, Canada

ROBERT L. ESKAY
Section of Neurochemistry
Laboratory of Clinical Studies
Division of Intramural, Clinical and Biologic Research
National Institute on Alcohol Abuse and Alcoholism
National Institutes of Health
Bethesda, Maryland 20892

RICHARD FAIRTILE
Veterans Administration Medical Center
Washington, D.C. 20422

JOHN M. FARAH, JR.
Experimental Therapeutics Branch
National Institute of Neurological and Communicative Disorders and Stroke
National Institutes of Health
Bethesda, Maryland 20205

GIORA FEUERSTEIN
Neurobiology Research Division
Department of Neurology
Uniformed Services University of the Health Sciences
Bethesda, Maryland 20814

MICHELLE FEVRE-MONTAGE
Clinique Endocrinologique Hôpital de l'Antiquaille
69321 Lyon Cedex 5, France

STEPHEN L. FOOTE
Department of Psychiatry
University of California, San Diego
La Jolla, California 92093

ROGER M. FREIDINGER
Merck Sharp and Dohme Research Laboratories
West Point, Pennsylvania 19486

DADIN FU
Veterans Administration Medical Center and Tulane University School of Medicine
New Orleans, Louisiana 70146

HAROLD GAINER
Laboratory of Neurochemistry and Neuroimmunology
National Institute of Child Health and Human Development
National Institutes of Health
Bethesda, Maryland 20892

JERRY D. GARDNER
Digestive Diseases Branch
National Institute of Arthritis, Diabetes, Digestive and Kidney Diseases
National Institutes of Health
Bethesda, Maryland 20205

D. R. GEHLERT
Department of Psychiatry
University of Utah
Salt Lake City, Utah 84142

M. GIGUERE
Departments of Medicine, Molecular Endocrinology and Urology
Laval University Medical Center
Quebec G1V 4G2, Canada

ALLAN GOLDSTEIN
The George Washington University School of Medicine and Health Sciences
Washington, D.C. 20892

K. GULYA
Department of Pharmacology
University of Arizona
Tucson, Arizona 85724
Permanent address:
Central Research Laboratory
Medical University
Szeged 6720, Hungary

MAC E. HADLEY
Department of Anatomy
University of Arizona
Tucson, Arizona 85724

NICHOLAS R. HALL
The George Washington University School of Medicine and Health Sciences
Washington, D.C. 20892

J. HAM
National Institute for Medical Research
The Ridgeway
Mill Hill, London NW7 1AA, England

CELESTE HART
Diabetes Branch
National Institute of Arthritis, Diabetes, Digestive and Kidney Diseases
National Institutes of Health
Bethesda, Maryland 20892

CAROL HERRON
Department of Biochemistry
The George Washington University School of Medicine and Health Sciences
Washington, D.C. 20037

MARION T. HIEROWSKI
Veterans Administration Medical Center and Tulane University School of Medicine
New Orleans, Louisiana 70146

TERESA L. HO
Institutes of Bio-Organic Chemistry and Biological Sciences
Syntex Research
Palo Alto, California 94304

HENRY H. HOLZGREFE
Lipids Research
The Upjohn Company
Kalamazoo, Michigan 49001

VIVIAN Y. H. HOOK
Laboratory of Cell Biology
National Institute of Mental Health
National Institutes of Health
Bethesda, Maryland 20205
Present address:
Department of Biochemistry
Uniformed Services University of the Health Sciences
Bethesda, Maryland 20814

JOEL HORWITZ
Department of Physiology and Biophysics
University of Illinois College of Medicine
Chicago, Illinois 60680

VICTOR J. HRUBY
Department of Chemistry
University of Arizona
Tucson, Arizona 85721

DAVID M. JACOBOWITZ
Clinical Neuroscience Branch
National Institute of Mental Health
National Institutes of Health
Bethesda, Maryland 20205

ROBERT T. JENSEN
Digestive Diseases Branch
National Institute of Arthritis, Diabetes, and Digestive and Kidney Diseases
National Institutes of Health
Bethesda, Maryland 20205

JEAN-CLAUDE JOB
Hôpital St. Vincent de Paul
Paris, France.

S. JOHNSTON
School of Psychology
University of Ottawa
Ottawa, Ontario K1N 6N5, Canada

JAMES E. JUMBLATT
Department of Biochemistry
Tufts University School of Medicine
Boston, Massachusetts 02111

G. KAHALY
Department of Endocrinology
University of Mainz
Mainz, Federal Republic of Germany

NED H. KALIN
Department of Psychiatry
University of Wisconsin Madison, and Psychiatry Service
William S. Middleton Memorial Veterans Hospital
Madison, Wisconsin 53705

P. S. KALRA
Department of Obstetrics and Gynecology
University of Florida School of Medicine
Gainesville, Florida 32610

SATYA P. KALRA
Department of Obstetrics and Gynecology
University of Florida College of Medicine
Gainesville, Florida 32610

ABBA J. KASTIN
Veterans Administration Medical Center
University of New Orleans and Tulane University School of Medicine
New Orleans, Louisiana 70146

O. KHORRAM
Department of Physiology
The University of Texas Health Science Center at Dallas
Southwestern Medical School
Dallas, Texas 75235

CLIFFORD KING
Veterans Administration Medical Center
Washington, D.C. 20422

PATRICK KITABGI
Centre de Biochimie du CNRS
Faculté des Sciences
Parc Valrose
06034 Nice Cedex, France

STANISLAW KONTUREK
Medical Academy
Krakow, Poland

LOUIS Y. KORMAN
Section on Gastroenterology
Veterans Administration Medical Center
Washington, D.C. 20422

U. KRAUSE
Department of Endocrinology
University of Mainz
Mainz, Federal Republic of Germany

F. LABRIE
Departments of Medicine, Molecular Endocrinology and Urology
Laval University Medical Center
Quebec G1V 4G2, Canada

Y. LACOURCIERE
Departments of Medicine, Molecular Endocrinology and Urology
Laval University Medical Center
Quebec G1V 4G2, Canada

ROBERT A. LAHTI
CNS Diseases Research
The Upjohn Company
Kalamazoo, Michigan 49001

SARA FRYER LEIBOWITZ
The Rockefeller University
New York, New York 10021

CSABA LERANTH
Department of Obstetrics and Gynecology and Section of Neuroanatomy
Yale University School of Medicine
New Haven, Connecticut 06510

DEREK LEROITH
Diabetes Branch
National Institute of Arthritis, Diabetes, Digestive and Kidney Diseases
National Institutes of Health
Bethesda, Maryland 20892

MAXINE A. LESNIAK
Diabetes Branch
National Institute of Arthritis, Diabetes, Digestive and Kidney Diseases
National Institutes of Health
Bethesda, Maryland 20892

CHOH HAO LI
Laboratory of Molecular Endocrinology
University of California
San Francisco, California 94143

TEHMING LIANG
Merck Sharp and Dohme Research Laboratories
Rahway, New Jersey 07065

WILLIAM LOWE
Diabetes Branch
National Institute of Arthritis, Diabetes, Digestive and Kidney Diseases
National Institutes of Health
Bethesda, Maryland 20892

M. D. LUMPKIN
Department of Physiology
The University of Texas Health Science Center
at Dallas
Southwestern Medical School
Dallas, Texas 75235

NEIL J. MacLUSKY
Department of Obstetrics and Gynecology and
Section of Neuroanatomy
Yale University School of Medicine
New Haven, Connecticut 06510

C. MASER-GLUTH
Department of Pharmacology
University of Heidelberg
Heidelberg, Federal Republic of Germany

S. M. McCANN
Department of Physiology
The University of Texas Health Science Center
at Dallas
Southwestern Medical School
Dallas, Texas 75235

Z. MERALI
School of Psychology
University of Ottawa
Ottawa, Ontario K1N 6N5, Canada

J. MIZRAHI
Department of Pharmacology
Medical School
University of Sherbrooke
Sherbrooke, Quebec J1H 5N4, Canada

G. MONFETTE
Hôtel-Dieu Hospital
St. Jérôme, Quebec J7Y2T8, Canada

TERRY W. MOODY
Department of Biochemistry
The George Washington University
School of Medicine and Health Sciences
Washington, D.C. 20037

ROBERT L. MOSS
Department of Physiology
The University of Texas Health Science Center
at Dallas
Dallas, Texas 75235

FREDERICK NAFTOLIN
Department of Obstetrics and Gynecology and
Section of Neuroanatomy
Yale University School of Medicine
New Haven, Connecticut 06510

CHARLES B. NEMEROFF
Departments of Psychiatry and Pharmacology and
Center for Aging and Human Development
Duke University Medical Center
Durham, North Carolina 27710

JOHN J. NESTOR, JR.
Institutes of Bio-Organic Chemistry and
Biological Sciences
Syntex Research
Palo Alto, California 94304

THOMAS L. O'DONOHUE
Experimental Therapeutics Branch
National Institute of Neurological and
Communicative Disorders and Stroke
National Institutes of Health
Bethesda, Maryland 20205

RICHARD D. OLSON
Veterans Administration Medical Center
University of New Orleans and Tulane
University School of Medicine
New Orleans, Louisiana 70146

N. ONO
Department of Physiology
The University of Texas Health Science Center
at Dallas
Southwestern Medical School
Dallas, Texas 75235

W. ORMANNS
Department of Pathology
University of Würzburg
Würzburg, Federal Republic of Germany

D. C. PARISH
Laboratory of Neurochemistry and Neuroimmunology
National Institute of Child Health and Development
National Institutes of Health
Bethesda, Maryland 20205

P. PARMASHWAR
School of Psychology
University of Ottawa
Ottawa, Ontario K1N 6N5, Canada

J. T. PELTON
Department of Chemistry
University of Arizona
Tucson, Arizona 85724

JUAN C. PENHOS
Diabetes Branch
National Institute of Arthritis, Diabetes, Digestive and Kidney Diseases
National Institutes of Health
Bethesda, Maryland 20892

ROBERT L. PERLMAN
Department of Physiology and Biophysics
University of Illinois College of Medicine
Chicago, Illinois 60680

DEBRA S. PERLOW
Merck Sharp and Dohme Research Laboratories
West Point, Pennsylvania 19486

DAVID C. PERRY
Department of Pharmacology
George Washington University Medical Center
Washington, D.C. 20037

J. M. POLAK
Departments of Biochemistry and Medicine
Royal Postgraduate Medical School
Hammersmith Hospital
London W12 OHS, England

REBECCA M. PRUSS
Laboratory of Cell Biology
National Institute of Mental Health
National Institutes of Health
Bethesda, Maryland 20205

REMI QUIRION
Douglas Hospital Research Centre and Department of Psychiatry
Faculty of Medicine
McGill University
Verdun, Quebec H4H 1R3, Canada

FRANCOISE RAYNAUD
Hôpital St. Vincent de Paul
Paris, France

LILLIAN RECANT
Veterans Administration Medical Center
Washington, D.C. 20422

TOMMIE W. REDDING
Veterans Administration Medical Center and Tulane University School of Medicine
New Orleans, Louisiana 70146

D. REGOLI
Department of Pharmacology
Medical School
University of Sherbrooke
Sherbrooke, Quebec J1H 5N4, Canada

MARC ROGER
Fondation de Recherche en Hormonologie
94268 Fresnes Cedex, France

WILLIS K. SAMSON
Department of Physiology
The University of Texas Health Science Center at Dallas
Dallas, Texas 75235

TOMI K. SAWYER
Biopolymer Chemistry/Biotechnology
The Upjohn Company
Kalamazoo, Michigan 49001

ANDREW V. SCHALLY
Veterans Administration Medical Center and Tulane University School of Medicine
New Orleans, Louisiana 70146

ALAIN B. SCHREIBER
Institutes of Bio-Organic Chemistry and Biological Sciences
Syntex Research
Palo Alto 94304

PEGGY J. K. D. SCHREUR
CNS Diseases Research
The Upjohn Company
Kalamazoo, Michigan 49001

J. SCHREZENMEIR
Department of Endocrinology
University of Mainz
Mainz, Federal Republic of Germany

JOSHUA SHEMER
Diabetes Branch
National Institute of Arthritis, Diabetes, Digestive and Kidney Diseases
National Institutes of Health
Bethesda, Maryland 20892

JACK E. SHERMAN
Department of Psychiatry
University of Wisconsin Madison, and Psychiatry Service
William S. Middleton Memorial Veterans Hospital
Madison, Wisconsin 53705

JEAN SIMON
Diabetes Branch
National Institute of Arthritis, Diabetes, Digestive and Kidney Diseases
National Institutes of Health
Bethesda, Maryland 20892

GERHARD SKOFITSCH
Laboratory of Clinical Science
National Institute of Mental Health
National Institutes of Health
Bethesda, Maryland 20205

ERIC M. SMITH
Department of Microbiology
University of Texas Medical Branch
Galveston, Texas 77550

D. G. SMYTH
National Institute for Medical Research
The Ridgeway
Mill Hill, London NW7 1AA, England

BRYAN L. SPANGELO
The George Washington University School of Medicine and Health Sciences
Washington, D.C. 20892

B. GLENN STANLEY
The Rockefeller University
New York, New York 10021

R. ST.-ARNAUD
Departments of Medicine, Molecular Endocrinology and Urology
Laval University Medical Center
Quebec G1V 4G2, Canada

DOUGLAS J. STAPLES
Biopolymer Chemistry/Biotechnology
The Upjohn Company
Kalamazoo, Michigan 49001

SERGE A. ST-PIERRE
INRS-Santé
Institut National de la Recherche Scientifique
Pavillon Gamelin
Montreal, Quebec H1N 3M5, Canada

JILL A. STIVERS
Clinical Neuroscience Branch
National Institute of Mental Health
National Institutes of Health
Bethesda, Maryland 20205

T. STRACK
Department of Endocrinology
University of Mainz
Mainz, Federal Republic of Germany

BALAZS SZOKE
Veterans Administration Medical Center and Tulane University School of Medicine
New Orleans, Louisiana 70146

JAMES P. TAM
The Rockefeller University
New York, New York 10021

KIM TIMMERS
Veterans Administration Medical Center
Washington, D.C. 20422

ARTHUR S. TISCHLER
Department of Pathology
Tufts University School of Medicine
Boston, Massachusetts 02111

ROBERT TOMBES
Department of Biochemistry
The George Washington University School of Medicine and Health Sciences
Washington, D.C. 20037

I. TORRES-ALEMAN
Veterans Administration Medical Center and Tulane University School of Medicine
New Orleans, Louisiana 70146

JEAN-EDMOND TOUBLANC
Hôpital St. Vincent de Paul
Paris, France

JACQUES TOURNIAIRE
Clinique Endocrinologique Hôpital de l'Antiquaille
69321 Lyon Cedex 5, France

SIDNEY UDENFRIEND
Roche Institute of Molecular Biology
Roche Research Center
Nutley, New Jersey 07110

GEORGE V. VAHOUNY
Department of Biochemistry
The George Washington University
School of Medicine and Health Sciences
Washington, D.C. 20037

RITA J. VALENTINO
Department of Pharmacology
George Washington University Medical Center
Washington, D.C. 20037

DANIEL F. VEBER
Merck Sharp and Dohme Research Laboratories
West Point, Pennsylvania 19486

P. VECSEI
Department of Pharmacology
University of Heidelberg
Heidelberg, Federal Republic of Germany

JEAN-PIERRE VINCENT
Centre de Biochimie du CNRS
Faculté des Sciences
Parc Valrose
06034 Nice Cedex, France

NANCY R. VOYLES
Veterans Administration Medical Center
Washington, D.C. 20422

ROBERT J. WALDBILLING
Armed Forces Radiobiology Research Institute
Bethesda, Maryland 20205

J. K. WAMSLEY
Department of Psychiatry
University of Utah
Salt Lake City, Utah 84142

MICHAEL WELLS
Veterans Administration Medical Center
Washington, D.C. 20422

MARK H. WHITNALL
Laboratory of Neurochemistry and Neuroimmunology
National Institute of Child Health and Human Development
National Institutes of Health
Bethesda, Maryland 20892

ANITA E. WILKERSON
CNS Diseases Research
The Upjohn Company
Kalamazoo, Michigan 49001

BRIAN WILLIAMS
Merck Sharp and Dohme Research Laboratories
Neuroscience Research Centre
Terlings Park, Essex, United Kingdom

ROSALYN S. YALOW
Solomon A. Berson Laboratory
Veterans Administration Medical Center
Bronx, New York 14068

H. I. YAMAMURA
Department of Pharmacology
University of Arizona
Tucson, Arizona 85724

JAMES E. ZADINA
Veterans Administration Medical Center
University of New Orleans and Tulane University School of Medicine
New Orleans, Louisiana 70146

Index